5G Networks

5G Networks

Fundamental Requirements, Enabling Technologies, and Operations Management

Anwer Al-Dulaimi, Xianbin Wang, and Chih-Lin I

IEEE PRESS

WILEY

Published by John Wiley & Sons, Inc., Hoboken, New Jersey.
Published simultaneously in Canada.

For general information on our other products and services or for technical support, please contact our Customer Care Department within the United States at (800) 762-2974, outside the United States at (317) 572-3993 or fax (317) 572-4002.

Wiley also publishes its books in a variety of electronic formats. Some content that appears in print may not be available in electronic formats. For more information about Wiley products, visit our web site at www.wiley.com.

Library of Congress Cataloging-in-Publication Data

ISBN: 9781119332732

Printed in the United States of America.

V10004142_082918

Contents

2 Waveform Design for 5G and Beyond *51*

Ali Fatih Demir, Mohamed Elkourdi, Mostafa Ibrahim, and Huseyin Arslan

3 Full-Duplex System Design for 5G Access *77*

Shu-ping Yeh, Jingwen Bai, Ping Wang, Feng Xue, Yang-seok Choi, Shilpa Talwar, Sung-en Chiu, and Vinod Kristem

Foreword

The fifth generation (5G) mobile networks is the first wireless systems that will allow ubiquitous Gigabit service to all connected users and appliances. This revolutionary telecommunications technology will not only improve the well-known performance metrics that we rely on to measure capacity or end-to-end delays but will - for the first time - also enable measuring and proactively influencing the user experience. The arrival of 5G will have conceptual, technical, economical, and social impacts that will change the human life to new digitized culture where everything is tracked, recorded, and logged into a file that is stored somewhere in cloud. Therefore, industrial and research communities are in a race to define all the key requirements, conducting analysis, implementing testbeds, and performing field trails. Alongside, the standardization bodies are finalizing new standards that shape the commercial-off-the-shelf products. The regulatory offices of governments are also trying to deliver new legislation that regulates 5G technologies and capabilities. In the mid of this outreaching change, we as engineers need to understand diverse 5G technologies and services to shape the vision towards networks of the future.

The transformation to 5G triggers the need to restructure the mobile network architecture. On the radio access technology (RAT) side, the New Radio (NR) comprises multiple-radio interfaces transmitting over different spectrum bands. This bridges pillars of spectrum and extends NR bandwidth beyond licensed Long-Term Evolution (LTE) bands to include millimeter waves above 6 GHz; and the trusted non-3GPP access allows for access of systems like WiFi too. However, a base station with multi-radio interfaces needs to support new schemes of interoperability for rerouting packets between different radio interfaces. The goal is to be able to access the spectrum at any available band using enhanced technologies such as non-orthogonal multiple access (NOMA), high-performance antennas, and massive MIMO.

The densification of macro and small cells in emerging ultradense networks (UDNs) is an architectural challenge at the Radio Access Network (RAN), in 3GPP referred to as (R)AN. A UDN requires new approaches to spectrum sharing and interference mitigation for friendly coexistence with other appliances.

This also requires supporting artificial intelligence (AI) at the base station and network levels with new information exchange platforms to distribute knowledge on the spectrum status between various network nodes. From previous generations, it is understood that RAN deployments will initiate the migration process towards the 5G network. Therefore, NRs will be attached to the fourth-generation (4G) core network to use the current entities of evolved packet core (EPC) and IP Multimedia Subsystem (IMS). The 3rd Generation Project (3GPP) has already drafted standards for 5G Core Network (5GC).

Nevertheless, during the transition stage, the NR base stations will be interfaced to the control and data planes through LTE RAN in a non-standalone mode. Once 5GC is deployed, the NR base stations will be disconnected from the 4G core and interfaced to 5GC in a standalone mode. At this point, 4G LTE base stations will also be disconnected from the 4G core and interfaced to NRs to access the 5GC. Driven by the need to optimize service and maintain higher quality of service (QoS), there is a need to verify the new suite of 5GC interfaces and define any subsequent changes to the call session initiation processes and applications storage. Once deployed, 5G networks will be able to provide unprecedented/ uninterrupted throughputs to support transmissions of ultrahigh definition (UHD) videos and Enhanced Voice Services (EVS).

The other major change in the 5G era is the employment of network function virtualization (NFV) concepts that will enhance the flexibility of deploying advanced concepts, such as edge-clouds. Moving away from proprietary hardware to use software applications running in orchestrated virtual machines requires dynamic resource scaling subject to processed traffic. The NFV will considerably alter the current network hierarchy by enabling on-fly instantiation of virtual function networks (VNFs) that virtually interconnect to each other. The software defined networking (SDN) is another feature that automates adaptation of routes between various network slices to avoid congested switches. SDN also provides the network with additional virtual layers to interconnect switches in data centers.

Provided with distributed cloud capabilities, SDN/NFV architectures enable 5G to connect new industries and empower unique service scenarios. This powerful ability to connect various users and appliances builds new alignment for stable access to individual components that run at vertical industries. Moreover, 5G is the networking technology for Internet of Things (IoT) backing remote access services, connected cars, and vehicle to everything (V2X) communications. Considering all these connections, there is a continuous need to investigate new green protocols and equipment for sustainable green communication networks.

With 5G designs running at full steam, there are a few pioneering projects in the world proving 5G viability via a real-world rollout. At the forefront is the UK 5GUK testbed that is the world first to deploy an end-to-end 5G system composed of heterogeneous vendor components. At the end of 2017, King

College London has already demonstrated the UK first 3.5GHz Massive MIMO as well as the world first fully softwarized/virtualized 5G end-to-end call.

This new IEEE/Wiley book entitled 5G Networks: Fundamental Requirements, Enabling Technologies, andOperationsManagement is thus a very timely publication. It is an exceptional milestone document that will educate audiences from academia and industry. The book editors are from leading research and development industries that are involved directly with 5G development as well as academia that has underpinned much of its research. This book has been written by the world-renowned experts who have studied, analyzed, and proposed new solutions for all 5G protocols, potential vertical industries, and standardization efforts.

The authors are researchers, professors, directors of standards, and chief technology officers of the world leading operators. The book has 20 chapters that look at all key aspects of 5G allowing readers to obtain academic, industrial, and regulatory knowledge. Bridging theory and application, I am confident that this book will become a reference document for our community for many years to come.

Chair Professor, Centre for Telecommunications
Research Prof. Mischa Dohler, FIEEE FREng FRSA
King's College London
London, UK

Preface

From research communities to industry world, the fifth generation (5G) development is gaining momentum driven by the market demand and business opportunities. The developers of 5G network adopts new ideas involving a variety of technologies and applications beyond the boundaries of past mobile generation. In this new network, business case models and user satisfaction are very important drivers to consider. This open space for exchanging ideas is also seen at the standardization bodies where partnerships and discussions started to draft new schemes incorporating different technologies and mechanisms. The 5G is not a collection of technologies that connects humans using machines, it is a whole new technology package that connects humans and machines in one pool. Therefore, the concepts for many of the technological advances that we consider to be a crucial part of the upcoming network are still need definition prior potential 5G deployment in 2020. Although the basic network requirements are well understood such as minimal latency, powerful processing features, unified communication bus, efficient interfacing and management of machine data, and gigabyte downloading capacities. This book energies the 5G technology development further through a carful investigation of potential network radio resources, technology integration, virtualization of appliances, and foreseen standards. The studied topics demonstrate that 5G is not just about a high speed mobile internet connectivity, in fact it is about influencing society and economy advances using associated vertical applications that considers type of requested service and user satisfaction rather than connected network interface.

This book provides a comprehensive and advanced analysis for all 5G network segments. It is written for all audience to discuss the state of art, technologies, vertical applications, and standards. The editors vision was to provide the research community with a single document that covers all aspects of 5G research. The book is meant to be an inclusive document for theoretical concepts and recent industry developments toward 5G. The authors are world leading experts from industry and academia who are engaged in 5G projects on day-to day basis. For academic research, this book explores a wide range of challenges for 5G networks such as radio resources management, waveform design, security, etc. It will be an excellent text book for students and researchers to learn mythology and identify methods to model their solutions. The given results throughout different chapters can also be used as a reference in comparative studies. For industry research, this book defines behavior of systems, solutions and technologies for deployments, review of standards, etc. This type of material will help researchers and developers in identifying development and testing plans and use this book as technical manual.

This book concludes the work of many teams and leading researchers from around the globe. It helps learners to build upon knowledge, develop ideas, and expand visions to network level solutions. The content is written and indexed to help step-by-step knwoldge increment process for formal classrooms and self-learning. Finally, this book also bridges the gap between academia and industry through a mixture of visions that allow both communities to learn from each other and motivate each other to enhance knowledge and improve communication systems.

Anwer Al-Dulaimi, EXFO Inc., Canada

Xianbin Wang, Western University, Canada

Chih-Lin I, China Mobile Research Institute, China

Author Bios

 *Anwer Al-Dulaimi* (M'11, SM'17) is a System Engineering Specialist in the R&D department at EXFO Inc., Toronto, Canada. Dr. Al-Dulaimi received his Ph.D. degree in electrical and computer engineering from Brunel University, London, U.K., in 2012 after receiving M.Sc. and B.Sc. honours degrees in communication engineering. He was a Postdoctoral Research Fellow in the department of electrical and computer engineering, University of Toronto, Canada. During his postdoctoral time, Dr. Al-Dulaimi contributed to the LTE research through project collaborations with Blackberry Advanced Research Team-Canada and Standardization Team, UK. He has been awarded many grants by the Wireless World Research Forum (WWRF), IEEE Standards Association (IEEE-SA), etc. He has published many academic papers and was awarded the best IEEE/WWRF Vehicular Technology Magazine paper for three times. His research interests include 5G wireless communications and network design and optimization, cloud networks, and Internet of Things. Dr. Al-Dulaimi is an IEEE Distinguished Lecturer. He is the chair of IEEE 1932.1 working group "Standard for Licensed/Unlicensed Spectrum Interoperability in Wireless Mobile Network". He is the editor of IEEE 5G Initiative Series in IEEE Vehicular Technology Magazine, editor of vehicular networking series in IEEE Communication Standards Magazine, associate editor of IEEE Communications Magazine, editor of IEEE 5G Tech Focus letter, and guest editor of many special issues in IEEE journals. He was the recipient of the 2013 Worldwide Universities Network Cognitive Communications Consortium best paper for outstanding research in cognitive communications for his edited book entitled "Self-Organization and Green Applications in Cognitive Radio Networks". Dr. Al-Dulaimi is an Associate Fellow of the British Higher Education Academy and registered as a Chartered Engineer by the British Engineering Council in 2010.

Xianbin Wang (S'98-M'99-SM'06-F'17) is a Professor and Canada Research Chair at Western University, Canada. He received his Ph.D. degree in electrical and computer engineering from National University of Singapore in 2001. Prior to joining Western, he was with Communications Research Centre Canada (CRC) as a Research Scientist/Senior Research Scientist between July 2002 and Dec. 2007. From Jan. 2001 to July 2002, he was a system designer at STMicroelectronics, where he was responsible for the system design of DSL and Gigabit Ethernet chipsets. His current research interests include 5G technologies, Internet-of-Things, communications security, and locationing technologies. Dr. Wang has over 300 peer-reviewed journal and conference papers, in addition to 26 granted and pending patents and several standard contributions. Dr. Wang is a Fellow of IEEE and an IEEE Distinguished Lecturer. He has received many awards and recognition, including Canada Research Chair, CRC President's Excellence Award, Canadian Federal Government Public Service Award, Ontario Early Researcher Award and five IEEE Best Paper Awards. He currently serves as an Editor/Associate Editor for IEEE Transactions on Communications, IEEE Transactions on Broadcasting, and IEEE Transactions on Vehicular Technology and He was also an Associate Editor for IEEE Transactions on Wireless Communications between 2007 and 2011, and *IEEE Wireless Communications Letters* between 2011 and 2016. Dr. Wang was involved in a number of IEEE conferences including GLOBECOM, ICC, VTC, PIMRC, WCNC and CWIT, in different roles such as symposium chair, tutorial instructor, track chair, session chair and TPC co-chair.

Chih-Lin I is the Chief Scientist of Wireless Technologies of China Mobile, in charge of advanced wireless communication R&D effort of China Mobile Research Institute (CMRI). She established the Green Communications Research Center of China Mobile, spearheading major initiatives including 5G Key Technologies R&D; high energy efficiency system architecture, technologies, and devices; green energy; C-RAN and soft base station. Dr. I received her Ph.D. degree in Electrical Engineering from Stanford University, and has more than 30 years experience in wireless communication technical domain. She has worked in various world-class companies and research institutes, including wireless communication fundamental research department of AT&T Bell Labs; Headquarter of AT&T, as the Director of Wireless Communications Infrastructure and Access Technology; ITRI of Taiwan, as the Director of Wireless Communication Technology; Hong Kong ASTRI, as the VP and the Founding GD of Communications Technology Domain. Dr. I received the Trans. COM Stephen Rice Best Paper Award, and is a winner of CCCP "National 1000 talent" program. She was an elected Board Member of IEEE ComSoc, Chair of ComSoc Meeting and Conference Board, and the Founding Chair of IEEE WCNC Steering Committee. She is currently the Chair of FuTURE Forum 5G SIG, an Executive Board Member of GreenTouch, a Network Operator Council Member of ETSI NFV, and an Adjunct Professor of BUPT. Dr. I has shown frequent presence in many important and high-level public occasions for speech delivery. She is often invited as the keynote speaker for diverse audience from academia, industry and governments. She is very active in many venues such as conferences, summits, workshops, panels and so on. This year she has delivered nearly 30 speeches in lots of events such as IEEE WCNC, IEEE ICC, IEEE VTC, IEEE PIMRC, Global Professional Services Forum and so on, which included a 3-hour-long tutorial on C-RAN in Cloud RAN Conference in Paris.

List of Contributors

Ahmad Shahidan Abdullah
Faculty of Electrical Engineering
Universiti Teknologi Malaysia
Johor, Malaysia

Anwer Al-Dulaimi
R&D Department
EXFO Inc.
Toronto, Canada

Saba Al-Rubaye
Instituto de Telecomunicações
Campus Universitário de Santiago
Aveiro - Portugal

Ahmad Alsharoa
Electrical and Computer
Engineering department
Iowa State University (ISU)
Ames, IA
USA

Huseyin Arslan
Department of Electrical Engineering
University of South Florida
Tampa, FL
USA
and
School of Engineering and
Natural Sciences
Istanbul Medipol University
Istanbul, Turkey

Jingwen Bai
Intel Corporation
Intel Lab
Santa Clara, CA
USA

Ioannis-Prodromos Belikaidis
R&D Department
WINGS ICT Solutions
Athens, Greece

Kui Cai
Science and Math Cluster
Singapore University of Technology
and Design
Singapore

Abdulkadir Celik
Computer, Electrical, Mathematical
Sciences & Engineering Division
King Abdullah University of
Science and Technology (KAUST)
Kingdom of Saudi Arabia

Batu K. Chalise
Department of Electrical and
Computer Engineering
New York Institute of Technology
Old Westbury, NY
USA

Kishor Chandra
Electrical Engineering, Mathematics
and Computer Science Department
Delft University of Technology
Delft, The Netherlands

Jun Cheng
Department of Intelligent
Information Engineering
and Sciences
Doshisha University
Kyoto, Japan

Yuhao Chi
State Key Laboratory of Integrated
Services Networks
Xidian University
Xi'an, China

Sung-en Chiu
University of California
Electrical and Computer Engineering
San Diego, CA
USA

Alex Jinsung Choi
Deutsche Telekom
T-Laboratories Innovation
Friedrich-Ebert-Allee 140,
53113 Bonn
Germany

Yang-seok Choi
Intel Corporation
Intel Lab
Santa Clara, CA
USA

John Cosmas
Department of Electronic and
Electrical Engineering
Brunel University London
Uxbridge, UK

Xavier Costa-Pérez
5G Networks R&D Group
NEC Laboratories Europe GmbH
Heidelberg, Germany

Linglong Dai
Department of Electronics Engineering
Tsinghua University
Beijing, China

Panagiotis Demestichas
R&D Department
WINGS ICT Solutions
Athens, Greece
and
Department of Digital Systems
University of Piraeus
Piraeus, Greece

Ali Fatih Demir
Department of Electrical Engineering
University of South Florida
Tampa, FL
USA

Zhiguo Ding
School of Electrical and Electronic
Engineering
The University of Manchester
Manchester, UK

Mohamed Elkourdi
Department of Electrical Engineering
University of South Florida
Tampa, FL
USA

Norsheila Fisal
Faculty of Electrical Engineering
Universiti Teknologi Malaysia
Johor, Malaysia

Frank H. P. Fitzek
5G Lab Germany and Technical
University Dresden
Dresden, Germany

Vassilis Foteinos
R&D Department
WINGS ICT Solutions
Athens, Greece

Renaud Di Francesco
Sony Europe Research and
Standardisation Department
Sony Mobile
Lund, Sweden

Maria Pia Galante
Technology Innovation Department
TIM
Torino, Italy

Caixia Gao
School of Electrical and
Electronic Engineering
North China Electric
Power University
Beijing, China

Andrés Garcia-Saavedra
5G Networks R&D Group
NEC Laboratories Europe GmbH
Heidelberg, Germany

Andreas Georgakopoulos
R&D Department
WINGS ICT Solutions
Athens, Greece

Fabio Giust
5G Networks R&D Group
NEC Laboratories Europe GmbH
Heidelberg, Germany

Abdul Hadi Fikri Abdul Hamid
Faculty of Electrical Engineering
Universiti Teknologi Malaysia
Johor, Malaysia

Shuangfeng Han
China Mobile Research Institute
China Mobile Communications
Corporation
Beijing, China

Syed Ali Hassan
School of Electrical Engineering
and Computer Science (SEECS)
National University of Sciences
and Technology (NUST)
Islamabad, Pakistan

Chih-Lin I
Wireless Technologies
China Mobile Research Institute
China Mobile Communications
Corporation
Beijing, China

Mostafa Ibrahim
School of Engineering and Natural
Sciences
Istanbul Medipol University
Istanbul, Turkey

Muhammad Ali Imran
School of Engineering
University of Glasgow
Glasgow, UK

Bruno Jacobfeuerborn
Deutsche Telekom AG
Berlin, Germany

Dushantha Nalin K. Jayakody
National Research Tomsk
Polytechnic University
Tomsk, Russia

Sangsoo Jeong
SK Telecom
Network Technology R&D Center
Hwangsaeul-ro, 258beon-gil,
Bundang-gu, Seongnam-si,
Gyeonggi-do
Korea

Ruicheng Jiao
Department of Electronics
Engineering Tsinghua University
Beijing, China

Sungho Jo
SK Telecom
Network Technology R&D Center
Hwangsaeul-ro, 258beon-gil,
Bundang-gu, Seongnam-si,
Gyeonggi-do
Korea

Ahmed E. Kamal
Electrical and Computer
Engineering department
Iowa State University (ISU)
Ames, IA
USA

Peter Karlsson
Sony Europe Research and
Standardisation Department
Sony Mobile
Lund, Sweden

Jouni Korhonen
Nordic Semiconductor
Espoo, Finland

Evangelos Kosmatos
R&D Department
WINGS ICT Solutions
Athens, Greece

Vinod Kristem
Intel Corporation
Intel Lab
Santa Clara, CA
USA

Rongpeng Li
College of Information Science and
Electronic Engineering
Zhejiang University
Hangzhou, China

Xi Li
5G Networks R&D Group
NEC Laboratories Europe GmbH
Heidelberg, Germany

Ying Li
State Key Laboratory of Integrated
Services Networks
Xidian University
Xi'an, China

Orestis-Andreas Liakopoulos
R&D Department
WINGS ICT Solutions
Athens, Greece

Marco Liebsch
5G Networks R&D Group
NEC Laboratories Europe GmbH
Heidelberg, Germany

Shahid Mumtaz
Instituto de Telecomunicações
Aveiro, Portugal

Akihiro Nakao
The Fifth Generation Mobile
Communications Promotion
Forum (5GMF)
Tokyo, Japan

Muhammad Shahmeer Omar
School of Electrical Engineering
and Computer Science (SEECS)
National University of Sciences and
Technology (NUST)
Islamabad, Pakistan

Jinhyo Park
SK Telecom
ICT R&D Center
SK T-Tower, 65, Eulji-ro, Jung-gu
Seoul, Korea

Anggrit Dewangkara Yudha Pinangkis
Electrical Engineering, Mathematics
and Computer Science Department
Delft University of Technology
Delft, The Netherlands

R. Venkatesha Prasad
Electrical Engineering, Mathematics
and Computer Science Department
Delft University of Technology
Delft, The Netherlands

Junaid Qadir
Information Technology University
Lahore, Pakistan

Chen Qi
College of Information Science and
Electronic Engineering
Zhejiang University
Hangzhou, China

Mohd Rozaini Abd Rahim
Faculty of Electrical Engineering
Universiti Teknologi Malaysia
Johor, Malaysia

Rozeha A. Rashid
Faculty of Electrical Engineering
Universiti Teknologi Malaysia
Johor, Malaysia

Ahmad M. Rateb
Faculty of Electrical Engineering
Universiti Teknologi Malaysia
Johor, Malaysia

Jonathan Rodriguez
Instituto de Telecomunicações
Aveiro, Portugal
and
University of South Wales
Pontypridd, UK

G. Romano
Technology Innovation Department
TIM
Torino, Italy

Mohd Adib Sarijari
Faculty of Electrical Engineering
Universiti Teknologi Malaysia
Johor, Malaysia

Kohei Satoh
The Fifth Generation Mobile
Communications Promotion
Forum (5GMF)
Tokyo, Japan

Hamdan Sayuti
Faculty of Electrical Engineering
Universiti Teknologi Malaysia
Johor, Malaysia

Vincenzo Sciancalepore
5G Networks R&D Group
NEC Laboratories Europe GmbH
Heidelberg, Germany

Takashi Shimizu
The Fifth Generation Mobile
Communications Promotion
Forum (5GMF)
Tokyo, Japan

Guanghui Song
Department of Intelligent
Information Engineering and Sciences
Doshisha University
Kyoto, Japan

Himal A. Suraweera
Department of Electrical
and Electronic Engineering
University of Peradeniya
Peradeniya, Sri Lanka

Shilpa Talwar
Intel Corporation
Intel Lab
Santa Clara, CA
USA

Stavroula Vassaki
R&D Department
WINGS ICT Solutions
Athens, Greece

Panagiotis Vlacheas
R&D Department
WINGS ICT Solutions
Athens, Greece

Bichai Wang
Department of Electronic
Engineering
Tsinghua University
Beijing, China

Ping Wang
Apple Inc.
Santa Clara, CA
USA

Xianbin Wang
Department of Electrical and
Computer Engineering
Western University London
ON, Canada

Risto Wichman
Department of Signal Processing
and Acoustics
Aalto University
Espoo, Finland

Chen Xu
School of Electrical and Electronic
Engineering
North China Electric Power University
Beijing, China

Feng Xue
Intel Corporation
Intel Lab
Santa Clara, CA
USA

Shu-ping Yeh
Intel Corporation
Intel Lab
Santa Clara, CA
USA

Zarrar Yousaf
5G Networks R&D Group
NEC Laboratories Europe GmbH
Heidelberg, Germany

Honggang Zhang
College of Information Science and
Electronic Engineering
Zhejiang University
Hangzhou, China

Zhifeng Zhao
College of Information Science and
Electronic Engineering
Zhejiang University
Hangzhou, China

Gan Zheng
Wolfson School of Mechanical
Electrical and Manufacturing
Engineering
Loughborough University
Leicestershire, United Kingdom

Zhenyu Zhou
School of Electrical
and Electronic Engineering
North China Electric
Power University
Beijing, China

List of Abbreviations

2G	Second-generation
3G	Third-generation
3GPP	Third-Generation Partnership Project
4G	Fourth-generation
5G	Fifth-generation
5GMF	Fifth-Generation Mobile Communication Promotion Forum
AAA	Authentication, authorization, accounting
AaaS	Analytics as a Service
AAL	Active-assisted living
ACI	Adjacent channel interference
ADC	Analog-to-digital converter
ADM	Alternative direction method
AF	Access function
AM	Active mode
AMF	Access and mobility management function
AMI	Advanced metering infrastructure
AN	Access network
AN	Active networks
API	Application programming interface
APOLLO	Analytics Platform for Intelligent Operation
AR	Augmented reality
ARP	Address Resolution Protocol
ARP	Allocation retention priority
ARQ	Automatic repeat query
ATM	Asynchronous transfer mode
AUSF	Authentication service

aWESoME	A web service middleware for ambient intelligence
AWGN	Additive white Gaussian noise
AxC	Antenna carrier flows
B2B	Business-to-business
B2C	Business-to-consumer
B2C/B2B	Business-to-consumer/business-to-business
BBU	Base band unit
BER	Bit error rate
BF	Basic frames
BH	BackHaul
BLER	Block error ratio
BLP	Binary linear programming
BP	Basic pursuit
BPM	Business process management
BPSO	Binary particle swarm optimization
BS	Base station
BSS	Business support system
BSS/OSS	Business support systems/operation supporting system
CAPEX	Capital expenditure
CCDF	Complementary cumulative distributive function
CCN	Content-centric networking
CCNF	Common Control Network Functions
CCP	Cloud computing platform
CDF	Cumulative distribution function
CDM	Code-division multiplexing
CDMA	Code-division multiple access
CEI	Customer Experience Index
CEM	Customer experience management
CIR	Channel impulse response
CM	Connectivity management
CMDP	Constrained Markov decision process
CMF	Context management function
cmWave	Centimeter wave
CN	Core network
CO	Central office

CoMP	Coordinated multipoint
COSMOS	Composable, open, scalable, mobile-oriented system
COTS	Commercial off-the-shelf
CP	Cyclic prefix
CP	Control plane
CP/UP	Control plane/user plane
CPRI	Common Public Radio Interface
CQI	Channel quality indicator
CRAN	Cloud radio access network
C-RAN	Centralized RAN
CRP	Chinese restaurant process
CRS	Cell-specific reference signal
CS	Compressive sensing
CS	Coordinated scheduling
CSCC	Common spectrum control channel
CSF	Centralized service functions
CSI	Channel state information
CSI-IM	Channel Station Interference Information Measurement
CSI-RS	Channel state information reference signal
CSMA	Carrier-sense multiple access
CTU	Contention transmission units
CU	Central unit
CUE	Cellular user equipment
D2D	Device-to-device
DA	Distribution automation
DAC	Digital-to-analog converters
DC	Data Center
DC	Dual connectivity
DCA	Dynamic channel assignment
DCF	Distributed coordination function
DCN	Dedicated core networks
DDL–OMP	Dynamic dictionary learning-orthogonal matching pursuit
DEA	Differential evolution algorithm
DEV	Devices
DF	Decision feedback

DFR	Dynamic functional recomposition
DFT	Digital Fourier transform
DL	Downlink
DMRS	Demodulation reference signal
DoD	Diagnostic on device
D-RAN	Distributed RAN
DS-CDMA	Direct sequence-CDMA
DSP	Digital signal processing
DU	Distributed unit
DU	Digital unit
E-UTRAN	Evolved Universal Terrestrial Radio Access Network
E2E	End-to-end
EE	Energy efficiency
EH	Energy harvesting
EM	Element managers
eMBB	Enhanced mobile broadband
EMS	Element management system
ENSD	End-to-end NSD
EO	Evolutionary optimization
ESP	Encapsulated security payload
eV2X	Enhanced vehicle-to-everything
EXIT	Extrinsic information transfer function
FBMC	Filter-bank multicarrier
FCC	Federal Communications Commission
FCOMMA	Fronthaul Enhancement CP/UP Separation, Open Architecture, MEC & M-CORD, Analytics(SON) Agent
FCS	Frame check sequence
FD	Full duplex
FDE	Frequency-domain equalization
FDMA	Frequency-division multiple access
FDS	Frequency-domain spreading
FDV	Frame delay variation
FEC	Forward error correction
FFR	Fractional frequency reuse

FH	FrontHaul
FLR	Frame loss ratio
FM	Flow management
FMSS	Flexible mobile service steering
FN	False negatives
FP	False positives
FPC	Fractional power control
FPGA	Field-programmable gate array
FRAND	Fair, reasonable, and nondiscriminatory
FRER	Frame replication and elimination for reliability
FSM-KW	Fourier series method with Kaiser Window
GAIA	Global access to the Internet for all
Gas	Genetic algorithms
GBR	Guaranteed bit rate
GEO	Geostationary
GFDM	Generalized frequency-division multiplexing
GG	Green grid
GM	Grand master
GNSS	Global Navigation Satellite System
GOCA	Group orthogonal coded access
GPP	General-purpose processors
GPRS	General Packet Radio Service
GPS	Global Positioning System
gRPC	Google Remote Procedure Call
GS	Gale–Shapley
GTP	GPRS Tunneling Protocol
GWCN	Gateway Core Network
HAN	Home area network
HAP	Higher altitude platforms
HARQ	Hybrid automatic retransmit request
HD FDD	Half-Duplex Frequency-Division Duplex
HetNet	Heterogeneous networks
HF	Hyperframes
HR	Human resources
HW/SW	Hardware/software
IA	Impact analysis

IaaS	Infrastructure-as-a-Service
IBFD	In-band full duplex
ICI	Intercarrier interference
ICN	Information-centric networks
ICT	Information and communication technology
IDMA	Interleave-division multiple access
IEC	International Electrotechnical Committee
IET	Interspersing Express Traffic
IETF	Internet Engineering Task Force
IFG	Interframe gap
IGMA	Interleave-grid multiple access
IM	Instantaneous messaging
IML	Individual message level
IntM	Interference mitigation
IoT	Internet of Things
IP	Internet Protocol
IPR	Intellectual Property Right
IRA	Irregular repeat–accumulate
ISD	Intersite distance
ISG	Industry Specification Group
ISI	Intersymbol interference
ISP	Internet Service Provider
ITU	International Telecommunications Union
IX	Internet Exchange
JMPA	Joint message passing algorithm
JP	Joint processing
KA	Keep alive
KASO	Knowledge-aware and service-oriented
KPI	Key Performance Indicator
LAA	Licensed-assisted access
LAP	Low-altitude platform
LCM	Life cycle management
LCRS	Low-code rate spreading
LDM	Layered-division multiplexing
LDP	Label Distribution Protocol
LDPC	Low-density parity check

LDS	Low-density signature
LDS-CDMA	Low-density spreading–code-division multiple access
LDS-SVE	Low-density spreading–signature vector extension
LeaPS	Lean Packet System
LEO	Low Earth orbit
LINP	Logically isolated network partitions
LLR	Log-likelihood ratio
LNA	Low-noise amplifier
LNSD	Local NSD
Log-MPA	Logarithmic-domain message passing algorithm
LoS	Line-of-sight
LPN	Low-power node
LSSA	Low code rate and signature-based shared access
LTE	Long-term evolution
LTE-A	Long-term evolution-advanced
M2M	Machine-to-machine
MAC	Media access control
MANO	Management and orchestration
MAP	Maximum a *posteriori*
MBB	Mobile broadband
MBER	Minimum bit error rate
MBH	Mobile BackHaul
MCC	Mission-critical communication
MCL	Maximum possible coupling loss
M-CORD	Mobile-Central Office rearchitected as a data center
MEC	Mobile edge computing
MECaaS	Mobile edge computing as a service
MEO	Medium Earth orbit
MF	Matched filter
MFH	Mobile FrontHaul
MIM	Mobile instantaneous messaging
MIMO	Multiple-input and multiple-output
ML	Machine learning
ML	Maximum-likelihood
MLE	Maximum-likelihood estimation
MM	Mobility management

MMC	Massive machine communication
MME	Mobility management entity
mMIMO	Massive MIMO
MMSE	Minimum mean-squared error
mMTC	Massive machine-type communications
mmWave	Millimeter wave
MNO	Mobile network operators
MOCN	Multioperator Core Network
MPA	Message passing algorithm
MPLS	Multiprotocol label switching
MPLS-TP	Multiprotocol label switching–transport profile
MRC	Maximum ratio combining
MBS	Microcell BS
MSC	Mobile switching centers
MSPP	Multiservice Provisioning Platform
MTA	Multitenancy application
MTBF	Mean time between failures
MUD	Multiuser detection
MU-MIMO	Multiple-user multiple-input multiple-output
MUSA	Multiuser shared access
MU-SCMA	Multiuser SCMA
MWC	Mobile World Congress
NAS	Network access stratum
NBI	Northbound Interface
NCMA	Nonorthogonal coded multiple access
NDN	Named Data Networking
NETCONF	Network Configuration Protocol
NFV	Network Function Virtualization
NFVO	Network Function Virtualization Orchestrator
NGC	Next-generation core
NGFI	Next-Generation Fronthaul Interface
NGMN	Next-Generation Mobile Networks
NG-OSS	Next-Generation Operations Support System
NLoS	Non-line-of-sight
NMS	Network Management System

NN	Neural networks
NOCA	Nonorthogonal coded access
NOMA	Nonorthogonal multiple access
NR	New radio
NS	Network Services
NSD	Network Services Descriptor
NSGA	Nondominated sorting genetic algorithm
NSIL	Network slice instance layer
NSI	Network slice instances
NSSAI	Network slice selection assistance information
NVS	Network virtualization substrate
OAM	Operations, administration, and maintenance
OBSAI	Open-Base Station Architecture Initiative
OCP	Open Compute Project
OEM	Original equipment manufacturers
OFDM	Orthogonal frequency-division multiplexing
OFDMA	Orthogonal frequency-division multiple access
OLPC	Open-loop power control
OMA	Orthogonal multiple access
OMP	Orthogonal matching pursuit
OOBE	Out-of-band emissions
OPEX	Operating expenditure
OSF	Operation supporting functions
OSS	Operation Supporting System
OTN	Optical Transport Networks
OTT	Over the top
PA	Power amplifier
PAPR	Peak-to-average power ratio
PBB-TE	Provider Backbone Bridge Traffic Engineering
PCF	Policy control
PCRF	Policy and charging rules function
PDCP	Packet Data Convergence Protocol
PDF	Probabilistic density function
PDMA	Pattern-division multiple access
PECP	Path Computation Element Protocol
PEP	Policy enforcement points
PHB	Per-hop behavior

PIC	Parallel interference cancellation
PLE	Path loss exponent
PMI	Precoding matrix indicator
PNC	Piconet controller
PNF	Physical network function
POaaS	Policy as a Service
PoC	Proof of concept
POF	Protocol oblivious forwarding
PON	Passive Optical Network
POTN	Packet Optical Transport Network
PPIC	Partial parallel interference cancellation
PRB	Physical resource block
PRTC	Primary Reference Time Clock
PSO	Particle swarm optimization
PTN	Packet transport networks
PTS	Packet transport system
PW	Pseudowire
QAM	Quadrature amplitude modulation
QCI	QoS class identifier
QoE	Quality of experience
QoS	Quality of service
QPSK	Quadrature phase shift keying
RAM	Random-access memory
RAN	Radio access network
RAT	Radio access terminal
RAT	Radio access technologies
RB	Resource block
RCA	Root cause analysis
RDMA	Repetition division multiple access
RE	Renewable energy
REC	Radio equipment controller
RF	Radio frequency
RFID	Radio frequency identification
RI	Rank indicator
RLNC	Random linear network coding

RM	Risk management
RMSE	Root mean square error
RNL	Radio network layer
ROADM	Reconfigurable optical add drop multiplexing
RoE	Radio over Ethernet
RRC	Radio resource control
RRH	Remote radio head
RRM	Radio resource management
RRU	Remote radio unit
RS	Reference signal
RSMA	Resource spread multiple access
RU	Radio Unit
RWBS	Repeated weighted boosting search
RX	Receiver
SAF	Store-and-forward
SAM	Security and AAA management
SAMA	Successive interference cancellation amenable multiple access
SAN	Storage Area Network
SBI	Southbound interface
SBS	Small cell BS
SCADA	Supervisory control and data acquisition
SCMA	Sparse code multiple access
SD	Sphere detector
SDC	Software-defined computing
SDDC	Software-defined data center
SDK	Software development tool kit
SDMA	Spatial-division multiple access
SDN	Software-defined networking
SDO	Standards developing organizations
SDR	Software-defined radio
SDRA	Software-defined resource allocation
SDRAN	Software-defined radio access network
SDS	Software-defined storage

SDT	Software-defined topology
SE	Spectrum efficiency
SFC	Service function chaining
SGSN	Serving GPRS switching nodes
SIC	Successive interference cancellation
SIC	Self-interference cancellation
SIL	Service instance layer
SIM	Subscriber identity module
SINR	Signal to interference and noise ratio
SISO	Single-input single-output
SLA	Service-level agreements
SM	Sleep mode
SMARTER	New Services and Markets Technology Enablers
SMF	Session management function
SMS	Short message service
SoHAN	Service-oriented architecture for home area network
SON	Self-organized networking
SONAC	Service-oriented network autocreation
SQMO	Slice QoS/QoE MANO
SQMon	Service quality monitoring
SRS	Sounding reference signal
STA	Station
STDFE	Space-time decision feedback equalization
STE	Space-time equalizer
SVM	Support vector machine
SYLPH	Services layer over light physical device
TA	Timing advance
TAE	Time alignment error
TAS	Time-aware shapers
TC	Transparent clock
TC	Traffic class field
TCO	Total cost of ownership
TCP	Transmission control protocol
TDD	Time-division duplex

TDM	Time-division multiplexing
TDMA	Time-division multiple access
Telco	Telecommunication Company
TETRA	Terrestrial Trunked Radio
TG	Traditional grid
TH	Turbo Hadamard
TiaaS	Telco Infrastructure as a Service
TIP	Telecom Infrastructure Project
TN	True negatives
TNL	Transport network layer
TP	True positives
T-PANI	T-Packet Analysis and Network Intelligence
T-SDN	Transport SDN
TSN	Time-sensitive networking
TTI	Transmission time interval
TTM	Time to market
TVWS	TV band white space
TX	Transmitter
UA	Universal access
UAV	Unmanned areal vehicles
UCA	Uniform cylindrical array
UCTN	Unified and converged transport network
UDM	Unified data management
UHDTV	Ultrahigh-definition television
UI	User interface
UICC	Universal integrated circuit card
UL	Uplink
ULPC	Uplink power control
Unified-O	Unified orchestration
UP	User plane
URC	Unity rate code
URLLC	Ultrareliable and low-latency communications
USIM	Universal Subscriber Identity Module
UWB	ultra-wideband
UX	User experience

V2X	Vehicular-to-anything communication
vCore	Virtualized core
vEPC	Virtualized evolved packet core
VI	Virtual infrastructures
vIDS	Virtualized intrusion detection system
VIM	Virtual infrastructure manager
vIPS	Virtualized intrusion prevention system
VM	Virtual machine
VNE	Virtual network embedding
VNF	Virtual network function
VNFC	Virtualized network function domain
VNFD	Virtualized network function descriptor
VNF-FG	VNF forwarding graphs
VNFM	VNF manager
VNFM	Virtualized network function manager
VR	Virtual reality
V-RAN	Virtualized radio access network
VTN	Virtual tenant network
WCDMA	Wideband code-division multiple access
WDM	Wavelength-division multiplexing
WER	Word error rate
WiLD	Wi-Fi over long distance
WLAN	Wireless local area networks
WPAN	Wireless personal area networks
WRC	World Radio Conference
XaaS	Anything as a Service

Introduction

Anwer Al-Dulaimi,[1] Xianbin Wang,[2] and Chih-Lin I[3]

[1] R&D Department, EXFO Inc., Toronto, Canada
[2] Department of Electrical and Computer Engineering, Western University London, ON, Canada
[3] Wireless Technologies, China Mobile Research Institute, China Mobile Communications Corporation, Beijing, China

The fifth generation (5G) mobile network is a new generation of wireless systems that is intended to connect users faster and more reliable than any previous generation. The industry is expecting 5G to support 1000-fold gains in capacity to meet consumer demand driven by ultra high definition (UHD) videos and cloud-based applications. The 5G will deliver the underlaying architecture that connects all machinery and human-type users with large bandwidth considering the type of requested service. In addition, 5G needs to support ultra-low latencies for new use cases, such as the virtual reality in vehicle-type communications. The enabling techniques for such services include millimeter waves, network identification with small cells, massive MIMO, beam-forming, and full duplex. The concept of spectrum extensions also motivated new areas of research to extend the cellular spectrum to unlicensed band above 6 GHz. Moreover, the rapid changes in radio access network can also be seen in 5G core network (5GC) by introducing service-based interfaces (SBI) for slicing end-to-end architecture. It is well understood that many other variations will be implemented at different layers and network segments for fully adaptive network that can steer traffic to corresponding users or processing entities. In reality, 5G is a new suite of technology that continue to evolve with very specific requirements considering capacity and performance. Therefore, it is crucial to define the trends for the 5G evolution and how to abstract related directions for research. In this book, we divide the 5G challenges into five parts to investigate: physical layer for 5G radio interface technologies, radio access technology for 5G networks, 5G network

5G Networks: Fundamental Requirements, Enabling Technologies, and Operations Management, First Edition.
Anwer Al-Dulaimi, Xianbin Wang, and Chih-Lin I.

interworking and core network advancements, vertical 5G applications, and R&D and 5G standardization. We further detail the technical challenges and given solutions in this chapter.

I.1 Motivation and Directions

I.1.1 Changes in Mobile Market and Trends of Future Services

The tremendous increase in connected users and downloaded applications is already driving mobile operators to the border edges considering supporting technology and available radio resources. Therefore, mobile operators are adopting new radio access technologies that can provide service anywhere and through any interface of connectivity. The 5G networks will employ massive cloud storage to facilitate their computational operations. The business model for any type of connectivity will be deployed in the form of orchestrated services such as cloud central utility management and control for machinery. These kinds of service had never been available in previous generations causing a significant increase in the number of end-users. Therefore, the challenge is not only to meet the load demand and users expectations but also to devise innovative technologies provided with radio resources. Considering the ongoing development plans and technology evolution, we identify the following directions as the most significant trends in 5G domain:

1) The UHD and super high definition (SHD) videos, tactile Internet, and rich social network services are some examples of emerging services that require additional spectrum resources. Therefore, one of the main challenges for 5G is the spectrum extension using multi-radio interfaces that incorporate multibands in any single transmission. A 5G base station (BS) may incorporate Wi-Fi, LTE, and millimeter-wave radio access interfaces. The end-users will be connected to multi-radio interfaces that may transmit concurrently different portions of the same data. Also, there are new technologies that will improve network accessibility to spectrum such as massive MIMO antennas, NOMA, and so on. It is well understood that no single enabling technologies will be able to deal with the spectrum challenge, but it is end-to-end solutions that slice the network architecture end-to-end (E2E).

2) The recent advances in computer designs and cloud-based operating systems allowed mobile operators to start restructuring their networks using network function virtualization (NFV) and software-defined networking (SDN) technologies. This transition will provide 5G the ability to drive computational resources towards high traffic demands. However, this indicates that current route selection procedures and network performance statistics may not be applicable in the next generation due to dynamic route changes. Therefore,

interoperability in 5G means a network that can talk and exchange information with any user (human or machine) without being limited by the medium of communications or data path. Nevertheless, most of data will be stored in the cloud that raise a major challenge of having data centers and packets content-aware machines installed within the network backbone to balance the load and avoid network failure or service interruption.

3) Wireless connectivity will be the main port for communications between any human-to-human (H2H), human-to-machine (H2M), machine-to-human (M2H), machine-to-machine (M2M), and vehicle-to-X (V2X). This unprecedented number of connected users and appliances requires developing new adaptive network that monitors users status, identify resources available, and launch the catalog of relevant service. These different technologies provide the necessary framework for the Internet of things (IoT).

There is also a need to look at the evolution of this technology and how they are defined by the standardization bodies. We have all these standards coming separately from 3GPP, Wi-Fi alliance, IEEE, ETSI, and so on, but 5G network will integrate almost all known communication technology standards. Therefore, it is the time to ask if we will really need a new forum that gather all specialists and researchers to shape the road for the future. In this book, we will not just study the 5G enabling technologies but also investigate the relevant vertical applications and ongoing standardization activities.

I.2   5G Requirements

The 5G requirements reflect the end-user requirements and the upper band availability in both spectrum and technology. Considering user experience, the 5G will offer new user experience packages targeting user requirements rather than guaranteeing the quality of service for available services. Subsequently, 5G will employ a variety of radio interface technologies operating at different spectrum bands to allow ubiquitous connectivity for both humans and appliances. The trend is to design a new network that can accommodate the growth in device and applications for years to come. To achieve such goals, 5G needs to provide the following.

I.2.1   New Spectrum Bands

The current cellular networks are consuming all assigned channels and have exhausted their licensed resources. Moreover the dynamic adaption between channels within the licensed band is not generating any additional radio resources. Therefore, meeting the growing demand can happen only by extending the spectrum accessibility for mobile operators. This triggered investigations over the significant free spectrum available in the unlicensed band (specially the 5 GHz). However, that raised concerns for the Wi-Fi alliance because of uncertainty on how much that will impact the Wi-Fi. There is also more spectrum available in spectrum bands in the (6-100) gigahertz range. These additional band extensions are extremely necessary for the 5G to be able to meet the demand for high capacity and data rates. Supporting unlicensed bands will require significant changes to network at all layers to provide the backward compatibility between enabling technologies.

I.2.2 New Radio Access Technologies for Ultradense Deployments

The continuous developments in the massive MIMO and beamforming techniques allow accessing the millimeter wave spectrum to support ultradense high-capacity scenarios. To this end, base stations should be able to divert transmission patterns toward users in such a sharp beam that is almost point-to-point link. This requires new radio interfaces that can take advantage of ultradense deployed antennas surrounding an end user. The new RATs should be able to interact and communicate with multiple MIMO units in/around its coverage to have the suitable sharp pensile beam. These flexible air interfaces should support different characteristics inherent in large frequency ranges. Provided by carrier aggregation, the advantage of such new RATs is that it provides contiguous carrier bandwidths of 1-2 GHz.

I.2.3 Physical Layer Designs

There is a need to optimize the bandwidth and power in the current systems to meet the new vehicular to everything (V2X) communications and also to provide the necessary framework for future applications such as the tactile Internet. These new technologies combined with industry automation and transport systems to health care, education, and gaming demand require a minimal latency. Therefore, achieving the targeted 5G radio latency value of less than 1 ms will require extremely small radio subframe lengths. This means that future 5G will employ a flexible dynamic time division duplex (TDD) model that can adjust the frame size according to the requested service. Such an optimization will require significant changes to the LTE uplink/downlink configuration scheme, resource allocations, control signaling and synchronization intervals, sparse code, and transmitted power.

I.2.4 Architectural and Cloud Integration

The ultradense deployment of primary and small cells as planned in 5G will increase the complexity of data processing and operations management from an operator perspective. However, the virtualization of various network components will allow operators not just to increase operations efficiency using cloud-managed operations but also to process intrasite operations using mobile edge computing. Network cloudification spans deep into the 5G core network (5GC) design that operates on top of a virtualization software layer without any dependency on the underlaying hardware. The software-defined networking will be the other cloud component that facilitates network adaptations and traffic forwarding.

These technologies evolve 5G network architecture, information, and signaling schemes and will be investigated and analyzed in this book.

I.3   ORGANIZATION OF THE TEXT

To address different challenges and relevant solutions that support the development of 5G network from concept to standard, this book is organized into five parts: Physical Layer for 5G Radio Interface Technologies, Radio Access Technology for 5G Networks, 5G Network Interworking and Core Network Advancements, Vertical 5G Applications, and R&D and 5G Standardization. Each part has many chapters that provide the readers with enough detailed knowledge and solutions on studied topics.

Part I: Physical Layer for 5G Radio Interface Technologies

This section has seven chapters that investigate topics related to physical layer. In Chapter 1 "Emerging Technologies in Software, Hardware and Management Aspects towards the 5G Era: Trends and Challenges," the authors provide a review on the status and challenges for hardware and software development for 5G wireless communications. Although the authors extend their survey to media access control (MAC) layer, radio resource management (RRM), and machine learning, the chapter is more suitable for readers who are interested in obtaining basic understanding of 5G physical layer and advanced waveform technologies combined with coding and modulation algorithms. In Chapter 2 "Waveform Design for 5G and Beyond," the authors dive more into physical layer and provide a comprehensive review of potential 5G waveform technologies. Their chapter provides deep analysis on multicarrier schemes, subcarrier wise filtering, and subband-wise filtered multicarrier modulation (MCM) techniques. The variation in performance between such waveforms confirms that slicing the spectrum locally allows asynchronous transmission across adjacent subbands and improves coexistence with other waveforms. The results confirm that employing guard bands using filter bank multicarrier (FBMC) is one of the options that highlights the merits of efficient spectrum utilization for fifth generation (5G) communication systems. Chapter 3 "Full-Duplex System Design for 5G Access" provides a downlink-uplink scheduler design to mitigate intracell UE-to-UE interference. The authors use low-complexity open-loop UL power control and BS beamforming nulling for various deployment scenarios. In this way, authors propose two static/semistatic interference mitigation (IM) methods to control BS-to-BS interference: The first approach employs beam nulling to suppress the line-of-sight component of BS-to-BS interference. The second scheme, UL power control, is used to improve signal strength to combat BS-to-BS interference. The given analysis shows that a combination of the two proposed schemes can improve uplink data rate and achieve 1.9 throughput gain compared with half-duplex operation (HD-FDD) in LTE. In Chapter 4 "Non-Orthogonal Multiple Access for 5G," the authors provide a comprehensive survey on non-orthogonal multiple access (NOMA) technology that enables a new dimension in which signals can be separated and given access to a base station. The basic concepts of NOMA are investigated along with the selection criterion for choosing between NOMA and orthogonal multiple access (OMA), considering information theory. NOMA is categorized into two: power-domain and code-domain NOMA. The key features of both schemes are addressed by focusing on spectral efficiency, system performance, receiver complexity, and so on. It is expected that NOMA will be widely used in future 5G wireless communication systems to enable massive connectivity and low latency. Therefore, future directions in NOMA research are also summarized at the end of the chapter to lay down the ground for researchers in this area. Chapter 5 "Code Design for Multiuser MIMO" proposes a new coding scheme for the future 5G multiuser MIMO systems. The authors provide a low-complexity coding scheme, called multiuser repetition-aided irregular repeat-accumulate (IRA) code to enhance system reliability for multiuser MIMO. Compared with conventional coding scheme, this new scheme is using repetitions of each user's code word along with parity checks to reduce system complexity. The control of the fraction of repetitions in the code word provided with a degree distribution optimization of the IRA encoder allows constructing a low-rate code with near capacity performance. This will significantly increase the number of users and concurrent transmissions supporting more repetitions with very low encoding and decoding complexities.

In Chapter 6 "Physical Layer Approaches to Security in 5G Systems," the authors review the security challenge for 5G networks. The 5G supports connectivity to a large number of diverse industries beyond cellular customers especially with Internet of Things (IoT). There is another dimension for this topic considering new computational platforms such as cloud core, visualization, and software-defined networking that scales security challenge in 5G. This chapter reviews all these challenges opposed to 4G challenges. Authors admit the need to investigate more methods to secure 5G communications and models used to authenticate users to guarantee privacy in 5G network operation. The chapter also provides some analysis for several full-duplex-based physical layer security techniques that exhibit better secrecy performance compared to the conventional half-duplex modes. Chapter 7 "Codebook-Based Beamforming Protocols for 5G Millimeter Wave Communications" reviews beamforming protocols proposed for millimeter wave (mmWave) systems and the associated challenges. The mmWave communications are widely accepted as additional radio interface in 5G to support high-throughput in Gbps orders. However, the high path loss in mmWave bands is the barrier for a successful usage of this technology. The authors investigate antenna beamforming and relevant evolution and advancements for IEEE 802.15.3c and IEEE 802.11ad antennas. The authors expand the scope of discussion by including research challenges in mmWave beamforming systems.

Part II: Radio Access Technology for 5G Networks

This section explores the 5G technologies at RAN and above layers. In Chapter 8 "Universal Access in 5G Networks: Potential Challenges and Opportunities in Urban and Rural Environments," the authors review the potential RAN technologies to support 5G requirements in urban or the rural environments. There are many system-level architectures that can divert and control traffic at local and large network domains such as software-defined networking. Moreover, 5G is a combination of various techniques such as device-to-device (D2D) communications, MIMO, mmWave communications, full-duplex transmissions, and Internet of Things. Therefore, network designers need to develop a roadmap for deploying such technologies considering density of users at various network sites. This chapter helps develop a vision of RAN-wise technology for 5G and improves the understanding of enabling technologies reviewed throughout this work.

In Chapter 9 "Network Slicing for 5G Networks," the authors study network slicing for efficient local resources access to improve network capacity. The RAN is assumed to employ different self-contained network technologies that concurrently operate under same governing entities. To this end, the network slicing is triggered not only by spectrum or technology availability but also by application and network site. An efficient resource management can be achieved by using cloud-based network that can orchestrate network operations and instantiate various virtual network functions (VNFs) to process arrival requests. The efficient coordination of network slices is achieved by separating control and data planes to support higher network adaptations using software defined networking controller. The authors show how different service catalogs are managed by management and organization (MANO) entity that is the highest entity of virtualization platforms. Chapter 10 "The Evolution Toward Ethernet-Based Converged 5G RAN" provides a comprehensive study of underlaying infrastructure that interfaces various network components focusing on midhaul and fronthaul. The authors start by providing a comprehensive review of current LTE technology and how to evolve RAN architecture toward

5G. The study reviews the ongoing 3GPP work to define 5G fronthaul interfaces and interworking between cloud-based network layers. The Ethernet tools for time-sensitive networking are also studied with the Ethernet frame preemption. The chapter investigates the use cases of various Ethernet standards and their potential features and also deployment scenarios to interface 5G network components. In Chapter 11 "Energy-Efficient 5G Networks Using Joint Energy Harvesting and Scheduling," the authors study green 5G networks where all base stations can harvest energy from renewable energy sources (e.g., solar). The BS power consumption constraints are given as of binary BS sleeping status and user-cell association variables to formulate a binary linear programming problem. The scheme needs a prior knowledge of RF status to evaluate the required energy considering incoming traffic. Therefore, the study analyzes two cases: First, the BS has zero knowledge on future RE statistics; second, the BS has a perfect knowledge of the network statistics. The chapter includes some numerical results of the proposed system performance and future directions of research.

Part III: 5G Network Interworking and Core Network Advancements

This section explores the new changes at the core network and necessary changes to support interworking with other technologies. In Chapter 12 "Characterizing and Learning the Mobile Data Traffic in Heterogeneous Cellular Network," the authors study a learning mechanism to increase the network awareness of traffic fluctuations. The 5G is anticipated to process a significant amount of traffic with increasing numbers of connected users and downloaded applications. Therefore, traffic predictability is a key feature to develop a learning framework that exploits traffic characteristics. The proposed mechanism is analyzed and characterized for different service types (e.g., mobile service messaging, web browsing, video) for a selected BS and variable number of packets. This chapter will help initiate the discussion on creating a unified learning framework that predicts the future traffic loads for large-sized 5G networks with accuracy and no dependencies on past knowledge. Chapter 13 "Network Softwarization View of 5G Networks" studies the softwarization of the core network and the implementation of NFV concepts to improve backbone performance in handling end-to-end communications. A softwarized network allows more flexibility in managing life cycle of various functions and services by instantiating the necessary virtual entities upon need. This model of networking is redesign of the current network architecture for more flexible and agile operations management. The chapter also proposes the network slicing solution in a new framework that extends from logically isolated network partitions (LINP) to sliced software entities. The SDN technology provides the mechanism to compose and manage slices created on top of the infrastructure. This combination of network softwarization model and network slicing facilitate harmonization of enabling technologies for 5G networks at both core and RAN sides. Finally, the content of this work reflects the 5G roadmap in Japan as defined by the Fifth Generation Mobile Communications Promotion Forum (5GMF). In Chapter 14 "Machine-Type Communication in the 5G Era: Massive and Ultra-Reliable Connectivity Forces of Evolution, Revolution, and Complementarity," the authors review machine-type communications in 5G era. The supervisory control and data acquisition are key factors to achieve efficient connectivity and control in machine communications. Therefore, integrating the supervisory control and data acquisition (SCADA) system with 5G network is a critical challenge to support monitoring and remote control of connected machines. Typically, this requires rethinking about current industrial communications and applying the necessary changes for 5G considering ultrareliable low-latency communications (URLLC) metrics.

Part IV: Vertical 5G Applications

This section explores some of vertical applications of future 5G network. In Chapter 15 "Social Aware Content Delivery in Device to Device Underlay Networks," the authors review the content delivery using joint peer discovery and resource allocation targeting to maximize the system sum rate with guaranteed QoS of both cellular and D2D links. The focus of this chapter is social networking and solving the joint optimization problem of users selecting similar contents and spectrum resources. This chapter considers matching between D2D transmitters and receivers, and the matching between D2D pairs and resource blocks (RBs). Therefore, the authors propose a three-dimensional matching process to coordinate allocation of users, contents, and spectrum resources, considering social and physical layers information. The given results validate the novelty of the new iterative matching algorithm compared with random matching. Chapter 16 "Service-Oriented Architecture for IoT Home Area Network in 5G" investigates the massive number of IoT devices and subsequent generated traffic causing high latency, low energy efficiency, poor interoperability between various network nodes, and high operational cost. To comply with 5G network features, the authors propose service-oriented network architecture for IoT devices operating within a Home Area Network (HAN). This architecture incorporates a middleware layer that enables efficient utilization of node resources, supports interoperability between heterogeneous nodes within the HAN, and facilitates development of new applications. The given results show a significant improvement in performance and could allow seamless integration of legacy IoT devices with the coming 5G network. In Chapter 17 "Provisioning Unlicensed LAA Interface for Smart Grid Applications," the authors study the problem of high data demand and control signaling for power utilities that exceed the available resources of current smart grid networks. The 5G allows building new communication architectures that employ more wireless access with sliced UDN and extended spectrum bands. In this work, the authors propose to transfer utility messages using cellular communications through Licensed Assisted Access (LAA) technology. The LAA unlicensed radio interface is used to replace the aggregators at consumer sites and retransmit the obtained data to control center using licensed radio interface over LTE band. To solve the bottleneck of Wi-Fi unlicensed band access, a new spectrum utilization algorithm is proposed to increase the reliability of data interception in the smart grid communications. The interoperability is another key feature in LAA that allows transferring the data between licensed and unlicensed bands.

Part V: R&D and 5G Standardization

This section covers the ongoing industrial research and development and standardization activities in 5G areas.

In Chapter 18 "5G Communication System: A Network Operator Perspective," the authors review the state of the art of potential 5G systems and how they all contribute to highly dynamic topology changes, replacement of the end-to-end model by real mesh topologies, massive number of IoT devices, and a paradigm shift from ubiquitous content communication to control and steering communication. The focus of this chapter is network softwarization techniques such as SDN, NFV technologies, and the use of network coding as a service for mobile edge computing and distributed edge caching. The chapter describes holistic testbeds of the Deutsche Telekom and the 5G Lab Germany as a fundamental step toward creating an experimental model to emulate 5G communication systems. The chapter also provides more details about the compatibility of those testbeds with other industry proof-of-concept (PoC) testbeds to allow readers a better underrating on the current scope of 5G implementation. Chapter 19 "Toward All-IT 5G End-to-End Infrastructure" explains the ongoing work of South Korean wireless telecommunications operator (SK Telecom) to transform itself into a platform service company at the dawn of 5G age. The advances in SDN/NFV-based virtualization technology and cloud computing have triggered many innovative changes in SK architecture. The order of these changes is "Scalable, Cognitive, Automated, Lean, and End-to-End." SK Telecom use ATSCALE architecture with its key functions and implementation technologies to drive structural innovation detected at a Software-Defined Data Center (SDDC) using a SDx-based technology. The chapter also explains the POC testbed that incorporates Software-Defined RAN (SDRAN) with virtualized core components for end-to-end testing. In the last chapter, Standardization: The Road to 5G," the authors provide an overview of ongoing 5G standardization efforts by various standardization bodies. The authors are from RAN technical standards in Telecom Italia Mobile (TIM). The chapter reflects on the European Union vision that 5G should support a fully connected society by integrating all humans and machines within its communication infrastructure. The authors describe the standardization process phases as vision, technical specifications, and policy and profiling. The International Telecommunication Union-Radiocommunication Sector (ITU-R) and 3GPP are the main players in standard development and defining the technology for the new 5G network.

I.4 Summary

This book combines all aspects of 5G research from world experts into a single text that provides readers with all necessary information from theoretical concepts to industry testbeds. The editors believe that this book will stimulate new ideas between research community members to develop 5G communications. The challenges imposed by this new digitized world require us to develop a network that can bring humans and machines together as one pool of users for service processing. The 5G is just a new story in the history of humanity, but it will not be the last. The effort to improve life quality and preserve the world environment are the messages that we carry as engineers and we should all work to achieve these noble goals.

Part I

Physical Layer for 5G Radio Interface Technologies

1

Emerging Technologies in Software, Hardware, and Management Aspects Toward the 5G Era: Trends and Challenges

Ioannis-Prodromos Belikaidis,[1] Andreas Georgakopoulos,[1] Evangelos Kosmatos,[1] Stavroula Vassaki,[1] Orestis-Andreas Liakopoulos,[1] Vassilis Foteinos,[1] Panagiotis Vlacheas,[1] and Panagiotis Demestichas[1,2]

[1] R&D Department, WINGS ICT Solutions, Athens, Greece
[2] Department of Digital Systems, University of Piraeus, Piraeus, Greece

1.1 Introduction

As the number of smartphones and demand for higher data-rate connections keep explosively growing, the technology has to pursue this trend in order to be able to provide the suitable communication schemes. By 2020, it is calculated that the total global mobile wireless user devices will be over 10 billion, with the mobile data traffic growing more than 200 times over compared to 2010 numbers. It is foreseeable that 4G mobile communication system could no longer meet the need of users service requirements. Thus, the new 5G communication systems are promising a complete network structure with unlimited access to information, providing the requested service demands to users far beyond what the current 4G offers by supporting innovative new wireless technologies and network architecture to meet the extremely high-performance requirement (Table 1.1).

These features include support for new types and massive number of devices, for very high mobile traffic volumes, universal access for users, very high frequency reuse and spectrum reuse in wireless technologies, automated provisioning, configuration and management of a wide range of new network services, ultrareliable, ultralow latency, ultradensification, and even more. 5G networks would be a heterogeneous networks (HetNets), meaning that different networks will be integrated all together to a unified system, enabling

5G Networks: Fundamental Requirements, Enabling Technologies, and Operations Management, First Edition.
Anwer Al-Dulaimi, Xianbin Wang, and Chih-Lin I.

Table 1.1 Main comparison between 5G and previous generations.

5G	4G and Earlier
New waveforms based on filter banks and other novel technologies	Waveforms based mainly on OFDM and variations
Gbps performance	Up to hundreds Mbps
End-to-end latency of some milliseconds	End-to-end latency of hundreds of milliseconds
Support of massive MIMO	SISO and limited MIMO technologies
Support of mm-wave bands up to hundreds GHz	Operation mainly below 6 GHz bands
Efficient support of massive number of devices in ultradense environments	Support of limited number of devices in dense/congested areas

aggregation of multiple existing radio access technologies (RATs) such as LTE-A, Wi-Fi, D2D, and even lightly licensed. Delivering all 5G requirements in order to support the new features and services, a substantial change on the network architecture is inevitable.

Normally, in the past such a process would need deployment of specialized devices build for a specific application and with fixed functionalities. Thus, any development and transformation to follow the constantly increasing and heterogeneous market requirements demands a huge investment to change/deploy hardware. Nowadays, various technologies and architectures have been utilized in order to solve this problem and provide a faster introduction and adaptation of new technologies to the communications systems. One of these elements that offers reprogrammability of the network elements in order to solve new problems or to establish new more suitable functions is software defined networking (SDN). This architecture provides dynamic, manageable, cost-effective, and adaptable solutions, making it ideal for the high-bandwidth, dynamic nature of today's applications. SDN decouples the dependence of implementing instructions that are provided by multiple, vendor-specific devices and protocols, making it bound to an exact hardware. Such aspects are also related to the flexible deployment of functionality to hardware, software, or mixed. As Section 1.3 proposes, hardware has better execution performance, but software offers greater flexibility.

Networking, computing, and storage resources would be integrated into one programmable, unified, and flexible infrastructure that will be more cost-effective and with higher scalability at minimum cost. This unification will allow for an optimized and more dynamic utilization of various distributed resources, and the pooling of fixed, mobile, and broadcast services.

Also, 5G will be designed to be a viable, robust, and scalable technology. Another positive result with the introduction of 5G communication systems will be the drastic energy consumption reduction and energy harvesting that will help the industry to have an astounding usage growth. Since network services will rely progressively on software, the creation and growth will be further encouraged. In addition, the 5G infrastructures will provide network solutions and involve vertical markets such as automotive, energy, food and agriculture, city and buildings management, government, health care, manufacturing, and public transportation.

The rest of the chapter is structured as follows: Section 1.2 elaborates on the main 5G requirements, while Section 1.3 presents the status and challenges in hardware and software development. Section 1.4 elaborates on the status and challenges in 5G wireless communications by focusing on physical layer, MAC and RRM. Finally, Section 1.5 investigates the benefits of machine learning in 5G network management.

1.2 5G Requirements and Technology Trends

In the digital era, users and devices are becoming more dependent on various applications and services that involve the creation, access/ communication, processing, and storage of digital content. These developments have been tremendously accelerated by wireless/mobile technologies, which have offered unparalleled access/ communication opportunities to users. 5G is expected to be dominated mainly by the following application classes, including massive machine-type communications (mMTC), enhanced mobile broadband (eMBB), and ultrareliable and low latency communications (URLLC). These main classes will facilitate scenarios related to critical and demanding applications for the realization of smart cities, as well as the realization of applications for Industry 4.0 and automation aspects. Also, self-driving aspects are expected to pose strict requirements especially on the latency in order to ensure reliable and secure service with very high rate of availability. Figure 1.1 illustrates the main application areas of 5G, as mentioned before.

Also, 5G has to support tight quality-related requirements in order to provide enhanced user experience in more demanding environments compared to previous generations. As mentioned also in Reference [1], such requirements include the following:

- Support of very high bitrates of more than 1 Gbps in heterogeneous environments.
- Support of very high number of devices (massive machine communications) especially in ultradense environments.
- Support of very high mobility (e.g., up to 500km/h).
- Support of aircraft communications.

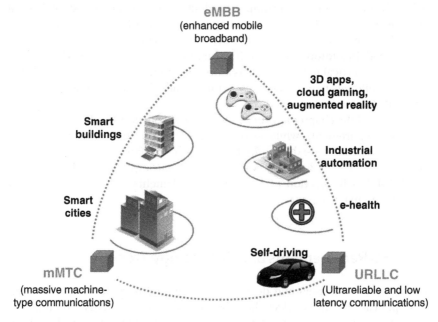

Figure 1.1 5G main application areas

- Support of mission critical communications with very low latency (down to 1ms).

For the successful realization of the aforementioned application areas, there are certain technological and networking trends that can lead toward the 5G direction. In terms of technology, trends are as follows:

- *Network Slicing:* Dynamic network slicing in 5G enables the design, deployment, customization, and optimization of different network slices running on a common network infrastructure. It leverages innovations in cloud mobile access and core [2].
- *Cloud and Fog/Mobile Edge Computing:* Cloud aspects can involve the allocation of resources to physical components in content servers, activation of the appropriate volume/type of functional components, and the determination of the interconnections and links between the physical elements. In this respect, they will manage in an aggregate manner all types of resources, namely, communication, computing, and storage, by taking advantage of significant processing powers/storage facilities associated with cloud platforms and virtualization through abstractions of resources and service components (which are pooled and universally accessible/sharable), in an on-demand, elastic, and scalable manner.

- *Separation of Control and User Plane:* Separation on control and user plane can potentially lead to more efficient usage of resources and energy efficiency as well. For example, the Green Touch initiative proposed that user plane (data) can be served mainly by small cells while a limited number of macro cells can serve as a signaling umbrella in order to handle control plane aspects.
- *Virtualization of Networking Functions:* Virtualization may include solutions based on software-defined networking (SDN) and network function virtualization (NFV) principles, which may be used for accelerating the application/service deployment times and flexibility (in general) in the network. For instance, through the standardized interfaces of the SDN model there can be instructions on how to handle new applications/services. Likewise, through NFV there can be an easier implementation of applications/services and networking intelligence (activation in cloud and instructions toward forwarding elements). The overall challenge is to evaluate the potential of these concepts, in terms of impact in the application/service deployment times, QoS/QoE.

In terms of wireless network trends there are the following (Figure 1.2):

- *Massive MIMO and Utilization of mmWave:* An important direction is related to the usage of massive multiple input–multiple output (MIMO) that can significantly enhance the spectral and energy efficiency of the wireless network. Moreover, the extra capacity that is needed by 5G networks for facilitating massive IoT, mobile broadband communications, and so on. can be provided through the utilization of extra spectrum by exploiting bands above 6 GHz. Mm-wave frequencies can be used for outdoor point-to-point backhaul links or for supporting indoor high-speed wireless applications (e.g., high-resolution multimedia streaming) [3]. Moreover, as the millimeter waves have a short wavelength, it becomes possible to pack a large number of antenna elements into a small area, which consequently helps realize massive MIMO at both the base stations and user devices [3].
- *Novel Multiple Access Schemes:* Novel multiple access schemes, such as nonorthogonal multiple access (NOMA), is one of the techniques being considered that uses cancellation techniques in order to remove the more powerful signal. Of course, well-known techniques such as orthogonal frequency division multiple access (OFDMA) can be exploited as well.
- *Ultradense Infrastructures:* Another direction that will continue to progress is the constant decrease in the cell size, at the expense of a corresponding constant increase in the number of cells that will be deployed. Cells of different sizes, characterized as macrocells, microcells, picocells, or femtocells will continue to be deployed. Specifically, hundreds of small cells are deployed per macrocell. The challenge with ultradense networks is to deploy and operate the appropriate set of cells, so as to carry data traffic, without severely increasing the signaling traffic (increases with the number of cells),

by minimizing the impact of mobility and radio conditions, and by achieving cost and energy efficiency.

- *New Waveform and Advanced Coding:* New candidate waveforms for 5G would be needed in order to serve specific service requirements, for example, higher or lower bandwidths, sporadic traffic, ultralow latency, and higher data rates compared to legacy technologies. Also, robustness toward distortion effects such as interference, RF impairments, and so on is important to be supported by new waveforms.

1.3 Status and Challenges in Hardware and Software Development

The increasing network demands require new network services, higher performance, increased bandwidth, lower energy consumption, and increased resilience. These demands imply a higher number of network devices and stations, increasing the cost and the energy consumption. The network centralization and functions virtualization become more significant as they enable better distribution of the available resources, less hardware utilization, and an easier to upgrade network as current devices and architectures are meeting their limits. New implementation techniques must be introduced in order to further increase reusability, flexibility along with performance and energy consumption at the same time. According to this approach, the network functions can be moved to software as much as possible, without affecting networks latency. The functions virtualization can be achieved at any level, using a partitioning technique between software and hardware functions, which takes into account the available resources. The current network systems introduce a static and customized functions virtualization. This manual and static partitioning may lead to high performance but it is not reusable and not reconfigurable, thus limiting

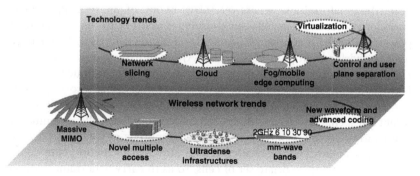

Figure 1.2 5G technology and wireless network trends.

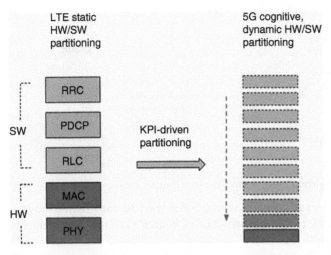

Figure 1.3 Cognitive, dynamic HW/SW partitioning in network stack layers.

network upgrades and resource allocation. Also, the processing power cannot be shared among nodes offering limited efficiency and spectrum capacity. Full virtualization is not always available as the devices of the underlying network might not be able of such a task, or virtualization might be limited according to available physical and computational resources. A cognitive and dynamic HW/SW partitioning can provide reconfigurable and flexible HW/SW partitioning to both device and network element architectures in 5G technologies, considering high performance and energy consumption reduction, according to the specified performance scenarios. The HW/SW partitioning is applicable for either inside a network stack layer and/or between multiple network stack layers, as shown in Figure 1.3.

1.3.1 Problem Statement

The cognitive, dynamic partitioning takes into account a set of given network functions, the KPIs that have to be optimized and the KPI constraints relative to the available resources. The technique result provides the HW or SW implementation decision for each given function, according to the given policies. These policies consist of the KPIs and their constraints. The result of the implementation has to change according to the policies alteration. The cognitive dynamic HW/SW partitioning task is to provide the best HW/SW partitioning of the 5G network stack functions, considering the given KPIs per scenario and the available HW/SW resources. The partitioning algorithm's result can be parsed to the management programs for further decision-making on the implementation part. The partitioning solution has to communicate with other programs to be aware of the available resources. Moreover, the partitioning

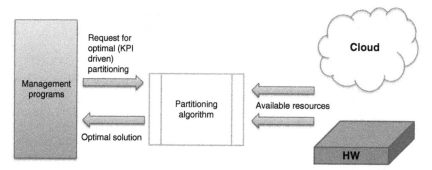

Figure 1.4 Partitioning algorithm functionality and communication.

must be aware of the performance scenario and the KPI constraints; thus the partitioning would interact with management programs and monitoring agents, as seen in Figure 1.4.

1.3.2 Solution

The cognitive dynamic HW/SW partitioning algorithm considers a multiple and diverse set of objectives, relative to 5G requirements. To this purpose, the current solution contains an evolutionary multiobjective algorithm, aiming to solve a specified optimization problem. This implies that the optimization problem formulation is one of the most significant parts of the solution, as it has to accurately address the requirements, the KPIs, and the objectives of a 5G network. The multiobjective algorithm receives as input the functional graph of the defined functions, along with the systems and functions KPIs/Constraints and optimization goals. Then the number of possible solutions is populated according to the optimization problem formulation. After having the set of solutions, a multiobjective algorithm will search for the best solution, according to the given KPIs, constraints, and optimization goals.

1.3.2.1 Functions Definition (LTE, 3GPP-Based PHY Functions)

The cognitive dynamic HW/SW partitioning algorithm is able to decide on device and network element functions, either inside a network stack layer and/or between multiple layers. Most of the current research and implementations are using functions from the LTE MAC layer or its reconfiguration while trying to introduce softwarization and reconfiguration of the LTE PHY layer functions. Some implementations are also able to simulate full LTE networks. The most notable among them are the LENA ns-3 [4] LTE simulator from CTTC, which provides full software implementation of virtual LTE networks, the OpenAirInterface [5], OpenLTE [6], and srsLTE [7]. In order to provide a first realization of the HW/SW partitioning challenges, the selected subset of functions used for the evaluation of the HW/SW partitioning is derived from the OpenLTE physical layer software implementation in Octave source code. The functions that compose the Octave code are identified, manually implemented in Verilog code, and characterized according to the KPIs. Specifically, the functions used for the evaluation of the algorithm untill now are the following:

- eNodeB LTE PHY [8] layer, downlink, OpenLTE, Octave (SW functions). The Octave source code that describes an LTE-based frequency domain downlink transmitter implementation in SW. This code contains the functions that are specified from the existing LTE, 3GPP standards.
- eNodeB LTE PHY layer function, broadcast channel:
  - Cyclic redundancy check (CRC), Verilog code (HW function)
  - Convolutional encoding, Verilog code (HW function)
  - Rate matching, Verilog code (HW function)
- eNodeB LTE PHY layer function, physical downlink control channel:
  - Cyclic redundancy check (CRC), Verilog code (HW function)
  - Convolutional encoding, Verilog code (HW function)
  - Rate matching, Verilog code (HW function)
  - Pseudorandom sequence generation, Verilog code (HW function).

The seven latter functions were manually implemented in Verilog code and evaluated in order to be bit accurate in conjunction to the corresponding SW functions. These functions are the first to be implemented in HW as they introduce a significant amount of delay in the SW execution time. Furthermore, while the implementation of broadcast channel and physical control channel, in Verilog, is similar, their HW mapping results in different power consumption that makes them perfect candidates to exercise the capabilities of the multiobjective algorithm.

1.3.2.2 Parameters (KPIs)/ Constraints Definition

The partitioning algorithmic solution considers the most critical KPIs for the 5G networks regarding partitioning and virtualization, so far. The considered KPIs are described in the following lines.

- Execution time
  - The measured execution time of the LTE network functions when implemented in SW.
  - The measured execution time of the LTE network functions when implemented in HW.
- Energy consumption of Verilog modules derived from power analysis using appropriate FPGA IDE.
- Measured SW memory utilization (e.g., RAM) of LTE functions when implemented in SW.
- Communication time, considering measured time of data transferring with respect to send and receive communication functions, between HW and SW implemented LTE functions.

- Reusability referring to the available HW resources, inversely relative to HW functions utilization.

1.3.2.3 Functional Graph (Dataflow Graph) Provision

The communication scheme between the utilized LTE functions forms a dataflow graph, including nodes that represent the functions implemented in HW or SW and edges that represent the link/interconnection between two functions. The communication overhead is also applied as weight to each corresponding link. The GUI accompanying the algorithmic solution provides the user with the ability to add or remove nodes/functions and edges. The partitioning algorithm receives this information and forms an array structure that includes the mapping of the interconnected components. The array items provide the multiobjective algorithm with the ability to find the optimal decision that also considers the communication overhead. An example is provided in Figure 1.5.

Figure 1.5 is generated by the user interface that controls the partitioning algorithm. The solution includes a predefined set of functions and interconnections, referring to the already implemented functions that form the graph in the figure. The user interface provides the user with the ability to add or

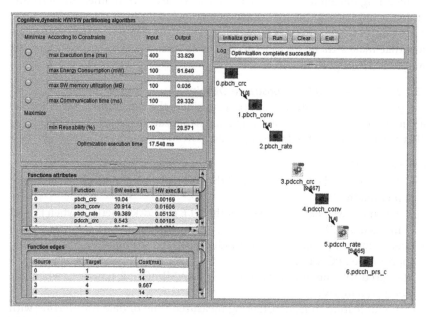

Figure 1.5 Graphical User Interface with generated data flow graph. The cyclic redundancy check function pdcch_crc and the rate matching encoding pdcch_rate functions of the physical downlink control channel are implemented in SW.

remove functions, interconnections, and their KPIs in order to create a graph that better suits his/her needs.

1.3.3 Optimization Problem Formulation

The core process of the optimization procedure is the optimization problem formulation as it guides the whole optimization process toward the goals of the problem. The main steps of the procedure are the following:

Input: This process receives as input the implemented or user provided functions along with their KPIs.

- N is the number of considered functions.
- M is the number of considered KPIs.
- An initial set of decision variables vector is created using random processes
  - $x = [x_1, \cdots x_N]$ is the decision variables vector, with x_n, $n = 1 \cdots N$, representing HW implementation ($x_n = 1$) or SW implementation ($x_n = 0$) of function n.

Process: When a KPI value is minimized, another KPI value is increased. In order to specify this relationship, the algorithm includes equations for each KPI with factors representing the relative difference. The objectives to be optimized are similar or even exact predefined KPIs, implemented by objective functions as described in the following paragraph– KPI-driven binary optimization:

- $f(x) = [f_1(x), \ldots f_M(x)]$ is the vector of objective functions to be minimized, each objective function f_m, $m = 1 \ldots M$, representing a corresponding KPI optimization.
- $f_m(x) = a_1 m x_1 + \cdots + a_N m x_N$, *where*, $a_n m$ is the difference of the KPI m value between HW implementation and SW implementation of function n. The factor $a_n m$ is used to make the algorithm aware of the inverse relationship between the KPIs.
- Each objective function f_m, $m = 1 \cdots M$, can have a corresponding constraint, for example, if m is the full model execution time-latency and it should be less than 80 s, this mean $f_m < = 80$.

A multiobjective algorithm will create a set of possible solutions (solution: $x = [x_1, \cdots, x_N]$) by measuring the objective function values for every solution. Then the algorithm evaluates the available solutions and sorts them considering the optimization target (specific KPI e.g., power).

Output: Finally, the algorithm selects the optimal solution ($x = [x_1, \cdots, x_N]$) representing HW implementation ($x_n = 1$) or SW implementation ($x_n = 0$) of function n) as described earlier.

The next section provides a description of the multiobjective algorithmic solution that is utilized toward this direction.

1.3.4 Evolutionary Multiobjective Algorithmic Solution

This section provides a description of the multiobjective algorithmic solution that is utilizing the optimization problem formulation. The multiobjective algorithm derives a subset of solutions from the set of all possible solutions according to the provided multiple objective functions, their KPIs, their constraints, and the optimization goal. The initial set of all possible solutions is provided from the objective functions that are considered in the optimization problem formulation. The functions and KPIs/Constraints are provided by the user utilizing the solutions user interface. The selected multiobjective algorithm for the HW/SW partitioning is the nondominated sorting genetic algorithm NSGA II [9]. The algorithm begins its search with an initial population of individual's solutions usually created at random within a specified lower and upper bound on each variable. Once the population in initialized, the population is sorted based on nondomination into each front. Once the nondominated sort is complete, the crowding distance (the Euclidian distance between each individual's solution) is assigned. After the population members are evaluated, the selection operator chooses better solutions with a larger probability to fill an intermediate mating pool. For this purpose, the tournament selection procedure takes place in which two solutions can be picked at random from the evaluated population and the better of the two can be picked. The crossover operator is to pick two or more solutions (parents) randomly from the mating pool and create one or more solutions by exchanging information among the parent solutions. Each child solution, created by the crossover operator, is then mutated with a mutation probability so that on an average one variable gets mutated per solution. In the context of real-parameter optimization, a simple Gaussian probability distribution with a predefined variance can be used with its mean at the child variable value. This operator allows an EO to search locally around a solution and is independent of the location of other solutions in the population. The elitism operator combines the old population with the newly created population and chooses to keep better solutions from the combined population. Such an operation makes sure that an algorithm has a monotonically nondegrading performance. Finally, the user of an EO needs to choose termination criteria. Often, a predetermined number of generations is used as a termination criterion. In most cases, this algorithm is able to find much better spread of solutions and better convergence near the true Pareto-optimal front compared to other multiobjective algorithms, on a diverse set of difficult test problems.

1.3.5 Testbed Setup

The current implementation of the cognitive dynamic HW/SW partitioning algorithm is tested considering the dynamic hot spot use case, resulting in the LTE eNodeB PHY layer reconfiguration, according to prespecified mea-

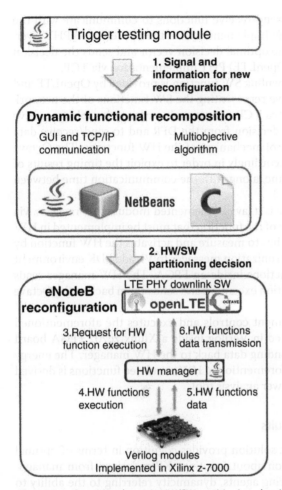

Figure 1.6 Cognitive dynamic HW/SW partitioning algorithm test-bed.

sured KPIs and optimization goals mostly referring to execution time or energy consumption reduction. The testbed of the demonstration includes the SW environments that will host the SW modules and a HW development environment along with an FPGA board, regarding the HW modules, as provided in Figure 1.6:

- The trigger module sends a signal that indicates a hot spot or not a hot spot condition. It is implemented in Java and it is connected to the dynamic functional recomposition module via TCP protocol.
- The dynamic functional recomposition module is the main module under test and contains the cognitive dynamic HW/SW partitioning algorithm

along with its GUI, TCP send/receive functions to communicate with the trigger module, and the SW implementation of the LTE eNodeB PHY layer. The algorithm provides the optimal decision vector and sends the decision to the SW LTE eNodeB (OpenLTE) PHY implementation via TCP.

- The LTE eNodeB PHY downlink SW function is provided by OpenLTE and it is implemented in Octave representing the SW functions of the network stack. It is facilitated by java TCP functions, in order to be able to receive the HW/SW partitioning decision from the DFR and to send/receive data to/from Vivado SDK control mechanism of the HW functions. The Octave code has been modified accordingly in order to exploit the timing results of the functions execution time, along with the communication time between HW and SW functions.

- The HW manager module is a java implemented module that receives, via TCP connection, the name of the function that must be implemented in HW along with the data that it has to measure and activates the HW function by sending appropriate synchronization messages to Vivado SDK environment that controls the HW functions inside an FPGA. The HW manager reads the results of the HW function execution and sends them back to the Octave OpenLTE module.

- The Vivado SDK environment controls and executes the aforementioned HW functions implemented in Verilog inside a Xilinx zc702 FPGA board and provides the corresponding data back to the HW manager. The energy consumption of the two aforementioned HW simulated functions is derived from the Xilinx Vivado power analysis tool.

1.3.6 Preliminary Test Results

The partitioning algorithmic solution provides cognition in terms of optimal decision based on information about functions KPIs; derived from management programs and monitoring agents, dynamicity referring to the ability to dynamically move functions from HW to SW implementation and vice versa according to the specified policies, considering also the communication overhead. Current results show improved overall performance (execution time, latency/communication overhead) by 70% and power consumption reduction by 50% with respect to the digital baseband processing of an LTE eNB PHY layer of Tx. Figure 1.7 illustrates the improvements in overall execution time of the addressed and tested physical layer functions, when HW/SW partitioning is performed targeting high-performance (a) and normal performance scenarios (b). In a high-performance scenario, more functions will be executed in HW while in normal performance more functions will move to SW in order to reduce power consumption.

The first set of results imply that the partitioning algorithm can achieve high performance with respect to algorithmic solution execution time, below thresh-

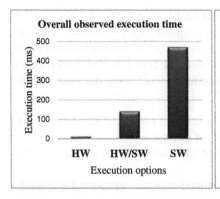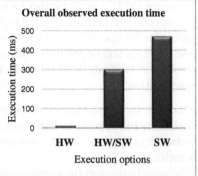

Figure 1.7 i) SW implementation, ii) HW/SW optimal partitioning, iii) HW implementation results in execution time (ms) for high-performance and normal performance scenarios using Octave

olds for LTE current handover time and user experience maintenance and algorithmic solution complexity. In addition, the cognitive, dynamic HW/SW partitioning algorithm provides reduction of the power consumption, limited memory overhead, enabling the interaction between algorithmic solution with management programs and monitoring agents. The proposed HW/SW partitioning will be further refined and upgraded considering improvements on performance, memory overhead, and portability of the current algorithmic solution. Furthermore, there is ongoing investigation for integration with the CTTCs LENA ns-3 simulator in order to have a full stack LTE network implementation to better clarify the benefits and challenges of HW/SW partitioning. Moreover, since the algorithmic solutions design enables KPIs and functions extensions future implementations and KPIs will be investigated in order to meet a diverse set of objectives such as user data rate and even capacity.

1.3.7 Status and Challenges in 5G Wireless Communications

This section provides useful insights on the advancements in physical layer, MAC, and RRM. Details on the aforementioned advancements are provided in the sections that follow.

1.3.7.1 Novel Physical Layer Aspects

One of the key aspects that will evolve with 5G has to do with the novel design of a unified, flexible air interface, its components, and procedures, so as to effectively deal with the issue of 5G requirements through such an adaptation. Therefore, it is important to develop a new spectrum agnostic 5G air interface for carrier frequencies below 6 GHz. This is motivated by the fact that todays licensed bands for cellular usage are all below 6 GHz, and the World Radio

Conference (WRC) in 2015 also focused on below 6 GHz spectrum among other aspects. Furthermore, even if higher frequency spectrum bands are made available for 5G operation in the future, having effective means for utilizing 5G below 6 GHz is still of relevance due to the more favorable radio propagation properties. A unified, flexible 5G air interface would have the following key characteristics:

- Flexibility to support the broad class of services with their associated KPIs
- Scalability to support the high number of devices
- Versatility to support the diverse device types and traffic/transmission characteristics
- Efficiency to support the requirements on energy consumption and resource utilization
- Future-proofness to support easy integration of new features.

The new air interface will meet the requirements on the 5G main KPIs (e.g., for increased throughput, reduced latency, etc.) with increased flexibility, reliability, future-proofness, and cost and energy efficiency. Notably, devices will be designed to support one common air interface for all services and more devices will be produced applying similar/common chip sets; thus achieving economy of scale for vendors. Also, the wireless system is expected to be more scalable and thus better suited to follow load variations between the services—both temporal and in different locations.

1.3.7.2 Novel Frame Design Based on Service Requirements

As previously mentioned, among the main objectives of a new, unified 5G air interface should be its flexibility to be adapted to the diverse requirements imposed nowadays by the heterogeneous service demands. This diversity of requirements creates a very challenging environment in terms of service-specific KPIs and channel characteristics as it should be flexible enough to satisfy these needs while in parallel optimize the resource utilization and minimize the overhead introduced by the multiservice support functionalities/mechanisms. The 5G services are foreseen to include mobile broadband (MBB) services, supporting high data rates and high coverage, massive machine communication (MMC) services, supporting small packet sizes and infrequent transmissions, mission critical communication (MCC) services with strict delay bounds and reliability factors, and vehicle-to-anything communication (V2X) services supporting both bs-to-device and device-to-device transmissions.

The current status of a stiff frame structure with, for example, a fixed transmission time interval (TTI) value either cannot satisfy the extremely strict requirements of specific services (e.g., delay requirements of MCC services) or results in underutilization and waste of resources due to inefficient resource management. Therefore, a flexible frame structure supporting the coexistence of different transmission time intervals (TTIs) is more than necessary in order to accomplish these diverse service requirements. The introduction of a flexible TTI supporting different TTI durations will both accomplish ultralow latency capabilities (e.g., in case of MCC services) by facilitating short TTI durations and high spectral efficiency gains (e.g., in case of MBB services) by utilizing long TTI durations. The selection of TTI scaling, which is the set of available TTI values, is of high importance, because it directly affects the effectiveness of the proposed flexible frame structure solution.

We proposed two methods for TTI scaling:

- 2^N *Scaling:* Definition of a set of TTI durations with double duration in each one.
- *Scaling Based On Service Classification:* Definition of a minimum set of TTI durations (mapped to a set of services)

In case of 2^N scaling, the minimum and maximum TTI length is defined and then a set of TTI lengths is generated based on the 2^N approach. For the 5G services mentioned already, reasonable values of TTI duration can be between 0.125 and 4 ms, therefore, the generated TTI values belong to the set (0.125, 0.25, 0.5, 1, 2 and 4ms). This approach is graphically depicted in Figure 1.8.

According to the second method, a set of predefined services (e.g., MBB, MMC, MCC, V2X) are analyzed based on their requirements in order for each service to estimate the TTI value that best reflects its characteristics, satisfy its requirements in terms of KPI values (e.g., average delay, max delay, throughput), and minimize the system resource overhead. Then a set of TTI lengths are generated that are mapped to the aforementioned services or group

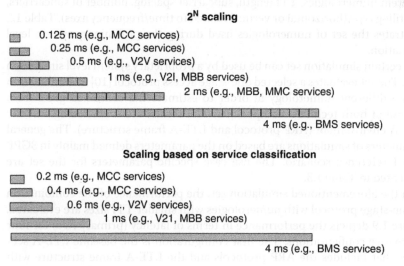

Figure 1.8 Flexible TTI - TTI Scaling

Table 1.2 Numerologies used in simulation scenarios.

RB Type	TTI (ms)	Subcarrier Spacing (kHz)	Number of Symbols per TTI
RB1 (LTE like)	1	15	14
RB2	0.5	15	7
RB3	0.25	30	7

of services. In Figure 1.8, an indicative set of TTI lengths is depicted based on an initial analysis and service classification. Each TTI length is mapped to one or several services. Regarding the comparison between the two scaling methods, the second method is better fitted to the service requirements as it is based on analysis of the special characteristics and optimum TTI selection for each service type. However, in case of absence of resource partitions in the spectrum, because of the different multiply factors between TTI values, this method cannot succeed high multiplexing gain between different services in the frequency domain as resource gaps may emerge for a set of selected TTI values. The first method may not be optimally, fitted to the services, although it eliminates the gaps of the spectrum (in case of no recourse partitions), while it occupies the minimum length in the packet header (for eight different TTI values, 3 bits are required).

1.3.7.3 Support of Different Numerologies

The following parameters are made available for configuration in order to create different numerologies: TTI length, subcarrier spacing, number of subcarriers, and tiling type (horizontal or vertical related to time/frequency axes). Table 1.2 illustrates the set of numerologies used during the preliminary system level evaluation.

A certain simulation set can be used by a proprietary system level simulation tool. The set evaluates a selected one-stage access protocol [10] in combination with a different numerology in order to estimate the combined gains from the use of both technical components. The results are evaluated against the LTE-A environment (ARP protocol and LTE-A frame structure). The general parameters of simulations are based on the parameters defined mainly in 3GPP case 1 reference scenario. The use case specific parameters for the set are presented in Table 1.3.

In the aforementioned simulation set, the performance of the combination of one-stage protocol with numerologies with smaller TTI sizes are evaluated. Figure 1.9 depicts the performance in terms of latency (primary KPI) for four alternative configurations. The first configuration is the baseline LTE-A scenario that includes the ARP protocols and the LTE-A frame structure with TTI length of 1 ms. In the other three configurations, the one-stage protocol is

Table 1.3 Simulation scenario parameters.

Parameters	Value
Reference scenario	3GPP case 1
Network topology	19 3-sectorized base stations (57 cells)
Intersite distance	500 m
Bandwidth	FDD 40MHz - 20MHz (uplink)
Request generation	Poisson
UE data traffic size	100 bytes / data report
PRACH allocation	1 PRACH allocation / TTI
Backoff window	Discrete uniform distribution (2,20) TTIs
Max connection attempts	4
Preambles	64

adopted in combinations with three different numerologies with TTI lengths of 1, 0.5, and 0.25 ms.

Figure 1.9 shows that the LTE-A (ARP, TTI = 1ms, PRACH per 1ms) has latency values between 5 and 7.5 ms, which is a performance far from the KPI target of 1 ms. The other configurations show improved latency values highly affected by the request rate. In detail, the numerology of TTI = 1 ms have relatively low-latency values only for low request volumes, while the numerologies of TTI = 0.5 and 0.25ms have relatively low-latency values for all

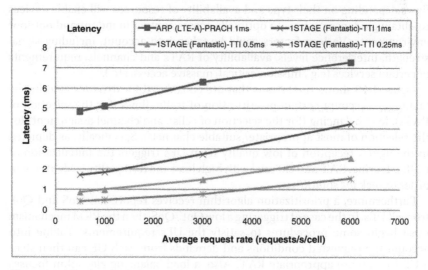

Figure 1.9 KPI 2 latency (primary KPI) - second set of simulation results.

the examined request volumes. Latency values below the KPI target are observed only for the numerologies of TTI = 0.5 and 0.25 ms and for low request rates.

The aforementioned latency results depict the latency from the UE to the BS (uplink), while no processing delay was taken into consideration.

1.3.8 Enhanced Radio Resource Management (RRM) and MAC Adaptation for 5G

In 5G networks it is important to proceed to the investigation and development of technologies that address the well-known challenges of predicted growth in mobile connections and traffic volume by successfully addressing the lack of dynamic control across wireless network resources that is leading to unbalanced spectrum loads and a perceived capacity bottleneck. Resource management with three degrees of freedom can be taken into consideration: (i) densification, (ii) rationalized traffic allocation over heterogeneous wireless technologies, and (iii) better load balancing across available spectrum bands in licensed, lightly licensed, and unlicensed spectrum portions. Moreover, the MAC has to be adapted in order to be able to support the deployment of various 5G services that call for increased reliability, reduced latency, and higher throughput in licensed, lightly licensed, and unlicensed bands.

Traffic steering provides operators with the necessary functionality in order to let them optimize resource utilization, QoS/QoE, and power consumption of cells and UEs by directing the traffic to the RAT or layer that is the most appropriate/suitable for a certain type of service. Steering of certain traffic flows depending on their type and availability of resources will enable devices to obtain guidance on how to optimally access content in indoor and outdoor environments. Various factors could be taken into account, including signal strength, interference levels, availability of RATs and channels, requirements of certain services (e.g., mission critical, massive access, etc.).

Figure 1.10 shows the actions that are considered for dynamic steering by taking into account certain prioritization of traffic flows (for the selection of RATs), load balancing (for the selection of cells), and channel assignment (for the selection of most appropriate/ suitable channels). Specifically, as depicted in the figure, reception of low quality from UEs triggers the initialization of traffic steering. RAT and channel with high load and low quality is identified in order to seek for solutions.

Furthermore, a prioritization algorithm receives reports on QoS and QoE from all UEs. In the case of triggering a low QoE/QoS, then the RRM mechanism must begin some procedures to satisfy the UEs requirements. Taking into account the requested conditions and demands from each UE can then steer a UE to its more appropriate RAT. Also, a load balancing algorithm focuses explicitly on achieving a good load balance between cells of the same RAT.

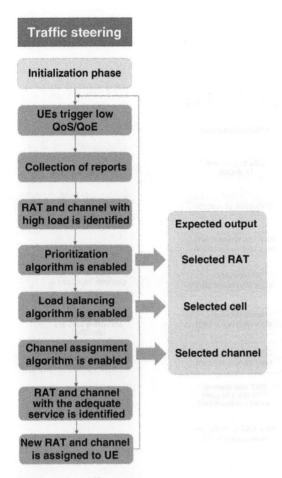

Figure 1.10 Traffic steering

Thus, based on sensing mechanism, the system is monitoring the loads and collects, the measurements from all cells of a particular RAT; an overloaded cell is identified and then under loaded cells that are in the vicinity are identified. An active and eligible UE can then be moved from the overloaded cell to the under loaded adjacent cell that meets its requirements in order to gradually arrive at the preferred load balance. In addition to the network load, the user experience (QoE) is also monitored throughout the sensing mechanism, and if needed an inter-RAT handover procedure is invoked to move the UE to a different RAT. Finally, at the time of connection establishment, the decision mechanism with the utilization of these algorithms and mechanisms is pursued to select the right RAT, cell, and channel to be used by the user throughout the connection.

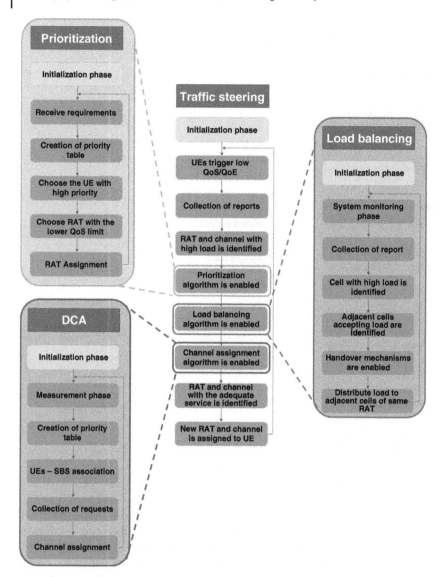

Figure 1.11 Dynamic Channel Assignment (DCA), prioritization, and load balancing as integral parts of traffic steering.

Figure 1.11 illustrates the integral parts of the Dynamic Channel Assignment (DCA); prioritization; load balancing, which have been previously discussed. Such concepts can be evaluated through system level simulations. Therefore, it is critically important to develop a system level simulator in a way to make a flexible, accurate, and efficient simulation of this complex heterogeneous net-

Figure 1.12 5G main application areas

works with multiple RAT, cells, and channels for 5G networks. Specifically, it should be possible to support heterogeneous networks consisting of LTE macrocells (licensed bands), small cells that utilize lightly licensed bands, and Wi-Fi with unlicensed bands, with BSs and APs densely deployed to provide users with seamless connectivity and demanded services, and finally, the radio resource management mechanism that assigns the most appropriate RAT, cell, and channel based on the requirements of each UE and at the same time keeping the system as load balanced as much as allowed. The particular simulator is a proprietary system-level simulation tool that is fully developed in Java with various capabilities and has been calibrated according to the 3GPP specifications. It takes into account various parameters such as traffic level, available infrastructure elements, available channels, and evaluates the various test cases. The calibration state of the proprietary simulator has been evaluated against the reference results of the 3GPP LTE calibration campaign [36.814] [11]. As a result, the cumulative distribution function (CDF) of coupling loss and downlink SINR have been examined in order to calibrate the tool with leading operators and vendors such as Nokia, Ericsson, DoCoMo, Huawei, and Telecom Italia.

The simulators playground is illustrated in Figure 1.12, where multiple macrocells and small cells were deployed in order to simulate a cellular environment as an example. The simulator is capable of creating various scenarios for sparse to dense and even ultradense deployment of various cells and technologies.

1.4   5G Network Management Aspects Enhanced with Machine Learning

1.4.1   Machine Learning for Service Classification in 5G Networks

A challenge for future wireless communication networks is the satisfaction of the diverse requirements coming from heterogeneous services. In 5G networks, the coexistence of different services like Mobile Broadband (MBB), massive machine-type communications (MMC), and mission critical communications (MCC) having various requirements in terms of both capacity and QoS will constitute a key prerequisite. Hence, one of the main issues that should be addressed by the 5G management system is the simultaneous provisioning of these services satisfying the corresponding requirements so as to optimize the network in order to be resource and energy efficient. A first step toward this direction is to be able to identify each service type in order to prioritize the services and be able to allocate efficiently the network resources.

Knowledge of QoS requirements per service flow could be provided by the higher layers as, for example, assumed in the HSPA and LTE, where sets of QoS parameters are available for RRM functionalities such as admission control and packet scheduling decisions. As an example from LTE, each data flow (bearer) is associated with a QoS profile consisting of the following downlink related parameters:

- Allocation retention priority (ARP)
- Guaranteed bit rate (GBR)
- QoS class identifier (QCI)

In particular, the QCI includes parameters like the layer-2 packet delay budget and packet loss rate. However, for the cases where detailed QoS parameters are not made available from the higher layers, the use of novel service classification techniques should be considered, in which the base stations monitor the traffic flows to extract more detailed service classification information and identify the service type providing this information input to packet scheduling algorithms and other RRM functionalities.

The support of fast and reliable traffic characterization is a necessary step in order to understand the network resource-usage and to provide differentiated and high QoS/QoE through prioritization targeting in increasing resource-usage and energy efficiency. In addition, the service classification process can interact with new services and procedures provided in 5G networks to support flexibility and adaptability to traffic variability.

In this section, the use of various machine learning mechanisms for the service classification problem is described and the performance of different algorithms is investigated. The considered classification methods reside in the area of statistical-based classification techniques and they are realized by

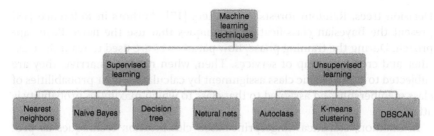

Figure 1.13 Overview of machine learning techniques.

exploiting several flow-level measurements (e.g., traffic volume, packet length, interpacket arrival time, and so forth) to characterize the traffic of different services. Then, to perform the actual classification, supervised machine learning techniques are applied to these measurements. It should be noted that in contrary to other methods of traffic classification, like payload-based classification, which need to analyze the packet payload or need to use deep packet inspection technologies, statistical-based classification techniques are usually very lightweight, as they do not access packet payload and can also leverage information from flow-level monitors.

1.4.2 State-of-the-Art Machine Learning Mechanisms for Traffic Classification

In the literature, there are a lot of studies that focus on application and service discrimination based on traffic classification learning techniques as presented in detail [12], [13], [14]. Various machine learning mechanisms are usually employed belonging to either unsupervised or supervised machine learning, as illustrated in Figure 1.13. It should be noted that recently semisupervised machine learning, mechanisms have also been proposed. However, considering that the majority of the existing works considers the other two categories, we focus only on supervised and unsupervised machine learning [15]. In the first case, clustering algorithms like K-Means, DBSCAN, and Autoclass [16] are investigated. The objective of these mechanisms is to group flows that have similar patterns into a set of disjoint clusters. The major advantage of these schemes is that they do not require a training phase like the supervised ones but they automatically discover the classes via the identification of specific patterns in the data set. However, the resulting clusters do not certainly map 1:1 to services as usually the number of clusters is greater than the number of service types and even in the case of 1:1: mapping, the clusters still need to be labeled in order to be mapped to the corresponding services.

Regarding the supervised machine learning techniques, which is also the approach that is analyzed in this section, there are various classification schemes that have been proposed for the traffic classification problem like Nave Bayes,

Decision trees, Random forests, and others [17]. Authors in Reference [18] present the Bayesian classification techniques that use the naive Bayes approach. During the training phase, flow parameters are used to train the classifier and create a group of services. Then, when new flows arrive, they are subjected to probabilistic class assignment by calculating their probabilities of class membership and assigned to that class to which maximum probability is attained.

In addition, statistical fingerprint-based classification techniques as presented in Reference [19] classify traffic based on a set of preselected parameters (e.g., packet size, interarrival time). During the training phase, a data set of flows from each service are used in order to analyze the data set and create the service fingerprint. This fingerprint is usually a PDF (probabilistic density function) vector used to identify the service. During the classification phase, the algorithm checks the behavior of a flow against the available set of PDF vectors. Also, support vector machine (SVM) techniques, first proposed in Reference [20], are binary supervised classification algorithms that transform a nonlinear classification problem in a linear one, by means of what is called a kernel trick.

Furthermore, artificial neural networks (NNs) consist of a collection of processing elements that are highly interconnected and transform a set of inputs to a set of desired outputs that is inspired by the way biological nervous systems works. In Reference [21], authors proposed a NN, in which a multilayer perception classification network is used for assigning probabilities to flows. A set of flow features are used as input to the first layer of network, while the output classifies flow into a set of traffic classes by calculating the probability density function of class membership. Also, decision tree algorithms, which are mentioned in Reference [22], represent a completely orthogonal approach to the classification problem, using a tree structure to map the observation input to a classification outcome. In these supervised classification algorithms, the data set is learnt and modeled, therefore, whenever a new data item is given for classification, it will be classified according to the previous data set. The problem of ML data set validation is discussed in Reference [23] that highlights three training issues that should be considered in ML classification. These issues refer to the algorithm's impact when training and testing data sets are collected from same/different network, when the real online traffic classes of the training data set are not presented, and finally the impact of the geographic place where the network traffic is captured. Real Internet traffic data sets collected from a campus network are used to study the traffic features and classification accuracy for each validation training issue, demonstrating the impact of each issue.

1.4.3 Classification Approach and Evaluation Metrics

In this section, the problem of service classification is investigated employing a set of different classification mechanisms that belong to the supervised ma-

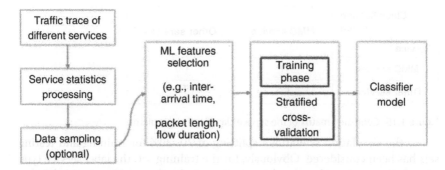

Figure 1.14 Proposed mechanism for the service classification process.

chine learning category. Before presenting the performance evaluation of each mechanism, the algorithmic procedure that has been followed is described and each step is explained in detail.

Figure 1.14 presents the algorithmic procedure that is followed for the classification mechanism. As can be seen, the first step refers to the collection of a number of traces from different services. For the considered simulation scenario, three service types are considered referring to MCC, MMC, and MBB communication while other services (like broadcast/multicast services) will also be considered in the future. The different traces of each service have been generated using specific traffic models. More specifically, the generation of different types of MCC/MMC traffic was following the traffic models presented in 802.16p [24], while for the generation of MBB traffic, video streaming traffic (YouTube) that follows the traffic models presented in Ref. [25] is assumed.

The second step refers to the statistical processing of these traces in order to separate them in flows. In particular, a flow is considered as a series of packets transmissions that have the same source and destination and for which the interarrival time is below a specific threshold. After this processing, a number of features for each flow is generated, including interarrival time statistics (mean value, standard deviation), packet size statistics (total value, mean/max/min value, standard deviation), and other flow characteristics like total number of packets, source, destination, and flow direction. Subsequently, some feature engineering tasks are performed before proceeding to the main classification mechanism. These tasks include the selection of the most representative features, the transformation of categorical features into numerical values, the normalization of features values and other tasks that guarantee a high data quality (e.g., replace missing values). Then, the implementation of machine learning mechanism follows including two main phases: the training phase and the cross-validation phase. It should be noted that stratification is applied in this case in order to randomly sample the flow data set in such a way that each service type is properly represented in both training and testing data sets.

Classification result / Service	MMC service	Other services
MMC service	TP	FN
Other services	FP	TN

Figure 1.15 Confusion matrix of the service classification problem.

For the simulation scenario, a splitting of 70–30% for training and testing sets has been considered. Obviously, for the training set, the label service type of each flow is considered as known whereas for the testing set, this label is considered as unknown and each flow is labeled using the classifier model. The outcome of the proposed mechanism is a classifier model that can be employed in unknown flows in order to recognize them and label them in an accurate way.

To evaluate the performance of the classification mechanisms, various metrics have been defined and can be used in the train/test sets to select the most adequate mechanism for the specific problem. To illustrate the relationship between the different evaluation metrics, a very useful tool that provides a holistic view of each algorithm's performance is the confusion matrix. The confusion matrix is actually a two-dimensional matrix, in which the horizontal axis represents the predicted class (outcome of the algorithm) whereas the vertical axis represents the true class. In Figure 1.15, the confusion matrix for the considered classification problem is presented, where FP, TP, FN, and TN stand for false positives, true positives, false negatives, and true negatives, respectively, and they are defined as follows:

- FP: The percentage of other services flow that are incorrectly classified as MMC service.
- TP: The percentage of MMC services flow that are correctly classified as MMC service.
- FN: The percentage of MMC services flow that are incorrectly classified as other services.
- TN: The percentage of other services flow that are correctly classified as other services.

Some of the most common evaluation metrics used for classification problems are the accuracy metric, the precision, the recall, and the F1 score. More specifically, these are defined as follows:

- Accuracy is defined as the percentage of correct predictions to the total number of predictions and is given by

$$\frac{(TP + TN)}{(TP + FP + TN + FN)}.$$

- Precision is defined as the percentage of the instances that were correctly predicted as belonging in a class among all the instances that were classified as belonging in this class and is given by

$$\frac{TP}{(TP + FP)}.$$

- Recall is defined as the percentage of the instances of a specific class that were correctly classified as belonging to this class and is given by

$$\frac{TP}{(TP + FP)}.$$

- F1 score is defined as the harmonic mean of the precision and recall and is given by

$$\frac{2 \times \text{Precision} \times \text{Recall}}{(\text{Precision} + \text{Recall})}.$$

- To be able to choose the best mechanism for a classification problem, the investigation of a single metric, like accuracy, is not always enough as the misclassification of a specific class instances may be more important than the correct classification of others. For this reason, other evaluation metrics have also to be applied to make the most appropriate choice depending on the problem's characteristics.

1.4.4 Evaluation Performance of Classification Mechanisms

In the considered simulation scenario, the performance of a set of different machine learning mechanisms has been investigated, including base classifiers such as naive Bayes classifier, support vector machines, tree classifier, k-nearest neighbor classifier, logistic regression as well as ensemble-based classifiers like random forest classifier. The goal of ensemble methods is to combine the predictions of several base estimators built with a given learning algorithm in order to improve generalizability and robustness over a single classifier. Usually, two families of ensemble methods are distinguished: the averaging methods (e.g., random forests [26]) in which several classifiers are developed independently and then the average of their predictions is used, and the boosting methods (e.g., AdaBoost– Adaptive Boosting) where base classifiers are built sequentially and one tries to reduce the bias of the combined estimator. To be able to compare the various machine learning mechanisms, in Table 1.4, the accuracy metric of each algorithm is presented, where a Dump classifier that classifies all the flows as type 0 (MMC service) is also considered resulting in 0.512 accuracy. From this table, it can be seen that Decision tree and the

Table 1.4 Accuracy score for each classification mechanism.

Classification Mechanism	Accuracy
Naive Bayers	0.808
Support vector machine	0.662
Decision tree	0.976
k-nearest neighbor classifier	0.952
Logistic regression	0.685
Random forest classifier	0.988

Random Forest algorithms lead to the highest accuracy values, outperforming the other machine learning algorithms.

However, to provide a more complete view of each classifiers performance, the corresponding confusion matrices are illustrated in Figure 1.16. The horizontal axis of this matrix represents the predicted class whereas the vertical axis represents the true class. It should be noted that Class 0, Class 1, and Class 2 refer to MMC, MCC, and MBB service types, respectively. In the considered scenario, considering that it is desired to eliminate the possibility that a MCC service is misclassified as another service type, the optimal model should have high values of recall whereas high accuracy values for the case of MMC and MBB services are required. The results of confusion matrix show that the Decision Tree and the Random Forest algorithms result in extremely good results as they misclassify only a few flows, also resulting in high values of recall and precision, as can be seen in Figure 1.17. Therefore, these two classification mechanisms can be selected for further consideration for the problem of service classification.

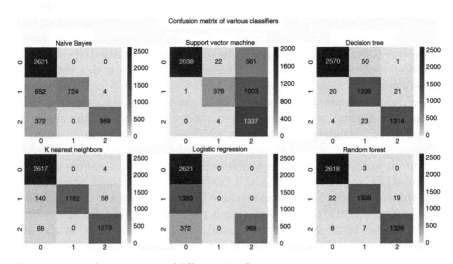

Figure 1.16 Confusion matrices of different classifiers.

Class\ metrics	Precision	Recall	F1 score	Class\ metrics	Precision	Recall	F1 score
MMC	0.99	1.00	0.99	MMC	0.99	0.97	0.98
MCC	0.99	0.97	0.98	MCC	0.94	0.97	0.96
MBB	0.99	0.99	0.99	MBB	0.98	0.99	0.99
Avg/total	0.99	0.99	0.99	Avg/total	0.98	0.98	0.98

Figure 1.17 Evaluation metrics for selected classification mechanisms.

1.5 Conclusion

5G is the next frontier of innovation for entire mobile industry. Consequently, the three major objectives for 5G are support of massive capacity and massive connectivity; support for an increasingly diverse set of services, applications, and users; and in addition flexible and efficient use of all available noncontiguous spectrum for widely different network deployment scenarios. Framed in this context, this chapter elaborated on the status and challenges in hardware/software development and in 5G wireless communications by focusing on physical layer, MAC, and RRM. Also, the benefits of machine learning in 5G network management were discussed. By taking into account the diversity of infrastructure, radio resources, and services that will be available in 5G, an adaptive network solution framework will become a necessity. Breakthrough developments in several RAN technologies will be required for realizing novel, 5G solutions. Such technologies include, among others, multiple access and advanced waveform technologies combined with coding and modulation algorithms, massive access protocols, massive MIMO, and virtualized and cloud-based radio access infrastructure.

Acknowledgment

Part of this work has been performed in the framework of the Horizon 2020 projects, SPEED-5G (ICT-671705), FANTASTIC-5G (ICT-671660), and Flex5Gware (ICT-671563) by receiving funds from the European Union.

References

1 NGMN, 5G White Paper. Available at www.ngmn.org (accessed April 2017).
2 J. Elliott and S. Sharma, Dynamic end-to-end network slicing unlocks 5G possibilities. Available at https://insight.nokia.com/dynamic-end-end-network-slicing-unlocks-5g-possibilities (accessed Nov. 2016).
3 T. E. Bogale and L. B. Le, "Massive MIMO and mmWave for 5G wireless HetNet: potential benefits and challenges," *IEEE Veh. Technol. Mag.*, vol. 11, no. 1, pp. 64–75, 2016.

4 LENA, LTE-EPC Network Simulator. Available at http://iptechwiki. cttc.es/LTE-EPC.

5 N. Nikaein, R. Knopp, F. Kaltenberger, L. Gauthier, C. Bonnet, D. Nussbaum, and R. Ghaddab, "OpenAirInterface 4G: an open LTE network in a PC, MOBI-COM 2014," *20th Annual International Conference on Mobile Computing and Networking*, 2014, Maui, Hawaii, Sept. 7–11, 2014.

6 Q. Zheng, H. Du, J. Li, W. Zhang, and Q. Li, "Open-LTE: an open LTE simulator for mobile video streaming," in *2014 IEEE International Conference on Multimedia and Expo Workshops (ICMEW)*, July 2014.

7 I. Gomez-Miguelez, A. Garcia-Saavedra, P. D. Sutton, P. Serrano, C. Cano, and D. J. Leith, *srsLTE: An Open-Source Platform for LTE Evolution and Experimentation*, Feb. 2016.

8 ETSI, LTE, Evolved Universal Terrestrial Radio Access (E-UTRA), Multiplexing and channel coding (3GPP TS 36.212 version 10.1.0 Release 10), 2011.

9 K. Deb, A. Pratap, S. Agarwal, and T. Meyarivan, "A fast and elitist multiobjective genetic algorithm: NSGA-II," *IEEE Trans. Evol. Comput.*, vol. 6, no. 2, pp. 182–197, April 2002.

10 S. Saur, A. Weber, and G. Schreiber, "Radio access protocols and preamble design for machine-type communications in 5G," in *2015 49th Asilomar Conference on Signals, Systems and Computers*, Nov 2015, pp. 3–7.

11 3GPP TR 36.814, Further advancements for E-UTRA physical layer aspects, March 2010.

12 N. Namdev, S. Agrawal, and S. Silkari, "recent advancement in machine learning based Internet traffic classification," *Procedia Comput. Sci.*, vol. 60, pp. 784–791, 2015.

13 S. Valenti et al. Reviewing traffic classification, in *Data Traffic Monitoring and Analysis*, Springer, 2013, pp.123–147.

14 P. Amaral, J. Dinis, P. Pinto, L. Bernardo, J. Tavares, and H. S. Mamede, "machine learning in software defined networks: data collection and traffic classification," *2016 IEEE 24th International Conference on Network Protocols (ICNP)*, Singapore, 2016, pp. 1–5.

15 T. Glennan, L. Christopher, and S. M. Erfani. "Improved classification of known and unknown network traffic flows using semi-supervised machine learning," *Australasian Conference on Information Security and Privacy*. Springer International Publishing, 2016.

16 T. Nguyen and G. Armitage, "A survey of techniques for internet traffic classification using machine learning," *IEEE Commun. Surv. Tutor.*, vol 10, no. 4, pp. 56-76, 2008.

17 J.S. Aafa et al. "A survey on network traffic classification techniques," *Int. J. Eng. Res. Technol.*, vol. 3, no. 3, 2014.

18 A. Moore and D. Zuev, "Internet traffic classification using Bayesian analysis techniques," in *ACM International Conference on Measurement and Modeling of Computer Systems (SIGMETRICS) 2005*, Banff, Alberta, Canada, June 2005.

19 M. Crotti, M. Dusi, F. Gringoli, and L. Salgarelli, "Traffic classification through simple statistical fingerprinting," *Comput. Commun. Rev.*, vol. 37, no. 1, pp. 516, 2007.

20 X. Peng, L. Qiong, and L. Sen, "Internet traffic classification using support vector machine [j]," *J. Comput. Res. Dev.*, vol. 3, p. 010, 2009.

21 T. Auld, A. W. Moore, and S. F. Gull, "Bayesian neural networks for Internet traffic classification," *IEEE Trans. Neural Netw.*, vol. 18, no. 1, pp. 223–239, 2007.

22 N. Williams, S. Zander, and G. Armitage, "A preliminary performance comparison of five machine learning algorithms for practical IP traffic flow classification," *Comput. Commun. Rev.* vol. 36, no. 5, pp. 5–16, 2006.

23 H. A. H. Ibrahim, O. R. Aqeel Al Zuobi, M. A. Al-Namari, G. MohamedAli, and A. A. A. Abdalla, "Internet traffic classification using machine learning approach: datasets validation issues," *2016 Conference of Basic Sciences and Engineering Studies (SGCAC)*, Khartoum, 2016, pp. 158–166.

24 IEEE 802.16, Machine to Machine (M2M) Evaluation Methodology Document (EMD), 2011.

25 P. Ameigeiras et al. "Analysis and modelling of YouTube traffic," *Trans. Emerg. Telecommun. Technol.*, vol. 23, no. 4, p. 360–377, 2012.

26 G. Biau, L. Devroye, and G. Lugosi, "Consistency of random forests and other averaging classifiers," *J. Mach. Learn. Res.*, vol. 9, pp. 2015–2033, 2008.

Mr. Ioannis-Prodromos Belikaidis is an electrical and computer engineer. He has been involved in routing and security of VANET (Vehicular Ad-hoc NETworks) projects and also in WiMAX propagation research. He also has extensive experience through 5G and energy-related projects that were conducted in the scope of his under- and post-graduate studies and internships in companies of the energy sector in Greece. He has knowledge on algorithms and procedures for complex networks, mobile, and pervasive, distributed computing, and currently he is active in 5G R&D.

Dr. Andreas Georgakopoulos is a solution architect and he has been involved in projects and R&D initiatives for almost a decade. Currently, he is active in 5G and Energy R&D activities and intelligent management of dynamic access networking paradigms such as opportunistic networking. He has also participated in the GreenTouch Initiative for energy-efficient networks (led by Bell Labs USA) and Wireless World Research Forum. He has also acted as work package and task leader in recent H2020 projects related to ICT and Energy.

Dr. Evangelos Kosmatos received his Dipl-Ing. and Ph.D. degrees from the school of Electrical and Computer Engineering (ICCS) of National Technical University of Athens (NTUA), Greece, in 2002 and 2008, respectively. He has participated in several EU projects (IST AQUILA, ENAMORADO, SMS, INCASA, STRONGEST, IDEALIST, CONFES, FANTASTIC-5G) as well as national projects (CONFES, WisePON). His research interests include 4G, 5G networks, optical networks (EPON, GPON, WDM-PON), Radio-over-Fiber (RoF) networks, protocols and algorithms supporting Quality of Service in heterogeneous networks, sensors/actuators networks (Internet of Things), middleware, and distributed technologies.

Dr. Stavroula Vassaki received the diploma degree in electrical and computer engineering (ECE) from the National Technical University of Athens (NTUA), Greece. Also, she received the M.Sc. degree in economics and administration of telecommunication networks from the National and Kapodistrian University of Athens, Greece and the Ph.D. degree from the NTUA focusing on the area of radio resource management in future telecommunication networks. Since 2006, she has been an associate researcher/ project engineer with the Mobile Radio Communications Laboratory, ECE, NTUA, participating in various research-oriented projects. From August 2009 to January 2011, she was an RF Engineer with the Non-Ionizing Radiation Office, Greek Atomic Energy Commission. She has been involved in various international and national research projects. Her research interests focus on wireless and satellite communications networks and include resource management mechanisms, optimization theory, game theory, and machine learning.

Mr. Orestis-Andreas Liakopoulos received the bachelor's degree in Physics from National and Kapodistrian University of Athens, Department of Electronics, Computers, Telecommunications and Control in 2012 and master's degree in electronic automation from national and Kapodistrian University of Athens, Faculties: Physics, Informatics and Telecommunications in Electronic Automation, Programming, Computer Science in 2014. In 2014, he participated in MIMO System Physical Layer Design and Implementation, in NKUOA cooperation with ANTCOR S.A. (ICT4 Growth, GSRT). Currently, he is a solution engineer at WINGS ICT Solutions. At the European level, he has been actively involved in the following international research and development programs: H2020-PHANTOM and H2020-Flex5Gware. His research interests are in the areas of structured programming, communication networks, parallel and distributed systems, object oriented programming, design and Use of database systems, and parallel computer systems technology.

Dr. Vassilis Foteinos graduated from the Department of Informatics & Telecommunications of the National and Kapodestrian University of Athens in 2008, received his M.Sc. diploma from the Department of Digital Systems of the University of Piraeus in 2011, and Ph.D. from the same department in 2015. He has conducted lab lessons for undergraduate students on the topic Computer Networks and Network Management in the Department of Digital Systems, University of Piraeus. Currently, he is participating in the H2020-Flex5Gware project. His main interests include the (co-)design of Hardware–Software systems and the development of innovative mechanisms for reconfigurable platforms in future 5G networks.

Dr. Panagiotis Vlacheas received the diploma and Ph.D. degree in electrical and computer engineering from the National Technical University of Athens (NTUA), School of Electrical and Computer Engineering, Athens, Greece, in 2003 and 2010, respectively. In 2004, he received the Ericssons awards of excellence in Telecommunications for his diploma thesis. Currently, he is a senior solution architect at WINGS ICT Solutions. At the European level, he has been actively involved in a number of international research and development programs, among others H2020-PHANTOM acting as Technical Manager, H2020-PROTEUS, H2020-Flex5Gware, H2020-iKaaS and MigraineNet project. His research interests are in the areas of cloud-IoT-Big Data integration, cognitive IoT, mobile edge computing, embedded systems, M2M communications, and reconfigurable and cognitive networks.

Prof. Panagiotis Demestichas is an experienced, senior electrical engineer with 25 years of experience in R&D and a professor in the University of Piraeus, Department of Digital Systems. He is active in the areas of 5G systems (services, environment conditions, ultradense infrastructures, spectrum aspects, virtualization, and softwarization), technoeconomic management and optimization, sustainability, simulation platforms, proof of concept, and validation by means of experiments involving environment condition generation and advanced analytics. He was also the project coordinator of ICT-FP7 OneFIT project and the technical manager of ICT-FP7 E3 project.

opment programs H2020 PHANTOM and H2020 FlexSMaps. His research interests are in the areas of structured programming, concurrent communications, parallel and distributed systems, object-oriented programming, design, and use of databases systems, and parallel computer systems technology.

Dr. Vassilis Vlachos graduated from the Department of Informatics & Telecommunications of the National and Kapodistrian University of Athens in 1998, received his M.Sc. degree from the Department of Digital Systems of the University of Piraeus in 2015, and Ph.D. from the same department in 2016. He has introduced lab courses for undergraduate students on the Computer Computer Networks and Network Management at the Department of Digital Systems, University of Piraeus. Currently, he is participating in the H2020 FlexOpware project. His main interests include the co-design of Hardware-Software systems and the development of innovative mechanisms for reconfigurable platforms in future 5G networks.

Dr. Panagiotis Vlacheas received the diploma and Ph.D. degree in electrical and computer engineering from the National Technical University of Athens (NTUA), School of Electrical and Computer Engineering, Athens, Greece, in 2003 and 2010, respectively. In 2004, he received the Ericsson award of excellence in Telecommunications for his diploma thesis. Currently, he is a senior solution architect at WINGS ICT Solutions. At the European level, he has been actively involved in a number of international research and development programs among others H2020 OneM2M technical alliance, H2020 PROFITS, H2020 FlexOpware, H2020 iKaaS and MazinNet project. His research interests are in the areas of cloud-IoT-Big Data integration, cognitive IoT, machine learning, embedded systems, M2M communications, and 5G mobile future networks.

Prof. Panagiotis Demestichas is an experienced senior electrical engineer with 20 years of experience in R&D and a professor in the University of Piraeus, Department of Digital Systems. He is active in the areas of 5G systems, services management, conditions, ultra-dense infrastructures, spectrum aspects, virtualization, and softwarization techniques, resource management and optimization, scalability, simulation platforms, proof of concept, and validation by means of experiments involving component condition generation and advanced analytics. He was also the project coordinator of ICT-FP7 OneFIT project and the technical manager of ICT-FP7 E3 project.

2

Waveform Design for 5G and Beyond

Ali Fatih Demir,[1] Mohamed Elkourdi,[1] Mostafa Ibrahim,[2] and Huseyin Arslan[1,2]

[1] Department of Electrical Engineering, University of South Florida, Tampa, FL, USA
[2] School of Engineering and Natural Sciences, Istanbul Medipol University, Istanbul, Turkey

2.1 Introduction

The standardization activities of wireless mobile telecommunications have begun with analog standards that were introduced in the 1980s, and a new generation develops almost every 10 years to meet the exponentially growing market demand. The leap from analog to digital started in second-generation (2G) systems, along with the use of mobile data services. The digital evolution with 3G enabled video calls and global positioning system (GPS) services on mobile devices. The 4G systems pushed the limits of data services further by better exploiting the time–frequency resources using orthogonal frequency-division multiple access (OFDMA) as an air interface [1]. Recently, the International Telecommunications Union (ITU) has defined the expectations for 5G [2], and the study of the next-generation wireless system is in progress with a harmony between academia, industry, and standardization entities to accomplish its first deployment in 2020.

5G is envisioned to improve major key performance indicators (KPIs), such as peak data rate, spectral efficiency, power consumption, complexity, connection density, latency, and mobility, significantly. Furthermore, the new standard should support a diverse range of services all under the same network [3]. The IMT-2020 vision defines the use cases into three main categories as enhanced mobile broadband (eMBB), massive machine-type communications (mMTC), and ultra-reliable low-latency communications (URLLC) featuring 20 Gb/s peak data rate, $10^6/km^2$ device density, and less than 1 ms latency,

5G Networks: Fundamental Requirements, Enabling Technologies, and Operations Management, First Edition.
Anwer Al-Dulaimi, Xianbin Wang, and Chih-Lin I.

respectively [4]. A flexible air interface is required to meet these different requirements. As a result, the waveform, which is the main component of any air interface, has to be designed precisely to facilitate such flexibility [5].

This chapter aims to provide a complete picture of the ongoing 5G waveform discussions and overviews the major candidates. The chapter is organized as follows: Section 2.2 provides a brief description of the waveform and reveals the 5G use cases and waveform design requirements. Also, this section presents the main features of CP-OFDM that is currently deployed in 4G LTE systems. CP-OFDM is the baseline of the 5G waveform discussions since the performance of a new waveform is usually compared with it. Section 2.3 examines the essential characteristics of the major waveform candidates along with the related advantages and disadvantages. Section 2.4 summarizes and compares the key features of the waveforms. Finally, Section 2.5 concludes the chapter.

2.2 Fundamentals of the 5G Waveform Design

2.2.1 Waveform Definition

The waveform defines the physical shape of the signal that carries the modulated information through a channel. The information is mapped from the message space to the signal space at the transmitter, and a reverse operation is performed at the receiver to recover the message in a communications system. The waveform, which defines the structure and shape of the information in the signal space, can be described by its fundamental elements: symbol, pulse shape, and lattice, as shown in Figure 2.1. The symbols constitute the random part of a waveform whereas the pulse shape and the lattice form the deterministic part.

- *Symbol:* A symbol is a set of complex numbers in the message space that is generated by grouping a number of bits together. The number of bits grouped within one symbol determines the modulation order that has a high impact on the throughput.
- *Pulse Shape:* The form of the symbols in the signal plane is defined by the pulse shaping filters. The shape of the filters determines how the energy is spread over the time and frequency domains and has an important effect on the signal characteristics.
- *Lattice:* The lattice is generated by sampling the time–frequency plane, and the locations of samples define the coordinates of the filters in the time–frequency grid. The lattice geometry might present different shapes such as rectangular and hexagonal according to the formation and distances between the samples. Furthermore, the lattice can be exploited by including additional dimensions such as space domain.

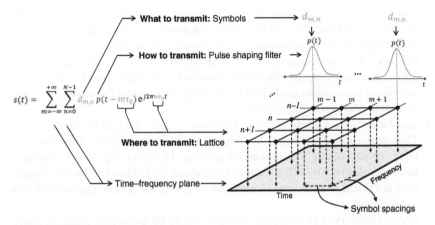

$$s(t) = \sum_{m=-\infty}^{+\infty} \sum_{n=0}^{N-1} d_{m,n}\, p(t - m\tau_0)\, e^{j2\pi n v_0 t}$$

Figure 2.1 Waveform definition [6].

2.2.2 5G Use Cases and Waveform Design Requirements

The new radio for 5G should support a wide range of services as discussed in Section 2.1. Primarily, the applications that require larger bandwidth and spectral efficiency falls into eMBB category, whereas the ones that have a tight requirement for device battery life falls into mMTC. Usually, the industrial smart sensors [7] or medical implants [8] operate several years without the demand for maintenance and hence low device complexity and high energy efficiency are crucial for these mMTC services. Furthermore, the mission-critical applications such as remote surgery [9] or self-driving vehicles [10] are represented in URLLC. The key requirements that are associated with each of these use cases are summarized in Table 2.1 and the following design criteria are essential to meet these requirements of 5G:

- *High Spectral Efficiency:* The modulation order, type of pulse shaping filters, and density of the lattice play an important role in determining the spectral

Table 2.1 The 5G use cases.

	eMBB	mMTC	URLLC
Critical Requirements	High Spectral Efficiency	Massive Asynchronous Transmission	High Reliability
	Low Latency	High Energy Efficiency	Low Latency
		Low Device Complexity	

efficiency. The guard units in time or frequency domains and other extra overheads decrease the spectral efficiency, which is critical especially for the eMBB type of communications. Furthermore, the multiantenna techniques [11] such as beamforming and massive MIMO is another crucial aspect to utilize the lattice more efficiently. However, the self-ISI and self-ICI of a waveform prevents to apply MIMO techniques in a straightforward way and increases complexity significantly.

- *Low Latency:* 5G targets a latency less than 1 ms for the URLLC applications. This goal can be managed by shortening the transmission time interval (TTI) or increasing the subcarrier spacing. However, the latter approach increases the relative CP overhead for a given TTI. Also, the localization in time is critical, and shorter filter/window durations are needed to fulfill this requirement.

- *High Reliability:* The reliability is evaluated by bit error rate (BER) or block error rate (BLER), and it is extremely important for mission-critical communications where errors are less tolerable. In addition, the retransmissions due to errors cause an increase in latency, and hence high reliable links are desirable to provide low latency as well.

- *Massive Asynchronous Transmission:* It is envisioned that there will be a huge number of nodes communicating over the 5G network for mMTC services. To maintain the synchronicity, excessive overhead is required for these applications. However, it decreases the spectral efficiency significantly. The waveforms that have strict synchronization requirements to achieve interference-free communications are not suitable for mMTC applications. Therefore, the waveforms that are well localized in the multiplexing domain are more suitable to relax the synchronization requirement for these type of applications.

- *Low Device Complexity:* The computational complexity is another critical metric of the waveform design and depends on the number of operations required at the transmitter or receiver. Additional windowing, filtering, and interference cancellation algorithms increase complexity substantially, and the system designer should consider it to design a cost-and energy-efficient transceivers.

- *High Energy Efficiency:* Low computational complexity and low peak-to-average power ratio (PAPR) provides high energy efficiency. PAPR is a statistical metric that is evaluated by complementary cumulative distributive function (CCDF) of the signal. Low PAPR is required to operate power amplifiers (PAs) efficiently, which are one of the most energy-hungry components in a transceiver.

2.2.3 The Baseline for 5G Waveform Discussion: CP-OFDM

Orthogonal frequency-division multiplexing (OFDM) is the most popular multicarrier modulation scheme that is currently being deployed in many standards such as the downlink of 4G LTE and the IEEE 802.11 family [12]. Its primary advantage over the single-carrier transmission schemes is its ability to cope with frequency selective channels for broadband communications. The data is divided into parallel streams, and each is modulated with a set of narrow subcarriers. The bandwidth of each subcarrier is set to be less than the coherence bandwidth of the channel. Hence, each subcarrier experiences a single-tap flat fading channel that can be equalized in the frequency domain with a simple multiplication operation. Also, OFDM systems utilize the spectrum in a very efficient manner due to the orthogonally overlapped subcarriers and allow flexible frequency assigning. A discrete OFDM signal on baseband is expressed as follows:

$$s_{\mathrm{OFDM}}[k] = \sum_{n=0}^{N-1} d_n e^{j2\pi k \frac{n}{N}}, \tag{2.1}$$

where d_n is the complex data symbol at subcarrier n, and N represents the total number of subcarriers. OFDM can easily be implemented by the inverse fast Fourier transform (IFFT) algorithm. Afterward, the cyclic prefix (CP) is added by copying the last part of the IFFT sequence and appending it to the beginning as a guard interval. The CP length is determined based on the maximum excess delay of the channel to alleviate the effect of intersymbol interference (ISI). However, it is hard-coded in 4G LTE and does not take into account the individual user's channel delay spread. As a result, the fixed guard interval leads to a degradation in the spectral efficiency. Furthermore, the CP yields to handling the interference in a multipath environment by ensuring circularity of the channel and by enabling easy frequency-domain equalization (FDE). OFDM partly diminishes the intercarrier interference (ICI) as well by setting the subcarrier spacing according to the maximum Doppler spread. A block diagram of conventional CP-OFDM transmitter and receiver is shown in Figure 2.2.

A major disadvantage of any multicarrier system, including CP-OFDM, is high peak-to-average power ratio (PAPR) due to the random addition of subcarriers in the time domain. For instance, consider the four sinusoidal signals as shown in Figure 2.3 with different frequencies and phase shifts [13]. The resulting signal envelope presents high peaks when the peak amplitudes of the different signals are aligned at the same time. As a result of such high peaks, the power amplifier at the transmitter operates in the nonlinear region causing a distortion and spectral spreading. In addition, as the number of subcarriers increases, the variance of the output power increases as well.

Another critical issue related to the CP-OFDM systems is its high out of band emissions (OOBE). The OFDM signal is well localized in the time domain with a rectangular pulse shape that results in a sinc shape in the frequency

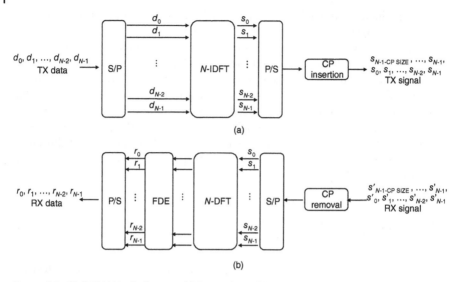

Figure 2.2 CP-OFDM block diagram. (a) Transmitter. (b) Receiver.

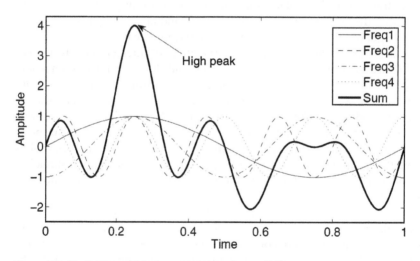

Figure 2.3 The PAPR problem in multicarrier schemes [13].

domain as shown in Figure 2.4. Especially, the sidelobes of the sincs at the edge carriers cause significant interference and should be reduced to avoid adjacent channel interference (ACI). Typically, OOBE is reduced by various windowing/filtering approaches along with the guard band allocation [14] to meet the spectral mask requirements of the various standards. 3GPP LTE uses 10% of total bandwidth as guard bands to handle this problem. However, fixed guard allocation decreases the spectral efficiency.

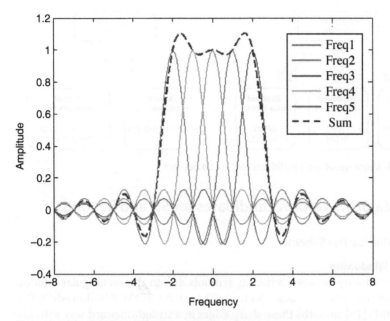

Figure 2.4 The OOBE problem in multicarrier schemes.

Furthermore, OFDM systems are more sensitive to synchronization errors than single carrier systems. As an example, if the orthogonality is lost due to the frequency offset, Doppler spread, or phase noise, the leakage from other subcarriers causes intercarrier interference (ICI). Similarly, the timing offset causes ISI or ICI when it occurs outside the guard interval.

Numerous waveforms are proposed considering all these disadvantages for the upcoming 5G standard [4,15–18]. Although backward compatibility, low implementation complexity, and easy multiple-input multiple output (MIMO) integration still make CP-OFDM an important candidate for the new standards, it seriously suffers from its limited flexibility and the unfriendly coexistence with different numerologies for various channel conditions and use cases in addition to the aforementioned problems. The proposed waveforms provide better flexibility and time–frequency localization using various filtering/windowing approaches and precoding strategies with certain trade-offs. Also, the external guard interval, that is CP, is suggested to be replaced with flexible internal guard interval to improve spectral efficiency further and to provide better performance. However, it is only being considered for the single-carrier schemes for practical reasons currently. The major waveform candidates for 5G and beyond are classified, as shown in Figure 2.5, and discussed thoroughly in the following sections.

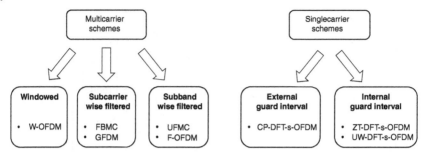

Figure 2.5 Major waveform candidates for 5G and beyond.

2.3 Major Waveform Candidates for 5G and Beyond

2.3.1 Multicarrier Schemes

2.3.1.1 Windowing

The discontinuity between adjacent symbols due to the rectangular window shape in the timedomain causes high OOBE for CP-OFDM. Windowed-OFDM (W-OFDM) [15] smooths these sharp edges in a straightforward way with low complexity. The baseband W-OFDM can be expressed as follows:

$$s_{\text{W--OFDM}}[k] = \sum_{m=-\infty}^{+\infty} \sum_{n=0}^{N-1} d_{m,n} g[k - m(N + L_{\text{CP}} + L_{\text{Ext}})] e^{j2\pi k \frac{n}{N}}, \quad (2.2)$$

where $d_{m,n}$ is the complex data transmitted on the nth subcarrier and m^{th} OFDM symbol, L_{CP} presents the CP length, L_{Ext} expresses the windowing extension, and $g[n]$ shows the windowing function. Several windowing functions have been evaluated in detail [19] with different trade-offs between the width of the main lobe and suppression of the side lobes. An illustration of windowing operation at the transmitter is shown in Figure 2.6.

Initially, the CP is further extended on both edges at the transmitter and the extended part from the beginning of the OFDM symbol is appended to the end. The windowing operation is applied symmetrically on both edges of the

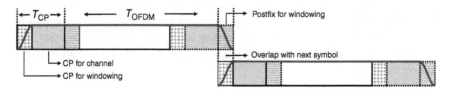

Figure 2.6 W-OFDM (Transmitter).

Edge windowing

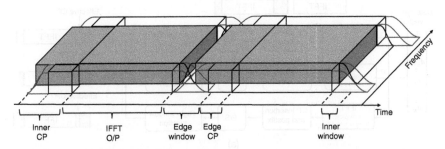

Figure 2.7 Edge windowing technique [20].

OFDM symbol, and the transitions parts (i.e., ramp-ups and ramp-downs) of adjacent symbols are overlapped to shorten the extra time domain overhead resulting from windowing. In addition, the windowing operation is performed at the receiver as well to reduce the interference from other users.

Edge windowing [20] is another approach to reduce the high OOBE of CP-OFDM. It is well known that the outer subcarriers have a higher influence on OOBE problem compared to the inner subcarriers. However, conventional windowing techniques apply the same window for all subcarriers within an OFDM symbol. As a result, the spectral efficiency decreases due to an extra windowing duration or the performance degrades since the effective CP length of channel shortens. The proposed approach borrows the CP duration of the channel to perform windowing and maintains the spectral efficiency (Figure 2.7). The longer windows that decrease the effective CP size are applied only to the edge subcarriers, as shown in Figure 2.8 and hence the OOBE is suppressed with a minimal performance loss. If the CP of the edge subcarriers is less than or equal to the maximum excess delay of the channel, neither ISI nor ICI is observed. Therefore, these edge subcarriers should be assigned to the user equipments that experience shorter delay spread. The edge windowed OFDM provide a better spectrum confinement with a low complexity and negligible performance loss [21].

Although windowing approaches present lower OOBE compared to CP-OFDM, the effect is limited and nonnegligible guard bands are still required. However, these methods can be applied along with the filtering approaches that are discussed in the following sections to provide better spectral confinement.

2.3.1.2   Subcarrier-Wise Filtering
2.3.1.2.1   FBMC
Filter bank multicarrier (FBMC) yields a good frequency-domain localization by extending the pulse duration in the time domain and using properly designed pulse shaping filters [6,19]. These flexible filters are applied at the

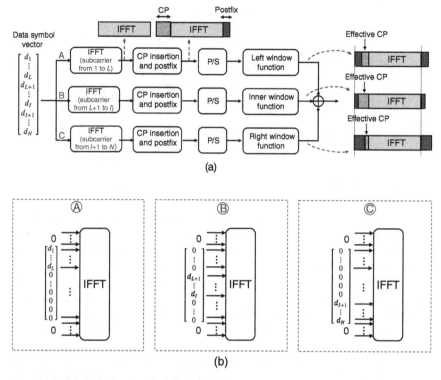

Figure 2.8 Edge windowing block diagram [21].

subcarrier level, and they enable adaption to various channel conditions and use cases. There are various ways to implement FBMC such as filtered multitone (FMT), cosine modulated multitone (CMT), and staggered modulated multitone (SMT). However, SMT, which is widely known as offset quadrature amplitude modulation OQAM–FBMC, is the focus of 5G waveform discussion due to its ability to handle interference while allowing dense symbol placement in the time–frequency lattice [4]. OQAM signaling provides staggering of "in-phase" and "quadrature-phase" components in both time and frequencydomains as shown in Figure 2.9 and hence, orthogonality is maintained within the real and imaginary domains separately.

The baseband FBMC/OQAM is expressed as follows:

$$s_{\text{FBMC–OQAM}}[k] = \sum_{m=-\infty}^{+\infty} \sum_{n=0}^{N-1} d_{m,n} g[k - m\frac{N}{2}] e^{j2\pi k\frac{n}{N}} e^{j\phi_{m,n}}, \tag{2.3}$$

where g represents the prototype filter, $\phi_{m,n}$ is an additional phase term at subcarrier n and symbol index m, which is expressed as $\frac{\pi}{2}(m + n)$. The $d_{m,n}$ is

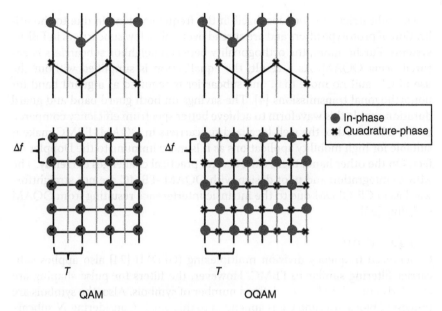

Figure 2.9 QAM signaling versus OQAM signaling [22].

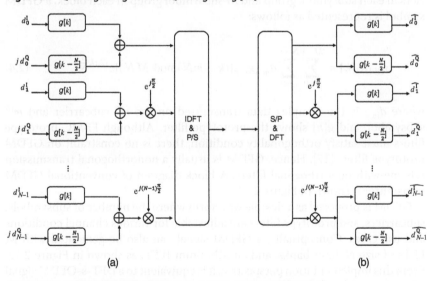

Figure 2.10 OQAM-FBMC block diagram. (a) Transmitter. (b) Receiver.

real valued since the real and imaginary parts are transmitted with a delay. Also, to address a perfect reconstruction of symbols, the prototype filter must satisfy the orthogonality condition [17]. A block diagram of conventional OQAM–FBMC transmitter and receiver is shown in Figure 2.10.

The subcarriers are well localized in the frequency domain due to the utilization of prototype filters and are spread over only a few subcarriers in FBMC systems. Furthermore, the orthogonality between neighbor subcarriers is ensured using OQAM. As a result, the equalization is simplified without the use of CP, and no more than one subcarrier is required as a guard band for nonorthogonal transmissions [4]. The savings on both guard band and guard duration enable this waveform to achieve better spectrum efficiency compared to CP-OFDM. Also, the well-localized subcarriers in OQAM–FBMC make it suitable for high mobility applications as it is more immune to the Doppler effect. On the other hand, there exist several practical challenges currently. The MIMO integration and pilot design with OQAM–FBMC are not straightforward as in CP-OFDM due to the intrinsic interference resulting from OQAM signaling [23].

2.3.1.2.2 GFDM

Generalized frequency division multiplexing (GFDM) [24] also applies subcarrier filtering similar to FBMC. However, the filters for pulse shaping are circularly convoluted over a defined number of symbols. Also, the symbols are processed blockwise, and CP is appended to this block. Considering N subcarriers in each subsymbol group and M subsymbol group in each block, a GFDM symbol is represented as follows:

$$s_{\text{GFDM}}[k] = \sum_{m=0}^{M-1} \sum_{n=0}^{N-1} d_{m,n} g_{m,n}[(k - mN)\text{mod}(MN)] e^{j2\pi k \frac{n}{N}}, \qquad (2.4)$$

where $d_{m,n}$ is the complex data transmitted on the nth subcarrier and m^{th} subsymbol, and $g[n]$ shows the prototype filter. Although FBMC prototype filters must satisfy orthogonality condition, there is no constraint on GFDM prototype filters [17]. Hence, GFDM is usually a nonorthogonal transmission scheme with nonorthogonal filters. A block diagram of conventional GFDM transmitter is shown in Figure 2.11.

GFDM is proposed as a flexible waveform where the number of subsymbols, subcarriers, and prototype filters are adjustable for various channel conditions and use cases. Conceptually, a GFDM signal can also be generated with M FFTs of size N, filter banks, and an MN-point IFFT, as shown in Figure 2.12. From this implementation perspective, it is equivalent to a DFT-s-OFDM signal when the rectangular function is used as a prototype filter, which also explains lower PAPR compared to CP-OFDM. This equivalency is further discussed in the following single-carrier waveform discussion. Also, it is equivalent to CP-OFDM when M equals to 1.

GFDM shares the well-frequency-localized characteristic with OQAM-FBMC. Hence, it is suitable for high-mobile scenarios and provides more

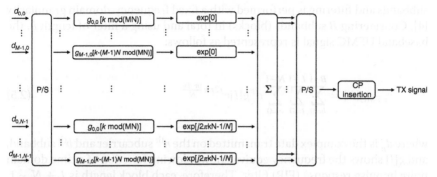

Figure 2.11 GFDM block diagram (Transmitter).

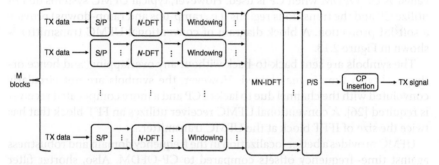

Figure 2.12 Equivalency of GFDM and DFT-s-OFDM (Transmitter).

immunity to synchronization errors. Although GFDM provides flexibility in the waveform, the nonorthogonal transmission scheme requires complex successive interference cancellation (SIC) algorithms at the receiver. Similar to OQAM-FBMC, pilot design and MIMO transmission is complicated. Furthermore, the block-wise transmission causes latency that makes it infeasible for mission-critical applications.

The waveforms that perform subcarrier-wise filtering, that is FBMC, and GFDM require a new transceiver design, and there is no backward compatibility with 4G LTE. The next section describes the subband-wise filtered multicarrier schemes and these operations are applicable to the current standards without significant changes.

2.3.1.3 Subband-Wise Filtered MCM
2.3.1.3.1 UFMC
Universal filtered multicarrier (UFMC) [25] applies subband-wise filtering to reduce OOBE. The subband-wise filtering is considered as a compromise between the whole band filtering and subcarrier-wise filtering. Hence, the filters are shorter compared to the FBMC, where the length of filters are much longer than the symbol duration. The total available bandwidth is partitioned into

subbands and filtering is performed with a fixed frequency-domain granularity [4]. Considering B subbands (blocks) in total and using a filter of length L, the baseband UFMC signal is represented as follows:

$$s_{\text{UFMC}}[k] = \sum_{b=0}^{B-1} \sum_{l=0}^{L-1} \sum_{n=0}^{N-1} d_n^b g[l] e^{j2\pi k \frac{(n-l)}{N}} \tag{2.5}$$

where d_n^b is the complex data transmitted on the nth subcarrier and b^{th} subband, and $g[l]$ shows the frequency equivalent windowing function of a time domain finite impulse response (FIR) filter. Therefore, each block length is $L + N - 1$. The use of CP is optional to provide better immunity against ISI, and it is also called as UF-OFDM when CP is used. However, typical UFMC systems do not utilize CP and the transitions regions (i.e., ramp-ups and ramp-downs) provide a soft ISI protection. A block diagram of conventional UFMC transmitter is shown in Figure 2.13.

The symbols are sent back-to-back without any overlapping, and hence orthogonality in time is maintained. However, the symbols are not circularly convoluted with the channel due to lack of CP and a more complicated receiver is required [26]. A conventional UFMC receiver utilizes an FFT block that has twice the size of IFFT block at the UFMC transmitter.

UFMC provides a better localization in the frequency domain and robustness against time–frequency offsets compared to CP-OFDM. Also, shorter filter lengths compared to subcarrier-wise filtering makes it more suitable for low-latency applications. On the other hand, these shorter filters offer limited OOBE suppression. Furthermore, increased complexity due to the lack of CP and complicated filtering operations should be dealt with intelligently to design practical communications systems.

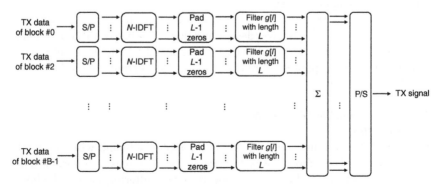

Figure 2.13 UFMC block diagram (Transmitter).

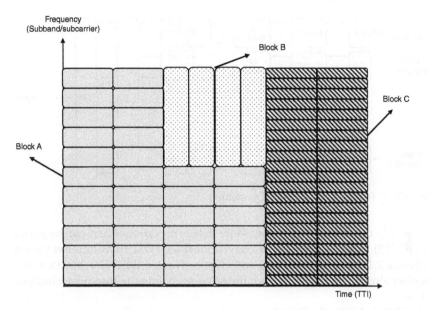

Figure 2.14 An example of flexible resource allocation enabled by f-OFDM.

2.3.1.3.2 f-OFDM

Filtered-OFDM (f-OFDM) [27] is another subband-wise filtered multicarrier scheme, but the filtering granularity is more flexible than UFMC. The partition in the time–frequency grid is adjusted based on the different channel conditions and use cases, as shown in Figure 2.14. This flexibility makes f-OFDM more suitable for the use of different numerologies (such as bandwidth, subcarrier spacing, CP duration, and transmission time interval) compared to UFMC with the cost of increased complexity. Considering B blocks in total, the baseband f-OFDM signal is represented as follows:

$$s_{\text{f-OFDM}}[k] = \sum_{b=0}^{B-1} \sum_{m=0}^{M-1} \sum_{l=0}^{L_b-1} \sum_{n=0}^{N-1} d_{m,n}^{b} g_b[l] e^{j2\pi k \frac{(n-l-mL_{\text{CP}})}{N}}, \tag{2.6}$$

where $d_{m,n}^{b}$ is the complex data transmitted on the bth block, n^{th} subcarrier, and m^{th} subsymbol, $g_b[l]$ shows the frequency equivalent windowing function of a time domain FIR filter on the bth block, and L_{CP} presents the CP size. As can be seen in the equation, f-OFDM maintains the CP in contrast to UFMC. Therefore, it is more immune to the ISI and needs less complex receiver. Ideally, the frequency-domain window $g_b[l]$ is desired to be a rectangle with a size of L_b. However, it corresponds to an infinite length sinc shape response in the time domain, and hence it is impractical. Therefore, windowed sinc functions

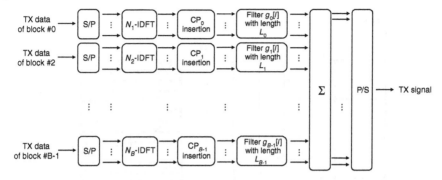

Figure 2.15 f-OFDM block diagram (Transmitter).

are used in the filtering operation. More details on various filters can be found in Ref. [19]. A block diagram of conventional f-OFDM transmitter is shown in Figure 2.15. Matched filtering and identically sized IFFT/FFT blocks at the receiver also differs f-OFDM from the other subband-wise filtering technique, UFMC.

F-OFDM shares all advantages of well frequency -localized waveforms such as low OOBE, allowing asynchronous transmission, being feasible for different numerologies, and requiring less number of guard tones. Although f-OFDM cannot provide low OOBE as subcarrier-wise filtered multicarrier schemes due to the use of shorter filter lengths, it is compatible with MIMO transmission scheme and does not require any successive interference cancellation (SIC) algorithm. However, complexity is still the main drawback of f-OFDM compared to CP-OFDM.

2.3.2 Single-Carrier Schemes

2.3.2.1 CP-DFT-s-OFDM

Discrete Fourier transform spread OFDM (DFT-s-OFDM) waveforms are proposed to mitigate the high PAPR problem in CP-OFDM while maintaining the useful characteristics of it. The data input can be modeled as independent and identically distributed random variables, and as a result, the corresponding output of IDFT in CP-OFDM has a high variance. Such high variance is reduced by providing correlation to the input data by performing a DFT operation before the IDFT process, as shown in Figure 2.16.

The utilization of CP in DFT-s-OFDM ensures circularity of the signal at the receiver and enables easy FDE to handle multipath channel effect. This waveform can be interpreted in two ways [18]. One interpretation is that it is a precoded CP-OFDM scheme, where PAPR is mitigated by DFT precoding. This interpretation provides to consider different precoding methods. The other interpretation is that it is a transmission scheme that upsamples the

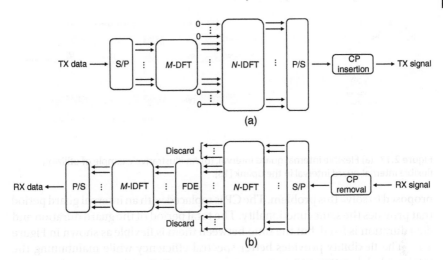

Figure 2.16 CP-DFT-s-OFDM block diagram. (a) Transmitter. (b) Receiver.

input data by the ratio of the IDFT and DFT block sizes (i.e., N/M where N>M), and performs a circular pulse shaping with a Dirichlet sinc function. This interpretation lets the designers consider different pulse shaping approaches to reduce the PAPR further and handle the OOBE.

DFT-s-OFDM is deployed in the uplink of the 4G LTE due to its lower PAPR feature that provides better power efficiency. The low complexity, support of dynamic spectrum access, and MIMO compatibility make CP-DFT-s-OFDM a significant candidate for the 5G, similar to the CP-OFDM. However, the spectral efficiency of this waveform is also comparable to CP-OFDM and suffers from high OOBE because of the discontinuity between adjacent symbols. Hence, similar windowing and filtering approaches as discussed in the multi-carrier schemes discussion that can be applied to improve spectral efficiency. In addition to that, internal guard concept for DFT-s-OFDM is being discussed for the 5G, which provides more flexibility in the system design as explained in the following sections.

2.3.2.2 ZT-DFT-s-OFDM

The guard interval is hard-coded in 4G LTE systems, and there exist only two options as normal and extended CP. However, the base station is preset to only one of these guard intervals because the use of different guard interval durations results in different symbol durations and consequently a different number of symbols per frame. This leads to the generation of mutual asynchronous interference even when the frames are aligned [16]. Hence, the users with two different CP durations do not coexist together in the same cell. As a result, the nonflexible guard interval penalizes the user equipments that experience better channel conditions. Zero-tail-DFT-spread-OFDM (ZT–DFT-s-OFDM) [28] is

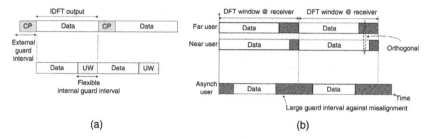

Figure 2.17 (a) Flexible internal guard interval. (b) An illustrative example of utilizing flexible internal guard interval in the uplink [18].

proposed to solve this problem. The CP is replaced with an internal guard period that provides the same functionality. The total period of the guard duration and data duration is fixed, but the ratio between them is flexible as shown in Figure 2.17. The flexibility provides better spectral efficiency while maintaining the total symbol duration.

Zero vectors with variable lengths are inserted into the head and tail of the data before DFT process in this approach. The tail length is set to be longer than the delay spread of the channel and, hence the leakage to the next symbol does not have significant power. Also, the zeros in the head, which are usually shorter than the zeros at the tail, provide a smoother transition and yield substantial OOBE reduction [16]. A block diagram of conventional ZT-DFT-s-OFDM transmitter and receiver is shown in Figure 2.18.

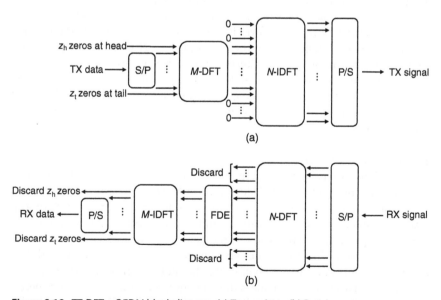

Figure 2.18 ZT-DFT-s-OFDM block diagram. (a) Transmitter. (b) Receiver.

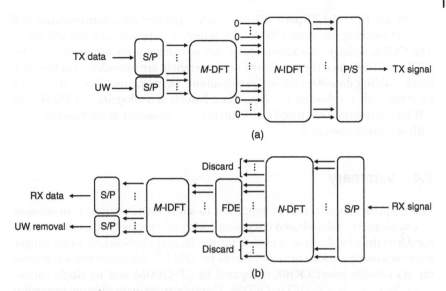

Figure 2.19 UW-DFT-s-OFDM block diagram. (a) Transmitter. (b) Receiver.

The fixed sequences (i.e., zero vectors) appended to each symbol ensure circularity at the receiver, and hence ZT-DFT-s-OFDM supports single-tap FDE. However, a residual energy of the data part in the last samples introduces a noncyclical leakage to the next symbol [16] and hence, internal guard interval approach do not provide perfect circularity as CP does. Furthermore, this leakage is a limiting factor in the link performance for the users utilizing high-order modulations in a multipath environment.

The PAPR and OOBE are low for ZT-DFT-s-OFDM, and spectral efficiency is increased due to flexible guard interval. The internal guard feature makes it suitable for different symbol durations without introducing mutual asynchronous interference. However, this flexibility causes extra overhead to track delay spread of the channel. Also, windowing can easily be applied to decrease OOBE without further extra guard duration.

2.3.2.3 UW-DFT-s-OFDM

Unique word DFT-spread OFDM (UW-DFT-s-OFDM) [29] is another single-carrier scheme that utilizes flexible internal guard band. The zero tails and heads of the ZT–DFT-s-OFDM are replaced with a fixed sequence that enhances cyclic properties of the signal in this approach. Since the fixed sequence is inserted before the DFT process as shown in Figure 2.19, the orthogonality is provided between the data and the unique word. The circularity is also ensured in this waveform and as a result simple FDE is supported. However, the leakage from the data part limits its link performance for the high-order modulations similar to ZT-DFT-s-OFDM.

Different from the ZT part, UW can also be exploited for synchronization and channel tracking purposes [30,31]. Therefore, it improves spectral efficiency. The OOBE leakage characteristics of this waveform are comparable to ZT–DFT-s-OFDM since the continuity of the symbols are provided with the DFT block. Adding data-dependent "perturbation" signal or modifying the kernel function with windowing the input data (which is analogous to GFDM with UW) mitigates the OOBE and PAPR further [18]. However, these benefits come with increased complexity.

2.4 Summary

The frequency localization is important to allow asynchronous transmission across adjacent subbands and coexistence with other waveforms. On the other hand, the time localization is critical for low latency applications where longer filter/window durations are not feasible for URLLC. All discussed waveforms for 5G provide lower OOBE compared to CP-OFDM and its single-carrier equivalent, that is, CP-DFT-s-OFDM. The subcarrier-wise filtering operation in FBMC results in the best frequency localization among the candidate waveforms due to the use of longer filter lengths. Although GFDM is another subcarrier-wise filtered waveform, the rectangular window shape in the time domain causes abrupt transitions and increases OOBE. However, windowing can be performed on this waveform, and W-GFDM presents a good spectral confinement as well. Furthermore, relatively shorter filters in the subband-wise filtered waveforms lead to a better time localization with a price of increasing the OOBE compared to the subcarrier-wise filtered waveforms.

Most multicarrier schemes suffer from high PAPR and are not suitable when high energy efficiency is required. However, GFDM exhibits a reduced PAPR characteristic due to its equivalency to DFT-spread waveforms, as discussed before. The single-carrier schemes are preferable in energy-limited use cases along with the use of flexible guard intervals that provide better spectral confinement and improved PAPR.

The spectral efficiency is another critical design criteria that is highly affected by the window/filter duration, the shape of filter, and extra overheads. Well-frequency-localized waveforms reduce the need for guard bands and hence leading to better efficiency in the frequency domain. On the other hand, the waveforms that do not utilize a guard interval, such as FBMC, are expected to have higher efficiency in the time domain. However, BER/BLER performance decreases substantially in a multipath fading channel due to lack of guard interval. As a result, complex receivers are required since an easy FDE is not possible. MIMO compatibility is also essential to achieve high throughput. The schemes that allow interference, such as FBMC and GFDM, cannot deploy straightforward MIMO algorithms.

Finally, the guard interval in the time domain makes a waveform more robust against ISI and time-offsets. In addition, the guard bands or the use of well-localized waveforms in the frequency domain make a waveform robust against carrier frequency offset and Doppler effects that reduce ICI and adjacent channel interference (ACI) in a multiple access environment. As a result, FBMC has the best immunity to ICI and is the most vulnerable to ISI.

A summary of the main advantages/disadvantages of these major 5G candidate waveforms is provided in Table 2.2. Moreover, the time –frequency grids, OOBE, and the pulse shapes of these waveforms are presented in Figures 2.20 and 2.21.

Table 2.2 The 5G waveform candidates.

Multicarrier Schemes		
Waveform	**Advantages**	**Disadvantages**
CP-OFDM	• Simple FDE • Easy MIMO integration • Flexible frequency assignment • Low implementation complexity	• High OOBE and PAPR • Strict synchronization requirement • Poor performance for high mobility applications • Hard-coded CP
W-OFDM	• All advantages belongs to CP-OFDM • Lower OOBE compared to CP-OFDM	• Either poor spectral efficiency or BER performance (depending on windowing type)
OQAM-FBMC	• Best frequency localization (i.e., lowest OOBE) • Good spectral efficiency (no guard band or CP) • Suitable for high-mobility applications • Convenient for asynchronous transmission	• Challenging MIMO integration and pilot design • No immunity to ISI due to lack of CP • High implementation complexity • Increased power consumption due to OQAM signaling
GFDM	• Flexible design • Good frequency localization • Reduced PAPR	• Higher latency due to block processing • Challenging MIMO integration and pilot design • High implementation complexity
UFMC	• Good frequency localization • Shorter filter length compared to subcarrier-wise operations (i.e., OQAM-FBMC and GFDM) • Compatible with MIMO	• No immunity to ISI due to lack of CP • High receiver complexity due to increased FFT size
F-OFDM	• Flexible filtering granularity • Better frequency localization • Shorter filter length compared to subcarrier-wise operations (i.e., OQAM-FBMC and GFDM) • Compatible with MIMO	• Very high implementation complexity
Single-carrier Schemes		
Waveform	**Advantages**	**Disadvantages**
CP-DFT-s-OFDM	• All advantages belongs to CP-OFDM • Low PAPR	• High OOBE • Strict synchronization requirement • Hard-coded CP
ZT-DFT-s-OFDM	• Flexible guard interval • Better spectral efficiency • Lower OOBE compared to CP-DFT-s-OFDM	• Strict synchronization requirement • Extra control signaling • Limited link performance for higher order modulation
UW-DFT-s-OFDM	• Flexible guard interval • Best spectral efficiency • Lowest OOBE and PAPR	• Strict synchronization requirement • Extra control signaling • Limited link performance for higher order modulation • High implementation complexity

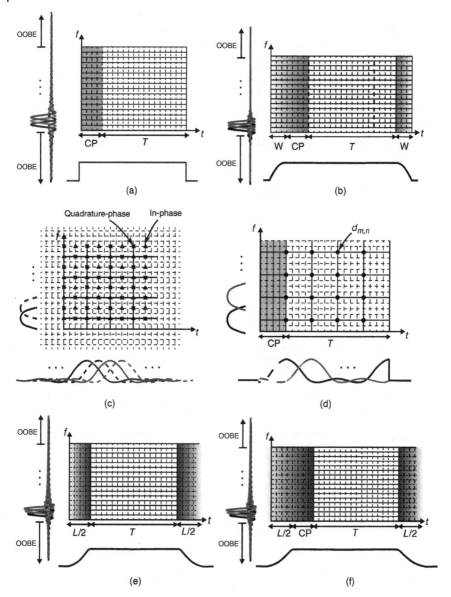

Figure 2.20 Comparison of multicarrier schemes. (a) CP-OFDM, (b) W-OFDM, (c) OQAM-FBMC, (d) GFDM, (e) UFMC, and (f) F-OFDM.

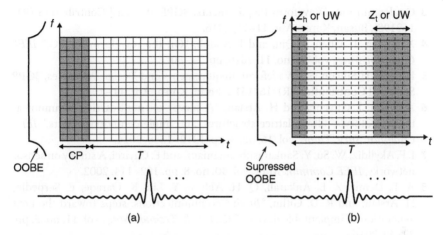

Figure 2.21 Comparison of single-carrier schemes. (a) CP-DFT-s-OFDM. (b) ZT-DFT-s-OFDM or UW-DFT-s-OFDM.

2.5 Conclusions

In this chapter, the ongoing new radio design discussions are summarized, and the major waveform candidates for 5G and beyond are presented. The main 5G use cases are identified along with the associated critical design requirements, and a brief description of CP-OFDM is provided as a baseline for the new waveform discussion. The candidate waveforms are classified considering the way the spectrum is utilized (i.e., single carrier versus multicarrier), the fashion in which the signal processing techniques are performed (i.e., windowing, subcarrier-wise filtering, and subband-wise filtering), and the guard interval type that is adopted (i.e., internal and external). The advantages and disadvantages of each waveform are discussed in detail. It could be concluded that there is no waveform that fits all requirements yet and the physical layer should be designed considering the specific use cases and requirements. Unlike the previous standards, the new generations will support high flexibility to fully exploit and increase further the potential of future communications systems.

References

1 M. Elkourdi, B. Pekoz, E. Guvenkaya, and H. Arslan, "Waveform design principles for 5G and beyond," in *2016 IEEE 17th Annual Wireless and Microwave Technology Conference (WAMICON)*, Apr. 2016, pp. 1–6.

2 ITU-R, IMT Vision - Framework and overall objectives of the future development of IMT for 2020 and beyond Technical Report M.2083-0, Sept. 2015.

3 Qualcomm Inc., Waveform Requirements, 3GPP Standard Contribution (R1-162198), Busan, Korea, Apr. 11–15, 2016.

4 X. Zhang, L. Chen, J. Qiu, and J. Abdoli, "On the waveform for 5G," *IEEE Commun. Mag.*, vol. 54, no. 11, 2016, pp. 74–80.

5 Huawei, HiSilicon, 5G Waveform: Requirements and Design Principles, 3GPP Standard Contribution (R1-162151), Busan, Korea, Apr. 11–15, 2016.

6 A. Sahin, I. Guvenc, and H. Arslan, "A survey on multicarrier communications: prototype filters, lattice structures, and implementation aspects," *IEEE Commun. Surv. Tutor.*, vol. 16, no. 3, pp. 1312–1338, 2014.

7 I. F. Akyildiz, W. Su, Y. Sankarasubramaniam, and E. Cayirci, A survey on sensor networks, *IEEE Commun. Mag.*, vol. 40, no. 8, pp. 102–114, 2002.

8 A. F. Demir, Z. E. Ankarali, Q. H. Abbasi, Y. Liu, K. Qaraqe, E. Serpedin, H. Arslan, and R. D. Gitlin, "*In vivo* communications: steps toward the next generation of implantable devices," *IEEE Veh. Technol. Mag.*, vol. 11, no. 2, pp. 32–42, 2016.

9 A. F. Demir, Q. Abbasi, Z. E. Ankarali, A. Alomainy, K. Qaraqe, E. Serpedin, and H. Arslan, "Anatomical region-specific *in vivo* wireless communication channel characterization," *IEEE J. Biomed. Health Inf.*, vol. 21, no. 5, pp. 1254–1262, 2017.

10 C. Urmson and W. Whittaker, "Self-driving cars and the urban challenge," *IEEE Intell. Syst.*, vol. 23, no. 2, pp. 66–68, 2008.

11 M. Hafez and H. Arslan, "On directional modulation: an analysis of transmission scheme with multiple directions," in *2015 IEEE International Conference on Communication Workshop (ICCW)*, June 2015, pp. 459–463.

12 T. Hwang, C. Yang, G. Wu, S. Li, and G. Y. Li, "ofdm and its wireless applications: a survey," *IEEE Trans. Veh. Technol.*, vol. 58, no. 4, pp. 1673–1694, 2009.

13 Y. Rahmatallah and S. Mohan, "Peak-to-average power ratio reduction in OFDM systems: a survey and taxonomy," *IEEE Commun. Surv. Tutor.*, vol. 15, no. 4, pp. 1567–1592, 2013.

14 A. F. Demir and H. Arslan, "The impact of adaptive guards for 5G and beyond," in *2017 IEEE 28th Annual International Symposium on Personal, Indoor, and Mobile Radio Communications (PIMRC)*, Oct. 2017, pp. 1–5.

15 Qualcomm Inc., Waveform Candidates, 3GPP Standard Contribution (R1-162199), Busan, Korea, Apr. 11–15, 2016.

16 G. Berardinelli, K. I. Pedersen, T. B. Sorensen, and P. Mogensen, "Generalized DFT-spread-OFDM as 5G waveform," *IEEE Commun. Mag.*, vol. 54, no. 11, pp. 99–105, 2016.

17 H. Lin and P. Siohan, Major 5G waveform candidates: overview and comparison. *Signal Processing for 5G*. Chichester: John Wiley & Sons, Ltd, 2016, pp. 169–188.

18 A. Sahin, R. Yang, E. Bala, M. C. Beluri, and R. L. Olesen, "Flexible DFT-S-OFDM: solutions and challenges," *IEEE Commun. Mag.*, vol. 54, no. 11, pp. 106–112, 2016.

19 B. Farhang-Boroujeny, "OFDM versus filter bank multicarrier," *IEEE Signal Process. Mag.*, vol. 28, no. 3, pp. 92–112, 2011.

20 A. Sahin and H. Arslan, "Edge windowing for OFDM based systems," *IEEE Commun. Lett.*, vol. 15, no. 11, pp. 1208–1211, 2011.

21 Samsung, Discussion on Multi-Window OFDM for NR Waveform, 3GPP Standard Contribution (R1-166746), Gothenburg, Sweden, Aug. 22–26, 2016.

22 Rohde & Schwarz, 5G Waveform Candidates, Technical Report, 2016.

23 H. Lin, "Flexible configured OFDM for 5G air interface," *IEEE Access*, vol. 3, pp. 1861–1870, 2015.

24 N. Michailow, M. Matthe, I. S. Gaspar, A. N. Caldevilla, L. L. Mendes, A. Festag, and G. Fettweis, "Generalized frequency division multiplexing for 5th generation cellular networks," *IEEE Trans. Commun.*, vol. 62, no. 9, pp. 3045–3061, 2014.

25 V. Vakilian, T. Wild, F. Schaich, S. t. Brink, and J. F. Frigon, "Universal-filtered multi-carrier technique for wireless systems beyond LTE," in *2013 IEEE Globecom Workshops (GC Wkshps)*, Dec. 2013, pp. 223–228.

26 F. Schaich, T. Wild, and Y. Chen, "Waveform contenders for 5G: suitability for short packet and low latency transmissions," in *2014 IEEE 79th Vehicular Technology Conference (VTC Spring)*, May 2014, pp. 1–5.

27 X. Zhang, M. Jia, L. Chen, J. Ma, and J. Qiu, "Filtered-OFDM: enabler for flexible waveform in the 5th generation cellular networks," in *2015 IEEE Global Communications Conference (GLOBECOM)*, Dec. 2015, pp. 1–6.

28 G. Berardinelli, F. M. L. Tavares, T. B. Sorensen, P. Mogensen, and K. Pajukoski, "Zero-tail DFT-spread-OFDM signals," in *2013 IEEE Globecom Workshops (GC Wkshps)*, Dec. 2013, pp. 229–234.

29 A. Sahin, R. Yang, M. Ghosh, and R. L. Olesen, "An improved unique word DFT-spread OFDM scheme for 5F systems," in *2015 IEEE Globecom Workshops (GC Wkshps)*, Dec. 2015, pp. 1–6.

30 J. Coon, M. Sandell, M. Beach, and J. McGeehan, "Channel and noise variance estimation and tracking algorithms for unique-word based single-carrier systems," *IEEE Trans. Wirel. Commun.*, vol. 5, no. 6, pp. 1488–1496, 2006.

31 M. Huemer, H. Witschnig, and J. Hausner, "Unique word based phase tracking algorithms for SC/FDE-systems," in *IEEE Global Telecommunications Conference, 2003. GLOBECOM '03*, vol. 1, Dec. 2003, pp. 70–74.

Ali Fatih Demir received the B.S. degree in electrical engineering from Yildiz Technical University, Istanbul, Turkey, in 2011, and the M.S. degree in electrical engineering and applied statistics from Syracuse University, Syracuse, NY, USA in 2013. He is currently pursuing the Ph.D. degree as a member of the Wireless Communication and Signal Processing (WCSP) Group in the Department of Electrical Engineering, University of South Florida, Tampa, FL, USA. His current research interests include waveform design, multicarrier systems, *in vivo* communications, and brain–computer interfaces. He is a student member of the IEEE.

Mohamed H. Elkourdi received the B.Sc. degree in telecommunication and electronics engineering with distinction from Applied Science University, Amman, Jordan, in 2010, and the M.S. degree in telecommunication and signal processing from New Jersey Institute of Technology, Newark, NJ, USA in 2013. He is currently pursuing the Ph.D. degree as a member of the Innovation in Wireless Information Networking Laboratory (iWINLAB) in the Department of Electrical Engineering, University of South Florida, Tampa, FL, USA. His current research interests include waveform design, multicarrier systems, multiple access techniques, and MIMO systems.

Mostafa Ibrahim received the B.Sc. degree from Ain Shams University, Cairo, Egypt in 2010, and the M.Sc. degree from Istanbul Medipol University, Istanbul, Turkey, in 2017, both in electronics and communication engineering. Prior to joining the Communications, Signal Processing, and Networking Center (CoSiNC) at Istanbul Medipol University in 2015, he was with the Egyptian Air Force, where he was a communication engineer officer (2011–2013), and with the Center for Nanoelectronics and Devices (CND) at the American University in Cairo, where he worked on on-chip energy harvesting systems optimization (2013–2014). His current research interests include air–ground channel modeling and waveform/modulation design beyond OFDMA.

Huseyin Arslan received the B.S. degree from Middle East Technical University, Ankara, Turkey, in 1992 and the M.S. and Ph.D. degrees from Southern Methodist University, Dallas, TX, USA, in 1994 and 1998, respectively. From January 1998 to August 2002, he was with the research group of Ericsson Inc., NC, USA, where he was involved with several projects related to 2G and 3G wireless communication systems. Since August 2002, he has been with the Department of Electrical Engineering, University of South Florida, Tampa, FL, USA, where he is a professor. In December 2013, he joined Istanbul Medipol University, Istanbul, Turkey, where he worked as the Dean of the School of Engineering and Natural Sciences. His current research interests include waveform design for 5G and beyond, physical layer security, dynamic spectrum access, cognitive radio, coexistence issues on heterogeneous networks, aeronautical (high-altitude platform) communications, and *in vivo* channel modeling and system design. He is currently a member of the editorial board of the *Sensors Journal* and the *IEEE Surveys and Tutorials*. He is a fellow of the IEEE.

3

Full-Duplex System Design for 5G Access

Shu-ping Yeh,[1] Jingwen Bai,[1] Ping Wang,[2] Feng Xue,[1] Yang-seok Choi,[1] Shilpa Talwar,[1] Sung-en Chiu,[3] and Vinod Kristem[1]

[1] *Intel Corporation, Intel Lab, Santa Clara, CA, USA*
[2] *Apple Inc., Santa Clara, CA, USA*
[3] *University of California, Electrical and Computer Engineering, San Diego, CA, USA*

3.1 Introduction

The next-generation wireless access, 5G, is expected to offer much higher capacity and performance beyond what the current 4G network can provide. In order to meet the new requirements of 5G networks and to overcome the limits in nowadays systems, it is anticipated that revolutionary technologies, including full-duplex (FD), should be introduced to existing and evolving systems, like LTE-A and Wi-Fi. In-band full-duplex (IBFD) techniques that support simultaneous transmit and receive in the same frequency at the same time can potentially double the spectrum efficiency. However, IBFD operation has long been considered as a daunting challenge due to the stringent requirement on self-interference cancellation (SIC) for suppressing the significant interference from transmitter (TX) to receiver (RX) chain in FD systems. Analyses from References [1–3] indicate that, in order to achieve IBFD, the power of self-interference from TX side to RX end is required to be attenuated by at least 100 dB.

Over the past several decades, many methods were developed to attenuate the self-interference. However, most of them approached the problem through physically "disconnecting" the leakage channel between the TX and RX sides. Since around 2010, new architectures have emerged taking advantage of the development in circuit and advanced signal processing. Researchers developed various SIC methods and demonstrated with prototyping efforts which

5G Networks: Fundamental Requirements, Enabling Technologies, and Operations Management, First Edition.
Anwer Al-Dulaimi, Xianbin Wang, and Chih-Lin I.
© 2018 by The Institute of Electrical and Electronics Engineers, Inc. Published 2018 by John Wiley & Sons, Inc.

achieved more than 120 dB leakage echo attenuation. Advances in both theory and practice [4–6] paved the way to enable potential application of FD operation in real wireless systems. Multiple full-duplex usage scenarios have been considered in the literatures, including full-duplex small cell access [7–9], full-duplex wireless backhaul, full-duplex relay [10], full-duplex Wi-Fi [11–13], and so on. In this chapter, we focus on system designs for full-duplex 5G small cell access. For other deployment scenarios, we briefly cover some of the challenges and related works in Section 3.6.2.

As the transmit power of small cell base station (BS) is generally lower (20–35 dBm in current LTE small cell products) than macro-BS, the SIC capability requirement for small cell BS is less stringent and can possibly be met with state-of-the-art technologies. However, unlike point-to-point FD transmission, interference environment in FD cellular network is more complicated. New interferences, including residual self-interference, BS-to-BS interference in the uplink (UL), and UE-to-UE interference in the downlink (DL), create challenges in system designs seeking to fully exploit the spectrum benefits from FD transmission. Recently, researchers [7,9,14], investigated the system performance and design challenges of FD small cell cellular systems. In Reference [7], the complicated interference scenarios introduced by FD operation were considered and they concluded that the FD performance gain can be mostly maintained with proper interference mitigation (IM). However, previous results are based on theoretical capacity approximation from long-term SINR where the effect of intelligent scheduling exploiting channel diversity was not evaluated. Also, there are limited discussions on the complexity of the proposed IM algorithms. Therefore, the focus of our research is to identify practical IM methods that efficiently address BS-to-BS and UE-to-UE interference in FD networks and to evaluate system-level performance via simulation tools that complies with 3GPP LTE evaluation methodology and captures realistic wireless environment models, including frequency-selective fading and fast fading.

The main contributions of this work are the following. (1) We propose and verify that BS-to-BS interference can be effectively mitigated via low-complexity open-loop UL power control and BS beamforming nulling. (2) The impact of UE-to-UE interference is evaluated and we propose joint DL–UL scheduler to manage intracell UE-to-UE interference when needed. (3) We identify and analyze practical design requirements to enable FD operation, including new reference signal and measurement procedure for UE-to-UE interference and low-overhead CQI feedback for the joint DL–UL scheduler. (4) We extensively simulate the performance of our proposed FD system designs under various deployment scenarios. Under full-buffer traffic where the interference environment is most severe, our proposed IM schemes for FD small cells can always achieve 1.7x–2x throughput gain over HD counterpart. Our results indicate that full-duplex is indeed a promising and practical technology for future small cell access.

The structure of this chapter is as follows. We first review the self-interference cancellation technologies that enable full-duplex operation in Section 3.2. Then, a general overview of system design challenges for full-duplex cellular systems is provided in Section 3.3, with an emphasis on interference complications for FD small cell access. Our proposals for FD small cell system design are described in Section 3.4. Specifically, we present practical interference mitigation schemes to treat BS-to-BS and UE-to-UE interferences, including null beamforming, uplink power control, joint intracell scheduler, CQI feedback, and reference signal design. Section 3.5 demonstrates the system performance with the proposed design in full-duplex system-level-simulator, by considering various indoor/outdoor scenarios using 3GPP models. Finally, we conclude the chapter and lay out future directions in Section 3.6.

3.2 Self-Interference Cancellation

The basis for enabling full-duplex access is the self-interference cancellation (SIC) capability that can substantially suppress TX echo in the RX chain. Taking advantage of the recent development in circuit and advanced signal processing, multiple research efforts have emerged with SIC methods and prototypes demonstrating more than 100 dB achievable attenuation of leakage signal. The advancement in SIC technology allows devices to transmit with IBFD and expands the possibilities of multiple full-duplex applications in the network. In this section, we provide a brief review of the key recent works that successfully achieved significant SIC (>100 dB), including our own Intel design and prototyping.

3.2.1 General SIC Architectures

The key requirement for self-interference cancellation is that the leakage signal from the TX chain should be suppressed sufficiently such that it has no impact on decoding RX signals. There are two critical constraints on the TX leakage signal power. First, the final TX leakage should not degrade the signal to interference and noise (SINR) level seen at the RX decoder, that is, the final TX leakage power should be much lower than the typical interference plus noise power observed at the RX receiver. Second, when converting the analog RX signal to the digital domain, the amplitude of the input, which consists of TX leakage plus RX signal and interference, to the analog-to-digital converter (ADC) should not exceed the ADC dynamic range. Typically, the gain of the low-noise amplifier (LNA) before the ADC is carefully chosen so the ADC dynamic range can best represent the RX signal with minimal quantization error. If there is no proper TX echo cancellation in the analog domain, the strong leakage signal would have saturated any practical LNA at the RX RF front end

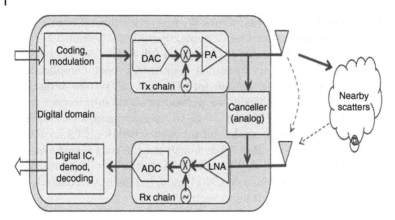

Figure 3.1 Two stages in canceling self-interference.

[1,3,6]. As a result, all contemporary SIC designs for full-duplex divide the SIC procedure into at least two parts, one is to attenuate TX leakage in the analog domain before the RX LNA, and the other is to further cancel the remaining echo in the digital domain after ADC.

There are several ways to achieve the cancellation in analog domain before RX ADC, and they are typically combined. One method is to enable better antenna separation with advanced circulator, different antenna polarization, or TX beamforming. Another method is to enable analog domain adaptive filtering, where the filters take the output of TX power amplifier (PA) as input and then apply adaptive filtering to cancel the leakage signal at the RX end. Both methods have achieved significant performance recently.

With respect to the self-interference cancellation in digital domain, the canceler takes the TX signal (baseband or signal past PA) as input, transforms to multiple kernel signals to model nonlinear distortion, and then applies adaptive filtering to cancel the leakage signal. Figure 3.1 is an illustration of the two stages and the corresponding methods in a high-level.

3.2.2 Self-Interference Cancellation State of the Art

Since 2010, researchers from MIT and Stanford [5] proposed architectures that can cancel self-interference for more than 110 dB. In their designs, a RF cancellation circuit in the analog domain cancels the leakage signal for more than 55 dB. Then in the digital domain, another 55 dB suppression is achieved. In the RF cancellation circuit, several delayed versions of the signal from the RX RF front end are prepared and summed up with certain weights. Adaptation algorithm (e.g., LMS) is applied to adjust the weights so that a good cancellation is achieved at the RX RF front end. The performance of this stage depends on the number of delayed signals and the delays themselves. Besides the analog cancellation, an isolator between the TX and RX antennas provides around 20 dB attenuation. In the digital cancellation block, both linear and nonlinear leakage signals are dealt with. This block takes the TX signal and transforms it to a set of kernels for canceling the linear and nonlinear distortions. Another adaptive filtering algorithm is applied to adjust the kernel weights. Based on the above architecture, it was demoed that more than 110 dB cancellation can be achieved for a 80 MHz Wi-Fi OFDM waveform, with 20 dBm TX power at 2.45 GHz [5]. In Mobile World Congress (MWC) 2015, it was demoed that full-duplex operation on LTE band can be achieved. Trials have been carried out with different operators (e.g., Deutsche Telekom [15]).

A team from Rice University started looking at simultaneous transmission and reception around 2010. In their design [6], in addition to analog and digital cancellation, passive suppression via MIMO antenna techniques were used. The MIMO techniques include the optimization on antenna placement, antenna nulling, and polarization. The prototyping in 2012 [6] showed the total cancellation is between 70 and 100 dB, with a median of 85 dB.

In 2013, researchers from Intel presented new techniques for the analog cancellation block [4]. Instead of aligning the delayed versions of carrier domain analog signal (after the PA) with adaptive weights, a phase shifter is introduced to the transmitted passband signal. With the new phase-shifted signal, new degree of freedom is introduced for the analog cancellation. Much smaller number of delayed signals are needed for the same cancellation target. Following this work, research and prototyping effort were carried out by Intel and TUT. In MWC 2014, more than 110 dB cancellation was shown publicly based on OFDM signals (Figure 3.2). As of now, with new techniques introduced, such as nice antenna separation via polarization, another 20 dB or more SIC is

Figure 3.2 Intel demo at MWC 2014 with 110 dB cancellation.

anticipated. In winter 2016, our 2×2 MIMO prototype demonstrating more than 110 dB cancellation verified that the architecture can be extended for multiantenna systems.

In 2014, researchers from Yonsei University and National Instruments used a dual polarization antenna to achieve 42 dB of isolation between the TX and RX antennas. Additional 60 dB cancellation is achieved by other analog methods. Overall, based on a SDR platform for a 40 MHz bandwidth, 60 dB cancellation is achieved in the analog domain, while 43 dB is achieved by digital domain method.

As more and more research efforts have been carried out for IBFD operation, various techniques have been improved or introduced. Examples include new circulator design [16], polarization techniques applied to mmWave system [17], massive MIMO, and so on, or new antenna design (Intel Labs is developing new cross-polarization antenna that provides 67 dB isolation in chamber and 45–50 dB isolation when hands are 10 cm close to the antennas).

3.3   FD System Design: Opportunities and Challenges

Though it is obvious that full-duplex can achieve $2 \times$ physical layer gain, the realizable gain from FD operation in cellular wireless networks is indefinite and depends heavily on the system design and the underlying network deployment characteristics. In this section, we highlight the key design challenges for FD cellular systems. The major challenge for FD wireless network design is the increasing interference arising from FD operation. We first describe the characteristics of new interferences in FD networks. For different types of new interferences, we briefly discuss potential methods for interference mitigation (IM) and review related past works. From practical implementation perspective, we further explain the challenges in designing efficient interference measurement for FD networks and elaborate the importance to reduce IM complexity.

3.3.1   New Interferences in FD Systems

The interference environment of FD cellular network is illustrated in Figure 3.3, where the black solid lines, black dash lines, and double black dash-dot lines indicate the signal link, the conventional interference link, and the new interferences from FD transmission, respectively. One of the three new interferences, I_{Echo} represents the self-interference introduced by FD and we have discussed the interference cancellation solutions in Section 3.2. In the following sections, we explain the characteristics of the other two new interferences in FD systems in more detail.

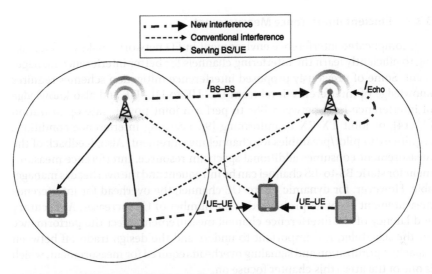

Figure 3.3 Interference in full-duplex cellular networks.

3.3.1.1 BS-to-BS Interference

In FD cellular system, DL transmission from neighboring BS may greatly impact UL reception of the serving BS. Since BS-to-BS channel is closer to line of sight with much smaller pathloss and also the transmit power and antenna gain at BS is much larger than those of UE, BS-to-BS interference ($I_{BS \to BS}$) can easily dominate desired weak UL signal if not properly addressed. Nonetheless, BS-to-BS interference channel is more static since BSs are not mobile and usually have steady traffic. Therefore, static interference mitigation schemes can be applied to suppress BS-to-BS interference, for example, elevation beam nulling [4] and semistatic UL/DL power control [18].

3.3.1.2 UE-to-UE Interference

For FD cellular systems, UL signal may create interference to downlink (DL) signal reception, especially to DL UE nearby; this is called UE-to-UE interference ($I_{UE \to UE}$). UE-to-UE interference is more dynamic since the scheduling decision can be varying and also the UE can be mobile and with nonpersistent traffic. UE-to-UE interference can be from UL UE served by the same BS (intracell UE-to-UE interference) or from UL UE in neighboring cells (intercell UE-to-UE interference). UE-to-UE interference mitigation can be done via clever scheduling that prevents strong interfering DL–UL UE pairs to transmit at the same time. Simple distance-based UE–UE pairing is suggested in References [8,9]. In References [4,7,14], power optimization was jointly done with scheduling to further reduce UE-to-UE interference. Network centralized cross-cell scheduling was also considered in References [7,14] to mitigate intercell UE-to-UE interference.

3.3.2 Efficient Interference Measurement

The complicated interference environment in FD networks makes it challenging to efficiently learn the interfering channels for better interference management. Some of previously proposed interference mitigation schemes requires knowledge of interference level from UL–UE to DL–UE and also knowledge of interference level between BSs to perform joint power–rate optimization [4,7,14], or joint TX/RX beamforming [9]. Learning interference conditions require extra pilot/preambles for channel measurement. Also, feedback of the measurement consumes additional spectrum resource. Interference measurement for static BS-to-BS channel can be infrequent and the overhead is manageable. However, for dynamic UE-to-UE channel, the overhead for interference measurement becomes prohibited as the number of UE increases. As accuracy and latency of the interference channel measurement affect the performance for the scheduler, it is important to understand the design trade-off between capacity optimization and signaling overhead required for measurement, which is one of the areas this chapter focuse on.

3.3.3 Complexity and Latency Consideration

Most past works focus on deriving the optimal solutions for best FD capacity gain without considering the practical implementation aspects. For example, dynamic power allocation requires extra signaling that reduces the expected throughput gain from power optimization. It is thus more suitable to consider semistatic power control for interference management. Another important design consideration is how much cross-cell coordination is required to effectively manage intercell interference. Tight cross-cell coordination that can potentially achieve optimal system throughput [7] requires large amount of signaling, low-latency backhaul, and intelligent central controller. In this chapter, we consider distributed IM where coordination is within the same cell and demonstrate distributed approach is sufficient in most small cell deployments. For scenarios with strong intercell interference, we envision semistatic frequency planning approaches such as fractional frequency reuse that can be practical solutions with good complexity-performance trade-off.

In the next section, we describe practical and efficient interference mitigation schemes addressing BS-to-BS and UE-to-UE interferences that can best exploit FD gain in small cell cellular systems. We also consider reference signal structure, interference measurement, and simplified CQI feedback signaling modified from existing LTE systems for FD small cell access system.

3.4 Designing the FD System

In this section, we describe in detail our solutions to address the system design challenges discussed in Section 3.3 for small cell FD access, including applying beamform nulling and uplink open-loop power control to handle the BS-to-BS interference, adopting joint intracell scheduler for UE-to-UE interference mitigation, efficient CQI feedback methods to enable joint scheduler, as well as reference signal structures and measurement methods for UE interference.

3.4.1 Overall Design for FD

Our FD system design target is to achieve the most system throughput gain from FD operation. In order to achieve this goal, we must carefully manage the additional interference introduced by FD transmission. Self-interference can be suppressed via various echo cancellation techniques described in Section 3.3. In this section, we focus on designs that address the two new system interferences in FD networks, BS-to-BS interference and UE-to-UE interference. Based on the characteristics of BS-to-BS interference and UE-to-UE interference, we propose to mitigate BS-to-BS interference with relatively static schemes and adopt dynamic interference mitigation (IM) approach to manage UE-to-UE interference.

We propose two static/semistatic IM methods to control BS-to-BS interference: The first approach is elevation beam nulling that can greatly suppress the line of sight component of BS-to-BS interference. The second scheme, UL power control, can further enhance UL signal strength to combat BS-to-BS interference. With a combination of both schemes, we can greatly improve uplink data rate and achieve 1.9× throughput gain compared with HD FDD system.

For UE-to-UE interference, we propose to manage interference from UL UE belonging to the same serving cell via intelligent scheduling decision at the BS. With knowledge of UE-to-UE interference level, the BS can schedule (DL UE, UL UE) pair with low UL UE to DL UE interference to transmit together. In Section 3.4.3, we describe how scheduling decision can be made. In addition, we investigate the required signaling procedure to obtain UE-to-UE interference level and propose practical design recommendation on reference signal and channel quality feedback. For UL UE interference from neighbor cells, we also briefly cover some potential approaches to manage intercell UE-to-UE interference, such as fractional frequency reuse, in Section 3.6.

3.4.2 Design to Mitigate BS-to-BS Interference

Figure 3.4 shows the cumulative distribution function (CDF) of the ratio of the conventional UL interference ($I_{UE \to BS}$) power to the BS-to-BS interference ($I_{BS \to BS}$) power. Detail definitions of the three deployment scenarios shown in the figure can be found in Section 3.5. From Figure 3.4, we observe that BS-to-BS interference is much stronger than conventional UL interference, especially in dense deployment scenario like the indoor test case for

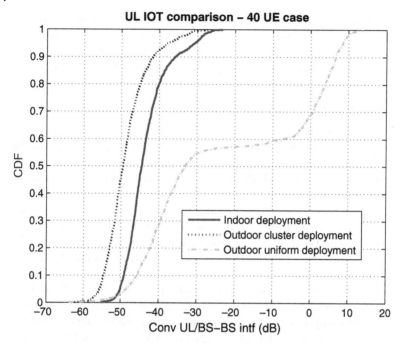

Figure 3.4 Interference in full-duplex cellular networks.

LAA evaluation, where over 90% of chances the BS-to-BS interference is 40 dB stronger than conventional UL interference. Such statistics indicate that, without proper BS-to-BS interference mitigation, full-duplex operation can significantly degrade uplink reception and hence provide no system throughput advantage.

As discussed in Section 3.3.1.1, one of the key characteristics of BS-to-BS interference is that the incoming direction and strength of the interference is relatively stable. The nature of BS-to-BS interference suggests that semistatic approach should be sufficient to mitigate BS-to-BS interference. Hence, we propose two semistatic IM schemes to manage BS-to-BS interference: elevation beam nulling and power control. The design concept of the two schemes are described in the following sections.

3.4.2.1 Elevation Beam Nulling

When transmitting with multiple antennas, base station can adjust the transmit power at the antennas to create different beam patterns. Horizontal and 3D beamforming/beam nulling has been widely used and researched in MIMO systems nowadays. In addition, most base stations deployed today employ elevation antenna down-tilting to optimize signal strength within their coverage area. The same concept can be revisited to help mitigate the BS-to-BS interference problem in FD networks.

In Reference [4], we take advantage of the fact that the altitudes of most BSs are fixed and propose to use elevation beam nulling to suppress BS-to-BS interference. When every BS has similar altitude, BS-to-BS interference can be greatly suppressed by simply forming nulls at the vicinity of 90° in elevation. For BSs with different antenna altitudes, wider null width is required to suppress BS-to-BS interference. Denote the wavelength of carrier frequency as λ. Consider a BS with four horizontal antennas each with 4λ height, each antenna can be split into eight $1/2\lambda$ elements stacked vertically, creating an equivalent two-dimensional 4×8 antenna array. We can then apply 3D beam forming/beam nulling techniques to form beam or null at arbitrary azimuth and elevation angle. For simplicity, we proposed in Reference [4] to separate horizontal and vertical beamforming and split beamforming weight of each antenna element into multiplication of horizontal weight and vertical weight. Elevation beam nulling will create a null at certain elevation angle over the entire range of azimuth angle. Horizontal weight can be designed based on conventional close-loop or open-loop MIMO techniques.

Assuming 15° antenna tilting, the vertical weights can be computed based on MMSE-based nulling. Antenna patterns for the cases with 4λ antenna height and 6λ antenna height are shown in Reference [4]. Suppose the null width requirement is within $\pm 1°$ (between 89° and 91°), the proposed beam nulling algorithm can guarantee 35 dB nulling for small cell BS and 40 dB null for large cell BS. Large cells require deeper null than small cells because the increase in transmit power for large cell BS is much higher than the extra pathloss from longer intersite distance (ISD). For a 25 m high BS antenna, a wide null from 89° to 91° can effectively suppress interference toward neighbor BSs at 500 m distance with antenna height variation from 16.3 m to 33.7 m. One drawback of the elevation nulling approach is that it may cause minor degradation in the cell edge UE receive power level and thus reduction in cell size may be required while applying elevation nulling to guarantee edge UE performance.

Note that, for enterprise deployment with BS deployed on the ceiling and antenna facing downward, the antenna beam-pattern is typically designed with 20–40 dB attenuation in 90° elevation. Therefore, even without additional design for FD–BS, substantial amount of BS-to-BS interference suppression can be achieved.

Since the same antenna pattern is applied for both transmit and receive, a total of 70 dB and 80 dB BS-to-BS interference suppression can be achieved for small cell and large cell deployment, respectively. Based on the CDF curve in Figure 3.4, BS-to-BS interference becomes negligible compared with conventional UL interference if the full 70 dB nulling gain can be achieved. However, for smaller cells with shorter ISD, wider null width is required to accommodate

BS antenna height variation. Also, in multipath wireless environment, there may be reflected beams from the main beam lobe causing strong BS-to-BS interference. Both conditions prevent from achieving the full nulling gain. Therefore, additional interference mitigation may be needed to combat the remaining BS-to-BS interference. In the next section, we describe our proposal to incorporate UL power control to further improve UL performance in FD cellular networks.

3.4.2.2 Uplink Power Control

In addition to elevation beamforming, BS-to-BS interference can be further reduced via DL or UL power control. Raising UL power can overpower BS-to-BS interference but with the drawback of rising power consumption. On the other hand, reducing BS-to-BS interference via decreasing DL power risks in shrinking coverage. Operators may adopt either or both approaches to manage BS-to-BS interference depending on deployments. In this chapter, we assume that DL power cannot be further reduced due to coverage constraints, and focus on uplink power control (ULPC) design.

Previously, ULPC algorithms were proposed to mitigate BS-to-BS or UE-to-UE interference in dynamic TDD [19,20] and FD [18] systems. However, prior designs tackled BS-to-BS and UE-to-UE interferences independently without jointly considering the impact of ULPC on both DL and UL performance. While increasing UL TX power improves UL SINR, the UE-to-UE interference is concurrently enhanced that degrades DL SINR. In this chapter, we design ULPC for FD systems based on UL open-loop power control (OLPC) used in LTE. Our design guidelines jointly consider both BS-to-BS and UE-to-UE interferences.

OLPC in existing LTE systems tunes the target received power level at BS, P_0, and the fractional power control (FPC) parameter, α_{FPC}, to control the distribution of uplink SNR experienced by serving cell UE. The UL transmit power can be computed via the following formula:

$$P_{tx} = \min\left\{P_{max}, P_0 + 10\log(\#PRB) + \alpha_{FPC} \cdot PL\right\}$$

The transmit power of the UE is denoted as P_{tx}, P_{max} denotes the maximum UE transmit power, #PRB is the number of physical resource blocks used for UL transmission, and PL is the pathloss of the link between UE and the serving BS.

To overcome BS-to-BS interference, we propose to boost P_0 to $B \cdot P_0$ for FD operation where B denotes the boosting factor. The optimal value of B depends on the distribution of $I_{BS \to BS}/I_{UE \to BS}$ and $I_{BS \to UE}/I_{UE \to UE}$, where $I_{x \to y}$ denotes interference from x to y. The boosting factor B can be uniform across all cells or be different for each cell based on $I_{BS \to BS}$ level. We describe both boosting strategies in the following and present the simulation results for BS-to-BS IM combining beamforming nulling [4] and the proposed OLPC schemes in Section 3.5.

3.4.2.2.1 Uniform Power Boosting

The goal for boosting factor adjustment is to minimize SINR degradation due to FD transmission. Below are the DL and UL SINR in HD transmission:

$$SINR_{HD-UL} = \frac{P_{rx,UE \to BS}}{I_{UE \to BS} + N_0},$$

$$SINR_{HD-DL} = \frac{P_{rx,BS \to UE}}{I_{BS \to UE} + N_0}.$$

$P_{rx,A \to B}$ denotes the received signal power from A to B. Here, $P_{rx,A \to B}$ and $I_{A \to B}$ are power level observed when there is no extra boosting, that is, using the default target P_0 for HD transmission in UL OLPC. When all cells adopt the same boosting factor B for UL OLPC, the FD SINR in DL and UL are as follows:

$$SINR_{FD-UL} = \frac{B \cdot P_{rx,UE \to BS}}{B \cdot I_{UE \to BS} + I_{BS \to BS} + N_0},$$

$$SINR_{FD-DL} = \frac{P_{rx,BS \to UE}}{I_{BS \to UE} + B \cdot I_{UE \to UE} + N_0}.$$

For UL, the BS experiencing stronger $I_{BS \to BS}$ suffers from more SINR degradation. Strong $I_{BS \to BS}$ suggests neighboring cells are closer and $I_{UE \to BS}$ is likely to be high as well. In such high-interference scenario, N_0 can be ignored in SINR computation and, by taking $B = I_{BS \to BS}/I_{UE \to BS}$, we can bound the UL SINR degradation within 3 dB. We define B_{UL} as the top x percentile of $I_{BS \to BS}/I_{UE \to BS}$ among all cells. When $B \geq B_{UL}$, we can guarantee $(100 - x)$ percent of BSs experience less than 3 dB SINR degradation. The value of B_{UL} can be found via checking the CDF plot like Figure 3.4 with $I_{BS \to BS}$ adjusted according to achievable elevation nulling.

Similarly, in the DL, more SINR degradation are observed for UE experiencing higher $I_{UE \to UE}$. Omitting N_0 in SINR computation for highly interfered UE, we can bound the DL SINR degradation within 3 dB by taking $B = I_{BS \to UE}/I_{UE \to UE}$. We define B_{DL} as the lowest x percentile of $I_{BS \to UE}/I_{UE \to UE}$ among all UE. When $B \leq B_{DL}$, we can guarantee $(100 - x)$ percent of UE experience less than 3 dB DL SINR degradation. Again, the value of B_{DL} can be found via checking the CDF plot of $I_{BS \to UE}/I_{UE \to UE}$. Two types of $I_{BS \to UE}/I_{UE \to UE}$ CDF curves are plotted in Figure 3.5. In Figure 3.5a, $I_{UE \to UE}$ includes both intercell and intracell UE-to-UE interference and less UL power boosting can be tolerated. For the case where intracell UE-to-UE interference can be managed, B_{DL} can be selected based on $I_{BS \to UE}/I_{intercellUE \to UE}$ as plotted in Figure 3.5b. Clearly, B_{UL} and B_{DL} can serve as reference to choose the

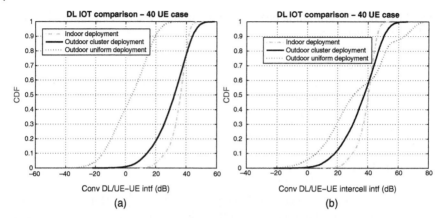

Figure 3.5 CDF of $I_{BS\rightarrow UE}/I_{UE\rightarrow UE}$. (a) Including both intercell and intracell UE-to-UE interference. (b) Intercell UE-to-UE interference only.

boosting factor B, to leverage between uplink and downlink performance. For instance, if $B_{DL} > B_{UL}$, setting B in the middle of (B_{DL}, B_{UL}) is satisfactory, in the sense that both the UE-to-UE interference and BS-to-BS interference can be sufficiently mitigated. If $B_{DL} < B_{UL}$, we shall choose $B_{DL} < B < B_{UL}$. For different selected B values, there is a trade-off between UL and DL performance. The simulation results for applying different boosting factors after beamforming nulling under various deployment scenarios are shown in Section 3.5.

3.4.2.2.2 Cell-Specific Power Boosting
The boosting factor can also be cell-specific based on the $I_{BS\rightarrow BS}$ level measured at each BS. The idea is that ULPC boosting factor can be reduced for cells experiencing low $I_{BS\rightarrow BS}$. Assuming boosting factors for neighboring cells are similar, the FD UL SINR formula previously derived can be reused to approximate the required boosting factor for each cell. Notice that when $B = 1 + I_{BS\rightarrow BS}/N_0$, FD UL SINR becomes the same as HD UL SINR. We thus propose, for BS experiencing low $I_{BS\rightarrow BS}$ level, boosting factor should be set to $B = 1 + I_{BS\rightarrow BS}/N_0$, and when $I_{BS\rightarrow BS}$ is high, boosting factor should be set based on the uniform power boosting rule previously derived. For example, if we choose to only protect UL SINR, the per-BS boosting factor can be set as follow:

$$B_{BS} = \min\left\{1 + \frac{I_{BS\rightarrow BS}}{N_0}, B_{UL}\right\}.$$

One advantage of cell-specific power boosting strategy is that UE served by BS experiencing low $I_{BS\rightarrow BS}$ can save power. Note that the boosting factor for cells with low $I_{BS\rightarrow BS}$ may be less than the boosting factors for their neighbor cells. In this case, the actual FD UL SINR becomes smaller than HD UL SINR for boosting factor $B = 1 + I_{BS \to BS}/N_0$. In most deployments, $I_{BS \to BS}$ observed at neighboring cells should be similar, thus, the boosting factor variation should be small and only minor SINR degradation is expected. Also, cells with very low $I_{BS \to BS}$ may be able to tolerate more SINR degradation as they operate most likely in noise-limited regime.

3.4.3 Design to Mitigate UE-to-UE Interference

In this section, we cover the design details to manage UE-to-UE interference in full-duplex cellular system, including joint downlink–uplink scheduler and the required channel quality indicator (CQI) feedback to enable joint scheduling. Corresponding signaling overheads introduced with various CQI feedback methods are discussed and their efficiency is compared with the baseline CQI feedback in current LTE systems.

3.4.3.1 Joint Downlink–Uplink Scheduler

UE-to-UE interference is very dynamic and can be handled by schedulers that carefully select best pairs of DL and UL UE for simultaneous transmission. In order to manage UE-to-UE interference from both intercell and intracell UE, inter cell cooperative schedulers that require extra signaling and coordination between multiple base stations can be designed. However, such cross-cell co-operation schemes can be highly complex and the throughput gain from joint optimization may diminish due to the signaling overhead. Therefore, in this chapter, we focus on distributed solutions that manage intracell UE-to-UE interference only. We propose that a BS can jointly schedule the DL and UL UE for transmission on the same resource block (RB) to minimize the impact of interference from the UL UE to the DL UE. We envision, in 5G, DL and UL will share the same multiplexing terminology and the same scheduling granularity will be used for DL and UL to enable flexible resource allocation. Thus, in the following of this chapter, we simply assume that both UL and DL are transmitted in OFDMA while describing our scheduler design. In the following of this section, we describe how the joint DL–UL scheduling metric is derived, how UE-to-UE interference impacts the joint scheduling metric, and how the joint scheduler can be simplified when scheduling FD-capable UE.

3.4.3.1.1 Joint DL–UL Scheduling Metric

In conventional half-duplex cellular system, the scheduler at a BS independently schedules uplink and downlink UE based on the optimization metric for the transmit direction of the traffic. In full-duplex cellular system, the same naive scheduler may accidentally select a UE pair with strong UL UE to DL UE interference for joint transmission. In order to avoid that UL UE interference degrades DL reception, we propose a joint intracell DL–UL scheduler, which selects the best pair of DL and UL UE in the serving cell based on a joint scheduling metric that optimizes both directions of traffic. Since UE pairs with strong UE-to-UE interference that severely degrades the joint scheduling metric would not be selected for joint transmission, UE-to-UE interference is automatically managed when the joint scheduling performance metric is optimized. In the following, we consider the scheduling metric for general α-fair schedulers and derive the corresponding joint DL–UL scheduling metric for FD transmission.

The α-fair throughput function, $f_\alpha(\mathrm{TP}_u)$, is defined as

$$
f_\alpha(\mathrm{TP}_u) = \begin{cases} \dfrac{\mathrm{TP}_u^{1-\alpha}}{1-\alpha}, & \text{for } \alpha \neq 1 \\[2mm] \log(\mathrm{TP}_u), & \text{for } \alpha = 1 \end{cases},
$$

where TP_u denote the throughput of user u.

A general α-fair scheduler targets to maximize the objective function of sum α-fair throughput: $\sum_u f_\alpha(\mathrm{TP}_u)$. The special case of $\alpha = 1$ is the proportional fair scheduler, which is generally used in current cellular system. Other special cases, such as $\alpha = 0$ and $\alpha \to \infty$, correspond to the max-sum-throughput scheduler and the max-min-throughput scheduler, respectively.

A full-duplex BS needs to schedule both DL and UL transmissions, and thus the objective function should consider the performance for both UL and DL directions. The joint scheduling metric can be computed depending on how the fairness for UL and DL traffic is defined. For instance, to guarantee fairness across all UE where DL and UL resource usage are separately considered, the corresponding objective function is as follows:

$$
w_{\mathrm{DL}} \cdot \sum_{u \in \mathcal{U}_{\mathrm{DL}}} f_\alpha(\mathrm{TP}_u^{\mathrm{DL}}) + w_{\mathrm{UL}} \cdot \sum_{u \in \mathcal{U}_{\mathrm{UL}}} f_\alpha(\mathrm{TP}_u^{\mathrm{UL}})
$$

The weighting factors w_{DL} and w_{UL} indicate how precious the DL and UL resources are, respectively. $\mathcal{U}_{\mathrm{DL}}$ denotes the set of UE with DL traffic and $\mathcal{U}_{\mathrm{UL}}$ is the set of UE with UL traffic.

We solve the above optimization problem and obtain the joint scheduling metric used by the scheduler for jointly selecting the (DL,UL) UE pairs. When making the scheduling decision for RB i, the scheduler can search across all potential UE pairs and select the (DL,UL) UE pair (u_1, u_2) with the highest scheduling metric as defined in the following equation:

$$
w_{\mathrm{DL}} \cdot \frac{R_{u_1,\mathrm{UE}_{\mathrm{UL}}=u_2}^{\mathrm{DL}}(\mathrm{RB}_i)}{(\mathrm{RTP}_{u_1}^{\mathrm{DL}})^\alpha} + w_{\mathrm{UL}} \cdot \frac{R_{u_2,\mathrm{UE}_{\mathrm{DL}}=u_1}^{\mathrm{UL}}(\mathrm{RB}_i)}{(\mathrm{RTP}_{u_2}^{\mathrm{UL}})^\alpha}, \tag{3.1}
$$

where $R^{DL}_{u_1,UE_{UL}=u_2}(RB_i)$ is the achievable data rate for DL UE u_1 on RB_i when u_2 is scheduled in the UL direction, and $R^{UL}_{u_2,UE_{DL}=u_1}(RB_i)$ denotes the achievable rate for UL UE u_2 on RB_i when u_1 is scheduled in the DL direction. $RTP^{DL}_{u_1}$ and $RTP^{UL}_{u_2}$ denotes the recent DL throughput for UE u_1 and recent UL throughput for UE u_2, respectively.

Depending on the complexity and performance requirement, the selection procedure can be achieved by either exhaustive search or optimization algorithms such as greedy search, dynamic programming, and Hungarian algorithm for maximum weight matching. In some scenarios, there can be implementation constraints that require the scheduler to select the UE for one transmit direction first then the other direction. For instance, current LTE system transmits the UL grant 4 ms in advance, that is, the scheduling decision for UL is determined 4 ms before the actual transmission. In this case, BS needs to first schedule the UL UE based on traditional UL scheduling metric, and then select the DL UE to be paired with that optimizes the joint scheduling metric defined in Eq. 3.1. In the following of this chapter, we only consider exhaustive search that simultaneously schedules both DL and UL direction to demonstrate the ultimate benefit of joint scheduler for FD transmission. In Section 3.5, we evaluate the proposed joint scheduler design via simulation and demonstrate the potential performance improvement.

3.4.3.1.2 Impact of UE-to-UE Interference on the Joint Scheduling Metric

The new joint DL–UL scheduling metric is a function of the achievable DL and UL data rates. Due to UE-to-UE interference, the achievable DL rate varies when the DL UE is paired with a different UL UE. $R^{DL}_{u_1,UE_{UL}=u_2}(RB_i)$ reflects how interference from UL UE u_2 degrades the DL rate performance of UE u_1. Similarly, in order to address UE-to-UE interference, the achievable UL rate can change when pairing with different DL UE. For example, we can adopt advanced UL power control schemes where UL UE adjust their transmit power specifically for their paired DL UE to manage intracell UE-to-UE interference level. In consequence, the achievable UL rate for UE u_2, $R^{UL}_{u_2,UE_{DL}=u_1}(RB_i)$, depends on how much UE u_2 back-off its transmit power to protect DL UE u_1. However, in typical cellular systems, such rapid UL power adjustment is prohibited by its significant signaling overhead requirement. In this chapter, we do not consider sophisticated power control schemes, but assume UL power settings are computed based on procedures developed in Section 3.4.2.2 where a UL UE transmits with the same power when pairing with different DL UE. Therefore, we can simplify the UL achievable rate in Eq. 3.1 to $R^{UL}_{u_2}(RB_i)$. In addition, conventional HD UL measurement and feedback mechanism can be reused to obtain the UL rate estimation for FD UL UE.

To understand how UE-to-UE interference impacts the DL UE data rate, we decompose the overall DL interference into three components:

$$I_{\text{overall}} = I_{\text{UE} \rightarrow \text{UE}_{\text{intra}}}(\text{UE}_{\text{DL}}, \text{UE}_{\text{UL-sched}})$$
$$+ I_{\text{UE} \rightarrow \text{UE}_{\text{inter}}}(\text{UE}_{\text{DL}}) + I_{\text{BS} \rightarrow \text{UE}_{\text{inter}}}(\text{UE}_{\text{DL}}), \tag{3.2}$$

which correspond to the interference value observed at the DL UE from (1) the paired UL UE within the same cell, (2) the UE-to-UE interference from neighboring cells, and (3) the conventional DL interference from neighboring BSs, respectively. We further combine interference from neighbor cell UE and BSs and denote it as $I_{\text{inter}}(\text{UE}_{\text{DL}})$, where $I_{\text{inter}}(\text{UE}_{\text{DL}}) = I_{\text{UE} \rightarrow \text{UE}_{\text{inter}}}(\text{UE}_{\text{DL}}) + I_{\text{BS} \rightarrow \text{UE}_{\text{inter}}}(\text{UE}_{\text{DL}})$.

The degree of FD gain a joint DL–UL scheduler can provide depends on whether the intracell UE-to-UE interference term in Eq. 3.2, $I_{\text{UE} \rightarrow \text{UE}_{\text{intra}}}(\text{UE}_{\text{DL}}, \text{UE}_{\text{UL-sched}})$, is strong or weak when compared with the latter two intercell interference components. When intercell interference dominate, naive scheduler that independently schedules DL and UL transmission with conventional DL/UL data rate estimation and feedback mechanism for HD transmission can provide plenty FD gain. However, as the contribution of intracell UE-to-UE interference grows, joint DL–UL scheduler that carefully avoid DL and UL UE with strong UE-to-UE interference being paired together can increasingly improve DL SINR. For the joint DL–UL scheduler to work, it is essential to acquire accurate DL data rate estimate that reflects the UE-to-UE interference from the potential pairing UL UE.

The DL UE data rate can be estimated based on feedback of rank indicator (RI), precoding matrix indicator (PMI), and channel quality indicator (CQI). UE-to-UE interference has less impact on the transmit direction and rank for BS-side beamforming. Therefore, the same RI and PMI feedback mechanism for HD transmission can be reused in FD networks. For RI report, we propose UE computes the wideband rank indicator based on overall interference, including intracell UE-to-UE interference, intercell UE-to-UE interference, and conventional intercell BS-to-UE interference. For PMI report, each DL UE calculates PMI based on the wideband RI and the subband serving channel information measured from serving cell reference signals. After computing wideband RI and subband PMIs, DL UE then feedbacks these information to the BS.

On the contrary, CQI reflects the received SINR of a DL UE and can be very sensitive to the UE-to-UE interference level. For every DL and UL UE pair, the joint scheduler should be aware of the corresponding DL SINR, where the interference part includes the UL UE to DL UE interference and overall intercell interference experienced by the DL UE. It is crucial for practical implementation to design efficient and accurate interference measurement and CQI feedback mechanisms to estimate DL UE SINR when pairing with different UL UE. In Sections 3.4.3.2 and 3.4.3.3, we address these design aspects in details.

3.4.3.1.3 Scheduling FD-Capable UE

When there are full-duplex capable UE with both DL and UL traffic, the joint scheduler can be further simplified. With sufficient SIC to suppress TX echo, the achievable DL rate when a UE transmitting in FD can be higher than when the UE is paired with another UL UE because of extra UE-to-UE interference in the latter case. Therefore, scheduler at the BS can choose to always bundle UL and DL transmission for such FD-capable UE, that is, when scheduled for transmission within one scheduling unit, a FD-UE transmits data in both DL and UL direction. In addition to complexity reduction, such *bundle scheduler* can reduce signaling overhead as there is no need to measure the impact of intracell UE-to-UE interference that is otherwise required if a FD–UE is paired with another UE transmitting in opposite direction.

Network with Pure FD-UE When all UE are full-duplex-capable UE with bidirectional traffic, the joint scheduling metric for a bundle scheduler can be simplified as

$$
w_{\text{DL}} \cdot \frac{R_u^{\text{DL}}(RB_i)}{(RT\, P_u^{\text{DL}})^\alpha} + w_{\text{UL}} \cdot \frac{R_u^{\text{UL}}(RB_i)}{(RT\, P_u^{\text{UL}})^\alpha},
$$

where $R_u^{\text{DL}}(RB_i) = R_{u,\text{UL}-\text{UE}=u}^{\text{DL}}(RB_i)$ and $R_u^{\text{UL}}(RB_i) = R_{u,\text{DL}-\text{UE}=u}^{UL}(RB_i)$.

As the bundle FD scheduler allocates one scheduling unit to the same FD UE for DL and UL transmission, it simply needs to compute the joint scheduling metric for each FD–UE and select the one with highest value to be scheduled.

Network with Mix of FD/HD UE When there are UE with only half-duplex capability deployed together with FD-UE, an efficient design is to combine the joint DL–UL scheduler with bundle FD scheduler. Similar to joint DL–UL scheduler, all possible (DL,UL) UE pairs of HD UE can be potential candidates to be scheduled. In addition, the bundle FD scheduler precludes the (DL, UL) UE pairs from two different FD UE to be scheduled in opposite directions. As for pairing between FD UE and HD UE, there can be two variations of scheduler design:

1) *Allow HD UE to be paired with FD UE:* A HD UE can be paired with another FD UE for concurrent transmission. This variation requires measurement and reporting of interference from a FD UE to a DL UE and interference from a UL UE to a FD UE. In this case, scheduling metrics of more (DL,UL) UE pair combinations are compared when selecting the best (DL,UL) pair to be scheduled.

2) *Disallow HD UE to be paired with FD UE:* Scheduler disregards FD UE when selecting UE to be paired with HD UE. Here, FD UE do not need to feed back interference information with respect to other UE. The bundle scheduler simply selects to schedule the (DL, UE) pair with the best scheduling metric from all possible (DL,UL) UE pairs of HD UE and all FD UE pairs where the (DL,UL) UE are the same FD UE.

Bundle scheduler reduces complexity and signaling overhead at the cost of not exploring all possible UE pair combinations. We evaluated with simulation for bundle scheduler testing both pure FD UE deployment and mix of FD/HD UE deployment. Simulation results in Section 3.5 show bundle scheduler, although sacrificing some user selection diversity can achieve most of the FD gain.

3.4.3.2 Channel Quality Indicator Feedback for Joint Scheduling

In Section 3.4.3.1, we briefly describe how CQI feedback affects the computation of joint scheduling metric. In this section, we provide a more thorough discussion on CQI feedback strategies for FD access.

UE-to-UE interference affects the achievable DL data rate when a FD BS schedules different UE for simultaneous DL and UL transmission. When pairing with different UL UE, the overall interference level experienced by a DL UE varies and it is essential for a joint DL and UL scheduler to know the corresponding achievable DL data rate to obtain a reliable joint scheduling metric. Therefore, new CQI feedback that indicates how UE-to-UE interference degrades DL rate may be needed in order to achieve better UE-to-UE interference mitigation.

In this section, we look into different feedback mechanisms and update rules that provide UE-to-UE interference information to assist joint DL and UL scheduling in FD cellular system. We start from reviewing the current CQI feedback mechanism in LTE networks to understand its limitation for estimating interference level with FD transmissions. Then, we investigate feedback schemes that can collect additional information regarding intracell UE-to-UE interference level. We first develop a feedback rule providing extensive intracell UE-to-UE interference information where a DL UE feedbacks the per subband CQI for each potential pairing UL UE to the joint scheduler. Such extensive feedback mechanism requires huge signaling overhead and is not practical for real-life implementation. Therefore, we later design two wideband feedback strategies that significantly reduce the overhead while preserving the critical UE-to-UE interference information. Both the wideband and subband feedback information can be exploited to trade-off complexity and performance at the FD BS for joint scheduling. Finally, we provide examples of joint scheduler designs that utilize different combinations of feedback strategies to estimate the joint scheduling metrics and analyze the corresponding signaling overhead for different designs.

3.4.3.2.1 Conventional CQI Feedback Schemes

In current half-duplex LTE system, the DL CQI feedback contains information (4 quantized bits) of the measured SINR where interference come from other BSs. LTE system usually groups several consecutive physical resource blocks (PRBs) together as one scheduling unit labeled as subband. The CQI feedback can report either measurements for individual subbands or a collective wideband SINR measurement. Subband CQI feedback is typically used to explore scheduling gain from frequency-selective fading.

In FD systems, conventional CQI feedback mechanism that reports one single SINR estimate of each subband for one DL UE can be reused. The CQI feedback can be the total SINR, including new UE-to-UE interference. Further is the DL SINR measured by UE u_{DL} while the UL UE scheduled in the measurement subband is $u_{UL-sched}$.

$$\text{SINR}^{(DL)}_{conv}(u_{DL}) = \frac{P_{rx}(u_{DL})}{I_{UE \rightarrow UE_{intra}}(u_{DL}, u_{UL-sched}) + I_{inter}(u_{DL}) + N}. \quad (3.3)$$

This CQI feedback mixes intracell and intercell interference together. Base station cannot distinguish how much intracell UE-to-UE interference further degrades the SINR. In addition, this feedback only includes UE-to-UE interference measurement from one UL UE. Joint scheduler has no information on UE-to-UE interference from other candidate UL UE on this subband.

Depending on the reference signal design, as described in Section 3.4.3.3, the UE may be able to distinguish intracell UE-to-UE interference from intercell interference. If that is the case, the UE may choose to feedback conventional CQI based on DL SINR with only intercell interference. We name this feedback scheme *conventional feedback variation 1*. The SINR reported by this kind of CQI feedback is the following:

$$\text{SINR}^{(DL)}_{conv1}(u_{DL}) = \frac{P_{rx}(u_{DL})}{I_{inter}(u_{DL}) + N}. \quad (3.4)$$

We can combine *conventional feedback variation 1* with new CQI feedback mechanisms described later to obtain more accurate per DL–UL pair SINR estimate.

3.4.3.2.2 Subband per DL–UL Pair CQI Feedback

The subband CQI feedback mechanism in LTE downlink can be reused to feedback subband per DL–UL pair CQI specific to FD system for joint scheduling.

New reference signal designs that will be described in Section 3.4.3.3 can be used to estimate the SINR for each DL–UL pair. DL UE u_{DL} computes per DL–UL pair CQI based on per pair SINR estimate, as shown further.

$$\text{SINR}^{(DL)}_{\text{pairwise}}(u_{DL}, u_{UL}) = \frac{P_{rx}(u_{DL})}{I_{UE \to UE_{\text{intra}}}(u_{DL}, u_{UL}) + I_{\text{inter}}(u_{DL}) + N}. \quad (3.5)$$

This feedback strategy provides full information on the intracell UE-to-UE interference, yet it requires huge signaling overhead. In next section, we will list a few alternatives for CQI feedback to leverage some performance for complexity reduction.

3.4.3.2.3 Wideband DL–UL Pairability CQI Feedback

The first overhead reduction proposal is to only provide BS with the information of whether a UL UE is an aggressor (interferer) or a nonaggressor (noninterferer) to a DL UE with 1-bit feedback. This 1-bit information indicates whether a DL UE can be paired with a UL UE or not. We describe in the following how to classify a DL–UL UE pair as pairable or not and how to update the pairable information maintained within the joint scheduler.

The pairable criteria can be set based on whether the ratio between intracell UE–UE interference and intercell UE–UE interference, $E(I_{\text{inter}-\text{ue2ue}}(\text{DL}-\text{UE}))$, exceeds a given threshold, that is,

$$\frac{I_{UE \to UE_{\text{intra}}}(\text{UE}_{DL}, \text{UE}_{UL})}{E[I_{UE \to UE_{\text{inter}}}(\text{UE}_{DL})]} \begin{cases} > \text{threshold}, & \text{aggressor}('0') \\ \text{otherwise}, & \text{nonaggressor}('1') \end{cases}.$$

Each DL UE measures the intracell UE-to-UE interference, and then, based on the knowledge of average intercell UE-to-UE interference, it bitmaps the ratio between the measured intracell UE–UE and average intercell UE–UE interference into 0 or 1 by comparing with a threshold, where "0" entry indicates that the DL–UL UE pair is not allowed to be scheduled in the same resources while "1" entry indicates that the UL and DL UE can be paired for joint scheduling. The value of the threshold can be set based on statistical knowledge of the ratio between the two interferences. Either a global threshold can be used across all BSs or individual threshold can be set for each BS:

1) *Global Threshold:* All BS adopt the same threshold for wideband UL–DL pair table based on overall network-wide long-term channel knowledge.
2) *Individual Threshold:* Each BS sets its threshold for wideband UL–DL pair table based on long-term channel knowledge of its own cell.

UE updates the table periodically based on its instantaneous and previous measurements or aperiodically upon BS requests. As the pairability status

Table 3.1 Example wide-band 0/1 DL-UL pair table in FD cellular system.

DL UE ID \ UL UE ID	1	2	...	N_U
1	0	1	...	1
⋮	⋮	⋮	⋱	⋮
N_D	1	0	...	1

changes more slowly than time-varying channel, the wideband feedback information can be updated periodically or aperiodically at relatively low frequency based on the BS configuration (e.g., every 40/80/160 ms) or upon request.

An example pairability table is illustrated in Table 3.1 where each entry of the table basically indicates if the UL UE is an aggressor to the DL UE or not for each UL/DL UE pair.

There can be multiple variants of the wideband 0/1 DL–UL pair table. For instance, instead of using the ratio between the measured intracell UE–UE and average intercell UE–UE interference, the table can also be obtained using intracell UE–UE versus average total (BS-UE + UE-UE) intercell interference. Another variant is that instead of 0/1 table using 1-bit feedback, we can also obtain, for example, 2-bit 0/1/2 UL–DL pair table, where we can have finer granularity to detect the strong aggressors (strong interferers), weak aggressors (weak interferers), and nonaggressors (noninterferers) for each DL UE with 2-bit feedback. The selection of the threshold for 0/1 or 0/1/2 UL–DL pair table depends on the actual deployment scenario, system requirement, and the schedulers used for the FD system.

The procedure for updating the wideband DL–UL pair table is as follows: The UE updates intracell UE–UE interference based on instantaneous measurement and intercell interference is estimated via moving average method. Based on the updated interference estimate, DL UE will bitmap the interference ratio according to the mechanism given already and feedback periodically or upon BS's request.

There can be a variant of the procedure to feedback the DL–UL pairability. In the variant update rule, the BS will initially set the UL–DL pair table as a all '1's table, meaning initially every UL UE can be paired with every DL UE in the cell. Next, the DL UE observes interference level for a certain period of time. Based on the observation, statistical signal processing techniques, such as maximum likelihood, K-means clustering, or other machine learning techniques, can be used to detect whether UL UE are aggressors to the DL UE. The DL UE only need to feedback the aggressor ID if a new aggressor is detected, and the UL–DL pair table at the BS will be updated accordingly. The BS can reset the UL–DL pair table periodically, and DL UE then repeat the same update process as described already to identify aggressors.

3.4.3.2.4 Wideband per DL–UL Pair Differential CQI Feedback

Another strategy to feedback UE-to-UE interference information with reduced signaling is to only indicate how much UE-to-UE interference degrades DL SINR. We propose to feedback a multibit wideband differential CQI, which can delineate per DL–UL pair UE–UE interference level at a finer scale with more bits than the 1-bit pairable feedback.

The multibit wideband differential CQI feedback will characterize the CQI degradation for each UL–DL pair for joint scheduling in the FD cellular system. We define the wideband CQI degradation $\mathtt{dfCQI}(\mathrm{UE_{DL}}, \mathrm{UE_{UL}})$, caused by UE-to-UE interference from $\mathrm{UE_{UL}}$ to $\mathrm{UE_{DL}}$, as the ratio between the average overall downlink interference plus noise over the average intercell interference plus noise,

$$
\mathtt{dfCQI}(\mathrm{UE_{DL}}, \mathrm{UE_{UL}}) = \frac{E[I_{\mathrm{UE} \rightarrow \mathrm{UE_{intra}}}(\mathrm{UE_{DL}}, \mathrm{UE_{UL}})] + E[I_{\mathtt{inter}}(\mathrm{UE_{DL}})] + N}{E[I_{\mathtt{inter}}(\mathrm{UE_{DL}})] + N}.
$$

This metric represents the ratio between the actual interference experienced under joint DL and UL transmission versus the interference level without intracell UE-to-UE interference. This feedback can be used as an adjustment indicator, on top of the conventional LTE subband feedback mechanism with new reference signal design described in Section 3.4.3.3 such that the CQI only reflects the interference without intracell UE-to-UE interference.

The average intra- and intercell interference can be computed using moving average method after measurement at DL UE side. We can quantize the CQI degradation from 0 to 2^{N-1} using N bits. Such wideband CQI differential values can be used at the BS to adjust CQI level for joint scheduling and rate adaption. We illustrate in the next paragraph how a joint scheduler can incorporate differential wideband CQI to derive more accurate per DL–UL pair SINR. The DL UE can feedback the wideband UL–DL per pair differential CQI periodically or aperiodically at relative low frequency, compared to conventional CQI feedback, based on the BS's request and configuration, for example, every 40/80/160 ms.

3.4.3.2.5 CQI Feedback Methods Comparison

In the following, we provide a few examples of how CQI feedback methods proposed in this section can be used in joint DL–UL scheduler and compare their corresponding signaling overhead with conventional subband CQI feedback for a 20 MHz LTE-based system as baseline. Typically, CQI is quantized in 4 bits. We denote the number of subbands as N_{sb}. For a 20 MHz LTE system with $N_{\mathrm{sb}} = 20$ and 5 ms CQI feedback periodicity, the overhead of subband

CQI feedback is 4-bit $\times N_{sb} = 80$ bits every 5 ms for each DL UE. We assume the same subband configuration for the following example schedulers.

Example 3.1 Naive Scheduler

We consider the simplest scheduler design that requires no modification to existing HD system and name it as naive scheduler. The naive scheduler schedules DL and UL UE independently based on conventional CQI feedback. As shown in Eq. 3.3, there can be mismatch between the intracell UE-to-UE interference measured through conventional CQI and the actual interference experienced during data transmission if different UL UE are scheduled during measurement and transmission periods. For scenarios where intracell UE-to-UE interference is weak compared with the intercell interference, the mismatch in CQI estimate is small and naive scheduler with conventional CQI metric can achieve good performance in FD networks. However, if intracell UE-to-UE interference is so strong that significantly degrades the SINR and achievable rate, naive scheduler may result in poor DL performance. The signaling overhead required for a naive scheduler is the same as conventional CQI feedback, which is 4-bit $\times N_{sb} = 80$ bits every 5 ms for each DL UE.

Example 3.2 Joint A Scheduler

The second example that we name as joint A scheduler is a design requiring complete interference information, that is, UE-to-UE interference between each DL–UL UE pair for all subbands. The joint A scheduler selects DL–UL UE pair based on the full subband per DL–UL pair CQI feedback that indicates the per pair SINR as shown in Eq. 3.5. In addition, the joint scheduler also considers the possibility that a DL UE is scheduled for half-duplex transmission and thus requires CQI feedback of DL SINR with only intercell interference. Table 3.2 illustrates the CQI information that DL UE should periodically feedback to the joint A scheduler for each subband.

Assuming 10 UL UE per cell, the overhead required for joint A scheduler for each DL UE is 4-bit $\times N_{sb} \times (N_U + 1) = 4 \times 20 \times 11 = 880$ bits every

Table 3.2 Full subband per UL–DL pair DL-CQI table.

DL UE ID \ UL UE ID	1	2	...	N_U	Empty
1	$CQI_{1,1}^{(DL)}$	$CQI_{1,2}^{(DL)}$	...	$CQI_{1,N_U}^{(DL)}$	$CQI_{1,\emptyset}^{(DL)}$
$\vdots$	$\vdots$	$\vdots$	$\ddots$	$\vdots$	$\vdots$
N_D	$CQI_{N_D,1}^{(DL)}$	$CQI_{N_D,2}^{(DL)}$	...	$CQI_{N_D,N_U}^{(DL)}$	$CQI_{N_D,\emptyset}^{(DL)}$

Table 3.3 Example of JointB-sCQI table.

Wideband, less frequent (e.g., every 160 ms)						Subband, every 5 ms	
DL UE ID ╲ UL UE ID	1	2	...	N_U		DL UE ID	Conventional
1	0	1	...	1	+	1	$CQI_{conv,1}^{(DL)}$
⋮	⋮	⋮	⋱	⋮		⋮	⋮
N_D	1	0	...	1		N_D	$CQI_{conv,N_D}^{(DL)}$

5 ms. We can see that the joint A scheduler, compared with the conventional LTE scheduler, introduces huge signaling overhead (880 bit versus 80 bit per DL UE in the example system). Next, we will introduce alternative CQI feedback approaches to reduce the complexity compared to joint A scheduler.

Example 3.3 Joint B-sCQI Scheduler
The first simplification is to use 1-bit wideband DL–UL pairability CQI feedback to supplement subband conventional CQI feedback. As indicated in Table 3.3, a wideband 0/1 pairability table updated with lower frequency can be combined with the subband conventional CQI feedback for making the scheduling decision.

Assume periodicity illustrated in Table 3.3 used with 20 subbands and 10 UL UE, the overhead of this scheme per DL UE is 4-bit $\times N_{sb} = 80$ bits per 5 ms + $N_U = 10$ bits per 160 ms, which is much lower than joint A scheduler. Compared with conventional LTE feedback, the additional complexity is only for the wideband 0/1 table with lower feedback frequency so the added overhead is minimal.

Example 3.4 Joint B-sCQI-WbDiff Scheduler
The second variation, Joint B-sCQI-WbDiff Scheduler, utilizes wideband per DL–UL pair differential CQI feedback to adjust the conventional subband CQI feedback variation 1. As conventional subband CQI feedback variation 1 only captures intercell interference as shown in Eq. 3.4, the wideband per DL–UL pair differential CQI can be used to update SINR estimate for (DL,UL) UE pair (UE_{DL}, UE_{UL}) as follows:

$$SINR_{pairwise}^{(DL)}(UE_{DL}, UE_{UL}) = \frac{SINR_{conv1}^{(DL)}(UE_{DL})}{dfCQI(UE_{DL}, UE_{UL})}$$

Table 3.4 Joint B-sCQI-WbDiff CQI tables.

Wide band, less frequent (e.g., every 160 ms)			Subband, every 5 ms	
UL UE ID / DL UE ID	1	... N_U	DL UE ID	Conventional
1	$\mathrm{dfCQI}_{1,1}$	... dfCQI_{1,N_U}	1	$\mathrm{CQI}^{(DL)}_{\mathrm{conv}1,1}$
⋮	⋮	⋱ ⋮	⋮	⋮
N_D	$\mathrm{dfCQI}_{N_D,1}$	... dfCQI_{N_D,N_U}	N_D	$\mathrm{CQI}^{(DL)}_{\mathrm{conv}1,N_D}$

We can translate the above relationship in dB scale:

$$\mathrm{SINR}^{(DL)}_{\mathrm{pairwise,in\ dB}}(\mathrm{UE_{DL}, UE_{UL}}) =$$
$$\mathrm{SINR}^{(DL)}_{\mathrm{conv1,in\ dB}}(\mathrm{UE_{DL}}) - \mathrm{dfCQI}_{\mathrm{in\ dB}}(\mathrm{UE_{DL}, UE_{UL}}). \tag{3.6}$$

Higher CQI differential value per UL and DL UE pair indicates stronger interference DL UE experience from the paired UL UE. Thus, such (UL UE, DL UE) pair will result in degraded subband CQI for DL UE for joint scheduling in the same resources. On the other hand, if CQI differential value is very low for (UL UE, DL UE) pair, the subband CQI value of DL UE will be not degraded for scheduling and rate adaption. The corresponding CQI tables used are illustrated in Table 3.4.

Assume periodicity illustrated in Table 3.4 is used, the overhead of this scheme per DL UE is 4 bits $\times N_{\mathrm{sb}}$ per 5 ms + 4 bits $\times N_U$ per 160 ms, which is much lower than joint A scheduler and slightly higher than JointB-sCQI.

Example 3.5 Joint B-mCQI Scheduler
This scheduler uses wideband 1-bit pairability CQI feedback combining with a subset of subband per pair CQI feedback as indicated in Table 3.5, where the subband per pair CQI is only needed when the pair is shown as non aggressor from the wideband 1-bit pairability table.

The overhead of this scheme per DL UE is 4 bits $\times N_{\mathrm{sb}} \times (N_U \times P_{\mathrm{nonaggressor}} +$ 1) per 5 ms+ N_U bits per 160 ms. $P_{\mathrm{nonaggressor}}$ is the probability that a UL UE is not an aggressor to certain DL UE.

Example 3.6 Joint B-mCQI Reduced Scheduler
This scheduler is similar to the one in Example 5, except it uses wideband 2 bit per pair table (0/1/2), instead of 0/1 table, to combine with a subset of subband per pair table as indicated in Table 3.5, where the subband per pair CQI is only needed when the pair is shown as weak aggressor from the wideband one. The table is similar to Table 3.5 except the subband CQI is only provided when the

Table 3.5 Example of JointB-mCQI table.

Wideband, less frequent
(e.g., every 160 ms)   Subband, every 5 ms

UL DL	1	...	N_U		UL DL	1	...	N_U	Empty
1	0	...	1	+	1	N/A	...	$\text{CQI}^{(DL)}_{1,N_U}$	$\text{CQI}^{(DL)}_{1,\varnothing}$
⋮	⋮	⋱	⋮		⋮	⋮	⋱	⋮	⋮
N_D	1	...	0		N_D	$\text{CQI}^{(DL)}_{N_D,1}$	...	N/A	$\text{CQI}^{(DL)}_{N_D,\varnothing}$

UL UE is a weak aggressor to the DL UE, that is, entry in the table is "1." The corresponding overhead is (4 bit $\times N_{sb} \times (N_U \times P_{\text{weakaggressor}} + 1)$ per 5 ms + $2 \times N_U$ per 160 ms. $P_{\text{weakaggressor}}$ is the probability of a UL UE being a weak aggressor.

The performance comparison of these CQI methods is demonstrated in Section 3.5.

So far we have discussed the scheduler scheme and the CQI feedback mechanisms for joint DL–UL full-duplex transmission. In next section, we will discuss how the measurement for the CQI feedback is obtained, which includes the reference signal design and interference measurement at DL UE.

3.4.3.3 Interference Measurement and Reference Signal Design

For downlink reference signals, there are cell-specific reference signal (CRS), UE-specific channel state information reference signal (CSI-RS), and demodulation reference signal (DMRS). For uplink, there are sounding reference signal (SRS) and demodulation reference signal (DMRS). All of them use dedicated resources (i.e., not overlapping with other channels). Reference signals mainly have two applications. On one hand, they are used to conduct channel estimation for demodulation and decoding in both uplink and downlink; on the other hand, for downlink, they are also used to estimate the downlink channel state information (CSI) and then feedback CSI to base station for scheduling. For uplink, they are also used to estimate the uplink CSI for base station to schedule uplink resource in next transmission opportunity. In addition, downlink CRS can be used for handover triggering (by estimating pathlosses between serving cell and neighbor cell) and finer synchronization.

3.4.3.4 IM-RS Signal

To facilitate full-duplex joint scheduler to optimize scheduling of UL and DL UE pair in FD cellular system, we suggest a new UE-to-UE interference measurement reference signal (UE–UE IM-RS) structure, for uplink to downlink per-pair interference measurement, where all data transmission is not allowed at such resource. The main purpose of this UE–UE IM-RS is to measure the interference coming from individuals of other intracell UL UE, intra-ue2ue(DL UE, UL UE), and/or total or individuals from other inter cell UL UE due to full-duplex operation. With such individual UE-to-UE interference information, especially individual pairs of intracell UE-to-UE interference, joint schedulers can utilize such interference information to jointly schedule uplink and downlink UE together.

The UE–UE IM-RS has one or combinations of the following features/structures that are designed specifically for UE-to-UE interference measurement in FD system:

1) The UE–UE IM-RS is transmitted by UE (potential interferers) that are configured by BS.
2) The UE-UE IM-RS resources are orthogonal to all the data transmission and other reference signals, and (quasi-)orthogonal among themselves that are used by different UE.
3) The IM-RS sequence set that are used within the same cell shall have low or zero correlation, whereas sequences between neighbor cells or cells more far away may allow more correlation.

Note that the third attribute is similar to the design of downlink CRS and uplink DMRS in the current LTE system, with the differences that IM-RS is configured by BS and then transmitted by UE. And each time UE may transmit different IM-RS sequences due to flexibility of BS configuration. On the other hand, downlink CRS is transmitted by BSs with fixed sequences based on its assigned cell id, and uplink DMRS is fixed for each UE by the UE ID. These differences facilitate the UE-to-UE interference measurement in FD system for different purposes.

3.4.3.4.1 Channel Station Interference Information Measurement (CSI-IM) Design

In addition, additional resource element can also be used to measure total UE-to-UE interference, which we call UE–UE CSI-IM. This CSI-IM can be used for multiple purposes and implemented in various ways: For instance, by muting all downlink transmission in UE–UE CSI-IM and allowing all UL UE transmit data in the UE–UE CSI-IM resource, overall UE-to-UE interference, $I_{UE \to UE}(UE_{DL})$, can be measured by each downlink UE. On the other hand, by muting downlink and intracell uplink transmission in UE–UE CSI-IM and allowing all UL UE in other cell to transmit in the UE–UE CSI-IM resource, intercell UE-to-UE interference, $I_{UE \to UE_{inter}}(UE_{DL})$, can be estimated. Different cells can take turns to mute its intracell transmission to measure its own $I_{UE \to UE_{inter}}(UE_{DL})$.

Besides allowing uplink data transmission in CSI-IM resource to measure UE-to-UE interference, one can also use such CSI-IM resource to transmit

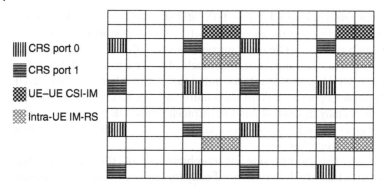

CRS port 0
CRS port 1
UE–UE CSI-IM
Intra-UE IM-RS

Figure 3.6 Example RS pattern with both UE–UE CSI-IM RS and UE–UE IM-RS in one PRB, based on LTE OFDM subframe and RS structure assuming two transmit antennas.

UE-specific reference signals. With reference signals transmitted in CSI-IM resource, one can, for example, obtain the intracell UE-to-UE interference from scheduled pair, $I_{UE \rightarrow UE_{intra}}(UE_{DL}, UE_{UL,sched})$, in addition to $I_{UE \rightarrow UE_{inter}}(UE_{DL})$, by using orthogonal sequences/resources for intracell and intercell reference signals. Such reference signal can use similar design as the downlink CSI RS in LTE advanced (3GPP release 10+), where code division multiplexing (CDM) and frequency/time division multiplexing(FDM/TDM) can be combined or individually used to support the various requirement of sequence pool size. $I_{UE \rightarrow UE_{intra}}(UE_{DL}, UE_{UL,sched})$ can be used for decoding purpose and hence it is expected to be measured in every subframe. Compared to the IM-RS we mentioned earlier (sequence length need to be the number of possible uplink scheduled UE if it is required to be orthogonal to distinguish each possible UL–DL pair), the number of orthogonal CSI-IM sequences can be much less since it only need to distinguish intracell and intercell interferences. Figure 3.6 illustrates an example of frame and reference signal structure for one PRB, where both UE–UE CSI-IM and UE–UE IM-RS presents. The design of the density of the IM-RS signal and CSI-IM should be based on the objective of good trade-off between the estimation accuracy and overhead. Please note that for illustration simplicity, Figure 3.6 shows aligned uplink transmission timing and DL reception timing. The actual position of the IM-RS transmission/reception depends on the individual timing advance (TA) of the uplink/downlink UE pairs as well as the propagation delay between the pair of UE. This requirement is similar as the requirement on the pilots between devices in device-to-device communications.

3.4.3.4.2 UE-to-UE Interference Measurement with UE–UE IM-RS

We summarize the steps to measure UE-to-UE interference as follows:

Step 0: Construct a fixed set of (quasi-)orthogonal UE–UE IM-RS sequences

$$\{X_{\text{imrs},j}(k)\}_{j=1}^{N}, k = 1, \ldots, K,$$

where K is the sequence length, and N is the total number of sequences. In an OFDM framework, k would be the resource element index and the overall sequence of length K spread one subband of a basic scheduling unit.

Step 1: BS configures a subset of uplink UE with different IM-RS indexes for reference signal generation with the mapping $s(j)$ from IM-RS index j to UE id $s(j)$, where $s(j) = -1$, a null id, if j is not configured to any uplink UE.

Step 2: Each of the configured uplink UE generates the reference signal according to its given IM-RS index j as $X_{\text{imrs},j}(k)$.

Step 3: For the downlink UE of interest, it receives

$$Y(k) = \sum_{j=1}^{N} \left[H_{s(j)}(k) X_{\text{imrs},j}(k) \right] + N(k),$$

where $H_{s(j)}$ is the UE–UE channel from user $s(j)$ to the downlink UE, with $H_{(-1)} = 0$, and $N(k)$ is the additive noise. It also receives from the base station index set $\{j\}$.

Step 4: The downlink UE measures the interference power of jth index as

$$P_j = \left\| \sum_{j=1}^{N} \left[Y(k) X_{\text{imrs},j}(k) \right] \right\|.$$

Once DL UE has the interference power of jth index, it can map to a specific SINR between this DL UE and uplink UE using jth sequence and feedback to eNb. eNb can use such information for joint scheduler.

Please note that if $H_{s(j)}(k)$ is relatively constant over the subband resource K, step 4 can be implemented by estimating $H_{s(j)}$ via algorithms like successive interference cancellation or sparse recovery algorithm, for example, L1 minimization/sparse Bayesian learning, and calculated the power as $P_j = |H_{s(j)}|^2$.

To summarize, with the proposed reference signal structures and interference measurement mechanism, one can flexibly obtain at least the following possible signal-to-interference ratios:

1) $\frac{P_{rx}(\text{UE}_{\text{DL}})}{I_{\text{BS}}(\text{UE}_{\text{DL}}) + N}$: SINR with conventional DL interference measured by conventional CRS.

2) $\frac{P_{rx}(\text{UE}_{\text{DL}})}{I_{\text{UE} \to \text{UE}}(\text{UE}_{\text{DL}}) + N}$: SINR with respect to total UE-to-UE interference, using conventional CRS/DMRS signal or CSI-IM proposed in this section.

3) $\text{SINR}_{\text{total}}(\text{UE}_{\text{DL}}) = \frac{P_{rx}(\text{UE}_{\text{DL}})}{I_{\text{BS}}(\text{UE}_{\text{DL}}) + I_{\text{UE} \to \text{UE}}(\text{UE}_{\text{DL}}) + N}$, that is, SINR w.r.t. total interference, which can be measured using conventional CRS signal or obtained from (1) and (2).

4) $\text{SINR}_{\text{conv1}}^{(\text{DL})}(\text{UE}_{\text{DL}}) = \frac{P_{rx}(\text{UE}_{\text{DL}})}{I_{\text{UE}\rightarrow\text{UE}_{\text{inter}}}(\text{UE}_{\text{DL}})+I_{\text{BS}}(\text{UE}_{\text{DL}})+N}$, that is, SINR w.r.t. intercell UE-to-UE interference, where $I_{\text{UE}\rightarrow\text{UE}_{\text{inter}}}(\text{UE}_{\text{DL}})$ can be derived from the total UE-to-UE interference and the intracell UE-to-UE interference with intra-UE IM-RS or directly measured by UE-to-UE CSI-IM.

5) $\text{SINR}_{\text{conv}}^{(\text{DL})}(\text{UE}_{\text{DL}}, \text{UE}_{\text{UL,sched}}) = \frac{P_{rx}(\text{UE}_{\text{DL}})}{I_{\text{UE}\rightarrow\text{UE}_{\text{intra}}}(\text{UE}_{\text{DL}},\text{UE}_{\text{UL,sched}})+I_{\text{UE}\rightarrow\text{UE}_{\text{inter}}}(\text{UE}_{\text{DL}})+I_{\text{BS}}(\text{UE}_{\text{DL}})+N}$: This is per DL–UL pair SINR with respect to interference between DL UE and the scheduled UL UE, plus all intercell UE-to-UE and conventional DL interferences, where
$I_{\text{UE}\rightarrow\text{UE}_{\text{intra}}}(\text{UE}_{\text{DL}}, \text{UE}_{\text{UL,sched}})$ can be obtained with reference signal CSI-IM transmitted. This SINR can be used for demodulation purpose as it reflects the actual SINR with the scheduled pair.

6) $\text{SINR}_{\text{pairwise}}^{(\text{DL})}(\text{UE}_{\text{DL}}, \text{UE}_{\text{UL,IM}}) = \frac{P_{rx}(\text{UE}_{\text{DL}})}{I_{\text{UE}\rightarrow\text{UE}_{\text{intra}}}(\text{UE}_{\text{DL}},\text{UE}_{\text{UL,IM}})+I_{\text{UE}\rightarrow\text{UE}_{\text{inter}}}(\text{UE}_{\text{DL}})+I_{\text{BS}}(\text{UE}_{\text{DL}})+N}$: This is per DL–UL pair SINR with respect to interference between DL UE and each possible UL UE plus all intercell UE-to-UE and conventional DL interferences, where
$I_{\text{UE}\rightarrow\text{UE}_{\text{intra}}}(\text{UE}_{\text{DL}}, \text{UE}_{\text{UL,IM}})$ can be obtained using the new IM-RS signal.

So with such reference signal structure, UE can obtain interference information, including $I_{\text{UE}\rightarrow\text{UE}_{\text{intra}}}(\text{UE}_{\text{DL}}, \text{UE}_{\text{UL,IM}})$ using IM-RS, and $I_{\text{UE}\rightarrow\text{UE}_{\text{inter}}}(\text{UE}_{\text{DL}})$ using CSI-IM, needed for joint scheduler proposed in Section 3.4.3.1. Once DL UE has these interference power, it can map to a specific SINR between this DL UE and the UL UE and feedback to the base station. BS can then use such information for joint scheduler.

The proposed reference signal would introduce additional signal overhead. Take a conventional 10 MHz FDD (20M total spectrum) LTE system as baseline and consider the illustrated new reference structure in Figure 3.4 for full-duplex, the total resource elements (RE) per PRB is $\#symbol \times \#subcarrier = 14 \times 12 = 168$; out of which the conventional LTE CRS occupies 16RE per PRB for 2-TX system; the new CSI-IM could occupy 4 RE per PRB (assumption), which is 4/168 = 2.38%; the intra-UE IM-RS could occupy 1 2 RE per PRB on every 40/80/160 ms periodicity, and so if we use 2 REs per PRB on every 80 ms interval, the overhead would be 2/80/168 = 0.015%. Therefore, by estimation, the overall efficiency with full-duplex system is roughly 88.1% whilst half-duplex is 90.5% with 2.4% additional overhead due to new the reference signal usage.

In Section 3.5, we will illustrate the performance and complexity of the proposed UE-to-UE interference mitigation schemes with simulation results.

3.5 System-Level Performance Analysis

In this section, we present the system-level simulation results for FD small cells. Specifically, we compare the system throughput performance between full-duplex small cells and half-duplex small cells. We first describe our simulation assumptions and evaluation methodology. Then, we demonstrate how BS-to-BS interference mitigation (IM) schemes developed in Section 3.4.2 can improve UL performance in FD networks and how UE-to-UE IM schemes proposed in Section 3.4.3 further improves FD DL throughput. Finally, we test our BS-to-BS and UE-to-UE IM methods under deployments with FD-capable UE and under different small cell and UE densities. From the simulation results, we verify that FD small cells can always provide close to 2× capacity gain under full buffer traffic.

3.5.1 General Simulation Methodology and Assumptions

To demonstrate the potential realizable full-duplex gain in real-world small cell deployments, we simulated FD and HD small cells based on LTE evaluation methodology. Specifically, we consider stand-alone small cell deployment where there is no macro-BS deployed or macro-BSs are operating on a different bandwidth from small cell BSs. Therefore, we can omit macro-UE from our evaluation and focus on the performance of small cell UE. For FD small cells, we assumed all the base stations can transmit in full-duplex while UE within the cell can be either conventional half-duplex UE or FD-capable UE.

We denote a small cell base station as a low power node (LPN) to match 3GPP terminology. To test how our IM schemes perform, we focus on evaluating the most severe interference environment where there are both DL and UL full buffer traffic within each small cell. Our simulation assumptions match closely with 3GPP evaluation methodologies.[1][2][3] The basic simulation settings are summarized in Table 3.6.

We consider the small cell deployment and channel models defined in 3GPP standards for dynamic TDD[3] and Licensed Assisted Access[4] evaluation. In the following, we explain in detail the deployment and channel models used for the simulations.

3.5.1.1 Deployment Models

Our simulation is based on hexagonal grid macrocell deployment with three sectors per cell site and 500 m site-to-site distance. Within every macrocell sector, 4 small cells are deployed with 10 DL UE and 10 UL UE served by each small cell, unless otherwise specified. We review multiple 3GPP

1 3gpp TR36.814, http://www.3gpp.org/dynareport/36814.htm.
2 3gpp TR 36.843, http://www.3gpp.org/dynareport/36843.htm.
3 3gpp TR 36.828, http://www.3gpp.org/dynareport/36828.htm.
4 3gpp TR 36.889, http://www.3gpp.org/dynareport/36889.htm.

Table 3.6 Basic simulation settings.

Parameter	Value
Carrier frequency	2 GHz
	20 MHz
System bandwidth	(FD or TDD: 20 MHz for both DL & UL;
	FDD: 10 MHz UL + 10 MHz DL)
LPN TX power	Indoor deployment: 24 dBm;
(No power control,	Outdoor cluster deployment: 30 dBm;
transmit at max pwr)	Outdoor uniform deployment: 24 dBm
UE TX power	23 dBm
UE UL power control	Open loop fractional
Noise figure	UE: 9 dB, LPN: 5dB (-174 dBm/Hz thermal noise)
Antenna gain	LPN: 5 dBi; UE: 0 dBi
Antenna pattern	2D Omni directional antenna for both LPN and UE
LPN antenna configuration	2TX, 2RX (codebook-based SU-MIMO)
UE antenna configuration	1TX, 2RX
Scheduler	Different FD schedulers based on proportional fair metric
Traffic model	full buffer
	Model large-scale fading for UE-to-UE and BS-to-BS interference
FD interference assumptions	BS-to-BS: total 40 dB suppression from elevation nulling
	Self-Echo: perfect cancellation
Receiver type	MMSE-IRC receiver
Uplink rate adaptation	Target 10% BLER based on SRS measurement
Downlink rate adaptation	Target 10% BLER based on CQI/PMI/RI feedback
DL CSI feedback	Subband CQI/PMI/RI error free feedback every 5 ms
UL CSI feedback	Subband SRS measurement every 10 ms
HARQ modeling	Synchronized 8 ms delay retransmission
HARQ retransmission scheme	CC with maximum 4 retransmissions
Control channel and RS overhead	DL: 3PDCCH + 2TXCRS UL: SRS + PUCCH

documentations and select three different small cell deployments to model typical small cell usage scenarios. They are based on deployment scenarios specified in LAA[4] and dynamic TDD[3] evaluation methodology. In the following, we explain in detail the three different deployment models.

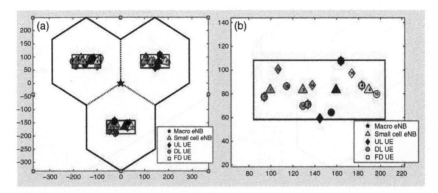

Figure 3.7 Illustration of indoor deployment.

1) *Indoor Deployment*:

The indoor deployment scenario is used for LAA evaluation. It models the planned small cell deployment for indoor enterprise usage case. Within each macrocell sector, there is one indoor building with LPNs deployed equally spaced inside. Each indoor building is of the size of 120 m × 50 m and is deployed at the center of each macrocell sector. UE are randomly dropped within the building in uniform distribution with the restriction that the number of UE served by each LPN is the same where cell association rule is based on maximum received signal strength. The indoor deployment is illustrated in Figure 3.7, where Figure 3.7a shows the enterprise building location within macrocells and Figure 3.7b zooms in to one enterprise building to illustrate small cell (4 LPNs per macro) and UE dropping within a building.

Detail specifications for indoor deployment are summarized in Table 3.7.

2) *Outdoor Cluster Deployment*:

The outdoor cluster deployment is used in both LAA and dynamic TDD evaluation methodology. It models the clustered small cell deployment for

Table 3.7 Indoor deployment details.

Deployment Specs	Value
Building dropping	Fixed placement (center of macro sector)
LPN dropping	Equally spaced inside the building
UE dropping	Within rectangular block of 120 m × 50 m
LPN-LPN dist	30 m
LPN-UE min dist	3 m
UE–UE min dist	3 m

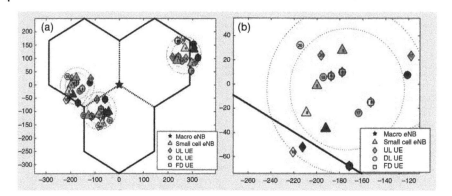

Figure 3.8 Illustration of outdoor cluster deployment.

outdoor hot spot usage case where the hot spot is larger than the coverage area for a single small cell. An example deployment of outdoor cluster scenario is illustrated in Figure 3.8, where Figure 3.8a shows the cluster locations within macrocells and Figure 3.8b zooms in to one cluster to illustrate LPNs and UE droppings.

Within each macrocell sector, one cluster is randomly dropped in uniform distribution. Within each cluster, four LPNs are randomly dropped within inner circle (inner dotted circle in Figure 3.8b) and UE are randomly dropped within outer circle (outer dotted circle in Figure 3.8b) both in uniform distribution.

Detail specifications for outdoor cluster deployment are summarized in Table 3.8.

Table 3.8 Outdoor cluster deployment details.

Deployment Specs	Value
Pico dropping radius	50 m
UE dropping radius	70 m
Macro-BS-LPN min dist	105 m
LPN–LPN dist	20 m
Macro-UE min dist	35 m
LPN–UE min dist	3 m
UE- UE min dist	3 m

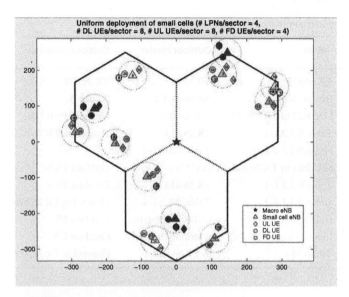

Figure 3.9 Illustration of outdoor uniform deployment.

3) *Outdoor Uniform Deployment*:

The outdoor uniform deployment is used in dynamic TDD evaluation. It models the small cell deployment for outdoor hot spot usage case where single small cell can sufficiently cover a hot spot and there are multiple such small hot spot within a macro cell. The outdoor uniform deployment is illustrated in Figure 3.9. The LPNs are randomly dropped based on uniform distribution. There are four LPNs per macrocell. Within the coverage area of the small cell, indicated by the dotted circle in the figure, UE are randomly dropped with uniform distribution.

Detail specifications for outdoor uniform deployment are summarized in Table 3.9.

Table 3.9 Outdoor uniform deployment details.

Deployment Specs	Value
Pico dropping	Entire macrocell sector
UE dropping radius	50 m
Macro-BS-LPN min dist	75 m
LPN–LPN dist	40 m
Macro-UE min dist	35 m
LPN–UE min dist	10 m
UE–UE min dist	N/A (3 m is used in our simulation)

Table 3.10 Channel model summary.

	Indoor	Outdoor cluster	Outdoor uniform
	TR 36.889	TR 36.889	TR 36.828
	Annex A.1.1	Annex A.1.2	Annex A
LPN↔UE link	ITU InH in TR 36.814	ITU Umi in	Outdoor Pico-UE
	Table B.1.2.1-1	TR 36.814	channel in TR 36.828
	and A.2.1.1.5-1	Table B.1.2.1-4	
LPN↔LPN link	ITU InH in TR 36.814	ITU Umi in	Outdoor Pico-
(Fast fading	Table B.1.2.1-1	TR 36.814	Outdoor Pico
not modeled)	and A.2.1.1.5-1	Table B.1.2.1-4	channel in TR 36.828
UE↔UE link	D2D Indoor-to-	D2D Outdoor-to-	Outdoor UE-
	Indoor channel in	Outdoor channel	Outdoor UE
	TR 36.843	in TR 36.843	channel in TR 36.828

3.5.1.2 Channel Models

Unlike synchronous half-duplex system where only the channel between LPN and UE needs to be modeled, we need to model the LPN–LPN and UE–UE channels as well in full-duplex networks since LPN and UE can transmit simultaneously. Wireless environment for LAA and dynamic TDD involves asynchronously DL and UL transmission among neighboring LPNs where LPN and UE can also concurrently transmit. Therefore, we adopt the channel models defined in LAA and dynamic TDD evaluation methodology[3,4] for our FD network simulation.

A high level summary of the channel models for the three deployment scenarios we previously defined is shown in Table 3.10. Detail explanation of the channel models can be found in 3GPP documentations[1,2].

All channels are assumed to be reciprocal. Note that the channel models for the outdoor cluster deployment are different in LAA and dynamic TDD evaluation. In general, dynamic TDD assumes much smaller UE-to-UE and BS-to-BS interference. Therefore, we only consider the LAA channel model for outdoor cluster deployment in our simulations since the LAA channel model is more challenging for FD operation.

3.5.2 Performance of BS-to-BS Interference Mitigation Schemes

In this section, we present the simulation results of different UL power control settings for mitigating BS-to-BS interference. As shown in Figure 3.4, the BS-to-BS interference can be significantly stronger than conventional UL interference ($I_{UE \rightarrow BS}$). Without proper BS-to-BS interference suppression, the FD UL SINR can be too weak to support acceptable UL transmission rate. Therefore, elevation beam nulling is essential to decrease BS-to-BS interference to tolerable level. Beam pattern in Reference [4] shows 70–80 dB nulling can be achieved when ISD is greater than 500 m. For small cell deployments described in Section 3.5.1.1, lower achievable nulling gain should be assumed due to shorter ISD and NLOS component in BS-to-BS link. In the following simulations, we assume BS-to-BS interference can be partially reduced via elevation beam nulling by at least 40 dB.

For UL OLPC evaluation, our baseline half-duplex system is operating in 20 MHz TDD. We evaluate only indoor and outdoor cluster deployment where BS-to-BS interference is more severe according to CDF Figure 3.4. Simulation resulting in this session assume no UE-to-UE interference mitigation where FD LPNs employ naive scheduler that independently schedules DL and UL traffic. For UL open-loop fractional power control, we assume $P_0 = -76\,\text{dBm}$ and $\alpha_{\text{FPC}} = 0.8$. The boosting factor B will boost target receive power from P_0 to $B \cdot P_0$.

We collect the per UE throughput statistics with different boosting factor settings and then plot the percentage of realizable FD throughput gain. Compared against HD LPN operating in TDD where DL and UL take turns to transmit, a FD-capable LPN can simultaneously communicate in both DL and UL. We define $\text{TP}_{\text{TDD}_{\text{DL}},\text{max}}$ as the throughput that can be achieved if the entire 20 MHz channel is scheduled for DL transmission only, and $\text{TP}_{\text{TDD}_{\text{UL}},\text{max}}$ represents the UL counterpart. Suppose that the TDD system allocates r_{DL} percent of time to DL and $r_{\text{UL}} = 1 - r_{\text{DL}}$ of time to UL, the actual throughputs for TDD_DL and TDD_UL are $\text{TP}_{\text{TDD}_{\text{DL}}} = r_{\text{DL}}\text{TP}_{\text{TDD}_{\text{DL}},\text{max}}$ and $\text{TP}_{\text{TDD}_{\text{UL}}} = r_{\text{UL}}\text{TP}_{\text{TDD}_{\text{UL}},\text{max}}$, respectively. The maximum achievable FD throughput is the sum of $\text{TP}_{\text{TDD}_{\text{DL}},\text{max}}$ and $\text{TP}_{\text{TDD}_{\text{UL}},\text{max}}$, that is, sum of the throughput from transmitting only in TDD-DL for the entire duration and the throughput from transmitting only in TDD-UL for the entire duration. The percentage of achievable FD gain is defined as $100 \times \frac{\text{TP}_{\text{FD}}}{(\text{TP}_{\text{TDD}_{\text{DL}},\text{max}}+\text{TP}_{\text{TDD}_{\text{UL}},\text{max}})}$. The percentages of achievable FD DL and FD UL gain are $100 \times \frac{\text{TP}_{\text{FD}_{\text{DL}}}}{\text{TP}_{\text{TDD}_{\text{DL}},\text{max}}}$ and $100 \times \frac{\text{TP}_{\text{FD}_{\text{UL}}}}{\text{TP}_{\text{TDD}_{\text{UL}},\text{max}}}$, respectively. The 5, 50, and 95% and mean UE throughput of FD small cells and TDD small cells are compared to derive the percentage of realizable FD gain with respect to TDD system. Simulation results are averaged over 10 trials with 1-tier (21 macrocell sectors) wraparound deployment where there are 4 small cells per macrocell sector and 10 DL UE + 10 UL UE per small cell.

Performance of different boosting factor settings for LAA indoor and outdoor deployment are shown in Figures 3.10 and 3.11.

Figures 3.10a and 3.11a plot the impact of boosting factor values on UL throughput performance. FD UL performance is severely degraded under small UL OLPC boosting factors. When boosting factors exceed B^{UL} selected based

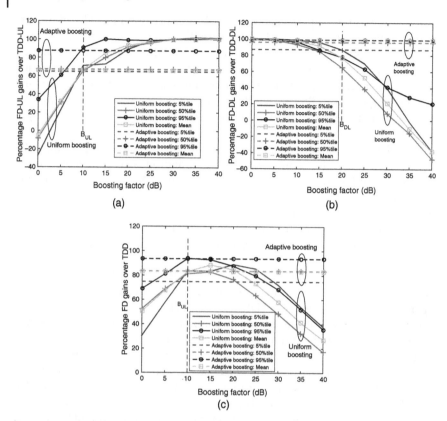

Figure 3.10 Percentage of realizable FD throughput gain with respect to TDD for indoor deployment with 10 DL UE + 10 UL UE per small cell. (a) Percentage of realizable FD-UL gain. (b) Percentage of realizable FD-UL gain. (c) Percentage of realizable total (DL + UL) FD gain.

on top 5% $I_{BS \to BS}/I_{UE \to BS}$, the FD UL throughput is at least 60% better than TDD UL. Difference in B^{UL} reflects that BS-to-BS interference is slightly more dominated in LAA outdoor than LAA indoor case. With very high boosting factor, optimal 100% FD UL throughput gain can be achieved.

Figures 3.10b and 3.11b show the impact of boosting factor values on DL throughput performance. We observe that small UL OLPC boosting factors can reach optimal 100% FD DL throughput gain while large UL OLPC boosting factors cause significant DL throughput degradation. Selecting B_{DL} based on the lowest 3% $I_{BS \to UE}/I_{UE \to UE}$, FD DL throughput is at least 65% higher than TDD DL for LAA indoor and 90% better for LAA outdoor deployment for boosting factors below B_{DL}.

Selection of boosting factor should balance between DL and UL throughput gain as the trends for DL and UL FD throughput gain with increasing boosting

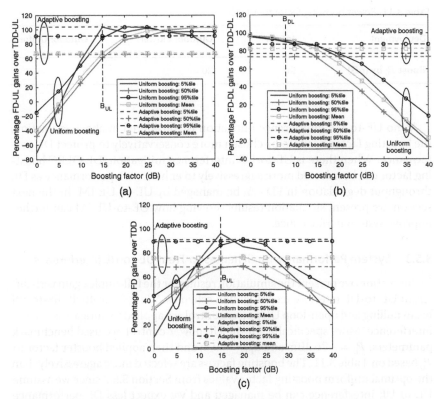

Figure 3.11 FD gain over TDD for outdoor cluster deployment with 10 DL UE + 10 UL UE per small cell. (a) UL UE throughput gain. (b) DL UE throughput gain. (c) DL + UL UE throughput gain.

factor are different. To jointly consider DL and UL throughput, Figures 3.10c and 3.11c plot the DL+UL FD gain over TDD. In LAA indoor deployment, B_{UL} is smaller than B_{DL}. By choose boosting factor B satisfying $B_{UL} < B < B_{DL}$, similar level of total DL+UL FD gain can be achieved. Optimal boosting factor achieves >80% FD gain over TDD. Horizontal lines in the figures denote the FD gain over TDD for cell-specific boosting factor (adaptive boosting). The performance of cell-specific boosting factor is close to the optimal uniform boosting factor and B_{UL}, except more degradation for cell-edge DL users. In LAA outdoor deployment, B_{UL} is larger than B_{DL}. Since less degradation was observed in DL performance, pushing boosting factor close to B_{UL} shows better DL+UL FD gain. Optimal uniform boosting factor achieves 75–80% FD gain over TDD for mean and 50% throughput and 90% FD over TDD gain for 5% and 95% throughput. The cell-specific boosting factor matches very well with the performance of optimal uniform boosting factor.

Table 3.11 Power boosting factor for ULPC.

Deployment	Indoor	Outdoor Cluster	Outdoor Uniform
Boost P_0 by (dB)	20	20	5

As no UE-to-UE interference mitigation is considered in above evaluation, the boosting factor needs to be chosen more conservatively to protect DL performance. When there is UE-to-UE interference mitigation, UL OLPC boosting factor can be selected more aggressively to enhance UL performance as DL throughput degradation in FD can be managed by UE-to-UE IM. In the next section, we present simulation results showing how UE-to-UE IM can further improve system performance.

3.5.3 System Performance for Schemes to Treat UE-to-UE Interference

In this section, we present the simulation results for the full-duplex gain with different UE-to-UE interference mitigation schemes. We assume 40 dB BS-to-BS beam nulling and open-loop UL power control are applied to mitigate BS-to-BS interference. More specifically, for the UL power control, we used benchmark parameters: $P_0 = -76$ dBm and $\alpha_{FPC} = 0.8$, and then applied booster factor to P_0 based on Table 3.11. The boosting factors are selected more aggressively than the optimal uniform boosting factor values from Section 3.5.2 since we assume UE-to-UE interference can be managed and we expect less DL performance degradation due to higher UE transmit power.

After UL power boosting, the DL interference comparison curves shown in Figure 3.5 should be moved toward the left as UE-to-UE interference increases. As expected, with power boosting, UE-to-UE interference is not trivial any more when comparing with conventional DL interference, especially for the outdoor uniform deployment when both intracell and intercell UE-to-UE interference are considered. The potential benefit for intracell UE-to-UE interference mitigation is shown is Figure 3.5b where conventional DL interference is compared against intercell UE-to-UE interference only. In the following, we demonstrate the exact throughput benefit by simulating the joint scheduling schemes proposed in Section 3.4.3.

3.5.3.1 System Performance of Joint Scheduler

The simulation parameters for results in this section are the same as settings summarized in Table 3.6. We compare the full-duplex results against FDD small cells with same bandwidth, that is, 20 MHz FD versus 10 MHz FDD DL + 10 MHz FDD UL. Our performance metric is defined as FD gain that is the per UE throughput gain of full-duplex system over FDD system. The deployment

Table 3.12 UL gain versus DL degradation with uplink power boosting, with 10 UE per LPN.

	P_0 boost (dB)	UL FD gain recovery (mean of TP)	DL FD gain degradation
Indoor	20	$0.95 \rightarrow 1.94$	$2.0 \rightarrow 1.78$
Outdoor cluster	20	$0.6 \rightarrow 1.91$	$2.0 \rightarrow 1.7$
Outdoor uniform	5	$1.89 \rightarrow 1.94$	$1.38 \rightarrow 1.21$

scenario for indoor and outdoor cluster deployments are the same as described in the previous section for BS-to-BS IM. For outdoor uniform scenario, we simulate only three macrocell sectors without wraparound to showcase the scenario where intracell UE-to-UE interference is more dominant to intercell interference.

We compared the performance of two scheduling strategies: *Naive scheduler* where DL and UL traffic are scheduled independently and *joint* DL–UL *scheduler* where a joint DL–UL proportional-fair scheduling metric is used for selecting the (DL, UL) UE pair for full-duplex transmission. Table 3.12 summarizes the performance results when there is no intracell UE-to-UE IM, that is, *naive scheduler* is used, and BS-to-BS interference is mitigated via beam nulling and UL power boosting. It can be seen that in sparse outdoor uniform deployment scenario, interference from UE within the same cell can cause much stronger interference than from neighbor cell BS, and the DL FD gain can be as low as 1.2× with boosted uplink power and thus smarter scheduler to handle UE-to-UE interference is needed.

We then apply the *joint DL–UL scheduling* approach described in Section 3.4.3.1 to mitigate UE-to-UE interference. Simulation results to compare the throughput performance with joint scheduler versus naive scheduler are summarized in Table 3.13. It shows that DL FD gain for outdoor uniform deployment can be improved to 1.74×. For dense deployments, the joint DL–UL scheduling approach also mitigates the DL performance degradation caused by UL power boosting and improves the DL FD gain from 1.78× and 1.7× to 1.9× and 1.81× for indoor and outdoor cluster deployments, respectively. This joint scheduler result utilizes full CQI feedback, that is, per subband per UL–DL UE pair SINR information section. The signaling overhead with this scheme is huge compared to half-duplex conventional LTE CQI feedback, that is, 880 bits versus 80 bits as discussed in Section 3.4.3.3. The performance for various CQI feedback schemes to trade-off performance with signaling complexity is presented next.

3.5.3.2 Performance of Various CQI Feedbacks
In Section 3.4.3.2, we presented various CQI feedback strategies for joint scheduler. In this section, we will compare the FD gain achievable via three of the

Table 3.13 Average full-duplex gain with proper interference mitigation.

Deployment Scenarios		Interference Mitigation Scheme		
		BS-nulling (40 dB) + Naive scheduler	BS-nulling + UL PwCtrl + Naive scheduler	BS-nulling + UL PwCtrl + Joint DL–UL scheduler
Indoor	DL	2.0×	1.78×	1.9V
	UL	0.95×	1.94×	1.96×
Outdoor	DL	2.0×	1.7×	1.81×
Cluster	UL	0.6×	1.91×	1.92×
Outdoor	DL	1.46×	1.25×	1.74×
Uniform	UL	1.77×	1.86×	1.86×

proposed CQI examples (1) wideband per UL–DL pair table (*JointB-sCQI*), (2) wideband per UL–DL pair differential CQI feedback (*JointB-sCQI-WbDiff*), and (3) full CQI with subband per UL–DL pair CQI feedback (*Joint A*). Our performance baseline is naive scheduler that uses only conventional LTE CQI feedback without UE-to-UE interference information.

For wideband per UL–DL pair table(*JointB-sCQI*), the threshold to categorize a UL UE as aggressor or nonaggressor to a DL UE can be set based on the statistical observation of the ratio between intracell UE-to-UE interference and intercell UE-to-UE interference. We observe the CDF curve for the ratio of these two interferences and choose threshold of 50 dB for outdoor uniform deployment and 0 dB for indoor and outdoor cluster deployment when generating the wideband 0/1 table.

We first look at the dense environments, that is, indoor and outdoor cluster deployments, where conventional DL interference is more dominant. We can see that both naive and joint-B-sCQI achieves good signaling and performance trade -off, and simple 1-bit table indicating pairable (UL,DL) pair is sufficient to assist the joint scheduler to make good scheduling decision. The performance is almost the same as the full-CQI (*Joint A*) case. We have similar observation for outdoor cluster deployment and here omitted the results for simplicity (Table 3.14).

For outdoor uniform deployment where intracell UE-to-UE interference is more dominant, the DL performance is not satisfactory with only 1-bit pair table feedback (*JointB-sCQI*). Table 3.15 shows only 1.5× FD gain was observed in the DL, which implies more information of UE-to-UE interference should be feedback to the joint scheduler. With the 4-bit differential CQI feedback for each (DL,UL) UE pair (*JointB-sCQI-WbDiff*), the DL FD can be improved to 1.77× that is about 95% of the gain achievable with full CQI feedback. The two efficient CQI feedback schemes only require 0.4–1.6% more feedback compared

Table 3.14 Performance of various CQI feedback methods for indoor deployment.

Mean FD Gain	Naive	JointB-sCQI	Joint A
DL	1.78×	1.9×	1.88×
UL	1.94×	1.96×	1.92×
Sum	1.85×	1.92×	1.9×
Overhead	+0%	+0.4%(10/32/80)	+1100%

Table 3.15 Mean FD gain of various CQI feedback methods for outdoor uniform deployment.

	Naive	JointB-sCQI	JointB-sCQI-WbDiff(4-bit)	JointA
DL	1.21×	1.5×	1.77×	1.83×
UL	1.94×	1.94×	1.94×	1.94×
Sum	1.51×	1.68×	1.84×	1.88×
Overhead	+0%	+0.4%	+1.6%(40/32/80)	+1100%

to current LTE CQI feedback. Simulation results demonstrated that the two efficient feedback methods can achieve almost all FD gain with about 1000× overhead saving over the full-CQI (joint A) case.

In summary, for dense deployment scenario such as LAA indoor/outdoor where conventional downlink interference is more dominant than intracell UE-to-UE interference, joint scheduler with very simplified CQI feedback, such as wideband per UL–DL pair table (*JointB-sCQI*), already achieves 1.9× performance gain and thus sufficient to be used. In sparse deployment scenario such as outdoor uniform deployment with more dominant intracell UE-to-UE interference, joint scheduler with simplified CQI feedback, such as wideband per UL–DL pair differential CQI feedback (*JointB-sCQI-WbDiff*), is sufficient to achieve 1.77× DL performance gain over HD system. The additional feedback overhead compared to conventional LTE CQI feedback is only 0.4%, 1.6% more, in the example system under consideration.

3.5.4 System Performance for Various Operation Regimes

In previous sections, we mainly consider deployments with 10 DL + 10 UL UE per small cell and 4 LPN per macrocell sector when demonstrating the system performance. In this section, we will present the results with various UE and LPN densities to validate the performance gain of full-duplex system for various operation regimes. Results for outdoor uniform deployment are presented in

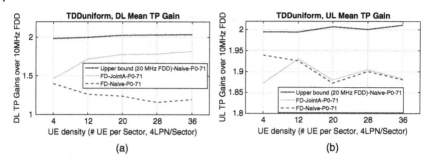

Figure 3.12 Throughput Performance gain with different HD UE Density in Outdoor Uniform Deployment (a) DL gain (b)UL gain.

this section to illustrate how multiuser diversity and density of LPNs affect the overall FD gain in the entire network.

3.5.4.1 Performance for Various UE Densities and Bundle Scheduler

For UE density study, we consider three types of UE composition: (1) small cells with pure half-duplex UE, (2) small cells with pure full-duplex UE, and (3) small cells with mix of HD and FD UE. For deployments with presence of FD UE, we consider bundle scheduler that schedules the same FD UE in both DL/UL direction for concurrent transmission, as discussed in Section 3.4.3.1. This simple bundle scheduler design that guarantees no intracell UE-to-UE interference with reduced signaling overhead is compared against the most complicated joint A scheduler.

3.5.4.1.1 Performance of UE Density for Pure HD UE

Figure 3.12 shows the performance gain of DL and UL throughput in outdoor uniform deployment, with number of DL/UL UE per cell increasing from 4 to 36. Note that the x-axis in Figure 3.12a and b represents the number of DL and UL UE per macrocell sector, respectively. As the number of UE increases, joint DL–UL scheduler can exploit multiuser diversity to achieve more FD DL gain. On the other hand, naive scheduler is unable to improve throughput performance as UE density increases. Furthermore, FD gain actually degrades since intracell UE-to-UE interference is more dominant in outdoor cluster deployment.

3.5.4.1.2 Performance of UE Density for Pure FD UE

For systems with pure FD UE, both joint A and bundle FD scheduler are evaluated by simulations. Bundle FD scheduler schedules UL/DL traffic from the same FD UE in the same resource block. Since there is no intracell UE-to-UE interference, no additional interference measurement or CQI feedback is required and conventional CQI feedback mechanism is used in our simulation.

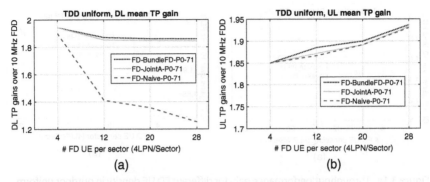

Figure 3.13 Throughput performance gain with different FD UE density in outdoor uniform deployment. (a) DL gain. (b)UL gain.

As we can see from simulation results in Figure 3.13, despite with less scheduling flexibility, the bundle scheduler performs close to joint A scheduler in all deployment scenarios. We even observe noticeable UL TP improvement for bundle scheduler in some deployment settings and the gain comes mainly from increasing UL transmission opportunity for cell-edge UE.

3.5.4.1.3 Performance of UE Density for Mixed FD and HD UE

Now we examine the performance when there are mixed FD and HD UE in the system. For fair comparison, the total number of UE with DL traffic is fixed to 28 and so is the total number of UE with UL traffic. We consider cases with 4, 12, 20, and 28 FD UE with bidirectional traffic, which corresponds to 24, 16, 8, and 0 pure DL UE plus 24, 16, 8, and 0 pure UL UE per sector. Bundle scheduler for mixed FD and HD (variation 1) mentioned in Section 3.4.3.1 is compared with joint A and naive scheduler. As we can see from Figure 3.14, the performance of bundle scheduler is comparable to joint A scheduler and much better than naive scheduler in outdoor uniform deployments.

3.5.4.2 Performance of Various LPN Densities

In this section, we will study the impact of LPN density on the performance of FD system with joint intracell scheduler and UL power control. We use dynamic TDD channel and outdoor uniform deployment models (Tier 0), assuming 10 DL UE + 10 UL UE per LPN to study the effect of LPN density.

With different LPN density, the BS-to-BS interference level varies. Therefore, instead of using a fixed uplink power for every LPN density, select different boosting factor power for different LPN densities such that the sum throughput can be maximized subject to the throughput of DL/UL gain is larger than a threshold(i.e., 5% gain is larger than 1.3x) and half-duplex throughput is within typical range. With reference $P_0 = -76$ dB, the power boost

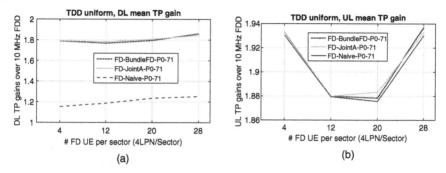

Figure 3.14 Throughput performance gain for different FD UE density in outdoor uniform deployment with mix of FD and HD UE. (a) DL gain. (b)UL gain.

Table 3.16 Power boost factors for different LPN densities.

Number of LPN per macrocell sector	1	4	9	17	25
Optimal boosting factor for joint A scheduler	0	5	5	10	10
Optimal boosting factor for naive scheduler	−5	0	5	10	10

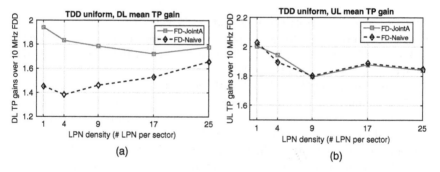

Figure 3.15 Throughput performance gain for different LPN density in outdoor uniform deployment with pure HD UE. (a) DL gain. (b) UL gain.

factors under various LPN densities used in the simulation are summarized in Table 3.16.

The performance of full duplex system under various LPN densities is shown in Figure 3.15. We can see that FD gain is a convex function of number of LPNs. When the number of LPN is small, intracell UE-to-UE interference is dominated and joint scheduler can effectively boost FD gain by managing intracell UE-to-UE interference. Under high LPN density, the intercell interference from neighbor cell BSs and UE becomes dominated that results in naive scheduler performance approaching joint scheduler when number of LPNs is large.

In summary, in this section, we investigate the FD performance in various operating regimes:

- For HD UE system, FD gain of joint scheduler grows with increasing UE density due to multiuser diversity.
- For systems with FD capable UE, bundle scheduler reduce signaling overhead and achieve close to joint scheduler in DL and outperforms joint scheduler in UL.
- For systems with various LPN density, the average sum FD gain is a convex function of LPN density and joint scheduler always achieves at least 1.8× gain over half-duplex system. In the high LPN density regime, naive performs close to joint scheduler.

3.6 Conclusions and Future Directions

The development of self-interference cancellation technology enables devices to communicate in full-duplex, which fosters new system design opportunities to double the spectrum efficiency. In this chapter, we first review existing self-interference cancellation techniques and their capability. With 120 dB+ self-interference suppression, small cell base stations are immune to TX echo and can operate in full-duplex. However, new system design challenges arise with simultaneous DL and UL transmissions coexisting in the network. After carefully studying the characteristics of the new BS-to-BS and UE-to-UE interferences, we identify practical interference mitigation (IM) schemes that can effectively manage the new interferences in the most efficient manner. Our proposed IM scheme can be summarized as follows.

- *BS-to-BS Interference Mitigation:*
  BS-to-BS interference can be significantly stronger than convention UL UE-to-BS interference and requires higher order interference suppression. However, as BS-to-BS interference is rather static for their fixed location and steady traffic loading, we can develop static or semistatic IM scheme to combat it. We propose to combine the below two methods for BS-to-BS IM:

  1) *Elevation Beam Nulling:* Neighboring small cells typically are with antennas of similar height in outdoor hot spot or indoor enterprise deployment. Therefore, forming a null at 90° elevation can effectively suppress interference from direct BS-to-BS link. Results in Reference [4] show 35–40 dB nulling can be achieved via careful vertical beam pattern design.
  2) *Uplink Open-Loop Power Control:* The UE can as well boost its transmit power to overcome BS-to-BS interference. Based on long term statistics of interference power, a boosting factor can be selected to

boost the target received power level for open-loop power control to improve UL SINR, as described in Section 3.4.2.2.

- *UE-to-UE Interference Mitigation:*
  In contrast to BS-to-BS interference, UE-to-UE interference is weaker yet more dynamic. For environments with strong BS-to-BS interference, UE may need to boost UL transmission power to improve UL SINR under full-duplex transmission. As UL power increases, the impact of UE-to-UE interference becomes more noticeable. To mitigate intracell UE-to-UE interference, we propose joint downlink-uplink scheduling scheme to select UE pairs with less interference in between for concurrent DL and UL transmission. We also studied practical design considerations for realizing joint scheduling, such as reference signal design and signaling feedback overhead, in Sections 3.4.3.3 and 3.4.3.2. Intercell UE-to-UE interference mitigation requires coordination between different cell sites. Though intercell UE-to-UE interference mitigation is not one of the focus area in this chapter, we will describe some potential solutions in Section 3.6.1.1.

Simulation results shown in Section 3.5 demonstrate that with proper interference mitigation, more than 1.7× spectrum efficiency gain can be achieved in FD networks. Furthermore, the gain can be realized with only marginal increase in signaling overhead. By testing over wide range of small cell deployments, we conclude that it is very promising to increase spectrum efficiency by more than 70% with FD transmission.

This study only covers limited perspective for FD system design. In the following, we list some future research direction to further improve current design and to understand the FD system more comprehensively.

3.6.1 Improvement to the Current Design

3.6.1.1 Intercell UE-to-UE Interference Handling

Throughout this chapter, the UE-to-UE interference scheme addresses intracell scheduler only. However, when intercell interference becomes stronger, such as when there is macro station (HetNet) with high transmission power nearby UE, or when the victim UE is at cell edge vulnerable to the uplink UE from neighbor cells, scheduler just for intracell UE-to-UE interference may not be sufficient. Some level of intercell coordination would be needed. This could include, but not limited to (1) methods to identify intercell victims/aggressors, (2) reference signal design for estimating intercell UE-to-UE interference level, (3) low latency, low-overhead protocol design for information sharing and signaling exchange between base stations, and (4) radio resource management (RRM) strategies for intercell coordination.

For intercell RRM, the frequency planning concept of fractional frequency reuse (FFR) can be explored where a base station can be assigned with a prioritized usage frequency band, which is orthogonal to the prioritized usage frequency bands of its neighboring base stations. By restricting the UE who suffer or cause strong intercell UE-to-UE interference to only transmit on the prioritized usage frequency band of its serving cells, we can orthogonalize transmission of major aggressor–victim pairs and minimize the performance degradation caused by intercell UE-to-UE interference. For UE that do not experience strong UE-to-UE interference, all bands are allowed for transmission. When UE are capable of detecting interference and distinguishing the aggressor/victim UE identity, more advanced strategy can be adopted such as allowing UE to access all frequency bands except for the priority usage bands of its major interfering neighbor cell UE. The concept of DL-centric zone and UL-centric zone can also be used if a UE can be classified as a pure aggressor or as a pure victim. Based on the fact that a UE is causing interference or receiving interference, the UE should avoid using the DL-centric zone for UL transmission if the UE is causing high interference and avoid using the UL-centric zone for DL transmission if the UE is receiving high interference. DL transmission is always allowed for a pure DL victim in the DL-centric zone and UL transmission is always allowed for a pure UL aggressor in the UL-centric zone if its major interfering UE does not belong to a cell utilizing current band as its priority usage band.

Note that backhaul communication latency can significantly affect the performance of intercell interference coordination. For a system with rapidly changing interference pattern caused by on-and-off user traffic, RRM strategy determined from past network observation may be outdated for managing current interference condition. Therefore, the intersite cooperation mechanism should consider system loading and traffic characteristics while determining the optimal IM strategy and the update frequency for RRM rules. The triggering criteria of different RRM rule should be carefully designed to maximize the potential gain from intercell RRM.

3.6.1.2 Traffic Asymmetry

In this chapter, simulation evaluation is conducted based on full buffer and symmetric traffic in downlink and uplink directions. For asymmetric traffic, the gain of full-duplex is expected to be lower. How to optimize the scheduler to explore the gain of full-duplex by considering non-full-buffer traffic shall be designed and evaluated.

3.6.1.3 MIMO Full-Duplex

Our work focuses on scenarios with only one or two antennas at the small cell base station. When the number of TX and RX antenna increases, scalable self-interference cancellation architecture will be required for full-duplex transmission as larger number of RX chains need to be protected from self-echo. In

Reference [26], a beamforming approach is used to suppress TX echo for full-duplex massive MIMO system. From the network performance perspective, FD scheduling and MIMO beamforming strategy should be jointly considered in order to maximize per user throughput.

3.6.2 More Scenarios and Future Work

This chapter is mainly focused on small cell access system with full-duplex-capable small cell in the network, extended scenarios, such as HetNet scenario [9], mixed FD, and HD small cells [21], are not addressed. In addition, full-duplex technology also attracts lots of attention in other wireless networks, including self-backhaul (relay), usage in unlicensed band for Wi-Fi and LAA. In this section, we will provide brief discussions on potential full-duplex system design directions in these areas.

3.6.2.1 Full-Duplex Self Backhaul (Relay)

Relay node can be deployed for boosting cell capacity and improve cell coverage. Half-duplex relay uses orthogonal resource between self-backhaul and relay-cell-access and thus impact spectrum efficiency greatly. For full-duplex relay, it relaxes the constraint of orthogonal resource allocation between self-backhaul and relay-cell-access links and thus system throughput can be largely improved. As a result, it draws lots of attention and some studies indicate it could be one of the first areas of practical usage of full-duplex techniques [15]. However, full-duplex self-backhaul introduces additional interferences to the whole network, and system design shall handle these interferences carefully [10].

In a network with half-duplex legacy macro station and half-duplex UE, full-duplex can be used on the relay node whilst macro station and UE remains on existing half-duplex LTE network. In such configurations, full-duplex brings three new types of interferences:

1) *Self-Interference:* Self-interference occurs from the transmission of backhaul signal to the access node reception at relay access node, and also transmission of access to backhaul reception at the relay backhaul node. These interferences can be addressed by self-interference cancellation methods as mentioned in Section 3.2.
2) *Interferences Cross Relay Nodes:* On uplink direction, uplink relay backhaul transmission may cause interference to the reception of another relay node on access link when UE transmits to the relay node; on downlink direction, downlink relay access transmission may cause interference to backhaul reception of neighbor relay node when the macro station transmits to the neighbor relay node.
3) *Interference due to Frequency Reuse:* On uplink direction, UL relay access transmission (UE->Relay) to UL relay backhaul reception (Relay->Macro),

and DL relay backhaul transmission (Macro->Relay) may cause interference to DL relay access (Relay->UE). This type of interference is similar to HETNET interferences.

For type 2 and 3 interferences, strategies for power control, resource allocation, beamforming, and so on shall be properly designed such that the self-backhaul system can enjoy close to 2× spectrum efficiency with high reliability.

3.6.2.2 Full-Duplex Wi-Fi System

In 802.11 WLANs, an AP can use full-duplex capability to simultaneously transmit/receive packets for downlink/uplink traffic. STA can be full-duplex-capable or half-duplex legacy devices. When the STA is FD capable, dual-link transmission can be set up between AP and STA to potentially double the system capability. When the STA is HD or only have one way traffic, the HD STA can be paired with another HD STA to set up a 3-node-form transmission (STA1->AP, AP->STA2) to utilize the full-duplex capability of AP. In this form of transmission, there could be STA-to-STA interference. MAC design shall consider choice of transmission modes and how to pair the uplink STA with downlink STA when it is a 3-node-form transmission to maximize the system gain. The MAC design can be categorized into two directions: (1) for legacy Wi-Fi system with half-duplex STAs, how to enable full-duplex transmission with minimal or without any change to the spec [13] and (2) for new Wi-Fi systems (802.11ax+), how to design new MAC [11,12] to maximize the gain of FD efficiently. Another direction is to investigate the hidden nodes/ exposed nodes problem for full-duplex Wi-Fi system. Full-duplex transmission can resolve the hidden node problem in CSMA systems [4] as it can send tones whilst it is receiving and other nodes can sense the transmission and avoid hidden node collision. On the other hand, full-duplex transmission may increase the exposed node problem in dense networks as the interference region increases with the additional reverse link transmission, and thus more nodes are exposed and prevented from simultaneous transmission. The trade-off between hidden node and exposed node and solutions [22] to mitigate the expose node problem in full-duplex systems are interesting topics for future study.

3.6.2.3 Full-Duplex Application in LAA

LTE unlicensed band solutions (LAA/LTE-U) are newly added in 3GPP to enable LTE in unlicensed spectrum and co-exist with Wi-Fi. Full-duplex is a promising technology to be used in such systems. On one hand, out-of-band full-duplex can be used to mitigate the interference between LTE channel and Wi-Fi adjacent channel to resolve the Wi-Fi/LTE-U coexistence issue; on the other hand, full-duplex can be used to boost the throughput if it is used on LTE or Wi-Fi or both side. System design has to account for the individual full-duplex capable network issues (on cellular or Wi-Fi as we discussed previously), and also account for new issues existing in LAA full-duplex systems, including how to maintain the frequency reuse of LTE network with CSMA mechanisms adapted for full-duplex transmission.

References

1 A. Sabharwal, P. Schniter, D. Guo, D. Bliss, S. Rangarajan, and R. Wichman, "In-band full-duplex wireless: challenges and opportunities," *IEEE J. Select. Area Commun.*, vol. 32, no. 9, pp. 1637–1652, 2014.

2 M. Chung, M. Sim, J. Kim, D. Kim, and C. Chae, Prototyping real-time full duplex radios. Available at http://arxiv.org/abs/1503.03013, 2015.

3 J. I. Choi, M. Jain, K. Srinivasan, P. Levis, and S. Katti, "Achieving single channel, full duplex wireless communications," in *Proc. ACM MobiCom*, 2010, pp. 1–12.

4 Y.-S. Choi and H. Shirani-Mehr, "Simultaneous Transmission and reception: algorithm, design and system level performance", *IEEE Trans. On Wirel. Commun.*, vol. 12, no. 12, pp. 5992–6010, 2013.

5 D. Bharadia, E. McMilin, and S. Katti, "Full duplex radios," in *SIGCOMM*, 2013.

6 M. Duarte and A. Sabharwal, "Full-duplex wireless communications using off-the-shelf radios: feasibility and first results," in *Proc. Asilomar Conference* on *Signals, Systems*, and *Computers*, 2010, pp. 1558–1562.

7 S. Goyal, P. Liu1, S. Panwar, R. A. DiFazio, R. Yang, E. Bala, "Full duplex cellular systems: will doubling interference prevent doubling capacity?," *IEEE Commun. Mag.*, vol. 53, no. 5, pp. 121–127, 2015.

8 R. Li, Y. Chen, G.Y. Li, G. Liu "Full-Duplex Cellular Networks: It Works!," *IEEE Commun, Mag.*, vol. 55, no. 4, pp. 184–191, 2017.

9 S. Han, C. Yang, P. Chen, "Full duplex asssted inter-cell interference cancellation in heterogeneous networks," *IEEE Trans. Commun.*, vol. 63, no. 12, pp. 5218–5234, 2015

10 R. Pitaval et al., "Full-duplex self-backhauling for small-cell 5G networks," *IEEE Wirel. Commun.*, vol. 22, no. 5, pp. 83–89, 2015.

11 N. Singh, D. Gunawardena, A. Proutiere, B. Radunovi, H. V. Balan and P. Key, "Efficient and fair MAC for wireless networks with self-interference cancellation," *in Proc. WiOpt 2011*, May 2011.

12 S. Goyal, P. Liu, O. Gurbuz, E. Erkip, and S. Panwar, "A distributed MAC protocol for full duplex radio," in *Forty Seventh Asilomar Conference on Signals, Systems and Computers (ASILOMAR)*, 2013.

13 A. Tang and X. Wang. "A-duplex: medium access control for efficient coexistence between full-duplex and half-duplex communications," *IEEE Trans. Wirel. Commun.*, vol. 14, no. 10, pp. 5871–5885, 2015.

14 DUPLO Project http://www.fp7-duplo.eu/.

15 News from Fiercewireless.com. Available at http://www.fiercewireless. com/tech/story/deutsche-telekom-completes-5g-full-duplex-field-trial-kumu-networks/2015-09-28.

16 R. Fleury, D. L. Sounas, and A. Alu, "Magnetless circulators for electromagnetic and acoustic waves," in *2016 European Conference on Antennas and Propagation*, Davos, Switzerland, April 10–15, 2016.

17 Dinc, et al, "A 60GHz same-channel full-duplex CMOS transceiver and link based on reconfigurable polarization-based antenna cancellation," in Radio Frequency Integrated Circuits Symposium (RFIC), 2015.

18 H. Shirani-Mehr, Y. Choi, R. Yang, and A. Papathanassiou, Method and apparatus for power control in full-duplex wireless systems with simultaneous transmission reception, U.S. Patent 20140078939 A1, Mar. 27, 2014.

19 A. Kini, M. Hosseinian, P. Sadeghi, and J. Stern-Berkowitz, "A dynamic subframe set power control scheme for interference mitigation in reconfigurable TD-LTE systems", in WOCC, 2014.

20 Q. Chen, H. Zhao, L.Li, H. Long, J Wang, and X. Hou "A Closed-Loop UL Power Control Scheme for Interference Mitigation in Dynamic TD-LTE Systems", in *IEEE 2015 Vehicular Technology Conference (VCT Spring)*, 2015.

21 S. Goyal, C. Galiotto, N. Marchetti, and S. Panwar, "Throughput and Coverage for a Mixed Full and Half Duplex Small Cell Network", in *International Conference on Communications (ICC 2016)*, 2016.

22 L. Wang, W. Kaishun, and M. Hamdi. "Combating hidden and exposed terminal problems in wireless networks." *IEEE Trans. Wirel. Commun.*, vol. 11, no. 11, pp. 4204–4213, 2012.

23 S. Han, C. I, Z. Xu, C. Pan, and Z. Pan "Full duplex: coming into reality in 2020?", IEEE Globecom, 2014.

24 Intel, Intel Ideation Meeting Report on wireless full duplex, Sept. 2015.

25 A. Goldsmith, *Wireless Communications*, Cambridge Press, 2005.

26 E. Everett, C. Shepard, L. Zhong, and A. Sabharwal, SoftNull: many-antenna full-duplex wireless via digital beamforming, *IEEE Trans. Wirel. Commun.*, vol. 15, no. 12, pp. 8077–8092, 2016.

Shu-ping Yeh received the B.S. degree from National Taiwan University in 2003, and the M.S. and Ph.D. degrees from Stanford University in 2005 and 2010, respectively, all in electrical engineering. She is currently a Senior Research Scientist with Intel Labs. She specializes in wireless technology development for cellular and local area network access. Her recent research focus includes advanced self-interference cancellation technology, full-duplex PHY/MAC system designs, and multitier multi-RAT heterogeneous networks.

Jingwen Bai received the B.E degree in electronic science and technology from the Beijing University of Posts and Telecommunications, Beijing, China, in 2011, and the M.S and Ph.D. degrees in electrical and computer engineering from Rice University, Houston, TX, USA, in 2013 and 2016, respectively. She is currently a Research Sc ientist with Intel labs, Santa Clara, CA, USA. Her research interest includes network information theory, interference management, and algorithm design for future wireless networks.

Ping Wang is currently an engineer in iOS wireless technologies in Apple Inc. She received her Ph.D. degree from McGill University, Canada and her M.Eng and B.Eng degree from Beijing University of Posts and Telecommunications, all in electrical engineering. Before joining Apple, she worked as a staff research scientist in the Wireless Communication Lab under Intel Labs, where her research was focused on system design in advanced cellular and Wi-Fi wireless networks, including interference mitigation for full-duplex systems. Prior to that, she also worked as a principal member of technical staff on radio access networks in CTO office of AT&T Labs and as a system engineer in Ericsson Canada in LTE R&D.

Feng Xue is a senior research scientist with Intel Labs. His current research interests are 4G/5G wireless communications, information theory, signal processing, and system optimization. He received the Ph.D. degree in electrical and computer engineering from the University of Illinois, Urbana-Champaign, in 2006, and the M.S. and B.S. degrees from the Institute of Systems Science, Chinese Academy of Sciences and Shandong University, respectively. Except a short 2 year stint at the Qualcomm Research Center in San Diego, he has been working at Intel Labs since 2006. He has published more than 40 papers in top journals and conferences, and he has filed more than 50 patents.

 *Yang-Seok Choi* received B.S. degree from Korea University, Seoul, South Korea in 1990, M.S.E.E. degree from Korea Advanced Institute of Science and Technology, Taejon, South Korea, in 1992, and Ph.D. degree from Polytechnic University, Brooklyn, NY, in 2000, all in electrical engineering. He has been in industry for 25 years in AT&T Labs-Research, Samsung, National semiconductor, and a start-up. He joined Intel in 2004 and led WiMax PHY Standards development team. In 2013, he joined Intel Labs and is a manger leading Wireless Interference Technology team. He has been focusing on future wireless communications in Intel Labs. He holds 60+ U.S. patents.

 *Shilpa Talwar* received the M.S. degree in electrical engineering and the Ph.D. degree in applied mathematics from Stanford University in 1996. She held several senior technical positions in wireless industry involved in a wide range of projects, including algorithm design for 3G/4G & WLAN chips, satellite communications, GPS, and others. She is currently the Director of Wireless Multicomm Systems and a senior principal architect with the Wireless Communications Laboratory, Intel, where she leads a research team focused on advancements in network architecture and technology innovations for 5G, and contributed to the IEEE and 3GPP standard bodies, including 802.16m and LTE-advanced. She is also coordinating several university collaborations on 5G, and leads Intel Strategic Research Alliance on 5G. She has authored over 50+ technical publications. She holds 33 patents (42 additional pending). Her research interests include heterogeneous networks, multiradio interworking, mm wave communications, advanced MIMO, and full-duplex and interference mitigation techniques.

 *Sung-En Chiu* received the B.S. and M.S. degrees in electronics engineering from National Chiao Tung University, Hsinchu, Taiwan, in 2008 and 2010, respectively. From 2011 to 2013, he served as a 3GPP RAN1 delegate for the Industrial Technology Research Institute, Hsinchu, Taiwan. He is currently working toward the Ph.D. degree in electrical and computer engineering at the University of California, San Diego, CA, USA. His Ph.D. research topics are on the sparse signal recovery and adaptive information processing using Bayesian models.

Vinod Kristem received his Bachelor of Technology degree in electronics and communications engineering from the National Institute of Technology (NIT), Warangal, in 2007. He received his Master of Engineering degree in telecommunications from the Department of Electrical Communication Engineering, Indian Institute of Science, Bangalore, India, in 2009. From 2009 to 2011, he was with Beceem Communications Pvt. Ltd., Bangalore, India (acquired by Broadcom Corp.), where he worked on channel estimation and physical layer measurements for WiMAX and LTE. He is currently working toward his Ph.D. degree in electrical engineering at University of Southern California, Los Angeles. His current research interests include the multiantenna systems, ultrawideband systems, and channel measurements and modeling.

4

Nonorthogonal Multiple Access for 5G

Linglong Dai,[1] Bichai Wang,[1] Ruicheng Jiao,[1] Zhiguo Ding,[2] Shuangfeng Han,[3] and Chih-Lin I[3]

[1] Department of Electronics Engineering, Tsinghua University, Beijing, China
[2] School of Electrical and Electronic Engineering, The University of Manchester, Manchester, UK
[3] Wireless Technologies, China Mobile Research Institute, China Mobile Communications Corporation, Beijing, China

4.1 Introduction

The rapid development of the mobile Internet and the Internet of Things (IoT) leads to challenging requirements for the fifth generation (5G) of wireless communication systems, which is fuelled by the prediction of 1000-fold data traffic increase by the year 2020 [1]. Specifically, the key performance indicators (KPI) advocated for 5G solutions can be summarized as follows [2]: (i) The spectral efficiency is expected to increase by a factor of 5–15 compared to 4G; (ii) To satisfy the demands of massive connectivity for IoT, the connectivity density target is 10 times higher than that of 4G, that is, at least $10^6/km^2$; (iii) 5G is also expected to satisfy the requirements of a low latency (radio latency $\leq$ 1 ms), low cost ($\geq$100 times the cost-effectiveness of 4G), and the support of diverse compelling services. In order to satisfy these stringent requirements, advanced solutions have to be conceived.

Over the past few decades, wireless communication systems have witnessed a "revolution" in terms of their multiple access techniques. Specifically, for 1G, 2G, 3G, and 4G wireless communication systems, frequency-division multiple access (FDMA), time-division multiple access (TDMA), code-division multiple access (CDMA), and orthogonal frequency-division multiple access (OFDMA) have been used as the corresponding key multiple access technologies, respectively [3], [4]. From the perspective of their design principles, these

5G Networks: Fundamental Requirements, Enabling Technologies, and Operations Management, First Edition.
Anwer Al-Dulaimi, Xianbin Wang, and Chih-Lin I.

multiple access schemes belong to the category of OMA, where the wireless resources are orthogonally allocated to multiple users in the time, frequency, and code domains or according to the fact based on their combinations. We might collectively refer to these domains as "resources." In this way, the users' information-bearing signals can be readily separated at a low complexity by employing relatively cost-effective receivers. However, the number of supported users is limited by the number of available orthogonal resources in OMA. Another problem is that despite using orthogonal time-, frequency-, or code-domain resources, the channel-induced impairments almost invariably destroy their orthogonality and hence typically high-complexity "orthogonality restoring measures," such as multiuser equalizers have to be invoked. Consequently, it remains a challenge for OMA to satisfy the radical spectral efficiency and massive connectivity requirements of 5G.

The innovative concept of NOMA has been proposed in order to support more users than the number of available orthogonal time-, frequency-, or code-domain resources. The basic idea of NOMA is to support nonorthogonal resource allocation among the users at the ultimate cost of increased receiver complexity, which is required for separating the nonorthogonal signals.

Recently, several NOMA solutions have been actively investigated [5,6], which can be basically divided into two main categories: power-domain NOMA [7–52] and code-domain NOMA [53–86], including multiple access solutions relying on low-density spreading (LDS) [53–62], sparse code multiple access (SCMA) [63–84], multiuser shared access (MUSA) [85], successive interference cancelation-amenable multiple access (SAMA) [86], and so on. Some other closely- related multiple access schemes such as spatial-division multiple access (SDMA) [87–100], pattern- division multiple access (PDMA) [101,102], and bit-division multiplexing (BDM) [103] have also been proposed. The milestones of multiple access techniques are summarized in Figure 4.1.

In this chapter, we will discuss the basic principles as well as the advantages of NOMA in Section 4.2. In Section 4.3, the design principles and key features of dominant power-domain NOMA schemes will be discussed in detail. Section 4.4 will introduce dominant code-domain NOMA schemes in terms of their basic principles and key features. Moreover, in Section 4.5, we will briefly introduce other NOMA schemes to present a thorough review of NOMA. A detailed comparison of the dominant NOMA schemes will be provided in Section 4.6. Besides, Section 4.7 will introduce some performance evaluations and transmission experiments of NOMA, so as to verify the advantage of this new technology. In Section 4.8, we will highlight the opportunities and future research trends of NOMA, in order to provide some insights into this promising field. Finally, our conclusions are offered in Section 4.9.

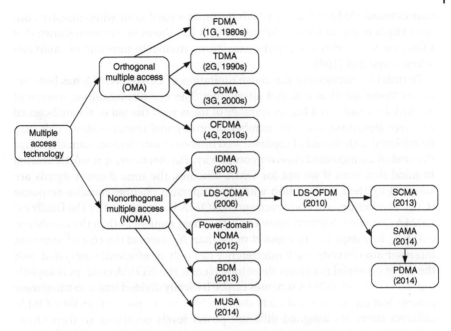

Figure 4.1 The milestones of multiple access technique developments.

4.2 Basic Principles and Advantages of NOMA

OFDMA is used for 1G, 2G, 3G, and 4G, respectively, multiple users in conventional OMA schemes, such as FDMA, TDMA, CDMA, and d orthogonal to radio resources in the time, frequency, and code- domain or to their combinations. More specifically, each user transmits a unique, user-specific signal over its unique frequency resource in FDMA; hence, the receiver can readily detect all users' data in their corresponding frequency bands, respectively. Similarly, in TDMA, an exclusive time slot is for different users' signals in the time domain. In CDMA, multiple users are allocated to each user; hence, it is easy to distinguish the same time-frequency resources they share, as well as the transmitted sequences, such as Walsh-Hadamard codes. Hence, a low-complexity decorrelation receiver can be used for multiuser detection (MUD). OFDMA can be regarded as a smart integration of FDMA and symbols of different users may be mapped to orthogonal spreading TDMA, where the radio resources are orthogonally partitioned in the time–frequency grid. Theoretically, as a benefit of orthogonal resource allocation, there is no interference among users in OMA systems; hence, low-complexity detectors can be used to separate the different users' signals. However, the maximum number of supportable users is rigidly restricted by the number of orthogonal resources available in conventional OMA schemes, which becomes a hard limit when massive connectivity is required for 5G. Additionally, it has theoretically been shown that OMA cannot always achieve the maximum attainable sum rate of multiuser wireless systems [104].

In order to circumvent the above limitation of OMA, NOMA has been recently investigated as a design alternative. The key distinguishing feature of NOMA is to support a higher number of users with the aid of nonorthogonal resource allocation, than the number of orthogonal resource slots. This may be achieved with the aid of sophisticated inter-user interference cancellation at the cost of an increased receiver complexity. Furthermore, it is worth bearing in mind that even if we opt for OMA schemes, the time domain signals are smeared by their convolution with the dispersive channel impulse response (CIR). In addition, a historic concept of NOMA is constituted by the family otf CDMA systems relying on more nonorthogonal sequences than the number of chips N_c in a sequence to support more than N_c users at the cost of imposing interuser interference. Such interference can only be efficiently mitigated with the aid of powerful multiuser detectors. Hence, the NOMA concept is appealing. The family of NOMA schemes can be basically divided into two categories: power-domain NOMA and code-domain NOMA. In power-domain NOMA, different users are assigned different power levels according to their channel quality, while the same time-frequency-code resources are shared among multiple users. At the receiver side, power-domain NOMA exploits the users' power difference, in order to distinguish different users based on successive interference cancellation (SIC). Code-domain NOMA is similar to CDMA or multicarrier CDMA (MC-CDMA), except for its preference for using low-density sequences or nonorthogonal sequences having a low cross-correlation.

4.2.1 Channel Capacity Comparison of OMA and NOMA

From the perspective of information theory, for the capacity of multiple access channels in both additive white Gaussian noise (AWGN) and fading scenarios, we have the following results for OMA and NOMA (applicable to both power-domain NOMA and code-domain NOMA):

- **AWGN Channel:** In the uplink of an AWGN channel supporting K users (K can be larger than 2), the capacity of the multiple access channel can be formulated as [105]

$$\sum_{i=1}^{K} R_i \leq W \log \left(1 + \frac{\sum_{i=1}^{K} P_i}{N_0 W} \right), \tag{4.1}$$

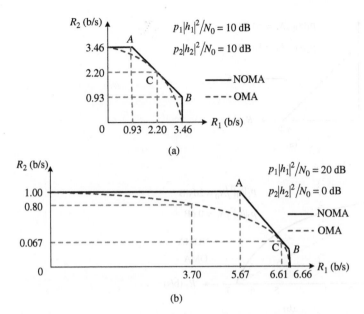

Figure 4.2 Channel capacity comparison of OMA and NOMA in the uplink AWGN channel. (a) Symmetric channel. (b) Asymmetric channel [5]. (Copyright 2016, IEICE, Permission No.: 18RB0010.)

where W is the bandwidth, P_i is the transmitted power, and N_0 is the power spectral density of Gaussian noise. More particularly, according to the capacity analysis found in the pioneering contribution of Tse and Viswanath [104], Figures 4.2 and 4.3 from Ref. [5] provided the channel capacity comparison of OMA and NOMA, where a pair of users communicating with a base station (BS) over an AWGN channel is considered as an example without loss of generality. Figure 4.2 showed that the uplink of NOMA is capable of achieving the capacity region, while OMA is suboptimal in general, except at one point. However, at this optimal point, rate fairness is not maintained, since the rate of the low-power user is much lower than that of the higher power user when the difference of the received powers of the two users is high. Note that the results for the simple two-user case can be extended to the general case of an arbitrary number of users [104]. Explicitly, it is shown in Ref. [104] that there are exactly $K!$ corner points when the K-user scenario is considered and the K-user NOMA system can achieve the same optimal sum rate at all of these $K!$ corner points.

In the downlink, the boundary of the capacity region is given by the rate tuples [106]:

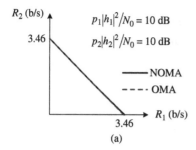

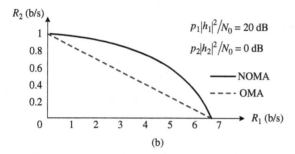

Figure 4.3 Channel capacity comparison of OMA and NOMA in the downlink AWGN channel. (a) Symmetric channel. (b) Asymmetric channel [5]. (Copyright 2016, IEICE, Permission No.: 18RB0010.)

$$R_k = W \log \left(1 + \frac{P_k |h_k|^2}{N_0 W + \left(\sum_{j=k+1}^{K} P_j \right) |h_k|^2} \right),$$ (4.2)

which is valid for all possible splits $P = \sum_{k=1}^{K} P_k$ of total power at the BS. The optimal points can be achieved by NOMA with the aid of superposition coding at the transmitter and SIC at each of the receivers [104]. More particularly, Figure 4.3 showed that the boundary of the rate pairs of NOMA is in general beyond that of OMA in asymmetric channels.

- **Fading Channels**: In fading channels, the sum capacity in the uplink–provided that the channel state information (CSI) is only known at the receiver–can be represented as

$$C_{\text{sum}} = E\left\{\log\left(1 + \frac{\sum_{k=1}^{K}|h_k|^2 P_{\text{ave}}}{N_0}\right)\right\}, \tag{4.3}$$

where we assume that each user has the same average power P_{ave}. In this way, OMA is strictly suboptimal in the uplink, while NOMA relying on MUD is optimal [104].

- **MIMO–NOMA and MIMO–OMA**: NOMA can also be extended to MIMO scenarios, where the BS has M antennas and each user is equipped with N antennas. Additionally, multiple users can be randomly grouped into M clusters with two users in each cluster. It has been shown in Ref. [107] that MIMO–NOMA is strictly better than MIMO–OMA in terms of the sum channel capacity (except for transmission to a single user in MIMO systems), that is, for any rate pair achieved by MIMO–OMA schemes, there is a power split for which MIMO–NOMA is capable of achieving rate pairs that are strictly higher.

4.2.2 Advantages of NOMA Compared to OMA

We can see from the capacity analysis that it is feasible for NOMA to achieve a higher transmission rate than OMA. Specifically, the main advantages of NOMA compared to the classical OMA can be summarized as follows:

- **Improved Spectral Efficiency and Cell-Edge Throughput**: The time-frequency resources are shared nonorthogonally among users both in the power-domain NOMA and in the code-domain NOMA. As already described, in the uplink of AWGN channels, although both OMA and NOMA are capable of achieving the maximum attainable sum capacity, NOMA supports a more equitable user fairness. Additionally, the capacity bound of NOMA is higher than that of OMA in the downlink of AWGN channels. In multipath fading channels subject to intersymbol interference (ISI), although OMA is indeed capable of achieving the maximum attainable sum capacity in the downlink, NOMA relying on MUD is optimal, while OMA remains suboptimal, if the CSI is only known at the downlink receiver.

- **Massive Connectivity**: Nonorthogonal resource allocation in NOMA indicates that the number of supportable users/devices is not strictly limited by the number of available orthogonal resources. Therefore, NOMA is capable of significantly increasing the number of simultaneous connections in rank-deficient scenarios, hence massive connectivity can be realized by NOMA.

- **Low Transmission Latency and Signaling Cost**: In conventional OMA relying on access-grant requests, a user has to first send a scheduling request to the base station (BS). Then, upon receiving this request, the BS schedules the user's uplink transmission by responding to a clear-to-send signal in the downlink channel. Thus, a high transmission latency and a high signaling overhead will be imposed, which becomes unacceptable in the case of massive 5G-style connectivity. Specifically, the access-grant procedure in LTE takes about 15.5 ms before the data are transmitted [108]. In this way, the radical requirement of maintaining a user delay below 1 ms cannot be readily satisfied [101]. By contrast, dynamic scheduling is not required in some of the uplink NOMA schemes. To elaborate a little further, in the uplink of a SCMA system, grant-free multiple access can be realized for users associated with preconfigured resources defined in the time and frequency domains, such as the codebooks, as well as the pilots. By contrast, at the receiver, blind detection and compressive sensing (CS) techniques can be used for performing joint activity and data detection [80]. Hence again, beneficial grant-free uplink transmission can be realized in NOMA, which is capable of significantly reducing both the transmission latency and the signaling overhead.

- **Relaxed Channel Feedback**: The requirement of channel feedback will be relaxed in power-domain NOMA, because the CSI feedback is only used for power allocation. Hence, there is no need for accurate instantaneous CSI knowledge. Therefore, regardless of whether fixed or mobile users are supported, having a limited-accuracy outdated channel feedback associated with a certain maximum inaccuracy and delay will not severely impair the attainable system performance, as long as the channel does not change rapidly.

Given the above prominent advantages, NOMA has been actively investigated, with a views for employment in 5G as a promising solution. Next, we will discuss and compare the dominant power-domain NOMA and code-domain NOMA solutions in Sections 1.3 and 1.4, respectively.

4.3 Power-Domain NOMA

In this section, we will discuss the first category of NOMA, namely, power-domain NOMA. In Refs [7]–[10], the concept and key features of power-domain NOMA have been described in detail. In contrast to the multiple access schemes relying on the time, frequency, code domain or on their combinations, NOMA can be realized in a recently emerged new domain, namely, in the power domain. At the transmitter, different signals generated by different users are directly superimposed on each other after classic channel coding and modulation. Multiple users share the same time-frequency resources, and then are detected at the receivers by MUD algorithms such as SIC. In this way, the spectral efficiency can be enhanced at the cost of an increased receiver complexity compared to conventional OMA. Additionally, it is widely recognized based on information theory that nonorthogonal multiplexing using superposition coding at the transmitter and SIC at the receiver not only outperforms classic orthogonal multiplexing but it is also optimal from the perspective of achieving the capacity region of the downlink broadcast channels [104].

Some practical considerations for power-domain NOMA, such as multiuser power allocation, signaling overhead, SIC error propagation and user mobility, were discussed in Ref. [7]. Furthermore, the benefits of power-domain NOMA over OMA were clarified in Refs [11,12] with the aid of experimental trials. To achieve a further enhancement of its spectral efficiency, the authors of Refs [7]–[10],[13]–[16] discussed the combination of NOMA with MIMO techniques. As a further development, the receiver design was discussed in Refs [17]–[19], while user grouping as well as resource allocation were investigated in Refs [20]–[31]. On the other hand, in order to increase the attainable performance of the SIC receiver, cooperative NOMA transmission has been proposed in Refs [32], [33]. A range of investigations related to multicell NOMA schemes were carried out in Ref. [34]. Moreover, since having an increased number of cell-edge users typically degrades the efficiency of coordinated multipoint (CoMP) transmissions, a promising NOMA solution was proposed for a CoMP system in Ref. [35]. Additionally, the performance of NOMA techniques supporting randomly distributed users was evaluated in Ref. [36]. These simulation results demonstrated that the outage performance of NOMA substantially depended on both the users' targeted data rates and on their allocated power. In Refs [37]–[46], the system-level performance of power-domain NOMA was evaluated, and the associated simulation results showed that both the overall cell throughput and the cell-edge user throughput, as well as the degree of proportional rate-fairness of NOMA, were superior to those of OMA. Furthermore, the impact of the residual interference imposed by realistic imperfect channel estimation on the achievable throughput performance was investigated in Refs [47]–[49].

Let us now elaborate on the power-domain NOMA techniques in this section. First, the basic principle of power-domain NOMA relying on a SIC receiver will be discussed. Then, a promising extension relying on integrating NOMA with MIMOs will be discussed for the sake of increasing its attainable spectral efficiency. Another compelling extension to a cooperative NOMA transmission scheme will also be presented. Finally, the networking aspects of NOMA solutions will be discussed.

4.3.1 Basic NOMA Relying on a SIC Receiver

First, we consider the family of single antenna systems relying on a single BS and K users.

In the downlink, the total power allocated to all K users is limited to P, and the BS transmits the signal x_i to the ith user subjected to the power-scaling coefficient p_i. In other words, the signals destined for different users are weighted by different power-scaling coefficients and then they are superimposed at the

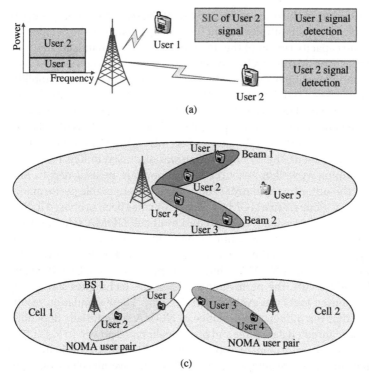

Figure 4.4 Power-domain NOMA. (a) Basic NOMA relying on a SIC receiver. (b) NOMA in MIMO systems. (c) Network-NOMA.

BS according to:

$$x = \sum_{i=1}^{K} \sqrt{p_i} x_i, \tag{4.4}$$

where $E[|x_i|^2] = 1$ $(i = 1, 2, \ldots, K)$ denotes the normalized power of the user signals, and we have $P = \sum_{i=1}^{K} p_i$. The received signal y_i of the ith user is

$$y_i = h_i x + v_i, \tag{4.5}$$

where h_i denotes the channel gain between the BS and the ith user, while v_i associated with the power density N_i represents the Gaussian noise plus the intercell interference.

At the receiver, SIC is used for MUD. The optimal order of SIC detection relies on detecting the strongest and hence least contaminated user to the weakest user (determined by $|h_i|^2/N_i$, $i = 1, 2, \ldots, K$). Based on this optimal

SIC-detection order, any user can detect its information without substantial interference–contamination imposed on the other users whose normalized channel gain is smaller than that of this user. It is intuitive that the users associated with small normalized channel gains should be allocated higher power levels in order to improve their received signal-to-interference and noise ratio (SINR), so that a high detection reliability can be guaranteed. Although the users having larger normalized channel gains require less power, they are capable of correctly detecting their data with a high probability, as a benefit of SIC. Without loss of generality, a descending-order-based power allocation $p_1 \geq p_2 \geq \cdots \geq p_K$ can be assumed. Assuming perfectly error-free decoding of the interfering signals, the achievable rate of user i ($i = 1, 2, \ldots, K$) can be written as

$$R_i = W \log \left(1 + \frac{p_i |h_i|^2}{N_i W + \left(\sum_{j=k+1}^{K} p_j \right) |h_i|^2} \right). \tag{4.6}$$

In the case of two users as shown in Figure 4.4a, we assume that the normalized channel gain of the second user is lower than that of the first one, that is, $|h_1|^2/N_1 > |h_2|^2/N_2$, and thus $p_1 < p_2$. The second user detects its signal by regarding the signal of the first user as interference. The first user first detects the signal of the second user, and then subtracts its remodulated version from the received signal, so that the first user can detect its own signal without interference from the second user.

Assuming that the transmission bandwidth is normalized to 1 Hz, the data rates of the first user and the second user can be represented as

$$R_1 = \log_2 \left(1 + \frac{p_1 |h_1|^2}{N_1} \right), \tag{4.7}$$

$$R_2 = \log_2 \left(1 + \frac{p_2 |h_2|^2}{p_1 |h_1|^2 + N_2} \right), \tag{4.8}$$

respectively. Thus, by tuning power allocation coefficients, the BS can adjust the data rate of each user. More particularly, it has been shown in Figure 4.5 [7] that this NOMA scheme is capable of achieving higher rates than OFDMA. On the other hand, this NOMA scheme makes a full use of the natural difference of channel gains among the users, which implies that the near–far effect is effectively harnessed to achieve higher spectral efficiency. As a result, both the attainable sum capacity and the cell-edge user data rate can be improved [8].

In the uplink, the signal received at the BS is given by

$$y = \sum_{i=1}^{K} h_i \sqrt{p_i} x_i + v,$$

(4.9)

where p_i and x_i are the transmit power and signal transmitted by the ith user, respectively. Furthermore, v associated with the power density N_0 represents the Gaussian noise plus the intercell interference at the BS. SIC is used for reliable signal detection at the BS. Without loss of generality, we assume that $p_1|h_1|^2 \geq p_2|h_2|^2 \geq \cdots \geq p_K|h_K|^2$, and accordingly the optimal decoding order for SIC is $x_1, x_2, \cdots, x_K$. Before the BS detects the ith user's signal, it decodes the jth $(j < i)$ user's signal first and then removes $(i-1)$ users' signals from the observation y. The remaining $(K-i)$ signals are regarded as interference. As a result, the achievable data rate of the ith user becomes

$$R_i = W \log \left(1 + \frac{p_i|h_i|^2}{N_0 W + \left(\sum_{j=k+1}^{K} p_j \right) |h_i|^2} \right).$$

(4.10)

As illustrated in Section 2.1, this NOMA scheme is capable of achieving the maximum attainable multiuser capacity in AWGN channels both in the uplink and in the downlink. Furthermore, this NOMA scheme has the potential of striking a more attractive trade-off between the spectral efficiency and user-fairness.

When the number of users is sufficiently high, the SIC-induced error propagation may have a severe effect on the error probability in the absence of preventative measures. However, some advanced user pairing and power allocation methods, as well as powerful channel coding schemes, can be used for reducing the error probability. Indeed, it has been shown that error propagation only has a modest impact on the NOMA performance even under the worst-case scenario [7], [38].

4.3.2 NOMA in MIMO Systems

Although the same time-frequency resources can be shared by multiple users in the basic NOMA employing SIC, the improvement of spectral efficiency still remains limited, hence may not satisfy the expected spectral efficiency improvements of 5G. An appealing solution is the extension of the basic NOMA using SIC by amalgamating it with advanced MIMO techniques [9],[10].

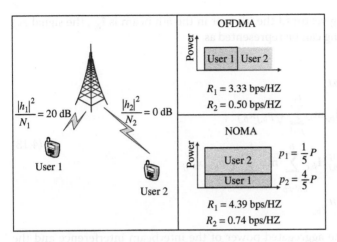

Figure 4.5 Comparison of NOMA and OMA [7].

As illustrated in Figure 4.4b, in downlink NOMA of MIMO systems, M_{BS} BS antennas are used for generating B different beams in the spatial domain with the aid of beamforming. Within each beam, the signals of multiple users may be transmitted by superimposing them, hence leading to the concept of intrabeam superposition modulation, which is similar to the basic NOMA using SIC, as discussed above. The bth ($1 \le b \le B$) transmitter beamforming vector is denoted as $\mathbf{m}_b$. Let us assume that the number of users in the bth beam is k_b, the transmitted symbol of the ith user in the bth beam is $x_{b,i}$, and the corresponding power-scaling coefficient is $p_{b,i}$. Then, by accumulating all signals of the B different beams, the M_{BS}-dimensional transmitted downlink signal vector at the BS can be formulated as

$$\mathbf{x}_0 = \sum_{b=1}^{B} \mathbf{m}_b \sum_{i=1}^{k_b} \sqrt{p_{b,i}} x_{b,i}. \tag{4.11}$$

Assuming that each user has N_r receiver antennas, the N_r-dimensional received signal vector of the ith user encapsulated in the bth beam can be represented as

$$\mathbf{y}_{b,i} = \mathbf{H}_{b,i} \mathbf{x}_0 + \mathbf{v}_{b,i}, \tag{4.12}$$

where $\mathbf{H}_{b,i}$ denotes the channel matrix of size $N_r \times M_{BS}$ between the BS and the ith user in the bth beam, and $\mathbf{v}_{b,i}$ denotes the Gaussian noise plus the intercell interference.

At the receiver, a pair of interference cancellation approaches are used for removing the inter-beam interference and the intrabeam interference, respectively. The interbeam interference can be suppressed by spatial filtering, which is similar to the signal detection algorithm of SDMA systems. Assuming that the spatial filtering vector of the ith user in the bth beam is $\mathbf{f}_{b,i}$, the signal $z_{b,i}$ after spatial filtering can be represented as

$$
\begin{aligned}
z_{b,i} &= \mathbf{f}_{b,i}^{H}\mathbf{y}_{b,i} \\
&= \mathbf{f}_{b,i}^{H}\mathbf{H}_{b,i}\mathbf{m}_{b}\sum_{j=1}^{k_{b}}\sqrt{p_{b,j}}x_{b,j} \\
&\quad + \mathbf{f}_{b,i}^{H}\mathbf{H}_{b,i}\sum_{\substack{b'=1\\b'\neq b}}^{B}\mathbf{m}_{b'}\sum_{j=1}^{k_{b}}\sqrt{p_{b',j}}x_{b',j} \\
&\quad + \mathbf{f}_{b,i}^{H}\mathbf{v}_{b,i}.
\end{aligned}
\tag{4.13}
$$

By normalizing the aggregated power of the interbeam interference and the receiver noise plus intercell interference to unity, we can rewrite (4.13) as

$$
z_{b,i} = \sqrt{a_{b,i}}\sum_{j=1}^{k_{b}}\sqrt{p_{b,j}}x_{b,j} + q_{b,i},
\tag{4.14}
$$

where $q_{b,i}$ is the normalized term representing the sum of the interbeam interference and receiver noise plus intercell interference, while $a_{b,i}$ is formulated as

$$
a_{b,i} = \frac{|\mathbf{f}_{b,i}\mathbf{H}_{b,i}\mathbf{m}_{b}|^{2}}{\left\{\begin{array}{c}\displaystyle\sum_{\substack{b'=1\\b'\neq b}}^{B}\sum_{j=1}^{k_{b}}p_{b',j}|\mathbf{f}_{b,i}\mathbf{H}_{b,i}\mathbf{m}_{b'}|^{2}\\[2mm]+\mathbf{f}_{b,i}^{H}E[\mathbf{v}_{b,i}\mathbf{v}_{b,i}^{H}]\mathbf{f}_{b,i}\end{array}\right\}}.
\tag{4.15}
$$

After spatial filtering, the system model (4.14) becomes similar to that of the basic NOMA combined with SIC, as described above. Therefore, the interbeam interference can be suppressed, and then intrabeam SIC is invoked for removing the interuser interference imposed by superposition coding within a beam.

Naturally, more users can also be simultaneously supported, because more than two users can share a single beamforming vector. To elaborate a little further, observe in Figure 4.6 [8] that both the basic NOMA combined with SIC and the extended NOMA relying on MIMO are capable of achieving a higher sum-rate than OFDMA. Furthermore, the sum-rate of NOMA in the context of MIMO systems is higher than that of the basic NOMA using a single antenna at the BS. Additionally, in this NOMA scheme, the number of reference signals required is equal to the number of transmitter antennas, regardless of

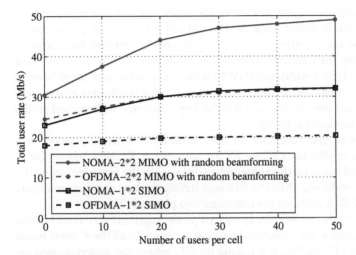

Figure 4.6 System-level performance for NOMA applying opportunistic random beamforming in downlink [8].

the number of nonorthogonal user signals. In this way, when the number of users is increased beyond the number of transmitter antennas, the number of orthogonal downlink reference signals is not increased beyond the number of transmit antennas.

In the uplink systems associated with K users, where the BS is equipped with M_{BS} antennas and each user has a single transmitter antenna, the signal received at the BS can be represented as

$$y = \sum_{i=1}^{K} \mathbf{h}_i \sqrt{p_i} x_i + \mathbf{v}, \tag{4.16}$$

where $\mathbf{h}_i$ is the M_{BS}-dimensional channel vector between user i and the BS. Furthermore, p_i and x_i are the transmit power and the signal transmitted by the ith user, respectively, while $\mathbf{v}$ represents the Gaussian noise plus the intercell interference vector.

At the receiver, MMSE-SIC receiver can be used to realize signal detection [5]. It has been shown in Ref. [5] that for any decoding order, the sum throughput for all K users is equal to the maximum of the total user throughput given the received signal vector in (4.16). In practice, the decoding order can be adjusted according to the actual requirements, such as user fairness.

4.3.3 Cooperative NOMA

Recently, a cooperative NOMA transmission scheme was proposed in Ref. [32]. Similar to the basic NOMA, cooperative NOMA also uses a SIC receiver for detecting the multiuser signal. Therefore, the users associated with better channel conditions can be relied upon as relays in order to improve the reception reliability of the users suffering from poor channel conditions. For cooperative transmission, for example, short-range communication techniques–such as Bluetooth and ultra-wideband (UWB) schemes–can be used for delivering signals from the users benefitting from better channel conditions to the users with poor channel conditions, which is the key difference with respect to the basic NOMA associated with SIC.

Without loss of generality, let us now consider a downlink cooperative NOMA system relying on a single BS for supporting K users, where the K users are ordered based on their channel qualities, with the first user having the worst channel condition, while the Kth user having the best channel condition. Cooperative NOMA relies on the following two phases [32].

The first phase, also termed the broadcast phase, represents the direct transmission. In this phase, the BS sends downlink messages to all the K users based on the principle of basic NOMA relying on SIC, where the superimposed information of the K users obeys the total power constraint. The SIC process is implemented at the user side. As a result, the users having better channel conditions have the knowledge of the signals intended to the users having poor channel conditions [109].

The second phase represents the cooperative transmission. During this phase, the cooperating users transmit their signals via their short-range communication channels, such as Bluetooth or UWB. Particularly, the second phase includes $(K - 1)$ time slots. In the first time slot, the Kth user broadcasts the superposition of the $(K - 1)$ signals destined for the remaining users. Then the SIC process is invoked again at these $(K - 1)$ users. The $(K - 1)$th user combines the signals received during both phases by using maximum ratio combining (MRC), and it detects its own information at a higher SNR than that of the traditional SIC. Similarly, in the kth time slot, where $1 \leq k \leq K - 1$, the $(K - k + 1)$st user also broadcasts the $(K - k)$ superposed signals for the remaining $(K - k)$ users, whose channel conditions are worse than that of this user. Then the $(K - k + 1)$st user combines the observations gleaned from both phases and it detects its own information at a higher SNR than the traditional SIC. Therefore, the employment of cooperation is indeed capable of enhancing the reception reliability. Note that it can only be invoked at low user loads, because the above-mentioned regime requires potentially excessive resources for cooperation. This might reduce the extra gain of NOMA.

To elaborate a little further, in practice the participation of all users in the cooperative NOMA cannot be readily realized due to the extra requirement of short-range communication resources as well as owing to the complex signal processing associated with a high-signaling overhead. A promising solution to this problem is to reduce the number of cooperating users. Without loss of generality, we consider the appealingly simple case of having only two cooperative users as an example. Let us assume that the users are sorted in the order of improving channel qualities and that the mth and nth users are paired together, where we have $m < n$. It has been shown in Ref. [32] that the worst choice of m and n is $n = m + 1$, while the optimal choice is to group two users experiencing significantly different channel qualities.

Again, the cooperative NOMA further exploits the specific feature that the users having better channel qualities have the knowledge of other users' signals, whose channel qualities are poor. In this way, the maximum diversity gain can be achieved for all users by transmitting the signals of those specific users who have better channel qualities to other users.

4.3.4 Network-NOMA

NOMA also exhibits its own benefits in multicell applications, leading to the concept of network-NOMA. However, by directly applying the single-cell NOMA scheme to multicell scenarios, network-NOMA may result in severe intercell interference between the NOMA user pairs of the adjacent cells. As an example, a downlink cellular system having two cells and four users is depicted in Figure 4.4c, where a two-user NOMA scheme is considered, with user 1 and user 2 being served by BS 1, while user 3 and user 4 being served by BS 2. However, there may exist a strong interference between user 1 and user 3 due to their close proximity, which potentially leads to a significant performance degradation of the network-NOMA scheme.

To mitigate the intercell interference, joint transmit precoding of all NOMA users' signals can be utilized. However, all users' data and channel information should be available at the BSs involved, and finding the optimal transmit precoder is not trivial. Moreover, multiuser precoding, which is applicable in a single-cell NOMA may not be feasible in network-NOMA, since a beam generated via geographically separated BS antennas may not be capable of covering more than one angularly separated user for intrabeam NOMA. By exploiting that the CIRs of different users is likely to be rather different in the multicell scenario, a reduced-complexity transmit-precoding scheme was proposed for network-NOMA [34], where the precoder is applied only to the signals of the cell edge users, such as users 1 and 3 of Figure 4.4c.

4.3.5 User Grouping and Resource Allocation

In power-domain NOMA, the channel gain difference among users is translated into different multiplexing gains [5]. Therefore, both user grouping and resource allocation have substantial effects on the achievable throughput. In order to optimize the user grouping and resource allocation, a reasonable optimization criterion has to be found first. Then, a compelling performance versus complexity trade-off has to be struck. Specifically, the classic proportional fair

(PF) scheduler is known to strike an attractive trade-off between capacity and fairness. Hence, it has been used in both the orthogonal 3G and 4G multiple access systems. Therefore, it has also been widely considered as a beneficial optimization criterion [5,23]–[26] in NOMA systems. Specifically, both multiuser scheduling and power allocation per frequency block can be realized by maximizing the product of the average user throughput of all the users within a cell [5], [23]. In the uplink, the transmission power is usually independently determined for all users, thus scheduling a user set can be configured at a given power level. In the downlink, under the constraint of a fixed total power, user scheduling and power allocation should be jointly optimized. More particularly, for a given scheduling user set, the iterative water-filling-based power allocation algorithm [5], [23] can be used, which achieves the maximum weighted sum of the user throughput, when exploiting the uplink–downlink duality. Given the optimal power allocation for each scheduling user set, the scheduling user set is selected by maximizing the optimization criterion.

Additionally, the max–min fairness criterion associated with instantaneous CSI knowledge, that is, maximizing the minimum user rate, and the min–max fairness criterion relying on the average-CSI knowledge, that is, minimizing the maximum outage probability, have been considered in Ref. [27] for deriving the power allocation. Low-complexity algorithms have also been developed for solving the associated nonconvex problems. The max–min fairness criterion has also been used in MIMO NOMA systems in Ref. [28], where a dynamic user allocation and power optimization problem was investigated. Specifically, a sub-optimal two-step method has been proposed. In the first step, the power allocation is optimized by fixing a specific combination of user allocation according to the max–min fairness, while the second step considered all the user allocation combinations. Furthermore, the joint power and channel allocation optimization has been shown to be NP-hard in Ref. [29], and an algorithm combining Lagrangian duality and dynamic programming was proposed for delivering a competitive suboptimal solution. Furthermore, in Ref. [30], power allocation has been conceived for the NOMA downlink supporting two users when 18 practical modulation schemes are employed. To elaborate a little further, the mutual information metric rather than Shannon-theoretic throughput metric has been used for deriving a more accurate result, and it has been shown that the power allocation problem formulated for maximizing the total mutual information depends on the modulation schemes employed.

Considering the intercell interference, fractional frequency reuse (FFR), which allows the users under different channel conditions to rely on different reuse factors, has been employed in Refs [23,29,31] for further enhancing the performance of the celledge users. FFR-based power allocation strikes a trade-off between the frequency bandwidth utilization per cell and the impact of intercell interference.

4.3.6   mmWave Communications and Power-Domain NOMA

With the development of wireless communication, spectrum resources at microwave frequencies have become more and more scarce. Fortunately, mmWave communication has access to a large unused spectrum, which makes it a very promising technology enabling multi-Gbps wireless access in 5G communications [2]. However, when applying mmWave communication to mobile cellular, its benefit will highly rely on multiple access strategies. To be specific, due to the diversity of users in 5G cellular, the requirements on data rate will be quite different between different users, and to allocate a whole resource block to a user with a low-data-rate requirement is a waste of resource. As a result, such inefficient OMA that is, TDMA/FDMA/CDMA, may offset the benefit of mmWave communication in the future 5G cellular [110], which indicates the necessity of adopting NOMA in mmWave system. Moreover, as mmWave could provide large bandwidth, and the use of NOMA promises a high spectral efficiency because of the nonorthogonal resource allocation, the combination of these two technologies will further improve the system throughput.

In addition to orders-of-magnitude larger bandwidths, the smaller wavelengths at mmWave allow more antennas in the same physical space, which enables massive multiple-input multiple-output (massive MIMO) to provide more multiplexing gain and beamforming gain. As a result, mmWave communication is always associated with large adaptive antenna arrays. However, the fundamental limit of existing massive MIMO system is that the number of supported users cannot be larger than the number of RF chains, which fails to meet the massive connectivity requirement in 5G [111]. Fortunately, by using NOMA in mmWave MIMO systems, we can take advantage of MIMO–NOMA presented before and break this limit by serving multiple users in one beam, and thus accommodate much more users than RF chains and further improve the spectral efficiency of the system.

Generally speaking, power-domain NOMA requires more transmit power compared to the OMA transmissions, which is caused by the increase in noise power attributed to a large bandwidth and the interference arising due to the nonorthogonal resource allocation [112]. Fortunately, by using the directional transmission capabilities offered by the mmWave large-scale antenna arrays, this problem can be largely mitigated. To be specific, as the wavelength of mmWave is much smaller, more antennas can be deployed at the transmitter, and thus narrower beams could be formed, which can effectively eliminate interbeam interference with the help of spatial filters.

4.3.7   Application of Power-Domain NOMA

Recently, the concept of power-domain has been successfully applied to ATSC 3.0 [113], which is a new next-generation broadcasting standard in the United

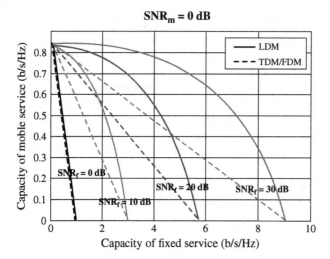

Figure 4.7 Channel capacity advantage of LDM versus TDM/FDM [113].

States, and this physical-layer nonorthogonal multiplexing technology is named layered division multiplexing (LDM).

Specifically, a two-layer LDM structure consists of the upper layer (UL) and the lower layer (LL) is accepted by ATSC 3.0 to improve spectral efficiency and provide more versatile broadcasting services. The UL with higher power allocation is used to deliver mobile services to indoor, portable, and hand-held receivers, while the LL is designed to deliver high data rate services, such as UHDTV or multiple HDTV services to fixed reception terminals, where the operational SNR is usually high due to the large and possibly directional receive antennas [113]. At the transmitter, the data of each layer are first processed by its own physical-layer signal processing modules, including channel encoding, interleaving, modulation, and so on, and then the signals from all layers are superimposed over the same time-frequency resources. At the receivers, to decode the UL signal, the lower power LL service is treated as an additional interference. To decode the LL signal, the receiver firstly needs to cancel the UL signal, which is referred to as SIC procedure in power-domain NOMA.

As shown in Figure 4.7 from Ref. [113], the channel capacity for the mobile and fixed services of the LDM and TDM/FDM systems is compared. It is observed that LDM offers better performance than TDM/FDM in all scenarios, and the higher the SNR threshold of the fixed service, the larger the advantage of the LDM systems.

4.4 Code-Domain NOMA

The NOMA schemes discussed in the previous section realize multiplexing in the power domain. By contrast, in this section, we introduce the other main category of NOMA schemes, which achieves multiplexing in the code domain. The concept of code-domain NOMA is inspired by the classic CDMA systems, in which multiple users share the same time-frequency resources, but adopt unique user-specific spreading sequences. However, the key difference compared to CDMA is that the spreading sequences are restricted to sparse sequences or nonorthogonal low cross-correlation sequences in NOMA. In this section, we first present the initial form of NOMA based on sparse spreading sequences, that is, LDS-CDMA [53]–[56]. Then, the family of LDS-aided multicarrier OFDM systems (LDS-OFDM) [57]–[62] will be discussed, which retains all the benefits of OFDM-based multicarrier transmissions in terms of its ISI avoidance, together with MUD-assisted LDS-CDMA operating at a lower complexity than that of the optimal maximum *a posteriori* probability (MAP) detector. Another important extension of LDS-CDMA is SCMA [63]–[84], which still enjoys the benefit of low-complexity reception, but has a better performance than LDS-CDMA. A suite of other improved schemes and special forms of CDMA, such as MUSA [85] and SAMA [86], will also be discussed in this section.

4.4.1 Low-Density Spreading CDMA (LDS-CDMA)

Developed from the classic concept of CDMA, LDS-CDMA is designed for limiting the amount of interference imposed on each chip of conventional CDMA systems by using LDS instead of conventional spreading sequences. The basic principle of LDS-CDMA has been discussed in Ref. [53]–[54]. Additionally, Refs[53], [54] also discussed the iterative MUD based on the message passing algorithm (MPA) imposing a lower complexity than that of the optimal MAP detector. Specifically, in Ref. [53], the performance of LDS-CDMA communicating over memoryless Gaussian channels using BPSK modulation was analyzed. The simulation results showed that the performance of LDS–CDMA is capable of approaching the single-user performance for a normalized user-load as high as 200%. However, the performance of LDS-CDMA operating in multipath fading channels is still under investigation at the time of writing. The challenge is that the multipath fading channels will destroy the original LDS structure. On the other hand, a structured approach of designing LDS codes for LDS-CDMA has been proposed in Ref. [55], where the basic idea is to map the signature constellation elements to the spreading matrix hosting the spreading sequences. Furthermore, the capacity region of LDS-CDMA was calculated using information theoretic analysis in Ref. [56], and the accompanying simulation results showed how the attainable capacity depended on the spreading sequence density factor as well as on the maximum number of users associated with each chip, which provided insightful theoretical guidelines for practical LDS system designs.

Let us now consider a classic synchronous CDMA system operating in the uplink and supporting K users with the aid of N_c chips (N_c equals to the number of observations at the receiver). The transmitted symbol x_k of user k is first generated by mapping a sequence of independent information bits to a constellation alphabet, that is, x_k is taken from a complex-valued constellation set $\mathbb{X}$. Then, the transmitted symbol x_k is mapped to a spreading sequence $\mathbf{s}_k$, such as the set of widely used PN sequences, which is unique for each user. The signal received during chip n can be represented by

$$y_n = \sum_{k=1}^{K} g_{n,k} s_{n,k} x_k + w_n, \tag{4.17}$$

where $s_{n,k}$ is the nth component of the spreading sequence $\mathbf{s}_k$, $g_{n,k}$ is the channel gain of user k on chip n, and w_n is a complex-valued Gaussian noise sample with a zero mean and a variance of σ^2. When we combine the signals received during all the N_c chips, the received signal vector $\mathbf{y} = \left[y_1, y_2, \ldots, y_{N_c}\right]^T$ is formulated as

$$\mathbf{y} = \mathbf{Hx} + \mathbf{w}, \tag{4.18}$$

where $\mathbf{x} = \left[x_1, x_2, \ldots, x_K\right]^T$, $\mathbf{H}$ is the channel matrix of size ($N_c \times K$), and the element $h_{n,k}$ in the nth row and the kth column of $\mathbf{H}$ is denoted by $g_{n,k} s_{n,k}$. Finally, $\mathbf{w} = \left[w_1, w_2, \cdots, w_{N_c}\right]^T$, and $\mathbf{w} \sim \mathcal{CN}\left(0, \sigma^2 \mathbf{I}\right)$.

In classic CDMA systems, the elements of the spreading sequences $\mathbf{s}_k (k = 1, \ldots, K)$ are usually nonzero, that is, the spreading sequences are not sparse. Consequently, the signals received from all the active users are overlaid on top of each other at each chip, and every user will be subjected to interuser interference imposed by all the other users. If the spreading sequences are orthogonal, it is straightforward to eliminate the interferences, hence the information of all users can be accurately detected by a low-complexity correlation receiver. However, the classical orthogonal spreading sequences can only support as many users as the number of chips. By contrast, the above-mentioned PN-sequence family has many more codes than the number of chips in a sequence, but since the codes are nonorthogonal, they impose interference even in the absence of non dispersive channels. Hence, they require more complex MUDs. Another natural idea, which leads to LDS-CDMA, is to use sparse spreading sequences instead of the classic "fully-populated" spreading sequences to support more users, where the number of nonzero elements in the spreading sequence is much lower than N_c for the sake of reducing the interference imposed on each

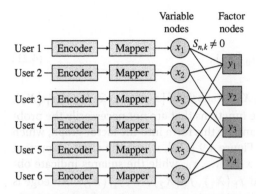

Figure 4.8 Illustration of LDS–CDMA: six users only employ four chips for transmission, which implies that a normalized user-load of 150can be achieved.

chip. Therefore, LDS-CDMA is potentially capable of improving the attainable system performance by using low-density spreading sequences [53], which is the key distinguishing feature between conventional CDMA and LDS-CDMA.

In LDS-CDMA, all transmitted symbols are modulated onto sparse spreading sequences. In this way, each user will only spread its data over a small number of chips, as shown in Figure 4.8. As a result, the number of the superimposed signals at each chip will be less than the number of active users, which means that the interference imposed on each chip will be efficiently reduced, hence mitigating the multiuser interference by carefully designing the spreading sequences. Therefore, the received signal at chip n in LDS-CDMA systems can be rewritten as

$$y_n = \sum_{k \in N(n)} g_{n,k} s_{n,k} x_k + w_n = \sum_{k \in N(n)} h_{n,k} x_k + w_n, \tag{4.19}$$

where $N(n)$ denotes the set of users whose sparse spreading sequences have a nonzero element at chip n, namely, $N(n) = \{k | s_{n,k} \neq 0\}$.

At the receiver, MUD based on message passing algorithm may be performed. Given the joint probability function $p(x_1, \ldots, x_E)$ for random variables $x_1, x_2, \ldots, x_E$, the message-passing algorithm is capable of simplifying the calculation of the marginal probability distribution for each variable as follows:

$$p(x_e) = \sum_{\sim \{x_e\}} p(x_1, x_2, \ldots, x_E), \tag{4.20}$$

where $\sim \{x_e\}$ represents all variables except for x_e. We assume that the joint probability function can be decomposed into the product of some positive functions, namely,

$$p\left(x_1, x_2, \ldots, x_E\right) = \frac{1}{Z} \prod_{d=1}^{D} f_d\left(X_d\right), \tag{4.21}$$

where Z is a normalized constant, X_d is a subset of $\{x_1, x_2, \ldots, x_E\}$, and $f_1\left(X_1\right)$, $f_2\left(X_2\right), \ldots, f_D\left(X_D\right)$ are positive functions that are not necessarily the probability functions. Then, we can translate this form into the factor graph, which is a bipartite graph, as shown in Figure 4.9 [114], where the circles represent variable nodes corresponding to $x_1, x_2, \ldots, x_E$, while the squares indicate observation nodes corresponding to $f_1\left(X_1\right), f_2\left(X_2\right), \ldots, f_D\left(X_D\right)$. An edge is present between a variable node x_e and an observation node $f_d\left(X_d\right)$ if and only if $x_e \in X_d$.

In general, the message passing algorithm relies on the factor graph representation of the problem as its input and returns the marginal distribution of all variable nodes. Messages can be passed between the variable node and the observation node through the edge between them, and the message can be interpreted as the soft value that represents the reliability of the variable associated with each edge. The marginal distribution of a variable node can be interpreted as a function of the messages received by that variable node. The iterative form of the message passing algorithm can be represented as

$$m_{d \to e}^{(t)}\left(x_e\right) \propto \sum_{\{x_i | i \in N(d) \setminus e\}} f_d\left(X_d\right) \prod_{i \in N(d) \setminus e} m_{i \to d}^{(t-1)}\left(x_i\right), \tag{4.22}$$

$$m_{e \to d}^{(t)}\left(x_e\right) \propto \prod_{i \in N(e) \setminus d} m_{i \to e}^{(t-1)}\left(x_e\right), \tag{4.23}$$

where $m_{d \to e}^{(t)}\left(x_e\right)$ denotes the message transmitted from the observation node $f_d\left(X_d\right)$ to the variable node x_e at the tth iteration. Similarly, $m_{e \to d}^{(t)}\left(x_e\right)$ presents the message transmitted from the variable node x_e to the observation node $f_d\left(X_d\right)$. If the maximum number of iterations is T, the marginal probability distribution for each variable can be finally calculated as

$$p\left(x_e\right) \propto \prod_{d \in N(e)} m_{d \to e}^{(T)}\left(x_e\right). \tag{4.24}$$

It has been theoretically shown that the marginal distribution can be accurately estimated with the aid of a limited number of iterations, provided that the factor graph does not have loops [114]. However, in many practical situations, the presence of loops cannot be avoided. Fortunately, the message passing algorithm is quite accurate for "locally tree-like" graphs, which implies that the length of the shortest loop is restricted to $\mathcal{O}(\log(E))$. Therefore, in

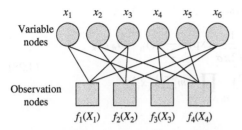

Figure 4.9 Factor graph representation of MPA.

most practical applications associated with a sparse structure, we can obtain an accurate marginal distribution estimate by an appropriate design of the factor graph.

In LDS-CDMA, the optimum MAP detection of $\mathbf{x}$ in (4.18) can be formulated as:

$$\hat{x}_k = \arg\max_{a \in \mathbb{X}} \sum_{\substack{\sim \{x_k\} \\ x_k = a}} p(\mathbf{x}|\mathbf{y}). \tag{4.25}$$

Without loss of generality, we assume that the transmitted symbols and noise are identically and independently distributed (i.i.d), and the transmitted symbols obey the uniform distribution. Then according to Bayes' rule, (4.25) can be reformulated as

$$\hat{x}_k = \arg\max_{a \in \mathbb{X}} \sum_{\substack{\sim \{x_k\} \\ x_k = a}} \prod_{n=1}^{N_c} p(y_n|\mathbf{x}_{[n]}), \tag{4.26}$$

where

$$p\left(y_n|\mathbf{x}_{[n]}\right) = \frac{1}{\sqrt{2\pi}\sigma} \exp\left\{-\frac{1}{2\sigma^2}(y_n - \sum_{k \in N(n)} h_{n,k}x_k)^2\right\}. \tag{4.27}$$

Observe that (4.26) represents a marginal process of $\prod_{n=1}^{N_c} p(y_n|\mathbf{x}_{[n]})$, which is similar to the form of the decomposable joint probability function of the message passing algorithm, apart from a normalization constant Z. To elaborate a little further, we can regard each term $p(y_n|\mathbf{x}_{[n]})$ as a positive function $f_d\left(X_d\right)$ in the message passing algorithm. Then, the factor graph shown in Figure 4.8 can be constructed just like that of Figure 4.9. Therefore, we can rewrite the iterative Eqs. 4.22 and 4.23 as follows:

$$
m_{n \to k}^{(t)}\left(x_k\right) \propto \sum_{\{x_i | i \in N(n) \backslash k\}} \frac{1}{\sqrt{2\pi}\sigma} \exp\{-\frac{1}{2\sigma^2}(y_n - h_{n,k}x_k
$$
$$
- \sum_{i \in N(n) \backslash k} h_{n,i}x_i)^2\} \prod_{i \in N(n) \backslash k} m_{i \to n}^{(t-1)}\left(x_i\right), \tag{4.28}
$$

$$
m_{k \to n}^{(t)}\left(x_k\right) \propto \prod_{i \in N(k) \backslash n} m_{i \to k}^{(t-1)}\left(x_k\right). \tag{4.29}
$$

Finally, the (approximate) marginal probability distribution of each variable after T iterations can be calculated by (4.24).

In the case of LDS, the number of edges in the factor graph is relatively low, hence less and longer loops can be expected based on a meritorious design of the factor graph based on a beneficial sparse spreading sequence design. Additionally, assuming that the maximum number of users superimposed on the same chip is w, the receiver complexity is on the order of $\mathcal{O}\left(|\mathbb{X}|^w\right)$ instead of $\mathcal{O}\left(|\mathbb{X}|^K\right)$ $(K > w)$ for conventional CDMA.

The performance of LDS-CDMA and direct sequence-CDMA (DS-CDMA), which is adopted by the 3G WCDMA systems, have been compared in Ref. [53]. More specifically, as shown in Figure 4.10 [53], LDS-CDMA outperforms DS-CDMA using the best-found spreading sequences, where the MMSE-based partial parallel interference cancellation (PPIC) receiver has been adopted by both schemes. Furthermore, when LDS-CDMA relies on the MPA receivers, its performance approaches the single user bound within a small margin of 1.17 dB at a BER of 10^{-4}.

4.4.2 Low-Density Spreading-Aided OFDM (LDS-OFDM)

OFDM and MC-CDMA are close relatives, especially when considering frequency-domain spreading, which spreads each user's symbols across all the OFDM subcarriers, provided that the number of spreading-code chips is identical to the number of subcarriers. Then multiple users may be supported by overlaying the unique, user-specific spreading sequences of all users on top of each other across all subcarriers. As always, the spreading sequences may be chosen to be orthogonal Walsh–Hadamard codes or nonorthogonal m-sequences, as well as LDSs, for example.

Hence, LDS-OFDM can be interpreted as an integrated version of LDS-CDMA and OFDM, where, for example, each user's symbol is spread across a carefully selected number of subcarriers and overlaid on top of each other in the frequency domain. To elaborate a little further, in the conventional OFDMA system, only a single symbol is mapped to a subcarrier, and different symbols

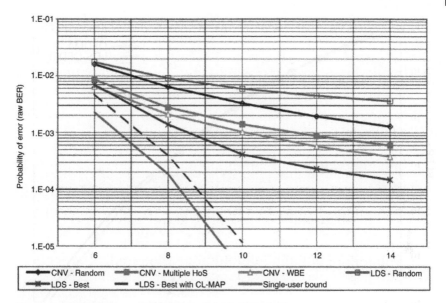

Figure 4.10 Comparison of LDS-CDMA and DS-CDMA [53].

are transmitted on different subcarriers, which are orthogonal and hence do not interfere with each other. Therefore, the total number of transmitted symbols is restricted by the number of orthogonal subcarriers. By contrast, in the LDS-OFDM system, the transmitted symbols are first multiplied with LDS sequences, whose length is equal to the number of subcarriers and the resultant chips are transmitted on different subcarriers. When using LDS spreading sequences, each original symbol is only spread to a specific fraction of the subcarriers. As a result, each subcarrier carries chips related to a fraction of the original symbols. Suffice to say that apart from the already accentuated benefits, frequency-domain spreading is particularly advantageous in strongly frequency-selective channels, which would often obliterate some of the subcarriers and their information, while in the presence of frequency-domain spreading they would only affect some of the chips conveying the original symbols. This is likely to allow us to still recover the original symbols. We note in closing that the family of MUDs designed using the message passing algorithm for LDS-CDMA can also be used for LDS-OFDM in order to separate the overlaid symbols at the receiver.

At the time of writing, a number of insightful LDS-OFDM investigations have already been disseminated in the literature. For example, the system model and properties of LDS-OFDM, including its frequency diversity order, receiver complexity, and its ability to operate under rank-deficient conditions in the presence of more users than chips have been presented in Ref. [57]. An upper

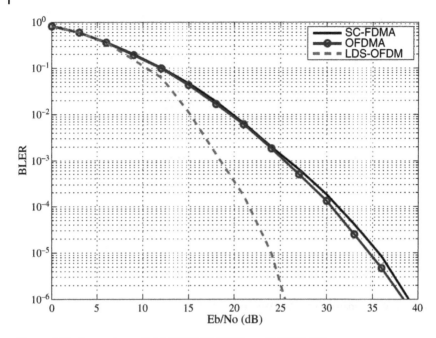

Figure 4.11 Comparison of LDS-OFDM, SC-FDMA, and OFDMA [59].

limit was imposed on the number of users per subcarrier, in order to control the receiver complexity in Ref. [58]. Additionally, in Refs [59], [60], the link-level performance comparison between LDS-OFDM, SC-FDMA, and OFDMA is presented in terms of the block error rate (BLER). It can be observed from Figure 4.11 that SC-FDMA and OFDMA have the same BLER performance, since only one RB is allocated to each user, and thus the same frequency diversity is achieved by both techniques. However, LDS-OFDM achieves lower BLER under the same E_b/N_0, due to the frequency diversity gained by spreading on more than one resource block. Furthermore, in order to improve the achievable performance of LDS-OFDM, a joint subcarrier and power allocation method was proposed in Ref. [61], with the objective of maximizing the weighted sum-rate using an efficient greedy algorithm. As a further result, a pair of PAPR reduction techniques have been proposed for LDS-OFDM in Ref. [62].

4.4.3 Sparse Code Multiple Access

The recently proposed SCMA technique constitutes another important NOMA scheme, which relies on code domain multiplexing developed from the basic LDS-CDMA scheme. In Ref. [63], SCMA was extensively discussed in terms of its transmission and multiplexing aspects, as well as in terms of its factor

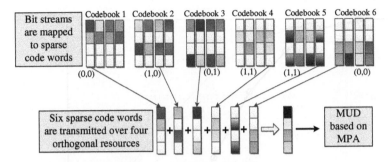

Figure 4.12 SCMA encoding and multiplexing.

graph representation and receiver architecture relying on the message passing algorithm. In contrast to the basic LDS-CDMA, as illustrated in Figure 4.12, the bit-to-constellation mapping and spreading operations in SCMA are intrinsically amalgamated, hence the original bit streams are directly mapped to different sparse code words, where each user has its own codebook. Without loss of generality, we assume that there are J codebooks, where each code book contains M code words of length K_l, and the number of nonzero elements in every code word is N_{nz}. For example, in Figure 4.12, we have $J = 6$, $M = 4$, $K_l = 4$, and $N_{nz} = 2$. We consider the rank-deficient scenario of $K_l < J$, which is capable of supporting the massive connectivity expected in 5G. All code words in the same codebook contain zeros in the same $(K_l - N_{nz})$ dimensions, and the positions of zeros in the different codebooks are unique and distinct for the sake of facilitating collision avoidance for any pair of users. Therefore, the maximum number of codebooks is restricted by the selection of K_l and N_{nz}, which is equal to $\begin{pmatrix} K_l \\ N_{nz} \end{pmatrix}$. For each user, $\log_2 M$ bits are mapped to a complex code word. The code words of all users are then multiplexed onto K_l shared orthogonal resources, such as the OFDM subcarriers. Due to the sparsity of code words, the signal received on subcarrier k can be represented by

$$y_k = \sum_{j \in N(k)} h_{kj} x_{kj} + w_k, \tag{4.30}$$

where x_{kj} is the kth component of the code word $\mathbf{x}_j$ for user j, h_{kj} is the channel gain of user j at the kth subcarrier, and w_k denotes the complex-valued Gaussian noise with zero mean and variance σ^2. Similar to LDS-CDMA, the message passing algorithm can also be used for MUD at the SCMA receiver. However, the receiver complexity may become excessive. To circumvent this problem, improved variants of the message passing algorithm have been proposed in Refs [64]–[72]. Specifically, a low-complexity logarithmic-domain message passing algorithm (Log-MPA) was proposed in Ref. [73]. The associated simulation

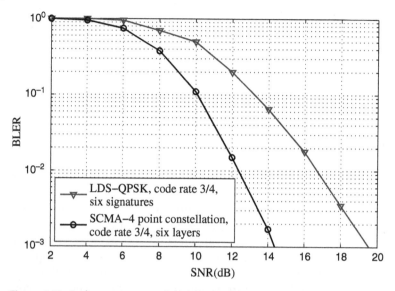

Figure 4.13 Performance comparison between SCMA and LDS [63].

results showed that the performance degradation of Log-MPA over the full-complexity MPA was negligible in practical applications, despite the fact that the Log-MPA achieved over 50% complexity reduction. For Log-MPA, the conditional channel probability calculation imposes up 60% of the total computational complexity in the whole decoding procedure. In Ref. [64], a dynamic search algorithm based on classic signal uncertainty theory was proposed for eliminating any unnecessary conditional channel probability calculation without degrading the decoding performance. On the other hand, in order to improve the BER performance of SCMA, the powerful turbo principle has been invoked in Ref. [74] for exchanging extrinsic information between the SCMA detector and the channel decoder. By contrast, in Ref. [75], a low-complexity turbo-like combination of iterative detection and iterative decoding was conceived for striking a compelling performance versus complexity balance.

Quantitatively, the performance of SCMA relying on a multidimensional constellation having four points was compared with that of LDS using QPSK modulation in Figure 4.13 [63]. These simulation results show that SCMA outperforms LDS in terms of its block error ratio (BLER). The key difference between LDS and SCMA is that SCMA relies on multidimensional constellations for generating its codebooks, which results in the so-called constellation shaping gain. [63] This gain is unavailable for other NOMA schemes. More explicitly, the "shaping gain" terminology represents the average symbol energy gain, when we change the shape of the modulation constellation. In general, the shaping gain is higher when the shape of the constellation becomes similar to a sphere. However, the SCMA codebook design is complex [63], since the different layers are multiplexed with the aid of different codebooks. However, the best design criterion to be used for solving the multidimensional constellation problem is unknown at the time of writing. Having said this, it is anticipated that using the powerful semianalytical tool of extrinsic information transfer-function (EXIT) charts for jointly designing the channel code and the constellation would lead to near-capacity performance. As a further solution, a multistage design approach has been proposed for finding a meritorious suboptimal solution in Ref. [63]. More details concerning the codebook design can be found in Ref. [76].

Specifically, in order to simplify the optimization problem of the multidimensional constellation design, a mother constellation can be generated first by minimizing the average alphabet energy for a given minimum Euclidian distance between any two constellation points. More particularly, an optimized design of the mother constellation based on the classic star-QAM signaling constellation has been proposed in Ref. [77]. The resultant simulation results showed that the star-QAM-based codebooks are capable of significantly enhancing the BER performance of the square-QAM-based codebooks. Once the mother constellation has been obtained, the codebook-specific operation can be applied to the mother constellation in order to obtain specific constellations for each codebook. More specifically, the codebook-specific operations, such as phase rotation, complex conjugation, and dimensional permutation, can be optimized for introducing correlation among the non zero elements of the code words, which is beneficial in terms of recovering the code words contaminated by the interference imposed by other tones. Additionally, different power can be assigned to the symbols superimposed over the same time-frequency index for ensuring that the message passing algorithm can operate more efficiently by mitigating the interferences between the paired layers. Furthermore, inspired by the family of irregular low-density parity check (LDPC) codes, an irregular SCMA structure has been proposed in Ref. [78], where the number of nonzero elements of the code words can be different for different users. In this way, users having different QoS requirements can be simultaneously served.

Again, in the uplink of a SCMA system, grant-free multiple access can be realized by carefully assigning the codebooks and the pilots to the users based on Ref. [79]. As mentioned in Section 4.2, a user does not have to send a scheduling request to the BS in the grant-free transmission scheme, thus a significant latency and signaling overhead reduction can be expected. As shown in Figure 4.14, the pre configured resource to be assigned to the users may be referred to as a contention transmission units (CTU). There are J codebooks

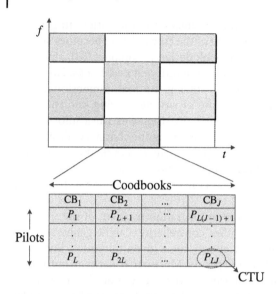

Figure 4.14 Definition of a contention transmission unit (CTU).

defined over a time-frequency resource, and L pilot sequences are associated with each codebook. The grant-free multiple access regime allows contentions to occur, when multiple users are assigned to the same CTU. The network detects the uplink packets by attempting their reception using all possible access codes assigned to the predefined contention region. Then a random back-off procedure can be invoked, when collisions occur. At the receiver, blind detection and compressive sensing (CS) techniques can be used for performing joint activity and data detection, for example, with the aid of the joint message passing algorithm (JMPA) of Ref. [80].

SCMA can also be used in the downlink in order to improve the system throughput, leading to the multiuser SCMA (MU-SCMA) concept [81]. Based on a limited knowledge of the channel conditions of different users, the BS simply pairs the users, where the transmit power is appropriately shared among multiple users. This regime is hence reminiscent of the NOMA scheme relying on the previously mentioned power-domain multiplexing. Compared to MU-MIMO, MU-SCMA is more robust to channel quality variations, and indeed, the provision of near-instantaneous CSI feedback is unnecessary for this open-loop multiple access scheme [81]. In Ref. [82], the concept of single-cell downlink MU-SCMA is extended to an open-loop downlink coordinated multipoint (CoMP) solution, which was termed as MU-SCMA-CoMP. In this scheme, the SCMA layers and transmits power are shared among multiple users within a

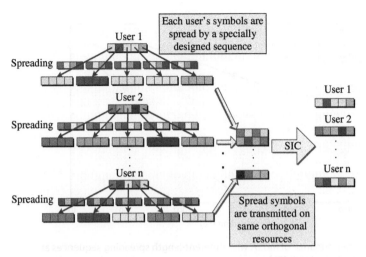

Figure 4.15 Uplink MUSA system.

CoMP cluster. The analysis and simulation results in Ref. [82] demonstrated its robustness to both user mobility and channel aging. Furthermore, the capacity of downlink massive MIMO MU-SCMA was analyzed in Ref. [83] based on random matrix theory and it has been shown that compared to massive MIMO MU-OFDMA systems, Massive MIMO MU-SCMA is capable of achieving a higher sum rate.

In a nutshell, the efficiency of SCMA has been verified by both simulations and real-time prototyping in Ref. [84]. Both the lab tests and the field tests demonstrated that SCMA is capable of supporting up to three times more users than the number of resource slots, while still maintaining a link integrity close to that of orthogonal transmissions.

4.4.4 Multi User Shared Access

MUSA is another NOMA scheme relying on code-domain multiplexing, which can be regarded as an improved CDMA-style scheme.

In the uplink of the MUSA system of Figure 4.15, all transmitted symbols of a specific user are multiplied with the same spreading sequence. (Note that different spreading sequences can also be used for different symbols of the same user, which results in beneficial interference averaging.) Then, all symbols after spreading are transmitted over the same time-frequency resources, such as OFDM subcarriers. Without any loss of generality, we assume that each user transmits a single symbol every time, and that there are K users as well as N

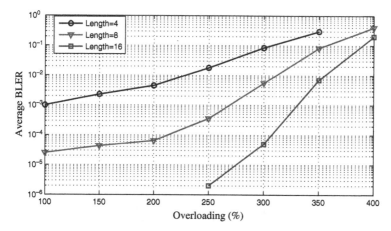

Figure 4.16 The performance of MUSA with different-length spreading sequences at different normalized user loads [85].

subcarriers. Rank-deficient scenarios can also be supported by MUSA, that is, $K > N$, which will impose interference among the users. At the receiver, linear processing and SIC are performed in order to separate the different users' data according to their channel conditions.

In the downlink of the MUSA system, the users are separated into G groups. In each group, the different users' symbols are weighted by different power-scaling coefficients and then they are superimposed. Orthogonal sequences of length G can be used as spreading sequences in order to spread the superimposed symbols from G groups. More specifically, the users from the same group employ the same spreading sequence, while the spreading sequences are orthogonal across the different groups. In this way, the inter group interferences can be removed at the receiver. Then, SIC can be used for carrying out intra group interference cancellation by exploiting the associated power difference.

In MUSA, the spreading sequences should have low cross-correlation in order to facilitate near-perfect interference cancellation at the receiver. The MUSA technique is capable of improving the downlink capacity, which is an explicit benefit of the associated SINR difference and SIC. As a further compelling benefit, MUSA is capable of guaranteeing fairness among the multiplexed users without any capacity loss. In a nutshell, with the advent of advanced spreading sequences and powerful state-of-the-art SIC techniques, substantial gains can be obtained by MUSA, even for a normalized user-load as high as say 300 %, which is shown in Figure 4.16 [85].

4.4.5 Successive Interference Cancellation Aided Multiple Access (SAMA)

Let us consider an uplink SAMA system supporting K users with the aid of N orthogonal OFDM subcarriers, where we have $K > N$, that is, when the system is rank deficient. The system model of SAMA is similar to that of MUSA, but in SAMA the non zero elements of any spreading sequence $\mathbf{b}_k$ for user k are equal to 1, and the spreading matrix $\mathbf{B} = (\mathbf{b}_1, \mathbf{b}_2, \ldots, \mathbf{b}_K)$ is designed based on the following principles [86]:

- The number of groups with different number of 1's in the spreading sequence should be maximized.
- The number of overlapped spreading sequences that have the same number of 1's should be minimized.

Then the maximum number of user supported with the aid of N orthogonal subcarriers can be calculated as

$$\binom{N}{1} + \binom{N}{2} + \cdots + \binom{N}{N} = 2^N - 1. \qquad (4.31)$$

For example, spreading matrices for $N = 2, K = 3, N = 3, K = 7$, and $N = 4, K = 15$ can be designed as follows:

$$\mathbf{B}_{2,3} = \begin{pmatrix} 1 & 1 & 0 \\ 1 & 0 & 1 \end{pmatrix}, \qquad (4.32)$$

$$\mathbf{B}_{3,7} = \begin{pmatrix} 1 & 1 & 0 & 1 & 1 & 0 & 0 \\ 1 & 1 & 1 & 0 & 0 & 1 & 0 \\ 1 & 0 & 1 & 1 & 0 & 0 & 1 \end{pmatrix}, \qquad (4.33)$$

$$\mathbf{B}_{4,15}$$
$$= \begin{pmatrix} 1 & 1 & 0 & 1 & 1 & 1 & 1 & 1 & 0 & 0 & 0 & 1 & 0 & 0 & 0 \\ 1 & 1 & 1 & 0 & 1 & 1 & 0 & 0 & 1 & 1 & 0 & 0 & 1 & 0 & 0 \\ 1 & 1 & 1 & 1 & 0 & 0 & 1 & 0 & 1 & 0 & 1 & 0 & 0 & 1 & 0 \\ 1 & 0 & 1 & 1 & 1 & 0 & 0 & 1 & 0 & 1 & 1 & 0 & 0 & 0 & 1 \end{pmatrix}. \qquad (4.34)$$

At the receiver, the message passing algorithm is invoked for separating the signals of the different users. The design objective used for determining the spreading matrices in SAMA is to facilitate convenient interference cancellation [86]. Consider $\mathbf{B}_{4,15}$, for example. The spreading sequence of the first user has four non zero elements, hence the resultant diversity order is 4. Thus, the first user's symbol is the most reliable one. Therefore, the first user's symbol can be readily determined in a few iterations, which is beneficial for the convergence of the symbol detection process of all the other users having lower diversity orders.

4.5 Other NOMA Schemes

Apart from the prominent power-domain NOMA and code-domain NOMA solutions discussed in Sections 1.3 and 1.4, respectively, a range of alternative multiple access schemes have also been investigated recently, which will be discussed in this section.

4.5.1 Spatial Division Multiple Access

SDMA is one of the powerful NOMA schemes, and the philosophy of SDMA may be deemed to be related to that of classic CDMA, based on the following philosophy. Even if orthogonal Walsh–Hadamard spreading sequences are employed for distinguishing the users, when they are transmitted over dispersive channels, their orthogonality is destroyed by their convolution with the CIR even in the absence of co-channel interference. Hence, we end up with a potentially infinite variety of received sequences. This leads to the appealing concept of simply using the unique, user-specific CIRs for distinguishing the users, instead of unique, user-specific spreading sequences. Naturally, when the users transmitting in the uplink are close to each other, their CIRs become quite similar, which aggravates the task of the MUD in separating their signals. The beneficial properties of this family of solutions have attracted substantial research efforts, as detailed in Refs [87]–[100].

To elaborate a little further, given the potentially infinite variety of CIRs, these sophisticated SDMA systems are capable of operating under highly rank-deficient conditions, namely, when the number of mobile users transmitting in the uplink is much higher than the number of BS uplink-receiver antennas. This would avoid the hard-limited user-load of the Walsh–Hadamard codes based CDMA systems, since the system performance would only gracefully decay upon increasing the number of users. The resultant SDMA systems tend to exhibit a performance similar to their rank-deficient CDMA counterparts relying on m-sequences, for example.

Since these SDMA systems rely on the CIR for distinguishing the users, they require accurate CIR estimation, which become extremely challenging, when the number of users is much higher than the number of BS receiver antennas. This logically leads to the concept of joint channel and data estimation, which attracted substantial research interest [91],[97]. These high-end solutions often rely on powerful nolinear bioinspired MUDs exchanging their soft information with the channel estimator.

Below we will elaborate further on a variety of powerful solutions in a little more detail. In Ref. [87], the family of minimum bit error rate (MBER) MUDs was shown to be capable of outperforming the classic minimum mean-squared-error (MMSE) MUD in terms of the achievable BER owing to directly minimizing the BER cost function. In this paper, genetic algorithms (GAs) were invoked for finding the optimum weight vectors of the MBER MUD in the context of multiple antenna-aided multi user OFDM. It was shown that the MBER MUD is capable of supporting more users than the number of receiver antennas available in highly rank-deficient scenarios.

A novel parallel interference cancellation (PIC)-based turbo space time equalizer (STE) structure was designed in Ref. [88] for multiple antenna-assisted uplink receivers. The proposed receiver structure allowed the employment of nonlinear type of detectors such as the Bayesian decision feedback (DF)-assisted turbo STE or the MAP STE, while operating at a moderate computational complexity. The powerful receivers based on the proposed structure tend to outperform the linear turbo detector benchmarker based on the classic MMSE criterion, even if the latter aims for jointly detecting all transmitters' signals. Additionally, the PIC-based receiver is also capable of equalizing nonlinear binary precoded channels. The performance difference between the presented algorithms was discussed using the powerful semi analytical tool of extrinsic information transfer function (EXIT) charts.

Wang et al. [89] demonstrated that the iterative exchange of extrinsic information between the K-best sphere detector (SD) and the channel decoder is appealing, since it is capable of achieving a near MAP performance at a moderate complexity. However, the computational complexity imposed by the K-best SD significantly increases when using a large value of K for the sake of maintaining a near-MAP performance in a high-throughput uplink SDMA/OFDM system supporting a large number of users and/or a high number of bits/symbols. This problem is further aggravated when the number of users/MSs exceeds that of the receive antennas at the BS, namely, in the challenging scenario of rank-deficient systems. It was demonstrated that the iterative decoding convergence of this two-stage system may be improved by incorporating a unity rate code (URC) having an infinite impulse response, which improves the efficiency of the extrinsic information exchange. Although this results in a slightly more complex three-stage system architecture, it allows us to use a low-complexity SD having a significantly reduced detection candidate list size. Alternatively, a reduced SNR is required. For example, given a target BER of 10^{-5} and a candidate list size of 32 for the SD, the three-stage receiver is capable of achieving a performance gain of 2.5 dB over its two-stage counterpart in a rank-deficient SDMA/OFDM 4-QAM system supporting eight co-channel users and employing for receive antennas at the BS, namely, in an (8×4) rank-deficient system having a normalized user-load of two. For the sake of further enhancing the three-stage concatenated receiver, the proposed iterative center-shifting SD scheme and the so-called irregular convolutional codes (IrCCs) were intrinsically amalgamated, which led to an additional performance gain of 2 dB.

In Ref. [90], Chen et al. proposed a space-time decision feedback equalization (ST-DFE)-assisted MUD scheme for multiple receiver antenna-aided SDMA systems. Again, a sophisticated MBER MUD design was invoked, which was shown to be capable of improving the achievable BER performance and enhancing the attainable system capacity over that of the standard MMSE design. An appealing adaptive implementation of the MBER ST-DFE assisted MUD was also proposed using a stochastic gradient-based least bit error rate algorithm, which was demonstrated to consistently outperform the classical least mean square (LMS) algorithm, while imposing a lower computational complexity than the LMS algorithm for the binary signalling scheme considered. It was demonstrated that the MBER ST-DFE assisted MUD is more robust to channel estimation errors as well as to potential error propagation imposed by decision feedback errors, than the MMSE ST-DFE-assisted MUD.

The development of evolutionary algorithms (EAs) [91], such as GA, repeated weighted boosting search (RWBS), particle swarm optimization (PSO), and differential evolution algorithms (DEAs) stimulated wide interests in the communication research community. However, the quantitative performance versus complexity comparison of GA, RWBS, PSO, and DEA techniques applied to the joint channel estimation and turbo MUD/decoding in the context of SDMA/OFDM systems is a challenging problem, which has to consider both the channel estimation problem formulated over a continuous search space and the MUD optimization problem defined over a discrete search space. Hence, the capability of the GA, RWBS, PSO, and DEA to achieve optimal solutions at an affordable complexity was investigated in this challenging application by Zhang and coworkers [91]. Their study demonstrated that the EA-assisted joint channel estimation and turbo MUD/decoder are capable of approaching both the Cramer-Rao lower bound of the optimal channel estimation and the BER performance of the idealized optimal maximum likelihood (ML) turbo MUD/decoder associated with perfect channel estimation, respectively, despite imposing only a fraction of the idealized turbo ML-MUD/decoder's complexity.

From the discussions above, we can see that the concept of NOMA has already existed in various systems, such as SDMA, where users are distinguished using the unique, user-specific CIRs. Actually these systems require accurate CIR estimation to successfully realize MUD, which becomes extremely challenging when the number of users is much higher than that of receiver antennas. Solving this CIR estimation problem logically leads to the concept of joint channel and data estimation, and these high-end solutions often rely on powerful nonlinear MUDs. In fact, most of the studies focus on MUD design, and a series of nonlinear MUD algorithms such as parallel interference cancellation (PIC) [88] and space-time decision feedback equalization (ST-DFE [90] have been proposed. Further, with the development of evolutionary algorithms, algorithms like genetic algorithm (GA) and particle swarm optimization (PSO) may be explored to acquire accurate CIR estimation [91]. In contrast, power-domain NOMA transmits the superposition of multiuser signals with different power-allocation coefficients, and usually SIC is used at the receiver to detect multiuser signals. As the channel gain difference among users is translated into different multiplexing gains [5], both user grouping and resource allocation have substantial effects on the achievable throughput. As a result, most of the studies concerning power-domain NOMA focus on user grouping, resource (power) allocation, and performance analysis. Recently, with the development of mmWave communication and massive MIMO, combining power-domain NOMA with mmWave and massive MIMO has become a promising technique [110]–[112], which is presented in detail in Sections 1.3.5 and 1.3.6.

4.5.2 Pattern Division Multiple Access

Apart from the SDMA already mentioned, the family of PDMA schemes [101],[102] constitutes another promising NOMA class that can be implemented in multiple domains. At the transmitter, PDMA employs nonorthogonal patterns, which are designed by maximizing the diversity and minimizing the overlaps among multiple users. Then, multiplexing can be realized in the code, power, or spatial domains, or in fact in their combinations. Multiplexing in the code domain is reminiscent of SAMA [86]. Multiplexing in the power domain has a system model similar to multiplexing in the code domain, but power scaling has to be considered under the constraint of a given total power. Multiplexing in the spatial domain leads to the concept of spatial PDMA, which relies on multiantenna-aided techniques. In contrast to MU-MIMO, spatial PDMA does not require joint precoding for realizing spatial orthogonality, which significantly reduces the system's design complexity. Additionally, multiple domains can be combined in PDMA to make full use of the various available wireless resources. The simulation results of Ref. [102] demonstrated that compared to LTE, PDMA is potentially capable of achieving a 200% normalized throughput in the uplink, and more than 50% throughput gain may be achieved in the downlink.

4.5.3 Signature-Based NOMA

Signature-based NOMA schemes are also proposed as promising candidates for 5G. Low code rate and signature-based shared access (LSSA) is one of them, and the transmitter structure for uplink massive machine-type communication

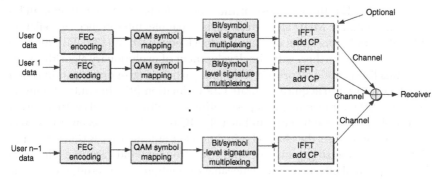

Figure 4.17 The LSSA transmitter structure.

(mMTC) is depicted in Figure 4.17 [115]. LSSA multiplexes each user's data in bit or symbol level with specific signature pattern, which consists of reference signal (RS), complex/binary sequence, and permutation pattern of a short length vector. All the users' signatures share the same short vector length, which can be chosen randomly from the mobile terminal or assigned to the user from the network. Moreover, LSSA can be optionally modified to have a multicarrier variant in order to exploit frequency diversity provided by wider bandwidth, and achieve lower latency. It can also support asynchronous uplink transmission, because the BS is able to distinguish/detect overlaid user signals by correlating with signature patterns, even if the transmission timing is different from each other.

Similar to LSSA, resource spread multiple access (RSMA) also assigns unique signatures to separate different users and spreads their signals over all the available time and frequency resources [116], [117]. The unique signatures can be power, spreading/scrambling codes with good correlation properties, interleaver, or their combinations, and interference–cancellation-type receivers are utilized. Depending on different application scenarios, RSMA can include the following [118]:

- **Single Carrier RSMA**: It is optimized for battery power consumption and link budget extension by utilizing single carrier waveforms and very low peak-to-average power ratio (PAPR) modulations. It allows grant-free transmission and potentially allows asynchronous access.
- **Multi carrier RSMA**: It is optimized for low-latency access and allows for grant-free transmissions.

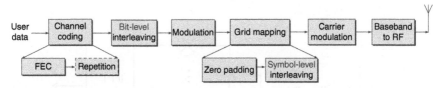

Figure 4.18 The schematic of IGMA transmitter.

4.5.4 Interleaver-Based NOMA

Interleave-grid multiple access (IGMA) is an interleaver-based multiple access scheme. It can distinguish different users based on different bit-level interleavers, different grid mapping patterns, or the combinations of these two techniques [119]. The typical transmitter system structure using IGMA is shown in Figure 4.18. Specifically, the channel coding process can be either simple repetition (spreading) of moderate coding rate forward error correction (FEC) or low coding rate FEC, and the grid mapping process may vary from sparse mapping based on zero padding to symbol-level interleaving, which could provide another dimension for user multiplexing. Compared to the need of well-designed code word or code sequences, the sufficient source of bit-level interleavers and/or grid mapping patterns not only provide enough scalability to support different connection densities but also flexibility to achieve good balance between gain from channel coding and benefit from sparse resource mapping. Moreover, the symbol-level interleaving randomizes the symbol sequence order, which may further bring benefit in terms of combating frequency-selective fading and leveraging intercell interference. Besides, the relatively low-complexity multiuser detector can be applied and sparse grid mapping pattern could further reduce detection complexity.

Another interleaver-based multiple access scheme, interleave-division multiple access (IDMA), has also been proposed. Explicitly, IDMA performs the interleaving of the chips after the symbols are multiplied by the spreading sequences. As shown in Ref. [120], compared to CDMA, IDMA is capable of achieving about 1 dB E_b/N_0 gain at a BER of 10^{-3} in highly loaded systems having a normalized userload of 200%. The gain is mostly attributable to the fact that chip interleaving results in an increased diversity gain compared to conventional bit interleaving.

4.5.5 Spreading-Based NOMA

There are also many other NOMA schemes based on spreading codes, which are consistent with the concept of aforementioned code domain NOMA, and nonorthogonal coded multiple access (NCMA) is one of them [121]. NCMA is based on resource spreading by using nonorthogonal spreading codes with

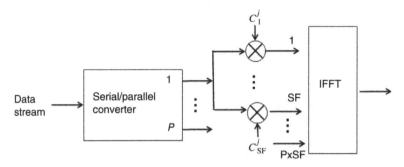

Figure 4.19 NOCA transmitter structure.

minimum correlation, which is obtained by Grassmannian line packing problem [122]. Using methods like exploiting additional layers through superposed symbol, it can provide additional throughput and improve connectivity with small loss of block error ratio (BLER). Besides, since the receiver of NCMA system adopts parallel interference cancellation (PIC), the multiuser detection can be implemented with low complexity. Consequently, NCMA is very suitable for scenarios like huge connections with small packet in massive machine-type communication (mMTC), or for reducing the collision probability in contention-based multiple access.

nonorthogonal coded access (NOCA) is also a spreading-based multiple access scheme [123]. Similar to other spreading-based schemes, the basic idea of NOCA is that the data symbols are spread using nonorthogonal sequences before transmission, which can be applied in frequency domain and/or time domain based on configuration. The basic transmitter structure is shown in Figure 4.19, where SF denotes the spreading factors and C_j is the spreading sequence of the jth user. Specifically, the original modulated data sequence is first converted into P parallel sequences, then each sequence is mapped onto SF subcarriers. In order to meet the requirements under different scenarios, the spreading factors can be adaptively changed.

Group orthogonal coded access (GOCA) belongs to sequence-based nonorthogonal multiple access[124]. GOCA sequences can be divided into different groups, and each group has same orthogonal sequence set and different nonorthogonal sequence sets. To keep orthogonality between orthogonal sequences in the same group, GOCA transmitter uses localized frequency domain repetition or time-domain repetition. Because of significant multiuser interference reduction due to the use of orthogonal sequences in the same group, GOCA with SIC receiver can achieve good performance in high overloading.

4.5.6 Bit Division Multiplexing

Other than the NOMA schemes mentioned above, BDM [103] is also proposed, which is particularly useful for downlink transmission. Its basic concept relies on hierarchical modulation, but the resources of multiplexed users are partitioned at the bit level instead of the symbol level. Strictly speaking, the resource allocation of BDM is orthogonal in the bit domain, but multiuser signals can share the same constellation, which implies that they are superimposed in the symbol domain.

4.5.7 CS-Based NOMA

As a further development, CS can be considered as an alternative technique in the detection of user activity and signal separation designed for NOMA schemes by exploiting either the user activity sparsity or data sparsity [125], [126]. A range of CS-based random access has been conceived recently, such as the family of asynchronous random access protocols [127] and compressive random access [128]. Additionally, in Ref. [129], random multiple access relying on CS was invoked for maximizing the system's total throughput. Furthermore, the attainable throughput associated with different amounts of channel knowledge was discussed, which provided useful insights into the quantitative benefits of CS in the context of throughput maximization in random multiple access schemes. Furthermore, the joint detection of both node activity and the received data was proposed in Ref. [130] for machine-type communication, which exploited the expected sporadic nature of communication. All in all, CS is expected to play an important role in NOMA schemes.

4.5.8 Miscellaneous NOMA Schemes

Apart from the NOMA schemes introduced above, there are five other schemes proposed by different companies during Rel-14 3GPP NR Study Item, which are listed in Table 4.1. Among these five NOMA schemes, three of them are spreading-based NOMA, namely, low-density spreading-signature vector extension (LDS-SVE) proposed by Fujitsu [132], frequency-domain spreading (FDS) [133], and low code rate spreading (LCRS) [133], both proposed by Intel. They spread user symbols on multiple RBs to get more diversity. Both repetition division multiple access (RDMA) from MTK [134] and interleaver division multiple access (IDMA) proposed by Nokia [135] are interleaver-based NOMA schemes. Further more, RDMA can separate different users' signals and utilize both time and frequency diversity by cyclic-shift repetition of the modulated symbols, and IDMA realizes MUD by exploring user-specific interleavers.

Table 4.1 Other 5 NOMA schemes proposed during Rel-14 3GPP NR Study Item.

NOMA schemes	Company	Full Name	Uplink (UL) Downlink (DL)
LDS-SVE [[132]]	Fujitsu	Low-density spreading-signature vector extension	UL/DL
FDS [[133]]	Intel	Frequency-domain spreading	UL
LCRS [[133]]	Intel	Low code rate spreading	UL
RDMA [[134]]	MTK	Repetition-division multiple access	UL
IDMA [[135]]	Nokia	Interleaver-division multiple access	UL

Table 4.2 Comparison of power-domain NOMA schemes.

NOMA Schemes	Basic NOMA with SIC	Power-domain NOMA		
		MIMO-NOMA	Cooperative NOMA	Network NOMA
Receiver	SIC	Spatial filter and SIC	Multi-stage SIC	SIC
Receiver complexity	Polynomial order	Polynomial order	Polynomial order	Polynomial order
Spreading gain	No	No	No	No
Shaping gain	No	No	No	No
Uplink (UL) or downlink (DL)	UL/DL	UL/DL	DL	DL
Characteristics	Basic NOMA	MIMO improving spectral efficiency	Strong users serving as relay	NOMA used in multi-cell
Design considerations	Power allocation; user pairing; interference cancellation			

4.6 Comparison and Trade-Off Analysis of NOMA Solutions

In this section, we provide a detailed trade-off analysis of different NOMA schemes mentioned above. Particularly, we first compare the schemes of power-domain NOMA and code-domain NOMA, respectively, and then analyze the pros and cons of these two families of NOMA, so as to provide valuable insights into this promising technology.

Table 4.3 Comparison of code-domain NOMA schemes.

			Code-domain NOMA		
NOMA schemes	**LDS-CDMA**	**LDS-OFDM**	**SCMA**	**MUSA**	**SAMA**
Receiver	MPA	MPA	MPA	SIC	MPA
Receiver complexity	Exponential order	Exponential order	Exponential order	Polynomial order	Exponential order
Spreading gain	Yes	Yes	Yes	Yes	Yes
Shaping gain	No	No	Yes	No	No
Uplink (UL) or downlink (DL)	UL/DL	UL/DL	UL/DL	UL/DL	UL/DL
Characteristic	Using LDS	Spreading symbols across subcarriers	Mapping bit streams to sparse code words	Code domain multiplexing	Special spreading matrix
Design considerations	Design of spreading sequences or codebooks; complexity reduction				

Specifically, Table 4.2 summarizes the comparison of existing power-domain NOMA schemes. In all the listed dominant power-domain NOMA schemes, NOMA with SIC is the basic one, while others are the evolutions of it. MIMO-NOMA takes advantage of multiple antennas, which can form multiple beams and prominently increase the spectral efficiency. In this scheme, the receiver of the system should first adopt the spatial filter to deal with the inter-beam interference, and then use SIC to detect multiuser signals within the same beam. Cooperative NOMA regards strong users as relays, since they have to detect weak users' signals first in the process of SIC, and thus can transmit those signals to weak users, where a multistage SIC is adopted at the receiver. Furthermore, Network NOMA refers to the multicell scenarios, and a reduced-complexity transmit-precoding scheme was proposed to deal with the interference of the cell-edge users. From Table 4.2, we can see that SIC is used in all power-domain NOMA schemes, and the complexity order is polynomial. In addition, power-domain NOMA schemes have neither shaping gain nor spreading gain, and the most important issues include power allocation, user pairing, and interference cancellation.

The comparison of dominant code-domain NOMA schemes are presented in Table 4.3. Developed from the classic concept of CDMA, LDS-CDMA is designed for limiting the amount of interference imposed on each chip by using low-density spreading sequences (LDS) instead of conventional spreading sequences used in CDMA systems. To fully explore the sparse nature of the

Table 4.4 Trade-off analysis of power-domain NOMA and code-domain NOMA.

NOMA Schemes	Power-Domain NOMA	Code-Domain NOMA
Receiver	SIC	MPA\SIC (MUSA)
Receiver complexity	Polynomial order	Exponential order
Spreading gain	No	Yes
Shaping gain	No	Yes (SCMA)
Signaling cost	Lower (a few-side information to signal the power assignment)	Higher (all the spreading sequences or codebooks have to be known)

LDS, a sub-optimal MPA method with the complexity of $\mathcal{O}(|\mathbb{X}|^w)$ is used at the receiver instead of the optimal maximum likelihood method, where $|\mathbb{X}|$ denotes the cardinality of the constellation set X, and w is the maximum number of non zero signals superimposed on each chip or subcarrier. Similarly, LDS-OFDM can be interpreted as an integrated version of LDS-CDMA and OFDM, where each user's symbol is spread across a carefully selected number of subcarriers. In contrast to the LDS-CDMA and LDS-OFDM, SCMA intrinsically amalgamates the mapping of bit streams to constellation and spreading operations, hence the original bit streams are directly mapped to different sparse code words. Besides, MUSA is another kind of NOMA scheme that multiplies all transmitted symbols of a specific user with the same spreading sequence and transmits symbols of different users over the same time-frequency resources, and SIC is used at the receiver. At last, SAMA has a very special spreading matrix discussed above, and the nonzero elements of any spreading sequence for each user are equal to one. From the theoretical perspective, all code-domain NOMA schemes are capable of achieving a beneficial "spreading gain" with the aid of using spreading sequences, which may also be termed as code words. As a matter of fact, achieving a "spreading gain" is an innate benefit of classic CDMA, which may also be viewed as a low-rate repetition-style channel coding scheme, where the code rate is given by the spreading factor. Moreover, SCMA intrinsically combines the mapping of bit streams to constellation and spreading operations, so it directly maps the original bit streams to different sparse code words and is capable of achieving an extra "shaping gain" due to the optimization of the associated multidimensional constellation [76].

Based on the discussions above, we can analyze the trade-off between power-domain NOMA and code-domain NOMA in details, as presented in Table 4.4. As mentioned above, all code-domain NOMA schemes enjoy the spreading gain, due to the use of spreading sequence, and SCMA is capable of achieving an extra shaping gain, while power-domain NOMAs cannot. However, the complexity of power-domain NOMA is significantly lower than that of code-domain NOMA, as the complexity order of the interference cancellation technique SIC used in power-domain NOMA is polynomial, while the complexity of MPA used in code-domain NOMA is exponential. For instance, the complexity of the SIC-MMSE is $\mathcal{O}(K^3)$, where K is the number of users supported, while that of MPA-based receiver is proportional to $\mathcal{O}(|\mathbb{X}|^w)$, where $|\mathbb{X}|$ denotes the cardinality of the constellation set $\mathbb{X}$, and w is the maximum number of nonzero signals superimposed on each chip or subcarrier. As a result, the complexity of code-domain NOMA will be very high in typical scenarios of massive connectivity. Moreover, the signaling overhead of power-domain NOMA is lower than that of code-domain NOMA, since only a few side information has to be transmitted to deliver the associated power assignment in power-domain NOMA, while all the spreading sequences or codebooks have to be known at the receiver in code-domain NOMA, which will increase the signalling cost, especially when the receiver does not know which users are active.

4.7 Performance Evaluations and Transmission Experiments of NOMA

We have provided some theoretical analyses of NOMA in the previous sections, which shows that NOMA yields better performance than traditional OMA schemes, and this makes it a promising candidate for the 5G wireless communication. In this part, we intend to present some performance evaluations and transmission experiments of NOMA, so as to verify the analytical results.

To assess the effectiveness of NOMA, NTT DOCOMO performed performance evaluations and transmission experiments using prototype equipment [136]– [139]. Specifically, in the experiment, the radio frame configuration was designed based on LTE Release 8, and the targets of these evaluations were Transmission Mode 3 (TM3) and Transmission Mode 4 (TM4), without and with feeding back a user Precoding Matrix Index to the base station, respectively [140].

Researchers also performed the experiment in an indoor radio-wave environment using prototype equipment, as shown in Figure 4.20. In this experiment, both UE1 and UE2 are stationary, and the former was near the base station, while the latter was at a point about 50 m from the base station (to the right outside the view in the photo). It has been shown that NOMA could obtain a

Figure 4.20 External view of NOMA prototype transmission equipment (indoor experiment environment) [140].

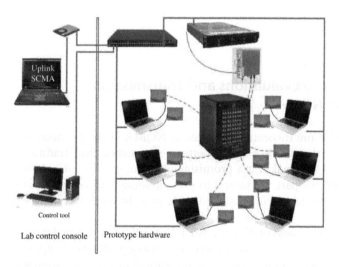

Figure 4.21 The setup of the uplink SCMA demo system [141].

gain of approximately 80% over OFDMA in terms of throughput, when 2×2 SU-MIMO is adopted.

Apart from NTT DOCOMO, Huawei Technologies developed SCMA-based multiuser uplink prototype to verify the advantages of SCMA technology in real communication systems [141]. The demo system consists of 1 base station with 2 antennas for diversity combined receiving, and 12 single-antenna users for uplink access and data transmission. The setup of the prototype is shown in Figure 4.21.

The basic system configurations of the demo are set to align with the current LTE TDD system. In particular, the researchers use LTE TDD configuration

Table 4.5 Specification for SCMA prototype [141].

Mode	Sparse Code Multiple Access
Number of active UEs	12 out of 14
UE transmit power	23 dBm (max) with open-loop power control
Basic waveform	OFDM / F-OFDM
MIMO mode	1 × 2 SIMO
Center frequency/bandwidth	2.6 GHz/20 MHz
Scheduled resource	48 RBs/4 RBs
Code rate	0.3–0.92
SCMA codebook	24 × 8, four points
Frame structure	TDD configuration 1, 4 subframes for PUSCH

Figure 4.22 Hardware setup of the in-lab SCMA demo system group [141].

1 and take the OFDMA as the baseline for performance comparison. The specification of the prototype system is shown in Table 4.5.

The in-lab prototype system is built with soft baseband, which means that all the baseband processing is realized by CPU instead of FPGA/DSP. At the base station side, one server (Huawei Tecal RH2288) is responsible for all the baseband processing and is connected with the standard commercial radio frequency components (Huawei product RRU3232). At the user side, the CPU of 1 laptop (MacBook Pro ME294CH/A) is used to model the processing of baseband for two users, which is connected to two mobile RF modules for testing. A user interface (UI) is developed to show the real-time throughput for each UE, supporting the real-time change of user status and system operation modes as well.

The prototype can run in either mode, OFDMA or SCMA, and support real-time switching from one to the other. To ensure a fair comparison, same data rate is kept for each user to guarantee the same quality of service. It is shown by the test that compared with the orthogonal multiple access baseline of 4G LTE, SCMA technology brings up to 300% overloading gain. For instance, 150% overloading gain can be observed from the fact that, when each user demands 12 physical resource blocks (RBs), a system with a total of 48 RBs can serve at most four users using orthogonal LTE OFDMA. However, with SCMA, the codebook design supports six users with the same amount of data to share the 48 RBs simultaneously, thus the equivalent delivered amount of data is actually 12 = 72 RBs other than 114 = 48 RBs, which results in the throughput gain of about 72/48 = 150%. The 300% gain is supported in a similar way, but needs a different codebook with larger spreading factor and larger number of data layers. In the prototype, a 24×8 SCMA codebook is used to allow 12 users (each with two data streams) to access and transmit simultaneously. For LTE OFDMA, however, only 4 users out of 12 can transmit data. The physical hardware of the in-lab prototype system is shown in Figure 4.22.

Besides fading simulator test in lab, SCMA prototype has been deployed to field trial to evaluate the performance. Specifically, four different test cases are designed, and UEs are deployed at the different locations to evaluate the performance of SCMA in different conditions. All the test cases shown in Figure 4.23 are as follows:

- **Case 1:** 12 UEs closely located in an area without mobility.
- **Case 2:** 12 UEs located in an area with distant separation but no mobility.
- **Case 3:** 12 UEs moving along a road about 120 m away from the BS (open-loop power control gives comparatively medium transmit power at UE).
- **Case 4:** 12 UEs moving along a road about 180 m away from the BS (open-loop power control gives comparatively high power at UE).

In all field trial tests, typical small packets of size 20 bytes (METIS definition) are used as payload for both LTE and SCMA, and the scheduling resources in whole system are limited to 4 RBSs in each subframe. The comparative testing results of OFDM and SCMA are shown in Figure 4.24, which indicates that SCMA achieves nearly 300% throughput gain compared with OFDM.

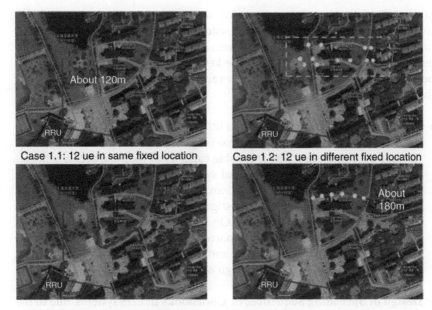

Figure 4.23 Four cases of SCMA field test [141].

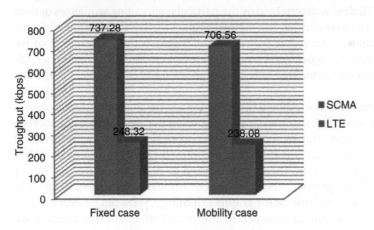

Figure 4.24 SCMA throughput gain over OFDM in field testing [141].

4.8 Opportunities and Future Research Trends

Existing NOMA schemes relying on either power-domain or code-domain multiplexing are capable of improving the spectral efficiency with the aid of nonorthogonal resource sharing. What's more, NOMA techniques are capable of operating in rank-deficient scenarios, which facilitates the support of massive connectivity. Therefore, NOMA solutions are considered as potentially promising 5G candidates. However, there are still numerous challenging problems to be solved. Hence, some of the key challenges of NOMA designs will be highlighted, along with opportunities and future research trends addressing these challenges.

- **Theoretical Analysis** To elaborate a little further, in-depth theoretical analysis is required to provide additional insights to guide and inform the associated system design. The attainable capacity of multiple access schemes constitutes one of the most essential system performance criteria. Specifically, the capacity bounds of code-domain NOMA relying on sophisticated spreading sequences has to be investigated. Similar methods and tools can also be conceived for MC-CDMA. On the other hand, the maximum normalized user load that may be supported is limited by both the achievable interference cancellation capability and the affordable receiver complexity, which is related to the specific design of both the spreading sequences and the receiver.
- **Design of Spreading Sequences or Codebooks** In LDS systems, due to the nonorthogonal resource allocation, there is mutual interference amongst the users. The maximum number of superimposed symbols at each orthogonal resource "index" is determined by the particular spreading sequences or code words of the users, which has a direct impact on the interference cancellation capability achieved at the receiver. Therefore, the factor graph of the message passing algorithm should be optimized to strike a compelling trade off between the normalized user-load supported and the receiver complexity imposed.

  In addition, it has been shown that the message passing algorithm is capable of determining the exact marginal distribution in case of an idealized cycle-free factor graph, and an accurate solution can be obtained with the aid of "locally tree-like" factor graphs, which implies that the cycle girth should be sufficiently high. Graph theory can be used to design cycle-free or "locally tree-like" factor graphs for NOMA without any loss of spectral efficiency. On the other hand, realistic factor graphs exhibiting cycles can be decomposed into cycle-free graphs in some practical applications. In this way, the message passing-based receiver is capable of attaining the optimal performance at the cost of a moderately increased receiver complexity. Additionally, the classic matrix design-principle and low-density parity-check (LDPC) code-based design methods can be invoked for constructing the factor graph of NOMA solutions.

  Apart from the challenge of factor graph design, we should also consider how to choose the nonzero values for each sequence. The nonzero values superimposed at the same resource indices should be distinct. A promising technique is to select different values from a complex-valued constellation for these nonzero elements in order to maintain the maximum possible Euclidean distance.

- **Receiver Design** The complexity of an MPA-based receiver may still become excessive for massive connectivity in 5G. Therefore, some approximate solutions of the MPA can be used for reducing receiver complexity, such as a Gaussian approximation of the interference, which models the interference-plus-noise as a Gaussian distribution. This approximation becomes more accurate, when the number of connections becomes high, as expected in 5G. Additionally, the MPA can be used to jointly detect and channel-decode the received symbols, where the constructed graph consists of variable nodes, observation nodes, and check nodes corresponding to the check equations of the LDPC code. In this way, extrinsic information can be more efficiently exchanged between the decoder and demodulator used at the receiver for improving the signal detection performance.

  For an SIC-based receiver, the associated error propagation may degrade the performance of some users. Therefore, at each stage of SIC, a high-performance nonlinear detection algorithm can be invoked for alleviating the influence of error propagation.

- **Grant-Free NOMA** As illustrated in Section 4.2, a high transmission latency and a high signaling overhead are encountered by an access-grant-based transmission scheme due to the uplink scheduling requests and downlink resource assignments required. It is expected that NOMA is capable of operating without grant-free transmissions at a low transmission latency, at a small signaling overhead, while supporting massive connectivity, especially in case transmitting short packets, as expected in 5G. Hence, contention-based NOMA schemes constitute a promising solution, in which one or more pre-configured resources are assigned to the contending users. On the other hand, integrated protocols – including random back-off schemes – can be considered as a technique of resolving nonorthogonal collisions, while reducing the packet dropping rates. Additionally, without relying on any access-grant procedure, the BS cannot obtain any information on the associated user activity, which however can be fortunately detected by CS-aided recovery algorithms due to the sparsity of user activity.

- **Resource Allocation** In power-domain NOMA, the interference cancellation capability of receivers is closely related to the accuracy of the power allocation scheme. On the other hand, the accuracy of allocating the power of each user directly affects the throughput of both power-domain NOMA and code-domain NOMA. By carefully adjusting the power allocation under a specific total power constraint, the BS becomes capable of flexibly controlling the overall throughput, the cell-edge throughput, and the rate-fairness of the users. The optimal resource allocation scheme has to search through the entire search space of legitimate solutions, and thus the complexity may become excessive. Both dynamic programming algorithms and greedy algorithms may be considered for realizing a near-optimal power allocation operating at a low complexity. Additionally, in order to support various applications, dynamic power allocation constitutes a promising research topic for future work.

- **Extension to MIMO** It is desirable to extend the existing NOMA schemes to their MIMO-aided counterparts, especially to large-scale MIMO systems, in order to further improve the attainable spectral efficiency by exploiting the spatial diversity gain and/or the multiplexing gain of MIMO systems. However, the design of MIMO-aided NOMA techniques is by no means trivial. Consider the power-domain NOMA as an example. Recall that the key idea of power-domain NOMA is to allocate the transmission power to users inversely proportionally to their channel conditions. For scenarios associated with single-antenna nodes, it is possible to compare the users' channel conditions, since channel gains/attenuations are scalars. However, in MIMO scenarios, the channels are represented by a matrix. Hence, it becomes difficult to decide which user's channel is better. This dilemma leads to implementational difficulties for NOMA solutions. This is still a promising open area at the time of writing, with very few solutions proposed in the open literature. A possible solution is to request the BS to form multiple beams, where NOMA techniques are invoked for supporting the users covered by the same directional beam and MIMO precoding/detection is used to cancel the inter beam interference [142]. Another possible solution is to assign different beams to different users individually, where the NOMA power allocation constraint has to be taken into consideration for the design of beamforming [143].

- **Cognitive Radio Inspired NOMA** The advantage of NOMA techniques can be simply illustrated by exploiting the concept of cognitive radio networks. Specifically, the user associated with poorer channel conditions in a NOMA system can be viewed as a primary user in the context of cognitive radio networks. If conventional OMA is used, the bandwidth resources assigned to this primary user, such as the time slots or frequency slots, are solely occupied by this user, and no other users can access these bandwidth resources, even if the primary user has a poor connection to the BS. The benefit of using NOMA is analogous to that of cognitive radio networks, where additional secondary users can be admitted into the specific beam occupied by the primary user. While these secondary users may impose performance degradations on the primary user, the overall system throughput can be significantly improved, particularly if the secondary users have better connections to the BS. By exploiting the appealing concept of cognitive radio networks, we can readily illustrate the performance gain of NOMA over conventional OMA, which significantly simplifies the design of NOMA systems [142]. For example, for the scenarios associated with MIMO schemes or in the presence of cochannel interference, the design of optimal power allocation is difficult, since it is challenging to decide the quality order of the users' channel conditions. The exploitation of cognitive radio networks may impose new constraints on the power allocation, which has to strike a similar throughput and fairness trade off as conventional NOMA.

4.9 Conclusions

In this chapter, we have discussed the key concepts and advantages of NOMA techniques, which constitute one of the promising technologies for future 5G systems. The dominant NOMA schemes have been introduced together with their comparison in terms of their operating principles, key features, receiver complexity, pros and cons, and so on. We also highlighted a range of key challenges, opportunities, and future research trends related to the design of NOMA, including the theoretical analysis, the design of spreading sequences or codebooks, their receiver design, the design issues of access grant-free NOMA, resource allocation schemes, extensions to large-scale MIMOs, and so on. It is expected that NOMA will play an important role in future 5G wireless communication systems supporting massive connectivity and low latency.

References

1 A. Osseiran, F. Boccardi, V. Braun, K. Kusume, P. Marsch, M. Maternia, O. Queseth, M. Schellmann, H. Schotten, H. Taoka, H. Tullberg, M. A. Uusitalo, B. Timus, and M. Fallgren, "Scenarios for 5G mobile and wireless communications: The the vision of the METIS project," *IEEE Commun. Mag.*, vol. 52, no. 5, pp. 26–35, May 2014.

2 F. Boccardi, R. W. Heath, Jr., A. Lozano, T. L. Marzetta, and P. Popovski, "Five disruptive technology directions for 5G," *IEEE Commun. Mag.*, vol. 52, no. 2, pp. 74–80, Feb. 2014.

3 A. W. Scott and R. Frobenius, "Multiple access techniques: FDMA, TDMA, and CDMA," *RF Measurements for Cellular Phones and Wireless Data Systems*, Wiley-Blackwell, pp. 413–429, Jan. 2008.

4 H. Li, G. Ru, Y. Kim, and H. Liu, "OFDMA capacity analysis in MIMO channels," *IEEE Trans. Inform. Theory*, vol. 56, no. 9, pp. 4438–4446, Sep. 2010.

5 K. Higuchi and A. Benjebbour, "Non-orthogonal multiple access (NOMA) with successive interference cancellation for future radio access," *IEICE Trans. Commun.*, vol. E98-B, no. 3, pp. 403–414, Mar. 2015.

6 L. Dai, B. Wang, Y. Yuan, S. Han, C.-L. I, and Z. Wang, "Non-orthogonal multiple access for 5G: Solutionssolutions, challenges, opportunities, and future research trends," *IEEE Commun. Mag.*, vol. 53, no. 9, pp. 74–81, Sep. 2015.

7 A. Benjebbour, Y. Saito, Y. Kishiyama, A. Li, A. Harada, and T. Nakamura, "Concept and practical considerations of non-orthogonal multiple access (NOMA) for future radio access," in *Proc. IEEE Intelligent Signal Processing and Communications Systems (IEEE ISPACS'13)*, Nov. 2013, pp. 770–774.

8 Y. Saito, Y. Kishiyama, A. Benjebbour, T. Nakamura, A. Li, and K. Higuchi, "Non-orthogonal multiple access (NOMA) for future radio access," in *Proc. IEEE Vehicular Technology Conference (IEEE VTC'13 Spring)*, Jun. 2013, pp. 1–5.

9 K. Higuchi and Y. Kishiyama, "Non-orthogonal access with random beamforming and intra-beam SIC for cellular MIMO downlink," in *Proc. IEEE Vehicular Technology Conference (IEEE VTC'13 Fall)*, Sep. 2013, pp. 1–5.

10 N. Nonaka, Y. Kishiyama, and K. Higuchi, "Non-orthogonal multiple access using intra-beam superposition coding and SIC in base station cooperative MIMO cellular downlink," in *Proc. IEEE Vehicular Technology Conference (IEEE VTC'14 Fall)*, Sep. 2014, pp. 1–5.

11 A. Benjebbour, K. Saito, A. Li, Y. Kishiyama, and T. Nakamura, "Non-orthogonal multiple access (NOMA): Conceptconcept, performance evaluation and experimental trials," in *Proc. IEEE International Conference on Wireless Networks and Mobile Communications (IEEE WINCOM'15)*, Oct. 2015, pp. 1–6.

12 A. Benjebbour, A. Li, K. Saito, Y. Saito, Y. Kishiyama, and T. Nakamura, "NOMA: From from concept to standardization," in *Proc. IEEE Conference on Standards for Communications and Networking (IEEE CSCN'15)*, Oct. 2015, pp. 18–23.

13 B. Kim, W. Chung, S. Lim, S. Suh, J. Kwun, S. Choi, and D. Hong, "Uplink NOMA with multi-antenna," in *Proc. IEEE Vehicular Technology Conference (IEEE VTC'15 Spring)*, May 2015, pp. 1–5.

14 Z. Ding, F. Adachi, and H. V. Poor, "The application of MIMO to non-orthogonal multiple access," *IEEE Trans. Wireless Wirel. Commun.*, vol. 15, no. 1, pp. 537–552, Jan. 2016.

15 Z. Ding, L. Dai, and H. V. Poor, "MIMO–NOMA design for small packet transmission in the internet Internet of thingsThings," *IEEE Access*, vol. 4, pp. 1393–1405, Apr. 2016.

16 Y. Lan, A. Benjebboiu, X. Chen, A. Li, and H. Jiang, "Considerations on downlink non-orthogonal multiple access (NOMA) combined with closed-loop SU-MIMO," in *Proc. IEEE Signal Processing and Communication Systems (IEEE ICSPCS'14)*, Dec. 2014, pp. 1–5.

17 C. Yan, A. Harada, A. Benjebbour, Y. Lan, A. Li, and H. Jiang, "Receiver design for downlink non-orthogonal multiple access (NOMA)," in *Proc. IEEE Vehicular Technology Conference (IEEE VTC'15 Spring)*, May 2015, pp. 1–5.

18 K. Saito, A. Benjebbour, Y. Kishiyama, Y. Okumura, and T. Nakamura, "Performance and design of SIC receiver for downlink NOMA with open-loop SU-MIMO," in *Proc. IEEE International Conference on Communication Workshop (IEEE ICCW'15)*, Jun. 2015, pp. 1161–1165.

19 X. Chen, A. Beiijebbour, A. Li, H. Jiang, and H. Kayama, "Consideration on successive interference canceller (SIC) receiver at cell-edge users for non-orthogonal multiple access (NOMA) with SU-MIMO," in *Proc. IEEE Annual International Symposium on Personal, Indoor, and Mobile Radio Communications (IEEE PIMRC'15)*, Aug. 2015, pp. 522–526.

20 Q. Sun, S. Han, C.-L. I, and Z. Pan, "On the ergodic capacity of MIMO NOMA systems," *IEEE Wirel.ess Commun. Lett.*, vol. 4, no. 4, pp. 405–408, Apr. 2015.

21 S. Timotheou and I. Krikidis, "Fairness for non-orthogonal multiple access in 5G systems," *IEEE Signal Process. Lett.*, vol. 22, no. 10, pp. 1647–1651, Mar. 2015.

22 K. Yakou, and K. Higuchi, "Downlink NOMA with SIC using unified user grouping for non-orthogonal user multiplexing and decoding order," in *Proc. IEEE International Symposium on Intelligent Signal Processing and Communication Systems (IEEE ISPACS'15)*, Nov. 2015, pp. 508–513.

23 J. Umehara, Y. Kishiyama, and K. Higuchi, "Enhancing user fairness in non-orthogonal access with successive interference cancellation for cellular downlink," in *Proc. IEEE International Conference on Communication Systems (IEEE ICCS'12)*, Nov. 2012, pp. 324–328.

24 N. Otao, Y. Kishiyama, and K. Higuchi, "Performance of non-orthogonal access with SIC in cellular downlink using proportional fairbased resource allocation," in *Proc. IEEE International Symposium on Wireless Communication Systems (IEEE ISWCS'12)*, Aug. 2012, pp.476–480.

25 X. Chen, A. Benjebbour, A. Li, and A. Harada, "Multi-user proportional fair scheduling for uplink non-orthogonal multiple access (NOMA)," in *Proc. IEEE Vehicular Technology Conference (IEEE VTC'14 Spring)*, May 2014, pp. 1–5.

26 F. Liu, P. Mähönen, and M. Petrova, "Proportional fairness-based user pairing and power allocation for non-orthogonal multiple access," in *Proc. IEEE 26th Annual International Symposium on Personal, Indoor, and Mobile Radio Communications (IEEE PIMRC'15)*, Aug. 2015, pp. 1127–1131.

27 S. Timotheou and I. Krikidis, "Fairness for non-orthogonal multiple access in 5G systems," *IEEE Signal Process. Lett.*, vol. 22, no. 10, pp. 1647–1651, Oct. 2015.

28 Y. Liu, M. Elkashlan, Z. Ding, and G. K. Karagiannidis, "Fairness of user clustering in MIMO non-orthogonal multiple access systems," *IEEE Commun. Lett.*, vol. 20, no. 7, pp. 1465–1468, Apr. 2016.

29 L. Lei, D. Yuan, C. K. Ho, and S. Sun, "Joint optimization of power and channel allocation with non-orthogonal multiple access for 5G cellular systems," in *Proc. IEEE Global Communications Conference (IEEE GLOBECOM'15)*, Dec. 2015, pp. 1–6.

30 J. Choi, "On the power allocation for a practical multiuser superposition scheme in NOMA systems," *IEEE Commun. Lett.*, vol. 20, no. 3, pp. 483–441, Jan. 2016.

31 Y. Hayashi, Y. Kishiyama, and K. Higuchi, "Investigations on power allocation among beams in non-orthogonal access with random beamforming and intra-beam SIC for cellular MIMO downlink," in *Proc. IEEE Vehicular Technology Conference (IEEE VTC'13 Fall)*, Sep. 2013, pp. 1–5.

32 Z. Ding, M. Peng, and H. V. Poor, "Cooperative non-orthogonal multiple access in 5G systems," *IEEE Commun. Lett.*, vol. 19, no. 8, pp. 1462–1465, Jun. 2015.

33 Z. Ding, H. Dai, and H. V. Poor, "Relay Selection selection for Cooperative cooperative NOMA," *IEEE Wireless Wirel. Commun. Lett.*, vol. 5, no. 4, pp. 416–419, Jun. 2016.

34 S. Han, C.-L. I, Z. Xu, and Q. Sun, "Energy efficiency and spectrum efficiency co-design: From from NOMA to network NOMA," *IEEE MMTC E-LetterLett.*, vol. 9, no. 5, pp. 21–24, Sep. 2014.

35 J. Choi, "Non-orthogonal multiple access in downlink coordinated two-point systems," *IEEE Commun. Lett.*, vol. 18, no. 2, pp. 313–316, Jan. 2014.

36 Z. Ding, Z. Yang, P. Fan, and H. V. Poor, "On the performance of non-orthogonal multiple access in 5G systems with randomly deployed users," *IEEE Signal Process. Lett.*, vol. 21, no. 12, pp. 1501–1505, Jul. 2014.

37 Y. Saito, A. Benjebbour, Y. Kishiyama, and T. Nakamura, "System-level performance evaluation of downlink non-orthogonal multiple access (NOMA)," in *Proc. IEEE Personal Indoor and Mobile Radio Communications (IEEE PIMRC'13)*, Sep. 2013, pp. 611–615.

38 A. Benjebbovu, A. Li, Y. Saito, Y. Kishiyama, A. Harada, and T. Nakamura, "System-level performance of downlink NOMA for future LTE enhancements," in *Proc. IEEE Global Communications Conference Workshops (IEEE Globecom Workshops'13)*, Dec. 2013, pp. 66–70.

39 Y. Saito, A. Benjebbour, A. Li, K. Takeda, Y. Kishiyama, and T. Nakamura, "System-level evaluation of downlink non-orthogonal multiple access (NOMA) for non-full buffer traffic model," in *Proc. IEEE Conference on Standards for Communications and Networking (IEEE CSCN'15)*, Oct. 2015, pp. 94–99.

40 A. Benjebbour, A. Li, Y. Kishiyama, H. Jiang, and T. Nakamura, "System-level performance of downlink NOMA combined with SU-MIMO for future LTE enhancements," in *Proc. IEEE Global Communications Conference Workshops (IEEE Globecom Workshops'14)*, Dec. 2014, pp. 706–710.

41 Y. Saito, A. Benjebbour, Y. Kishiyama, and T. Nakamura, "System-level performance of downlink non-orthogonal multiple access (NOMA) under various environments," in *Proc. IEEE Vehicular Technology Conference (IEEE VTC'15 Spring)*, May 2015, pp. 1–5.

42 M. Kimura and K. Higuchi, "System-level throughput of NOMA with SIC in cellular downlink under FTP traffic model," in *Proc. IEEE International*

Symposium on Wireless Communication Systems (IEEE ISWCS'15), Aug. 2015, pp. 1–5.

43 Y. Endo, Y. Kishiyama, and K. Higuchi, "Uplink non-orthogonal access with MMSE-SIC in the presence of inter-cell interference," in *Proc. IEEE International Symposium on Wireless Communication Systems (IEEE ISWCS'12)*, Aug. 2012, pp. 261–265.

44 P. Sedtheetorn and T. Chulajata, "Spectral efficiency evaluation for non-orthogonal multiple access in Rayleigh fading," in *Proc. IEEE International Conference on Advanced Communication Technology (IEEE ICACT'16)*, Jan. 2016, pp. 747–750.

45 Z. Ding, F. Adachi, and H. V. Poor, "Performance of MIMO–NOMA down-link transmissions," in *Proc. IEEE Global Communications Conference (IEEE GLOBECOM'15)*, Dec. 2015, pp. 1–6.

46 A. Li, A. Benjebbour, and A. Harada, "Performance evaluation of non-orthogonal multiple access combined with opportunistic beamforming," in *Proc. IEEE Vehicular Technology Conference (IEEE VTC'14 Spring)*, May 2014, pp. 1–5.

47 K. Yamamoto, Y. Saito, and K. Higuchi, "System-level throughput of non-orthogonal access with SIC in cellular downlink when channel estimation error exists," in *Proc. IEEE Vehicular Technology Conference (IEEE VTC'14 Spring)*, May 2014, pp. 1–5.

48 N. Nonaka, A. Benjebbour, and K. Higuchi, "System-level throughput of NOMA using intra-beam superposition coding and SIC in MIMO downlink when channel estimation error exists," in *Proc. IEEE International Conference On on Communication Systems (IEEE ICCS'14)*, Nov. 2014, pp. 202–206.

49 Z. Yang, Z. Ding, P. Fan, and G. K. Karagiannidis, "On the performance of non-orthogonal multiple access systems with partial channel information," *IEEE Trans. Commun.*, vol. 64, no. 2, pp. 654–667, Feb. 2016.

50 K. Higuchi and Y. Kishiyama, "Non-orthogonal access with successive interference cancellation for future radio access," *IEEE Vehicular Technology Society, Asia Pacific Wireless communications Communications symposium Symposium (IEEE APWCS'12)*, pp. 1–5, Aug. 2012.

51 N. Otao, Y. Kishiyama, and K. Higuchi, "Performance of non-orthogonal access with SIC in cellular downlink using proportional fair-based resource allocation," in *Proc. IEEE Wireless Communication Systems (IEEE ISWCS'12)*, Aug. 2012, pp. 476–480.

52 T. Takeda and K. Higuchi, "Enhanced user fairness using non-orthogonal access with SIC in cellular uplink," in *Proc. IEEE Vehicular Technology Conference (IEEE VTC'11 Fall)*, Sep. 2011, pp. 1–5.

53 R. Hoshyar, F. P. Wathan, and R. Tafazolli, "Novel low-density signature for synchronous CDMA systems over AWGN channel," *IEEE Trans. Signal Process.*, vol. 56, no. 4, pp. 1616–1626, Apr. 2008.

54 D. Guo and C.-C. Wang, "Multiuser detection of sparsely spread CDMA," *IEEE J. Sel. Areas Commun.*, vol. 26, no. 3, pp. 421–431, Apr. 2008.

55 J. Van De Beek and B. M. Popovic, "Multiple access with low-density signatures," in *Proc. IEEE Global Communications Conference (IEEE Globecom'09)*, Dec. 2009, pp. 1–6.

56 R. Razavi, R. Hoshyar, M. A. Imran, and Y. Wang, "Information theoretic analysis of LDS scheme," *IEEE Commun. Lett.*, vol. 15, no. 8, pp. 798–800, Jun. 2011.

57 R. Hoshyar, R. Razavi, and M. Al-Imari, "LDS-OFDM an efficient multiple access technique," in *Proc. IEEE Vehicular Technology Conference (IEEE VTC'10 Spring)*, May 2010, pp. 1–5.

58 M. Al-Imari, P. Xiao, M. A. Imran, and R. Tafazolli, "Uplink non-orthogonal multiple access for 5G wireless networks," in *Proc. 11th International Symposium on Wireless Communications Systems (IEEE ISWCS'14)*, Aug. 2014, pp. 781–785.

59 M. Al-Imari, M. A. Imran, R. Tafazolli, and D. Chen, "Performance evaluation of low density spreading multiple access," in *Proc. IEEE Wireless Communications and Mobile Computing Conference (IEEE IWCMC'12)*, Aug. 2012, pp. 383–388.

60 M. Al-Imari, M. A. Imran, and R. Tafazolli, "Low density spreading for next generation multicarrier cellular systems," in *Proc. IEEE Future Communication Networks (IEEE ICFCN'12)*, Apr. 2012, pp. 52–57.

61 M. Al-Imari, M. A. Imran, R. Tafazolli, and D. Chen, "Subcarrier and power allocation for LDS-OFDM system," in *Proc. IEEE Vehicular Technology Conference (IEEE VTC'11 Spring)*, May 2011, pp. 1–5.

62 M. Al-Imari and R. Hoshyar, "Reducing the peak to average power ratio of LDS-OFDM signals," in *Proc. IEEE Wireless Communication Systems (IEEE ISWCS'10)*, Sep. 2010, pp. 922–926.

63 H. Nikopour and H. Baligh, "Sparse code multiple access," in *Proc. IEEE 24th International Symposium on Personal Indoor and Mobile Radio Communications (IEEE PIMRC'13)*, Sep. 2013, pp. 332–336.

64 Y. Zhou, H. Luo, R. Li, and J. Wang, "A dynamic states reduction message passing algorithm for sparse code multiple access," in *Proc. IEEE Wireless Telecommunications Symposium (IEEE WTS'16)*, Apr. 2016, pp. 1–5.

65 Y. Du, B. Dong, Z. Chen, J. Fang, and X. Wang, "A fast convergence multiuser detection scheme for uplink SCMA systems," *IEEE Commun. Lett.*, vol. 5, no. 4, pp. 388–391, May 2016.

66 H. Mu, Z. Ma, M. Alhaji, P. Fan, and D. Chen, "A fixed low complexity message pass algorithm detector for uplink SCMA system," *IEEE Wirel. Commun. Lett.*, vol. 4, no. 6, pp. 585–588, Aug. 2015.

67 Z. Jia, Z. Hui, and L. Xing, "A low-complexity tree search based quasi-ML receiver for SCMA system," in *Proc. IEEE International Conference on Computer and Communications (IEEE ICCC'15)*, Oct. 2015, pp. 319-323.

68 Y. Liu, J. Zhong, P. Xiao, and M. Zhao, "A novel evidence theory based row message passing algorithm for LDS systems," in *Proc. IEEE International Conference on Wireless Communications & Signal Processing (IEEE WCSP'15)*, Oct. 2015, pp. 1–5.

69 D. Wei, Y. Han, S. Zhang, and L. Liu, "Weighted message passing algorithm for SCMA," in *Proc. IEEE International Conference on Wireless Communications & Signal Processing (IEEE WCSP'15)*, Oct. 2015, pp. 1–5.

70 K. Xiao, B. Xiao, S. Zhang, Z. Chen, and B. Xia, "Simplified multiuser detection for SCMA with sum-product algorithm," in *Proc. IEEE International Conference on Wireless Communications & Signal Processing (IEEE WCSP'15)*, Oct. 2015, pp. 1–5.

71 Y. Du, B. Dong, Z. Chen, J. Fang, and L. Yang, "Shuffled multiuser detection schemes for uplink sparse code multiple access systems," *IEEE Commun. Lett.*, vol. 20, no. 6, pp. 1231–1234, Jun. 2016.

72 A. Bayesteh, H. Nikopour, M. Taherzadeh, H. Baligh, and J. Ma, "Low complexity techniques for SCMA detection," in *Proc. IEEE Global Communications Conference Workshops (IEEE Globecom Workshops'15)*, Dec. 2015, pp. 1–6.

73 S. Zhang, X. Xu, L. Lu, Y. Wu, G. He, and Y. Chen, "Sparse code multiple access: An an energy efficient uplink approach for 5G wireless systems," in *Proc. IEEE Global Communications Conference(IEEE Globecom'14)*, Dec. 2014, pp. 4782–4787.

74 Y. Wu, S. Zhang, and Y. Chen, "Iterative multiuser receiver in sparse code multiple access systems," in *Proc. IEEE International Conference on Communications (IEEE ICC'15)*, Jun. 2015, pp. 2918–2923.

75 B. Xiao, K. Xiao, S. Zhang, Z. Chen, B. Xia, and H. Liu, "Iterative detection and decoding for SCMA systems with LDPC codes," in *Proc. IEEE International Conference on Wireless Communications & Signal Processing (IEEE WCSP'15)*, Oct. 2015, pp. 1–5.

76 M. Taherzadeh, H. Nikopour, A. Bayesteh, and H. Baligh, "SCMA codebook design," in *Proc. IEEE Vehicular Technology Conference (IEEE VTC'14 Fall)*, Sep. 2014, pp. 1–5.

77 L. Yu, X. Lei, P. Fan, and D. Chen, "An optimized design of SCMA codebook based on star-QAM signaling constellations," in *Proc. IEEE International Conference on Wireless Communications & Signal Processing (IEEE WCSP'15)*, Oct. 2015, pp. 1–5.

78 S. Zhang, B. Xiao, K. Xiao, Z. Chen, and B. Xia, "Design and analysis of irregular sparse code multiple access," in *Proc. IEEE International Conference on Wireless Communications & Signal Processing (IEEE WCSP'15)*, Oct. 2015, pp. 1–5.

79 K. Au, L. Zhang, H. Nikopour, E. Yi, A. Bayesteh, U. Vilaipornsawai, J. Ma, and P. Zhu, "Uplink contention based SCMA for 5G radio access," in *Proc. IEEE Global Communications Conference (IEEE Globecom'14)*, Dec. 2014, pp. 1–5.

80 A. Bayesteh, E. Yi, H. Nikopour, and H. Baligh, "Blind detection of SCMA for uplink grant-free multiple-access," in *Proc. IEEE Wireless Communications Systems (IEEE ISWCS'14)*, Aug. 2014, pp. 853–857.

81 H. Nikopour, E. Yi, A. Bayesteh, K. Au, M. Hawryluck, H. Baligh, and J. Ma, "SCMA for downlink multiple access of 5G wireless networks," in *Proc. IEEE Global Communications Conference (IEEE Globecom'14)*, Dec. 2014, pp. 1–5.

82 U. Vilaipornsawai, H. Nikopour, A. Bayesteh, and J. Ma, "SCMA for open-loop joint transmission CoMP," in *Proc. IEEE Vehicular Technology Conference (IEEE VTC'15 Fall)*, Sep. 2015, pp. 1–5.

83 T. Liu, X. Li, and L. Qiu, "Capacity for downlink massive MIMO MU-SCMA system," in *Proc. IEEE International Conference on Wireless Communications & Signal Processing (IEEE WCSP'15)*, Oct. 2015, pp. 1–5.

84 L. Lu, Y. Chen, W. Guo, H. Yang, Y. Wu, and S. Xing, "Prototype for 5G new air interface technology SCMA and performance evaluation," *China Commun.*, Supplement No. 1, pp. 38–48, Dec. 2015.

85 Z. Yuan, G. Yu, and W. Li, "Multi-user shared access for 5G," *Telecommunications Network Technology*, vol. 5, no. 5, pp. 28–30, May 2015.

86 X. Dai, S. Chen, S. Sun, S. Kang, Y. Wang, Z. Shen, and J. Xu, "Successive interference cancelation amenable multiple access (SAMA) for future wireless communications," in *Proc. IEEE International Conference on Communication Systems (IEEE ICCS'14)*, Nov. 2014, pp. 1–5.

87 M. Y. Alias, S. Chen, and L. Hanzo, "Multiple-antenna-aided OFDM employing genetic-algorithm-assisted minimum bit error rate multiuser detection," *IEEE Trans. Veh. Technol.*, vol. 54, no. 5, pp. 1713–1721, Sep. 2005.

88 A. Wolfgang, S. Chen, and L. Hanzo, "Parallel interference cancellation based turbo space-time equalization in the SDMA uplink," *IEEE Trans. Wirel. Commun.*, vol. 6, no. 2, pp. 609–616, Feb. 2007.

89 L. Wang, L. Xu, S. Chen, and L. Hanzo, "Three-stage irregular convolutional coded iterative center-shifting K-best sphere detection for soft-decision SDMA-OFDM," *IEEE Trans. Veh. Technol.*, vol. 58, no. 4, pp. 2103–2109, May 2009.

90 S. Chen, L. Hanzo, and A. Livingstone, "MBER space-time decision feedback equalization assisted multiuser detection for multiple antenna aided SDMA systems," *IEEE Trans. Signal Process.*, vol. 54, no. 8, pp. 3090–3098, Aug. 2006.

91 L. Hanzo, S. Chen, J. Zhang, and X. Mu, "Evolutionary algorithm assisted joint channel estimation and turbo multi-user detection/decoding for OFDM/SDMA," *IEEE Trans. Veh. Technol.*, vol. 63, no. 3, pp. 1204–1222, Mar. 2014.

92 S. Chen, A. Wolfgang, C. J. Harris, and L. Hanzo, "Symmetric RBF classifier for nonlinear detection in multiple-antenna-aided systems," *IEEE Trans. Neural Networks*, vol. 19, no. 5, pp. 737–745, May 2008.

93 A. Wolfgang, J. Akhtman, S. Chen, and L. Hanzo, "Reduced-complexity near-maximum-likelihood detection for decision feedback assisted space-time equalization," *IEEE Trans. Wirel. Commun.*, vol. 6, no. 7, pp. 2407–2411, Jul. 2007.

94 J. Akhtman, A. Wolfgang, S. Chen, and L. Hanzo, "An optimized-hierarchy-aided approximate Log-MAP detector for MIMO systems," *IEEE Trans. Wirel. Commun.*, vol. 6, no. 5, pp. 1900–1909, May 2007.

95 S. Chen, A. Livingstone, H. Q. Du, and L. Hanzo, "Adaptive minimum symbol error rate beamforming assisted detection for quadrature amplitude modulation," *IEEE Trans. Wirel. Commun.*, vol. 7, no. 4, pp. 1140–1145, Apr. 2008.

96 J. Zhang, S. Chen, X. Mu, and L. Hanzo, "Turbo multi-user detection for OFDM/SDMA systems relying on differential evolution aided iterative channel estimation," *IEEE Trans. Commun.*, vol. 60, no. 6, pp. 1621–1633, Jun. 2012.

97 J. Zhang, S. Chen, X. Mu, and L. Hanzo, —-, "Joint channel estimation and multi-user detection for SDMA/OFDM based on dual repeated weighted boosting search," *IEEE Trans. Veh. Technol.*, vol. 60, no. 7, pp. 3265–3275, Jun. 2011.

98 C.-Y. Wei, J. Akhtman, S.-X. Ng, and L. Hanzo, "Iterative near-maximum-likelihood detection in rank-deficient downlink SDMA systems," *IEEE Trans. Veh. Technol.*, vol. 57, no. 1, pp. 653–657, Jan. 2008.

99 A. Wolfgang, J. Akhtman, S. Chen, and L. Hanzo, "Iterative MIMO detection for rank-deficient systems," *IEEE Signal Process. Lett.*, vol. 13, no. 11, pp. 699–702, Nov. 2006.

100 L. Xu, S. Chen, and L. Hanzo, "EXIT chart analysis aided turbo MUD designs for the rank-deficient multiple antenna assisted OFDM uplink," *IEEE Trans. Wirel. Commun.*, vol. 7, no. 6, pp. 2039–2044, Jun. 2008.

101 "5G: Rethink mobile communications for 2020+," *FuTURE Mobile Communication Forum*, Nov. 2014.

102 S. Kang, X. Dai, and B. Ren, "Pattern division multiple access for 5G," *Telecommun. Netw. Technol.*, vol. 5, no. 5, pp. 43–47, May 2015.

103 J. Huang, K. Peng, C. Pan, F. Yang, and H. Jin, "Scalable video broadcasting using bit division multiplexing," *IEEE Trans. Broadcast.*, vol. 60, no. 4, pp. 701–706, Dec. 2014.

104 D. Tse and P. Viswanath, *Fundamentals of Wireless Communication*. 1em plus 0.5em minus 0.4em Cambridge: Cambridge University Press, 2005.

105 T. M. Cover and J. A. Thomas, *Elements of Information Theory*. 1em plus 0.5em minus 0.4em John Wiley & Sons, Inc., 2006.

106 P. P. Bergmans, "A simple converse for broadcast channels with additive white Gaussian noise," *IEEE Trans. Inf. Theory*, vol. 20, no. 2, pp. 279–280, Mar. 1974.

107 Y. Liu, G. Pan, H. Zhang, and M. Song, "On the capacity comparison between MIMO-NOMA and MIMO-OMA," *IEEE Access*, vol. 4, pp. 2123–2129, May 2016.

108 L. Wang, X. Xu, Y. Wu, S. Xing, and Y. Chen, "Sparse code multiple access-: Towards towards massive connectivity and low latency 5G communications," *Telecommun. Netw. Technol.*, vol. 5, no. 5, pp. 6–15, May 2015.

109 L. Li, L. Wang, and L. Hanzo, "Differential interference suppression aided three-stage concatenated successive relaying," *IEEE Trans. Commun.*, vol. 60, no. 8, pp. 2146–2155, May 2012.

110 Z. Xiao, L. Dai, P. Xia, J. Choi, and X. Xia, "Millimeter-Wave wave communication with non-orthogonal multiple access for 5G," submitted to *IEEE Wirel. Commun.* doi: arXiv:1709.07980.

111 B. Wang, L. Dai, Z. Wang, N. Ge, and S. Zhou, "Spectrum and energy efficient beamspace MIMO-NOMA for millimeter-wave communications using lens antenna array," *IEEE J. Sel. Areas Commun.*, vol. 35, no. 10, pp. 2370–2382, Oct 2017.

112 A. Marcano, and H. L. Christiansen, "Performance of non-orthogonal multiple access (NOMA) in mmWave wireless communications for 5G networks," in *Proc. IEEE International Conference on Computing, Networking and Communications (IEEE ICNC' 17)*, Jan. 2017, pp. 26–29.

113 L. Zhang, W. Li, Y. Wu, X. Wang, S-I. Park, H. M. Kim, J-Y. Lee, P. Angueira, and J. Montalban, "Layered-division-multiplexing: Theory theory and practice," *IEEE Trans. Broadcast.*, vol. 62, no. 1, pp. 216–232, Mar. 2016.

114 F. R. Kschischang, B. J. Frey, and H.-A. Loeliger, "Factor graphs and the sum-product algorithm," *IEEE Trans. Inf. Theory*, vol. 47, no. 2, pp. 498–519, Feb. 2001.

115 3GPP, "Low code rate and signature based multiple access scheme for New new Radioradio," TSG RAN1 #85, Nanjing, China, May 23rd"–27th, May 2016.

116 3GPP, "Discussion on multiple access for new radio interface," TSG RAN WG1 #84bis, Busan, Korea, April 11–15, 11th-15th, Apr. 2016.

117 3GPP, "Initial views and evaluation results on non-orthogonal multiple access for NR uplink," TSG RAN WG1 #84bis, Busan, Korea, April 11th–15th, Apr., 2016.

118 3GPP, "Candidate NR multiple access schemes," TSG RAN WG1 #84b, Busan, Korea, April 11th–15th, Apr., 2016.

119 3GPP, "Non-orthogonal multiple access candidate for NR," TSG RAN WG1 #85, Nanjing, China, May 23rd–27th, May 2016.

120 K. Kusume, G. Bauch, and W. Utschick, "IDMA vs. CDMA: Analysis analysis and comparison of two multiple access schemes," *IEEE Trans. Wirel. Commun.*, vol. 11, no. 1, pp. 78–87, Jan. 2012.

121 3GPP, "Considerations on DL/UL multiple access for NR," TSG RAN WG1 #84bis, Busan, Korea, April 11th–15th, Apr. 2016.

122 A. Medra and T. N. Davidson, "Flexible codebook design for limited feedback systems via sequential smooth optimization on the grassmannian manifold," *IEEE Trans. Signal Process.*, vol. 62, no. 5, pp. 1305–1318, Mar. 2014.

123 3GPP, "Non-orthogonal multiple access for nNew Radioradio," TSG RAN WG1 #85, Nanjing, China, May 23 –27th, May 2016.

124 3GPP, "New uplink non-orthogonal multiple access schemes for NR," TSG RAN WG1 #86, Gothenburg, Sweden, Aug. 22nd–26th, Aug., 2016.

125 B. Wang, L. Dai, Y. Yuan, and Z. Wang, "Compressive sensing based multi-user detection for uplink grant-free non-orthogonal multiple access," in *Proc. IEEE Vehicular Technology Conference (IEEE VTC'15 Fall)*, Sep. 2015, pp. 1–5.

126 B. Wang, L. Dai, T. Mir, and Z. Wang, "Joint user activity and data detection based on structured compressive sensing for NOMA," *IEEE Commun. Lett.*, vol. 20, no. 7, pp. 1473–1476, Jul. 2016.

127 V. Shah-Mansouri, S. Duan, L.-H. Chang, V. W. Wong, and J.-Y. Wu, "Compressive sensing based asynchronous random access for wireless networks," in *Proc. IEEE Wireless Communications and Networking Conference (IEEE WCNC'13)*, Apr. 2013, pp. 884–888.

128 G. Wunder, P. Jung, and C. Wang, "Compressive random access for post-LTE systems," in *Proc. IEEE International Conference on Communications Workshops (IEEE ICC'14)*, June. 2014, pp. 539–544.

129 J.-P. Hong, W. Choi, and B. D. Rao, "Sparsity controlled random multiple access with compressed sensing," *IEEE Trans. Wirel. Commun.*, vol. 14, no. 2, pp. 998–1010, Feb. 2015.

130 "Components of a new air interface -: building blocks and performance," *Mobile and Wireless Communications Enablers for the Twenty-Twenty Information Society METIS*, Mar. 2014.

131 B. Wang, K. Wang, Z. Lu, T. Xie, and J. Quan, "Comparison study of non-orthogonal multiple access schemes for 5G," in *Proc. IEEE International Symposium on Broadband Multimedia Systems and Broadcasting (IEEE BMSB'15)*, Jun. 2015, pp. 1–5.

132 3GPP, "Initial LLS results for UL non-orthogonal multiple access," TSG RAN WG1 #85, Nanjing, China, May 23rd-27th, May, 2016.

133 3GPP, "Multiple access schemes for new radio interface," TSG RAN WG1 #84bis, Busan, South Korea, 11th-15th, Apr. 2016.

134 3GPP, "New uplink non-orthogonal multiple access schemes for NR," TSG RAN WG1 #86, Gothenburg, Sweden, 22nd-26th, Aug. 2016.

135 3GPP, "Performance of Interleave Division Multiple Access (IDMA) in Combination with OFDM Family Waveforms," TSG RAN WG1 #84bis, Busan, South Korea, 11th–15th, Apr., 2016.

136 Y. Saito, A. Benjebbour, Y. Kishiyama, and T. Nakamura, "System-level performance evaluation of downlink non-orthogonal multiple access (NOMA)," in *Proc. IEEE Annu. Symp. PIMRC*, London, U.K., Sep. 2013, pp. 611–615.

137 Y. Saito, A. Benjebbour, Y. Kishiyama, and T. Nakamura, "System-level performance evaluation of downlink non-orthogonal multiple access (NOMA) under various environments," in *Proc. IEEE Vehicular Technology Conference (IEEE VTC-Spring' 15)*, May 2015, pp. 1–5.

138 K. Saito, A. Benjebbour, A. Harada, Y. Kishiyama, and T. Nakamura, "Link-level performance evaluation of downlink NOMA with SIC receiver considering error vector magnitude," in *Proc. IEEE Vehicular Technology Conference (IEEE VTC-Spring' 15)*, May 2015, pp. 1–5.

139 K. Saito, A. Benjebbour, Y. Kishiyama, Y. Okumura, and T. Nakamura, "Performance and design of SIC receiver for downlink NOMA with open-Loop SU-MIMO," in *Proc. IEEE International Conference on Communication Workshop (IEEE ICCW' 15)*, June. 2015, pp. 1161–1165.

140 A. Benjebbour, K. Saito, Y. Saito, and Y. Kishiyama, "5G radio access technology," *NTT DOCOMO Technical. J.*, vol. 17, no. 4, pp. 16–28, 2015.

141 L. Lu, Y. Chen, W. Guo, H. Yang, Y. Wu, and S. Xing, "Prototype for 5G new air interface technology SCMA and performance evaluation", *China Commun.*, vol. 12, no. supplement, pp. 38–48, Dec. 2015.

142 Z. Ding, F. Adachi, and H. V. Poor, "The application of MIMO to non-orthogonal multiple access," *IEEE Trans. Wirel. Commun.*, vol. 15, no. 1, pp. 537–552, Sep. 2015.

143 M. F. Hanif, Z. Ding, T. Ratnarajah, and G. K. Karagiannidis, "A minorization-maximization method for optimizing sum rate in non-orthogonal multiple access systems," *IEEE Trans. Signal Process.*, vol. 64, no. 1, pp. 76–88, Sep. 2015.

Linglong Dai received his B.S. degree from Zhejiang University in 2003, M.S. degree (with the highest honor) from the China Academy of Telecommunications Technology (CATT) in 2006, and Ph.D. degree (with the highest honor) from Tsinghua University, Beijing, China, in 2011. From 2011 to 2013, he was a Postdoctoral Research Fellow with the Department of Electronic Engineering, Tsinghua University, where he has been an Assistant Professor since July 2013 and then an Associate Professor since June 2016. His current research interests include massive MIMO, millimeter-wave communications, multiple access, and machine learning. He has published over 60 IEEE journal papers and over 40 IEEE conference papers. He also holds 15 granted patents. He has coauthored the book *mmWave Massive MIMO: A Paradigm for 5G* (Academic Press, Elsevier, 2016). He has received five conference Best Paper

Awards at IEEE ICC 2013, IEEE ICC 2014, IEEE ICC 2017, IEEE VTC 2017-Fall, and IEEE ICC 2018. He has also received the Outstanding Ph.D. Graduate of Tsinghua University Award in 2011, the Excellent Doctoral Dissertation of Beijing Award in 2012, the National Excellent Doctoral Dissertation Nomination Award in 2013, the URSI Young Scientist Award in 2014, the IEEE Transactions on Broadcasting Best Paper Award in 2015, the Second Prize of Science and Technology Award of China Institute of Communications in 2016, the Electronics Letters Best Paper Award in 2016, the IEEE Communications Letters Exemplary Editor Award in 2017, the National Natural Science Foundation of China for Outstanding Young Scholars in 2017, and the IEEE ComSoc Asia-Pacific Outstanding Young Researcher Award in 2017. He currently serves as Editor of *IEEE Transactions on Communications, IEEE Transactions on Vehicular Technology*, and *IEEE Communications Letters*. He also serves as Guest Editor of *IEEE Journal on Selected Areas in Communications, IEEE Journal of Selected Topics in Signal Processing*, and *IEEE Wireless Communications*. He is an IEEE Senior Member. Particularly, he is dedicated to reproducible research and has made a large amount of simulation code publicly available.

Bichai Wang received her B.S. degree in Electronic Engineering from Tsinghua University, Beijing, China, in 2015. She is currently working towards her Ph.D. degree in the Department of Electronic Engineering, Tsinghua University. Her research interests are in wireless communications, with emphasis on new multiple access techniques mmWave massive MIMO, and machine learning for wireless communications. She received the Freshman Scholarship of Tsinghua University in 2011, the Academic Merit Scholarships of Tsinghua University in 2012, 2013, and 2014, respectively, the Excellent Thesis Award of Tsinghua University in 2015, the National Scholarship of China in 2016, the IEEE VTC 2017-Fall Best Student Paper Award in 2017, and the IEEE Transactions on Communications Exemplary Reviewer Award in 2017.

Ruicheng Jiao received his B.S. degree in Physics from Tsinghua University in 2016. He is currently working towards his Ph.D. degree in the Department of Electronic Engineering, Tsinghua University. His research interests are in wireless communications, especially new multiple access techniques, such as nonorthogonal multiple access (NOMA), and machine learning for wireless communications. He received the IEEE Communications Letters Exemplary Reviewer Award in 2017.

Zhiguo Ding received his B.Eng degree in Electrical Engineering from the Beijing University of Posts and Telecommunications in 2000, and Ph.D. degree in Electrical Engineering from Imperial College London in 2005. From July 2005 to April 2018, he was working in Queen's University Belfast, Imperial College, Newcastle University and Lancaster University. Since April 2018, he has been with the University of Manchester as a Professor in Communications. From October 2012 to September 2018, he has also been an academic visitor in Princeton University. Prof. Ding's research interests are 5G networks, game theory, cooperative and energy harvesting networks, and statistical signal processing. He is serving as an Editor for *IEEE Transactions on Communications, IEEE Transactions on Vehicular Technology, and Journal of Wireless Communications and Mobile Computing and was an Editor for IEEE Wireless Communication Letters, IEEE Communication Letters from 2013 to 2016.* He received the best paper award in IET ICWMC-2009 and IEEE WCSP-2014, the EU Marie Curie Fellowship 2012-2014, the Top IEEE TVT Editor 2017 and IEEE Heinrich Hertz Award 2018.

Shuangfeng Han received his M.S. and Ph.D. degrees in Electrical Engineering from Tsinghua University, in 2002 and 2006, respectively. He joined Samsung Electronics as a Senior Engineer in 2006, where he was involved in MIMO and multi-BS MIMO. Since 2012, he has been a Senior Project Manager with the Green Communication Research Center, China Mobile Research Institute. His research interests are green 5G, massive MIMO, full duplex, nonorthogonal multiple access, and EE-SE co-design. He is currently the Vice Chair of the Wireless Technology Work Group in the China's IMT-2020 (5G) Promotion Group.

Chih-Lin I received her Ph.D. degree in electrical engineering from Stanford University. She has been with multiple world-class companies and research institutes leading research and development, including AT&T Bell Laboratories; Director of AT&T HQ; Director of ITRI Taiwan; and VPGD of ASTRI Hong Kong. Her current research interests center around green, soft, and open. She received the IEEE Transactions on Communications Stephen Rice Best Paper Award, was a winner of the CCCP National 1000 Talent Program, and received the 2015 Industrial Innovation

Award of the IEEE Communication Society for Leadership and Innovation in Next-Generation Cellular Wireless Networks. In 2011, she joined China Mobile as its Chief Scientist of Wireless Technologies, established the Green Communications Research Center, and launched the 5G Key Technologies Research and Development. She was a professor with National Chiao Tung University, an adjunct professor at Nanyang Technological University, and is currently an adjunct professor with the Beijing University of Posts and Telecommunications. She is the Chair of the FuTURE 5G SIG, an Executive Board Member of Green-Touch, a Network Operator Council Founding Member of ETSI NFV, a Steering Board Member of WWRF, a Steering Committee Member and the Publication Chair of the IEEE 5G Initiative, a member of the IEEE ComSoc SDB, SPC, and CSCN-SC, and a Scientific Advisory Board Member of Singapore NRF. She is spearheading major initiatives, including 5G, C-RAN, high energy efficiency system architectures, technologies and devices, and green energy. She was an area editor of the IEEE/ACM Transactions on Networking, an Elected Board Member of IEEE ComSoc, the Chair of the ComSoc Meetings and Conferences Board, and the Founding Chair of the IEEE WCNC Steering Committee.

5

Code Design for Multiuser MIMO

Guanghui Song,[1] Yuhao Chi,[2] Kui Cai,[3] Ying Li,[2] and Jun Cheng[1]

[1] *Department of Intelligent Information Engineering and Sciences, Doshisha University, Kyoto, Japan*
[2] *State Key Laboratory of Integrated Services Networks, Xidian University, Xi'an, China*
[3] *Science and Math Cluster, Singapore University of Technology and Design, Singapore*

Multiuser multiple-input multiple-output (MIMO) is a crucial technique to increase the channel capacity and improve communication qualities in the fifth-generation (5G) wireless communication systems. Its capacity gain relays on permitting a great deal of signals from multiple users to be multiplexed and providing a large multiantenna receive power gain. In this chapter, a low-complexity coding scheme, called multiuser repetition-aided irregular repeat-accumulate (IRA) code, and its code design are proposed for multiuser MIMO systems to further enhance their system reliability. A major difference between conventional coding scheme is that not only parity checks, which are generated by an IRA encoder, but also repetitions are used in each user's codeword to reduce the coding and decoding complexities. Repetition is a simple way to construct a low-rate code and is shown to be beneficial for multiuser decoding iteration. With a deliberate control of the fraction of repetitions in the code-word and a degree distribution optimization of the IRA encoder, near capacity performance is achieved. It is shown that as the number of users or transmission antenna increases, more repetitions can be used, therefore, very low encoding and decoding complexities are required.

5G Networks: Fundamental Requirements, Enabling Technologies, and Operations Management, First Edition.
Anwer Al-Dulaimi, Xianbin Wang, and Chih-Lin I.

5.1   Introduction

Future mobile networks are expected to achieve significantly increased capacity and reduced latency to support the rapid growth of mobile data traffic. New spectral efficient techniques, such as nonorthogonal multiple access (NOMA) in combination with multiuser multiple-input multiple-output (MIMO) [1–4], are considered as a cost-effective solution to fulfill these stringent requirements in the fifth-generation (5G) wireless networks. In the multiuser MIMO systems, multiple mobile users simultaneously transmit messages to the base station via multiple antennas and their signals are modulated by the same frequency carrier wave. Each user's message is separately encoded according to its own encoder at a certain coding rate. The coded signal are transmitted in a multiplexed way through multiple transmit antennas. The base station receives multiple superimposed signals from the users with Gaussian noise via multiple receive antennas, based on which a multiuser detection and decoding are performed to eliminate the Gaussian noise and recover each user's message.

Proper design of the codes for all the mobile users is critical in enhancing the system performance in the means of increasing the communication rate as well as elevating the reliability, which is so called approaching the multiuser MIMO channel capacity. In a conventional way, users are modulated with different power levels. The base station can first decode the users with high-power level by regarding the low-power users as noise. If the high-power user is decoded successfully, its signal can be reconstructed and removed from the receive antennas. Then the user with lower power could be decoded. In this way, all the users could be decoded successively. This decoding scheme is called successive interference cancellation (SIC). In SIC, each user can use a point-to-point channel capacity approaching code to achieve optimal. However, the SIC approach needs a perfect power control for each user, thus, a channel state information should be aware at the transmitter. Moreover, the decoding performance is very sensitive to the channel estimation errors [1][5]. In practice, SIC has other problems, such as error propagation [6]. As a result, more robust joint multiuser decoding scheme and joint multiuser code design are required. This motivates us to design good multiuser code for joint multiuser decoding.

Conventional multiuser code constructions, when the number of users and transmit antennas are large, are based on spreading, that is, a repetition code is serially concatenated on a channel code for each user [7–9]. Usually, the received signal-to-interference-plus-noise (SINR) for each user at the base station is very low due to a mass of interferences from the other users. The repetition code can provide multiple receive observations for each coded bits. After a maximum-ratio combining (MRC), the SINR can be significantly improved so that the channel decoding can be validated. However, theoretical analysis [10] shows that spreading in the means of serial concatenation between channel code and repetition may lead to capacity loss.

On the other hand, interleave-division multiple-access (IDMA) gives a very simplified multiuser coding scheme that each user can use the same low-rate code followed by a user-specific interleaving [11][12]. In References [12–18] good code for this structure were designed and showed that near capacity performance can be achieved under iterative decoding. However, a practical problem when constructing low-rate code is that both of the encoding and decoding complexities increase significantly as the code rate decreases, since low rate means large number of parity checks according to conventional coding principle, which inherently leads to high complexity.

In this chapter, we consider code design for multiuser MIMO system. We propose an IDMA-based multiuser repetition-aided irregular repeat-accumulate (IRA) code for this system. In our coding scheme, each user employs a channel code named repetition-aided IRA code followed by an interleaving. Coded bits in a repetition-aided IRA codeword are either repetitions or parity checks, generated by a conventional IRA encoder, of the original message bits. Repetition is an efficient way to construct low-rate code and scarcely encoding and decoding complexity increases. Moreover, we show that repetition is useful for multiuser iterative decoding, that is, it can aid the IRA component decoder to converge.

A main difference between our coding scheme and the conventional channel code serially concatenated with spreading schemes [7–9] is that, rather than spreading for each symbol, we only partially introduce repetitions in our codeword in a controllable manner. With a deliberate control of the fraction of repetitions in the codeword and a degree distribution optimization for the IRA component encoder, near capacity performance is achieved. We show that as the number of users or transmit antennas increase, more repetitions can be used in the codeword, therefore, very low encoding and decoding complexities are required. We give numerical comparisons to demonstrate advantages of our codes in terms of both performance and decoding complexity over previous low-rate constructions.

5.2 Multiuser Repetition-Aided IRA Coding Scheme

Figure 5.1 illustrates a K-user repetition-aided IRA coded MIMO system. Based on the IDMA scheme, the message data of user k is first encoded by a repetition-aided IRA encoder followed by a user-specific interleaver π^k, $k = 1, ..., K$. The repetition-aided IRA code is a low-rate channel code, which is described by a factor graph in Figure 5.2. In the factor graph, variable nodes, labeled by $u_1^k, ..., u_m^k$, denote message data bits. Code length of the repetition-aided IRA

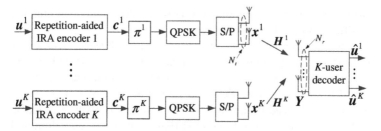

Figure 5.1 K-user repetition-aided IRA coded MIMO system.

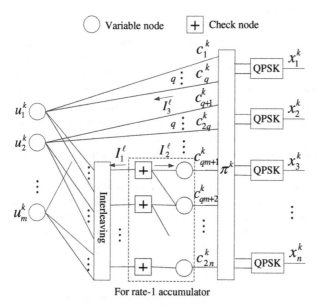

Figure 5.2 Factor graph of repetition-aided IRA code of user k. The graph in the dashed box is corresponding to the rate-1 accumulator.

code illustrated here is $2n$. Coded bits consist two components: $c_1^k, ..., c_{qm}^k$ are the direct repetitions of the original message data bits, that is, each message bit is repeated q times, and $c_{qm+1}^k, ..., c_{2n}^k$ are parity-check outputs from an IRA encoding, which consists an irregular message repetition, an interleaving, and an accumulation. The degree distribution of the irregular repetition is $\lambda(x) = \sum_i \lambda_i x^{i-1}$, which means the fraction of edges connected to a message node with i repetitions is λ_i. Here, the degree distribution is only considered for the edges shed to the interleaving, and the upper qm edges for the repetition component are not involved. Both of the repetition component output and the IRA component come into the user interleaver π^k, after which a QPSK modulation is applied to generate the output signal. Letus return to Figure 5.1. The QPSK signal vector of each user is transmitted parallely via N_t transmit antennas. If the base station has N_r receive antennas, the received signal $y_t \in \mathbb{C}^{N_r \times 1}$ at time t is a length-N_r complex signal vector. Each element of y_t is a superposition of K users' signals:

$$y_t = \sum_{k=1}^{K} H_t^k x_t^k + z_t, \quad t = 1, ..., \frac{n}{N_t}, \tag{5.1}$$

where $x_t^k = (x_{1,t}^k, x_{2,t}^k, ..., x_{N_t,t}^k)^T$, $x_{j,t}^k \in \{a + b\sqrt{-1} | a, b = \pm 1\}$, $j = 1, ..., N_t$, and $z_t \in \mathbb{C}^{N_r \times 1}$ is a complex Gaussian noise vector, that is $z_t \sim \mathcal{CN}(0, N_0)$. We assume Rayleigh fading $H_t^k \in \mathbb{C}^{N_r \times N_t}$, whose entries obey $\mathcal{CN}(0, 1)$, $k = 1, ..., K, t = 1, ..., \frac{n}{N_t}$.

The single-user rate of the repetition-aided IRA code is $R = \frac{\sum_i \lambda_{i/i}}{q \sum_i \lambda_{i/i+1}}$. The communication sum rate of the K users is $R_c^{\text{sum}} = 2KN_tR$.

The novelty of our K user repetition-aided IRA code comparing with conventional coding scheme is that both repetitions and parity checks are used in the codeword, where the repetitions are used for the following reasons:

a) Due to the interuser and interantenna interferences in multiuser MIMO systems, the signal-to-interference-and-noise ratio (SINR) of each user is usually very low. Repetition code can provide a large SINR amplification at the decoder to aid IRA decoding.

b) Repetition increases the encoding and decoding complexities very little, which is crucial for its application in systems with many users.

The main difference between our repetition-aided coding scheme and the conventional spreading scheme [7–9] is that the repetition is partially introduced in the codewords in a flexible manner rather than spreading for each symbol. This allows us to control the fraction of repetition in our codewords so that we can avoid the capacity loss caused by repetition.

5.3 Iterative Decoding and EXIT Analysis

Based on received signal $Y = (y_1, ..., y_{\frac{n}{N_t}})$, a multiuser decoding is performed to recover message vectors $u^k, k = 1, ..., K$. Two low-complexity belief propagation decoding schemes, low density parity check (LDPC)-like and turbo-like decodings, are considered here. In the following, we give a detail description for each decoding scheme and for the aim of code design, its extrinsic information transfer analysis is also given.

Both of the LDPC-like and turbo-like decoding schemes are accomplished by iterative processing between a multiuser detection (MUD) and multiple single-user decodings. The MUD first gives an initial estimation for the signal of each user, based on which single-user decodings are performed in parallel, whose outputs are fed back to MUD for the next iteration of processing. After a preset maximum number of iterations, hard decisions are made at the single-user decoders to obtain the estimated message vectors.

5.3.1 MUD

The MUD gives an initial estimation for each user's signal based on the superimposed signal Y. We use a very low-complexity detection scheme based on Gaussian approximation. Specifically, consider the estimation of $x_{j,t}^k, j = 1, ..., N_t, t = 1, ..., \frac{n}{N_t}, k = 1, ..., K$. Let $L_{A,j}^{k,Re}$ and $L_{A,j}^{k,Im}$ denote *a priori* LLRs of the real part $Re(x_{j,t}^k)$ and imaginary part $Im(x_{j,t}^k)$ of $x_{j,t}^k$, which are from the single-user decoding. At the beginning of iteration, $L_{A,j}^{k,Re} = L_{A,j}^{k,Im} = 0$. The mean and variance for $Re(x_{d,t}^k)$ are given by

$$E[Re(x_{j,t}^k)] = \tanh(L_{A,j}^{k,Re}/2),$$
$$Var[Re(x_{j,t}^k)] = 1 - \tanh(L_{A,j}^{k,Re}/2)^2.$$

Similarly, we can also obtain $E[Im(x_{j,t}^k)]$, $Var[Im(x_{j,t}^k)]$, $E[Re(x_{j,t}^m)]$, $Var[Re(x_{j,t}^m)]$, $E[Im(x_{j,t}^m)]$, and $Var[Im(x_{j,t}^m)]$. Let $\boldsymbol{h}_{j,t}^k$ be the jth column of H_t^k. By regarding the interference signal as complex Gaussian, MUD output as extrinsic LLR about $Re(x_{j,t}^k)$, and $Im(x_{j,t}^k)$ to the single-user decoder [12]

$$L_{E,j}^{k,Re} = \frac{2Re(\overline{\boldsymbol{h}_{j,t}^k}\boldsymbol{y}_t) - M_R(j,t)}{V_R(j,t) + |\boldsymbol{h}_{j,t}^k|^2 \frac{N_0}{2}}, \tag{5.2}$$

$$L_{E,j}^{k,Im} = \frac{2Im(\overline{\boldsymbol{h}_{j,t}^k}\boldsymbol{y}_t) - M_I(j,t)}{V_I(j,t) + |\boldsymbol{h}_{j,t}^k|^2 \frac{N_0}{2}}, \tag{5.3}$$

$$M_R(j,t) = \sum_{\substack{d=1 \\ d \neq j}}^{N_t} \left(Re(\overline{\boldsymbol{h}_{j,t}^k}\boldsymbol{h}_{d,t}^k)E[Re(x_{d,t}^k)] - Im(\overline{\boldsymbol{h}_{j,t}^k}\boldsymbol{h}_{d,t}^k)E[Im(x_{d,t}^k)] \right)$$

$$+ \sum_{\substack{m=1 \\ m \neq k}}^{K} \sum_{b=1}^{N_t} \left(Re(\overline{\boldsymbol{h}_{j,t}^k}\boldsymbol{h}_{b,t}^m)E[Re(x_{b,t}^m)] - Im(\overline{\boldsymbol{h}_{j,t}^k}\boldsymbol{h}_{b,t}^m)E[Im(x_{b,t}^m)] \right),$$

$$M_I(j,t) = \sum_{\substack{d=1 \\ d \neq j}}^{N_t} \left(\text{Re}(\overline{h_{j,t}^k} h_{d,t}^k) E[\text{Im}(x_{d,t}^k)] + \text{Im}(\overline{h_{j,t}^k} h_{d,t}^k) E[\text{Re}(x_{d,t}^k)] \right)$$

$$+ \sum_{\substack{m=1 \\ m \neq k}}^{K} \sum_{b=1}^{N_t} \left(\text{Re}(\overline{h_{j,t}^k} h_{b,t}^m) E[\text{Im}(x_{b,t}^m)] + \text{Im}(\overline{h_{j,t}^k} h_{b,t}^m) E[\text{Re}(x_{b,t}^m)] \right),$$

$$V_R(j,t) = \sum_{\substack{d=1 \\ d \neq j}}^{N_t} \left(\text{Re}(\overline{h_{j,t}^k} h_{d,t}^k)^2 \text{Var}[\text{Re}(x_{d,t}^k)] + \text{Im}(\overline{h_{j,t}^k} h_{d,t}^k)^2 \text{Var}[\text{Im}(x_{d,t}^k)] \right)$$

$$+ \sum_{\substack{m=1 \\ m \neq k}}^{K} \sum_{b=1}^{N_t} \left(\text{Re}(\overline{h_{j,t}^k} h_{b,t}^m)^2 \text{Var}[\text{Re}(x_{b,t}^m)] + \text{Im}(\overline{h_{j,t}^k} h_{b,t}^m)^2 \text{Var}[\text{Im}(x_{b,t}^m)] \right),$$

$$V_I(j,t) = \sum_{\substack{d=1 \\ d \neq j}}^{N_t} \left(\text{Re}(\overline{h_{j,t}^k} h_{d,t}^k)^2 \text{Var}[\text{Im}(x_{d,t}^k)] + \text{Im}(\overline{h_{j,t}^k} h_{d,t}^k)^2 \text{Var}[\text{Re}(x_{d,t}^k)] \right) +$$

$$\sum_{\substack{m=1 \\ m \neq k}}^{K} \sum_{b=1}^{N_t} \left(\text{Re}(\overline{h_{j,t}^k} h_{b,t}^m)^2 \text{Var}[\text{Im}(x_{b,t}^m)] + \text{Im}(\overline{h_{j,t}^k} h_{b,t}^m)^2 \text{Var}[\text{Re}(x_{b,t}^m)] \right),$$

where $\overline{h_{j,t}^k}$ and $|h_{j,t}^k|$ are the conjugate transpose and modulus of $h_{j,t}^k$.

For AWGN with $h_t^k = 1$, the computation complexity of $V_R(j)$ and $V_I(j)$ is $O(K)$. When h_t^k is a general fading, the complexity is $O((KN_t)^2)$. In the following, we further simplify MUD by using the average variance $(V_R(j) + V_I(j))/2$ to replace the accurate variances $V_R(j)$ and $V_I(j)$ of the real and imaginary parts in Eqs. 5.2 and 5.3. We consider the multiuser MIMO system with $N_t = N_r = 1$ but the principle applies to a general multiuser MIMO system. The MUD is approximated as

$$L_E^{k,\text{Re}} \approx \frac{4\text{Re}\left(\overline{h_t^k} y_t - \overline{h_t^k} \sum_{m \neq k} h_t^m E[x_t^m] \right)}{\sum_{m \neq k} |h_t^m|^2 (\text{Var}[\text{Re}(x_t^k)] + \text{Var}[\text{Im}(x_t^k)]) + N_0}, \tag{5.4}$$

$$L_E^{k,\text{Im}} \approx \frac{4\text{Im}\left(\overline{h_t^k} y_t - \overline{h_t^k} \sum_{m \neq k} h_t^m E[x_t^m] \right)}{\sum_{m \neq k} |h_t^m|^2 (\text{Var}[\text{Re}(x_t^k)] + \text{Var}[\text{Im}(x_t^k)]) + N_0}, \tag{5.5}$$

which can be computed in the manner of $O(K)$.

Now, we carry out to calculate the EXIT function of the MUD. Since the statistical EXIT function of detections of both the real and imaginary parts in Eqs. 5.2 and 5.3 are the same due to their symmetry, we formulate the EXIT function based on real part detection (Eq. 5.2). We first derive a conditional EXIT function for a given channel realization of $H_t^1, ..., H_t^K$. Then the ergodic EXIT function is derived by taking expectation of it over all the possible channel realizations.

Since the EXIT function is derived based on Gaussian approximation [19], we assume $L_{A,j}^{k,Re}$, $L_{A,j}^{k,Im}$, and $L_{E,j}^{k,Re}$ as Gaussian variables with means $Re(x_{j,t}^k)m_A$, $Im(x_{j,t}^k)m_A$, $Re(x_{j,t}^k)m_E$ and variances $2m_A$, $2m_A$, $2m_E$, respectively. For given $H_t^1, ..., H_t^K$, since $x_{d,t}^k, d \neq j, x_{b,t}^m, m \neq k$, and z_t are independent of $x_{j,t}^k$,

$$m_E = E[Re(x_t^k)L_{E,k}^{Re}]$$

$$= E[\frac{2|h_k|^2}{\sum_{j \neq k} \left(Re(h_j)^2 Var[Re(x_t^k)] + Im(h_j)^2 Var[Im(x_t^k)] \right) + \frac{N_0}{2}}]$$

$$= E[\frac{2|h_k|^2}{\sum_{j \neq k} \left(Re(h_j)^2(1 - tanh(\frac{L_{A,j}^{Re}}{2})^2) + Im(h_j)^2(1 - tanh(\frac{L_{A,j}^{Im}}{2})^2) \right) + \frac{N_0}{2}}],$$

which is in fact a function of m_A. The above expectation can be approximately obtained by Monte Carlo simulation. Using relation $m_A = J^{-1}(I_A)^2/2$, we obtain conditional EXIT function $T_s(I_A, K, N_0|H_t^1, ..., H_t^K) = J(\sqrt{2m_E})$ as a function of I_A. Then the EXIT function of ESE is expectation $T_s(I_A, K, N_0) = E[T_s(I_A, K, N_0|H_t^1, ..., H_t^K)]$ over matrix $H_t^1, ..., H_t^K$ whose entries obey Rayleigh($\frac{N_0}{2}$). Thus, $T_s(I_A, K, N_0)$ can be approximately obtained by Monte Carlo simulation [19][20]. Actually, the function is determined by the number of users K, transmit antennas N_t, receive antennas N_r, and noise spectrum density N_0.

We consider MUDs for multiuser system with $K = 10$, $N_t = N_r = 1$ over both AWGN and Rayleigh fading channel, and give the EXIT curves in Figure 5.3. The noise density spectra are set to be $N_0 = 5, 10, 15, 20$, and 25. It shows that EXIT curves over Rayleigh fading channel are always below that over AWGN, but the difference is only obvious at a large a priori input I_A. As the noise level increases, the EXIT functions over the Rayleigh fading channel approach that over AWGN, and both approach the single-user case with $K = 1$, which is a horizontal line. This indicates that when the system is very noisy, low-rate codes designed for a single-user AWGN channel, such as turbo-Hadamard code [12–14] and serial concatenated nonbinary LDPC and nonbinary repetition code [15][16], also perform well over multiuser channel. The EXIT functions also predict that when the channel is less noisy, codes work well for

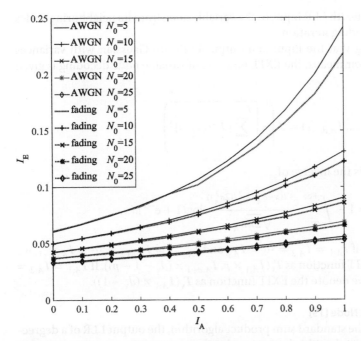

Figure 5.3 EXIT curves of MUD over the AWGN and Rayleigh fading multiuser channels with $K = 10$, $N_t = N_r = 1$, $N_0 = 5, 10, 15, 20$, and 25.

multiuser system that should be quite different from those for a single-user (point-to-point) channel. As the noise level decreases, the difference between good code over AWGN and Rayleigh fading channels should also be more obvious.

5.3.2 LDPC-Like Decoding

In the LDPC-like decoding scheme, the single-user decoding is performed on the code's factor graph, where in each decoding iteration, a local processing at each node is required (Figure 5.2).

5.3.2.1 Variable Node [19]
The local decoding at a variable node with degree d_v is a log-likelihood ratio (LLR) combination. That is, the output of a variable node on the jth edge is

$$L_{j,E} = \sum_{i=1, i \neq j}^{d_v} L_{i,A},$$

(5.6)

where $L_{i,A}$ is the ith LLR input to the variable node from its neighboring nodes in the last decoding iteration.

By assuming that the input and output LLRs are Gaussian with variances being twice their means, the EXIT function of variable node decoding is given by

$$T_v(I_{A,1}, ..., I_{A,d_v-1}) = J\left(\sqrt{\sum_{i=1}^{d_v-1} [J^{-1}(I_{A,i})]^2}\right)$$

where $J^{-1}(*)$ is the inverse of

$$J(\sigma_A) = 1 - \int_{-\infty}^{+\infty} \frac{e^{-((x-\sigma_A^2/2)^2/(2\sigma_A^2))}}{\sqrt{2\pi}\sigma_A} \log(1 + e^{-x})dx.$$

For simplicity, if $I_{A,1} = I_{A,2} = \cdots = I_{A,p}$ and $I_{A,p+1} = I_{A,p+2} = \cdots = I_{A,d_v-1}$, we denote this EXIT function as $T_v(I_{A,1} \times p, I_{A,p+1} \times (d_v - 1 - p))$. If $I_{A,1} = I_{A,2} = \cdots = I_{A,d_v-1}$, we denote the EXIT function as $T_v(I_{A,1} \times (d_v - 1))$.

5.3.2.2 Check Node [19]

According to the standard sum product algorithm, the output LLR of a degree-d_c check node on the jth edge is given by

$$L_{j,E} = 2\tanh^{-1}\left(\prod_{i=1,i\neq j}^{d_c} \tanh\left(\frac{L_{i,A}}{2}\right)\right), \tag{5.7}$$

where $L_{i,A}$ is the ith input LLR from its neighboring nodes in the last iteration.

Also, based on the Gaussian assumption and using a duality property between variable and parity check node decodings, the EXIT function of the check node decoding is

$$T_c(I_{A,1}, ..., I_{A,d_c-1}) \approx 1 - T_v(1 - I_{A,1}, ..., 1 - I_{A,d_c-1}).$$

Similarly, if $I_{A,1} = \cdots = I_{A,d_c-1}$, we denote the EXIT function as $T_c(I_{A,1} \times (d_c - 1))$.

5.3.3 Turbo-Like Decoding

The turbo-like decoding is similar as the LDPC-like decoding except that the local processings in the dashed box in Figure 5.2, corresponding to the deaccumulation, is replaced by a posterior probability processing based on the BCJR algorithm [21]. We refer to it as BCJR processor in this chapter.

We first illustrated the two-state trellis of accumulator in Figure 5.4, where $s_i \in \{0, 1\}$ is the encoder state at time i. Let $L_{i,A}^u$ and $L_{i,A}^c$, $i = 1, ..., 2n - qm$, be

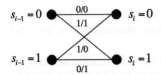

Figure 5.4 Trellis of rate-1 accumulator $\frac{1}{1+D}$. Branch labels are the input message/coded bits corresponding to the state transition.

the left and right LLR inputs to the BCJR processor, that is $L_{i,A}^u$ and $L_{i,A}^c$ are a priori LLRs of the ith input and output bits of the accumulator, respectively. The extrinsic LLR output about the ith information input and output bits are calculated using the standard BCJR algorithm [21, pp. 180–184]

$$L_{i,E}^u = \log \frac{\alpha_{i-1}(0)\gamma_i(0,0)\beta_i(0) + \alpha_{i-1}(1)\gamma_i(1,1)\beta_i(1)}{\alpha_{i-1}(0)\gamma_i(0,1)\beta_i(1) + \alpha_{i-1}(1)\gamma_i(1,0)\beta_i(0)} - L_{i,A}^u, \tag{5.8}$$

$$L_{i,E}^c = \log \frac{\alpha_{i-1}(0)\gamma_i(0,0)\beta_i(0) + \alpha_{i-1}(1)\gamma_i(1,0)\beta_i(0)}{\alpha_{i-1}(0)\gamma_i(0,1)\beta_i(1) + \alpha_{i-1}(1)\gamma_i(1,1)\beta_i(1)} - L_{i,A}^c. \tag{5.9}$$

Forward recursion:

$$\alpha_i(0) = \alpha_{i-1}(0)\gamma_i(0,0) + \alpha_{i-1}(1)\gamma_i(1,0), \tag{5.10}$$

$$\alpha_i(1) = \alpha_{i-1}(0)\gamma_i(0,1) + \alpha_{i-1}(1)\gamma_i(1,1), \tag{5.11}$$

$$\alpha_0(0) = 1, \alpha_0(1) = 0.$$

Backward recursion:

$$\beta_{i-1}(0) = \gamma_i(0,0)\beta_i(0) + \gamma_i(0,1)\beta_i(1), \tag{5.12}$$

$$\beta_{i-1}(1) = \gamma_i(1,0)\beta_i(0) + \gamma_i(1,1)\beta_i(1), \tag{5.13}$$

$$\beta_{2n-qm}(0) = \beta_{2n-qm}(1) = 0.5.$$

We calculate $\gamma_i(s_{i-1}, s_i)$ based on the trellis of Figure 5.4 and a priori LLRs $L_{i,A}^u, L_{i,A}^c$. According to its definition, $\gamma_i(s_{i-1}, s_i)$ is the probability of the state transition (s_{i-1}, s_i). For example, $\gamma_i(1,1)$ is the probability of the ith message and coded bits are 0, 1 according to Figure 5.4, which is the product of $\frac{e^{L_{i,A}^u}}{1+e^{L_{i,A}^u}}$ and $\frac{1}{1+e^{L_{i,A}^c}}$ using the relation between LLR and probability. Similarly, we have

$$\gamma_i(0,0) = \frac{e^{L_{i,A}^u} e^{L_{i,A}^c}}{(1+e^{L_{i,A}^u})(1+e^{L_{i,A}^c})}, \gamma_i(0,1) = \frac{1}{(1+e^{L_{i,A}^u})(1+e^{L_{i,A}^c})}, \tag{5.14}$$

$$\gamma_i(1,0) = \frac{e^{L_{i,A}^u}}{(1+e^{L_{i,A}^u})(1+e^{L_{i,A}^c})}, \gamma_i(1,1) = \frac{e^{L_{i,A}^u}}{(1+e^{L_{i,A}^u})(1+e^{L_{i,A}^c})}. \tag{5.15}$$

and s_i is the encoder state at time i, U_i^+ is the set of pairs (s_{i-1}, s_i) for possible state transition corresponding to that the ith input message bit is +1, and U_i^{-1} and $C_i^{\pm 1}$ are similarly defined.

Note that the statistical LLR transfer functions of the information and output coded bits in the BCJR processor are different, we need two EXIT functions to analyze the BCJR processor. Both EXIT functions can be obtained by Monte Carlo simulation as in Reference [22].

5.3.4 Decoding Complexity Computation

Our single-user decoding complexity is mainly caused by processings of parity-check bits. Directly calculating (Eq. 5.6) for a degree-d_v variable node involves $d_v(d_v - 2)$ numbers of addition/subtraction (+/−) operations in each iteration processing. The complexity is small when d_v is small but becomes very high when d_v is large. Alternatively, when $d_v > 3$, we first calculate a sum of all the d_v input LLRs, then subtract each LLR to get the extrinsic output on each edge. This method only needs $2d_v - 1$ numbers of (+/−) operations to calculate (Eq. 5.6). Each degree-3 check node processing of (Eq. 5.7) can be efficiently realized with 6 +/−, multiplication/division (×/÷), and exponent/logarithm (exp / log) operations [21, pp. 224–225]. Since there are $\frac{m\lambda_j/j}{\sum_i \lambda_i/i}$ degree-$(q + j)$ and $m(\frac{1}{R} - q)$ degree-3 variable nodes and $m(\frac{1}{R} - q)$ degree-3 check nodes in the factor graph of Figure 5.2, the average number of +/− operations per iteration per information bit in LDPC-like decoding is

$$\sum_j \frac{\lambda_j/j}{\sum_i \lambda_i/i}(2(q + j) - 1) + 9(\frac{1}{R} - q) = \frac{11}{R} - 9q - 1,$$

where we assume $q + j > 3$ in our code. Similarly, the numbers of ×/÷ and exp / log operations in LDPC-like decoding are $6(\frac{1}{R} - q)$ and $6(\frac{1}{R} - q)$, respectively, as shown in Table 5.1.

An iteration in turbo-like decoding involves the processing of message variable nodes and a BCJR processor. Our evaluation of BCJR processor complexity is based on the following facts: Eqs. 5.14 and 5.15 involve 2 +/−, 5 ×/÷, 2 exp / log Eqs. 5.10–5.13 involve 4 +/−, 8 ×/÷; and Eqs. 5.8 and 5.9 involve 6 +/−, 10 ×/÷, and 2 exp / log operations per trellis section. By adding the message variable node processing, we obtain the complexity of turbo-like decoding per iteration per information bit in Table 5.1. Overall for a fixed code rate, each kind of operation decreases linearly as repetition number q increases under both the LDPC and turbo-like decodings.

For our later comparison, we also evaluated the decoding complexities of turbo-Hadamard code [12–14] and 64-ary PCC [17] in Table 5.1. In the turbo-Hadamard code, the Hadamard order is 5 and number of convolutional-

Table 5.1 Number of operations per iteration per information bit in a single-user decoding of repetition-aided IRA under LDPC and turbo-like decodings, turbo-Hadamard code (Hadamard order 5, convolutional-Hadamard component number M) [23], and 64-ary PCC (repetition number q', accumulation component number M') [17].

Codes	+/−	×/÷	exp / log
Aided: LDPC-like	$\frac{11}{R} - 9q - 1$	$6(\frac{1}{R} - q)$	$6(\frac{1}{R} - q)$
Aided: turbo-like	$\frac{14}{R} - 12q - 1$	$23(\frac{1}{R} - q)$	$4(\frac{1}{R} - q)$
Turbo-Hadamard	$86.8M$	$17M$	$13.8M$
64-ary PCC	$460M' + 44.5q'$	$226M' + 87.3q' - 10.7$	$2(M' + q')$

Hadamard components is M. We always assume that the simplest 2-state convolutional code (accumulation) is employed in each convolutional-Hadamard component. The complexity in Table 5.1 is based on Algorithm 2 in Reference [23]. In the 64-ary PCC, q' is the number of 64-ary repetitions and M' is the number of 64-ary accumulation components. We assume fast Fourier transform-based nonbinary sum–product decoding with probability-based message updating is employed. The complexities in Table 5.1 are based on the following evaluations: (a) each degree-d_v 2^s-ary variable node processing involves $d_v(2^s - 1)$ of +/− and $(3d_v - 1)2^s$ of ×/÷ operations including normalization, (b) each degree-d_c check node processing involves sd_c2^{s+1} of +/− and $(2d_c - 1)2^s$ of ×/÷ operations, and (c) bit/symbol LLR interconversions involve $s(2 + 2^{s-1})$ of +/−, $s(2 + 2^s) - 2^s$ of ×/÷, and $2s$ of exp / log operations in total.

5.4 Code Optimization Procedure

For a given channel noise level, we find the code with the optimal repetition number q and message node degree distribution sequence $\lambda(x) = \sum_i \lambda_i x^{i-1}$ that give maximum sum rate R_c^{sum} under asymptotic error-free decoding. We give a finite search method based on the differential evolution algorithm to find these optimal parameters.

The decoding error performance can be predicted by the EXIT functions, given in Section 5.3, of the employed decoder. Since the EXIT functions of LDPC-like decoding is computation tractable, we optimize the code for this decoding scheme, while the decoding performance of the optimized codes will be verified by Mont Carlo simulation. Let I_1^ℓ, I_2^ℓ, and I_3^ℓ be mutual information outputs at the check and sum nodes at the ℓth iteration, as labeled in the factor graph of Figure 5.2. Using the EXIT functions formulated in Section 5.3, we get the following recursion formulas:

$$I_1^{\ell+1} = T_c\left(T_v(I_2^\ell, I_3^\ell) \times 2\right),$$

$$I_2^{\ell+1} = T_c\left(\sum_i \lambda_i T_v(I_1^\ell \times (i-1), I_3^\ell \times q), T_v(I_2^\ell, I_3^\ell)\right),$$

$$I_3^{\ell+1} = T_s\left(P\sum_i \lambda_i T_v(I_1^\ell \times i, I_3^\ell \times (q-1)) + (1-P)T_v(I_2^\ell \times 2), K, N_0\right),$$

where $P = \frac{q\sum_i \lambda_i/i}{q\sum_i \lambda_i/i+1}$, and the last equation reflects the fact that with probabilities P and $1 - P$, the input to a sum node is $\sum_i \lambda_i T_v(I_1^\ell \times i, I_3^\ell \times (q-1))$ and $T_v(I_2^\ell \times 2)$. The decoding error rate approaches 0 if $I_i^{\ell_{\max}} = 1, i = 1, 2$, where $\ell_{\max}$ is the maximum iteration number.

For a given number of users K and noise level N_0, $I_i^{\ell_{\max}}, i = 1, 2$, are only determined by the repetition number q and the degree sequence of the IRA encoder. Code optimization solves the following maximization:

$$\text{maximize} \quad R_c^{\text{sum}} = \frac{2KN_t\sum_i \lambda_i/i}{q\sum_i \lambda_i/i + 1}, \qquad (5.16)$$

$$\text{subject to} \quad I_i^{\ell_{\max}} = 1, i = 1, 2, \quad \sum_{i\in\varphi}\lambda_i = 1, 0 \leq \lambda_i \leq 1$$

where φ is a given set of degree types for optimization. Here, we only consider a limited number of degree types from a given finite set φ and find the optimal allocations for each type. To further simplify the search procedure, for each $q = 1, 2, ..., q^{\max}$, we find a local optimal degree sequence that maximizes Eq. 5.16 using a standard differential evolution algorithm [24]. Then the global optimal sequence is found from these local optimal sequences. If the global maximum sum rate is achieved by several different local sequences, we choose the sequence associated with maximum repetition number q since it also gives the minimum coding and decoding complexities (see Table 5.1).

5.5 Numerical Results and Comparisons

In this section, we provide numerical results for our code optimizations for multiuser systems over both AWGN and Rayleigh fading channels. We consider both code design for multiuser systems with single transmission and receive antenna $N_t = N_r = 1$ and multiple transmission and receive antennas. We compare our code with previous low-rate constructions, such as the turbo-Hadamard code and multiuser PCC.

Table 5.2 Optimal repetition number q and message node degree sequences for repetition-aided IRA codes under LDPC-like decoding over AWGN 10-user channel.

R_c^{sum}	1	1.2	1.6	2
q	5	5	5	5
λ_2			0.063021	0.162290
λ_3	0.140694	0.193123	0.228288	0.300876
λ_{10}	0.055662	0.047006	0.111951	
λ_{30}	0.269286	0.385269	0.226877	0.536834
λ_{50}			0.369864	
λ_{100}	0.534358	0.374602		
$\left(\frac{E_b}{N_0}\right)^*_{dB}$	0.13	0.44	1.07	1.84
S.B.	0.00	0.34	1.04	1.76

S.B. denotes Shannon bound.

5.5.1 AWGN Channel

In this section, we consider multiuser systems with $N_t = N_r = 1$ over the AWGN channel. For $K = 10$, message node degree types in $\varphi^*=\{3, 10, 30, 50, 80, 100\}$ or $\varphi=\{2\}\bigcup\varphi^*$ are used in our code optimization. The use of degree-2 message nodes will improve the code threshold although it might slightly deteriorate the word error rate (WER) performance [26]. By carefully setting the noise level N_0 in the optimization procedure, we can obtain the optimal code (with optimal repetition number q and the corresponding optimal degree sequence) under LDPC-like decoding with sum rates $R_c^{sum} = 1, 1.2, 1.6, 2$ in Table 5.2. Note that for sum rates $R_c^{sum} = 1, 1.2$, we only use degree types in φ^*, which is enough to find capacity approaching code, while for $R_c^{sum} = 1.6, 2$, we need to use the degree types in φ, so that all of their decoding thresholds $\left(\frac{E_b}{N_0}\right)^*_{dB}$ of the obtained codes are close to the corresponding capacities of AWGN multiuser channel [25].

To see the multiuser iterative behavior of our code more explicitly, we give an EXIT chart illustration of the decoding in Figure 5.5. We first show the EXIT curve of our rate-$(R = 1/20)$, corresponding to $R_c^{sum} = 1$, repetition-aided IRA code obtained in Table 5.2, which almost matches that of the 10-user MUD (ESE) EXIT curve with $N_0 = 19.4$, corresponding to $\left(\frac{E_b}{N_0}\right)_{dB} = 0.13$. Since our code is a parallel concatenation between a rate-1/5 repetition code and a rate-1/15 nonsystematic IRA code, we also illustrated the EXIT curves of these two components to see how these two components effect the decoding. The component of nonsystematic IRA decoding can provide a significant message

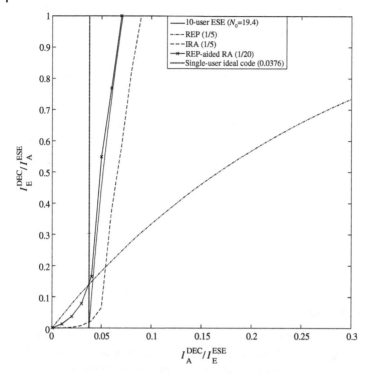

Figure 5.5 EXIT curves of repetition-aided IRA (rate= $1/20$), repetition (REP) ($q = 5$), non-systematic IRA (rate= $1/15$), ideal single-user code (rate= 0.0376), and 10-user ESE over AWGN multiuser channel ($N_0 = 19.4$).

amplification when a priori input I_A^{DEC} is large, but has extremely poor performance when I_A^{DEC} is near 0 (which is usually the area that code works in high multiuser interference environment). On the contrary, the repetition component decoding has a good message amplification at I_A^{DEC} near 0, which compensates the weakness of IRA decoding. Our code utilizes the characteristics of both repetition and IRA codes to achieve successful decoding.

Figure 5.5 also unveils an interesting fact in multiuser decoding, which is also previously observed in the coded CDMA systems [27]: A good multiuser iteration decoding does not require the code of each user being very strong. Weak code may be better for joint decoding. To see this more clearly, let us consider the extreme case of employing a single-user ideal (strong) code, whose EXIT curve is a vertical line, for each user. To iterate successfully in the 10-user system, the vertical line should be on the left of the 10-user ESE EXIT curve, as illustrated in Figure 5.5. Its code rate corresponding to the value of the abscissa at that point is exactly $T_s(I_A^{ESE} = 0, K, N_0) = 0.0376$, which is lower

Table 5.3 Optimal message node degree sequences of repetition-aided IRA codes at sum rate $R_c^{\text{sum}} = 1.5, 1.8$ under LDPC-like decoding over 15 and 20-user channels.

K	15		20	
R_c^{sum}	1.5		1.8	
q	7	1	14	1
λ_2			0.042658	
λ_3	0.188000	0.122982	0.252781	0.113208
λ_{30}	0.265001	0.233257	0.349902	0.122413
λ_{50}				0.109886
λ_{80}			0.354659	0.014119
λ_{100}	0.546999	0.321081		0.207254
λ_{500}		0.322680		0.433121
$\left(\frac{E_b}{N_0}\right)^*_{\text{dB}}$	0.90	1.02	1.58	1.74
S.B.	0.86		1.40	

than $1/20 = 0.05$ achieved by our repetition-aided IRA coding scheme. This is also the mutual information output of ESE when no a priori message is known about the other $K - 1$ users. In other words, there is no iteration improvement between single-user decoding and ESE and each user's decoding treats the other users signal as noise.

To see the impact of the repetition number on the code optimization, in Figure 5.6 we fix the value of $q = 1, 2, ..., 14$ to see the maximum achievable sum rate by optimizing the message node degree distribution. We deliberately fix the noise level at the four thresholds in Table 5.3.4 with $N_0 = 19.4, 15.05, 9.75, 6.55$. The maximum sum rate almost keeps a maximum value when $q \leq 5$ and decreases as $q > 5$. Note that empirically the sum rate for $N_0 = 19.4$ with $q = 1, 2$ can be improved by using more degree types. Therefore, $q = 5$ is a cutoff point, below which capacity of multiuser channel is approached by optimizing the degree distribution. For $q > 5$, there is a sum rate loss since too many repetitions in the codeword make the code too weak as in the conventional spreading scheme. Since the coding and decoding complexities decreases as q increases, we use $q = 5$ as our optimal code parameter. Note that for $N_0 = 19.4$ and 15.05, the maximum sum rates decrease very slowly as q increases, therefore, if one does not mind such an insignificant rate loss, many low-complexity codes are available.

In Figure 5.6, we also compared our code with the conventional spreading scheme, a systematic IRA code serially concatenated with a length-q spreading under LDPC-like decoding. The degree distribution of the systematic IRA code is also optimized using the differential evolution method. For this coding

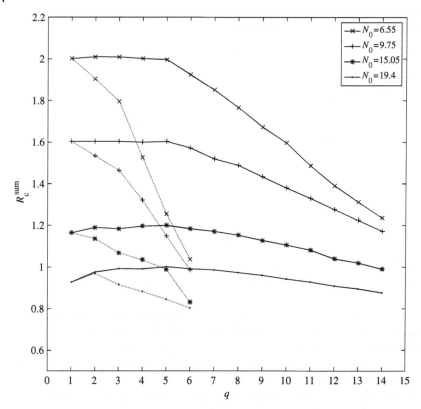

Figure 5.6 Achievable sum rates of optimized 10-user repetition-aided IRA codes (solid lines) and IRA coded (dashed lines) with message node degrees from $\{2, 3, 10, 30, 50, 80, 100\}$ under LDPC-like decoding over AWGN multiuser channels.

scheme, the maximum sum rate is achieved at $q = 1$ in most cases (the sum rate for $N_0 = 19.4$, $q = 1$ can also be improved using more degree types), which means that spreading always causes sum rate loss since spreading is performed for each coded symbol that makes the code too weak to overcome Gaussian noise. However, in our coding scheme we partially introduce repetitions in the codeword in a controllable manner that avoids such rate loss.

In fact, when the user number K increases we can use more repetitions in the codeword so that the decoding complexity for each user scarcely increases. For $K = 15$ and 20, we find near capacity degree distributions in Table 5.3 with $q = 7$ and 14 that achieve sum rates $R_c^{sum} = 1.5$ and 1.8, respectively. The gaps between their decoding thresholds and channel capacities are only 0.04 and 0.18 dB. For comparison, we fix $q = 1$ and find the optimal low-rate IRA codes in Table 5.3 using our code optimization procedure. Although their decoding thresholds also near the channel capacity, their graph consists very

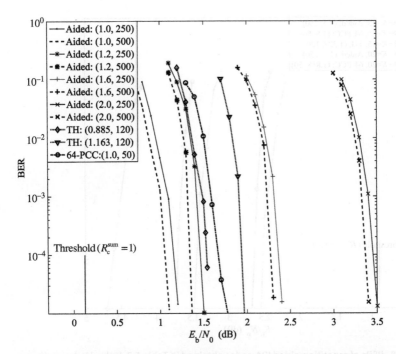

Figure 5.7 BERs of repetition-aided IRA codes obtained in Table 5.2, turbo-Hadamard codes [12–14] and 64-ary PCC [17] over AWGN 10-user channels. Parameters labeled for each curve are code: (sum rate, iteration number). In the turbo-Hadamard codes, Hadamard order is 5, numbers of convolutional-Hadamard components are $M = 3$ for sum rate 1.163 and $M = 4$ for sum rate 0.885. In the 64-ary PCC, numbers of accumulation components and repetitions over GF(64) are $M' = 3$ and $q' = 17$, respectively.

high degree-500 message nodes and moreover, their encoding and decoding complexities are much higher due to a large number of parity checks being used.

Since the above code optimizations are based on the assumption of infinite code length, to verify its practicability in predicting finite length code performance, in Figures. 5.7 and 5.8, we illustrate simulated bit error rate (BER) performances for our codes obtained in Tables 5.2 and 5.3 with finite code length under LDPC-like decoding. We consider a random interleaving for each user and use the information bit length of 4096 for each user. The gaps between the BER curves of decoding with 250 iterations and the estimated thresholds are about 1 dB at 10^{-4} for the code with $R_c^{\text{sum}} = 1$ of 10-user system (Figure 5.7) and 1.6 dB for the code with $R_c^{\text{sum}} = 1.5$ of 15-user system (Figure 5.8). The gap becomes larger for a code with higher sum rate because in our simulations we fixed the information length so that a higher sum rate leads to a shorter code length.

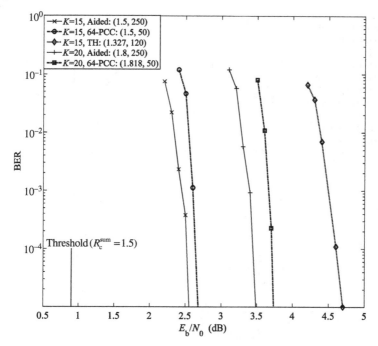

Figure 5.8 BERs of repetition-aided IRA codes obtained in Table 5.3, turbo-Hadamard codes, and 64-ary PCC over 15 and 20-user AWGN channels. Parameters labeled for each curve are number of user, code: (sum rate, iteration number). In the turbo-Hadamard codes, Hadamard order is 5, numbers of convolutional-Hadamard components are $M = 4$. In the 64-ary PCC, numbers of accumulation components and repetitions are $M' = 3, q' = 17$ for 15-user channels and $M' = 3, q' = 19$ for 20-user channels.

Furthermore, we compare the performance of our code with previously proposed low-rate codes, turbo-Hadamard code [12–14], and a 64-ary PCC [17], over 10 and 15-user channels. We perform 120 and 50 iterations, respectively, for turbo-Hadamard and the 64-ary PCC decodings, which are adequate for their convergence. Our code at sum rates 1 and 1.2 in Figure 5.7 and sum rate 1.5 in Figure 5.8 are much better than the turbo-Hadamard codes at lower sum rates 0.885, 1.163, and 1.327, respectively. Here the Hadamard order is 5, the numbers of the convolutional-Hadamard components are $M = 3$ for sum rate 1.163 and $M = 4$ for sum rates 0.885, 1.327. The BERs of the higher rate turbo-Hadamard codes are not illustrated since they are much poorer over the given multiuser channels. The reason is that the original low-rate turbo-Hadamard code is a strong code, which is designed for single-user AWGN channel, as we stated in Section 5.3.1. Also, our codes have a gain of 0.1–0.5 dB over the 64-ary PCCs at similar sum rates. Here the 64-ary PCCs have repetition number $q' = 3$ and accumulation components $M' = 17$ for sum rates 1, 1.5, and $M' = 19$ for

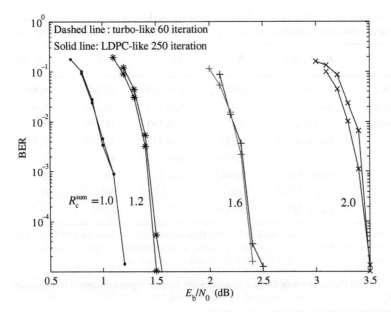

Figure 5.9 BERs of repetition-aided IRA codes obtained in Table 5.2 under LDPC-like (solid line) and turbo-like (dashed line) decodings over 10-user AWGN channel. The number of iterations for LDPC and turbo-like decodings are 250 and 60, respectively.

sum rate 1.818, which are the optimal 64-ary PCCs for the corresponding sum rates [17].

In Figure 5.9, we compared the performances of two decoding schemes of LDPC-like and turbo-like decodings given in Section 5.3 for the codes obtained in Table 5.2. It shows that the optimal codes for LDPC-like decoding also perform well under turbo-like decoding, which converges even faster due to a posterior probability processing for the accumulation. Turbo-like decoding with 60 iterations has similar performance as LDPC-like decoding with 250 iterations.

Table 5.4 illustrates the decoding complexities per iteration of our repetition-aided IRA obtained in Tables 5.2 and 5.3 under LDPC and turbo-like decodings, where the first number corresponds to the number of computing operations of LDPC decoding and the second number corresponds to the number of computing operations of turbo-like decoding. Since the turbo-like decoding with 60 iterations has similar performance as the LDPC-like decoding with 250 iterations, turbo-like decoding has lower complexity in terms of total number of computing operations, even though its single-iteration complexity is higher. It is interesting that due to the use of repetition, the decoding complexity of our code even decreases as the user number increases. For comparison with conventional coding schemes, we also illustrate decoding complexities of turbo-Hadamard

Table 5.4 Number of operations per iteration per information bit in a single-user decoding of repetition-aided IRA codes obtained in Tables 5.2 and 5.3 under LDPC, turbo-like decodings.

Codes	K, M, q, r	R	$+/-$	$\times/\div$	exp / log
Aided IRA	$K = 10, q = 5$	0.05	$174, 219$	$90, 345$	$90, 60$
(LDPC, turbo)	$K = 10, q = 5$	0.06	$138, 173$	$70, 269$	$70, 47$
-like decoding	$K = 15, q = 7$	0.05	$156, 195$	$78, 299$	$78, 52$
	$K = 20, q = 14$	0.045	$118, 143$	$49, 190$	$49, 33$
TH [23]	$r = 5, M = 3$	0.0581	260	51	41
	$r = 5, M = 4$	0.0442	347	68	55
64-PCC [17]	$q' = 17, M' = 3$	0.05	2137	2151	40
	$q' = 19, M' = 3$	0.0455	2226	2326	44

Decoding complexities of turbo-Hadamard (TH) code (Hadamard order 5, convolutional-Hadamard component number M) and 64-ary PCC (repetition number q', accumulation component number M') are also illustrated for comparison. $R = R_c^{\text{sum}}/(2KN_t)$ is code rate of each user.

Table 5.5 Optimal message node degree sequence for 10-user repetition-aided IRA code at sum rate $R_c^{\text{sum}} = 1, 1.2, 1.6, 2$ over Rayleigh fading channel.

R_c^{sum}	1	1.2	1.6	2
q	5	5	5	5
λ_2			0.068364	0.171579
λ_3	0.145577	0.196835	0.216802	0.284322
λ_{10}	0.031780	0.035279	0.129308	0.030637
λ_{30}	0.291608	0.386443	0.266998	0.513463
λ_{50}			0.138178	
λ_{80}			0.180351	
λ_{100}	0.531035	0.381443		
$\left(\frac{E_b}{N_0}\right)^*_{\text{dB}}$	0.38	0.75	1.43	2.36
S.B.	0.11	0.46	1.17	1.93

code and 64-ary PCC, which are calculated based on Table 5.1. Our code for a 20-user system under turbo-like decoding has much lower complexity than the similar rate turbo-Hadamard code in terms of total number of computing operations, and both have much lower decoding complexities than 64-ary PCCs.

5.5.2 Rayleigh Fading Channel

In this section, we present the numerical results of the code design for Rayleigh fading channel. For $K = 10$, $N_t = N_r = 1$, by fixing $q = 5$ we obtain the optimal

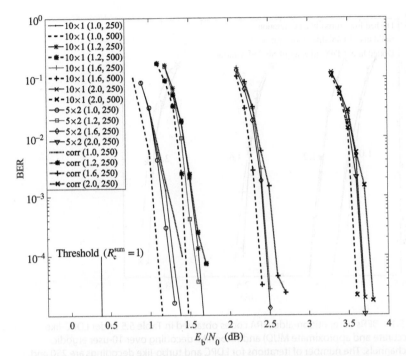

Figure 5.10 BERs of repetition-aided IRA codes over Rayleigh fading multiuser channels with receivers that perfectly know the channel. Parameters labeled for each curve are [user number] × [transmit antenna number] (sum rate, iteration number) for ergodic fading multiuser channel and corr (sum rate, iteration number) for a 10-user correlated fading channel. In the correlated fading multiuser channel, channel gain remains unchanged for every 100 symbols for every user.

message node degree distributions for the IRA components that achieve sum rate $R_c^{sum} = 1$, 1.2, 1.6, 2 under LDPC-like decoding in Table 5.5. Similarly as in AWGN channel, we used degree types from φ^* and φ for $R_c^{sum} = 1$, 1.2 and $R_c^{sum} = 1.6$, 2, respectively. All of their decoding thresholds are close to the corresponding capacities of multiuser channel [25]. Note that the optimal degree sequences for codes at $R_c^{sum} = 1$ and 1.2 are similar as that for AWGN channel in Table 5.2 since for low-rate code (works at high noise level), the EXIT functions of ESE over a Rayleigh fading channel are almost the same as that over the AWGN multiuser channel revealed in Section 5.3.1.

In Figure 5.10, we give our finite length BER simulations for the codes obtained in Table 5.5 under LDPC-like decoding, where the fading coefficient is generated independently for each transmitted symbol. As in the simulation in AWGN channel, we use random interleaving and information bit length 4096. To compare single and multiple transmitter antenna models, we simulated our

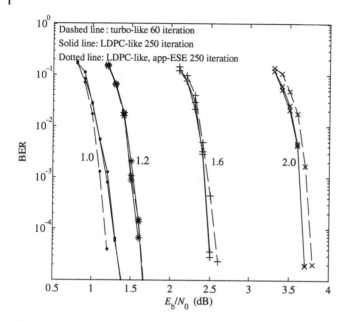

Figure 5.11 BERs of repetition-aided IRA codes obtained in Table 5.5 under LDPC-like (with accurate and approximate MUD) and turbo-like decoding over 10-user ergodic fading channels. The number of iterations for LDPC and turbo-like decodings are 250 and 60. The solid line is for LDPC-like decodings with the accurate MUD in Eqs. 5.2 and 5.3. The dotted line is for LDPC-like decoding with the approximate MUD in Eqs. 5.4 and 5.5. The dashed line is for turbo-like decoding.

codes for both a 10-user system with single transmission antenna $N_t = N_r = 1$ and a 5-user MIMO with double transmission antennas $N_t = 2, N_r = 1$. Note that in the latter, each user's codeword is separated into two equal length vectors that are transmitted by two antennas, respectively. The BER curves for these two cases are very close. At $R_c^{sum} = 1$, 1.2, the 5-user double-antenna case slightly outperforms the 10-user single-antenna case because we considered the same transmission length for each antenna, and the code length of each user is doubled in the former case. The gap between the BER curve at 10^{-4} with iteration number 250 for $R_c^{sum} = 1$ and the estimated threshold is about 0.9 dB.

Note that the above code design for the ergodic fading model is suitable if many independent channel realizations appear, and the receiver can accurately estimate them. In Figure 5.10, we also illustrated the BERs over 10-user correlated fading channel, where the channel gain remains unchanged for every 100 symbols for every user. Comparing the BERs over the ergodic fading channel with the same iteration number, a performance loss of less than 0.1 dB at 10^{-4}

was observed for each code. This demonstrates the applicability of our code to a block fading channel.

In Figure 5.11, we compared the LDPC-like (with accurate and approximate ESE) and turbo-like decoding over 10-user ergodic fading channels. We consider the codes obtained in Table 5.5 and show that the LDPC-like decoding with 250 iterations performs slightly worse than the turbo-like decodings with 60 iterations at low sum rate 1.0, but at high rates 1.6, 2.0, the former is slightly better. LDPC-like decodings with the accurate ESE in Eqs. 5.2 and 5.3 have almost the same performance as our proposed approximate ESE in Eqs. 5.4 and 5.5 whose complexity just increases linearly with the number of users.

Moreover, we consider multiple receiver antenna models with $K = 10$, $N_t = 1$, $N_r = 2, 4$ and obtain the corresponding optimal repetition number and degree sequence under LDPC-like decoding in Tables 5.6 and 5.7. In Figure 5.12, we give the finite length BER simulations for these codes over a 10-user Rayleigh fading MIMOs. Their decoding thresholds are also close to the channel capacity and the gaps between the BER curves at 10^{-4} and the estimated thresholds are about 0.9 dB for the code with $R_c^{sum} = 1$ over 10-user Rayleigh fading MIMO with $N_r = 2$ and 1.1 dB for the code with $R_c^{sum} = 1$ over 10-user Rayleigh fading MIMO with $N_r = 4$.

Table 5.6 Optimal message node degree sequence for 10-user repetition-aided IRA code at sum rate $R_c^{sum} = 1, 1.2, 1.6, 2$ over Rayleigh fading multiuser MIMO with $N_t = 1, N_r = 2$.

R_c^{sum}	1	1.2	1.6	2
q	3	3	3	3
λ_2			0.028748	0.058984
λ_3	0.110117	0.142135	0.172168	0.223082
λ_{10}	0.08113	0.102045	0.183508	0.237999
λ_{30}	0.204257	0.185187	0.270712	0.290983
λ_{50}	0.101873	0.312919	0.286429	0.188952
λ_{80}				
λ_{100}	0.502623	0.257714	0.058435	
$\left(\frac{E_b}{N_0}\right)^*_{dB}$	-3.47	-3.27	-2.87	-2.42
S.B.	-3.7	-3.52	-3.15	-2.8

Table 5.7 Optimal message node degree sequence for 10-user repetition-aided IRA code at sum rate $R_c^{sum} = 1, 1.2, 1.6, 2$ over Rayleigh fading multiuser MIMO with $N_t = 1, N_r = 4$.

R_c^{sum}	1	1.2	1.6	2
q	3	3	3	3
λ_2			0.021643	0.047414
λ_3	0.104111	0.129569	0.165659	0.215932
λ_{10}	0.091926	0.131282	0.246779	0.404276
λ_{30}	0.283793	0.274269	0.289242	0.195233
λ_{50}		0.191999	0.121544	0.054732
λ_{80}	0.297463	0.272882	0.155132	0.082412
λ_{100}	0.222707			
$\left(\frac{E_b}{N_0}\right)^*_{dB}$	−6.84	−6.79	−6.34	−5.97
S.B.	−7.09	−6.99	−6.69	−6.58

5.6 Conclusion

In this chapter, we proposed a very practical coding scheme, very efficient encoding and decoding for the future 5G multiuser MIMO systems. We give a detail code optimization procedure based on multiuser iterative decoding theory and EXIT analysis. Near capacity communication rate are achieved by the designed code for different user number and antenna numbers.

Although we used the IRA code for our parity-check component, it can also be realized by another code, such as an LDPC code or a spatial-coupled code [28]. Designing repetition-aided codes with other forms of parity-check component is an interesting future work.

We considered multiuser system with equal-power and equal-rate user transmission, which is in fact the most difficult scenario for code design. For unequal-power or unequal-rate multiuser system, we can always use the successive interference cancellation decoding that will make the code design much simpler. Related works are given in References [1–3,29].

Acknowledgement

This work was supported in part by three research fundings of the Japan Society for the Promotion of Science through the Grant-in-Aid for Scientific Research (C) under Grant 16K06373, the Ministry of Education, Culture, Sports, Science and Technology through the Strategic Research Foundation at Private Univer-

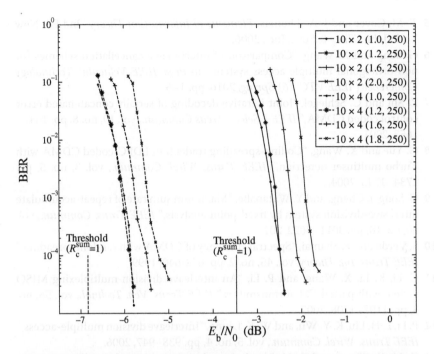

Figure 5.12 BERs of repetition-aided IRA codes over Rayleigh fading multiuser MIMOs with multiple receiver antennas and receives that perfectly know the channel. Parameters labeled for each curve are [user number] × [receive antenna number] (sum rate, iteration number) for ergodic fading multiuser MIMO.

sities (2014-2018) under Grant S1411030, and Singapore Ministry of Education Academic Research Fund Tier 2 MOE2016-T2-2-054.

References

1 Y. Saito, Y. Kishiyama, A. Benjebbour, T. Nakamura, A. Li, and K. Higuchi, "Non-orthogonal multiple access (NOMA) for cellular future radio access," in *Proc. IEEE Vehicular Technology Conference (IEEE VTC' 13)-Spring*, 2013, pp. 1–5.

2 Y. Liu, Z. Ding, M. Elkashlan, and J. Yuan, "Non-orthogonal multiple access in large-scale underlay cognitive radio networks," *IEEE Trans. Veh. Technol.*, vol. 65, no. 12, pp. 10152–10157, 2016.

3 Z. Ding, P. Fan, and H. V. Poor, "Impact of user pairing on 5G nonorthogonal multiple access," *IEEE Trans. Veh. Technol.*, vol. 65, no. 8, pp. 6010–6023, 2015.

4 H. Nikopour and H. Baligh, "Sparse code multiple access," in *Proc. IEEE International Symposium on Personal, Indoor and Mobile Radio Communications (IEEE PIMRC' 13)*, 2013, pp. 332–336.

5 T. M. Cover and J. A. Thomas, *Elements of Information Theory*, 2nd Ed. New York: John Wiley & Sons, Inc., 2006.

6 G. Song and X. Wang, "Comparison of interference cancellation schemes for non-orthogonal multiple access system," in *Proc. IEEE Vehicular Technology Conference (IEEE VTC' 16)-Spring*, 2016, pp. 1–5.

7 Z. Shi and C. Schlegel, "Joint iterative decoding of serially concatenated error control coded CDMA," *IEEE J. Select. Areas Commun.*, vol. 19, no. 8, pp. 1646–1653, 2001.

8 G. Yue and X. Wang, "Coding-spreading tradeoff in LDPC-coded CDMA with Turbo multiuser detection," *IEEE Trans. Wirel. Commun.*, vol. 3, no. 5, pp. 1734–1745, 2004.

9 G. Song, J. Cheng, and Y. Watanabe, "Maximum sum rate of repeat-accumulate interleave-division system by fixed-point analysis," *IEEE Trans. Commun.*, vol. 60, no. 10, pp. 3011–3022, 2012.

10 S. Verdu and S. Shamai, "Spectral efficiency of CDMA with random spreading," *IEEE Trans. Inf. Theory*, vol. 45, no. 2, pp. 622-640, 1999.

11 C. Li, K. Li, X. Wang, and P. Li, "An interleave-division-multiplexing MISO system with partial CSI at transmitter," *IEEE Trans. Veh. Technol.*, vol. 56, no. 3, pp. 1197–1208, 2007.

12 P. Li, L. H. Liu, K. Y. Wu, and W. K. Leung, "Interleave division multiple-access," *IEEE Trans. Wirel. Commun.*, vol. 5, no. 4, pp. 938–947, 2006.

13 K. Li, X. Wang, and P. Li, "Analysis and optimization of interleave-division multiple-access communication systems," *IEEE Trans. Wirel. Commun.*, vol. 6, no. 5, pp. 1973–1983, 2007.

14 P. Li, L. Liu, K. Wu, and W. Leung, "Approaching the capacity of multiple access channels using interleaved low-rate codes," *IEEE Commun. Lett.*, vol. 8, no. 1, pp. 4–6, 2004.

15 Y. Tsujii, G. Song, J. Cheng, and Y. Watanabe, "Approaching multiple-access channel capacity by nonbinary coding-spreading," in *Proc. IEEE International Symposium on Information Theory (IEEE ISIT' 13)*, 2013, pp. 2820–2824.

16 G. Song, Y. Tsujii, J. Cheng, and Y. Watanabe, "Finite field spreading for multiple-access channel," *IEEE Trans. Commun.*, vol. 62, no. 3, pp.1001–1010, 2014.

17 H. Han, G. Song, M. Yoshida, and J. Cheng, " -user nonbinary parallel concatenated code for Gaussian multiple-access channel," in *Proc. IEEE International Conference on Communications (IEEE ICC' 14)*, 2014, pp. 2064–2069.

18 F. J. Vazquez-Araujo, M. Gonzalez-Lopez, L. Castedo, and J. Garcia-Frias, "Interleave-division multiple access (IDMA) using low-rate layered LDGM codes," *Wirel. Commun. Mob. Comput.*, vol. 12, no. 14, pp. 1276–1283, 2012.

19 S. ten Brink, G. Kramer, and A. Ashikhmin, "Design of low-density parity-check codes for modulation and detection," *IEEE Trans. Commun.*, vol. 52, no. 4, pp. 670–678, 2001.

20 S. ten Brink and B. M. Hochwald, "Detection thresholds of iterative MIMO processing," in *Proc. IEEE International Symposium on Information Theory (IEEE ISIT' 02)*, 2002, p. 22.

21 W. E. Ryan and S. Lin, *Channel Codes: Classical and Modern*. Cambridge: Cambridge University Press, 2009.

22 S. ten Brink, "Convergence behavior of iteratively decoded parallel concatenated codes," *IEEE Trans. Commun.*, vol. 49, no. 10, pp. 1727–1737, 2001.

23 P. Li, W. Leung, and K. Wu, "Low-rate turbo-Hadamard codes," *IEEE Trans. Inf. Theory*, vol. 49, no. 12, pp. 3213–3224, 2003.

24 R. Storn and K. Price, "Differential evolution: a simple and efficient heuristic for global optimization over continuous spaces," *J. Glob. Optim.*, vol. 11, no. 4, pp. 341–359, 1997.

25 D. Tse and P. Viswanath, *Fundamentals of Wireless Communication*. Cambridge University Press, 2005.

26 S. ten Brink and G. Kramer, "Design of repeat-accumulate codes for iterative detection and decoding," *IEEE Trans. Signal Process.*, vol. 51, no. 11, pp. 2764–2772, 2003.

27 C. Schlegel, Z. Shi, and M. Burnashev, "Optimal power/rate allocation and code selection for iterative joint detection of coded random CDMA," *IEEE Trans. Inf. Theory*, vol. 52, no. 9, pp. 4286–4294, 2006.

28 S. Kudekar, T. Richardson, and R. Urbanke, "Threshold saturation via spatial coupling: why convolutional LDPC ensembles perform so well over the BEC," *IEEE Trans. Inf. Theory*, vol. 57, no. 2, pp. 803–834, 2011.

29 M. He, G. Song, and J. Cheng, "Rate optimization for repeat-accumulate interleave-division system by fixed-point analysis," in *Proc. IEEE International Conference on Communications (IEEE ICC' 14)*, 2014, pp. 2099–2105.

Guanghui Song received the B.S. degree in communication and engineering from Henan Normal University, Xinxiang, China, in 2006. He received the M.S. degree in telecommunications engineering from Xidian University, Xi'an, China, in 2009 and the Ph.D. degree from the Department of Intelligent Information Engineering and Sciences, Doshisha University, Kyoto, Japan, in 2012. From April 2013 to March 2015, he was a research assistant in Doshisha University. From April 2015 to March 2016, he was a postdoctoral fellow in the Department of Electrical and Computer Engineering, University of Western Ontario, London, Canada. Currently, he is a postdoctoral researcher in Singapore University of Technology and Design, Singapore. His research interests are in the areas of channel coding theory, multiuser coding, and coding for storage systems.

Yuhao Chi received the B.S. degree in electronic and information engineering from Shaanxi University of Science & Technology, Xi'an, China, in 2012. He received the Ph.D. degree in communication and information systems from Xidian University, Xi'an, China, in 2018. From 2016 to 2017, he received the state scholarship fund from China scholarship council to be an exchange Ph.D. student with Nanyang Technological University, Singapore, and a visiting student with the Singapore University of Technology and Design, Singapore. His current research interests include coding theory, multiuser coding and detection, and message passing algorithm.

Kui Cai received the B.E. degree in information and control engineering from Shanghai Jiao Tong University, Shanghai, China, the M.Eng. degree in electrical engineering from the National University of Singapore, and the joint Ph.D. degree in electrical engineering from the Technical University of Eindhoven, The Netherlands, and the National University of Singapore. She has been with Data Storage Institute, Singapore, since 1999, where she was the Program Leader of Non-Volatile Memory Coding and Signal Processing. She is currently an associate professor with the Singapore University of Technology and Design. She is the Vice-Chair (Academia) of the IEEE Communications Society, Data Storage Technical Committee. She is a recipient of the 2008 IEEE Communications Society Best Paper Award in coding and signal processing for data storage. Her research interests include coding theory, communication theory, and signal processing for various data storage systems and digital communications.

Ying Li received the B.S. degree in telecommunication engineering and the Ph.D. degree in communication and information systems from Xidian University, Xi'an, China, in 1995 and 2005, respectively. From 2011 to 2012, she was with the University of California, Davis, CA, USA, as a visiting scholar. She is currently a professor with the State Key Laboratory of Intergreated Services Networks, Xidian University. His current research interests are on design and analysis for wireless systems, including channel coding, wireless network communications, interference processing, relay transmission, and MIMO techniques.

Jun Cheng received the B.S. and M.S. degrees in telecommunications engineering from Xidian University, Xi'an, China, in 1984 and 1987, respectively, and the Ph.D. degree in electrical engineering from Doshisha University, Kyoto, Japan, in 2000. From 1987 to 1994, he was an Assistant Professor and Lecturer in the Department of Information Engineering, Xidian University. From 1995 to 1996, he was an Associate Professor in the National Key Laboratory on Integrated Service Network, Xidian University. In April 2000, he joined ATR Adaptive Communications Research Laboratories, Kyoto, Japan, where he was a visiting researcher. From August 2002 to June 2003, he was working as a Staff Engineer at the R&D Center, Panasonic Mobile Communications Co., Ltd. (formerly Wireless Solution Labs., Matsushita Communication Industrial Co., Ltd.), Yokosuka, Japan. From July 2003 to March 2004, he was a Staff Engineer at Next-Generation Mobile Communications Development Center, Matsushita Electric Industrial Co., Ltd., Yokosuka, Japan. In April 2004, he joined Doshisha University, Kyoto, Japan. Currently, he is a professor in the Department of Intelligent Information Engineering and Sciences, Doshisha University, Kyoto, Japan. His research interests are in the areas of communications theory, information theory, coding theory, array signal processing, and radio communication systems.

6

Physical Layer Techniques for 5G Wireless Security

Batu K. Chalise,[1] Himal A. Suraweera,[2] Gan Zheng,[3] and Risto Wichman[4]

[1] Department of Electrical and Computer Engineering, New York Institute of Technology, Old Westbury, NY, USA
[2] Department of Electrical and Electronic Engineering, University of Peradeniya, Peradeniya, Sri Lanka
[3] Wolfson School of Mechanical, Electrical and Manufacturing Engineering, Loughborough University, Leicestershire, United Kingdom
[4] Department of Signal Processing and Acoustics, Aalto University, Espoo, Finland

6.1 Introduction

Wireless communications and cellular networks are vulnerable to various security threats due to several reasons [1,2]. Wireless transmissions, by their very broadcasting nature propagate everywhere, thus a potential adversary can easily intercept wireless messages. Moreover, limited bandwidth, architectural complexity, and limited device computational power can contribute toward poor security guarantees in wireless systems. As such, development of second-generation (2G) cellular networks saw the introduction of several security functions, including encryption of the radio interface, authentication, and subscriber privacy. Encryption helped to protect voice signals, data, and signaling information against eavesdropping activity. 2G cellular also introduced a subscriber identity module (SIM) card as a method of adding strong authentication of network users.

With the introduction of third-generation (3G) cellular systems, security was further improved. For example, mutual authentication was implemented to prevent attacks by rogue networks while sophisticated security procedures were also introduced for network access, wired network, and user and application domains. Security evolution has continued with the fourth-generation (4G) Long Term Evolution (LTE) technology [3]. Built upon the lessons learned from previous generation networks, LTE security architecture has added features such as a tamper resistant universal integrated circuit card (UICC), new

5G Networks: Fundamental Requirements, Enabling Technologies, and Operations Management, First Edition.
Anwer Al-Dulaimi, Xianbin Wang, and Chih-Lin I.

cryptographic algorithms and a different key structure, secure signaling between the device and mobility management entity (MME), and security mechanisms for interworking with trusted Wi-Fi access.

Beyond 4G evolution in mobile communication are the fifth-generation (5G) systems [4]. It is anticipated that 5G would create a networked society at large and support a countless number of applications, industries, and devices. In addition, 5G will accommodate high throughput, high reliability, and ultralow latency requirements of new use cases, for example, in connected transportation, e-health, smart grids, and manufacturing. The need to support a large number of diverse industries beyond cellular customers and functions, such as cloud access, visualization, and software-defined networking, makes security in 5G a crucial issue. To make matters even more complicated, 5G will have to offer differentiated security to different services as required. For example, e-health will demand strong security while Internet of Things (IoT) can be based on lightweight security at most [5]. Hence, there is a need to research on secure communications, user authentication methods, and privacy guarantees under 5G network operation.

5G security should be designed with more options beyond node-to-node and end-to-end security available in 4G mobile communication systems. Moreover, designed protocols and key exchange solutions must provide better security than in 4G and also facilitate extremely low-latency use cases. To this end, other than the cryptographic security mechanisms that operate at upper layers, physical layer security can be used to safeguard 5G wireless communications [6]. Physical layer security techniques exploit wireless channel characteristics such as the noise, fading, and interference to degrade the received signal at the eavesdropper [7]. Physical layer security methods do not rely on computational complexity and it is difficult to compromise data even if malicious devices would have powerful computational capabilities. Furthermore, physical layer techniques and technologies, such as massive multiple-input multiple-output (MIMO), millimeter communications, cognitive radio, and full-duplex wireless, can be easily integrated to implement comprehensive 5G wireless security solutions.

Notations: Matrices (vectors) are denoted by bold-faced capital (lowercase letters) and the superscripts T, *, and H denote transpose, complex conjugate, and Hermitian transpose, respectively. $\text{tr}()$ stands for matrix trace, $\|\cdot\|$ denotes Euclidean norm, the absolute value is denoted by $|\cdot|$, $\mathbb{E}[.]$ denotes expectation, and the space of $M \times M$ matrix with complex entries is denoted by $\mathbb{C}^{M \times M}$. $\mathcal{CN}(.,.)$ and $\mathbf{I}$ denote complex Gaussian distribution and identity matrix, respectively.

6.1.1 Information Theoretic Security

Now we setup notation and terminology for the security defined from an information theoretic sense to support the discussion in the following sections.

Wireless communications has become a widespread technology used for many different applications in the modern world. Hence, in the last few years, security for wireless systems has become an important issue that has already attracted considerable research interest. Traditionally, in communication networks, security has been realized at the higher layers. However, in emerging applications, such as radio-frequency identification (RFID) networks and 5G IoT, use of low complexity devices make data encryption hard to implement. For this reason, there has been significant interest in developing physical layer security methods based on the properties of the wireless fading channel.

Today, there is a rich body of fundamental results for physical layer security due to information theoretic characterization of secrecy. Some of the pioneering and notable contributions made to the field of secure communication include Claude Shannon's work on noiseless cipher systems [8], Aaron Wyner's work on the wiretap channel model [9], and Csiszar and Korner's work on the broadcast channel with common messages [10]. In particular, Shannon assumed a case where both receiver and eavesdropper receive the same signal encrypted by a key only known to the transmitter and receiver but not to the eavesdropper. On the other hand, in the wiretap model assumed by Wyner, Alice (transmitter) wishes to communicate a secret message M by encrypting it into a code word X^n to Bob (receiver), while keeping the message confidential from Eve (eavesdropper) in the wiretap channel. Based on this model, Wyner defined the concept of *weak secrecy* that requires the normalized conditional entropy of the secret message given the channel output at Eve to be arbitrarily close to the entropy rate of the message. This definition of secrecy has shortcomings [11] and can be replaced with the notion of *strong secrecy* that requires the amount of information leaked to Eve vanish as the code word length, $n \to \infty$.

According to secure communications, Alice must encode the message into a code word useful to be decoded reliably at Bob. At the same time, for security the code word must be useless at Eve. The maximum rate at which both are possible is the secrecy capacity. A fundamental result reported in the literature is that if Eve has a worse channel than Bob, achieving a non-zero secrecy capacity is possible. In particular, several works have investigated the secrecy capacity for practical wireless channels. Consider the basic Gaussian wiretap model, where the received signals at Bob and Eve at the ith channel use are given by

$$Y_{B,i} = hX_i + N_{B,i}, \tag{6.1}$$

and

$$Y_{E,i} = gX_i + N_{E,i}, \tag{6.2}$$

where X_i is the signal transmitted with power P, and the constant channel coefficients between Alice and Bob and between Alice and Eve are h and g, respectively. Furthermore, $N_{B,i}$ and $N_{E,i}$ are the zero mean Gaussian noises at Bob and Eve with variances σ_B^2 and σ_E^2, respectively. For this channel, the secrecy capacity is given by [12]

$$C_s = \log\left(1 + \frac{P|h|^2}{\sigma_B^2}\right) - \log\left(1 + \frac{P|g|^2}{\sigma_E^2}\right). \tag{6.3}$$

Furthermore, we see that secure communication in this channel is possible only if the signal-to-noise ratio (SNR) of the Alice–Bob main channel is higher than the SNR of the eavesdropper channel, that is, $\frac{P|h|^2}{\sigma_B^2} > \frac{P|g|^2}{\sigma_E^2}$.

A limitation of the above Gaussian wiretap mode is that the channel conditions are assumed to be fixed. However, fading conditions of the Alice–Bob and Alice–Eve channels will change for different channel uses in practice. Under such multipath fading conditions, when Bob's and Eve's channel state information (CSI) is available at Alice, an important secrecy metric to study is the so-called ergodic secrecy capacity given by [11]

$$C_s = \max_{P(\cdot,\cdot) \in \mathcal{P}} \mathbb{E}\left[\log\left(1 + \frac{P(h,g)|h|^2}{\sigma_B^2}\right) - \log\left(1 + \frac{P(h,g)|g|^2}{\sigma_E^2}\right)\right], \tag{6.4}$$

where $P(h,g)$ is the power allocation and $\mathcal{P}$ is the set of power allocation functions that satisfy an average power constraint, P, and the expectation $\mathbb{E}[\cdot]$ is taken over all $(h,g) \in A$ where

$$A = \left\{(h,g) : \frac{|h|^2}{\sigma_B^2} > \frac{|g|^2}{\sigma_E^2}\right\}.$$

Beyond work on basic systems, there have been continuous efforts to obtain generalized results for complex models applicable in modern wireless networks. There is a sizable and growing body of work on secrecy capacity or achievable secrecy rates with optimum power allocation schemes for multiple access channels, interference channels, systems with multiple transmit and receive antennas, multicarrier systems, and relay systems. Implementation of well-integrated security solutions for 5G applications based on physical layer security techniques and traditional encryption methods will benefit from this ongoing work aimed at understanding the fundamental limits of wireless secrecy.

6.1.2 Organization of the Chapter

In this chapter, we will discuss physical layer security techniques for 5G wireless communications. Since in-band full-duplex is one of the promising techniques for enhancing capacity of 5G networks, we mainly focus on physical layer security techniques for systems that employ full-duplex transceivers.

In Section 6.1.2, we introduce full-duplex communications and transceiver architecture, and discuss prior art on full-duplex physical layer security schemes reported for the bidirectional, base station, and relay topologies. In Section 6.1.3, we exploit full-duplex receiver to transmit optimum jamming signals and confuse the eavesdropper. Moreover, optimum beamforming designs for full-duplex wireless security solutions are considered for multiantenna bidirectional and relay communication systems in Sections 6.1.4 and 6.1.5, respectively. All of these approaches are shown to have much better performance than the conventional secure half-duplex communication systems. In Section 6.1.6, we emphasize future challenges and open issues for the implementation of full-duplex secure transmissions in 5G systems. Finally, we conclude in Section 6.1.7 with a summary of main results and outline the key role that full-duplex will play for providing physical layer-based security solutions.

6.2 5G Physical Layer Architecture

The bold goal of 5G is to provide ubiquitous connectivity for a massive number of heterogeneous devices. 5G networks promise to deliver broadband, low latency, ultrareliable connectivity, and narrow-band low data rate connections for device-to-device (D2D) communications and IoT. These diverse requirements call for a portfolio of different radio access and connectivity solutions and demand flexibility and scalability from the system to adapt to various communication needs.

At physical layer, the technical solutions in 5G networks include large bandwidths at mm-waves, massive multiantenna transceivers, small cells and network densification, D2D communications, flexible numerology of orthogonal frequency division multiplexing (OFDM) waveforms, flexible frame structure, spectrum sharing, and advanced channel coding schemes, to name a few.

According to Cooper's law, wireless capacity has increased with the factor of one million since 1957 [13]. This has mainly been due to the deployment of ever smaller cells – the direction 5G is also taking. Usage of more bandwidth and frequencies, like 5G's expansion to mm-wave region, accounts for 1.5% of the capacity increase. Development at physical layer amounts to a mere 0.3% of the total capacity increase. However, physical layer lays the foundation for the operation of the whole communication system, and improvements at physical layer are multiplied in the upper layers of the communication stack.

Transmitting and receiving simultaneously at the same time–frequency resource, that is, in-band full-duplex communications, is a physical layer technique that can theoretically increase the throughput by 100% compared to half-duplex communications [14]. This is a significant achievement in view of typical advances in the physical layer. In practice, three issues decrease the theoretical 100% gain of the full-duplex communications, though. To double the throughput, both nodes of a point-to-point link must have some data to transmit. Cellular network operators' statistics show that typical uplink–downlink traffic ratios are on the order of 1:7-8 [15], so implementing full-duplex capability in a handset will not bring in the maximum gain. Instead, full-duplex access points and base stations serving multiple users simultaneously and full-duplex relays are not constrained by the asymmetric uplink and downlink traffic ratios. Second, if all nodes in the network communicate simultaneously, co-channel interference levels increase that again decreases the overall system capacity from the theoretical maximum. As usual, the interference can be controlled by dynamic scheduling and radio resource management. Third, the transmitted signal leaks inevitably into the receiver chain of the full-duplex transceiver. This self-interference (SI) seriously limits the gains from full-duplex operation unless it can be mitigated close to the noise floor of the receiver. Nevertheless, several prototypes have shown that it is possible to suppress SI down to the noise level [16,17], so full-duplex technology appears feasible.

To leverage the potential of full-duplex technology, 5G networks must support the suitable frame and control structures to enable the operation of full-duplex nodes. In 5G networks, this is facilitated by time division duplexing (TDD) between uplink and downlink together with flexible duplexing. The dynamic traffic variations expected in very dense network deployments call for the ability to dynamically assign transmission resources (time slots) to different transmission directions (uplink and downlink) to allow efficient utilization of the available spectrum. In very dense deployments and high frequencies with low-power nodes, the TDD-specific interference scenarios (base station-to-base station and D2D interference) will be similar to the base station-to-device and device-to-base station interference in frequency division duplex that has dominated the deployments of cellular networks this far. When the network allows dynamic allocation of uplink and downlink slots in TDD, some network nodes may operate in full-duplex mode to further increase the spectral efficiency or physical layer security of the 5G physical layer.

6.2.1 Full-Duplex Communications

In addition to improving the throughput, full-duplex transceiver may improve the secrecy rate by simultaneously receiving the signal of interest and sending a jamming signal to distract the eavesdropper [18]. At the same time, the jamming signal causes SI to the transceiver itself. Not surprisingly, SI is one of the key challenges in the implementation of full-duplex transceivers. Nevertheless, full-duplex transceivers have been utilized in wireless communication systems since a long time, in continuous-wave radars, and on-channel repeaters in broadcast and cellular systems; so the SI problem appears to be manageable.

Typically, the transmitter and receiver antennas in full-duplex transceivers are close to each other, or the transmitter and receiver chains may even share the same antennas. Therefore, the path loss of the SI signal is much less than that of the signal of interest, and without physical isolation and cancellation, strong SI signal would completely mask the signal of interest in the receiver chain. Thus, SI cancellation is an essential part of full-duplex transceivers.

As an example of the dynamics of the received signal, suppose the transmitter power is 23 dBm like in LTE user equipment. To detect the signal of interest, its level should be larger than the noise floor given by the noise figure, which is determined by receiver electronics and the signal bandwidth. Typical assumptions for the noise figure, for example, in 4G systems, are 5 dB in base station and 7 dB in handsets. Assuming, for example, 10 MHz signal bandwidth and 7 dB noise figure, the noise floor becomes −97 dBm and there is a 120 dB difference between the noise floor and the transmitted signal level within the same full-duplex transceiver. The goal of the SI cancellation is to reduce the SI close to the noise floor, preferably even below it, to keep performance degradation reasonable [19]. Therefore, SI should be suppressed by 120 dB, which is far from trivial.

SI cancellation techniques can be classified into passive and active ones [16,20,21]. The former refers to spatial separation, polarization, directivity, antenna design, and geometry of transmit and receive antenna arrays to minimize over-the-air leakage of the transmitted signal to the receiver chain [22,23]. Active cancellation techniques can be further divided into radio frequency (RF) or analog cancellation and digital baseband cancellation. The basic principle of the SI cancellation in analog domain is the same as in digital baseband: A replica of the interfering signal is created and then subtracted from the received signal. This calls for active electronic circuits that can cancel a time-varying multipath SI signal.

Figure 6.1 shows an example of a full-duplex transceiver architecture implementing RF cancellation and digital baseband cancellation. Transmit and receive antennas are separated spatially providing some passive isolation. The channel model in digital baseband is more accurate than the one in RF domain,

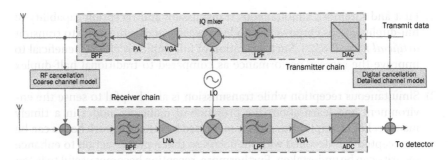

Figure 6.1 Full-duplex transceiver architecture.

because digital baseband allows more extensive signal processing. The cancellation signal is tapped at the output of the transmitter power amplifier, which ensures that the cancellation signal contains all RF impairments so that most of those imperfections are canceled, depending on the accuracy of the channel model.

At the receiver side, cancellation occurs at the input of the low-noise amplifier (LNA), and therefore, the dynamic range of the entire receiver chain can be relaxed. If the analog cancellation signal was injected to the receiver chain after the LNA, the SI signal after passive isolation should fit to the dynamic range of the LNA.

The signal of interest together with the remaining SI is converted into digital baseband. The power of the remaining SI signal after passive isolation and RF cancellation should be sufficiently small to avoid saturating analog-to-digital converter. Finally, the contribution of the transmitted digital baseband signal is subtracted from the received signal in the digital cancellation and the signal of interest is detected and decoded. If the transceiver is equipped with multiple transmit and receive antennas, SI can be mitigated using spatial suppression as well [24,25]. In any case, SI cannot be removed completely, and some residual SI remains after spatial isolation and analog and digital cancellation.

In analytical studies, like in this chapter, the effect of SI cancellation is often modeled by an attenuation factor operating on the SI channel. Changing the attenuation and consequently the power of the residual SI then allows evaluating performance bounds and determining the requirements for SI cancellation.

6.2.2 Security in Full-Duplex Communications

So far, majority of the work on physical layer security has assumed half-duplex operation. It is only recently that security schemes, their analysis, and optimization exploiting full-duplex operation have surfaced.

Full-duplex operation provides several important features useful for physical layer security implementation.

1) First and foremost, simultaneous transmission and reception capability of full-duplex radios empowers a node to receive and at the same to transmit *artificial noise* [26,27]. Such operation of full-duplex is clearly beneficial to improve the secrecy performance as compared to traditional half-duplex operation.

2) Simultaneous reception while transmission is also helpful to sense the environment and learn about the presence of malicious nodes in a timely manner. For example, messages transmitted by other network nodes can be intercepted and decoded without delays so that a proper action to enhance security can be undertaken. Furthermore, reception becomes useful to listen to pilot sequences transmitted by base stations and to acquire timely CSI.

Note that the CSI plays a pivotal role for designing beamforming schemes in multiple antenna systems and such systems can significantly improve the secrecy performance as compared to systems with single antennas.

3) By design, full-duplex communications adopt power allocation to control the harmful effect of SI at a node input [28]. Such power control is also useful for secrecy communications, since low transmit power and attenuation encountered in the wireless environment with high probability will weaken the received signal at the eavesdropper.

4) Full-duplex radios provide attractive medium access layer solutions to reduce delays/latency, improve collision and detection, and mitigate hidden/exposed node issues. It is clear that these features also influence the security performance.

5) Full-duplex operation offers complementary benefits to be integrated with many promising 5G technological components. For example, technologies, such as massive MIMO, millimeter wave communications, non-orthogonal multiple access (NOMA), cognitive radio, and cooperative communications, can be integrated with full-duplex operation to further improve the secrecy performance.

6.2.3 Prior Art

In this section we will introduce some of the current literature available on full-duplex security schemes operating at the physical layer.

In the current literature, three full-duplex use cases identified as (a) bidirectional topology, (b) relay topology, and (c) base station topology are promoted [17]. Under the bidirectional topology, nodes can transmit and receive data at the same time and frequency band. In the relay topology, a full-duplex relay performs retransmission of received signals from one node (source) to another (destination). The scenario where a base station serves an uplink user and transmits data to a downlink user at the same time and frequency band is referred to as the base station topology. The bidirectional topology is useful for communication in machine-type and IoT devices. An application envision for the relay topology is self-backhauling where the relay can be used as an intermediary to exchange signals between the core and access links of the network. The base station topology is applicable in cellular mobile networks and wireless local area networks to schedule simultaneous uplink and downlink transmissions. Several papers have proposed and studied physical layer-based security schemes applicable for the above topologies.

6.2.3.1 Bidirectional Topology

In Reference [29], optimization of a full-duplex malicious node, in terms of jamming covariance matrix design to minimize the secrecy rate of multiple antenna transmission between Alice (transmitter) and Bob (receiver), is presented. The authors in Reference [30] have considered transmit/receive antenna assignment at Bob and beamforming and power allocation design at Alice, in order to maximize the secrecy rate of secure full-duplex MIMO transmission. In Reference [31], a point-to-point system with two full-duplex legitimate users and an eavesdropper is assumed. Specifically, by designing artificial noise to act against the eavesdropper, it has been demonstrated that full-duplex operation can improve the system throughput, while maintaining secrecy and guaranteeing the quality-of-service (QoS). In Reference [32], degrees of freedom of a secure transmitter–receiver system have been studied, where the receiver operates in the full-duplex mode to transmit jamming signals to degrade Eve's channel. In Reference [33], full-duplex receiver jamming performance in a two-tier decentralized network has been analyzed under a stochastic geometry framework. In Reference [34], physical layer security in a wireless *ad hoc* network has been studied by assuming a hybrid full-/half-duplex receiver deployment strategy. Specifically, under this strategy, a fraction of nodes operate in full-duplex mode to send jamming signals, while other nodes operate in half-duplex node with the objective of choosing the fraction of full-duplex receivers for achieving the best security performance. Recent works have also considered extensions by assuming multiple receivers and eavesdroppers. Assuming a multiuser multiple-input single-output network, full-duplex receiver jamming has been used against eavesdroppers to provide secure communications in Reference [35]. In Reference [36], a communication system that consists of two full-duplex source nodes equipped with multiple antennas and a single antenna eavesdropper has been considered to address a robust secure beamforming problem with the objective of maximizing worst case sum secrecy rate under imperfect CSI conditions. The point-to-point Alice–Bob model has been extended to cognitive radio in Reference [37]. Assuming a single antenna Alice, multiple antenna Bob and Eve, and a primary user equipped with a single antenna each, in Reference [37], exact and asymptotic secrecy performance of zero-forcing (ZF) beamforming and two antenna reception schemes have been presented.

6.2.3.2 Base Station Topology

Secrecy design in cellular networks has been considered in Reference [38] by exploiting a model comprising of a full-duplex base station and multiple half-duplex uplink/downlink mobile users. The authors of Reference [38] have maximized the downlink secrecy rate using semidefinite programming (SDP) relaxation. Further more, in Reference [39] optimal and suboptimal zero-forcing beamforming schemes have been presented to provide physical layer security. In Reference [40], a power efficient resource allocation algorithm has been reported to enable secure multiuser MIMO communications with a full-duplex base station. The full-duplex base station has also been extended to the case of secure simultaneous wireless information and power transfer (SWIPT) in Reference [41]. Specifically, the system model of Reference [41] assumes an additional ideal user that harvests energy from the RF signals transmitted by the base station. For this system, authors in Reference [41] have solved a weighted uplink/downlink secrecy rate maximization problem by designing covariance matrices under a sum transmit power of the full-duplex base station and minimum harvested energy of the idle user constraint.

6.2.3.3 Relay Topology

Some papers have studied the secrecy performance of full-duplex relay systems [42–50]. Most of them consider the case of half-duplex source and destination nodes assisted by a full-duplex relay. In Reference [42], secrecy performance of a full-duplex relay system in the presence of source-Eve link and relay-Eve link has been studied. In Reference [43], one-way secure relay communication is considered, where full-duplex transmission and full-duplex jamming modes of the relay have been analyzed. In Reference [44], the achievable secrecy rate of a multihop relay network, where the relays operate in the decode-and-forward (DF) mode and transmit jamming signals to the eavesdropper, has been analyzed. Assuming Nakagami-m fading conditions and imperfect SI cancellation, in Reference [45], ergodic secrecy rate of full-duplex bidirectional communication has been analyzed. For the considered bidirectional system model, it has been reported in Reference [45] that a linear growth in the ergodic secrecy rate can be achieved with the log of SNR of the channel between the transceiver pair. In Reference [46], the secrecy performance of a transmitter/receiver pair (the relay of this system can operate in either full-duplex or half-duplex mode) has been compared to show the advantages of full-duplex operation. A buffer-aided full-duplex relay that switches between full-duplex/half-duplex operation has been considered in Reference [47] to enhance the secrecy performance of a cooperative communication system. Inspired by the fact that full-duplex or half-duplex mode has its own disadvantages, a relay mode switching scheme has been proposed in Reference [48] where it has been shown to significantly outperform half-duplex relay selection. In Reference [49], rank-two relay beamforming matrices based on Alamouti code and artificial noise covariances have been optimized for full-duplex relay systems. The work in Reference [50] considered a full-duplex heterogeneous network with multiple cognitive radio eavesdroppers for secrecy performance evaluation. To this end, secrecy outage probability of three relay selection schemes, namely, (a) partial relay selection, (b) optimal relay selection, and (c) minimum SI relay selection have been studied.

6.3   Secure Full-Duplex Receiver Jamming

Physical layer jamming is an efficient way to increase the secrecy rate of wireless systems to degrade the decoding capability of the eavesdroppers by introducing controlled jamming signals. There are several ways to transmit jamming signals. First, jamming signals can be embedded in the intended signals at the transmitter, which can be beamformed to avoid the legitimate receiver and only affect the eavesdroppers. This is normally referred as the *artificial noise approach* and requires multiple antennas at the transmitter [26,27]. Second, when the transmitter has only a single antenna and artificial noise is not possible, full-duplex wireless becomes a key technology to enable the receiver to decode the required signals while sending jamming signals at the same time to confuse the eavesdroppers. This method requires the receiver to cancel the strong SI to a satisfactory level. Third, when the receiver cannot deal with the SI, external full-duplex relays can be employed to send jamming signals to the eavesdroppers while receiving intended signals from the source simultaneously.

This section introduces a full-duplex multiantenna receiver as an alternative to transmit jamming signals to the eavesdropper. To be specific, we will maximize the secrecy rate in a wireless network with a single-antenna source, a multiantenna destination, and a multiantenna eavesdropper. Conventional wireless node works in the half-duplex mode and the achievable secrecy rate saturates as the transmit power increases. The operation, which allows simultaneous transmission and reception on the same frequency band has emerged as an attractive solution to improve the spectral efficiency. The full-duplex operation has also shown great potential to improve physical layer security, especially in the aforementioned unfavored situation that the transmitter has a single antenna and cannot jam the eavesdropper.

As shown in Figure 6.2, the main concept is that the intended receiver sends jamming signals to degrade the eavesdropper's channel and protect its own reception [18], which is termed receive jamming. The receive jamming will

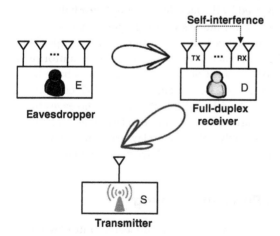

Figure 6.2 System model with full-duplex receiver jamming.

interfere the eavesdropper but also the receiver because of the SI, but if the SI can be well suppressed, it could effectively jam the eavesdropper without much degradation in the quality of the legitimate reception. When the receiver has multiple transmit or receive antennas, it can employ joint transmit and receive beamforming for simultaneous signal detection, SI suppression, and jamming emission. We further simplify the beamforming design to an efficient 1-D search. This scheme provides a self-protection against eavesdroppers without external helpers. Numerical results verify that the secrecy rate increases unbounded as the transmit power goes up, which is in contrast to the half-duplex case without receive jamming.

6.3.1 System Model

As in Reference [18], consider a secure communication system composed of one single-antenna source S, one destination (legitimate receiver) D, and one passive eavesdropper E, with M and M_e antennas, respectively. The receiver D has a total of M antennas divided to M_r receive antennas and M_t transmit antennas such that $M = M_t + M_r$.

Let $\mathbf{h}_{sd} \in \mathbb{C}^{M_r \times 1}$, $\mathbf{h}_{se} \in \mathbb{C}^{M_e \times 1}$, and $\mathbf{H}_{ed} \in \mathbb{C}^{M_e \times M_t}$ denote the S–D, S–E, and between D–E channels, respectively. $\mathbf{n}_D \sim \mathcal{CN}(0, \mathbf{I})$ and $\mathbf{n}_E \sim \mathcal{CN}(0, \mathbf{I})$ represent zero-mean i.i.d. complex Gaussian noise (with covariance $\mathbf{I}$) at D and E, respectively. The transmit signal s is assumed to be a zero mean complex Gaussian random variable with a power P_s.

The proposed receive jamming is illustrated in Figure 6.2. The receiver D protects itself by operating in the full-duplex mode to degrade the quality of the eavesdropper's link to S. To be specific, the receiver transmits a jamming signal while it simultaneously receives the source signal enabled by the full-duplex mode. This creates a feedback loop channel from the receiver output to the receiver input through the effective SI channel $\mathbf{H}_{si} = \sqrt{\rho}\mathbf{H} \in \mathbb{C}^{M_r \times M_t}$, where $\mathbf{H}$ is a fading loop channel [16,24,28]. Without loss of generality, we introduce a variable ρ with $0 \le \rho \le 1$ to quantify the effect of passive SI suppression such as antenna isolation [24]. Therefore, $\rho = 0$ refers to the ideal case of no SI while $0 < \rho \le 1$ corresponds to different SI suppression capabilities.

The receiver D transmits the Gaussian jamming signal with zero mean and a covariance matrix $\mathbf{Q}$. The covariance matrix $\mathbf{Q}$ has general rank and will be designed to maximize the secrecy rate with the power constraint $\text{tr}(\mathbf{Q}) \le P_d$. E is not aware of the full-duplex operation of D, and simply uses a maximum ratio combining (MRC) receiver, $\mathbf{h}_{se}^H$, for overhearing the legitimate transmission. The received signal at E is $\mathbf{y}_E = \mathbf{h}_{se}s + \mathbf{H}_{ed}\mathbf{n} + \mathbf{n}_E$. After applying MRC receiver, the data estimate at E is

$$\hat{s}_e = \frac{\mathbf{h}_{se}^H}{\|\mathbf{h}_{se}\|}(\mathbf{h}_{se}s + \mathbf{H}_{ed}\mathbf{n} + \mathbf{n}_E). \tag{6.5}$$

The destination D employs a linear receiver $\mathbf{r}$ on its received signal, $\mathbf{y}_D = \mathbf{h}_{sd}s + \sqrt{\rho}\mathbf{H}\mathbf{n} + \mathbf{n}_D$, to obtain the data estimate

$$\hat{s}_d = \mathbf{r}^H(\mathbf{h}_{sd}s + \sqrt{\rho}\mathbf{H}\mathbf{n} + \mathbf{n}_D), \text{ with } \|\mathbf{r}\| = 1. \tag{6.6}$$

The achievable secrecy rate can be written as [51]

$$R_S = \max\left\{ 0, \log_2\left(1 + \frac{P_s|\mathbf{r}^H\mathbf{h}_{sd}|^2}{1 + \rho\mathbf{r}^H\mathbf{H}\mathbf{Q}\mathbf{H}^H\mathbf{r}}\right) \right. \tag{6.7}$$

$$\left. - \log_2\left(1 + \frac{P_s\|\mathbf{h}_{se}\|^2}{1 + \frac{\mathbf{h}_{se}^H\mathbf{H}_{ed}\mathbf{Q}\mathbf{H}_{ed}^H\mathbf{h}_{se}}{\|\mathbf{h}_{se}\|^2}}\right) \right\}.$$

An interesting analogy can be made between the full-duplex receive jamming and the external jamming. The above secrecy rate can also be achieved with the help of an external jammer helper with M_t antennas, whose channels to D and E are $\rho\mathbf{H}$ and $\mathbf{H}_{ed}$, respectively. The difference in our proposed receive jamming is that D needs to perform the SI cancellation to achieve that rate and does not need the external assistance. In this section, since D has multiple antennas, we consider active SI mitigation in the spatial domain, in order to keep the complexity low, which is widely adopted in the literature [24,25,55].

6.3.2 Transmit and Receive Designs for SI Cancellation and Jamming

In this section, we study the design of the optimal linear receiver $\mathbf{r}$ and the optimal transmit covariance $\mathbf{Q}$ at D in order to effectively suppress the SI and generate jamming signals to E. Based on Eq. 6.7, we first formulate the secrecy rate maximization problem with a power constraint at D, that is,

$$\max_{\mathbf{Q}, \|\mathbf{r}\|=1} \frac{1 + \frac{P_s|\mathbf{r}^H\mathbf{h}_{sd}|^2}{1 + \rho\mathbf{r}^H\mathbf{H}\mathbf{Q}\mathbf{H}^H\mathbf{r}}}{1 + \frac{P_s\|\mathbf{h}_{se}\|^2}{1 + \frac{\mathbf{h}_{se}^H\mathbf{H}_{ed}\mathbf{Q}\mathbf{H}_{ed}^H\mathbf{h}_{se}}{\|\mathbf{h}_{se}\|^2}}} \tag{6.8}$$

$$\text{s.t. } \mathbf{Q} \succeq 0, \ \text{tr}(\mathbf{Q}) \leq P_d.$$

The problem (6.8) is non-convex and difficult to tackle in general. In the following we will first present a useful lemma about the properties of the optimal $\mathbf{Q}^+$ to simplify its optimization.

Lemma 6.1   *The problem (6.8) has an optimal rank-1 solution* $\mathbf{Q}^+$.

Proof: Let $(\mathbf{Q}^+, \mathbf{r}^+)$ be one optimal solution to the problem (6.8). If $\mathbf{Q}^+$ is rank-1, the desired result is obtained. If $\mathbf{Q}^+$ is not rank-1, let $\mathbf{r}^H \mathbf{H} \mathbf{Q}^+ \mathbf{H}^H \mathbf{r} = x$, $\mathbf{h}_{se}^H \mathbf{H}_{ed} \mathbf{Q}^+ \mathbf{H}_{ed}^H \mathbf{h}_{se} = y$ and consider the following problem

$$\min_{\mathbf{Q}} \ \mathbf{r}^H \mathbf{H} \mathbf{Q} \mathbf{H}^H \mathbf{r}, \quad \text{s.t.} \quad \text{tr}(\mathbf{Q} \mathbf{H}_{ed}^H \mathbf{h}_{se} \mathbf{h}_{se}^H \mathbf{H}_{ed}) = y, \ \text{tr}(\mathbf{Q}) \le P_d. \tag{6.9}$$

Obviously, $\mathbf{Q}^+$ is feasible for the above problem. We want to prove that $\mathbf{Q}^+$ is also the optimal solution to Eq. 6.9. Let $\mathbf{Q}'$ be any solution to the above problem (6.9). Then it holds that $\mathbf{r}^H \mathbf{H} \mathbf{Q}' \mathbf{H}^H \mathbf{r} = x$. This is because if $\mathbf{r}^H \mathbf{H} \mathbf{Q}' \mathbf{H}^H \mathbf{r} < x$, then $\mathbf{Q}'$ is feasible for the problem of Eq. 6.8 but achieves a strictly larger objective value than $\mathbf{Q}^+$ in Eq. 6.8. But this contradicts the fact that $\mathbf{Q}^+$ is the optimal solution to Eq. 6.8, so $\mathbf{Q}^+$ is also the optimal solution to Eq. 6.9.

Problem (6.9) is a homogeneous quadratically constrained quadratic program (QCQP) with two constraints. According to the results in Reference [54], it has a rank-1 solution. Thus, it follows that the problem of (6.8) has a rank-1 solution $\mathbf{Q}^+$. This completes the proof.

Since $\mathbf{Q}$ can be rank-1, we introduce a new variable $\mathbf{q}$ such that $\mathbf{Q} = P_d \mathbf{q} \mathbf{q}^H$ and $\|\mathbf{q}\| = 1$, and problem (6.8) will be simplified to

$$\max_{\|\mathbf{q}\|=1, \|\mathbf{r}\|=1} \frac{1 + \dfrac{P_s |\mathbf{r}^H \mathbf{h}_{sd}|^2}{1 + \rho P_d \mathbf{r}^H \mathbf{H} \mathbf{q} \mathbf{q}^H \mathbf{H}^H \mathbf{r}}}{1 + \dfrac{P_s \|\mathbf{h}_{se}\|^2}{1 + \dfrac{P_d \mathbf{h}_{se}^H \mathbf{H}_{ed} \mathbf{q} \mathbf{q}^H \mathbf{H}_{ed}^H \mathbf{h}_{se}}{\|\mathbf{h}_{se}\|^2}}}. \tag{6.10}$$

Next, we will focus on the equivalent problem (6.10), and we aim to jointly optimize $\mathbf{q}$ as well as the receiver design at D. From Eq. 6.7, the optimal linear receiver that maximizes the received signal-to-interference-plus-noise ratio (SINR) at D is given by the following adaptive minimum mean square error (MMSE):

$$\mathbf{r} = \frac{\left(\rho \mathbf{H} \mathbf{Q} \mathbf{H}^H + \mathbf{I}\right)^{-1} \mathbf{h}_{sd}}{\left\| \left(\rho \mathbf{H} \mathbf{Q} \mathbf{H}^H + \mathbf{I}\right)^{-1} \mathbf{h}_{sd} \right\|}. \tag{6.11}$$

Then the achievable secrecy rate is expressed as

$$
R_S = \max\left(0, \log_2\left(1 + P_s \mathbf{h}_{sd}^H \left(\rho \mathbf{H}\mathbf{Q}\mathbf{H}^H + \mathbf{I}\right)^{-1} \mathbf{h}_{sd}\right)\right.
$$

$$
\left. - \log_2\left(1 + \frac{P_s \|\mathbf{h}_{se}\|^2}{1 + \frac{\mathbf{h}_{se}^H \mathbf{H}_{ed} \mathbf{Q}\mathbf{H}_{ed}^H \mathbf{h}_{se}}{\|\mathbf{h}_{se}\|^2}}\right)\right)
$$

$$
= \max\left(0, \log_2\left(1 + P_s \|\mathbf{h}_{sd}\|^2 - \frac{\rho P_s P_d |\mathbf{h}_{sd}^H \mathbf{H}\mathbf{q}|^2}{1 + \rho P_d \mathbf{q}^H \mathbf{H}^H \mathbf{H}\mathbf{q}}\right)\right.
$$

$$
\left. - \log_2\left(1 + \frac{P_s \|\mathbf{h}_{se}\|^2}{1 + \frac{P_d |\mathbf{h}_{se}^H \mathbf{H}_{ed} \mathbf{q}|^2}{\|\mathbf{h}_{se}\|^2}}\right)\right), \tag{6.12}
$$

where we have used the matrix inversion lemma.

The secrecy rate expression R_S in Eq. 6.12 is still complicated. To tackle it, we introduce a parameter t and study the problem below:

$$
\max_{\mathbf{q}} \quad |\mathbf{h}_{se}^H \mathbf{H}_{ed} \mathbf{q}|^2 \tag{6.13}
$$

$$
\text{s.t.} \quad \frac{|\mathbf{h}_{sd}^H \mathbf{H}\mathbf{q}|^2}{1 + \rho P_d \mathbf{q}^H \mathbf{H}^H \mathbf{H}\mathbf{q}} = t, \quad \|\mathbf{q}\|^2 = 1,
$$

which is a non-convex quadratic optimization problem and still difficult to solve. To tackle this difficulty, we first study a modified but equivalent problem below by introducing a new rank-1 matrix variable $\tilde{\mathbf{Q}} = \mathbf{q}\mathbf{q}^H$:

$$
\max_{\tilde{\mathbf{Q}}} \quad \mathrm{tr}(\mathbf{h}_{se}^H \mathbf{H}_{ed} \tilde{\mathbf{Q}} \mathbf{H}_{ed}^H \mathbf{h}_{se}) \tag{6.14}
$$

$$
\text{s.t.} \quad \mathrm{tr}(\tilde{\mathbf{Q}}(\mathbf{H}^H \mathbf{h}_{sd} \mathbf{h}_{sd}^H \mathbf{H} - t\rho P_d \mathbf{H}^H \mathbf{H})) = t,
$$

$$
\tilde{\mathbf{Q}} \geq \mathbf{0}, \quad \mathrm{tr}(\tilde{\mathbf{Q}}) = 1.
$$

Equation 6.14 is a SDP problem and can be solved by standard convex optimization techniques [52]. Note that in Reference [54], it has been shown that the optimal $\tilde{\mathbf{Q}}$ should be rank-1, so Eqs. 6.13 and 6.14 are equivalent in the sense that given the optimal solution $\tilde{\mathbf{Q}}^+$ to Eq. 6.14, the optimal $\mathbf{q}^+$ to solve Eq. 6.13 can be extracted via $\tilde{\mathbf{Q}}^+ = \mathbf{q}^+ \mathbf{q}^{+H}$. Denote its optimal objective value as $R(t)$, the secrecy rate maximization problem can be formulated as

$$\max_{t \geq 0} \quad R(t) \triangleq \log \left(\frac{1 + P_s \|\mathbf{h}_{sd}\|^2 - \rho P_s P_d t}{1 + \frac{P_s \|\mathbf{h}_{se}\|^2}{1 + \frac{P_d h(t)}{\|\mathbf{h}_{se}\|^2}}} \right). \tag{6.15}$$

Thus, the maximum of $R(t)$ can be found via a 1-D search.

6.3.3 Results and Discussion

We now evaluate the performance of the proposed full-duplex receive jamming scheme using computer simulations. All channel entries are independently drawn from the Gaussian distribution $\mathcal{CN}(0, 1)$. Unless otherwise specified, it is assumed that E is equipped with the same number of receive antennas as D, that is, $M_e = M_r$, $M_t = M_r = 2$ and $\rho = 0.5$. The total transmit SNR, P_T, is used as power metric, and it is assumed that $P_S = \frac{1}{3} P_T$, $P_d = \frac{2}{3} P_T$.

To demonstrate the advantage of the secrecy rate performance of the proposed full-duplex scheme, we compare it with two benchmark schemes:

- The baseline half-duplex system, in which the achievable secrecy capacity is expressed as:

$$C_{S,HD} = \max \left(0, \log_2(1 + P_T \|\mathbf{h}_{sd,HD}\|^2) - \log_2(1 + P_T \|\mathbf{h}_{se}\|^2) \right), \tag{6.16}$$

where $\mathbf{h}_{sd,HD}$ denotes the channel between S and all antennas at D in the half-duplex mode. For fairness of comparison, we keep the same total power for the half-duplex and the full-duplex systems.

- Fixed receive beamforming for $\mathbf{r}$. We consider the following MMSE receiver that aims to suppress the SI and noise:

$$\mathbf{r} = \frac{\left(\rho \mathbf{H} \mathbf{H}^H + \mathbf{I} \right)^{-1} \mathbf{h}_{sd}}{\| \left(\rho \mathbf{H} \mathbf{H}^H + \mathbf{I} \right)^{-1} \mathbf{h}_{sd} \|}. \tag{6.17}$$

In Figure 6.3, the achievable secrecy rate is shown against total transmit SNR. First, it is observed that for the half-duplex mode, the secrecy rate saturates from very low SNR, because E has the same number of antennas as D, while S has a single antenna and cannot provide any countermeasure against the eavesdropper. On the other hand, for both full-duplex schemes, the secrecy rate keeps increasing as the transmit SNR increases; this is due to the fact that multiple transmit antennas at D help mitigate SI and generate jamming signal to E. It can also be seen that the optimal linear receiver at D can achieve 1 bits per channel use (bpcu) higher secrecy rate than the benchmark scheme with a fixed MMSE receiver.

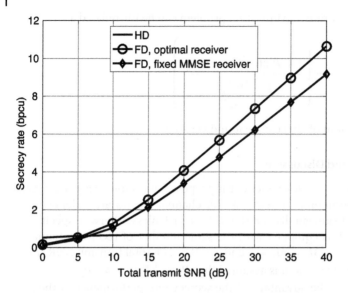

Figure 6.3 Secrecy rate versus total transmit power for the multiantenna case.

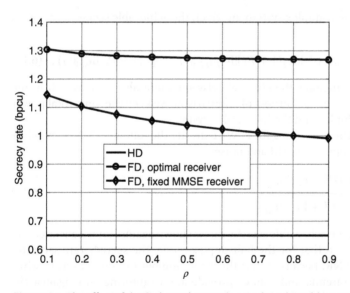

Figure 6.4 The effect of the SI channel strength ρ on the achievable secrecy rate.

In Figure 6.4, the impact of residual SI, denoted by ρ, is examined when the total transmit SNR is 15 dB. The half-duplex is not affected by the SI, therefore, the secrecy rate remains constant. The secrecy rates of both full-duplex schemes decrease as ρ increases. Both full-duplex schemes outperform the half-duplex scheme, even when at high ρ. It can also be seen that the performance gap between the optimal linear receiver and the fixed MMSE receiver widens at higher ρ. This is because the optimal linear receiver is more effective in dealing with the SI.

The above results have shown substantial performance gains over the conventional half-duplex operation at the destination, but with the assumption that the eavesdropper is unaware of the full-duplex receive jamming. An interesting future direction is to consider more sophisticated scenarios, where the eavesdropper knows the full-duplex strategy employed at the destination, and performs a similar full-duplex operation. In this case, non-cooperative game is a useful tool to study the interaction between the legitimate receiver and the eavesdropper.

6.4 Secure Full-Duplex Bidirectional Communications

In this section, secure bidirectional communication between a multiantenna access point (AP) and a user equipment (UE) is considered in the presence of a single passive eavesdropper. Future 5G systems will support direct bidirectional exchange of data between devices. To this end, full-duplex can support some main attributes of D2D communications such as low transmit power and symmetric traffic conditions between the nodes. In addition, as we show in this section, full-duplex operation can be utilized to enhance the secrecy performance of bidirectional communications.

6.4.1 System Model

Figure 6.5 shows a bidirectional communication system model consisting of one AP and one UE. The AP uses M_r antennas for receiving and M_t antennas for transmitting. Considering the constraints on the size and cost of the UE, we assume that it has two antennas, that is, one antenna for transmitting and the other for receiving. Both the UE and AP operate in the full-duplex mode. The eavesdropper is a single-antenna node and is mainly interested for decoding messages sent by the AP to the UE.

Let $\mathbf{h}_U \in \mathbb{C}^{M_r \times 1}$ be the uplink (UE-AP) channel, whereas $\mathbf{h}_D \in \mathbb{C}^{M_t \times 1}$ be the downlink (AP-UE) channel. The SI channels at the AP and UE are denoted by $\mathbf{H}_{rr} \in \mathbb{C}^{M_r \times M_t}$ and $h_m \in \mathbb{C}$, respectively. Let $\mathbf{g}_D \in \mathbb{C}^{M_t \times 1}$ be the AP-eavesdropper channel, whereas $g_U \in \mathbb{C}$ be the UE-eavesdropper channel. Then, the information rate at the AP can be expressed as

$$R_A = \log_2 \left(1 + \frac{P_U |\mathbf{w}_r^H \mathbf{h}_U|^2}{P_A |\mathbf{w}_r^H \mathbf{H}_{rr} \mathbf{w}_t|^2 + \sigma_A^2} \right),$$ (6.18)

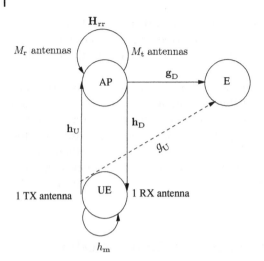

Figure 6.5 A full-duplex bidirectional communication system with an eavesdropper.

where $\mathbf{w}_r \in \mathbb{C}^{M_r \times 1}$ and $\mathbf{w}_t \in \mathbb{C}^{M_t \times 1}$ are the receive and transmit beamformers at the AP, respectively, P_U is the maximum transmit power of the UE, and P_A is that of the AP. The variance of the zero-mean AWGN at the AP is denoted by σ_A^2. Without loss of generality, it is assumed that $||\mathbf{w}_r|| = ||\mathbf{w}_t|| = 1$. Similarly, the information rate at the UE is given by

$$R_U = \log_2 \left(1 + \frac{P_A |\mathbf{h}_D^T \mathbf{w}_t|^2}{P_U |h_m|^2 + \sigma_U^2} \right), \tag{6.19}$$

where σ_U^2 is the variance of noise at the UE. Assuming that the eavesdropper is interested at decoding the signal broadcast by the AP, the eavesdropper information rate is given by

$$R_E = \log_2 \left(1 + \frac{P_A |\mathbf{g}_D^T \mathbf{w}_t|^2}{P_U |g_U|^2 + \sigma_E^2} \right), \tag{6.20}$$

where σ_E^2 denotes variance of noise at the eavesdropper. For notational simplicity, the effects of distance attenuation are included in the relevant channels. In this model, the secrecy rate is given by $C_s = \max(0, R_U - R_E)$.

6.4.2 Optimization for Secure Bidirectional Communications

The objective is to maximize the secrecy rate while satisfying that the AP information rate is above a threshold value, r_t. By changing this threshold value, the boundary of the secrecy-rate versus AP-rate region can be obtained.

The optimization problem is expressed as

$$\max_{||\mathbf{w}_r||=||\mathbf{w}_t||=1} C_s \text{ s.t. } R_A \geq r_t.$$

(6.21)

In this optimization problem, only R_A depends on $\mathbf{w}_r$. The optimum $\mathbf{w}_r$ is obtained from the optimization problem $\max_{||\mathbf{w}_r||=1} \dfrac{\mathbf{w}_r^H \mathbf{h}_U \mathbf{h}_U^H \mathbf{w}_r}{\mathbf{w}_r^H \left[P_A \mathbf{H}_{rr} \mathbf{w}_t \mathbf{w}_t^H \mathbf{H}_{rr}^H + \sigma_A^2 \mathbf{I} \right] \mathbf{w}_r}$. As such,

the solution of $\mathbf{w}_r$ is given by $\mathbf{w}_r = \dfrac{\mathbf{A}^{-1}\mathbf{h}_U}{||\mathbf{A}^{-1}\mathbf{h}_U||}$, where $\mathbf{A} \triangleq P_A \mathbf{H}_{rr} \mathbf{w}_t \mathbf{w}_t^H \mathbf{H}_{rr}^H + \sigma_A^2 \mathbf{I}$.

Substituting this $\mathbf{w}_r$ into R_A, we obtain $R_A = \log_2(1 + P_U \mathbf{h}_U^H \mathbf{A}^{-1} \mathbf{h}_U)$. Substituting R_U and R_E into Eq. 6.21, the optimization problem can be expressed as

$$\max_{||\mathbf{w}_t||=1} \frac{c_2}{c_1} \frac{|\mathbf{h}_D^T \mathbf{w}_t|^2 + c_1}{|\mathbf{g}_D^T \mathbf{w}_t|^2 + c_2}$$

$$\text{s.t. } \log_2 \left(1 + P_U \mathbf{h}_U^H \left[P_A \mathbf{H}_{rr} \mathbf{w}_t \mathbf{w}_t^H \mathbf{H}_{rr}^H + \sigma_A^2 \mathbf{I} \right]^{-1} \mathbf{h}_U \right) \geq r_t,$$

(6.22)

where $c_1 \triangleq \dfrac{P_U |h_m|^2 + \sigma_U^2}{P_A}$ and $c_2 \triangleq \dfrac{P_U |g_U|^2 + \sigma_E^2}{P_A}$. Since

$$\mathbf{h}_U^H \left[P_A \mathbf{H}_{rr} \mathbf{w}_t \mathbf{w}_t^H \mathbf{H}_{rr}^H + \sigma_A^2 \mathbf{I} \right]^{-1} \mathbf{h}_U = \frac{1}{\sigma_A^2} \left[\mathbf{h}_U^H \mathbf{h}_U - \frac{P_A |\mathbf{h}_U^H \mathbf{H}_{rr} \mathbf{w}_t|^2}{\sigma_A^2 + P_A ||\mathbf{H}_{rr} \mathbf{w}_t||^2} \right],$$

Eq. 6.22 can be further expressed as

$$\max_{||\mathbf{w}_t||=1} \frac{\mathbf{w}_t^H (\mathbf{h}_D^* \mathbf{h}_D^T + c_1 \mathbf{I}) \mathbf{w}_t}{\mathbf{w}_t^H (\mathbf{g}_D^* \mathbf{g}_D^T + c_2 \mathbf{I}) \mathbf{w}_t}$$

$$\text{s.t. } f \mathbf{w}_t^H (\sigma_A^2 \mathbf{I} + P_A \mathbf{H}_{rr}^H \mathbf{H}_{rr}) \mathbf{w}_t \geq \mathbf{h}_U^H \mathbf{H}_{rr} \mathbf{w}_t \mathbf{w}_t^H \mathbf{H}_{rr}^H \mathbf{h}_U,$$

(6.23)

where $f \triangleq \dfrac{1}{P_A} \left[||\mathbf{h}_U||^2 - \dfrac{(2^{r_t} - 1)\sigma_A^2}{P_U} \right]$. The objective function in Eq. 6.23 is

in Rayleigh quotient form. Let $\mathbf{B}$ and $\mathbf{C}$ be defined as $\mathbf{B} \triangleq \mathbf{h}_D^* \mathbf{h}_D^T + c_1 \mathbf{I}$ and $\mathbf{C} \triangleq \mathbf{g}_D^* \mathbf{g}_D^T + c_2 \mathbf{I}$, respectively. Then, $\max_{\mathbf{w}_t} \dfrac{\mathbf{w}_t^H \mathbf{B} \mathbf{w}_t}{\mathbf{w}_t^H \mathbf{C} \mathbf{w}_t}$ is equivalent to

$\max_{\mathbf{w}_t} \mathbf{w}_t^H \mathbf{C}^{-\frac{1}{2}} \mathbf{B} \mathbf{C}^{-\frac{1}{2}} \mathbf{w}_t$, whereas the constraint in Eq. 6.23 changes to

$$f \mathbf{w}_t^H \mathbf{C}^{-\frac{1}{2}} (\sigma_A^2 \mathbf{I} + P_A \mathbf{H}_{rr}^H \mathbf{H}_{rr}) \mathbf{C}^{-\frac{1}{2}} \mathbf{w}_t \geq \mathbf{h}_U^H \mathbf{H}_{rr} \mathbf{C}^{-\frac{1}{2}} \mathbf{w}_t \mathbf{w}_t^H \mathbf{C}^{-\frac{1}{2}} \mathbf{H}_{rr}^H \mathbf{h}_U.$$

(6.24)

However, the resulting optimization problem will be still non-convex. Thus, introducing a matrix auxiliary variable $\mathbf{W}_t = \mathbf{w}_t \mathbf{w}_t^H$ and relaxing rank-1 constraint on $\mathbf{W}_t$, the constrained secrecy-rate maximization is finally expressed

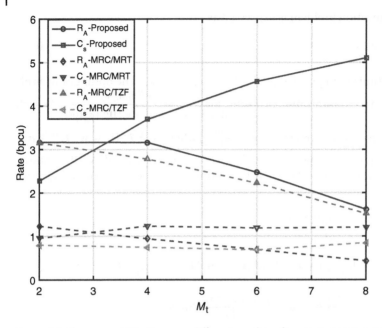

Figure 6.6 Secrecy and AP rate versus different number of transmit antennas at the AP.

as

$$\max_{\mathbf{W}_t} \operatorname{tr}(\mathbf{W}_t \mathbf{C}^{-\frac{1}{2}} \mathbf{B} \mathbf{C}^{-\frac{1}{2}})$$

$$\text{s.t.} \quad f \operatorname{tr}(\mathbf{W}_t \mathbf{C}^{-\frac{1}{2}} (\sigma_A^2 \mathbf{I} + P_A \mathbf{H}_{rr}^H \mathbf{H}_{rr}) \mathbf{C}^{-\frac{1}{2}}) \geq \operatorname{tr}(\mathbf{W}_t \mathbf{C}^{-\frac{1}{2}} \mathbf{H}_{rr}^H \mathbf{h}_U \mathbf{h}_U^H \mathbf{H}_{rr} \mathbf{C}^{-\frac{1}{2}}),$$

$$\operatorname{tr}(\mathbf{W}_t) = 1, \ \mathbf{W}_t \succeq 0. \tag{6.25}$$

This is a standard SDR problem with a single inequality constraint. As such, it can be shown that there exists a rank-1 optimum solution of $\mathbf{W}_t$ [53]. This means that the relaxation does not affect the optimality of Eq. 6.25.

6.4.3 Results and Discussion

In Figure 6.6, we demonstrate the effectiveness of the proposed secure full-duplex bidirectional communication method. To this end, the proposed optimum scheme is compared with different suboptimum schemes, such as MRC/maximum ratio transmission (MRT) and MRC/transmit ZF (TZF) schemes [55]. In the MRC/MRT scheme, the receive and transmit beamformers at the AP are given by $\mathbf{w}_r = \frac{\mathbf{h}_U}{||\mathbf{h}_U||}$ and $\mathbf{w}_t = \frac{\mathbf{h}_D^*}{||\mathbf{h}_D||}$, respectively. For MRC/TZF scheme, the same $\mathbf{w}_r$ is taken, whereas $\mathbf{w}_t$ is obtained by solving the following optimization problem

$$\max_{||\mathbf{w}_t||=1} \quad \mathbf{w}_t^H \mathbf{C}^{-\frac{1}{2}} \mathbf{B} \mathbf{C}^{-\frac{1}{2}} \mathbf{w}_t$$

$$\text{s.t.} \quad \mathbf{h}_U^H \mathbf{H}_{rr} \mathbf{w}_t = 0, \tag{6.26}$$

where the equality constraint represents the TZF constraint. Using similar approach as in Reference [53], the optimum TZF solution is given by

$$\mathbf{w}_t^{\mathrm{TZF}} = \mathbf{v}_{\max}\left(\mathbf{P}\mathbf{C}^{-\frac{1}{2}}\mathbf{B}\mathbf{C}^{-\frac{1}{2}}\mathbf{P}\right), \tag{6.27}$$

$\mathbf{v}_{\max}(\mathbf{X})$ denotes the eigenvector corresponding to the largest eigenvalue of $\mathbf{X}$, and $\mathbf{P}$ is the projection matrix given by $\mathbf{P} = \mathbf{I} - \frac{\mathbf{H}_{rr}^H \mathbf{h}_U \mathbf{h}_U^H \mathbf{H}_{rr}}{||\mathbf{h}_U^H \mathbf{H}_{rr}||^2}$. In Figure 6.6, we take $M = 10$, $\sigma_A^2 = \sigma_U^2 = \sigma_E^2 = 30$ dBm, $P_A = 40$ dBm, $P_U = 30$ dBm, and power of SI at both AP and UE as 30 dBm. It can be observed from this figure that the proposed method provides significantly higher secrecy rate than the MRC/MRT and MRC/TZF schemes. Moreover, the highest value of R_A is achieved with the proposed method. In this method, when M_t increases, the secrecy rate increases, whereas R_A decreases. The degradation in R_A is due to a decrease in M_r, which results in decreasing SI cancellation capability of the AP. When M_t increases, secrecy rate increases, since the AP has increasing degrees of freedom to supress signal leaked to the eavesdropper.

6.5 Secure Full-Duplex Relay Communications

In this section, we consider the problem of optimizing the receive and transmit relay beamformers in a secure full-duplex DF relay system [56]. First, the underlying optimization problem is equivalently expressed in terms of the relay transmit beamformer. Although the resulting optimization is non-convex and complicated to solve, we then equivalently reformulate it to a suitable form so that the optimum solution is obtained by efficiently solving semidefinite relaxation (SDR) problem, in conjunction with a line search.

6.5.1 System Model

As shown in Figure 6.7, we consider a DF relay communication system with a source S, a relay R, a destination (legitimate receiver) D, and a passive eavesdropper E. The source, destination, and eavesdropper are all equipped with a single antenna, whereas the FD relay is equipped with M antennas. M_r antennas of the relay are used for reception and the remaining $M_t \triangleq M - M_r$ are used for transmission. We assume that the relay estimates the S–R and R–D channels during the training phase. The eavesdropper is assumed to be an

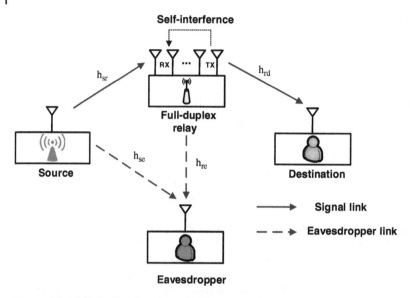

Figure 6.7 A full-duplex relay channel with an eavesdropper.

active user belonging to the legitimate network. Hence, the relay can estimate the E–R channel that is used as an estimate for the R–E channel using channel reciprocity. All channels are assumed to be flat fading and the corresponding channel estimation errors are negligible. This means that the obtained secrecy rate is an upper bound for the practical designs.

Let $\mathbf{h}_{sr} \in \mathbb{C}^{M_r \times 1}$, $\mathbf{h}_{rd} \in \mathbb{C}^{M_t \times 1}$, $h_{se} \in \mathbb{C}$, and $\mathbf{h}_{re} \in \mathbb{C}^{M_t \times 1}$ denote the S–R, R–D, S–E, and R–E channels, respectively. The signal received at the relay for the n-th time instant is expressed as

$$\mathbf{y}_r[n] = \sqrt{\frac{P_S}{d_1^\tau}} \mathbf{h}_{sr} s[n] + \mathbf{H}_{rr} \mathbf{r}[n] + \mathbf{n}_r[n], \tag{6.28}$$

where $s[n]$ is a zero-mean signal transmitted by the source with the variance $E[|s[n]|^2] = 1$, $\mathbf{H}_{rr} \in \mathbb{C}^{M_r \times M_t}$ is the residual SI channel at the relay, $\mathbf{r}[n] \in \mathbb{C}^{M_t \times 1}$ is the signal transmitted by the relay, and $\mathbf{n}_r[n] \in \mathbb{C}^{M_r \times 1}$ is the additive white Gaussian noise (AWGN) at the relay, having zero mean and the covariance matrix $E[\mathbf{n}_r[n]\mathbf{n}_r^H[n]] = \sigma_r^2 \mathbf{I}$. Moreover, d_1 is the source-relay distance, τ is the path loss exponent, and P_S is the transmit power of the source. The relay applies a beamformer, $\mathbf{w}_r \in \mathbb{C}^{M_r \times 1}$, to the received signal. The estimate of the relay is given by

$$\hat{s}[n] = \mathbf{w}_r^H \mathbf{y}_r[n] = \sqrt{\frac{P_S}{d_1^\tau}} \mathbf{w}_r^H \mathbf{h}_{sr} s[n] + \mathbf{w}_r^H \mathbf{H}_{rr} \mathbf{r}[n] + \mathbf{w}_r^H \mathbf{n}_r[n]. \tag{6.29}$$

Without loss of generality, we consider that $||\mathbf{w}_r|| = 1$. The relay applies a transmit beamformer, $\mathbf{w}_t \in \mathbb{C}^{M_t \times 1}$, to the estimated signal $\hat{s}[n]$. Thus, the relay's output is expressed as $\mathbf{r}[n] = \sqrt{P_R}\mathbf{w}_t\hat{s}[n - \delta]$, where δ denotes the processing delay of the relay, P_R is its transmit power, and without loss of generality, we assume that $||\mathbf{w}_t|| = 1$. Substituting $\mathbf{r}[n]$ into Eq. 6.29, the SNR for the S–R channel is given by

$$\gamma_{sr} = \frac{\frac{\kappa_1}{d_1^\tau}|\mathbf{w}_r^H \mathbf{h}_{sr}|^2}{\kappa_2|\mathbf{w}_r^H \mathbf{H}_{rr}\mathbf{w}_t|^2 + 1}, \tag{6.30}$$

where $\kappa_1 \triangleq \frac{P_S}{\sigma_r^2}$ and $\kappa_2 \triangleq \frac{P_R}{\sigma_r^2}$. Similarly, the SNR corresponding to the R–D link can be expressed as

$$\gamma_{rd} = \frac{\kappa_3}{d_2^\tau}|\mathbf{h}_{rd}^T \mathbf{w}_t|^2, \tag{6.31}$$

where d_2 is the relay-destination distance, $\kappa_3 \triangleq \frac{P_R}{\sigma_d^2}$, and σ_d^2 is the variance of zero-mean AWGN at the destination. Since the relay employs DF protocol, the end-to-end information rate from the source to the destination is given by

$$C_{df} = \log_2\left(1 + \min(\gamma_{sr}, \gamma_{rd})\right). \tag{6.32}$$

On the other hand, the signal received by the eavesdropper is

$$y_e[n] = \sqrt{\frac{P_S}{d_3^\tau}}h_{se}s[n] + \sqrt{\frac{P_R}{d_4^\tau}}\mathbf{h}_{re}^T\mathbf{w}_t\hat{s}[n - \delta] + n_e[n], \tag{6.33}$$

where d_3 and d_4 are the respective distances from the eavesdropper to the source and relay, and $n_e[n]$ is zero-mean AWGN at the eavesdropper, having the variance σ_e^2. Since $\delta \neq 0$, the S–E and R–E channels form an intersymbol interference channel. Assuming $\delta = 1$, the received signal of two time instants can be expressed as $\mathbf{y}_e = \bar{\mathbf{D}}\mathbf{x}_s + \mathbf{n}_e$, where $\mathbf{y}_e = [y_e[1], y_e[2]]^T$, $\mathbf{x}_s = [s[1], s[2]]^T$, $\mathbf{n}_e = [n_e[1], n_e[2]]^T$, and

$$\bar{\mathbf{D}} = \begin{pmatrix} \sqrt{\frac{P_S}{d_3^\tau}}h_{se} & 0 \\ \sqrt{\frac{P_R}{d_4^\tau}}\mathbf{h}_{re}^T\mathbf{w}_t & \sqrt{\frac{P_S}{d_3^\tau}}h_{se} \end{pmatrix}. \tag{6.34}$$

After some manipulations, the information rate for the eavesdropper can be expressed as

$$C_e = \frac{1}{2}\log_2\left(e^2 + \mathbf{w}_t^H \mathbf{D}\mathbf{w}_t\right), \tag{6.35}$$

where $e \triangleq 1 + \frac{\kappa_4 |h_{se}|^2}{d_3^\tau}$, $\mathbf{D} \triangleq \frac{\kappa_5 \mathbf{h}_{re}^* \mathbf{h}_{re}^T}{d_4^\tau}$, $\kappa_4 \triangleq \frac{P_S}{\sigma_e^2}$, and $\kappa_5 \triangleq \frac{P_R}{\sigma_e^2}$. Consequently, the secrecy rate, $C_s = \max \{0, C_{df} - C_e\}$, is expressed as

$$C_s = \max \left\{ 0, \log_2 \left(1 + \min \{\gamma_{sr}, \gamma_{rd}\}\right) - \frac{1}{2} \log_2(\gamma_e) \right\}, \tag{6.36}$$

where $\gamma_e \triangleq e^2 + \mathbf{w}_t^H \mathbf{D} \mathbf{w}_t$. The secrecy rate is a function of $\mathbf{w}_r$ and $\mathbf{w}_t$. In the following, an optimization method for maximizing C_s is proposed.

6.5.2 Proposed Optimization Solution

The objective is to maximize the information secrecy rate or equivalently $C_{df} - C_e$, which is expressed as

$$\max_{\{\|\mathbf{w}_r\|=1, \|\mathbf{w}_t\|=1\}} \left\{ \log_2 \left(1 + \min \left(\frac{\frac{\kappa_1}{d_1^\tau} |\mathbf{w}_r^H \mathbf{h}_{sr}|^2}{\kappa_2 |\mathbf{w}_r^H \mathbf{H}_{rr} \mathbf{w}_t|^2 + 1}, \frac{\kappa_3}{d_2^\tau} |\mathbf{h}_{rd}^T \mathbf{w}_t|^2 \right) \right) \right.$$
$$\left. - \frac{1}{2} \log_2 \left(e^2 + \mathbf{w}_t^H \mathbf{D} \mathbf{w}_t \right) \right\}. \tag{6.37}$$

In this optimization problem, only the SNR of the S—R channel depends on $\mathbf{w}_r$. Thus, the optimum $\mathbf{w}_r$ can be obtained by solving the following problem

$$\max_{\|\mathbf{w}_r\|=1} \frac{\mathbf{w}_r^H \mathbf{h}_{sr} \mathbf{h}_{sr}^H \mathbf{w}_r}{\mathbf{w}_r^H (\kappa_2 \mathbf{H}_{rr} \mathbf{w}_t \mathbf{w}_t^H \mathbf{H}_{rr}^H + \mathbf{I}) \mathbf{w}_r}. \tag{6.38}$$

The optimum solution of $\mathbf{w}_r$ can be expressed in terms of $\mathbf{w}_t$ as

$$\mathbf{w}_r = \frac{(\kappa_2 \mathbf{H}_{rr} \mathbf{w}_t \mathbf{w}_t^H \mathbf{H}_{rr}^H + \mathbf{I})^{-1} \mathbf{h}_{sr}}{\|(\kappa_2 \mathbf{H}_{rr} \mathbf{w}_t \mathbf{w}_t^H \mathbf{H}_{rr}^H + \mathbf{I})^{-1} \mathbf{h}_{sr}\|}. \tag{6.39}$$

Substituting Eq. 6.39 into Eq. 6.37, the resulting optimization problem turns to be a function of $\mathbf{w}_t$. Moreover, since $\max_x \log_2(f_1(x)) - \frac{1}{2} \log_2(f_2(x))$ is equivalent to $\max_x \frac{f_1(x)}{(f_2(x))^{\frac{1}{2}}}$ for positive-valued functions $f_1(x)$ and $f_2(x)$, the optimization problem is expressed as

$$\max_{\{\|\mathbf{w}_t\|=1\}} \frac{1 + \min \left(\frac{\kappa_1}{d_1^\tau} \mathbf{h}_{sr}^H (\kappa_2 \mathbf{H}_{rr} \mathbf{w}_t \mathbf{w}_t^H \mathbf{H}_{rr}^H + \mathbf{I})^{-1} \mathbf{h}_{sr}, \frac{\kappa_3}{d_2^\tau} |\mathbf{h}_{rd}^T \mathbf{w}_t|^2 \right)}{(e^2 + \mathbf{w}_t^H \mathbf{D} \mathbf{w}_t)^{\frac{1}{2}}}. \tag{6.40}$$

However, Eq. 6.40 is a complicated optimization problem and intractable. We demonstrate that this problem can be efficiently solved under a framework of convex optimization problems. By introducing an auxiliary variable u, where $u = \min\left(1 + \frac{\kappa_1}{d_1^\tau}\mathbf{h}_{sr}^H(\kappa_2\mathbf{H}_{rr}\mathbf{w}_t\mathbf{w}_t^H\mathbf{H}_{rr}^H + \mathbf{I})^{-1}\mathbf{h}_{sr}, 1 + \frac{\kappa_3}{d_2^\tau}|\mathbf{h}_{rd}^T\mathbf{w}_t|^2\right)$, Eq. 6.40 is equivalently expressed as

$$
\max_{\{u \geq 1, \|\mathbf{w}_t\|=1\}} \frac{u}{(e^2 + \mathbf{w}_t^H\mathbf{D}\mathbf{w}_t)^{\frac{1}{2}}}
$$

$$
\text{s.t.} \quad u \leq 1 + \frac{\kappa_1}{d_1^\tau}\mathbf{h}_{sr}^H(\kappa_2\mathbf{H}_{rr}\mathbf{w}_t\mathbf{w}_t^H\mathbf{H}_{rr}^H + \mathbf{I})^{-1}\mathbf{h}_{sr}, \tag{6.41}
$$

$$
u \leq 1 + \frac{\kappa_3}{d_2^\tau}|\mathbf{h}_{rd}^T\mathbf{w}_t|^2.
$$

Applying Sherman–Morrison formula to $r \triangleq \mathbf{h}_{sr}^H(\kappa_2\mathbf{H}_{rr}\mathbf{w}_t\mathbf{w}_t^H\mathbf{H}_{rr}^H + \mathbf{I})^{-1}\mathbf{h}_{sr}$, the first constraint in Eq. 6.41 can be expressed as

$$
\mathbf{w}_t^H\mathbf{H}_{rr}^H\mathbf{h}_{sr}\mathbf{h}_{sr}^H\mathbf{H}_{rr}\mathbf{w}_t \leq q(u)\left[1 + \kappa_2\mathbf{w}_t^H\mathbf{H}_{rr}^H\mathbf{H}_{rr}\mathbf{w}_t\right], \tag{6.42}
$$

where $q(u) \triangleq \frac{\|\mathbf{h}_{sr}\|^2}{\kappa_2} - \frac{d_1^\tau}{\kappa_1\kappa_2}(u - 1)$. After applying Eq. 6.42 in Eq. 6.41, it is evident that the joint optimization problem over u and $\mathbf{w}_t$ is still non-convex. However, for a given u, Eq. 6.41 can be expressed as

$$
\min_{\{\|\mathbf{w}_t\|=1\}} \mathbf{w}_t^H\mathbf{D}\mathbf{w}_t
$$

$$
\text{s.t.} \quad \mathbf{w}_t^H\mathbf{H}_{rr}^H\mathbf{h}_{sr}\mathbf{h}_{sr}^H\mathbf{H}_{rr}\mathbf{w}_t \leq q(u)\left[1 + \kappa_2\mathbf{w}_t^H\mathbf{H}_{rr}^H\mathbf{H}_{rr}\mathbf{w}_t\right],
$$

$$
\frac{d_2^\tau(u - 1)}{\kappa_3} \leq \mathbf{w}_t^H\mathbf{h}_{rd}^*\mathbf{h}_{rd}^T\mathbf{w}_t, \tag{6.43}
$$

which consists of non-convex quadratic inequality constraints. Introducing $\mathbf{W}_t \triangleq \mathbf{w}_t\mathbf{w}_t^H$ and relaxing the rank-1 constraint of $\mathbf{W}_t$, Eq. 6.43 can be expressed as the following SDR problem

$$
\min_{\mathbf{W}_t} \text{tr}\left(\mathbf{W}_t\mathbf{D}\right)
$$

$$
\text{s.t.} \quad \text{tr}\left(\mathbf{W}_t\mathbf{H}_{rr}^H\mathbf{h}_{sr}\mathbf{h}_{sr}^H\mathbf{H}_{rr}\right) \leq q(u)\left[1 + \kappa_2\text{tr}\left(\mathbf{W}_t\mathbf{H}_{rr}^H\mathbf{H}_{rr}\right)\right],
$$

$$
\frac{d_2^\tau(u - 1)}{\kappa_3} \leq \text{tr}\left(\mathbf{W}_t\mathbf{h}_{rd}^*\mathbf{h}_{rd}^T\right), \tag{6.44}
$$

$$
\text{tr}(\mathbf{W}_t) = 1, \ \mathbf{W}_t \succeq \mathbf{0}.
$$

Using a similar approach as in the optimization problem of Reference [53], it can be shown that $\mathbf{W}_t$ in Eq. 6.44 is rank-1, that is, the equality constraint $\mathbf{W}_t = \mathbf{w}_t\mathbf{w}_t^H$ holds true. Let $g(u)$ be the minimum objective function achieved in Eq. 6.44 for a given u. Then, the maximum secrecy rate is obtained from

$$\max_{u} \frac{u}{\left(e^2 + g(u)\right)^{\frac{1}{2}}}, \tag{6.45}$$

which can be solved by using a line search over u. Since $r \le ||\mathbf{h}_{sr}||^2$ and $|\mathbf{h}_{rd}^T \mathbf{w}_t|^2 \le ||\mathbf{h}_{rd}||^2$, the line search over u is confined to the following interval

$$u \in \left\{ 1, \min \left\{ 1 + \frac{\kappa_1 ||\mathbf{h}_{sr}||^2}{d_1^{\tau}}, 1 + \frac{\kappa_3 ||\mathbf{h}_{rd}||^2}{d_2^{\tau}} \right\} \right\}. \tag{6.46}$$

The proposed secure full-duplex relaying method is compared with the half-duplex one under *antenna conserved* scheme [22]. In the half-duplex method, the information rate of the two-hop relay link and the eavesdropper are, respectively, given by

$$C_{df,H} = \frac{1}{2} \log_2 \left(1 + \min \left(\frac{\kappa_1}{d_1^{\tau}} ||\tilde{\mathbf{h}}_{sr}||^2, \frac{\kappa_3}{d_2^{\tau}} |\tilde{\mathbf{h}}_{rd}^T \tilde{\mathbf{w}}_t|^2 \right) \right),$$

$$C_{e,H} = \frac{1}{2} \log_2 \left(1 + \frac{\kappa_4}{d_3^{\tau}} |h_{se}|^2 + \frac{\kappa_5}{d_4^{\tau}} |\tilde{\mathbf{h}}_{re}^T \tilde{\mathbf{w}}_t|^2 \right), \tag{6.47}$$

where, in contrast to the full-duplex mode, $\tilde{\mathbf{h}}_{sr}$, $\tilde{\mathbf{h}}_{rd}$, and $\tilde{\mathbf{h}}_{re}$ are $M \times 1$ channel vectors, and $\tilde{\mathbf{w}}_t$ is the $M \times 1$ beamforming vector. After some manipulations, the secrecy rate, $C_{s,H} = \max(0, C_{df,H} - C_{e,H})$, is given by

$$C_{s,H} = \max \left(0, \frac{1}{2} \log_2 \frac{\min \left(1 + \frac{\kappa_1}{d_1^{\tau}} ||\tilde{\mathbf{h}}_{sr}||^2, \tilde{\mathbf{w}}_t^H \mathbf{C} \tilde{\mathbf{w}}_t \right)}{\tilde{\mathbf{w}}_t^H \mathbf{A} \tilde{\mathbf{w}}_t} \right), \tag{6.48}$$

where $\mathbf{C} \triangleq \left(\mathbf{I} + \frac{\kappa_3}{d_2^{\tau}} \tilde{\mathbf{h}}_{rd}^* \tilde{\mathbf{h}}_{rd}^T \right)$ and $\mathbf{A} \triangleq \left(\left(1 + \frac{\kappa_4}{d_3^{\tau}} |h_{se}|^2 \right) \mathbf{I} + \frac{\kappa_5}{d_4^{\tau}} \tilde{\mathbf{h}}_{re}^* \tilde{\mathbf{h}}_{re}^T \right)$. It is obvious that $C_{s,H}$ is maximized by maximizing $\tilde{\mathbf{w}}_t^H \mathbf{C} \tilde{\mathbf{w}}_t$ and minimizing $\tilde{\mathbf{w}}_t^H \mathbf{A} \tilde{\mathbf{w}}_t$, which can be achieved by maximizing $\frac{\tilde{\mathbf{w}}_t^H \mathbf{C} \tilde{\mathbf{w}}_t}{\tilde{\mathbf{w}}_t^H \mathbf{A} \tilde{\mathbf{w}}_t}$. Thus, the optimum solution of $\tilde{\mathbf{w}}_t$ is given by $\tilde{\mathbf{w}}_t = \mathbf{v}_{max} \left(\mathbf{A}^{-1} \mathbf{C} \right)$, where $\mathbf{v}_{max}$ is the eigenvector corresponding to the maximum eigenvalue of $\mathbf{A}^{-1} \mathbf{C}$.

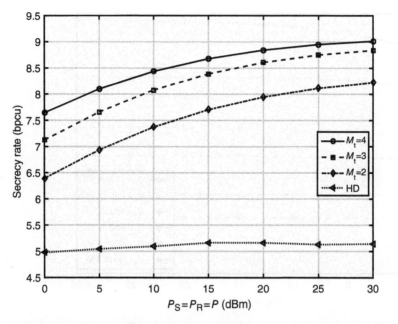

Figure 6.8 Secrecy rate versus P for different number of transmit antennas at the relay.

6.5.3 Results and Discussion

In this section, the effectiveness of the optimization method is demonstrated with computer simulations. The performance of the proposed full-duplex scheme is compared with the half-duplex one. For all simulations, we set $d_1 = d_2 = 40$ m, $M = 8$, $\sigma_r^2 = \sigma_d^2 = \sigma_e^2 = -80$ dBm, $P_S = P_R = P$, and $\tau = 3$. All results are obtained by averaging secrecy rate over 100 independent channel realizations.

Figure 6.8 shows the secrecy rate versus P for the proposed full-duplex scheme with different M_t. We take $d_3 = 200$ m and $d_4 = 50$ m in this figure. The secrecy rate increases with P. This result is worth noting, mainly due to the fact that the increasing source power can degrade secrecy rate since the source transmits omnidirectionally and only the relay has spatial degrees of freedom to minimize the power leaked to the eavesdropper. In fact, the result shows that the design is robust despite the source leaks power to the eavesdropper. On the other hand, the secrecy rate decreases when M_t decreases. This is due to relay's decreased signal suppression capability towards the eavesdropper. The performance of the half-duplex method is also displayed, where $M_r = M_t = 8$ due to the employment of the *antenna conserved* approach. The secrecy rate of the half-duplex scheme is approximately half of the full-duplex scheme. It is worth mentioning that the

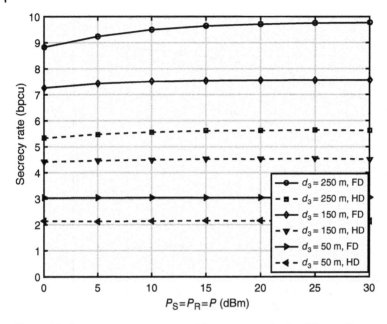

Figure 6.9 Secrecy rate versus power of the source or relay for different distances with the eavesdropper.

half-duplex mode uses twice the RF chains used in the full-duplex mode. Since the RF chains are more expensive than the antennas, the performance gain of the full-duplex mode over its half-duplex counterpart is achieved at a lower cost.

The secrecy rate versus P is demonstrated in Figure 6.9 for different values of d_3, where we take $d_4 = 50$ m and $M_t = 5$. It can be observed from this figure that the secrecy rate of the proposed full-duplex scheme decreases when d_3 decreases. This is due to the increased decoding capability of the eavesdropper that observes stronger S-E link. On the other hand, since the source is a single-antenna node, it does not have enough degrees of freedom to suppress the signal leaked to the eavesdropper.

6.6 Future Directions and Open Issues

Despite the fundamental results established and progress made with physical layer security techniques such as the ones described in this chapter, still much needs to be done to incorporate physical layer security solutions using full-duplex operation into practical system designs. Therefore, we now describe some open research issues that could establish future directions in the field.

Due to simultaneous transmissions, full-duplex operation in cellular networks can introduce increased levels of co-channel interference [57]. Therefore, in order to achieve performance gains, careful resource allocation is crucial. When physical layer solutions are to be created, the role played by the increased interference is not properly understood so far. For example, increased interference can lower the SINR at both useful node and eavesdropper nodes. Moreover, the amount of interference and mitigation at each node will be influenced by factors such as propagation characteristics, traffic conditions, and system configuration. Therefore, it remains an open issue to study the impact of co-channel interference and propose effective solutions to improve the secrecy performance.

Another important topic related to physical layer security is the secret key generation [58]. Specifically, one could exploit the temporal variation, channel reciprocity, and spatial decorrelation of a wireless channel to generate secret keys and exchange secure information. In the literature, such key generation methods have largely been studied assuming half-duplex operation. So far, the use of full-duplex operation for the generation of secret keys has not been thoroughly investigated.

5G wireless communications will include several important technological components such as massive MIMO, millimeter wave communications, NOMA, cognitive radio, and dense network operation. Combining full-duplex wireless with them is complementary and network performance gains can be improved. Moreover, they also provide several features for improving the security performance. For example, with massive MIMO pencil-like beams can be formulated with increased spatial resolution such that the leakage signal power in the direction of malicious users can be minimized. Similarly, millimeter wave transmissions experience high attenuation and thus signal strength at malicious user locations are reduced. If full-duplex operation is applied, further security schemes in combination with these systems can be implemented.

The schemes described in the previous sections are based on ideal assumptions such as perfect knowledge of CSI. However, in practice, available CSI is imperfect due to channel estimation errors as well as outdated due to user mobility and feedback delays. A terminal can leverage on full-duplex transmission to acquire accurate and timely CSI. To do so, a base station can transmit pilot sequences in the frequency used for uplink reception while SI cancellation can be used at the user to accurately estimate the channel. In secure communications with multiple antennas, accurate CSI is required for beamforming design. Specifically, beamforming designs must improve one's own signal strength while transmitting artificial noise aimed at the eavesdroppers and suppressing SI at the terminal. Clearly, such beamforming schemes are heavily dependent on the CSI accuracy and its timeliness. Therefore, it is important to design and optimize schemes robust against imperfect and outdated CSI conditions.

In full-duplex systems, SI severely reduces the performance. Although effective analog and digital SI cancellation schemes are available, they often require complex hardware that is out of the range for devices with low cost, low implementation complexity, and low power consumption. Therefore, in order to provide security solutions for 5G scenarios such as IoT, low-complexity SI cancellation schemes must be developed. To this end, recent advancements in miniaturized electronics, antenna design, and signal processing techniques would come useful.

Received SNR conditions dictate the performance achievable with half-duplex or full-duplex operation. As such, literature recommends hybrid operation of nodes where switching between the half-duplex and full-duplex modes of operation is performed by the network to realize the best performance. In the context of guaranteeing wireless security, there is a scope for work to be conducted to understand the operational regions where one mode of operation would yield significantly better performance than the other.

6.7 Conclusion

5G systems are expected to support a large number of users and devices with high throughput, low latency, and ultrareliable communications, which makes security a very important issue to address. To this end, physical layer security techniques based on the random features of the wireless fading channel are promising to implement secure data transmission. Moreover, physical layer security techniques can be combined with traditional encryption methods to provide well-integrated security solutions for 5G networks. Among various novel physical layer techniques introduced to 5G, full-duplex technology provides one of the most promising development paths to improve wireless security.

In this chapter, we studied several full-duplex-based physical layer security techniques that exhibit better secrecy performance compared to the conventional half-duplex operation. In particular, assuming multi-antenna bidirectional and relay communications, optimum beamforming designs for full-duplex wireless security solutions were proposed. Moreover, we also considered an alternative solution to conventional design of transmit jamming signals via multi-antenna full-duplex receiver that exploits its full-duplex operation mode to transmit jamming signals and confuse the eavesdroppers. We also highlighted future challenges and open issues for full-duplex secure transmission that need to be addressed when moving into 5G system implementation.

Acknowledgment

This work was supported in part by the Academy of Finland under Grant 288249, by the UK EPSRC under Grant EP/N007840/1, and by the European Commission H2020-MSCA-RISE-2015 under Grant 690750.

References

1 Y. Zou, J. Zhu, X. Wang, and L. Hanzo, "A survey on wireless security: technical challenges, recent advances, and future trends, *Proc. IEEE*, vol. 104, no. 9, pp. 1727–1765, 2016.

2 Ericsson, "5G security – scenarios and solutions," June 2015. Available at https://www.ericsson.com/assets/local/publications/white-papers/wp-5g-security.pdf.

3 J. Cao, M. Ma, H. Li, Y. Zhang, and Z. Luo, "A survey on security aspects for LTE and LTE-A networks," *IEEE Commun. Surv. and Tutor.*, vol. 16, no. 1, pp. 283–302, 2014.

4 J. G. Andrews, S. Buzzi, W. Choi, S. V. Hanly, A. Lozano, A. C. K. Soong, and J. C. Zhang, "What will 5G be?," *IEEE J. Select. Areas Commun.*, vol. 32, no. 6, pp. 1065–1082, 2014.

5 Z. Ling, J. Luo, Y. Xu, C. Gao, K. Wu, and X. Fu, "Security vulnerabilities of Internet of Things: a case study of the smart plug system," *IEEE Internet Things J.*, vol. 4, no. 6, pp.1899–1909, 2017.

6 N. Yang, L. Wang, G. Geraci, M. Elkashlan, J. Yuan, and M. D. Renzo, "Safeguarding 5G wireless communication networks using physical layer security," *IEEE Commun. Mag.*, vol. 53, no. 4, pp. 20–27, 2015.

7 M. Bloch and J. Barros, *Physical-Layer Security: From Information Theory to Security Engineering.* New York, NY: Cambridge University Press, 2011.

8 C. E. Shannon, "Communication theory of secrecy systems," *Bell Syst. Tech. J.*, vol. 28, no. 4, pp. 656–715, 1949.

9 A. D. Wyner, "The wire-tap channel," *Bell Syst. Tech. J.*, vol. 54, no. 8, pp. 1355–1387, 1975.

10 I. Csiszar and J. Korner, "Broadcast channels with confidential messages," *IEEE Trans. Inf. Theory*, vol. 24, no. 3, pp. 339–348, 1978.

11 H. V. Poor and R. F. Schaefer, "'Wireless physical layer security," *Proc. Natl. Acad. Sci. USA*, vol. 114, no. 1, pp. 19–26, 2017.

12 S. K. Leung-Yan-Cheong and M. E. Hellman, "The Gaussian wire-tap channel," *IEEE Trans. Inf. Theory*, vol. 24, no. 4, pp. 451–456, 1978.

13 M. Dohler, R. W. Heath, A. Lozano, C. B. Papadias, and R. A. Valenzuela, "Is the PHY layer dead?," *IEEE Commun. Mag.*, vol. 49, no. 4, pp. 159–165, 2011.

14 J. I. Choi, M. Jain, K. Srinivasan, P. Levis, and S. Katti, "Achieving single channel, full duplex wireless communication," in *Proc. 16th Annual International*

Conference on Mobile Computing and Networking (ACM MobiCom' 10), Chicago, IL, Sept. 2010, pp. 1–12.

15 International Telecommunications Union, "IMT traffic estimates for the years 2020 to 2030," 2015. Available at https://www.itu.int/dms_pub/itu-r/opb/rep/R-REP-M.2370-2015-PDF-E.pdf

16 M. Duarte, C. Dick, and A. Sabharwal, "Experiment-driven characterization of full-duplex wireless systems," *IEEE Trans. Wirel. Commun.*, vol. 11, no. 12, pp. 4296–4307, 2012.

17 A. Sabharwal, P. Schniter, D. Guo, D. W. Bliss, S. Rangarajan, and R. Wichman, "In-band full-duplex wireless: challenges and opportunities," *IEEE J. Select. Areas Commun.*, vol. 32, no. 9, pp. 1637–1652, 2014.

18 G. Zheng, I. Krikidis, J. Li, A. P. Petropulu, and B. Ottersten, "Improving physical layer secrecy using full-duplex jamming receivers," *IEEE Trans. Signal Process.*, vol. 61, no. 20, pp. 4962–4974, 2013.

19 D. Korpi, T. Riihonen, V. Syrjala, L. Anttila, M. Valkama, and R. Wichman, "Full-duplex transceiver system calculations: analysis of ADC and linearity challenges," *IEEE Trans. Wirel. Commun.*, vol. 13, no. 7, pp. 3821–3836, 2014.

20 E. Everett, A. Sahai, and A. Sabharwal, "Passive self-interference suppression for full-duplex infrastructure nodes," *IEEE Trans. Wirel. Commun.*, vol. 13, no. 2, pp. 680–694, 2014.

21 M. Heino, D. Korpi, T. Huusari, E. Antonio-Rodriguez, S. Venkatasubramanian, T. Riihonen, L. Anttila, C. Icheln, K. Haneda, R. Wichman, and M. Valkama, "Recent advances in antenna design and interference cancellation algorithms for in-band full duplex relays," *IEEE Commun. Mag.*, vol. 53, no. 5, pp. 91–101, 2015.

22 E. Aryafar, M. A. Khojastepour, K. Sundaresan, S. Rangarajan, and M. Chiang, "MIDU: enabling MIMO full duplex," in *Proc. 18th Annual International Conference on Mobile Computing and Networking (ACM MobiCom' 12)*, Istanbul, Turkey, Sept. 2012, pp. 257–268.

23 D. Bharadia, E. McMilin, and S. Katti, "Full duplex radios," in *Proc. ACM SIGCOMM' 13*, Hong Kong, China, Aug. 2013, pp. 375–386.

24 T. Riihonen, S. Werner, and R. Wichman, "Mitigation of loopback self-interference in full-duplex MIMO relays," *IEEE Trans. Signal Process.*, vol. 59, no. 12, pp. 5983–5993, 2011.

25 H. Q. Ngo, H. A. Suraweera, M. Matthaiou, and E. G. Larsson, "Multipair full-duplex relaying with massive arrays and linear processing," *IEEE J. Select. Areas Commun.*, vol. 32, no. 9, pp. 1721–1737, 2014.

26 S. Goel and R. Negi, "Guaranteeing secrecy using artificial noise," *IEEE Trans. Wirel. Commun.*, vol. 7, no. 6, pp. 2180–2189, 2008.

27 A. L. Swindlehurst, "Fixed SINR solutions for the MIMO wiretap channel," in *Proc. IEEE International Conference on Acoustics, Speech and Signal Processing (ICASSP)*, Taipei, Taiwan, Apr. 2009, pp. 2437–2440.

28 T. Riihonen, S. Werner, and R. Wichman, "Hybrid full-duplex/half-duplex relaying with transmit power adaptation," *IEEE Trans. Wirel. Commun.*, vol. 10, no. 9, pp. 3074–3085, 2011.

29 A. Mukherjee and A. L. Swindlehurst, "A full-duplex active eavesdropper in MIMO wiretap channels: construction and countermeasures," in *Proc. Asilomar Conference on Signal, Systems and Computers*, Pacific Grove, CA, Nov. 2011, pp. 265–269.

30 Y. Zhou, Z. Z. Xiang, Y. Zhu, and Z. Xue, "Application of full-duplex wireless technique into secure MIMO communication: achievable secrecy rate based optimization," *IEEE Signal Process. Lett.*, vol. 21, no. 7, pp. 804–808, 2014.

31 O. Cepheli, S. Tedik, and G. Karabulut Kurt, "A high data rate wireless communication system with improved secrecy: full duplex beamforming," *IEEE Commun. Lett.*, vol. 18, no. 6, pp. 1075–1078, 2014.

32 L. Li, Z. Chen, D. Zhang, and J. Fang, "A full-duplex Bob in the MIMO Gaussian wiretap channel: scheme and performance," *IEEE Signal Process. Lett.*, vol. 23, no. 1, pp. 107–111, 2016.

33 T. X. Zheng, H. M. Wang, Q. Yang, and M. H. Lee, "Safeguarding decentralized wireless networks using full-duplex jamming receivers," *IEEE Trans. Wirel. Commun.*, vol. 16, no. 1, pp. 278–292, 2017.

34 T. X. Zheng, H. M. Wang, J. Yuan, Z. Han, and M. H. Lee, "Physical layer security in wireless ad hoc networks under a hybrid full-/half-duplex receiver deployment strategy," *IEEE Trans. Wirel. Commun.*, vol. 16, no. 6, pp. 3827–3839, 2017.

35 B. Akgun, O. O. Koyluoglu, and M. Krunz, "Exploiting full-duplex receivers for achieving secret communications in multiuser MISO networks," *IEEE Trans. Commun.*, vol. 65, no. 2, pp. 956–968, 2017.

36 R. Feng, Q. Li, Q. Zhang, and J. Qin, "Robust secure beamforming in MISO full-duplex two-way secure communications," *IEEE Trans. Veh. Technol.*, vol. 65, no. 1, pp. 408–414, 2016.

37 T. Zhang, Y. Cai, Y. Huang, T. Q. Duong, and W. Yang, "Secure full-duplex spectrum-sharing wiretap networks with different antenna reception schemes," *IEEE Trans. Commun.*, vol. 65, no. 1, pp. 335–346, 2017.

38 F. Zhu, F. Gao, M. Yao, and H. Zou, "Joint information-and jamming-beamforming for physical layer security with full duplex base station," *IEEE Trans. Signal Process.*, vol. 62, no. 24, pp. 6391–6401, 2014.

39 F. Zhu, F. Gao, T. Zhang, K. Sun, and M. Yao, "Physical-layer security for full duplex communications with self-interference mitigation," *IEEE Trans. Wirel. Commun.*, vol. 15, no. 1, pp. 329–340, 2016.

40 Y. Sun, D. W. K. Ng, J. Zhu, and R. Schober, "Multi-objective optimization for robust power efficient and secure full-duplex wireless communication systems," *IEEE Trans. Wirel. Commun.*, vol. 15, no. 8, pp. 5511–5526, 2016.

41 Y. Wang, R. Sun, and X. Wang, "Transceiver design to maximize the weighted sum secrecy rate in full-duplex SWIPT systems," *IEEE Signal Process. Lett.*, vol. 23, no. 6, pp. 883–887, 2016.

42 G. Chen, Y. Gong, P. Xiao, and J. A. Chambers, "Physical layer network security in the full-duplex relay system," *IEEE Trans. Inf. Forensic Secur.*, vol. 10, no. 3, pp. 574–583, 2015.

43 S. Parsaeefard and T. Le-Ngoc, "Improving wireless secrecy rate via full-duplex relay-assisted protocols," *IEEE Trans. Inf. Forensic Secur.*, vol. 10, no. 10, pp. 2095–2107, 2015.

44 J. H. Lee, "Full-duplex relay for enhancing physical layer security in multi-hop relaying systems," *IEEE Commun. Lett.*, vol. 19, no. 4, pp. 525–528, 2015.

45 N. H. Mahmood, I. S. Ansari, P. Mogensen, and K. A. Qaraqe, "On the ergodic secrecy capacity with full duplex communication," in *Proc. IEEE International Conference on Communication (ICC' 17)*, Paris, France, May 2017, pp. 1–7.

46 H. Alves, G. Brante, R. Demo Souza, D. B. da Costa, and M. Latva-aho, "On the performance of full-duplex relaying under phy security constraints," in *Proc. IEEE International Conference on Acoustics, Speech and Signal Processing (ICASSP)*, Florence, Italy, May 2014, pp. 3978–3981.

47 A. El Shafie, A. Sultan, and N. Al-Dhahir, "Physical-layer security of a buffer-aided full-duplex relaying system," *IEEE Commun. Lett.*, vol. 20, no. 9, pp. 1856–1859, 2016.

48 H. He, P. Ren, Q. Du, and L. Sun, "Full-duplex or half-duplex? Hybrid relay selection for physical layer secrecy," in *IEEE Vehicular Technology Conference (VTC Spring)*, Nanjing, China, May 2016, pp. 1–5.

49 Q. Li, W.-K. Ma, and D. Han, "Sum secrecy rate maximization for full-duplex two-way relay networks using Alamouti-based rank-two beamforming," *IEEE J. Select. Top. Signal Process.*, vol. 10, no. 8, pp. 1359–1374, 2016.

50 N. P. Nguyen, C. Kundu, H. Q. Ngo, T. Q. Duong, and B. Canberk, "Secure full-duplex small-cell networks in a spectrum sharing environment," *IEEE Access*, vol. 4, pp. 3087–3099, 2016.

51 J. Vilela, M. Bloch, J. Barros, and S. W. McLaughlin, "Wireless secrecy regions with friendly jamming," *IEEE Trans. Inf. Forensic Secur.*, vol. 6, no. 2, pp. 256–266, 2011.

52 S. Boyd and L. Vandenberghe, *Convex Optimization*. Cambridge, U.K.: Cambridge University Press, 2004.

53 M. Mohammadi, B. K. Chalise, H. A. Suraweera. C. Zhong, G. Zhend, and I. Kirkidis, "Throughput analysis and optimization of wireless-powered multiple antenna full-duplex relay systems," *IEEE Trans. Commun.*, vol. 64, no. 4, pp. 1769–1785, 2016.

54 Y. Huang and S. Zhang, "Complex matrix decomposition and quadratic programming," *Math. Oper. Res.*, vol. 32, no. 3, pp. 758–768, 2007.

55 H. A. Suraweera, I. Krikidis, G. Zheng, C. Yuen, and P. J. Smith, "Low-complexity end-to-end performance optimization in MIMO full-duplex relay systems,"*IEEE Trans. Wirel. Commun.*, vol. 13, no. 2, pp. 913–927, 2014.

56 B. K. Chalise, W.-K. Ma, H. A. Suraweera, and Q. Li, "Secrecy rate maximization in full-duplex multiantenna decode-and-forward relay systems," in *Proc. IEEE International Workshop on Signal Processing Advances in Wireless Communications (SPAWC 2017)*, Sapporo, Japan, July 2017.

57 S. Goyal, P. Liu, S. S. Panwar, R. A. Difazio, R. Yang, and E. Bala, "Full duplex cellular systems: will doubling interference prevent doubling capacity?," *IEEE Commun. Mag.*, vol. 53, no. 5, pp. 121–127, 2015.

58 J. Zhang, T. Q. Duong, A. Marshall, and R. Woods, "Key generation from wireless channels: a review," *IEEE Access*, vol. 4, pp. 614–626, 2016.

Batu K. Chalise received the M.S. and Ph.D. degrees in electrical engineering from the University of Duisburg-Essen, Germany. He is currently an assistant professor in the Department of Electrical and Computer Engineering, New York Institute of Technology, NY, USA. He was a visiting assistant professor with the Department of Electrical Engineering and Computer Science, Cleveland State University, OH from 2015 to 2017. He was a wireless systems engineer with ArrayComm, San Jose, CA from 2013 to 2015 and a postdoctoral research fellow with the Center for Advanced Communications, Villanova University, PA, from 2010 to 2013. He has also held various research and teaching positions with the Catholic University of Louvain, Belgium, and University of Duisburg-Essen, Germany.

His research interests include signal processing for wireless and radar communications, wireless sensor networks, smart systems, and machine learning. He was a recipient of the U.S. Air Force Laboratory Summer Faculty Research Fellowship in 2016.

Himal A. Suraweera received the B.Sc. Eng. degree (Hons.) from the University of Peradeniya, Sri Lanka, in 2001, and the Ph.D. degree from Monash University, Australia, in 2007. He is currently a senior lecturer with the Department of Electrical and Electronic Engineering, University of Peradeniya. His research interests include relay networks, energy harvesting wireless communications, physical layer security, full-duplex and multiple input-multiple-output systems.

He is the recipient of the 2017 IEEE ComSoc Leonard G. Abraham Prize, the IEEE ComSoc Asia-Pacific Outstanding Young Researcher Award in 2011, the WCSP Best Paper Award in 2013, and the SigTelCom Best Paper Award in 2017. He was an editor of the *IEEE Journal on Selected Areas in Communications* (Series on Green Communications and Networking) from 2015 to 2016 and the *IEEE Communications Letters* from 2010 to 2015. He is currently serving as an editor of the *IEEE Transactions on Wireless Communications*, the *IEEE Transactions on Communications*, and the *IEEE Transactions on Green Communications and Networking*.

Gan Zheng received the B.Eng. and the M.Eng. from Tianjin University, Tianjin, China, in 2002 and 2004, respectively, both in electronic and information engineering, and the Ph.D. degree in electrical and electronic engineering from The University of Hong Kong in 2008. He is currently a senior lecturer in the Wolfson School of Mechanical, Electrical and Manufacturing Engineering, Loughborough University, UK. His research interests include edge caching, full-duplex radio, wireless power transfer, cooperative communications, cognitive radio, and physical-layer security. He is the first recipient of the 2013 IEEE Signal Processing Letters Best Paper Award, and he also received 2015 GLOBECOM Best Paper Award.

Risto Wichman received his M.Sc. and D.Sc. (Tech) degrees in digital signal processing from Tampere University of Technology, Tampere, Finland, in 1990 and 1995, respectively. From 1995 to 2001, he worked at Nokia Research Center as a senior research engineer. In 2002, he joined Department of Signal Processing and Acoustics, Aalto University School of Electrical Engineering, Finland, where he is a professor since 2008. His research interests include signal processing techniques for wireless communication systems.

7

Codebook-Based Beamforming Protocols for 5G Millimeter Wave Communications

Anggrit Dewangkara Yudha Pinangkis, Kishor Chandra, and R. Venkatesha Prasad

Electrical Engineering, Mathematics and Computer Science Department, Delft University of Technology, Delft, The Netherlands

7.1 Introduction

The explosive proliferation of mobile devices and ever increasing usage of data-hungry applications have resulted in the need of multi-Gbps wireless access. According to Cisco report [1], in 2015, wired networks still dominated 51% of IP traffic while Wi-Fi and mobile networks accounted for 48%. In 2020, as the number of mobile devices increase, it is predicted that the traffics from Wi-Fi and mobile device will grow significantly and will account for 66% of the global data traffic while the wired networks will only account for 34%. The increasing use of video applications that contribute to 75% of the mobile data traffic is one of the main contributing factors to this unprecedented surge in data traffic. The Next Generation Mobile Networks(NGMN) Alliance has identified several high data rate scenarios for 5G communications such as mass data download from a kiosk, 8K ultra-high definition wireless video transfer, augmented and virtual reality applications, video on demand systems in crowded public spaces, and in-vehicle environments requiring data rates ranging from tens of Mbps to multi-Gbps [2]. Furthermore, the mobile offloading and wireless fronthauling and backhauling would require enormous data rates. For example, extremely high quality (8K) video conferencing and gaming would require a per-user data rate of 300 Mbps in the downlink (DL) and 50 Mbps in the uplink (UL). Considering the mobile broadband access in a highly dense urban environment with 2500 connections/km^2, the resulting DL and UL traffic density would be around 750 Gbps/km^2 and 125 Gbps/km^2, respectively [3]. On the other hand,

5G Networks: Fundamental Requirements, Enabling Technologies, and Operations Management, First Edition.
Anwer Al-Dulaimi, Xianbin Wang, and Chih-Lin I.

current 4G system technology, that is, Long Term Evolution (LTE), can only provide a peak data rate of 100 Mbps per user with a DL traffic density 0.77 Gbps/km^2 [4]. Thus, it is evident that a huge gap exists between the future data traffic projections and the capacity of current mobile communication systems. To address this predicted surge in the mobile data usage, 5G networks target a 1000× increase in the existing network capacity [2].

Traditionally, reducing the cell size has been the main driver behind the network capacity growth from 2G to 4G cellular systems. However, the network densification resulting due to the closely spaced small cells in the sub-6 GHz frequency bands is interference limited. Although several techniques of coordinated transmission have been proposed to avoid the interference in small cells, the capacity is still limited by the intercell interference in dense small cell environments. Currently, Wi-Fi (IEEE 802.11b/g/n/ac) operating over 2.4/5 GHz dominates the indoor wireless space. Since its inception, Wi-Fi technology has gone through several amendments to meet the data rate requirements. Despite using very sophisticated physical (PHY) and medium access control (MAC) layer techniques, such as multiple user multiple-input multiple-output (MU-MIMO), higher order modulation and coding, channel bonding and frame aggregation, it is hard to improve the 2.4/5 GHz Wi-Fi data rates any further. For example, the IEEE 802.11ac uses channel bonding and multiuser MIMO schemes but it can only provide a peak data rate of around 1 Gbps because of the limited available bandwidth in the 5 GHz frequency band.

To achieve the targeted 1000× increase in the network capacity, many disruptive approaches are being pursued for 5G communications. These include dense small cell deployment, massive MIMO, millimeter wave (mmWave) radio access and the cloud radio access network (CRAN) architecture, and so on. Millimeter wave frequency band (30–300 GHz) has a large bandwidth availability in both licensed as well as unlicensed bands. This massive available bandwidth can support very high data rates in the order of multi-Gbps. This is why mmWave wireless access has become one of the most preferable air interface for supporting high data rate applications in 5G communications [5–7]. Because of the availability of large bandwidth, radio access and fronthauling and backhauling in the mmWave band (30–300 GHz) has emerged as a key candidate for the multi-Gbps wireless connectivity in the 5G communications [8]. The large frequency chunks are available in 27.5–29.5 GHz, 38.6–40 GHz, 57–66 GHz, 71–86 GHz, and 81–86 GHz bands comprising of both the licensed and unlicensed spectrum (see Figure 7.1). These frequency bands are being investigated for wireless personal area networks (WPANs), wireless local area networks (WLANs), mobile broadband access, and small cell fronthaul and backhaul connectivity in 5G networks. The unlicensed frequency band in 60 GHz band (57–66 GHz) has received most attention for short range high data rate communication resulting in standards such as IEEE 802.15.3c [9] and ECMA-387 [10] for WPAN applications and IEEE 802.11ad [11] for WLAN applications.

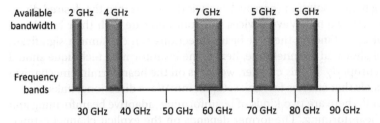

Figure 7.1 Available bandwidths in millimeter wave bands below 90 GHz [8].

The initial standardization efforts such as IEEE 802.15.3c and IEEE 802.11ad in the 60 GHz mmWave bands have mainly focused on WLAN/WPAN operations. However, recent measurement studies have supported the feasibility of mmWave-based mobile communication [12,13]. A further amendment to the 60 GHz WLAN standard IEEE 802.11ad is underway through the recently setup IEEE 802.11ay working group targeting a peak data rate of 20 Gbps using MU-MIMO and channel bonding in unlicensed mmWave bands above 45 GHz. Apart from the WLAN applications, IEEE 802.11ay targets new usage scenarios and applications, including broadband access in crowded public spaces, wireless connectivity in data centers, and fronthaul and backhaul communications in mmWave bands. There are several 5G-PPP projects under the Horizon 2020 Framework Program of the European Commission, such as mmMagic [14], FLEX5GWARE [15], and METIS [16] that are investigating mobile communications at mmWave frequencies.

mmWave-based small cells can provide much needed capacity gain due to the availability of large bandwidth. However, the mmWave signal propagation is significantly different from the sub-6 GHz signal propagations leading to several unique characteristics. As the free-space path loss is proportional to the square of carrier frequency, mmWave signals experience a significantly higher path loss as compared to the 2.4/5 GHz signals. To overcome the path loss, beamforming is used in the mmWave communication [17,18]. Thanks to the short wavelength in mmWave band making it feasible to pack a large number of antenna elements to enable high-gain beamforming. Beamforming results in additional antenna gain to overcome the path loss by focusing the signal power in the desired directions. Apart from the high path loss, an important propagation characteristic of mmWave signals is their limited ability to diffract around obstacles due to the short wavelengths. Furthermore, mmWave signals cannot penetrate through solid materials such as walls and metals. These properties make mmWave links highly susceptible to blockage from obstacles [19]. The highly directional beams that can be steered in the desired direction are much sought in mmWave communications to compensate for the high path loss as well as to circumvent the blockage by obstacles in order to find alternate paths.

Steering high-resolution (narrow beamwidth) beams into the desired direction such that two mmWave devices can communicate with their best beam pair is not easy. Finding the best beam directions can consume a significant amount of allocated channel time, hence, the beamforming technique should be chosen properly. In this chapter, we focus on the beamforming mechanism and protocols that enables fast link setup in mmWave directional links.

In general, there are two kinds of beamforming: adaptive beamforming and switched beamforming. The former depends on the explicit channel estimation that leads to computation and communication overheads, especially in mmWave communications with a large number of antenna elements implementing MIMO [20]. On the other hand, switched beamforming uses fixed predefined antenna sectors and employs MAC layer procedures to find the best beam direction. Antenna sectors are determined using predefined antenna weight vectors (amplitude and phase configurations) that are also called beam codebook. The codebook-based beamforming employs beam searching protocols at MAC layer that find the transmit and receive antenna orientations to find the best beam directions. IEEE 802.15.3c and 802.11ad have adopted codebook-based beamforming employing beam searching protocols at the MAC layer. These beamforming protocols rely on the exhaustive search where a considerable fraction of the allocated time slot is wasted in searching the best direction thus resulting in a very high beam searching overhead. Specially, in the dynamic channel conditions, frequent beam searching would be triggered that will significantly hamper the transmission capacity. Therefore, efficient MAC layer beamforming protocols are required that can find the appropriate beam directions with minimum overhead. Before discussing the beam searching algorithm, we provide a brief introduction of the underlying beamforming architectures as follows.

7.2 Beamforming Architecture

There are three kinds of beamforming architectures, namely, analog beamforming, digital beamforming, and the hybrid beamforming. Each beamforming architecture has its own trade-off between the complexity and the beamforming performance as explained in the following section.

7.2.1 Analog Beamforming

Analog beamforming is the simplest beamforming architecture. This architecture only uses a single radio frequency (RF) chain consisting of an analog to digital converter (ADC) and a mixer (up/down converter), as can be seen in the Figure 7.2. All antenna element weighting factors are applied in the analog domain. Although the analog beamforming is simple, controlling the beam in

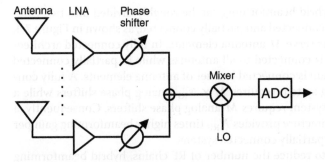

Figure 7.2 Analog beamforming architecture.

analog domain is not flexible. The phase shifts cannot be set to any arbitrary value. As a result, the beam performance (i.e., gain) depends on the resolution of the phase shifters that are attached at each antenna element.

7.2.2 Digital Beamforming

Digital beamforming uses digital signal processor to implement the beamforming algorithm. Since the antenna weighting factors are applied in the digital domain, the phase shift and the amplitude changes can be done flexibly. Each antenna element needs one RF chain that consists of an ADC and an up/down converter, as seen in the Figure 7.3. When a large number of antenna elements are used, the RF chain consumes a lot of power and it also increases the system cost.

7.2.3 Hybrid Beamforming

Hybrid beamforming can leverage the benefits of both analog and digital beamforming. It offers performance close to digital beamforming but with less num-

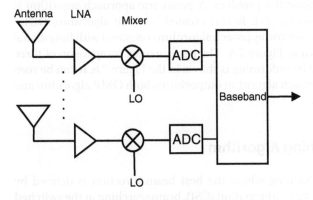

Figure 7.3 Digital beamforming architecture.

ber of RF chains. Hybrid beamforming can be roughly divided into two categories, namely, fully connected and partially connected, as shown in Figure 7.4, where N_{RF} RF chains serve M antenna elements. In fully connected architecture, each RF chain is connected to all antennas while in partially connected systems, each RF chain is connected to a set of antenna elements. A fully connected beamforming system requires $M \times N_{RF}$ analog phase shifters while a partially connected system requires M analog phase shifters. Consequently, a fully connected architecture provides N_{RF} times higher beamforming gain per transceiver than the partially connected system.

Due to its ability to reduce the number of RF chains, hybrid beamforming has emerged as a preferable solution in the 5G mmWave communications that exploits MIMO technology [21,22]. In particular, it is highly suitable for the beamforming employing massive MIMO technology where antenna arrays would consist of hundreds of antenna elements [23].

Hybrid beamforming approaches the ideal codebook (w) by multiplying predefined RF codebook (w_{RF}) and predefined baseband codebook (w_{BB}). In this case, w_{RF} is restricted by the set of feasible RF codebook $w_{RF_{dict}}$. Minimizing RMSE in $||w - w_{RF}w_{BB}||_2$ means that the beam in the hybrid architecture will be closer to the one in the fully digital architecture.

$$(w_{RF}^{opt}, w_{BB}^{opt}) = \underset{w_{RF}, w_{BB}}{\arg \min} || w - w_{RF}w_{BB} ||_2$$

$$\text{s.t.} [w_{RF}] \in w_{RF_{dict}} \tag{7.1}$$

$$|| w_{RF}w_{BB} ||_2^2 = 1.$$

There are some proposed algorithms to minimize the difference between the fully digital codebook and the hybrid codebook. To find the solution of Eq. 7.1, References [24] and [25] employ orthogonal matching pursuit (OMP) algorithm. In Reference [26], orthogonal matching pursuit with dynamic dictionary learning (DDL–OMP) algorithm, which is a modified version of OMP algorithm, is used to solve the problem. A geometric approach algorithm is also proposed in Reference [27]. Beams created in OMP algorithm, DDL–OMP algorithm, and geometric approach algorithm compared with fully digital beamforming are shown in Figure 7.5. The performance comparison of three algorithm in the hybrid beamforming is shown in the Figure 7.6. It can be seen that the geometric approach algorithm outperforms both OMP algorithm and DDL–OMP algorithm.

7.3 Beam Searching Algorithm

Unlike adaptive beamforming where the best beam direction is defined by measuring full channel state information (CSI), beam searching in the switched

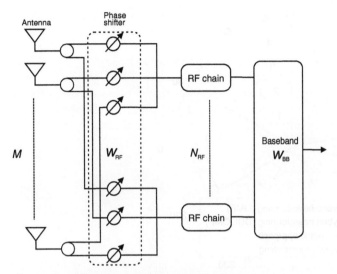

(a) Fully connected hybrid beam forming.

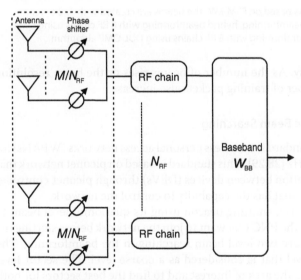

(b) Partially connected hybrid beam forming.

Figure 7.4 Hybrid beamforming architectures.

beamforming only depends on the measured signal quality of each predefined beams. Transmitter and receiver devices exchange their training packet to measure the channel quality for each beam candidate. Nevertheless, finding the best beam pair in the mmWave communication that exploits very narrow

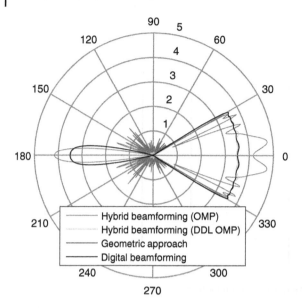

Figure 7.5 Beam patterns based on FSM-KW. The beams are created with 128 antenna elements in fully digital beamforming, hybrid beamforming with 8 RF chains using OMP algorithm, and hybrid beamforming with 8 RF chains using DDL OMP algorithm.

beamwidth is not easy. As the number of beams rises or the beam resolution gets higher, the number of training packets also increase.

7.3.1 IEEE 802.15.3c Beam Searching

IEEE 802.15.3c is a standard for wireless personal access networks (WPANs) for high data rate in 60 GHz [28,29]. This standard is based on piconet network that organizes communication between devices (DEVs) through piconet controller (PNC). PNC is a DEV that has the capability to control the network.

The PNC will start transmitting beacon using its quasiomni level beam to identify DEVs around the PNC that want to join the network before starting the beamforming. There are two level beam searching in the beamforming. The first level is sector level that is considered as a coarse level. The sector level beam is needed to find the area of interest and to find the best sectors for both DEVs. It will be followed by beam level where some beams will be covered by one sector. The beam level will select the best pair beams covered in the best selected sector. The beam searching, both in sector level and beam level, will be done in brute force search. Therefore, the number of training packets will be

$$N = \alpha(N_1^s \times N_2^s + N_1^b \times N_2^b), \qquad (7.2)$$

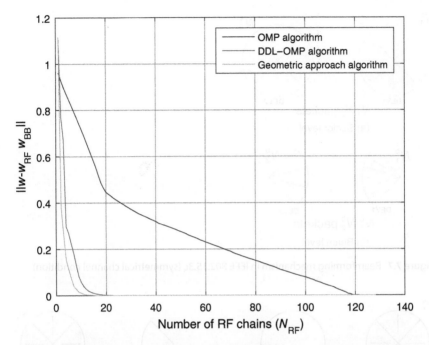

Figure 7.6 Comparison of root mean square error (RMSE), $||w - w_{RF}w_{BB}||_2$, with various number of RF chains in OMP algorithm, DDL–OMP algorithm and geometric approach algorithm. In this case, the beam created is based on FSM-KW beamforming, $A = 40$ and $M = 128$.

where $\alpha = 1$ when the channel is symmetrical and $\alpha = 2$ when the channel is asymmetrical. N_1^s and N_2^s denote number of sector candidates in DEV1 and DEV2, respectively, while N_1^b and N_2^b denote the number of beams covered by one sector in DEV1 and DEV2, respectively.

7.3.2 IEEE 802.11.ad Beam Searching

Instead of calling the device as DEV as in IEEE 802.15.3c, IEEE 802.11ad calls the device as station or STA. Before starting beamforming, STA1 will transmit beacons to initiate communication with STA2. The beacons are transmitted through its quasiomni beam (Figure 7.7).

Similar to IEEE 802.15.3c, IEEE 802.11ad also exploits two level of beam-training mechanism [30]. However, instead of using exhaustive beam searching at each level of beam searching as in IEEE 802.153c, IEEE 802.11ad uses its best sector or its best beam to receive the training packets. The beam-training mechanism is shown in Figure 7.8. By doing sector level beamforming and beam level beamforming, the total training packets will be

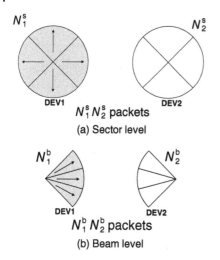

$N_1^s N_2^s$ packets

(a) Sector level

$N_1^b N_2^b$ packets

(b) Beam level

Figure 7.7 Beamforming mechanism in IEEE 802.15.3c (symmetrical channel condition).

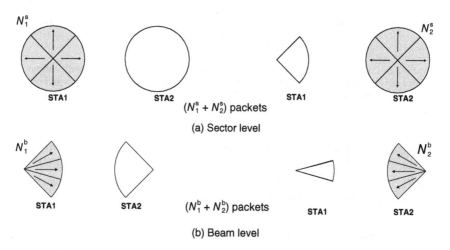

$(N_1^s + N_2^s)$ packets

(a) Sector level

$(N_1^b + N_2^b)$ packets

(b) Beam level

Figure 7.8 Beam searching mechanism in IEEE 802.11ad (symmetrical channel condition).

$$N = \alpha(N_1^s \times N_2^s + N_1^b \times N_2^b), \qquad (7.3)$$

where $\alpha = 1$ when the channel is symmetrical and $\alpha = 2$ when the channel is asymmetrical. N_1^s and N_2^s represent the number of sectors in STA1 and STA2, respectively. N_1^b and N_2^b represent the number of beams covered by one sectors in STA1 and STA2, respectively.

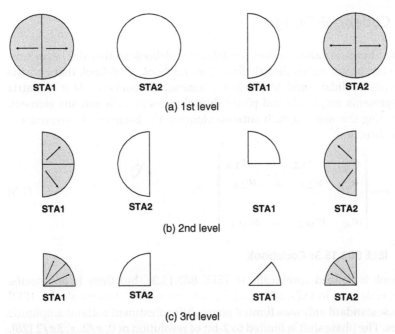

Figure 7.9 Illustration of beam searching mechanism in decrease-and-conquer (symmetrical channel condition).

7.3.3 Hierarchical Beam Searching

Instead of using only two-level beam searching as proposed in IEEE 802.15.3c and IEEE 802.11ad, hierarchical beam searching consists of multilevels. Hierarchical beam searching is proposed in References [26,31–35]. In order to minimize the beam training overhead, hierarchical beamforming generally uses two beam candidates at each level as can be seen in Figure 7.9. Therefore, the searching will follow a binary search-like algorithm.

By doing binary search-like beam searching, the total number of training packets will be

$$N = \alpha(2\log_2 N_1 + 2\log_2 N_2), \tag{7.4}$$

where N_1 is the number of possible finest beams in STA1, N_2 is the number of possible finest beams in STA2, $\alpha = 1$ when the channel is symmetrical and $\alpha = 2$ when the channel is asymmetrical. If N_1 and N_2 is associated with N_1^s, N_1^b, N_2^s and N_2^b as in IEEE 802.15.3c and IEEE 802.11ad, they will be equivalent with $N_1 = N_1^s \times N_1^b$ and $N_2 = N_2^s \times N_2^b$.

7.4 Codebook Design

Switched beamforming requires predefined codebook so that the beam candidates can be generated directly from the codebook. Codebook that consists of K beam candidates and M number of antenna elements is a $M \times K$ matrix that represents amplitude and phase shift changes at each antenna element. By adjusting the signal at each antenna element, the beam can be steered at a specific direction.

$$w = \begin{bmatrix} w_{1,1} & w_{1,2} & \cdots & w_{1,K} \\ w_{2,1} & w_{2,2} & \cdots & w_{2,K} \\ \cdots & \cdots & \cdots & \cdots \\ w_{M,1} & w_{M,2} & \cdots & w_{M,K} \end{bmatrix} \tag{7.5}$$

7.4.1 IEEE 802.15.3c Codebook

Codebook is defined specifically in IEEE 802.15.3c, but there is no specific codebook defined in IEEE 802.11ad. In order to reduce the complexity, IEEE 802.15.3c standard only uses limited phase shift adjustment without amplitude changes. The phase shift is limited to 2-bit of resolution or $0, \pi/2, \pi, 3\pi/2$ [28].

When the number of possible beams is more than the number of antenna elements, $K > M$, the codebook is defined as

$$w(m, k) = j^{\text{fix}\{ \frac{m \times \text{mod} \, [k + (K/2), K]}{K/4} \}} \quad \text{for} \quad m = 0 : M - 1 \text{ and } k = 0 : K - 1, \tag{7.6}$$

for a special case when $K = M/2$

$$w(m, k) = \begin{cases} (-j)^{\text{mod}(m,k)}, & \text{for} \quad m = 0 : M - 1 \text{ and } k = 0 \\ (-1)^{\text{fix}\{ \frac{m \times \text{mod}[k + (K/2), K]}{K/4} \}} & \text{for} \quad m = 0 : M - 1 \text{ and } k = 1 : K - 1 \end{cases} \tag{7.7}$$

Beam candidates with $M = 2$ $K = 4$ and $M = 4$ $K = 8$ are shown in Figure 7.10. It can be seen that due to limited phase shifter resolution, some beams cannot reach maximum gain. Moreover, there is a gap between adjacent beams that can be counted as loss. Therefore, to reduce the cusp loss, the number of antenna elements is set equal to twice the possible beams [36].

7.4.2 N-Phase Beamforming

N-phase beamforming is similar to IEEE 802.15.3c. However, this beamforming is designed to accommodate the availability of higher phase shift resolution.

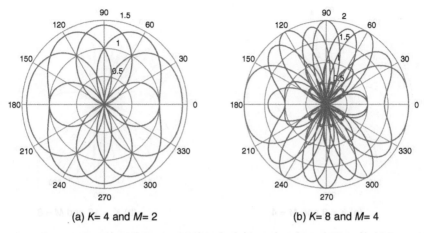

(a) $K = 4$ and $M = 2$ (b) $K = 8$ and $M = 4$

Figure 7.10 IEEE 802.15.3c beamforming with various number of beams.

Instead of using only 4-phase or 2-bits of resolution, this beamforming uses N-phase [37]. The codebook is defined as

$$w(m, k) = j^{\frac{2\pi}{N} \text{fix}\left\{ \frac{m \times \text{mod}[k+(K/2),K]}{K/N} \right\}} \quad \text{for} \quad m = 0 : M - 1 \text{ and } k = 0 : K - 1, \tag{7.8}$$

where M is number of antenna elements and K is number of possible beams.

Since the phase shifter resolution is higher than the one in IEEE 802.15.3c codebook, this beamforming offers better beamforming as depicted in Figure 7.11 where eight beams are generated with $N = 4$. All the beam candidates can reach maximum gain, unlike in IEEE 802.15.3c where some beams have lower array factor than the others. Nevertheless, for higher beam resolution higher phase resolution is needed.

7.4.3 DFT-Based Beamforming

Digital Fourier Transmorm (DFT)-based beamforming gives more flexible beams where each beam can reach the same maximum gain. However, this beamforming requires more flexible phase shift to compensate the flexibility of the beams.

DFT-based codebook can be formulated as

$$w(m, k) = e^{-j2\pi mk/K} \quad \text{for} \quad m = 0 : M - 1 \text{ and } k = 0 : K - 1, \tag{7.9}$$

where M is number of antenna elements and K is number of possible beams (Figure 7.12).

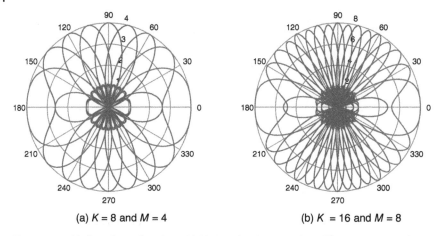

(a) $K = 8$ and $M = 4$ (b) $K = 16$ and $M = 8$

Figure 7.11 N-phase beamforming with N=8 and various number of beams generated.

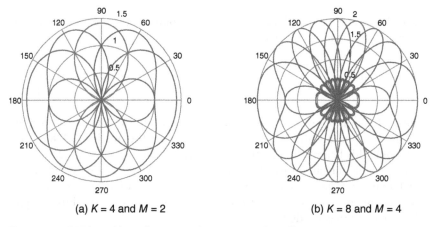

(a) $K = 4$ and $M = 2$ (b) $K = 8$ and $M = 4$

Figure 7.12 DFT-based beamforming with various number of beams.

From Eq. 7.9, it can be seen that the codebook requires higher phase shift resolution as K increases. The phase shift resolution in this codebook is $b = \log_2(K)$.

Compared to IEEE 802.15.3c, with the cost of higher phase shift resolution DFT-based codebook gives better beamforming performance where all of its possible beams can reach maximum gain. Beamforming in IEEE 802.15.3c when $K = 32$ suffers cusp loss 1.98 dB or losing about 36% of its power. If this beamforming is used in both DEV1 and DEV2, it is potential to loose 3.96 dB or 60% of its power. It can be seen that DFT codebook can minimize the losses where there is only 0.91 dB of loss. However, if both DEV1 and DEV2 suffers maximum cusp loss, it still have 1.92 dB of losses or about 35% of its power in total.

7.4.4 Fourier Series Method with Kaiser Window (FSM-KW) Beamforming

FSM-KW beamforming can be seen as a bandpass filter in the angle space domain [38]. This beamforming is intended to have uniform array factor in the desired beamwidth. Kaiser window is chosen in Reference [26] since the trade-off between sidelobe attenuation and beam transition width can be adjusted.

Beamforming codebook in FSM-KW can be defined as

$$w(m) = w_{\text{window}}(m)e^{-i\beta\psi_o}\frac{\sin(\beta\psi_b)}{\pi\beta} \text{ for } m = 0, 1, ..., M-1, \qquad (7.10)$$

where $w_{\text{window}}(m)$ are the Kaiser window samples and $\beta = m - M/2$ while ψ_o and ψ_b are given as

$$\psi_o = \frac{\pi}{2}(\cos\theta_1 + \cos\theta_2), \qquad (7.11)$$

$$\psi_b = \pi\sin(\theta_c)\sin\left(\frac{\Delta\theta}{2}\right) + \frac{\pi D}{M-1}. \qquad (7.12)$$

The D-factor can be defined from the sidelobe attenuation A (dB)

$$D = \begin{cases} \frac{A-7.95}{14.36}, & \text{if } A > 21 \\ 0.922, & \text{if } A \le 21. \end{cases} \qquad (7.13)$$

The γ factor can be calculated as

$$\gamma = \begin{cases} 0.11(A-8.7), & \text{if } A \geqslant 50 \\ 0.58(A-21)^{0.4} + 0.079(A-21), & \text{if } 21 < A < 5 \\ 0, & \text{if } A \le 21. \end{cases} \qquad (7.14)$$

The Kaiser window samples itself are given as

$$w_{\text{window}}(m) = \frac{I_0(\gamma\sqrt{1-4\beta^2})}{I_{0\times2}(\gamma)}, \qquad (7.15)$$

where I_0 is the zeroth order modified Bessel function of the first kind.

Unlike the IEEE 802.15.3c beamforming, N-phase beamforming, DFT-based beamforming, and FSM-KW beamforming can generate almost uniform array factor in the desired beamwidth with only small fraction of beamforming. Since the beamwidth and the beam direction can be set flexibly, this beamforming can be used in the hierarchical beam searching algorithm where the beam searching

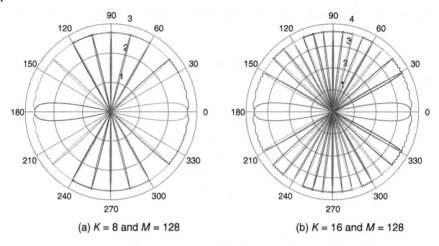

(a) $K = 8$ and $M = 128$ (b) $K = 16$ and $M = 128$

Figure 7.13 Beamforming based on FSM-KW beamforming that is used in Reference [26] and implemented on fully digital beamforming.

area can be divided into two areas at each beam searching level. The resulting beams using the FSM-KW beamforming can be seen in Figure 7.13.

7.5 Beamforming Evaluation

Comparing the three beam searching algorithms mentioned in Section 7.3, hierarchical beam searching has the lowest number of training packets if the hierarchical process is done in binary-like search. The comparison of training packets for some arbitrary number of beams is shown in Table 7.1. It can be concluded that hierarchical beam searching performance is better than the other mentioned beam searching methods.

Discussing how to design the beams, FSM-KW beamforming can accommodate hierarchical beam searching with low beam intersection and almost flat array factor in the 360° coverage. Although it costs more antenna elements compared to other beamforming, the complexity can be reduced with hybrid beamforming. Moreover, hybrid beamforming also offers ability to generate multiple beams that will be useful in the MIMO system.

Nevertheless, in general codebook design in IEEE 802.15.3c standard N-phase beamforming, and DFT-based beamforming, FSM-KW beamforming do not take into account the beam pattern of element antenna. The element pattern is assumed to have an isotropic beamforming that has the same gain in all angle direction. Therefore, the beam pattern is assumed to be the same as the array factor in the array antenna.

Table 7.1 Illustration of comparison of training packets in IEEE 802.15.3c, IEEE 802.11ad, and decrease-and-conquer beamforming, taking assumption that the channel is symmetric and both devices have the same number of sectors and beams.

Number of Sectors	Number of Beams per Sector	Number of Training Packets (Symmetrical Channel)		
		IEEE 802.15.3c	IEEE 802.11ad	Hierarchical (2 Beam Candidates)
4	2	20	12	12
4	4	32	16	16
8	8	128	32	24
8	16	320	48	28

In reality, it is impractical to steer the beam at low angle direction. A patch antenna that is usually used as an antenna element in the array antenna has limited beamwidth that will lead to scanning angle limitation, which is usually until $\pm 60°$ from the broadside direction [39,40]. Beamforming in more realistic condition when considering that a patch antenna has sinusoid beam pattern, $E(\theta) = \sin(\theta)$, as shown in Figure 7.14.

A patch antenna is usually preferred as an array antenna component due to its compactness, for example, patch antenna designed in Reference [41] has size of 1.7 mm $\times$ 2 mm. If this factor is not being considered such that 360° is only covered with one array antenna system, the beam at low-angle direction will have very low gain. Therefore, it is suggested to limit the beamforming scanning angle $\pm 60°$ from the broadside direction so that there are at least three antenna system to cover 360°.

7.6 Conclusion

In the context of 5G, mmWave communication has established itself as the main candidate for the multi-Gbps wireless connectivity in WPAN/WLANs and cellular access, backhaul and fronthaul in dense small cells. The small wavelength of mmWave signals allows close packing of a large number of antenna elements to form highly directional antenna arrays that can compensate for the high path loss observed at mmWave frequencies. However, forming a mmWave link using narrow beams is not easy as the receive and transmit antenna arrays need to find each other. The time spent to find the best beam directions has a direct consequence on the medium access utilization [42] as well as the initial access procedures [43] to discover an access point or base station. Generally, beamforming is performed using two approaches, namely, adaptive beamforming and switched beamforming. Switched beamforming outperforms adaptive

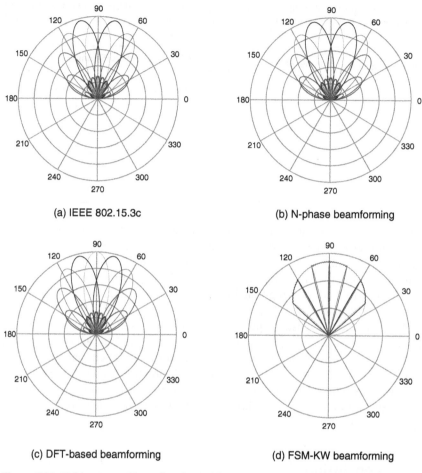

(a) IEEE 802.15.3c

(b) N-phase beamforming

(c) DFT-based beamforming

(d) FSM-KW beamforming

Figure 7.14 Field pattern of beamforming in (a) IEEE 802.15.3c, (b) N-phase, (c) DFT based, and (d) FSM-KW when the antenna element pattern is considered (taking assumption $E(\theta) = \sin(\theta)$).

beamforming in terms of computational complexity that has a direct impact on the baseband power consumption that is an important factor at mmWave frequencies. However, switched beamforming has limited flexibility where the beams have discrete angle directions. The discretization of beam directions introduces intersection loss (cusp loss) between the adjacent beam candidates. A compromising approach called hybrid beamforming that uses both the digital and analog beamforming has emerged as an attractive alternative to realize the mmWave beamforming. It is power efficient than the digital beamforming but can also provide satisfactory performance in terms of flexibility. Although hybrid beamforming is more complex than analog beamforming, the complexity cost gives the ability to transmit multiple simultaneous beams that can be utilized in mmWave MIMO systems in 5G communications to further increase the data rates.

References

1 Cisco, "Cisco Visual Networking Index: Global Mobile Data Traffic Forecast Update 2015–2020," *Cisco Public Information*, Feb. 2016.

2 NGMN Alliance, " 5G White Paper," 2015. Available at https://www.ngmn. org/uploads/media/NGMN 5G White Paper V1 0.pdf

3 5G PPP, " 5G PPP use cases and performance evaluation models," Version: 1.0, 2016. Available at https://5g-ppp.eu/wp-content/uploads/2014/02/5G-PPP-use-cases-and-performance-evaluation-modeling_v1.0.pdf

4 mmMAGIC, "Use case characterization, KPIs and preferred suitable frequency ranges for future 5G systems between 6 Ghz and 100 Ghz," ICT-671650-mmMAGIC/D1.1, Deliverable D1.1, 2015. Available at https://bscw.5g-mmmagic.eu/pub/bscw.cgi/d54427/mmMAGIC_D1.1.pdf

5 K. Chandra, Z. Cao, T. M. Bruintjes, R. V. Prasad, G. Karagiannis, E. Tangdiongga, H. P. A. van den Boom, and A. B. J. Kokkeler, "mCRAN: A radio access network architecture for 5G indoor communications," in *IEEE ICC 2015 – Workshop on Fiber-Wireless Integrated Technologies, Systems and Networks (ICC'15 – Workshops 09)*, 2015.

6 F. Boccardi, R. W. Heath, Jr., A. Lozano, T. L. Marzetta, and P. Popovski, "Five disruptive technology directions for 5G," *IEEE. Commun. Mag.*, vol. 52, no. 2, pp. 74–80, 2014.

7 K. Chandra, R. V. Prasad, and I. Niemegeers, "An architectural framework for 5G indoor communications," in *IEEE 2015 International Wireless Communications and Mobile Computing Conference (IWCMC)*. 2015, pp. 1144–1149.

8 A. Ghosh, T. A. Thomas, M. C. Cudak, R. Ratasuk, P. Moorut, F. W. Vook, T. S. Rappaport, G. R. MacCartney, S. Sun, and S. Nie, "Millimeter-wave enhanced local area systems: a high-data-rate approach for future wireless networks," *IEEE J. Select. Areas Commun.*, vol. 32, no. 6, pp. 1152–1163, 2014.

9 IEEE Standard for Information Technology, Telecommunications and Information Exchange between Systems Local and Metropolitan Area Networks – Specific Requirements. Part 15.3: Wireless Medium Access Control (MAC) and Physical Layer (PHY) Specifications for High Rate Wireless Personal Area Networks (WPANs) Amendment 2: Millimeter-Wave-Based Alternative Physical Layer Extension, Report, pp. 1–187, Apr. 2009.

10 Standard ECMA-387, High Rate 60 GHz PHY, MAC and HDMI PALs, Dec. 2010.

11 IEEE Standards Association, Draft Standard – Part 11:Wireless LAN Medium Access Control (MAC) and Physical Layer (PHY) Specifications – Amend-

ment 4: Enhancements for Very High Throughput in the 60 GHz Band, IEEE P802.11adTM/D9.0, July 2012.

12 T. Rappaport, S. Sun, R. Mayzus, H. Zhao, Y. Azar, K. Wang, G. Wong, J. Schulz, M. Samimi, and F. Gutierrez, "Millimeter wave mobile communications for 5G cellular: it will work!" *IEEE Access*, vol. 1, pp. 335–349, 2013.

13 C. Dehos, J. Gonzalez, A. De Domenico, D. Ktenas, and L. Dussopt, "Millimeter-wave access and backhauling: the solution to the exponential data traffic increase in 5G mobile communications systems?" *IEEE Commun. Mag.*, vol. 52, no. 9, pp. 88–95, 2014.

14 M. Tercero, P. von Wrycza, A. Amah, J. Widmer, M. Fresia, V. Frascolla, J. Lorca, T. Svensson, M.-H. Hamon, S. Destouet Roblot et al., "5G systems: the mmMAGIC project perspective on use cases and challenges between 6–100 Ghz," in *2016 IEEE Wireless Communications and Networking Conference*, 2016.

15 Flex5Gware, Flexible and efficient hardware/software platforms for 5G network elements and devices, 2015. Available at http://www.flex5gware.eu/

16 A. Osseiran, F. Boccardi, V. Braun, K. Kusume, P. Marsch, M. Maternia, O. Queseth, M. Schellmann, H. Schotten, H. Taoka, H. Tullberg, M. A. Uusitalo, B. Timus, and M. Fallgren, "Scenarios for 5G mobile and wireless communications: the vision of the METIS project," *IEEE Commun. Mag.*, vol. 52, no. 5, pp. 26–35, 2014.

17 W. Roh, J.-Y. Seol, J. Park, B. Lee, J. Lee, Y. Kim, J. Cho, K. Cheun, and F. Aryanfar, "Millimeter-wave beamforming as an enabling technology for 5G cellular communications: theoretical feasibility and prototype results," *IEEE Commun. Mag.*, vol. 52, no. 2, pp. 106–113, 2014.

18 K. Chandra, R. V. Prasad, B. Quang, and I. G. M. M. Niemegeers, "Cog-cell:cognitive interplay between 60 Ghz picocells and 2.4/5 Ghz hotspots in the 5G era," *IEEE Communications Magazine, Special issue on Emerging Applications, Services and Engineering for Cognitive Cellular Systems (EASE4CCS)*, 2015.

19 M. Jacob, S. Priebe, R. Dickhoff, T. Kleine-Ostmann, T. Schrader, and T. Kurner, "Diffraction in mm and sub-mm wave indoor propagation channels," *IEEE Trans. Microw. Theory Tech.*, vol. 60, no. 3, pp. 833–844, 2012.

20 K. Ramachandran, N. Prasad, K. Hosoya, K. Maruhashi, and S. Rangarajan, "Adaptive beamforming for 60 GHz radios: challenges and preliminary solutions," in *Proceedings of the 2010 ACM International Workshop on mmWave Communications: From Circuits to Networks*. ACM, 2010, pp. 33–38.

21 C. Kim, T. Kim, and J.-Y. Seol, "Multi-beam transmission diversity with hybrid beamforming for MIMO-OFDM systems," in *Globecom Workshops (GC Wkshps), 2013 IEEE*. 2013, pp. 61–65.

22 F. Sohrabi and W. Yu, "Hybrid digital and analog beamforming design for large-scale MIMO systems," in *2015 IEEE International Conference on Acoustics, Speech and Signal Processing (ICASSP)*, 2015, pp. 2929–2933.

23 T. L. Marzetta, "Noncooperative cellular wireless with unlimited numbers of base station antennas," *IEEE Trans. Wirel. Commun.*, vol. 9, no. 11, pp. 3590–3600, 2010.

24 O. El Ayach, R. W. Heath, S. Abu-Surra, S. Rajagopal, and Z. Pi, "Low complexity precoding for large millimeter wave MIMO systems," in *2012 IEEE International Conference on Communications (ICC)*, 2012, pp. 3724–3729.

25 O. El Ayach, S. Rajagopal, S. Abu-Surra, Z. Pi, and R. W. Heath, "Spatially sparse precoding in millimeter wave MIMO systems," *IEEE Trans. Wirel. Commun.*, vol. 13, no. 3, pp. 1499–1513, 2014.

26 D. De Donno, J. Palacios, D. Giustiniano, and J. Widmer, "Hybrid analog-digital beam training for mmWave systems with low-resolution RF phase shifters," in *2016 IEEE International Conference on Communications Workshops (ICC)*, 2016.

27 J. Palacios, D. De Donno, and J. Widmer, "Lightweight and effective sector beam pattern synthesis with uniform linear antenna arrays," *IEEE Antennas Wirel. Propag. Lett.*, vol. 16, pp. 605–608, 2016.

28 IEEE Standards Association, IEEE Std 802.15.3c - 2009, Part 15.3: Wireless Medium Access Control (MAC) and Physical Layer (PHY) Specifications for High Rate Wireless Personal Area Networks (WPANs), IEEE Std, 2009.

29 K. Chandra, A. Doff, Z. Cao, R. V. Prasad, and I. Niemegeers, "60 Ghz MAC standardization: progress and way forward," in *2015 12th Annual IEEE Consumer Communications and Networking Conference (CCNC)*, 2015, pp. 182–187.

30 I. C. Society, 802.11ad-2012 – IEEE Standard for Information Technology –Telecommunications and Information Exchange Between Systems – Local and Metropolitan Area Networks – Specific Requirements – Part 11: Wireless LAN Medium Access Control (MAC) and Physical Layer (PHY) Specifications Amendment 3: Enhancements for Very High Throughput in the 60 GHz Band. IEEE, 2012.

31 A. Alkhateeb, O. El Ayach, G. Leus, and R. W. Heath, "Channel estimation and hybrid precoding for millimeter wave cellular systems," *IEEE J. Select. Top. Signal Process.*, vol. 8, no. 5, pp. 831–846, 2014.

32 L. Chen, Y. Yang, X. Chen, and W. Wang, "Multi-stage beamforming codebook for 60GHz WPAN," in *2011 6th International ICST Conference on Communications and Networking in China (CHINACOM)*, 2011, pp. 361–365.

33 T. He and Z. Xiao, "Suboptimal beam search algorithm and codebook design for millimeter-wave communications," *Mob. Netw. Appl.*, vol. 20, no. 1, pp. 86–97, 2015.

34 S. Hur, T. Kim, D. J. Love, J. V. Krogmeier, T. A. Thomas, and A. Ghosh, "Millimeter wave beamforming for wireless backhaul and access in small cell networks," *IEEE Trans. Commun.*, vol. 61, no. 10, pp. 4391–4403, 2013.

35 Z. Xiao, T. He, P. Xia, and X.-G. Xia, "Hierarchical codebook design for beamforming training in millimeter-wave communication," *IEEE Trans. Wirel. Commun.*, vol. 15, no. 5, pp. 3380–3392, 2016.

36 J. Wang, Z. Lan, C.-S. Sum, C.-W. Pyo, J. Gao, T. Baykas, A. Rahman, R. Funada, F. Kojima, I. Lakkis et al., "Beamforming codebook design and performance evaluation for 60GHz wideband WPANs," in *2009 IEEE 70th Vehicular Technology Conference Fall (VTC 2009-Fall)*, 2009, pp. 1–6.

37 W. Zou, Z. Cui, B. Li, Z. Zhou, and Y. Hu, "N-phases based beamforming codebook design scheme for 60 Ghz wireless communication," *J. Beijing Univ. Posts Telecommun.*, vol. 35, no. 3, pp. 1–5, 2012.

38 S. J. Orfanidis, *Electromagnetic Waves and Antennas*. New Brunswick, NJ: Rutgers University, 2002.

39 R. J. Mailloux, *Phased Array Antenna Handbook*, vol. 2. Boston: Artech House, 2005.

40 T. C. Cheston and J. Frank, Phased Array Radar Antennas. *Radar Handbook*, McGraw Hill, 1990, pp. 7–1.

41 D. J. Joaquin, A new low-cost microstrip antenna array for 60 GHz applications, Ph.D. dissertation, Utah State University, 2016.

42 K. Chandra, R. V. Prasad, I. G. M. M. Niemegeers, and A. R. Biswas, "Adaptive beamwidth selection for contention based access periods in millimeter wave WLANS," in *2014 IEEE 11th Consumer Communications and Networking Conference (CCNC)* 2014, pp. 458–464.

43 M. Giordani, M. Mezzavilla, C. N. Barati, S. Rangan, and M. Zorzi, "Comparative analysis of initial access techniques in 5G mmwave cellular networks," in *IEEE 2016 Annual Conference on Information Science and Systems (CISS)*, 2016, pp. 268–273.

Anggrit Dewangkara Yudha Pinangkis is a master's student in electrical engineering at TU Delft. He obtained his bachelor's degree in telecommunication engineering from Institut Teknologi Bandung (ITB) in 2013. From 2013-2014, he was a research assistant at Radio Communication and Microwave Laboratory, ITB. His research interests include antenna beamforming, antenna design, and radar.

Kishor Chandra is a postdoctoral researcher at Delft University of Technology, The Netherlands, where he also received his Ph.D. in 2017. From 2009-2011, he was a research engineer with the Center for Development of Telemetics (CDOT), New Delhi, India, working on IP Multimedia Subsystems. He received his Master of Technology (M.Tech) in signal processing from Indian Institute of Technology (IIT) at Guwahati, India, in 2009 and his Bachelors of Engineering (B.E.) degree in electronics and communications engineering from K.E.C. Dwarahat, India, in 2007. He is awarded the prestigious Marie Curie Postdoctoral Individual fellowship (MSCA-IF-2017) for a 2-year research stay at CNRS/CentraleSupelec, Paris, France. He is also a recipient of ERCIM Alain Bensoussan Postdoctoral Fellowship. His research interests are in the area of 5G millimeter Wave Communications/Networking and Tactile Internet.

R. Venkatesha Prasad is an assistant professor at the Delft University of Technology, The Netherlands. He received the B.E. degree in electronics and communication engineering and M.Tech. degree in industrial electronics from the University of Mysore, Mysore, India, in 1991 and 1994, respectively, and the Ph.D. degree from the Indian Institute of Science, Bangalore, India, in 2003. He is a distinguished lecturer of IEEE and a senior member of IEEE and ACM.

Part II

Radio Access Technology for 5G Networks

8

Universal Access in 5G Networks: Potential Challenges and Opportunities for Urban and Rural Environments

Syed Ali Hassan,[1] Muhammad Shahmeer Omar,[1] Muhammad Ali Imran,[2] Junaid Qadir,[3] and Dushantha Nalin K. Jayakody[4]

[1] School of Electrical Engineering and Computer Science (SEECS), National University of Sciences and Technology (NUST), Islamabad, Pakistan
[2] School of Engineering, University of Glasgow, Glasgow, UK
[3] Information Technology University, Lahore, Pakistan
[4] National Research Tomsk Polytechnic University, Tomsk, Russia

8.1 Introduction

The demands for high data rates and ultra reliable coverage have led the researchers to pave the way for future wireless networks under the umbrella of 5G communications. Whereas 5G communications is an amalgamation of a multitude of technologies ranging from device-level algorithms such as low-power transmissions to system-level architectures such as software-defined networking (SDN), the challenges posed by each of these techniques are critical. 5G is known to work at the intersection of various techniques such as device-to-device (D2D) communications, massive multiple-input multiple-output (MIMO), millimeter wave (mmWave) communications, full-duplex transmissions, and Internet of Things (IoT) to name a few. However, an important challenge is to fill the gap between the rural and urban coverage in the context of *universal access* (UA). By UA, we mean that the advantages of all the technologies under 5G communications should be equally visible on every nook and corner of the globe, however, this stringent condition is not easy to meet.

Keeping in view of the above, we, in this chapter, aim to highlight the two divisions of the human populations living in either the urban or the rural

5G Networks: Fundamental Requirements, Enabling Technologies, and Operations Management, First Edition.
Anwer Al-Dulaimi, Xianbin Wang, and Chih-Lin I.

environments and state the technologies beneficial for these areas in terms of providing all promised benefits of 5G communications. Although the mmWave systems and massive MIMO techniques can be useful in dense and urban areas, the use of unmanned aerial vehicles (UAV) and drone communications with free-space optical or satellite links can be a way to provide 5G services to all areas of the globe.

The chapter is divided into two main streams: the first stream discusses the potential techniques for urban area coverages, which mainly include mmWave and massive MIMO systems, while the second portion discusses more on the rural areas and the technologies that can benefit the use of 5G in those areas.

8.2 Access for Urban Environments

This section will cover technologies and strategies proposed in literature to increase peak data rates while maintaining a coverage quality of service (QoS). Among the many techniques in literature, we focus on two core techniques of massive MIMO and mmWave systems. While both of these can be used effectively in urban and dense environments, they offer serious challenges to be deployed. Some of these opportunities and challenges are given further.

8.2.1 Massive MIMO

5G wireless communication links are anticipated to fortify emerging services with rigorous requisites on data rates, latency, and reliability. A new feature of the 5G communications includes the ability to serve a dense crowd of devices calling for novel ways of connecting to the network especially in urban environments. There are several situations where huge number of users gather in a limited area such as a shopping center. Macrocells that cover large spatial areas remain important to deliver continuous availability of enhanced mobile broadband services in dense urban situations [1].

The technology known as massive MIMO (mMIMO) is a succession from conventional MIMO when the number of transmit antennas at the base station (BS) exceeds a predefined number. It is a multiuser technology in which each base station is equipped with an array of M number of antenna elements and the BS utilizes them to communicate with K single-antenna terminals over the same time and frequency band [2]. By the use of random matrix theory, the authors in Reference [2] demonstrate that the adverse effects offered by noise and small-scale fading are diminished and that the required transmitted energy per bit is inversely proportional to the number of antennas in a MIMO cell. Furthermore, the fading effects can be completely turned down using the mMIMO technique. In general, as indicated in Reference [2], mMIMO can achieve 17 Mbps for each of 40 users in a 20 MHz channel in uplink and

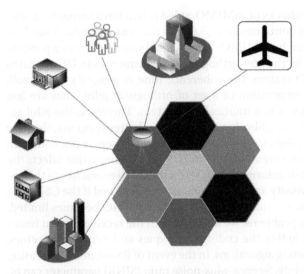

Figure 8.1 Massive MIMO in urban environment.

downlink and this rate will grow for cooperative scenarios. In Figure 8.1, we depict a mMIMO for an environment where multiuser scenario is assumed.

With a mMIMO system, a base station armed with a large number of antenna elements concurrently aids multiple single-antenna users in the same time–frequency resource. Some practical works done in Reference [3] used real-time wireless channels to demonstrate the efficacy of mMIMO systems. Through the channel measurements (2.6 GHz using a virtual uniform linear array (ULA) and a practical uniform cylindrical array (UCA) that is more dense in size) with 128 antenna terminals, it is determined that the theoretical rewards of this mMIMO can also be realized in practical situations. Therefore, with regard to this, the multiuser massive MIMO (MU-Massive MIMO) is identified as a potential candidate for urban environment. That can provide a reasonable quality of experience/ quality of service (QoE/QoS) for extremely demanding urban environments. One of the main necessities of 5G systems is to be able to aid many users in a dense setting. Massive MIMO can be seen as a promising solution to improve the data rates in densely populated urban areas.

By significantly increasing the spectral efficiency via large-scale antenna system in mMIMO, the support for universal access, especially, for urban environment is seen viable. When it comes to ultradensification, cost and interference have been recognized as the main encounter in urban placement [4]. In particular, massive MIMO is a suitable candidate for an efficient backhaul wireless link; a large number of user equipments (UEs) can be accommodated simultaneously as in the urban environments. This multiuser mMIMO set up amalgamates many base stations.

One of the main drawbacks of mMIMO-aided urban environment for universal access is the pilot contamination [5,6] and this sets an upper bound for spectral efficiency offered by mMIMO. The channel estimates from a particular UE is contaminated by the non-orthogonal pilot sequences of UE-allocated channels. In a mMIMO system, the coherence time is generally very small, which gives rise to the generation of a set of orthogonal pilots that are less than the number of devices in a multicellular system. Therefore, the pilot sequences are reused in the neighboring cells. When pilot reuse occurs, the base station gets interfering pilots from the neighboring cell users, a phenomenon commonly known as pilot contamination [6,7]. Pilot contamination affects the estimation of channel state information (CSI) on the receiver and hence the optimal detection. Full diversity and array gains are not achieved if the CSI is not perfect and hence the motivation behind using the mMIMO becomes limited. In addition to the above problems, we also deal with the receiver design issues in massive MIMOs, including the coding techniques and iterative detectors, precoding and beamforming algorithms. In the event of BS antennas increasing to infinity, the signal-to-interference-plus-noise ratio (SINR) parameter can be treated as an upper bound with a finite value relevant to the interference of pilot contamination [8,9]. This excessive interference signal can be treated as a useful energy source for massive MIMO-enabled base stations. This also helps to achieve green communication goals set up in the 5G technologies via radio frequency (RF) energy harvesting.

As a step for advance MU-Massive MIMO, some work has been done to use the spatial multiplexing techniques; this enables moving users like vehicles nodes [10] to help in boosting the capacity and enhancing the energy efficiency of the system. This new scheme combing with small cells improves the spectral efficiency. Use of small cells together with MU-Massive MIMO require more careful network planning for urban environments. This can also deploy in indoor environment. It helps in overcoming the high-penetration loss of outer to indoor (10–25 dB). Another important consideration in the urban mMIMO is the impact of large-scale shadowing, which generally does not average out with increasing the number of BS antennas. For instance, a natural model to study composite fading is the so-called K-fading that jointly captures the effects of multipath fading and shadowing. One of its parameters controls the shadowing intensity. The authors in Reference [11] have studied the impacts and showed that shadowing poses a stringent challenge for effective mMIMO operation. Figure 8.2 shows that increasing the BS antennas in a heavily shadowed environment does not increase the capacity of the system in a mMIMO scenario operating in K-fading. Here L represents the number of cochannel cells introducing pilot contaminations and v_0 and v_1 show the shadowing intensities of desired as well as interfering links, respectively. A large value of v shows low shadow intensity and vice versa. It can be seen that when the shadow intensity is

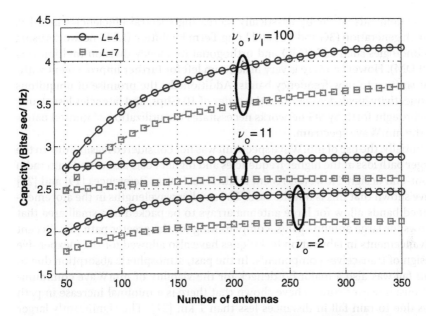

Figure 8.2 Capacity of mMIMO system in the presence of composite fading–shadowing. (Reproduced with permission of Reference 34. Copyright, 2012, IEEE.)

small, an elevation in the capacity can be observed showing that the multipath fading can be averaged out by increasing the BS antennas.

Many methods have been suggested to alleviate the impact of pilot contamination as this is especially harmful for urban environment. In References [**?**], a technique employing time shifting is proposed. Using channel correlation, a simple technique is proposed in References [14] and [15] to combat the pilot contamination among other approaches presented in the literature.

As already explained, the need of many antenna elements to deploy MU-Massive MIMO system for urban environment will present an issue of antenna size with many elements, which boils down to answering the question that which frequencies should be used by this scheme. Certainly, centimeter wave (cmWave) or millimeter wave (mmWave) bands can be used and are useful to reduce the array size [18]. Use of patch antennas and array antennas in the gigantic building in today's urban areas helps our targeted goals in 5G. However, the use of mmWave spectrum poses additional challenges that are discussed in the next section.

8.2.2 Millimeter Wave Technologies

8.2.2.1 Introduction and Background

The widespread use of smart mobile devices has seen with it an exponential increase in data requirements. Service providers have done well to meet these demands so far by drastically increasing spectral efficiencies in third, fourth generation (3G and 4G) and Long Term Evolution (LTE) networks using technologies such as MIMO and orthogonal frequency division multiplexing (OFDM). However, there is very little room left for further improvement without shifting to new frequency bands. Additionally, the promise of ubiquitous connectivity, peak data rates on the order of 10 Gbps, and latencies less than 1 ms brought forth by 5G networks necessitates the utilization of unused bands in the mmWave spectrum.

Initially, the use of mmWave spectrum was not considered feasible due to the larger path loss at the higher frequencies and higher costs associated with radio front-end design. However, recent studies, such as in References [19] and [20] have shown that this is not the case. The smaller wavelengths in the aforementioned bands allow for large antenna arrays to be packed in a small area that allows for large antenna gains, compensating for the greater path loss. Recent advancements in fabrication techniques have also allowed for the inexpensive design of transceiver components. In the past, atmospheric absorption due to rain fall was also a major bottleneck for the viability of mmWave communications; however studies have shown that there is a minimal increase in path loss due to rain fall in distances less than 1 km [21]. The significantly larger bandwidths available in the mmWave band directly translates to better data rates, which is one of the principal requirements of 5G networks. Highly directional beamforming antennas also enable deployment of mmWave networks in urban settings [22]. Combining mmWave transmissions with the different technologies already present in conventional microwave networks to enhance spectral efficiency promises to enable even greater data rates that provide an additional incentive to make the transition.

The mmWave spectrum is largely divided into five bands, each one of which has been allocated by the Federal Communications Commission (FCC) for a specific purpose. Nonetheless, these bands can still be used for mobile networks. Table 8.1 briefly describes the licenses assigned to each band and some of their prominent characteristics.

8.2.2.2 Analysis of mmWave Communication

8.2.2.2.1 Propagation Characteristics

One of the main concerns of using mmWaves in cellular communication was the larger path loss at the higher frequencies but studies, like References [19],[22] and [23] have shown that directional beamforming antennas have made mmWave communication viable for cell radii of up to 200 m, which is also the average size of small cells in current 4G and LTE systems. At such

Table 8.1 Different mmWave bands and their utilization and characteristics [25].

Frequency Band	Characteristics
28 GHz (27.5–29.5 GHz range)	The frequency ranges of 27.5–28.35 GHz and 29.1–29.25 GHz operate under FCC rules for fixed microwave services such as the fixed-satellite service (FSS), geostationary orbit (GSO), non-local television transmissions service (NON-LTTS), mobile satellite service (MSS), and non-geostationary orbit (NGSO)
38 GHz (36–40 GHz range)	A portion of this band is used for fixed microwave services, specifically point to point microwave links for backhaul
60 GHz (57–64 GHz range or V-band)	This band is currently unlicensed and shows high oxygen absorption characteristics but can still be used for mobile communications specifically in small cells. The standard 802.11ad is also expected to run in this band
70 GHz and 80 GHz (71–76 GHz and 81–86 GHz ranges, respectively, or the E-band)	This band falls under a lightly licensed paradigm as per FCC 101, however, high rain attenuation makes it suitable only for small cells with ranges up to 200m
90 GHz (92–95 GHz range or W-band)	This band is unlicensed for only indoor communications

distance, absorption loss due to rainfall is also minimal. A loss of 1 dB was recorded for a transmitter–receiver separation of 100 m [19].

Rigorous measurements were carried out in References [19] and [22] to determine the path loss of mmWave communications using channel sounding equipment in downtown New York, USA, and Austin, USA. The results showed that mmWaves have distinct propagation characteristics in line-of-sight (LoS) and non-line-of-sight (NLoS) environments. For example, the path loss exponent (PLE) in New York, at a carrier frequency of 28 GHz, was recorded to be 2.55 and 5.76 for LoS and NLoS paths, respectively. The best LoS PLE was 1.68 and the best NLoS PLE was recorded to be 4.58. Similarly, measurements showed that the LoS and NLoS PLEs in Austin, measured using a carrier frequency of 38 GHz, were 2.30 and 3.86, respectively. Subsequently, it can be inferred from the results that the PLEs also vary with the density of the scattering environments, that is LoS and NLoS PLEs are both higher in downtown New York than in Austin. This occurs because New York, being a major urban city, has a more dense scattering environment than the mildly urban region of Austin.

The channel measurements in Reference [19] also showed that there were significantly larger penetration losses at mmWave frequencies. This makes outdoor–indoor communications considerably harder. Typical construction materials such as bricks and tinted glass windows show penetration losses of 40.1 and 28.3 dB, respectively, at a distance of 5 m. This loss effectively isolates indoor and outdoor networks.

Small-scale measurements conducted in Reference [24] presented an interesting observation. The findings showed that wideband mmWave NLoS channels no longer exhibited characteristics of Rayleigh fading. The NLoS envelope followed a Rician distribution with a K factor between 5 and 8 dB. The LoS envelope also followed a Rician distribution, but with a K factor between 9 and 15 dB. This indicates that the individual or the sum of a few multipath components are adequately resolved at the receiver. Another interesting result of the experiment showed that the multipath components are spatially uncorrelated at distances greater than 2 and 5 wavelengths for LoS and NLoS environments, respectively. This has considerable implications for multiantenna diversity schemes in next-generation mmWave systems.

8.2.2.2.2 Blockage Models

The disparity in LoS and NLoS PLEs and relatively smaller cell radii in mmWave communications results in the need to effectively model blockages in a propagation environment. While simulations can be run without employing a blockage model, most tractable analysis use stochastic geometry to derive closed-form expressions for rate and coverage probability. Several blockage models have been proposed in current literature and each is briefly discussed. In all of these models, the probability of a link being a LoS link is given as a decreasing function of the transmitter–receiver separation distance. This is fairly easy to understand, as the greater distance, the less likely it is for the transmitter and receiver to be in LoS.

- 3GPP urban outdoor microcellular model [26]:

$$P_L(x) = \min(18/x, 1)(1 - e^{-x/36}) + e^{-x/36}, \tag{8.1}$$

where x is the distance (in meters) and P_L is the probability that a link is LoS.
- Exponential decay model [27]: This model is applicable to a scenario with outdoor users and both indoor and outdoor base stations (BSs). It is used to model blockages from buildings, using environmental statistics such as the percentage area covered by buildings (κ), the average building parameter (ρ), and the average building area (A).

$$P_L(x) = e^{-\beta x}, \tag{8.2}$$

where

$$\beta = \frac{-\rho \ln(1 - \kappa)}{\pi A} \tag{8.3}$$

- LoS ball model: In this model,

$$P_L(x) = \begin{cases} 1 & x \le D_0 \\ 0 & \text{otherwise,} \end{cases} \qquad (8.4)$$

where D_0 can be determined using moment matching techniques as outlined in Reference [28].
- Extended exponential decay model: This model extends the exponential decay model in Eq. 8.2 to incorporate blockages due to foliage in addition to the ones due to buildings, and is proposed in Reference [25].

$$P_L(x) = \begin{cases} e^{-\beta x + \eta}(1 - \min(ax + b, c)) & x < \frac{\eta}{\beta} \\ 1 - \min(ax + b, c) & \text{otherwise,} \end{cases} \qquad (8.5)$$

where β is given in Eq. 8.3, $\eta = \ln(1 - \kappa)$, and a, b, and c are environment variables, given in References [25] and [29].

The blockage models in Eqs. 8.1–8.4 are compared with simulation results using actual building locations in Reference [30]. The results show that the exponential decay model in Eq. 8.2 is the closest fit to simulation results for dense urban environments.

8.2.2.2.3 Coverage and Rate Performance of mmWave Networks

One of the defining characteristics of mmWave communications is that it is noise limited rather than interference limited as is the case for conventional microwave networks [30]. This means that there is very little or negligible interference in mmWave networks. This occurs due to the higher path loss and reduced LoS probability for interferers. The directional antennas proposed for mmWave networks is another reason for the reduced interference. Users not falling in the main lobe of transmitter antennas receive negligible signal strength from interfering BSs. This gives rise to new possibilities in spatial multiplexing.

The greater path loss in mmWave networks limits the signal-to-interference-plus-noise ratio (SINR) coverage probability in mmWave networks. This is evident from a common interpretation of the Frii's path loss equation for mmWave links used in literature [30], given by

$$\frac{P_r}{P_t} = \left(\frac{\lambda_c}{4\pi}\right)^2 R^{-\alpha}, \qquad (8.6)$$

where P_r is the received power, P_t is the transmit power, λ_c is the carrier wavelength, R is the link distance, and α is the path loss exponent, which is

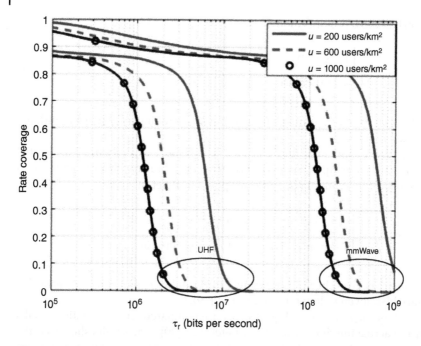

Figure 8.3 Rate coverage for mmWave and UHF networks for different user densities as given in Reference [31]. (Reproduced with permission of Reference 31. Copyright, 2016, IEEE.)

dependent on scattering environment and whether a link is LoS or NLoS in mmWave networks.

Assume a downlink in a mmWave network with a single user and a BS deployment process Φ. In such a network the SINR is calculated as

$$\text{SINR} = \frac{P_{\text{r},0}}{\sum_{k \in \Phi \setminus 0} P_{\text{r},k} + \sigma^2},\tag{8.7}$$

where $P_{\text{r},0}$ is the desired signal strength, $P_{\text{r},k}$ is the interference power from the kth interferer, and σ^2 is the noise. It is obvious from the equation that increasing path loss results in decreasing desired signal strength and decreasing interference power. This combined with the larger bandwidth makes mmWave networks noise limited. In spite of this fact, the received signal strength is also weaker that results in a loss in SINR. In order to improve coverage, a dense BS deployment scheme is required [28] that mandates greater infrastructure and the overall power consumption will also increase.

The greater bandwidth available at the mmWave frequencies (200 MHz at a carrier frequency of 28 GHz and 2 Ghz at a carrier frequency of 73 GHz), combined with improved spectral efficiencies in dense deployments offers a significant improvement in data rates over conventional microwave networks. This is verified by analytical and simulation results in References [28] and [31].

Figure 8.3 is a comparative study of the network rates in microwave and mmWave networks. In order to calculate the rate for the graphs, the authors use the relation

$$R = \frac{B}{N} \log_2(1 + \text{SINR}), \tag{8.8}$$

where B is the signal bandwidth and N is the number of users associated to a BS at a given moment in time. The graphs show that the mmWave rates are orders higher than the rates offered by microwave networks. It is pertinent to note that in this figure, the authors assumed a bandwidth of 2 GHz for mmWave and 20 MHz for microwave links.

Dense BS deployment, while an attractive solution to the mmWave network coverage problem has many limitations. First and foremost is the increased power consumption and infrastructure costs associated with dense network deployments. Second, overhauling the current network to support mmWave is not viable for 5G communication. A more effective solution would be to introduce hybrid networks that consist of both mmWave and microwave BSs. In this way a heterogeneous network can be set up with small cells operating in the mmWave band and macrocells operating to eliminate any coverage dead zones. Results in References [31]–[33] show that hybrid networks can significantly boost coverage with a minor loss in data rates. Hybrid networks can also serve to facilitate the transition to mmWave networks.

8.2.2.3 mmWave as a New Paradigm in Communications

There is a wide array of literature that suggests that mmWave networks will substantially increase data rates at little or no cost to coverage. Aside from high data rates, mmWave networks also promise improved spectral efficiency and opportunities for spatial multiplexing that can further increase data rates. This is without utilizing some of the existing techniques that are already used in 4G and LTE systems such as MIMO.

The small-scale fading characteristics determined in Reference [24] show how multiantenna diversity schemes can be used to substantially improve network performance in mmWave systems due to uncorrelated multipath components. However, there are several problems that need to be considered when designing mmWave MIMO systems. Such problems include hardware constraints due to high frequency and bandwidth and the use of large antenna arrays at both transmitters and receivers, as outlined in [28]. The use of separate RF chains for each antenna is difficult in mmWave communications, since the antennas are packed so close together. Analog-to-digital converters

(ADCs) and digital-to-analog converters (DACs) consume a large amount of energy particularly at mmWave frequencies, due to the larger bandwidths and sampling rates. Hence, a lot of research is happening into dividing the signal processing among the digital and analog phases of the RF chain so that low resolution or a lower number of ADCs/DACs can be used. Similarly, hybrid precoding and combining schemes are also being designed for mmWave communications. Distributed antenna diversity schemes, such as cooperative transmissions, are also expected to play a major role in next-generation mmWave networks.

The use of mmWave systems in 5G networks is essential if it is to produce on the promises of higher data rates, lower latencies, and ubiquitous connectivity. In urban cities it can be used to improve peak and edge data rates considerably. Not only does mmWave have applications in mobile communications, but it can also be used as an enabling technology for the Internet of Things (IoT) especially in a rich urban setting.

8.3 Providing Access to Rural Areas

8.3.1 Why Traditional Approaches Do Not Work for Rural Areas?

Mobile cellular technology, which reaches 80% of the human beings worldwide, can transform the lives of the "bottom billions" if we can achieve universal service access. Despite the unprecedented success of mobile technology, traditional cellular technology (2G, 3G, and 4G) still suffers from coverage problems, which have precluded universal service provisioning. In particular, mobile services have generally been limited to dense areas where revenues can compensate the capital expenditure (CAPEX) and their operating expenditure (OPEX) expended by mobile operators that are primarily incentivized by profit making. The lack of economic profitability has meant that operators have been loathe to bring their services to rural and remote sparsely populated areas. Another reason for the relatively poor penetration of cellular technology in rural areas is related to the affordability of services. Even though approximately 80–90% of the world's population live within the range of 2G/3G mobile signals, the reachability of mobile broadband is much more limited (only an estimated 10% of the world's population) [34].

8.3.2 Motivation for Aiming at Coverage in Rural Areas through 5G

...people lack many things: jobs, shelter, food, health care and drinkable water. Today, being cut off from basic telecommunications services is a hardship almost as acute as these other deprivations, and may indeed reduce the chances of finding remedies to them. –Kofi Annan, UN Secretary General.

We live in a world in which there is a great disparity between the lives of the rich and the poor. Currently, there are approximately 2 billion people who do not have access to mobile phone/ service. The *digital divide*–the differentiated opportunities available to various people on this planet due to their varying ability to access digital technology–is self-enforcing leading to increasing disparities between the rich and the poor. Although information and communication technology (ICT) has great potential in bridging this gap through the use of telecommunications, the benefits of ICT will only trickle down to the digitally deprived when our technologies are universally accessible and affordable.

Mobile telephony and the Internet has revolutionized all aspects of modern life. In the parlance of economists, access to mobile telephony has a *multiplier effect* on economic growth and employment. In particular, ICT can facilitate development in three important ways: (1) *inclusion* through which information barriers are overcome and opportunities are created and markets can be expanded; (2) *efficiency* that helps streamline economics transaction costs; and (3) *scale* that can be used to automate and facilitate services yielding economies to scale and positive network effects. For example, previous research shows that–everything else being the same–access to Internet and ICT can help improve a country's GDP by about 1% [35]. It is clear that to bring the benefits of telecommunications and the Internet to all, there is a need to prioritize the agenda of universal coverage for 5G.

While commissioning advanced 5G services that are envisioned for urban environments, which include high-bandwidth applications such as high-definition (HD) video streaming as well as very-low-latency applications such as tactile Internet and the Internet of Things (IoT) in rural environments, in a cost-effective fashion may be very challenging, there must be an effort to at least commission a "good-enough" 5G service that is sufficient for the more modest needs of rural users. This may be guaranteed by aiming for a more modest standard (e.g., a restricted but still contextually appropriate throughput to rural users) for the users in rural settings.

While the previous mobile technologies have largely failed to realize universal service provisioning (due to various technical, socioeconomical, and prioritization-related issues), 5G mobile cellular technology, which is currently in a formative phase and expected to be finalized around 2020, offers a clean slate to technologists and policy makers through which they can design for universal service provisioning. Keeping in mind the great social benefits of democratizing communications and through it access to the digital economy and the Internet the authors believe that universal service provisioning should be considered as one of the most important goal on the 5G research agenda.

8.3.3   5G Technologies Thrusts and Universal Coverage

"The digital divide is widening and is arguably of much larger concern than a local tenfold capacity increase in downtown Manhattan or in the streets of Tokyo." [36]

Since cellular technologies have traditionally not prioritized coverage as a primary goal, many of the adopted design choices are not perfectly in sync with the needs of universal mobile coverage and global access to the Internet for All (GAIA). To breakaway with the persistent universal coverage problems that have plagued every generation of the mobile standards, it is imperative that we critically analyze all aspects of the traditional mobile architectures.

Traditionally, cellular systems have had an overbearing focus on the peak data rate and have underemphasized universal coverage. It is conspicuous in the current inchoate 5G research that the overbearing focus of most of the stakeholders is not much different. Even though 5G is aiming for ambitious performance targets of 1000 increase in capacity; an edge rate of 100 Mbps; a peak data rate on the order of tens of Gbps; round trip latency of 1 ms, the coverage goals put forward have been modest (the 2015 Next Generation Mobile Networks Alliance (NGMN) white paper a relatively modest target of service 95% of the time at 95% of the locations–instead of aiming for universal coverage).

5G research is dominated by the three important trends of ultradensification, millimeter wave communication (mmWave), and massive MIMO (mMIMO). All of these three techniques are primed toward increasing the peak data rate, which may be useful in urban setting, however, are not ideal for universal coverage.

With ultradensification, we aim to provide more cells in a given area using technologies such as pico- and femtocells. The main market for ultradensification is mainly urban development whereas the coverage issues are mainly relevant for sparse rural environments. In terms of ultradensification, as the number of cells increase, cost and interference become important challenges. The cost issue related to ultradensification becomes even more pronounced when we consider rural/low-ARPU regions, which make the trend of ultradensification ill-suited for GAIA. In contrast to ultradensification, some researchers are exploring ultralarge cells (hundreds of kilometers in diameter) and macrocells [37]. The issues of reuse of existing infrastructure and the use of efficient energy resources also become important for the goal of universal service.

With millimeter wave communication (mmWave), the aim is to leverage more bandwidth by moving to the mmWave spectrum. Despite the great advantages that can come with mmWave as explained in the previous sections, it is important to note that the mmWave frequencies are not best suited for broad coverage due to the higher path loss in the mmWave band. The use of mmWave for wider coverage will entail more cost in terms of CAPEX (either due to the requirement of more mmWave base stations to achieve coverage, or due to the need of implementing high-complexity beamforming to overcome the attenuation). mmWave is being used by Facebook for their terrestrial connectivity system called Terragraph, which is a 60 GHz multinode system. But it is aimed at bringing high-speed Internet to dense urban areas. Terragraph uses the WiGig standard to provide in-room high-bandwidth communications. Since the 60 GHz band has poor propagation properties, Terragraph nodes are placed across the city at a distance of 200 to 250 m. The problem of interference in such settings is automatically ameliorated due to the vastness of available bandwidth and the signal absorbing nature of the band, this significantly simplifies the task of network planning. Thus, we conclude that while mmWave can have applications for universal coverage, it will be mostly restricted to bringing high-speed service to everyone in urban areas. The mmWave technology does not seem suited to rural areas in which relatively less number of users are spread out over large areas.

With massive MIMO (mMIMO), we can increase spectral efficiency via large-scale antenna systems. Even though mMIMO can provide great gains in rural/low-ARPU settings, the computational burden associated with it makes it unaffordable to the rural/low-ARPU region. Some researchers are working on ameliorating the computational burden of mMIMO since mMIMO, despite the computational challenges, has promised for universal coverage. It is worth pointing out that some research groups are exploring mMIMO technology for large coverage gains that can transfer to cost gains (fewer base stations per unit area). Researchers are also working on using mMIMO for energy reduction and for deploying ultralarge cell sizes (over 50 km) through the judicious use of massive antenna arrays. mMIMO is also being used by Facebook for their terrestrial connectivity systems called Project ARIES (Antenna Radio Integration for Efficiency in Spectrum). mMIMO uses spatial multiplexing and an antenna array at the base station to support multiple autonomous user terminals on the same time–frequency response. mMIMO techniques are versatile over a wide range of system parameters. In particular, the beamforming gain of mMIMO can be leveraged at lower frequencies to provide wide-coverage connectivity to rural areas.

8.3.4 Backhaul/Access Solutions for Rural Areas

For the operation of 5G to be sustainable in rural areas, there are many important challenges.

8.3.4.1 Terrestrial 5G Backhaul Solutions
The main backhaul challenges related to 5G can be seen in Reference [38]. If we restrict ourselves to wireless backhaul solutions, since wired backhaul solutions are often more expensive, there are two main approaches. The first is to use

a terrestrial wireless technology using technology such as Wi-Fi, WiMAX, or TVWS to develop long-distance backhaul links from rural 5G cells.

The use of Wi-Fi is seen in the so-called Wi-Fi over long distance links (WiLD) networks. WiLD links are usually point-to-point line-of-sight (LOS) wireless links that use high-gain directional antennas and communicate over long distances (10–100 km). The use of Wi-Fi is motivated by the commoditization of Wi-Fi that has resulted in the extensive availability of IEEE 802.11 hardware at a very cheap cost. Furthermore, Wi-Fi is license free, which adds to the appeal since it reduces administrative hurdles in setting up a rural setup.

The use of WiMAX is also promising for rural networks since they can also provide cheap long-distance communication. WiMAX technology also used line-of-sight links and directional antennas. Since WiMAX has a base-station-based network architecture, it requires a sufficient number of users to amortize the provisioning of expensive base stations, which makes the deployment of WiMAX in rural areas economically challenging. Compared to Wi-Fi, WiMAX is not as established and thus we anticipate that it will be more expensive than Wi-Fi due to licensing and equipment costs in the foreseeable future.

TV band whitespace (TVWS) is another upcoming technology that can provide a more competitive solution than Wi-Fi or WiMAX. The reason is that Wi-Fi and WiMAX operate at higher frequencies, where for the same equivalent isotropically radiated power, the communication range is much less than what is possible at the lower TV-band frequencies. This makes Wi-Fi and WiMAX ill-suited for long-distance wide-area wireless coverage. TVWS communication can benefit from the much better propagation characteristics at the lower frequencies and a host of other desirable features such as support for non-line-of-sight (NLOS) communication, including transmission through vegetation and small obstacles, which makes them highly suitable for various terrain configurations. These properties make TVWS very attractive for rural wireless networks. Broadly speaking, there are four key reasons for the wide interest in the use of TVWS in rural areas: (1) Low risk regulation: Since TVWS relies on a DSA model, the regulator does not face the problem of allocating the spectrum for years or to engage in auctions. (2) Ample availability of TV spectrum: In contrast to typical urban areas in North America and Europe, there are often wide unused spaces in developing regions (such as Sub Saharan African region) especially in the rural areas where there is an acute need of connectivity. (3) Suitability for rural environments: The use of UHF spectrum offers excellent propagation characteristics with the transmissions able to reach further, penetrate obstacles such as walls, and not require line of sight (this helps in reducing the infrastructural cost as the network can work now with fewer, more simpler, towers). (4) Opportunities for entrepreneurship: TVWS democratizes the wireless service market, which was earlier limited only to wealthy companies that could pay for licensed spectrum, and offers entrepreneurs to offer local competitive broadband services.

8.3.4.2 Airborne 5G Backhaul Solutions

Apart from terrestrial wireless technologies, rural networks can also make use of airborne infrastructure (such as satellites and aerial platforms including UAVs, drones, solar planes, etc.). The fact that satellite technology can provide much broader wireless coverage with lesser terrestrial infrastructure makes it well suited to rural networks. Satellite technology comes in various varieties, including those that make use of (1) geostationary (GEO) satellites; (2) medium Earth orbit (MEO) satellites; and (3) low Earth orbit (LEO) satellites. All these configurations have their own pros and cons. GEO satellites are typically expensive and require bulky user equipment; communication through GEO satellites also entails a high latency. LEO and MEO satellites, on the other hand, increase system complexity through the fact that these satellites do not remain with respect to the user on the ground.

In recent times, other aerial networking technologies are emerging such as those that make use of small UAV drones, balloons, low-altitude platforms (LAPs), and higher altitude platforms (HAPs). As can be expected, these technologies have their own bandwidth/ latency/ price/ performance trade-offs.

While satellites and aerial platforms cannot match cost effectively the performance of terrestrial 5G networks in terms of area spectral efficiency, these platforms have an advantage in terms of coverage. The wide coverage properties of satellites and aerial platforms make them an important solution that must be considered in any plan for 5G-based rural coverage. High-throughput satellites and aerial platforms can also be used in an adjunct capacity to augment other solutions. This stems from the fact that there will probably be no one-size-fits-solution that will meet all needs of 5G service. In areas where providing a terrestrial backhaul solution is difficult or cost-ineffective, satellite and aerial platforms can provide a viable alternative.

8.3.4.3 Joint Optimization of Access and Backhaul

A vast body of literature has investigated the optimization of radio access and the backhaul network individually and now there is a flurry of outputs that specifically deal with the joint optimization of the two interconnected networks. One important concept that relates the two parts of the end-to-end connectivity is the so-called functional split between the remote radio unit (RRU) and the base band unit (BBU). Depending on the positioning of the functional split the cost of the RRU and BBU changes and the corresponding costs of operating the resulting RRU/BBU unit is also affected. In addition to this, the requirements (both in terms of delay and throughput) of the backhaul are also dependent on the choice of the functional split. As a consequence, a designer needs to rely on an intricate model that captures all the interplay of these dependent variables to identify most suitable functional split given the available budget for the capital, operational, and backhaul cost to deploy a rural network. Details of such a tool can be found in Reference [39].

8.3.4.4 Application-Specific Design for Rural Coverage

The demand for different services in rural areas can be more compromising compared to the urban locations. This is owing to the fact that the only alternative to a compromised service is no coverage or no connectivity, which is not acceptable in this digital age. Due to this compromised expectation of many users in the rural area, the technology (both at radio access and the backhaul) can be adapted to reduce the operational as well as the capital expenditures to make the overall operation more cost-effective. In this regard, a viable solution is to use the older legacy technologies (3G, 2G, etc.) alongside the new but more economically available technologies like Wi-Fi for the connectivity of selected areas or hotspots in rural areas. The key performance indicators that can be adjusted with lower and more compromised targets include coverage (targeting only public places), latency/delay (using satellite backhaul if needed), throughput (using GPRS and older technologies for data), reliability (using wireless for the backhaul), and availability (using the delay tolerant services and UAVs for providing scheduled services). These compromises suit different applications individually. As an example, an email can easily tolerate the delay/latency; a text message will not be affected by throughput constraint, weather updates can cope with reliability if frequent updates are being transmitted.

If heterogeneous technologies are available in either a rural or an urban setting, it becomes necessary to utilize these depending on the offered performance by these technologies and the demanded requirements of the specific application we intend to serve. As an example, if an important health-related parameter needs to be monitored in real time, we would need to select the technologies that are reliable and have low latency. However, all radio access and backhaul technologies are not suitable for these performance metrics. So, a user with a specific application demand may need to select both the radio access as well as the backhaul in order to satisfy the requirements of this specific service. This will change the current mechanisms of selection of a best serving access point to adopt a more holistic approach of end-to-end connectivity, including both radio access and the backhaul link. The details of these innovative radio access point selection methods are presented in many recent publications, including Reference [41].

8.3.5 Cost-Effective Solutions to Enable Rural 5G

8.3.5.1 How to Reduce CAPEX

One important way CAPEX can be reduced is to explore how the need of infrastructure can be alleviated. This is the approach being taken by many solutions that propose airborne base stations (as in drones, balloons, or UAVs). For example, Google is exploring the use of solar-powered balloons that are controlled by winds of the stratosphere as a mechanism for providing cellular connectivity. Facebook, on the other hand, is using UAVs to deploy aerial base stations that beam their signals on the ground that can be used by end customers. CAPEX can also be reduced by avoiding duplication of infrastructure by using methods devised for resource sharing and resource pooling. In this regard, new methods of virtualization and resource sharing through software-defined networks (SDN), network functions virtualization (NFV), and cloud technology can be useful. Last, CAPEX can also be reigned in by encouraging the use of commodity devices, even if they have to be hacked according to the context (e.g., Wi-Fi, which is originally a local area networking technology has been effectively used as a low-cost long-distance technology).

8.3.5.2 How to Reduce OPEX

Along with the imperative of reducing CAPEX, there is also a need of reducing the recurring OPEX costs. This is a multidimensional issue solving which requires efforts in all the operations of 5G. One way to reduce the operating costs is to reduce the energy consumed through the use of energy-efficient methods, for example, this can be achieved by using different power states (such as full power, low power, or off) in different settings according to the context of the users, application, and the operator. Another way is to consider and develop new architectures—such as the cloud radio access network (CRAN) architecture—that are more energy efficient. One can also consider the use of renewable energy sources as solar power and wind power, this is especially important in rural areas in developing countries that have limited and intermittent power supply.

8.3.5.3 How to Jointly Optimize the CAPEX and OPEX

As discussed in the previous section, the CAPEX and OPEX of a cellular network not only depends on the choice of the density of the access points, the capability and technology on the radio access, and the capability and technology of the backhaul/fronthaul but also on several other design choices (like the functional split choice between the RRU and BBU). Sometimes, a costlier investment as a hardware functionality can reduce the long-term running cost of the network. Hence, the overall comparison for feasibility should also incorporate an estimate of the running life of a deployment and should be able to identify the break-even point considering the payback time for any investment in terms of more expensive hardware that can reduce the running cost. One example is the energy cost of an old generation RRU/BBU equipment compared to their capital cost when compared to a new generation equivalent units. Although the capital cost may be significantly higher for a new generation device but the amount of power saving will easily justify the higher capital cost by recovering the difference even in a couple of years. Hence, a more holistic framework of cost comparisons is needed to make such decisions. An example of such a framework is given in Reference [42].

8.3.5.4 Use of Self-Organized Networking for Rural Coverage

One of the most costly expenses in running a cellular network is the need for trained human resources (HR) to configure, optimize, and fix a running mobile network. This cost becomes even higher for a remote location since it involves the travel and related expenses for frequently sending or relocating the required human resource. For the more demanding future generations where the reconfigurations are needed more frequently, this model becomes completely infeasible if we solely rely on trained human resource. Self-organized networking is an emerging trend that offers possible solutions to this dilemma. Using the knowledge base of trained human resources and the more powerful capability of artificial learning-enabled cellular networks to acquire and strengthen the knowledge base, we can design a futuristic cellular network relying on self organised networking (SON) [44]. SON-enabled networks will be able to configure the devices and nodes when deployed [43], [44], optimize the parameters when the surrounding environments or the traffic demands change [43], heal the basic and more frequently occurring faults in the network [45] that require software-based updates or report the hardware faults after self-diagnosis. All these functionalities significantly reduce the required costs of commissioning and running a cellular network making it feasible for deployment in rural areas. It also opens the doors for community-based networking efforts with communities who do not have trained manpower in their own communities.

8.4 Conclusions

While 5G promises a paradigm shift for the future wireless systems, there are many challenges for the upcoming 5G network to operate globally without losing the desired quality of service. The rural–urban divide poses a critical challenge for 5G designers to choose and implement technologies that are specifically designed for particular situations. This opens up a plethora of opportunities for the researchers to delve into various aspects of communications theory and to come up with elegant solutions of providing universal access. Keeping in view, this chapter provided an overview, opportunities, and challenges of various techniques that are candidate solutions for providing access to urban as well as rural environments and it has been established that a single technology cannot be used universally owing to disparity in user densities and requirements. Massive MIMO and mmWave techniques are better suited for dense deployments. On the other hand, terrestrial and airborne services can be used in rural settings for providing reliable coverage. Besides all, the joint optimization of CAPEX and OPEX remains a big challenge.

References

1 M. Fallgren, B. Timus, et al. , Deliverable D1.1: Scenarios, requirements and KPIs for 5G mobile and wireless system, in *Mobile and Wireless Communications Enablers for the Twenty-Twenty Information Society*, 2013.

2 T. L. Marzetta, "Noncooperative cellular wireless with unlimited numbers of base station antennas," *IEEE Trans. Wirel. Commun.*, vol. 9,no. 11, pp. 3590–3600, 2010.

3 X. Gao, O. Edfors, F. Rusek, F. Tufvesson, "Massive MIMO performance evaluation based on measured propagation data," *IEEE Trans. Wirel. Commun.*, vol. 14, no. 7, pp. 3899–3911, 2016.

4 J. F. Monserrat et al., "METIS research advances towards the 5G mobile and wireless system definition," *EURASIP. J. Wirel. Commun. Netw.*, vol. 53, 2015. doi: https://doi.org/10.1186/s13638-015-0302-9.

5 F. Fernandes, A. Ashikhmin, and T. L. Marzetta, "Inter-cell interference in non-cooperative TDD large scale antenna systems," *IEEE J. Select. Areas Commun.*, vol. 31, no. 2, pp. 192–201, 2013.

6 M. Filippou, D. Gesbert, and H. F. Yin, "Decontaminating pilots in cognitive massive MIMO networks," in *Proc. IEEE ISWCS*, Aug. 2012, pp. 816–820.

7 Z. Mulk, S. A. Hassan, "On achievable rates in Massive MIMO-based hexagonal cellular system with pilot contamination," in *81st IEEE Vehicular Technology Conference (VTC-Spring)*, Glasgow, U.K., May 2015.

8 D. M. Wang, C. Ji, X. Q. Gao, S. H. Sun, and X. H. You, "Uplink sum-rate analysis of multi-cell multi-user massive MIMO system," in Proc. IEEE ICC, Jun. 2013, pp. 5404–5408.

9 N. Krishnan, R. D. Yates, and N. B. Mandayam, "Cellular systems with many antennas: large system analysis under pilot contamination," in *Proc. Annual Allerton Conference on Communication, Control, and Computing*, Oct. 2012, pp. 1220–1224.

10 V. Jungnickel, K. Manolakis, W. Zirwas, B. Panzner, V. Braun, M. Lossow, M. Sternad, R. Apelfrojd, T. Svensson, "The role of small cells, coordinated multipoint, and massive MIMO in 5G," in *IEEE Commun. Mag.*, vol. 52, no. 5, pp. 44–51, 2014.

11 T. Mushtaq, S. A. Hassan, and D. N. K. Jayakody, "Ergodic rate analysis of massive MIMO systems in K-fading environment," in *IEEE Vehicular Technology Conference (VTC-Fall)*, Canada, Sept. 2016.

12 R. Muharar and J. Evans, "Downlink beamforming with transmit-side channel correlation: a large system analysis," in *Proc. IEEE International Conference on Communication (ICC)*, June 2011.

13 S. Wagner, R. Couillet, M. Debbah, and D. T. M. Slock, "Large system analysis of linear precoding in MISO broadcast channels with limited feedback," *IEEE Trans. Inf. Theory*, vol. 58, no. 7, pp. 4506–4537, 2012.

14 R. Zakhour and S. V. Hanly, "Base station cooperation on the downlink: large system analysis," *IEEE Trans. Inf. Theory*, vol. 58, no. 4, pp. 2079–2106, 2012.

15 H. Huh, S.-H. Moon, Y.-T. Kim, I. Lee, and G. Caire, "Multi-cell MIMO downlink with cell cooperation and fair scheduling: a large system limit analysis," in *IEEE Trans. Inf. Theory*, vol. 57, no. 12, pp. 7771–7786, 2011.

16 H. Huh, A.M. Tulino, and G. Caire, "Network MIMO with linear zeroforcing beamforming: large system analysis, impact of channel estimation, and reduced-complexity scheduling," *IEEE Trans. Inf. Theory*, vol. 58, no. 5, pp. 2911–2934, 2012.

17 H. Huh, G. Caire, H. C. Papadopoulos, and S. A. Ramprashad, "Achieving massive MIMO spectral efficiency with a not-so-large number of antennas," in *IEEE Trans. Wirel. Commun.*, vol. 11, no. 9, pp. 3226–3239, 2012.

18 W. Roh, S. Ji-Yun, P. Jeongho, L. Byunghwan, L. Jaekon, K. Yungsoo, C. Jaeweon, C. Kyungwhoon, and F. Aryanfar, "Millimeter-wave beamforming as an enabling technology for 5G cellular communications: theoretical feasibility and prototype results," *IEEE Commun. Mag.*, vol 2, no. 2, pp. 106–113, 2014.

19 T. S. Rappaport, S. Sun, R. Mayzus, H. Zhao, Y. Azar, K. Wang, G. N. Wong, J. K. Schulz, M. Samimi, and F. Gutierrez, "Millimeter wave mobile communications for 5G cellular: it will work!," *IEEE Access*, vol. 1, pp. 335–349, 2013.

20 Z. Pi and F. Khan, "An introduction to millimeter-wave mobile broadband systems," *IEEE Commun. Mag.*, vol. 49, no. 6, pp. 101–107, 2011.

21 Z. Qingling and J. Li, "Rain attenuation in millimeter wave ranges," in *Proc. IEEE International Symposium on Antennas, Propagation and EM Theory*, Oct. 2006, pp. 1–4.

22 T. S. Rappaport, F. Gutierrez, E. Ben-Dor, J. N. Murdock, Y. Qiao, and J. I. Tamir, "Broadband millimeter-wave propagation measurements and models using adaptive-beam antennas for outdoor urban cellular communications," in *IEEE Trans. Antennas Propag.*, vol. 61, no. 4, pp. 1850–1859, 2013.

23 S. Rangan, T. S. Rappaport, and E. Erkip, "Millimeter wave cellular wireless networks: potentials and challenges," *Proc. IEEE*, vol. 102, no. 3, pp. 366–385, 2014.

24 M. K. Samimi, G. R. MacCartney, S. Sun, and T. S. Rappaport,"28 GHz millimeter-wave ultrawideband small-scale fading models in wireless channels," in *Proc. IEEE Vehicular Technology Conference (VTC Spring)*, 2016, pp. 1–6.

25 A. Ghosh, T. A. Thomas, M. C. Cudak, R. Ratasuk, P. Moorut, F. W. Vook, T. S. Rappaport, G. R. MacCartney, S. Sun, and S. Nie, "Millimeter wave enhanced local area systems: a high data rate approach for future wireless networks," *IEEE J. Select. Areas Commun.*, vol. 32, no. 6, pp. 1152–1163, 2014.

26 3GPP, "Evolved Universal Terrestrial Radio Access (E-UTRA): Further advancements for E-UTRA physical layer aspects (Release 9)," TR 36.814, 2010.

27 T. Bai, R. Vaze, and R. W. Heath, "Analysis of blockage effects on urban cellular networks," *IEEE Trans. Wirel. Commun.*, vol. 13, no. 9, pp. 5070–5083, 2014.

28 T. Bai and R. W. Heath, "Coverage and rate analysis for millimeter-wave cellular networks," in *IEEE Trans. Wirel. Commun.*, vol. 14, no. 2, pp.1100–1114, 2015.

29 M. N. Kulkarni, T. A. Thomas, F. W. Vook, A. Ghosh, and E. Visotsky, "Coverage and rate trends in moderate and high bandwidth 5G networks," in *IEEE Globecom Workshops (GC Wkshps)*, Dec. 2014, pp. 422–426.

30 M. N. Kulkarni, S. Singh, J. G. Andrews, "Coverage and rate trends in dense urban mmWave cellular networks," in *2014 IEEE Global Communications Conference (GLOBECOM)*, pp. 3809–3814, Dec. 2014.

31 M. S. Omar, M. A. Anjum, S. A. Hassan, H. Pervaiz, and Q. Niv, "Performance analysis of hybrid 5G cellular networks exploiting mmWave capabilities in suburban areas," in *IEEE International Conference on Communication (ICC)*, May 2016, pp. 1–6.

32 H. Munir, S. A. Hassan, H. B. Parveiz, L. Musavian, and Q. Ni, "Energy efficient resource allocation in 5G hybrid heterogeneous networks: a game theoretic approach," *84th IEEE Vehicular Technology Conference (VTC-Fall)*, Montreal, Canada, Sep. 2016.

33 S. A. R. Naqvi, S. A. Hassan, and Z. ul Mulk, "Pilot reuse and sum rate analysis of mmWave and UHF-based massive MIMO systems," in *IEEE Vehicular Technology Conference (VTC Spring)*, 2016, pp. 1–5.

34 O. Onireti, M. A. Imran, J. Qadir, and A. Sathiaseelan, "Will 5G see its blind side? Evolving 5G for universal internet access," in *ACM SIGCOMM Conference Proceedings*, 2016.

35 L. Waverman, M. Meschi, and M. Fuss, "The impact of telecoms on economic growth in developing countries," *The Vodafone Policy Paper Series*, vol. 2, no. 3, pp. 10–24, 2015.

36 M. Eriksson and J. van de Beek, "Is anyone out there? 5G, rural coverage and the next 1 billion," *IEEE ComSoc Technology News (CTN)*, 2015.

37 L. Chiaraviglio, N. Blefari-Melazzi, W. Liu, , J. A. Gutierrez, J. Van De Beek, R. Birke, et al., "5G in rural and low-income areas: are we ready?" ITU Kaleidoscope, Bangkok, Thailand, 2016.

38 X. Ge, H. Cheng, M. Guizani, and T. Han, "5G wireless backhaul networks: challenges and research advances," *IEEE Netw.*, vol. 28, no. 6, pp. 6–11, 2014.

39 M. Jaber, M. Imran, R. Tafazolli, and A. Tukmanov. "An adaptive backhaul-aware cell range extension approach." In *IEEE International Conference on Communication Workshop (ICCW)*, 2015, pp. 74–79.

40 M. Jaber, M. A. Imran, R. Tafazolli, and A. Tukmanov, "5G backhaul challenges and emerging research directions: a survey.," in *IEEE Access*, vol. 4, pp. 1743–1766, 2016.

41 M. Jaber, M. Imran, R. Tafazolli, and A. Tukmanov, "A distributed SON-based user-centric backhaul provisioning scheme," In *IEEE Access*, vol. 4, pp 2314–2330, 2016.

42 M. Jaber, D. Owens, M. A. Imran, R. Tafazolli, and A. Tukmanov, "A joint backhaul and RAN perspective on the benefits of centralised RAN functions," in *IEEE Intterational Conference on Communication Workshops (ICC)*, 2016, pp. 226–231.

43 A. Imran, M. A. Imran, and R. Tafazolli, "A novel self organizing framework for adaptive frequency reuse and deployment in future cellular networks." in *21st Annual IEEE International Symposium on Personal, Indoor and Mobile Radio Communications*, 2010, pp. 2354–2359.

44 O. G. Aliu, A. Imran, M. A. Imran, B. Evans, "A survey of self organisation in future cellular networks," *IEEE Commun. Surv. Tutor.*, vol. 15, no. 1, pp. 336–361, 2013.

45 O. Onireti, A. Zoha, J. Moysen, A. Imran, L. Giupponi, M. A. Imran, and A. Abu-Dayya, "A cell outage management framework for dense heterogeneous networks," In *IEEE Trans. on Veh. Tech.*, vol. 65, no. 4, pp. 2097–2113, 2016.

46 E. Bjornson, E. G. Larsson, and T. L. Marzetta, "Massive MIMO: ten myths and one critical question," *IEEE Commun. Mag.*, vol. 54, no. 2, pp.114-123, 2016.

Syed Ali Hassan received the B.E. degree in electrical engineering with highest honors from the National University of Sciences and Technology (NUST), Islamabad, Pakistan, in 2004, the M.S. degree in electrical engineering from the University of Stuttgart, Stuttgart, Germany, in 2007, the M.S. degree in mathematics from Georgia Institute of Technology (Georgia Tech in 2011), Atlanta, GA, USA, and the Ph.D. degree in electrical engineering from the Georgia Tech in 2011.

Currently, he is working as an assistant professor in the School of Electrical Engineering and Computer Science (SEECS), NUST, where he is the Director of the Information Processing and Transmission research group, which focuses on various aspects of theoretical communications. Prior to joining SEECS, he was a research associate in Cisco Systems Inc., San Jose, CA, USA. He has authored or coauthored more than 100 publications in international conferences and journals. His broader area of research is signal processing for communications. He was a Technical Committee Member and a Symposium Chair for various conferences/workshops as well as Guest Editor for journals.

Muhammad Shahmeer Omar received the B. Eng. degree in electrical engineering from the National University of Sciences and Technology, Islamabad, Pakistan, in 2015, where he worked in the Information Processing and Transmissions Laboratory. Currently, he is pursuing his master's degree in electrical engineering from Georgia Tech, USA. His research interests include fifth generation mobile networks and physical layer techniques for wireless sensor networks.

Muhammad Ali Imran is the Vice Dean Glasgow College UESTC and Professor of Communication Systems in the School of Engineering at the University of Glasgow. He was awarded his M.Sc. (Distinction) and Ph.D. degrees from Imperial College, London, U.K., in 2002 and 2007, respectively. He is an affiliate professor at the University of Oklahoma, USA, and a visiting Professor at 5G Innovation Centre, University of Surrey, UK. He has over 18 years of combined academic and industry experience, working primarily in the research areas of cellular communication systems. He has been awarded 15 patents, has authored/coauthored over 300 journals and conference publications, and has been principal/coprincipal investigator on over £6 million in sponsored research grants and contracts. He has supervised 30+ successful Ph.D. graduates. He has an award of excellence in recognition of his academic achievements, conferred by the President of Pakistan. He was also awarded IEEE Comsoc's Fred Ellersick award 2014, FEPS Learning and Teaching award 2014, and Sentinel of Science Award 2016. He was twice nominated for Tony Jean's Inspirational Teaching award. He is a shortlisted finalist for The Wharton-QS Stars Awards 2014, QS Stars Reimagine Education Award 2016 for innovative teaching, and VC's learning and teaching award in University of Surrey. He is a senior member of IEEE and a senior fellow of Higher Education Academy (SFHEA), UK.

Junaid Qadir is an associate professor at the Information Technology University (ITU) Punjab in Lahore, Pakistan. He is the Director of the IHSAN Lab at ITU that focuses on deploying ICT for development, and is engaged in systems and networking research. His research interests include the application of algorithmic, machine learning, and optimization techniques in networks. In particular, he is interested in the broad areas of wireless networks, cognitive networking, software-defined networks, and cloud computing. He serves as an associate editor for *IEEE Access, IEEE Communications Magazine*, and *Big Data Analytics*. He is a member of ACM and a senior member of IEEE.

Dushantha Nalin K. Jayakody received the B. Eng. degree with first-class honors from Pakistan and was ranked as the merit position holder of the University (under SAARC Scholarship). He received his M.Sc. degree in electronics and communications engineering from Eastern Mediterranean University, Cyprus (under the University full graduate scholarship) and was ranked as the first merit position holder of the department. He received the Ph. D. degree in electronics and communications engineering from the University College Dublin, Ireland. From 2014 to 2016, he was a postdoctoral fellow at the University of Tartu, Estonia and University of Bergen, Norway. Since 2016, he is an Associate Professor in the Institute of Cybernetics, National Research Tomsk Polytechnic University, Russia where he also serves as the Director of Tomsk Infocomm Lab. He has received the best paper award from the IEEE International Conference on Communication, Management and Information Technology (ICCMIT), Warsaw, Poland in April 2017. He is a member of IEEE and he has served as session chair or technical program committee member for various international conferences, such as IEEE PIMRC 2013/2014, IEEE WCNC 2014/2016, IEEE VTC 2015. He currently serves as a lead guest editor for the Elsevier *Physical Communications* journal and MDPI *Information* journal. Also, he serves as a reviewer for various IEEE transactions and other journals.

9

Network Slicing for 5G Networks

Xavier Costa-Pérez, Andrés Garcia-Saavedra, Fabio Giust,
Vincenzo Sciancalepore, Xi Li, Zarrar Yousaf, and Marco Liebsch

5G Networks R&D Group, NEC Laboratories Europe GmbH, Heidelberg Germany

9.1 Introduction

Mobile networks are a key element of today's society, enabling communication, access, and information sharing. Moreover, traffic forecasts predict that the demand for capacity will grow exponentially over the next years, mainly due to video services. However, as cellular networks move from being voice-centric to data-centric, operators' revenues are not able to keep pace with the predicted increase in traffic volume. Such pressure on operators' return on investment has pushed research efforts toward designing for 5G novel mobile network solutions able to open the door for new revenue sources. In this context, the *network slicing* paradigm has emerged as a key *5G disruptive technology* addressing this challenge.

Network slicing for 5G allows mobile network operators (MNOs) to open their physical network infrastructure platform to the concurrent deployment of multiple logical self-contained networks, orchestrated in different ways according to their specific service requirements; such network slices are then (temporarily) owned by tenants. The availability of this vertical market multiplies the monetization opportunities of the network infrastructure as (i) new players may come into play (e.g., automotive industry, e-health) and (ii) a higher infrastructure capacity utilization can be achieved by admitting network slice requests and exploiting multiplexing gains.

5G Networks: Fundamental Requirements, Enabling Technologies, and Operations Management, First Edition.
Anwer Al-Dulaimi, Xianbin Wang, and Chih-Lin I.

With network slicing for 5G networks, different services (e.g., automotive, mobile broadband, or haptic Internet) can be provided by different network slice instances. Each of these instances consists of a set of virtual network functions that run on the same infrastructure with a tailored orchestration. In this way, very heterogeneous requirements can be provided on the same infrastructure, as different network slice instances can be orchestrated and configured separately according to their specific requirements. Additionally, this is performed in a cost-efficient manner as the different network slice tenants share the same physical infrastructure.

While the 5G network slicing concept has been proposed recently [1], it has already attracted substantial attention. 3GPP has started working on the definition of requirements for network slicing [2], whereas NGMN identified network sharing among slices as one of the key 5G issues [3].

A *network slice* is defined by NGMN as "a set of network functions, and resources to run these network functions, forming a complete instantiated logical network to meet certain network characteristics required by the Service Instance(s)."

According to NGMN, the concept of network slicing involves three layers, namely, (i) service instance layer, (ii) network slice instance layer, and (iii) resource layer. The service instance layer (SIL) represents the end user and/or business services provided by the operator or the third-party service providers, which are supported by the network slice instance layer (NSIL). The NISL is in turn supported by the resource layer, which may consist of the organic resources such as compute, network, memory, storage, etc., or it may be more comprehensive as being a network infrastructure, or it may be more complex as network functions. Figure 9.1 depicts this concept where the resources at the resource layer are dimensioned to create several sub network instances, and network slice instances are formed that may use none, one, or multiple sub network instances.

9.2 End-to-End Network Slicing

The end goal of network slicing in 5G mobile networks is to be able to realize end-to-end (E2E) network slices starting from the mobile edge, continuing through the mobile transport (fronthaul (FH)/backhaul (BH)), and up until the core network (CN). The allocation of a slice involves the selection of the required functions, their constrained placement, the composition of the underlying infrastructure, and the allocation of the resources to fulfill the services' requirements, for example, bandwidth, latency, processing, resiliency.

We consider two main network slicing services that enable different degrees of explicit control and are characterized by different levels of automation of the mobile network slices management:

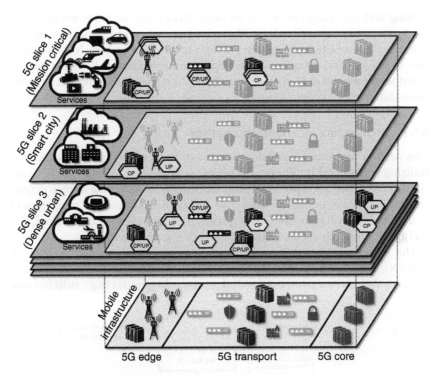

Figure 9.1 Network slicing in 5G networks.

1) The provision of *virtual infrastructures* (VI) under the control and operation of different tenants – in line with an Infrastructure-as-a-Service (IaaS) model, that is, creation of a *network slice instance*.
2) The provision of tenant's owned *network services* (NS), that is, creation of a *service instance*.

In the former service, the deployment of a mobile network deals with the allocation and deallocation of VIs. The logical entities within a VI encompassing a set of compute and storage resources are interconnected by a virtual, logical network (i.e., virtual nodes are interconnected by virtual links over the substrate network). The VIs can be operated by the tenant via different SDN control models. In the latter, NS are instantiated directly over a shared infrastructure, and as a set of interrelated virtual network functions (VNFs) connected through one or more VNF forwarding graphs (VNF-FGs).

Multi tenancy is an orthogonal characteristic that can be applied to both kinds of services, guaranteeing separation, isolation, and independence between different slices coupled with the efficient sharing of the underlying resources for both VI and NS concepts. In this context, a *tenant* is a logical entity owning and operating either one or more VIs or one or more network services, ultimately controlling their life cycle. A tenant can be associated with an administrative entity (e.g., mobile (virtual) network operators, M(V)NOs) or user of a given service (e.g., over-the-top service providers, OTTs) and implies a notion of ownership of one or more service instances and isolation between these instances.

9.2.1 Architecture for End-to-End Network Slicing

This section presents a novel architecture of the network slicing system for realizing end-to-end (E2E) network slices, as depicted in Figure 9.2. This architecture unifies the aspects of resource virtualization, virtual infrastructure, and network service management. The design follows the SDN principles of (*i*) data and control plane fully decoupled, (*ii*) control logically centralized, and (*iii*) applications having an abstracted view of resources and states.

The *data plane* is the infrastructure layer (referred as resource layer in NGMN), including mobile edge, mobile transport, and core. The infrastructure is a logical construct composed of links, forwarding nodes (e.g., switches and

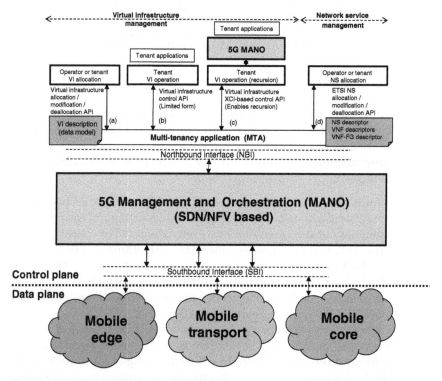

Figure 9.2 Architecture for network slicing

routers), cloud nodes (e.g., data centers), and so on, comprising a set of network, computing, and storage resources.

The *control plane* is divided into two layers: an application layer at the top and the 5G management and orchestration (MANO) platform below. The design of the MANO is based on the ETSI NFV management and network orchestration (MANO) framework [4] with integrated SDN-based control. The MANO provides an abstracted view of available resources and states and control, and management functions to an ecosystem of applications, via a *northbound interface* (NBI). On the other hand, the MANO is connected to the data plane elements via a *southbound interface* (SBI) to execute control and management functions (e.g., OpenFlow, SNMP, OVSDB) on the actual hardware components.

To offer end-to-end management through different network domains, the MANO system considers two different implementation alternatives:

- *Hierarchical Structure:* A global orchestration engine controls the MANO of each domain (edge, transport, and core) via a north/southbound interface.
- *Peer-to-Peer Structure:* MANOs per domain interact with each other via an east/westbound interface.

Parts of the architecture described are already under development on open-source projects for the implementation of the MANO, for example, Open-DayLight and ONOS for the SDN controllers, and OPNFV, ONAP, OSM, OpenStack, and OpenMANO for the ETSI NFV MANO.

With respect to the *multi-tenancy application (MTA)*, it implements the multi-tenancy support by coordinating and managing tenants access to a shared infrastructure, performing resource isolation between instances assigned to different tenants, and delivering multi-tenancy-related services, such as the allocation and operation of VIs, by means of dedicated APIs[1] in cooperation with the data plane, enforcing this logical separation. As shown in Figure 9.2, such APIs depend on the actual service: for the control of a VI or NS lifetime, instantiation, modification, and deletion (API classes (a) and (d) in Figure 9.2), and for the control of the VI in its limited or full-featured form (API classes (b) and (c), respectively). MTA relies on the multitenant capabilities and support of the underlying MANO elements if available or implements them as a software layer.

9.2.2 Deployment of Virtual Infrastructure

This service involves the dynamic allocation of a virtual infrastructure (VI), its operation, and, subsequently, deallocation. The actual realization of a VI combines, for example, aspects such as the partitioning and book-keeping of

---

1 In the considered model, a single tenant entity owns one or more instances of each service in a 1:N relationship.

resources or the instantiation of connections supporting virtual links characterized in terms of, for example, unreserved bandwidth or latency. The provisioning of a VI also commonly requires direct hardware element support, or its emulation through software (e.g., software-based hypervisors) for multiplexing over the shared infrastructure.

The allocation of a VI can be triggered by a tenant (such as a MVNO), either directly consuming the MTA API – Figure 9.2 API class (a) – or via the intervention of the infrastructure operator in a less dynamic environment, after an off-line service-level agreement (SLA). The VI concept is quite generic and could be extended to incorporate infrastructure elements beyond the ones considered herein. As part of the deployment of the VI, network, computing, and storage resources need to be partitioned and aggregated, eventually recursively if a hierarchy is enabled. This partitioning can be committed in full at the time of instantiation (hard allocation) or reflected in terms of predefined quotas that are enforced at the time of use (soft allocation). For example, the partitioning of storage resources relies on storage controllers, and existing technology to partition and aggregate volumes is sufficiently flexible to be used from a virtual infrastructure perspective. Computing resources have a containment relationship with, for example, virtual machines (VMs) or containers, and their partitioning is straightforward. The partitioning of network resources clearly involves interface cards and link bandwidth management. A common partitioning involves, for example, instantiating software nodes and binding physical (and logical) ports to instances, and relying on the asynchronous multiplexing associated with packet switching. This results in one or more virtual switches for that tenant or group of tenants attached to (a subset of) the physical switch ports.

It is noteworthy that VI allocation follows an IaaS model, so the actual use of the VI (including the functions and related business logic) is defined by the tenant. The infrastructure owner is agnostic to the VI end use. Once a given VI has been allocated, the MTA empowers the tenants with different degrees of control to be exerted over it, with different operational models of control and management. In simple terms, the control and management is restricted to the operational management. The operation of VI is mostly autonomous, with limited involvement of the tenant, such as monitoring and SLA validation. Each tenant is free to deploy their choice of the infrastructure operating system and control plane, allowing the optimization of the resource usage within each VI.

The former model involves the MTA offering an API that enables the tenant to have a limited form of control over the (abstracted) elements that constitute the VI – Figure 9.2 API class (b) – including a set of operations and policies that can be applied (e.g., retrieve an aggregated view of the virtual infrastructure topology and resource state and apply rules that affect element configuration and behavior). Low-level operations such as the actual configuration and monitoring of individual flows at the nodes may not be allowed. The latter implies per-tenant controller – Figure 9.2 API class (c) – or per-tenant MANO including, most importantly, the ability to offer network services over its allocated virtual infrastructure. This approach ultimately enables recursion.

9.2.3 Deployment of Network Services

The allocation of a network service (NS) extends and complements the concept of VI deployment – Figure 9.2 API class (d) – to deliver isolated chains of virtual services composed of specific VNFs in an automated manner, and exploiting the sharing of a common physical infrastructure with computing, storage, and network resources. The tenant request usually specifies the type of VNFs (i.e., the desired virtual application components) in the NS descriptor, their capabilities and dimensions through one or more VNF descriptors, and how they must be interconnected through a VNF-FG descriptor. Templates for the unified description of these information elements are currently under standardization process in the ETSI NFV ISG and in OASIS TOSCA standards [5].

When deploying a NS, the tenant is only interested in operating the applications that run in these virtual resources and expects that the needed level of resource capacity is seamlessly available in real time without any further configuration effort. The tenant has access to application-level interfaces only and the NS provisioning API follows an "intent-based" modeling approach where the tenant asks just for the composition of some network functions, without caring about how they should be deployed and delivered.

In this case, the MTA is responsible for maintaining and coordinating the logical mapping between tenants and their assigned services, in terms of NS and VNFs instances and underlying virtual resources, in compliance with the SLAs established. Multi-tenancy can be handled at different levels: at the lower level, a tenant has assigned physical and/or virtual resources in the domain of a Virtual Infrastructure Manager (VIM); at the upper levels, tenants have assigned VNFs and NSs. These different kinds of tenants can overlap and be merged in a single entity or be mapped over separate entities. For example, an MVNO can further virtualize the rented VI to serve different kinds of business customers, like CDN providers, delivering dedicated VNFs and NSs. The management of these tenants' relationships, together with the correlated authorization and SLA validation and assurance procedures are under the responsibility of the MTA. Moreover, NSs are not built directly on top of physical resources, but over virtual infrastructures through the allocation of VNFs and VNF-FGs in virtual machines and virtual network nodes, following a recursive approach. This involves the operation of multiple MTA instances deployed at different levels and requires the mediation of MANO components deployed over the VI itself.

9.2.4   E2E Network Slicing Implementations

Some early trials have been conducted demonstrating network slicing with cross-industry collaboration among operators, vendors, and vertical industries. Two specific examples are given here.

(1) Deutsche Telekom and Huawei demonstration of E2E autonomous network slicing. In this demo, enhanced mobile broadband (eMBB), massive machine-type communication (mMTC) and ultrareliable lowlatency communication (uRLLC) are envisaged as network classes that could be built as slices. E2E network slicing included not only the core network and RAN, but also interconnecting transport networks. The demo implements E2E network slicing automation based on service oriented network auto creation (SONAC). It uses software-defined topology (SDT), software-defined protocol (SDP), and software-defined resource allocation (SDRA) to ensure the automatic implementation of slice management, service deployment, resource scheduling, and fault recovery.

(2) SK Telecom, Deutsche Telekom (DT), and Ericsson have jointly built and demonstrated a trial network on federated network slicing for roaming, making SK Telecom and DT network slices available in each operators footprint, connecting South Korea and Germany. The demonstration was hosted at DTs corporate R&D center in Bonn, Germany and SK Telecoms 5G testbed at Yeongjongdo (the BMW driving center) in Korea. The demo featured an industrial maintenance use case involving a repair worker communicating via augmented reality (AR) with support colleagues in a visited network. The scenario used local breakout and edge cloud to enable the best service experience in terms of latency and throughput for the AR repairman.

9.3   Network Slicing MANO

This section highlights the challenges and requirements for the management and orchestration (MANO) of network slices based on some architectural and topological use cases of network slices and their possible configurations. More importantly it provides insights into the challenges and requirements of a MANO system for performing QoS/QoE-aware network slice orchestration.

Conceptually, a network slice is composed of VNFs that are chained in a specific order over virtual links (VL) in order to provide a specific network service. A single NFV infrastructure (NFVI), such as a data center, may support concurrent instances of multiple network slices, where each network slice is composed to support one or more service vertical, for example, mobile broadband services such as 4K video streaming or ultra reliable services such as e-health or autonomous driving applications. Provisioning of a credible MANO framework becomes a challenging proposition when concurrent network slices have to be managed and their respective resources orchestrated in order to ensure end-to-end service integrity. The challenge becomes more complex when the network slice(s) traverse over different NFVI-PoPs belonging to different administrative domains. The MANO for network slices can be broadly identified into two categories:

1) *Intra slice MANO:* The MANO framework supports the life cycle management (LCM) of a single slice independent of the other slices that may share the same NFVI domain.
2) *Inter slice MANO:* The MANO framework supports the LCM of multiple network slices in relation to each other. This may be due to the dependence of the respective network slices either due to some functional and/or resource sharing.

In view of the above two categories, 3GPP SA5 has identified several use cases, for network slice management [6] depending on the slice topology and configuration. Based on this, there are two main slice deployment possibilities:

Isolated Network Slices with Shared Access Network: In this use case, the operator creates two or more network slices that are totally isolated from each other in terms of the constituent resources, that is, the underlying compute, network, memory, and even the VNFs/PNFs. Such isolated slices do not interfere or impact on the performance of other slices as each slice utilizes only their respective allotted resources. Each network slice manages its own service requirements. For example, an operator can deploy one network slice for mobile broadband services while the other for machine-to-machine services. Each network slice will then manage its respective traffic without impacting on the performance of the other. If one network slice malfunctions or is taken down, it will not have any impact on the services being supported by other network slice(s) sharing the same infrastructure. However, the slice management and orchestration challenge becomes complex when a single slice instance may support more than one service instance, in which case each service will have its own QoS/QoE requirements. The MANO system must ensure compliance to the service QoS/QoE requirements.

Network Slices with Shared Resources: In this use case, two or more network slice instances may share one or more network functions whose functions/features may be required by multiple slices. For example, it could be that an operator instantiates two vEPC slices for two different MVNOs, but both the slices use the same instance of HSS/PCRF function. This use case makes the management of network slices more complex as any orchestration action taken on one network slice will have an impact on the other slice(s) that are sharing the same common network function. In case, the common network function becomes a performance bottleneck will adversely impact on all other network slices sharing the bottleneck network function. It should be noted that

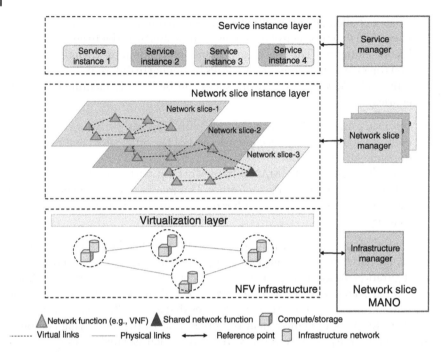

Figure 9.3 Network slice management and orchestration (MANO) overview.

the common network function may be a physical network function (PNF) or it can be a VNF. Thus, any MANO action taken on the common network function will have an impact on the network slices that are sharing the network function. The notion of network slices sharing a common network function is depicted in Figure 9.3. Moreover, the MANO of slices becomes more challenging when the slices traverse different administrative domains. In the following section, we will describe the common MANO tasks and give an overview of the ETSI NFV MANO system in terms of QoS/QoE-aware management and orchestration of network slices.

9.3.1 Management and Orchestration Architecture

As mentioned earlier, the 5G network is mainly composed of three layers, namely, the resource layer, the network slice instance layer, and the service instance layer. Each of these respective layers needs to be managed in coordination with other layers. Figure 9.3 provides an overview of this situation, which illustrates the three essential layers of a 5G network infrastructure where each layer has a management plane.

A MANO framework is thus required in order to provide a uniform and coherent management and orchestration of resources across these three tiers

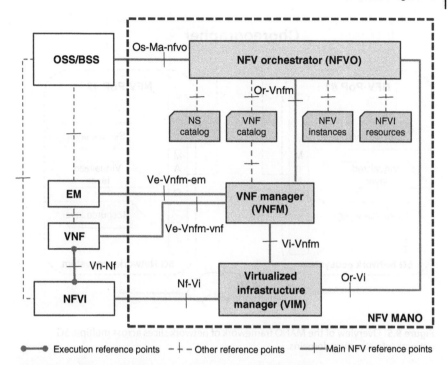

●━━● Execution reference points ─ ┼ ─ Other reference points ━┼━Main NFV reference points

Figure 9.4 ETSI NFV management and orchestration (MANO) framework.

of resources. It is necessary that the management plane of each layer is able to communicate with those of the other layers. In this regard ETSI ISG NFV is in the process of specifying a MANO system for an NFV environment that is referred to as NFV MANO. Its functional overview is depicted in Figure 9.4.

The ETSI NFV MANO framework has three main functional blocks, namely, the virtualized infrastructure manager (VIM), the VNF manager (VNFM), and the NFV orchestrator (NFVO). These three functional blocks are interconnected via specific reference points. There are additional data repositories that may contain necessary information about network service (NS), VNF, NFV, and NFVI that will enable the NFVO to perform its tasks. The MANO architecture also defines reference points for interfacing the MANO system with external entities like NFVI, OSS/BSS, VNFs, and element managers (EM) for delivering a unified management and orchestration of a VNF system. It should be noted that the VIM manages the physical/virtualized resources of the NFVI, whereas VNFM is responsible for the LCM of the individual VNF(s) and there could be multiple instances of VNFM. NFVO, on the other hand, manages the network service (NS) by performing service orchestration and resource orchestration

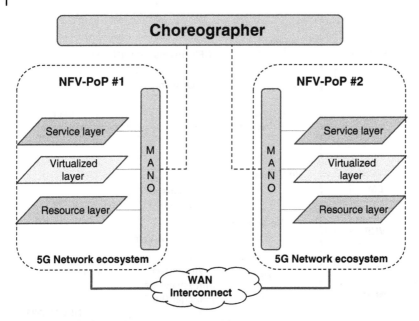

Figure 9.5 Overview of the MANO framework of network slices across multiple 5G network infrastructure domains.

in order to ensure end-to-end service integrity that is formed by one or more VNFs interconnected over virtual links.

When compared to Figure 9.3, the VIM corresponds to the infrastructure manager, the VNFM corresponds to the network slice manager while the NFVO corresponds to the service instance layer. It can thus be inferred that the ETSI NFV MANO system has the required building blocks for providing a MANO framework for the 5G network slices.

The MANO paradigm becomes more complex when the network slice(s) span across multiple 5G network infrastructure PoPs, for example, data centers. This is illustrated in Figure 9.5, where each NFV-PoP has its own MANO entity for the management of local resource/service instances. In case of the management and orchestration of network slices that span across multiple NFV-PoPs, belonging to different administrative domains and interconnected over some WAN infrastructure, an overarching entity is required, which is referred to as a choreographer in Figure 9.5. The choreographer shall ensure to coordinate across the individual MANO entities to ensure the overall end-to-end integrity of the service(s) utilizing the cross-domain network slice(s). In essence, the choreographer is like a management broker that shall coordinate between the MANO instances at different admin domains to ensure efficient and effective management and orchestration of network slices that spans across different domains. The choreographer is aware of the resource status in each admin domain. ETSI NFV recognized this use case and refers to the choreographer as an umbrella NFVO in Reference [7]. However, the role and functions of the choreographer are still an open research issue that is under investigation and its interfaces and operations have not yet been standardized.

9.3.2 Network Slicing MANO Tasks

A MANO system is supposed to orchestrate multiple complex management tasks in order to ensure the provisioning of a network slice service. A MANO framework for 5G virtualized networks infrastructure is designed to go beyond providing the traditional FCAPS (fault, configuration, accounting, performance, and security) management into providing additional management tasks. Some of the additional management functions, besides FCAPS are as follows:

1) Software image management
2) Service reliability management
3) Policy management mobility management
4) Bandwidth and latency management
5) QoS/QoE management
6) Mobility management
7) Energy management
8) Charging and billing management
9) Network slice update/upgrade
10) VNF life cycle management
11) Virtualized infrastructure management, that is, management of resource capacity, performance, fault, isolation, and so on.

As mentioned earlier, the basic building block of a network slice at the virtualization layer is the VNF. The MANO entity performs the life cycle management (LCM) of a network slice by managing the individual VNFs that are part of the network slice. Figure 9.6 provides an overview of the different phases of the LCM of the network slice that the MANO system has to orchestrate.

Preparation Phase: In this phase, the blueprint of the network slice is provisioned and is parsed to verify and prepare the resources required for the function and operation of the network slice. A slice blueprint is a descriptor file that provides all the information regarding the type/amount of physical/virtualized resources required for the network slice, its functional and operational performances, QoS, QoE, reliability, connectivity, VNFs, security requirements, and so on. The network slice blueprint is similar to the ETSI NFV network service descriptor (NSD) file. Based on this blueprint, or service template or descriptor file, the MANO system shall ensure the provisioning and allocation of the required resources, including the relevant VNFs required for the provisioning

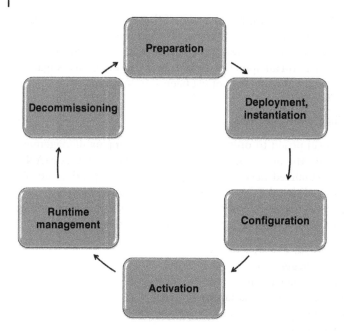

Figure 9.6 Management phases of the network slice life cycle.

of the network slice. The resources may be CPU, network, memory, storage database, VNFs, PNFs, and so on.

Deployment Phase: This phase is also referred to as the instantiation phase. In this phase all the resources that are stipulated in the slice blueprint are deployed and instantiated. For example, it will instantiate the relevant VNFs with their respective underlying resources specified by the VNF descriptor file.

Configuration Phase: In this phase, the deployed VNFs are configured based on the configuration requirement specified in the VNF descriptor files of the individual VNFs. This phase will also involve creating virtual links with specified bandwidth between the various VNFs forming a network slice.

Activation Phase: This phase includes any action that makes the network slice instance active, for example, attaching users to it, accepting connection, diverting traffic to it. It is at this phase when the k slice is fully operational and ready to provide its respective service.

Run time Management: In this phase, the MANO system performs run time management of the active network slice(s) by performing tasks such as network slice instance(s) scaling or migration in view of load conditions and network performance conditions. In slice migration, the MANO system may relocate a network slice, or part of it, for example, certain slice VNFs, to another host within the same infrastructure or in a different infrastructure. Similarly, the MANO system can scale-in or scale-out a network slice.

Decommissioning Phase: In this phase, the network slice instance is taken out of active service by removing the configurations and associated states and releasing all of the resources associated with the decommissioned network slice. After this phase, the network slice stands deleted and a new instance needs to be prepared before it can be deployed and instantiated.

9.3.3 Run Time Management of Network Slices

Each of the management phases discussed already have their own requirements and are a subject of individual research. However, in this section we will focus on a generic algorithmic approach for the network slice MANO system to perform a QoS/QoE-aware management and orchestration of network slices and their respective resources. The main objective of such an approach is to enable the MANO system to make accurate decisions on actions at run time in order to ensure long-term integrity of the network slice as per the service QoS/QoE requirements.

9.3.3.1 Generic QoS/QoE Slice MANO Algorithm

Post deployment and activation of network slices, the challenge is to ensure the QoS/QoE compliance of the service(s) that are being supported by the network slice(s). With reference to the ETSI-NFV MANO framework (see Figure 9.4), this compliance is being managed and enforced by the NFVO with the help of VNFM and VIM functional elements. Each time the NFVO detects service degradation it needs to derive appropriate orchestration actions in order to address the causes of degradation and maintain the service QoS/QoE within prescribed limits. The challenge is for the NFVO to determine the correct orchestration action, and this can be achieved if the NFVO is able to diagnose the actual causes of service degradation. It is important for the NFVO to derive the most appropriate orchestration action in order to minimize service disruption that may occur during the application of the management action and to stabilize the service with minimum disruption time. This challenge is made more difficult owing to the network slice topology and configuration as discussed earlier in the use cases for the slice management.

In view of those complexities and challenges, effective methods need to be devised and architectural extensions proposed that may enable the NFVO to determine the exact cause of the QoS/QoE degradation in order to accurately derive and implement appropriate management actions. In Figure 9.7, we provide a functional overview of the extensions to the NFV-MANO architecture in order to support QoS/QoE-aware slice management and orchestration.

Before we describe the key architectural extensions, it should be noted that the run time slice QoS/QoE orchestration has more diverse and stringent requirements than those of individual service instances. This being due to the fact that a single network slice may support multiple service instances and thus

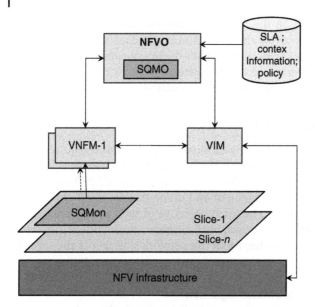

Figure 9.7 Architectural extension for QoS/QoE-aware slice management and orchestration.

the algorithmic impact may be felt across them, with the potential of affecting thousands of users and tens of thousands of service instances in the case of a mobile network slice. The main requirements of QoS/QoE-aware slice MANO system are as follows:

1) The QoS/QoE management system should be pre emptive (i.e., predictive) rather than reactive.
2) The QoS/QoE management system should be able to accurately determine and quantify the performance bottleneck that would enable the NFVO to derive a more suitable MANO action in order to ensure long-term sustainability and survivability of the network slice.

In perspective of these main requirements, the two essential components proposed in Figure 9.7 are as follows:

Service Quality Monitoring (SQMon) Each slice instance is proposed to have a SQMon component, which is responsible for the *QoS monitoring* and *QoE mapping* of the service instance(s) that is being supported on the respective slice. It can also monitor the resource utilization by individual VNF(s) that make up the slice instance. The SQMon is a multi parametric and configurable entity that can be configured and tuned to reflect each service functional and performance peculiarities, and based on them it is able to monitor the service's QoS and map it to QoE. The SQMon will periodically keep the VNFM updated of the network slice QoS/QoE status, which will determine any QoS/QoE degradation event. The VNFM will provide the NFVO with the network slice QoS/QoE performance reports.

Slice QoS/QoE MANO (SQMO) It is proposed to be a part of the NFVO, which is responsible to derive appropriate MANO decisions on orchestration actions (e.g., scale in/out/up/down, migrate, update/upgrade) on respective network slices. The unique feature of the SQMO is that in addition to the slice QoS/QoE reports from the VNFM/VIM, it relies on a variety of other inputs, such as SLA, user/device/service context information, physical/virtual resource utilization of the infrastructure, thereby enabling it to make an accurate determination of the reasons for QoS/QOE degradation. This informational variety enables the NFVO to make accurate MANO decisions on actions.

9.4 Network Slicing at the Mobile Edge

Mobile edge slicing presents a number of challenges that must be considered and properly addressed to achieve a slicing solution.

Spectrum and Radio Resource Constraints Mobile edge slicing operations must cope with the scarcity of radio resources. Dedicated carriers assigned to individual slices, which apparently may solve such a resource limitation problem, might impair the overall RAN slicing process efficiency as the potential multiplexing gain of sharing radio resources might not be fully exploited. A slicer controller should be developed to trade-off flexible resources sharing versus guaranteed isolated resources, for example, by adopting dynamic Transmission Time Interval (TTI) based on slice service requirements, as shown in Figure 9.8. Eventually, the time variability of the channel conditions, due to the user mobility, the presence of obstacles, and the channel fast-fading might also severely exacerbate the slicing process efficiency while requiring advanced mechanisms to mitigate the interference effect in ultra dense scenarios [8].

Multiple Radio Access Technology (RAT) 5G mobile edge integrates heterogeneous RATs and air interfaces, such as millimeter- or micro waves technologies. Each of these is characterized by different capabilities and specific requirements making the radio access planning not trivial. Matching agreed slice requirements together with RAT specifications could present several challenges, carefully addressing while accommodating mobile edge slice requests. For example, ultra reliable low-latency communication (URLLC) applications may require particular RAT features to satisfy stringent service requirements.

Security Issues while Exposing Sharing Information The novel 5G virtualized mobile edge can be designed considering different sharing options based on network elements shared between tenants, such as physical layer elements, MAC functions, or even the whole RAN facilities. However, such sharing option information is essential when slicing operations take place. This information exposure may lead to critical security issues, as confidential information is

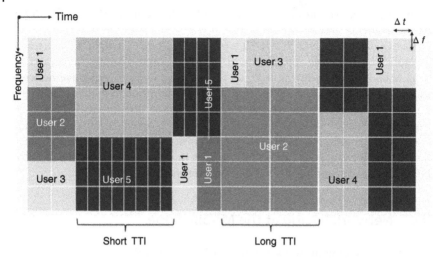

Figure 9.8 Radio tiles as enablers for mobile edge slicing. (From Reference [9].)

processed on network elements shared between two direct competitors, or to data market troubles, as the infrastructure provider can directly access to private data. Therefore, additional constraints preventing specific mobile edge slicing operations must be taken into account.

Mobility and User Attachment User mobility plays a key role in advancing the novel mobile edge network when network slicing concepts are applied. In particular, the major challenge raised is how users shall manage network slice attachment procedures. On the one hand, UEs seamlessly attach to available mobile edge slices: The network management decides based on slice service requirements, ongoing traffic flows, and current network conditions. This option fairly prevents users from simultaneously attaching to multiple slices. On the other hand, the user selfishly selects the most appropriate available slice, also being able to select multiple slices based on contractual agreements. However, the network may still guide the UE–slice pairing, for example, for load balancing purposes. Whereas in the former case, the mobility is completely and efficiently managed by the network itself, in the latter, advanced mobility mechanisms must be deployed to reduce vertical handover signaling burden.

Exploiting Spatial Slicing Virtualized RAN elements may implement different slicing configurations. Based on particular tenant requirements, an additional degree of freedom comes into play–the spatial domain. A particular mobile edge slice may be available only for a set of RAN elements. This would significantly add complexity in the slicing decisions but, at the same time, add more flexibility while tailoring specific spatial areas to required services.

Mobile Edge Slice Brokerage The new 5G ecosystem allows tenants to directly request network slices for a fixed time window. Such network extension automatically opens the network to new business scenarios, wherein various entities, that is, tenants, can directly acquire mobile edge resources on demand. While this novel aspect could add value, it requires an automatic resource brokerage in charge of (*i*) collecting on-demand mobile edge slice requests, (*ii*) monitoring current RAN resource usages, (*iii*) processing slice requirements to be accommodated into the mobile edge network, (*iv*) evaluating economic advantages of admitting/rejecting mobile edge slice requests. In addition to new slice requests, an automatic slice broker also needs to optimize/monetize the unused resource capacity in order to not only accommodate new requests but also to accommodate high traffic requests and/or load on existing slices. Thus, practical and efficient algorithms must be designed to perform admission control operations with the objective of maximizing the overall revenue, that is, minimizing unused mobile edge resources.

9.4.1 Enabling Solutions for Mobile Edge Slicing

A number of works have been produced to cover the above mentioned open issues. It is worth mentioning the former idea of RAN sharing in 3GPP standards as a limited shared mobile edge approach based on contractual agreements. Two active network sharing architectures have been defined, the Multi operator Core Network (MOCN) that allows each operator to share base stations connected on a separate core network and the Gateway Core Network (GWCN) allowing operators to share additionally the mobility management entity (MME) [10]. Then in Reference [11], a complementary network sharing management is specified enabling MVNOs to directly control the allocated resources.

There is huge body of work in the literature focusing on the spectrum sharing topic. A performance comparison study is provided in Reference [12] wherein different sharing scenarios are described. Authors in Reference [13] propose a controller providing wireless services to external applications assuming a three-dimensional resource grid (e.g., space, time, and frequency) that is promptly sliced and efficiently shared. NetShare [14] proposes a network-wide radio resource management framework that provides an efficient RAN sharing solution adopting the well-known proportional fairness criteria.

The most simplified network slice concept application is represented by a pure RAN element's reservation to particular services as an "isolated slice." However, this concept has been recently extended with the advent of the network virtualization paradigm: The higher the flexibility degree, the more the multiplexing gain while accommodating slice requests. The network virtualization substrate (NVS) concept was introduced in References [15] and [16] that allows the infrastructure provider to directly control resource allocation policies toward base station virtual instances as well as to provide decision power to the infrastructure tenants by customizing their own resource scheduling policies.

The NVS solution and the others are mostly oriented to MAC scheduler approaches which usually require MAC scheduler amendments. Conversely, in Reference [17], a gateway-based approach is proposed to remotely accommodate RAN slicing at the base stations without any MAC modification. This kind of approach might further be improved with the abstraction of the underlying mobile edge network, as reported in Reference [18].

Mobile edge slicing is also the main focus of different European 5G-PPP projects. METIS-II [19] discusses technological slicing solutions shedding the light on synchronous functions, such as multidimensional resource management, dynamic traffic steering, and resource abstraction. In 5G-NORMA project [20], a novel architecture has been introduced and deeply explored for enabling 5G networks to RAN slicing aspects by looking at control/data plane functional split for a dynamic slice composition.

Last, a broker entity in charge of monitoring and assigning portion of available capacity has been first introduced by 3GPP and extensively analyzed in Reference [21] with the objective of enabling the on-demand mobile edge slice provisioning. In addition, Reference [22] discusses a dynamic slicing scheme that flexibly schedules radio resources based on the requested service level agreements (SLAs) while maximizing the user rate and applying fairness criteria. Slicing brokering challenges are optimally faced in Reference [23] and [24]. The former focuses on admission control mechanisms to prevent SLA violation while trying to maximize the overall mobile edge resource utilization. The latter focuses on the economic issues related to mobile edge slice requests aiming at maximizing the overall revenue for the infrastructure provider.

9.4.2 Slice Requests Brokering

Mobile edge slice brokerage is the pivotal point of the network slicing development in new business scenarios. If not properly designed, the infrastructure provider may incur losses due to SLA violation in case of sudden high-traffic demands or mobile edge resource under-utilization. An architectural proposal for mobile edge slice brokering is depicted in Figure 9.9.

The mobile edge slice brokering process is performed at the network management level, which directly negotiates with tenant applications through dedicated northbound interfaces (following the SDN paradigm) or, in case of sharing operator network manager, by means of standardized interfaces, namely, Type-5. Authentication and service management services are provided while exchanging slice requests information. The network management layer is also able to exploit 3GPP conventional monitoring procedures in order to gather network key performance indicators (KPIs), for example, global network load measurements, and assist brokering procedures that may require real-time RAN element information. The mobile edge slice broker acts as a mediator converting incoming SLA slice requests into physical resource allocation, thereby com-

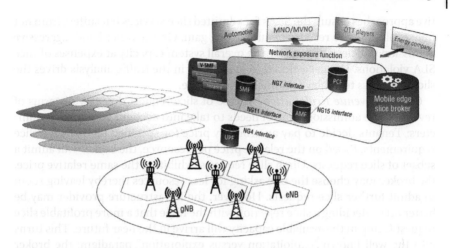

Figure 9.9 Virtualized RAN architecture for supporting network slicing.

posing a mobile edge slice. Those instructions are conveyed through the Itf-N interface to specific RAN elements, based on the slice service requirements.

The mobile edge slice request might consist, and not be limited to, the following parameters: (*i*) slice time duration, (*ii*) explicit physical resource amount in terms of resource blocks (PRBs), (*iii*) specific spatial areas directly affected by the service, (*iv*) starting time or time window within which the slice starting time shall fall, (*v*) heterogeneous service requirements, (*vi*) the slice's purpose for preventing security issues, and (*vii*) RATs selection. Type-5 and Itf-N interfaces are enhanced accordingly to enable different tenant applications to easily inquire such mobile edge slice request parameters. Therefore, an efficient slice allocation is obtained through *ad hoc* admission control algorithms. Such a rich combination of parameters enable the slice broker to perform accurate resource dimensioning and thus effectively handle high traffic requests.

Two different targets are identified while devising admission control strategies and hereinafter described: (*i*) Mobile edge resource utilization and (*ii*) overall revenue.

Mobile Edge Resource Utilization While attaining the first objective, brokering procedures take care of combining and dovetailing different RAN slice requirements in the overall system capacity. An accurate analysis of traffic demands associated with a specific mobile edge slice and future traffic predictions may benefit the overall slicing allocation process resulting in a boosted system capacity as allocated slice resources might have been over estimated. However, traffic demand forecasting failures may seriously impair the slice quality of service and bring additional penalties to the infrastructure provider. In this case, the challenge lies on finding a trade-off between conservative admission control policies and proactive slicing allocation decisions. On the one hand, conservative approaches ensure the SLAs for admitted slice services but suffers from not fully exploiting the resource multiplexing gain. On the other hand, aggressive admission policies can saturate the overall system capacity at expenses of slice SLA violations. Hence, a closed-loop solution in the traffic analysis drives the slicing process to achieve optimal performance.

Overall Revenue Exploitation. In case of slice request diversity in terms of revenues, the admission control needs to take into account additional parameters: Tenants decide to pay an arbitrary price for specified mobile edge slice requirements. Based on the relative price per resource, the broker will admit a subset of slice requests. Upon slice requests exhibiting the same relative price, the broker may choose the one asking for less resources thereby leaving room to admit further slice requests. However, the infrastructure provider may be better off by deciding a slice rejection with the hope that a more profitable slice request fitting in the available capacity will arrive in the near future. This turns into the well-known "exploitation versus exploration" paradigm: the broker may decide to keep accepting slice requests received by known tenants, marked as profitable due to the highest relative price willing to pay, whereas declining some other requests based on a rejection probability. Appropriate decision strategies based on monitored pricing values lead the broker to easily find the optimal strategy for admitting the slice requests while maximizing the overall revenue.

9.4.3   Managing Mobile Edge Slice Resources

Multiple mobile edge slices can share functions and resources. A centralized (SDN) controller can be designed to optimally manage the resource assignment process while attaining high resource efficiency and guaranteeing individual slice SLAs. The controller takes care of network functions, either virtual or physical that the network slices rely on. Interestingly, flexible and reliable slicing solutions leverage on detailed functional splitting configurations: Mobile edge slices are properly tailored on specific service requirements. Although a fixed splitting configuration of shared network functions and resources is beneficial for ensuring high quality of service, it may drastically degrade the network utilization. Conversely, dynamic configuration adjustments might bring multiplexing gains at the expense of (short) service disruptions as well as less determinism. Therefore, the main objective of the controller is to properly find the trade-off between flexible and static resource assignments accounting for sharing policies issued by the service provider.

An example of shared resources is the system spectrum divided into several fixed resource blocks (RBs) and variable length TTI blocks fully managed by the slice controller, as shown in Figure 9.8. Based on the slice requirements, the controller optimally selects a set of TTIs as a "masked" view of the shared resource pool assigned to the mobile edge slice. The advantage of such a solution relies on the instantaneous channel conditions monitoring as well as on the subsequent resource adjustment in order to handle fast channel dynamics. Other options may prefer a dedicated resource scheduler per slice that directly interacts with the northbound interface, of the SDN controller as a specified application. The scheduling application automatically retrieves the selected slice resource mask available on the northbound interface, thereby scheduling users only on granted resource blocks. In addition, the slice resource scheduler may implement own scheduling policies while preserving slice isolation constraints and confidentiality properties. Please note that controller decisions significantly affect the overall slicing efficiency as it might give more priority to particular mobile edge slices: Different objectives can be achieved such as fairness, interference mitigation, or spectral efficiency maximization.

9.5 Network Slicing at the Mobile Transport

Architectural flatness and decentralization – *pushing intelligence out to the edge* – has traditionally been an axiomatic criterion to design 3G/4G systems with affordable topological flexibility and high capacity; we refer to this architecture as distributed radio access network (D-RAN). More recently, an opposing paradigm, called Cloud-RAN (C-RAN), has gained momentum and holds itself out as a promising solution for 5G. In its purest form, the functionality of a base station (BS) is fully decoupled from the remote radio head (RRH) and is virtualized into a centralized cloud computing platform (CCP) [25]. (Virtual) BSs/CCPs are connected to the mobile core (responsible for charging, management, providing gateways to the Internet) via a backhaul (BH) network. On the other end, RRHs exchange digitized radio samples with CCPs through a high-capacity fronthaul (FH) network, typically using Common Public Radio Interface (CPRI) or Open Base Station Architecture Initiative (OBSAI) interface.

The C-RAN approach has been shown to improve spectrum efficiency and reduce costs (pooling gains) in certain setups [26,27]. However, its advantages become questionable when considering larger-scale deployments. This is due to the very stringent requirements of the FH traffic, which can only be met in practice by installing dark fibers that are very costly in terms of time and money. In addition, nowadays FH architectures have the following limitations [28]:

- Bandwidth usage is constant and independent of user load, that is, no statistical multiplexing
- Data rate requirements scale linearly with the number of antennas, which renders massive MIMO unfeasible
- Low (or none) path diversity between RRHs and CCPs (poor resilience, high inefficiency)

- FH and BH are incompatible in terms of physical interfaces, data, or control planes (i.e., no infrastructure reuse)

In light of the above, a redesign of the FH is being considered [28,29]. Notably, the IEEE 1914 WG has recently been set up to standardize the next-generation fronthaul interface (NGFI) tackling the aforementioned limitations. Two pivotal paradigms steer its design. First, NGFI shall be based on a simpler packet-based transport protocol[2] that would enable statistical multiplexing of traffic, infrastructure reuse, and routing with higher degrees of freedom. Second, NGFI proposes a flexible split of the RAN functionality. The idea is to divide a classic BS into a set of functions that can either be processed co located with the RRH or offloaded into a centralized CCP, depending on the transport requirements and centralization needs. In this way, we can relax the tough requirements of fully C-RAN centralization for the transport networks, allowing for the use of heterogeneous wired (e.g., optical, copper) and wireless transport technologies (e.g., mmWave, μ Wave) to reduce the overall transport costs. Each split option poses different requirements in terms of bandwidths and delays (analyzed in the Small Cell Forum [31]), it requires a good trade-off to balance the cost/performance gains (the more is the centralization, the higher the gains) and the requirements (the less is the centralization, the relaxer the transport network requirements). In principle, retaining as much centralization as possible would be desirable for achieving high-performance gains, when full centralization is not feasible due to the transport network constraints.

To allow for various degrees of radio access network (RAN) centralization varying from no centralization D-RAN to fully C-RAN, the division between BH and FH will blur as varying portions of functionality of the base stations need to be moved flexibly across the transport network as required for cost-efficiency/performance reasons. This leads to the convergence of FH and BH in the same transport network, which is envisioned for the next generation transport for 5G. In fact, the integration of FH and BH toward a common packet-based transport network is recently being studied under the umbrella of the 5G public–private partnership (5G-PPP) projects in Europe, referred to as *Crosshaul* in 5G-Crosshaul Project [32], or *X-haul* in 5G-Xhaul Project [33], or named as Fronthaul/Midhaul/Backhaul in iCirrus Project [34]. In this book, the integrated FH and BH is referred as the *Mobile Crosshaul*.

Although multi tenancy in its wide sense is a concept that has been developed since long in many contexts, its applicability and benefits within transport networks have been addressed more recently, with specific research work regarding, for example, network virtualization in projects such as GEYSERS and STRAUSS [35,36]. For the integrated FH and BH, the 5G-PPP projects, such as

---

2 Indeed Ethernet with some enhancements, frame preemption (IEEE 802.1Qbu) and scheduling (802.1Qbv), has been shown to be promising [30].

5G-Crosshaul and 5G-Xhaul projects, are still looking for ways to support multitenancy in the future 5G transport networks, allowing for dynamic network resource management and infrastructure sharing.

The concept of multitenancy for the mobile crosshaul is illustrated in Figure 9.10, where the owner of the physical infrastructure allocates virtual infrastructures over its substrate network, providing multiple network slices to offer to different tenants. Each tenant, for example, a MNO or MVNO owns a network slice and operates the allocated virtual infrastructure. In this example, tenant A, C, and D owns the network slice 1, 2, and 3, respectively. Moreover, tenant A itself is an MNO who also owns the physical infrastructure that can be shared by other MVNOs. The MVNO tenants can further deploy their own NS or allow multiple third-party tenants (e.g., OTTs) to instantiate their NS on top of the virtual infrastructure, for example, tenant B deploying its NS over the VI of tenant A. It is possible to instantiate a VI on top of another one following a recursive approach, by applying the same principles and operational procedures, for example, the VI of tenant D is instantiated over the one of tenant C.

The provision of network slices in the mobile crosshaul requires enabling technologies to be in place in the data plane (i.e., transport technologies), control and management plane (i.e., MANO), and application plane. For the transport in the data plane, traffic separation and isolation and resource partitioning techniques are needed. For the MANO to manage and orchestrate the mobile crosshaul, SDN and NFV are the key technologies that enable networks to be broken out from their underlying physical infrastructures so that they can, programmatically, provide connectivity as a service. In terms of application plane, novel SDN/NFV applications are needed to address the new challenges of integrated fronthaul/backhaul networks, and new requirements of the 5G tenant users requesting a slice for deployment of their own services or virtual infrastructures. In the following section, we will briefly sketch the main aspects at the data, control, and application layers to enable network slicing, and how requirements such as isolation and traffic separation are realized.

9.5.1 Enabling Mobile Transport Slicing Technologies

The mobile crosshaul transport network is comprised of forwarding elements and processing units. The forwarding elements are switching units, based on packet or circuit technologies, that interconnect a broad set of link and PHY technologies (e.g., optical, microwave, mmWave) using a common frame to transport both backhaul and fronthaul traffic. The processing units take care of most of the computational burden in the mobile crosshaul, including BBUs or MAC processors, VNFs, and other virtualized services (e.g., CDN-based transcoding). To this aim, the data plane makes use of a NFV infrastructure (NFVI) relying on generalized hardware components.

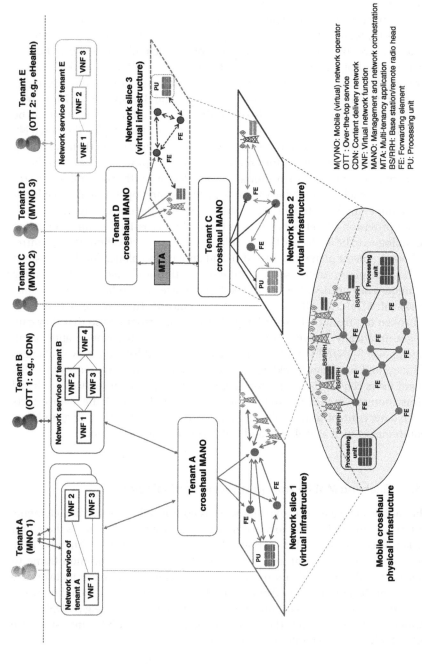

Figure 9.10 Network slicing in mobile crosshaul for multi-tenancy support.

When carrying the data of several tenants through the network, several requirements have to be considered:

- Traffic *Separation*. One tenant should not be able to listen to the traffic of other tenants or of the network provider.
- Traffic *Isolation*. The network has to provide guaranteed QoS to traffic of different tenants. Traffic of one tenant should not impact the QoS of the traffic of other tenants.
- Traffic *Differentiation*. The traffic of different tenants may be forwarded differently, even when entering or exiting the network at the same points of attachment.
- *Statistical Multiplexing*. Multiplexing gains should be possible among the traffic of different tenants.

The crosshaul forwarding elements may support circuit- or packet-switched forwarding or both. For circuit-switched forwarding, traffic separation and isolation can be achieved by creating different circuits per tenant. Although this is beneficial to achieve low and deterministic latency, for example, it does not provide statistical multiplexing gains among the traffic of different tenants. For packet-switched forwarding, the requirements are supported by using a common frame format across the fronthaul and backhaul network and different transmission technologies.

In terms of the design for the crosshaul common frame, we can leverage on existing approaches. One of the baseline approaches is to apply Provider Backbone Bridge Traffic Engineering (PBB-TE) [37] as common format to encapsulate the tenants' traffic. The forwarding elements can make use of the additional fields in the PBB-TE header (see Figure 9.11) to achieve the requirements.

Packet forwarding is harmonized across the network, thanks to the adoption of the crosshaul forwarding elements as switch and a forwarding abstraction model common to all the forwarding elements, either circuit or packet based. Such models are defined by the southbound protocols that define the interaction between the data and control planes. We propose the use of OpenFlow protocols suite as southbound interface for controlling the forwarding of the crosshaul common frames. OpenFlow defines a rich set of operations that can be applied to incoming packets for differentiating the forwarding behavior.

The backbone VLAN ID (B-VID) and the service ID (I-SID) allow identifying the traffic of different tenants by using unique identifiers per tenant or even per service of the tenants. This allows creating different virtual networks and keeping the traffic separate in the forwarding elements. The forwarding elements can also consider these identifiers to take different forwarding decisions per tenant.

To achieve the traffic separation regarding QoS, the three priority code point bits within the header can be used to distinguish different types of service

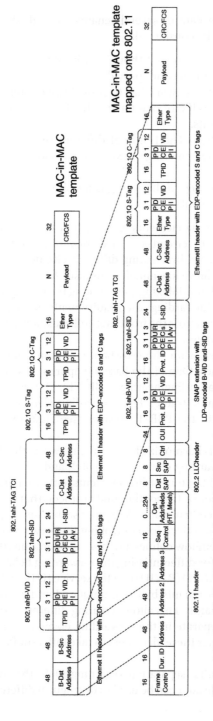

Figure 9.11 Provider Backbone Bridge Traffic Engineering (PBB-TE) header

within the network and to schedule the packets for forwarding based on this priority information. At the ingress of the network, this priority has to be set appropriately and consistently across the different tenants to simplify the rules within the network.

An alternative approach that is favored by the mobile network operators is Multi protocol Label Switching - Transport Profile (MPLS-TP). MPLS-TP is intended to provide next-generation converged packet transport networks tying together service routing and transport platforms, enabling a true carrier grade packet transport. Its major advantages are consistent operations and operations, administration, and maintenance (OAM) functions across the different network layers and the seamless interworking with IP/MPLS networks. MPLS-TP is highly scalable due to its multiplexing capability that can be used to create a network with multiple hierarchical layers. MPLS-TP also supports a large variety of services that are encapsulated into pseudowires and it can be carried over the existing and evolving transport network infrastructure.

MPLS-TP is based on MPLS but differs from the latter since the configuration is provided via management commands and not via routing protocols such as the Label Distribution Protocol (LDP). Typically, nodes are connected via either point-to-point pseudowires or multipoint-to-multipoint networks, implementing a mesh of pseudowires. MPLS-TP has a typical frame structure with an outer label for a label switched path (LSP), inner label for a pseudo wire (PW), and an optional PW control word. ITU-T and IETF standards allow the possibility of an unlimited number of indented labels, corresponding to an unlimited LSP hierarchy.

Figure 9.12 shows an MPLS-TP header using Ethernet as data link layer technology. The label in the LSP label can be used for distinguishing different tenants. Prioritization of different services can be based on the Traffic Class field (TC bits), while the PW can be used to transport different services, both packet and circuit oriented. According to IETF RFC 3270, MPLS-TP can differentiate traffic based on EXP-inferred PHB scheduling class LSP (E-LSP) or label-only-inferred PHB scheduling class LSP (L-LSP). In the first case, the TC field is used to determine the per-hop behaviour (PHB) to be applied to the packet and it supports a maximum number of eight scheduling classes, in the other case PHB scheduling class is explicitly assigned at the time of label establishment supporting an unlimited number of scheduling classes. Different LSPs can be provisioned to transport different flows belonging to different tenants. Edge nodes are in charge of classifying the incoming traffic and assigning it to the correct LSP while intermediate nodes need to analyze only the LSP label of the incoming packets and use the forwarding behaviour associated with the E-LSP or L-LSP. As such, multiple tenants can be supported over the same infrastructure without interfering.

MPLS-TP Frame

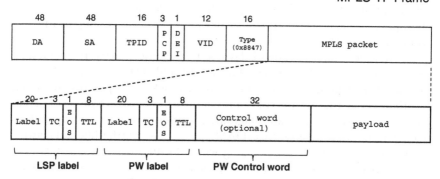

Figure 9.12 Multi-Protocol Label Switching - Transport Profile (MPLS-TP) header

9.5.2 Enabling Slicing Technologies for the Crosshaul MANO

The mobile crosshaul management and orchestration (MANO) can be designed based on the ETSI NFV MANO framework [4]. It may consist of a NFV Orchestrator (NFV-O), one or multiple VNF Managers (VNFMs), a Virtual Infrastructure Manager (VIM) responsible for controlling and managing the NFVI computing (via computing controllers), storage (via storage controllers), and network resources (via SDN controllers). Support of multitenancy has a strong requirement on the crosshaul MANO components, from the SDN controller to the VIM and MANO components, for the orchestration and deployment of virtual infrastructure and network services, as described in Section 9.2.

The SDN controller must support the provisioning of isolated virtual network infrastructures with a given set of capabilities. Traffic isolation can be achieved through the creation of tagged network connections, configuring the flows at the forwarding elements making use of the multi tenant features of the crosshaul common frame in the mobile crosshaul data plane. Traffic encapsulation and decapsulation (e.g., using PBB-TE or MPLS-TP headers, as described in Section 9.5.1) with tenant-specific tags must be enforced at the network edges in order to guarantee the proper isolation. At the northbound level, the SDN controller needs to support a service for the creation and operation of virtual networks assigned to specific tenants, which can be specified following intent-based network models; QoS requirements can be handled through the usage of queues or meters. An example of SDN application for provisioning of virtual network infrastructures with multitenancy support is the OpenDaylight Virtual Tenant Network (VTN) project [38]. The VTN application allows a tenant to request a virtual network composed of virtual bridges, routers, tunnels, tunnel end points, virtual interfaces, and virtual links. The mapping between network packets exchanged between OpenFlow switches at the data plane level and instances of virtual networks defined at the logical level is based on ports and/or VLANs.

At the VIM level, multi tenancy is handled through the modeling of the tenant concept, where each tenant has its own view of the VIM capacity, policies to regulate the access to the resources (e.g., a quota of dedicated resources), and, optionally, custom resource flavors and VM images. Requests for new VI must be authenticated and authorized, and they are evaluated based on the resources still available in the tenant's quota. Finally, the access to the instantiated VI is strictly limited to the tenant owing the specific instance. Most of the cloud computing platforms (e.g., VMware, OpenStack) support multitenancy.

A similar approach, based on per-tenant profiles and policies, needs to be adopted at the NFV Orchestration level, extending the virtual resources concept to VNF and NS entities. Each tenant must have the view and the control on its own VNFs and NSs only; they must be maintained fully isolated from other entities belonging to different tenants in order to guarantee their security and their desired KPI level independently on the load of other VNFs. New service requests must be granted depending on the tenants profile, in combination with the tenant-related policies at the VIM level, which may have an important role in the VIM selection. Currently, this implies extending the functions of the NFVOs such as Open Source Mano[3] and OpenBaton[4] to have the tenant separation and identify the mapping of a tenant to a NS.

9.5.3 Multi-tenancy Application for Slice Management and Orchestration

A coherent management and orchestration of multi tenancy is required horizontally for unifying the concepts of infrastructure virtualization and multitenancy in all involved segments and resources. The multi tenancy application (MTA) at the application level provides such management, becoming the logical decision entity and serving as an optimizer to decide the constrained allocation, operation, and deallocation of virtual infrastructures (VI) and/or of network service (NS) instances, in compliance with the desired degree of detail and control. The MTA acts as a front end between the tenants and the tenant's service instances. Essentially, for mobile crosshaul, this application is in charge of deciding an optimum subset of nodes (node mapping) and links (link mapping) in the substrate network to build a virtual infrastructure for a tenant that satisfies its resource demand and service-level agreements (SLAs), by solving classical

---

3 https://osm.etsi.org/
4 http://openbaton.github.io/

virtual network embedding (VNE) problems [39–44]. In our context, the VNE problem consists of the following aspects[5]:

- Slice Creation: Finding a subset of network, cloud nodes, and links, and allocation of the corresponding infrastructure resources (networking, storage, computing) in the substrate network to satisfy each slice request.
- Service Management: Optimize the placement of virtual network functions and even chaining them to provide end-to-end services.

To deploy and enforce the computed mapping, the MTA needs to interact/coordinate with several functional entities inside the crosshaul MANO, namely, the SDN controller, the NFVO, and the VIM, to collect information as well as to give commands. The MTA covers both network- and computing related functions. For the network-related services, the MTA first collects the information on physical topology, traffic paths, and link load through the control plane, then computes the optimum allocation of networking resources and commands the control plane to perform the required configuration. This may involve direct requests to the SDN controller (to provision network paths and/or to allocate virtual nodes providing the desired mapping between physical and virtual ports). For the computing-related services, the MTA may ask the NFVO to provide a virtual infrastructure topology specifying where the VNFs must be placed or instruct directly the VIM to enforce the mapping between virtual infrastructures and corresponding physical resource. The VIM itself will in turn request the SDN controller for the provisioning of required network paths and related node configurations.

9.6 Network Slicing at the Mobile Cloud

The definition of a functional architecture for a network slice comprises the design of network functions and the interfaces among them to support the realization by end users of the service realized through the network slice. The core network is traditionally the part of the mobile network where such network functions reside, namely, for the authentication and authorization of the user, the session and mobility management, QoS provisioning, charging, lawful interception, and so on, and these network functions are split into control plane and user plane functions.

The system architecture working groups within the 3rd Generation Partnership Project[6] (3GPP) have specified the mobile core for the 3G and 4G systems, therefore, they are also the primary actors for what concerns the specification

---

5 We note that this mathematical framework can also be applied to the core segment.
6 www.3gpp.org

of network slicing support in the core network. Although the 4G system was not specifically designed to cater for network slicing, it supports network sharing as per the architectures defined in Reference [11] (see Section 9.4.1). From release 13, 3GPP also introduced the Architectural Enhancement for Dedicated Core Networks (DCN) that can be seen as an alternative to support slicing in LTE networks [45,46]. Such enhancements enable a mobile network operator to deploy one or more dedicated core networks within a PLMN, each of those dedicated to specific types of subscribers. Differently from routing traffic through specific PGWs by appropriately configuring dedicated APNs in a group of UEs, DCN can result in using a whole dedicated set of EPC entities both in the control and data plane. In fact, DCN consists in instantiating a set of physical or virtualized core entities (MME, SGW, etc.) connected to the same RAN (i.e., announcing the same PLMN identifier). This way, an operator can instantiate a set of core resources for a given tenant and partition the subscribers in order to serve them through a dedicated core instance. For instance, M2M users can be isolated from the rest by being rerouted to a dedicated core network that they connect to using a shared RAN.

Nevertheless, different terminals and subscribers may have fundamentally different communication characteristics and therefore, different requirements on the underlying service. The most obvious example comes from comparing a modern smartphone and a static metering sensor. Although the former makes use of a disparate set of communication services (voice, video, web, etc.) even when in motion, the latter only generates small amount of data at periodic intervals from a fixed location. In such scenario, current EPC operations produce much more signaling overhead rather than actual data transmission, and some of such control plane procedures (e.g., related to paging, session, and mobility management) are not necessary at all. Based on this and other use cases, it has already been identified that there is a need to redesign the core network architecture and control protocols taking into account more light-weight operations and the tailoring of core functions to the terminal characteristics. Under these assumptions, slices in a 5G environment can be designed to cater only for specific functionalities and supported features that are required by classes of users (i.e., end devices).

The concepts already mentioned are encompassed in 3GPP SA1's "Feasibility Study on New Services and Markets Technology Enablers" (also known as "SMARTER" [47]). SMARTER has identified network slicing as a key enabler to support future service scenarios, and the requirements of which have been derived and documented in Reference [48].

Based on such requirements 3GPP members have drafted potential solutions documented in the "Study on Architecture for Next Generation System" [2]. The solutions therein mostly focus on the slice's architectural aspects (e.g., functions common to multiple slices versus dedicated functions in a slice) and operational aspects related to how UEs can (re-)select a slice, connect to multiple slices, and roaming support. The proposed solutions look at aspects that address separation between core network instances for improved slice isolation, as well as function reuse for better efficiency. One can therefore design an architecture for network slicing where all network functions are dedicated to a slice, or, conversely, where some functions are common to multiple slices, for instance, having control plane handling common between the slices, while the user planes are handled as different network slices.

The evaluation of the different solutions lead to the following "interim agreements," that are recognized by the 3GPP participants as common starting design objectives on which the specifications of the next generation system should be based upon:

- The network slice is a logical network (providing telecommunication services and network capabilities) including access network (AN) and core network (CN).
  - AN can be common to multiple network slices.
  - Network slices may differ for features supported and network functions optimizations use cases.
  - Networks may deploy multiple network slice instances delivering exactly the same optimizations and features - but dedicated to different groups of UEs, for example, as they deliver a different committed service and/or because they may be dedicated to a customer.
- A UE may provide network slice selection assistance information (NSSAI) consisting of a set of parameters to the network to select the set of RAN and CN part of the network slice instances (NSIs) for the UE.
- If a network deploys network slicing, then it may use UE-provided network slice selection assistance information to select a network slice. In addition, the UE capabilities and UE subscription data may be used.
- A UE may access multiple slices simultaneously via a single RAN. In such case, those slices share some control plane functions, for example, mobility management function and authentication function. These common functions are collectively identified as CCNF (Common Control Network Functions).
- The CN part of network slice instance(s) serving a UE is selected by the CN, not by the RAN.
- It shall be possible to handover a UE from a slice in next-generation core (NGC) to a DCN in EPC. There is not necessarily a 1:1 mapping between slice and DCN.
- The UE needs to be able to associate an application with one out of multiple parallel established PDU sessions. Different PDU sessions may belong to different slices.

In a subsequent step, 3GPP started to document the system architecture for a 5G system in a technical specification [49]. The current system architecture includes the control plane network functions for Core Access and Mobility

Management (AMF), Session Management (SMF), Policy Control (PCF), Authentication Service (AUSF), and Unified Data Management (UDM). Since network slices can differ in the supported features, multiple instances of a network slice may differ in the type of services and accordingly in the types of deployed network functions. Although most control plane functions can be associated with an individual network slice, the AMF represents a common instance for all network slices serving a UE, and takes a particular role in connecting a network slice in the core network with the access network. The core network part of a single or multiple instances of a network slice serving a UE is selected by the core network.

During its attachment procedure, a UE may provide NSSAI, which can be used by the access network to select a suitable AMF. The selected AMF can use the requested NSSAI, which comprises Single NSSAI (S-NSSI) corresponding to the network slices a UE intents to connect to, as well as information about the data network, to which the UE wants to connect, to select a network slice and the associated SMF.

In case the UE does not provide NSSAI, the access network can forward signaling associated with the registration procedure to a default AMF. The AMF can refer to and route the registration request to a different AMF in case the default AMF is not suitable to serve the UE's request. The procedure associated with a 5G system and initial specification on how the procedure applies to a sliced network are specified in Reference [50].

9.6.1 Control Plane Modularization to Support Network Slicing

Following the approach of having common network functions shared among slices, the concept of *architecture modularization* applied to the control plane is regarded as an essential design principle to build a flexible network architecture natively supporting slicing. According to this concept, conventional monolithic network functions, often corresponding to physical network elements in the existing systems, are split into basic building blocks defined with the proper granularity. This allows for defining different logical architectures (i.e., different network slices) by composing building blocks in a way that is seen as the most appropriate to fulfill the functional requirements to that particular slice. This approach goes along the philosophy that a slice is created to fulfill the requirements of a specific service, and a tailored functional architecture for that slice is the most suitable solution as opposed to the one-fits-all model.

The categorization and division of network functions into building blocks to assemble a network slice is not univocal. A formalized methodology is therefore necessary to define the basic set of network building blocks, upon which access agnostic end-to-end architectures can be designed. Such methodology may focus on CN and AN functional blocks altogether, making a tight coupling between them in order to optimize the overall procedures and protocols.

Nevertheless, this approach may lead to a (radio) access technology-specific architecture, thus hindering flexible deployment of those core functions to other (R)ATs, including those that are not yet available. It is hence considered as a key design objective to define AN abstractions and expose them to CN building blocks in order to be integrated in the E2E architecture. Moreover, the defined set of functions needs to allow tailored network architectures to be instantiated as independent network slices, yet enabling 5G devices to discover and connect to multiple network slices, allowing inter slice operations and optimization for specific use cases.

An exemplary modularization of network functions into building blocks is presented further and is a result of the 5G CONFIG project[7], a private industry research initiative working on concepts for the upcoming 5G core network architecture. A methodology for modularization as well as a set of basic building blocks serving as modules for a 5G network architecture are introduced in Reference [51]. It collects and categorizes the network functions that are currently defined by 3GPP specifications [46], as well as including additional features proposed for a 5G system:

- The *Access Function (AF)* is responsible for the last hop connectivity and terminates the interface that conveys the appropriate abstractions for radio connection management independently from the radio technology. Such interface should be common to all ANs connected to the core, but the AN may use access-specific protocols toward the AN. The AF is the converging point for CN-related access control management procedures, such as initial attachment, change of point of attachment, and detachment of end devices, allowing controlling entities to be aware of the latest network end point to which devices were connected.
- The *Connectivity Management (CM)* terminates the interfaces from the AN and the 5G end devices that carry the Network Access Stratum (NAS) signaling. The CM selects the user-plane functions that shall support the services for a specific device, and manages the device's session. For this reason, it plays a key role for the per-slice customization in the control plane as well.
- The *Flow Management (FM)* handles data flows in the user plane, by controlling how functions are traversed in the core toward access domains and other domains such as the service hosting infrastructures. It is involved in establishing and reconfiguring user plane paths and enforcing QoS and other policies to the user data flows. Together with the CM, the FM can be seen as the control plane functions that drive the operation of the user plane architecture described in the next section, provided the appropriate abstractions and interface to program the entities defined therein.

---

7 https://www.5g-control-plane.eu/

- The *Mobility Management (MM)* handles the connection issues that arise upon mobility events. This building block may be deployed in different flavors, for example, it can be enriched with slice-specific features (e.g., mobility support for multicast/broadcast communications, mobility pattern-based schemes, etc.) or even removed from a slice if not necessary (e.g., for fixed devices).
- The *Security and AAA Management (SAM)* is responsible to perform authentication and security-related functions. It may evolve rapidly and independently because of the introduction of new authentication and security methods, as 5G encompasses a wide range and type of end devices, which pose different requirements in terms of security.
- The *Context Management Function (CMF)* generates, stores, and distributes context that can be utilized by both data plane and control plane building blocks, which subscribe to it. The context refers to the information that describes the situation of an entity, which can be a device, user, network, radio environment, or an external application.

9.6.2 User Plane Simplification for Lean Packet Slices

The 3GPP EPC adopts the bearer concept to establish logical connections between a mobile device and its assigned packet data network gateway. In support of a more lightweight user plane with less per-packet overhead, no single point to aggregate traffic forwarding and treatment policies but optimized routing in between a mobile device's location and its one or multiple correspondent services, this section summarizes the concept of a simplified user plane to support slices associated with a more lean packet system. The design objective of the lean packet system (LeaPS) is to provide all devices with IP connectivity as base layer service, and to incrementally offer additional services like session management, QoS support, mobility management, and so on. In this way, connectivity serves as a platform to build services tailored to the requirements of classes of terminals and/or subscribers (that is, the end users of a slice). In LeaPS, a common transport infrastructure connects multiple access networks made of heterogeneous access technologies to distributed data centers where services are hosted, such data centers might be remotely located in the Internet, or in proximity to the users in edge clouds, as illustrated in Figure 9.13.

Instead of selection of a single packet data network gateway, and setting up a logical connection between that gateway and the mobile device, the control plane can select suitable Policy Enforcement Points (PEP) at the network edge (edge PEP) and in the proximity of the device's used correspondent services (root PEP). The control plane enforces policies for traffic selection and treatment in the edge PEP and one or multiple root PEPs to enable direct traffic forwarding between a device and its correspondent services, as well as to classify traffic for prioritized forwarding and metering. As depicted in Figure 9.13,

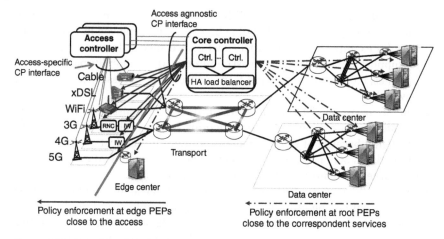

Figure 9.13 Overview of the lean packet system.

the control plane is logically separated from the user plane and can be represented by a tailored selection of control plane functions per the modularized architecture described in Section 9.6.1. The control plane communicates the traffic rules through well defined interfaces terminated by the PEPs: By traversing the pair of Edge and Root PEPs, a user traffic flow undergoes the policies that are defined by the control plane. Thus, the role of the control plane is to *(i)* determine the policies that are required for a particular slice, *(ii)* associate traffic flows to slices, and *(iii)* propagate the policies in the form of traffic rules to the appropriate PEPs. Such policies are related to mobility support and associated traffic steering, QoS, charging, and so on, and are implemented by PEPs through match/action rules in accordance with the SDN paradigm (e.g., based on the Openflow protocol).

This architecture leverages the control-/user plane separations to allow for independent scaling and flexible slice configuration over a common transport infrastructure being the network functions associated with a slice implemented by a set of forwarding devices through the policies defined for that slice. As exemplarily depicted in Figure 9.14, the control plane may have an abstract view of the user plane and associated edge PEPs and root PEPs. Such abstract view can represent a particular user plane slice instance. The mapping of the selected PEPs per the control plane's view of the abstract user plane slice to physical resources, such as routers or switches, can be accomplished by a network controller, for example, an SDN Controller.

9.7 Acknowledgment

The work presented in this chapter has been partially funded by the EU H2020 TRANSFORMER project (Grant agreement No.761536). The authors would

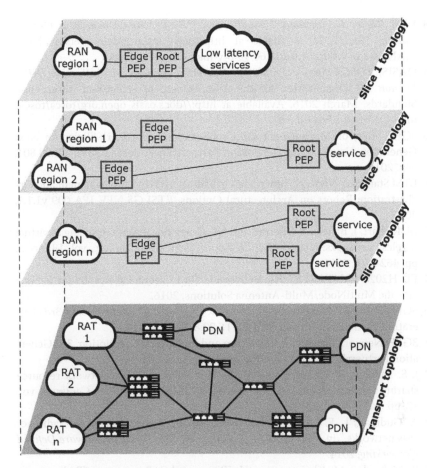

Figure 9.14 User plane abstractions for different slice instances.

like to acknowledge the contributions of their project colleagues, although the views expressed here are those of the authors and do not necessarily represent the project.

References

1 NGMN Alliance, *Description of Network Slicing Concept*, Public Deliverable, 2016.
2 3GPP, *Study on Architecture for Next generation System*, 3rd Generation Partnership Project (3GPP), TR 23.799, Sep. 2016.
3 NGMN Alliance, 5G White Paper, Public Deliverable, 2015.

4 ETSI, *Network Functions Virtualisation (NFV); Management and Orchestration*, Dec. 2014. Available at http://www.etsi.org/deliver/etsi_gs/NFV-MAN/001_099/001/01.01.01_60/gs_nfv-man001v010101p.pdf

5 OASIS, *TOSCA Simple Profile for Network Functions Virtualization (NFV) Version 1.0*, Organization for the Advancement of Structured Information Standards, March 2016. Available at http://docs.oasis-open.org/tosca/tosca-nfv/v1.0/tosca-nfv-v1.0.pdf

6 3GPP, Study on Management and Orchestration of Network Slicing for Next Generation Network, 3rd Generation Partnership Project (3GPP), TR 28.801, Sep. 2016.

7 ETSI Standard, Network Functions Virtualisation (NFV); Management and Orchestration; Report on Architectural Options, ETSI GS NFV-IFA 009 v1.1.1, 2016.

8 M. Richart, J. Baliosian, J. Serrat, and J. L. Gorricho, "Resource slicing in virtual wireless networks: a survey," *IEEE Trans. Netw. Serv. Manag.*, vol. 13, no. 3, pp. 462–476, 2016.

9 EU H2020 FANTASTIC 5G, Deliverable D4.1, Technical Results for Service Specific MultiNode/Multi-Antenna Solutions, 2016.

10 3GPP, Study on Radio Access Network (RAN) Sharing enhancements, 3rd Generation Partnership Project (3GPP), TS 22.852, Sep. 2014.

11 3GPP, Network Sharing; Architecture and Functional Description, 3rd Generation Partnership Project (3GPP), TS 23.251, June 2016.

12 J. S. Panchal, R. D. Yates, and M. M. Buddhikot, "Mobile network resource sharing options: performance comparisons," *IEEE Trans. Wirel. Commun.*, vol. 12, pp. 4470–4482, 2013.

13 A. Gudipati, L. E. Li, and S. Katti, "Radiovisor: a slicing plane for radio access networks," in *Proc. of ACM Workshop on Hot Topics in Software Defined Networking*, 2014.

14 R. Mahindra, M. Khojastepour, H. Zhang, and S. Rangarajan, "Radio access network sharing in cellular networks," in *Proceedings of IEEE International Conference on Network Protocols (ICNP'13)*, 2013.

15 R. Kokku, R. Mahindra, H. Zhang, and S. Rangarajan, "NVS: a substrate for virtualizing wireless resources in cellular networks," *IEEE/ACM Trans. Netw.*, vol. 20, no. 5, pp. 1333–1346, 2012.

16 X. Costa-Perez, J. Swetina, T. Guo, R. Mahindra, and S. Rangarajan, "Radio access network (RAN) virtualization for future carrier networks," *IEEE Commun. Mag.*, vol. 51, no. 7, pp. 27–35, 2013.

17 R. Kokku, R. Mahindra, H. Zhang, and S. Rangarajan, "CellSlice: Cellular wireless resource slicing for active ran sharing," in *5th International Conference on Communication Systems and Networks (COMSNETS'13)*, 2013.

18 J. He and W. Song, "App RAN: Application-oriented radio access network sharing in mobile networks," in *Proc. of the International Conference on Communications (ICC)*, 2015.

19 EU H2020 METIS-II, Deliverable D5.1, Draft Synchronous Control Functions and Resource Abstraction Considerations, 2016.

20 EU H2020 NORMA, Deliverable D5.1, Functional Network Architecture and Security Requirements, 2015.

21 K. Samdanis, X. Costa-Perez, and V. Sciancalepore, "From network sharing to multi-tenancy: the 5G network slice broker," *IEEE Commun. Mag.*, vol. 54, no. 7, pp. 32–39, 2016.

22 M. I. Kamel, L. B. Le, and A. Girard, "LTE wireless network virtualization: dynamic slicing via flexible scheduling," in *IEEE 80th Vehicular Technology Conference (VTC2014-Fall)*, Sept 2014.

23 V. Sciancalepore, K. Samdanis, X. Costa-PØrez, D. Bega, M. Gramaglia, and A. Banchs, "Mobile traffic forecasting for maximizing 5G network slicing resource utilization," in *Proc. of the IEEE International Conference on Computer Communications (INFOCOM)*, May 2017.

24 D. Bega, M. Gramaglia, A. Banchs, V. Sciancalepore, K. Samdanis, and X. Costa-PØrez, "Optimising 5G infrastructure markets: the business of network slicing," in *Proc. of the IEEE International Conference on Computer Communications (INFOCOM)*, May 2017.

25 Y. Lin et al., "Wireless network cloud: architecture and system requirements," *IBM J. Res. Dev.*, vol. 54, no. 1, pp. 4:1–4:12, 2010.

26 I. Chih-Lin et al., "Toward green and soft: a 5G perspective," *IEEE Comm. Mag.*, vol. 52, no. 2, pp. 66–73, 2014.

27 V. Suryaprakash, P. Rost, and G. Fettweis, "Are heterogeneous cloud-based radio access networks cost effective?" *J. Select. Areas Commun.*, vol. 33, no. 10, pp. 2239–2251, Oct 2015.

28 China Mobile, 'Next Generation Fronthaul Interface,' White paper, Alcatel-Lucent, Nokia, ZTE, Broadcom, Intel, June 2015.

29 I. Chih-Lin et al., "Rethink fronthaul for soft RAN," *IEEE Comm. Mag.*, vol. 53, no. 9, pp. 82–88, 2015.

30 T. Wan and P. Ashwood-Smith, "A performance study of CPRI over Ethernet with IEEE 802.1Qbu and 802.1Qbv enhancements," in *GLOBECOM*, Dec 2015, pp. 1–6.

31 Small Cell Forum, Release 6.0. Small cell virtualization functional splits and use cases, Jan. 2016.

32 X. Costa-Perez, A. Garcia-Saavedra, X. Li et al., "5G-Crosshaul: an SDN/NFV integrated fronthaul/backhaul architecture for 5G networks," *IEEE Wirel. Commun. Mag.*, vol. 24, no. 1, pp. 38–45, 2017.

33 5G-Xhaul, EU H2020 Project, Dynamically Reconfigurable Optical-Wireless Backhaul/Fronthaul with Cognitive Control Plane for Small Cells and Cloud-RANs." Available at http://www.5g-xhaul-project.eu/

34 iCirrus, EU H2020 Project, intelligent Converged network consolidating Radio and optical access aRound USer equipment. Available at http://www.icirrus-5gnet.eu/

35 GEYSERS FP7 project, A Generalised Architecture for Dynamic Infrastructure Services System. Available at http://www.geysers.eu/

36 STRAUSS FP7 EU-JP project, Scalable and efficient orchestration of Ethernet services using software-defined and flexible optical networks. Available at http://www.ict-strauss.eu/en/

37 IEEE 802.1 Task Group, IEEE 802.1ah-2008 - IEEE Standard for Local and metropolitan area networks – Virtual Bridged Local Area Networks Amendment 7: Provider Backbone Bridges, 2008.

38 OpenDaylight, OpenDaylight Virtual Tenant Network (VNT). Available at https://wiki.opendaylight.org/view/VTN:Main

39 Y. Zhu and M. Ammar, "Algorithms for assigning substrate network resources to virtual network components," in Proc. of the *25th IEEE International Conference on Computer Communications (INFOCOM 06)*, April 2006, pp. 1–12.

40 M. Yu, Y. Yi, J. Rexford, and M. Chiang, "Rethinking virtual network embedding: substrate support for path splitting and migration," *SIGCOMM Comput. Commun. Rev.*, vol. 38, no. 2, pp. 17–29, 2008.

41 N. M. M. K. Chowdhury, M. R. Rahman, and R. Boutaba, "Virtual network embedding with coordinated node and link mapping," in *IEEE INFOCOM 2009*, April 2009, pp. 783–791.

42 X. Cheng, S. Su, Z. Zhang, H. Wang, F. Yang, Y. Luo, and J. Wang, "Virtual network embedding through topology-aware node ranking," *SIGCOMM Comput. Commun. Rev.*, vol. 41, no. 2, pp. 38–47, 2011.

43 S. Zhang, Z. Qian, J. Wu, and S. Lu, "An opportunistic resource sharing and topology-aware mapping framework for virtual networks," in *Proc. IEEE INFOCOM 2012*, March 2012, pp. 2408–2416.

44 L. Gong, Y. Wen, Z. Zhu, and T. Lee, "Toward profit-seeking virtual network embedding algorithm via global resource capacity," in *Proc. IEEE INFOCOM 2014*, April 2014, pp. 1–9.

45 3GPP, Architecture Enhancements for Dedicated Core Networks: Stage 2, 3rd Generation Partnership Project (3GPP), TR 23.707, Dec. 2014.

46 3GPP, General Packet Radio Service (GPRS) enhancements for Evolved Universal Terrestrial Radio Access Network (E-UTRAN) access, 3rd Generation Partnership Project (3GPP), TS 23.401, May 2015.

47 3GPP, Feasibility Study on New Services and Markets Technology Enablers: Stage 1, 3rd Generation Partnership Project (3GPP), TR 22.891, Sep. 2016.

48 3GPP, Feasibility Study on New Services and Markets Technology Enablers – Network Operation: Stage 1, 3rd Generation Partnership Project (3GPP), TR 22.864, Sep. 2016.

49 3GPP, System Architecture for the 5G System: Stage 2, 3rd Generation Partnership Project (3GPP), TS 23.501, April 2017.

50 3GPP, Procedures for the 5G System: Stage 2, 3rd Generation Partnership Project (3GPP), TS 23.502, April 2017.

51 X. An, R. Trivisonno, H. Einsiedler, D. von Hugo, K. Haensge, X. Huang, Q. Shen, D. Carujo, K. Mahmood, D. Trossen, M. Liebsch, F. L. ao, C. Phan, and F.

Klamm, "Architecture modularisation for next generation mobile networks," in *Proc. of the European Conference on Networks and Communications (EuCNC)*, June 2017.

Xavier Costa-Perez is head of 5G Networks R&D at NEC Laboratories, Europe, where he manages several projects focused on 5G mobile core, backhaul/fronthaul, and radio access networks. The 5G Networks team contributes to projects for NEC products roadmap evolution and to European Commission research collaborative projects and related standardization bodies. He received his M.Sc. and Ph.D. in telecommunications from the Polytechnic University of Catalonia (UPC-BarcelonaTech).

Andres Garcia-Saavedra received his M.Sc and Ph.D. from University Carlos III of Madrid in 2010 and 2013, respectively. He then joined the Hamilton Institute in Trinity College Dublin, Ireland, as a research fellow in 2014. Since July 2015 he is working as a research scientist at NEC Laboratories Europe. His research interests lie in the application of fundamental mathematics to real-life computer communications systems and the design and prototyping of wireless communication systems and protocols.

Fabio Giust received his M.Sc. degree in telecommunications engineering at University of Padova, Italy, and a Ph.D. in telematics engineering at University Carlos III of Madrid, Spain. He joined NEC Laboratories Europe in 2015, and is currently working as a research scientist. His research interests cover 5G mobile networks, distributed mobility management, mobile edge computing, and network virtualization.

Vincenzo Sciancalepore received the M.Sc. degree in telecommunications engineering and telematics engineering in 2011 and 2012, respectively, and the double Ph.D. degree in 2015. From 2011 to 2015, he was a research assistant with IMDEA Networks, where he was involved in intercell coordinated scheduling for LTE-Advanced networks and device-to-device communication. He is currently a research scientist with NEC Europe Ltd., Heidelberg, where he is involved in network virtualization and network slicing challenges.

Xi Li received her M.Sc. in 2002 from TU Dresden and Ph.D. in 2009 from University of Bremen, Germany. From 2003 to 2014, she worked as a research fellow and lecturer in Communication Networks Group at University of Bremen, leading several industrial and European research projects on mobile networks. In 2014, she worked as a solution designer in Telefonica, Germany. In 2015 she joined NEC Laboratories Europe, and is working as a senior researcher on 5G Networks.

Zarrar Yousaf is a senior researcher at NEC Laboratories Europe, Germany. His current area of research is NFV/SDN-based systems in the context of 5G networks with specific focus on NFV MANO system and architecture. He is also actively involved in the ETSI NFV standards activities where he is a rapporteur of two work items and has made several contributions to more than eight published standard documents. He received his B.Sc. (1996) and M.Sc. (1999) in electrical engineering from University of Engineering and Technology, Peshawar, Pakistan. He has also received a M.Sc. in telecommunication and computers from George Washington University, USA (2001). He completed his Ph.D. from Dortmund University of Technology (TU Dortmund), Germany.

Marco Liebsch is chief researcher at NEC Laboratories Europe and is working in the area of mobility management, mobile content distribution, mobile cloud networking, and software-defined networking. He worked in different EU research projects and is contributing to standards in the IETF and 3GPP. For his thesis on paging and power saving in IP-based mobile communication networks, he received a Ph.D. degree from University of Karlsruhe, Germany, in 2007. He has a long record of IETF contributions as well as RFC, journal, and conference publications.

10

The Evolution Toward Ethernet-Based Converged 5G RAN

Jouni Korhonen

Nordic Semiconductor, Espoo, Finland

This chapter presents the evolution of the radio access network (RAN), specifically in 3GPP context, from the current fourth-generation (4G) toward the fifth-generation (5G) cellular network systems. The emphasis is on the RAN transport network evolution. The 5G network has ambitious goals when it comes to requirements for the new cellular system [1]. Specifically, requirements related to a new radio interface and overall system scalability are always at least a magnitude greater or stricter than the state-of-the-art features of the current generation. For instance, the peak bandwidth to a user should be 10–20 times more in 4G, end-to-end latency for ultralow latency and mission critical applications should be within a 1 ms boundary, and the number of connected devices should be on the order of a thousand more to address the emerging needs of the Internet of Things (IoT) market. All these wireless requirements are enabled by the advances in the new flexible, reconfigurable, and highly coordinated 5G radio interface, vastly increased availability of spectrum, and the softwarization of networking functions. One of the chief challenges for a successful deployment of 5G is overcoming the scalability and quality assurance challenges required in the 5G RAN transport network [2,3]. Knowing that the RAN transport network is already struggling with the current 4G radio, a new architecture and a paradigm shift are essential to meet desired targets of the 5G RAN network [4–6]. A good and detailed introduction to 5G and its RAN networking topics can be found in Reference [7].

5G Networks: Fundamental Requirements, Enabling Technologies, and Operations Management, First Edition.
Anwer Al-Dulaimi, Xianbin Wang, and Chih-Lin I.
© 2018 by The Institute of Electrical and Electronics Engineers, Inc. Published 2018 by John Wiley & Sons, Inc.

10.1   Introduction to RAN Transport Network

In order to understand the evolution toward the 5G radio access network (RAN), understanding the current Long-Term Evolution (LTE) RAN and its architecture is essential. The generations preceding LTE are of lesser interest because the 3GPP-defined 5G will only maintain backward compatibility with LTE (see Section 10.2 for further discussion).

LTE, or rather the Evolved Universal Terrestrial Radio Access Network (E-UTRAN) [8,9], is the access network part of the 3GPP Evolved Packet System (EPS) [10]. The other part is called Evolved Packet Core (EPC), which forms the core network and gateways part of the 3GPP EPS. E-UTRAN and LTE were introduced in 3GPP Release-8. The terms E-UTRAN and RAN are essentially the same and are used interchangeably through the rest of this chapter. E-UTRAN consists of base stations (BTS), or rather evolved Node-Bs (eNB), as LTE-capable base stations are called since 3GPP Release-8. E-UTRAN provides both user plane and control plane terminations toward user equipment (UE), that is, mobile devices. Base stations connect to the packet core network via an S1 interface. There are both control and user plane flavors of S1 called S1-MME and S1-U. The RAN internal interface between base stations is called X2. Similarly, there are both control plane and user plane flavors of X2 called X2-AP and X2-U. X2-AP and X2-U are the main interfaces in E-UTRAN. Figure 10.1 illustrates a very high-level EPS architecture where

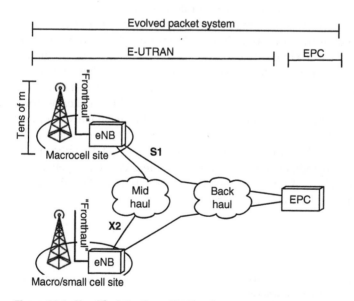

Figure 10.1 Simplified distributed RAN architecture in EPS.

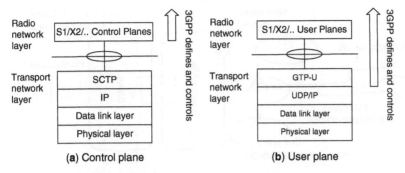

(a) Control plane   **(b)** User plane

Figure 10.2 E-UTRAN protocol stacks.

E-UTRAN is the part of the network from the EPC (core network) toward cell sites.

The E-UTRAN protocol stack is layered into a radio network layer (RNL) and a transport network layer (TNL). Whether the used protocols are over the S1 or X2 interfaces, they follow more or less the same layering (in this chapter, example references are for S1 only) [11]. Figure 10.2a and b illustrates both control (CP) and user planes (CP), respectively. 3GPP has always been brief describing and defining the details of the transport network player. Specifically, this concerns the data link and the physical layers. The main interest from the 3GPP side has been controlling the radio network layer with the exception of the GPRS tunneling protocol (GTP). The specifications merely list the requirements for the data link layer (*Any data link protocol that fulfills the requirements toward the upper layer may be used* [12]). The physical layer (layer 1) is equivalently briefly specified [13]. However, the layer 1 has specific interesting requirements that are essential when it comes to network synchronization in the RAN: line clock extraction capability and synchronization source management.

Figure 10.1 also names logical network parts in the RAN. For instance, the network segment between the base station and the core network is called a (mobile) *backhaul* [14]. The network segment between base stations is called a *midhaul*, and finally the "network" from a base station (or bottom of a mast) to a remote radio head (RRH) (typically at the top of a mast) is called a *fronthaul*. These network segment names are well established and also partly specified outside 3GPP [15]. Figure 10.3 illustrates how the RAN could look like from the transport network layer point of view. Note that here the fronthaul is already expanded to cover a deployment case where remote radio heads are not located at the immediate proximity of the base station. Interestingly, 3GPP "acknowledged" the existence of the fronthaul as late as in Release-14. The fronthaul became apparent as a part of the 3GPP new radio (NR) and *functional split* studies [3]. The entire discussion around the fronthaul has been very opinionated and controversial. We will discuss this topic further in Section

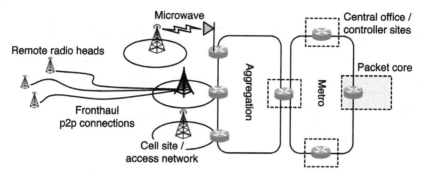

Figure 10.3 Typical cellular network deployment from the transport network point of view.

10.1.3. The other "hauls" are discussed further in Sections 10.1.1 and 10.1.2, respectively.

10.1.1 Backhaul Network

A *backhaul* network is a generic name for a network between a (packet) core and base station [15,16]. The transport network layer data link and physical links use a wide variety of transport solutions. However, Ethernet and IP/MPLS have established themselves as the de facto transport network layer technologies [17]. The trend is gearing toward bringing IP/MPLS as deep into RAN as possible. The IP traffic in backhaul comprises cellular user and control plane traffic (both S1 and X2 as of E-UTRAN) and other management traffic including time synchronization distribution.

PseudoWires (PW) [18] is a widely deployed solution replacing backhaul transport technologies such as asynchronous transfer mode (ATM) and time division multiplexing (TDM) transports that were popular prior to Ethernet and IP/MPLS. Of physical medias, microwave is still strong, with around 50% of the market share, but fiber (gray optics or wavelength-division multiplexing (WDM), etc) is predicted to bypass microwave links as the most popular transport media by the commercial deployments of the 5G. This development is purely driven by the exceedingly increasing bandwidth demand on the RAN transport networks.

Time synchronization distribution and network synchronization accuracy as a single selected technology area is becoming a key building block in Ethernet and IP-based backhaul networks. Specifically, enhancements in LTE radio co-ordination features and the forthcoming 5G NR require ever increasing time synchronization accuracy. Sending and receiving radio signals have always required frequency synchronization, and since the introduction of time division duplexing (TDD) and various combined multipoint (CoMP) flavors, a phase synchronization is required as well. [19]. We will discuss network synchronization in RAN further in Section 10.1.4. Other than network synchronization operating a mobile backhaul network is close to any well-managed Internet Service Provider (ISP) network for large enterprise customers without any particularly tight latency, frame delay variation, or reliability requirements. In the mobile backhaul, maximum allowed latencies between the base stations and the core networks are measured up to tens of milliseconds.

Mobile (i.e., backhaul as well) traffic volumes are increasing by a factor of 10 from 2015 to 2020.[1] This will pose a significant pressure on mobile operators to deploy more capacity into their backhaul transport infrastructure. The positive aspect is that backhaul has a manageable overhead, which consists of GTP and IPsec encapsulated security payload (ESP) tunneling imposed overhead, and radio and tunnel management-related control plane traffic. This totals around 10–25% overhead [20,21] depending on the deployment.

Another notable characteristic of the backhaul traffic is that the "busy hour" does not necessarily cause peak traffic volumes but the "quiet period," where one or very few mobile devices with good radio conditions are able to take all available spectrum out of a cell site (and its backhaul connection). Coming up with a proper transport capacity planning for such can be a nontrivial task, and some kind of a compromise is typically done between selecting the busy hour means and possible peaks. Anyway, since the backhaul traffic is mostly "user IP traffic" occasional packet losses or congestion are not fatal to the operation of the radio system.

10.1.2 Midhaul Network

A *midhaul* network is a generic name for a network between base stations (the X2 interface), and between base stations and small cells [15]. The midhaul can be considered as a flavor of a backhaul, and within 3GPP it is actually called as a backhaul. There are flavors of it: ideal and nonideal (backhaul). The difference mainly being the maximum allowed latency between base stations or a base station and a small cell [22]. The maximum latency of the ideal backhaul is as low as 2.5 μs. Actually, ideal backhaul more closely resembles a fronthaul than a backhaul.

10.1.3 Fronthaul Network

A *fronthaul* (network) is a generic name for the connectivity between the central unit (CU) and a distributed unit (DU). Both DU and CU are known by multiple names, such as remote radio head (RRH) or radio equipment (RE) in the case

---

1 http://www.cisco.com/c/en/us/solutions/collateral/service-provider/visual-networking-index-vni/mobile-white-paper-c11-520862.html

of DU. Similarly, CU is often called a baseband unit (BBU) or radio equipment controller (REC). As noted in Section 10.1, the fronthaul network was originally planned for a very short distance, from the bottom of a mast to the top of the mast, typically using coax cabling and analog radio signal transport. The fronthaul has then gradually been replaced by a fiber connectivity transporting radio signals in a digital domain.

The popular fronthaul protocols are the Common Public Radio Interface (CPRI) [23] and the Open Base Station Architecture Initiative (OBSAI) [24]. CPRI is clearly the dominant deployed solution and typically used as a synonym for the fronthaul. This section is no exception on that matter. CPRI defines the functional roles for both RE (i.e., RRH/DU) and RUC (i.e., BBU/CU). The RE does rather simple processing, such as channel filtering, digital-to-analog and vice versa conversions, and up and down conversions, carrier (de)multiplexing, power amplification, RF filtering, and TDD switching. The REC, on the other hand, is responsible for all digital signal processing (DSP) heavy lifting there is in the cellular protocol stack.

Compared to other "hauls," the fronthaul has very different characteristics. First, CPRI is essentially an *always on symmetric TDM stream* and typically run over a dedicated point to point dark fiber connection. In practice, proper networking/switching does not exist in CPRI. In some deployments daisy chaining is used to attach multiple RRHs into one CPRI connection. CPRI bundles everything into its founding framing structure, the *basic frame*, which contains control and management (C&M), synchronization, and user payload (typically in-phase and quadrature (I/Q) samples), all in one package. Other distinct characteristics include the following:

- Stringent latency, jitter, and synchronization requirements: numbers like 2 ppb frequency synchronization accuracy, ≈8 ns one way delay variation per link hop, and 100 μs maximum end-to-end latency are good examples.
- Bandwidth expansion is over tenfold to that at the user plane.
- In-band synchronization solution.
- In-band C&M always reserves the same amount of link bandwidth whether it is used or not.
- There is no interoperability across vendor equipment. This is intentional in order to force vendor lock between the RRH and the BBU.

These requirements might sound unreasonable, but one has to always remember the origin of the fronthaul, and for that original application the solution works actually extremely well. It is also important to understand that the 3GPP radio system has well-defined synchronization accuracy and latency budgets (see Section 10.1.4) [19]. Any additional synchronization error or jitter contributed by the fronthaul is directly away from that available budget and immediately reflected on the air interface. For example, in the 3GPP LTE radio system, the time alignment error (TAE) of the LTE signals is measured at the base station transmitter antenna port(s). Now, if the fronthaul contributes an unbounded synchronization error, the base station has no way to take that error into account in its processing.

The current fronthaul network exists on the available leftover time after the radio has completed the baseband processing out of what has been reserved for the Hybrid Automatic Retransmit reQuest (HARQ) protocol loop [25]. The critical time allocation unit in this context is the radio frame level transmission time interval (TTI), which for 4G LTE is 1 ms (the length of one LTE radio subframe and a single HARQ process slot and the granularity of the radio scheduler). The TTI length in the 5G NR is likely to be less (in order to reduce overall air interface round-trip time) [26]. In 4G LTE, the leftover time can be hundreds of microseconds, for example, a maximum close to 250 μs out of the total 3 ms HARQ loop reserved time [27,28]. This would allow fronthaul networks to span tens of kilometers between the RRHs and the BBU (assuming 5 μs per kilometer fiber delay). In practice, the available latency budget for the fronthaul has to be further divided into various latency components, as opposed to merely looking at the link propagation delay. Each networking node on the path and processing at the CU and DU contributes to the total latency. Equation 10.1 gives an example of fronthaul latency components on a high level. Figure 10.4 illustrates how the fronthaul actually became possible.

$$\text{Latency} = \sum_{m=0}^{\text{hops}} t_{\text{switch}_m} + \sum_{n=0}^{\text{hops}+1} t_{\text{propagation}_n} + t_{\text{DUproc}} + t_{\text{CUproc}}, \qquad (10.1)$$

where

$$t_{\text{switch}} = t_{\text{SaF}} + t_{\text{SelfQueuing}} + t_{\text{Queuing}} + t_{\text{MaxPacketSize}}, \qquad (10.2)$$

where

- t_{SaF} is the store and forward delay for a given packet P.
- t_{Queuing} is the switch queuing time for a packet P.
- $t_{\text{SelfQueuing}}$ is the worst case self-queuing time (that is, dependent on number of ingress ports) for a packet P.
- $t_{\text{propagation}_n}$ is the link propagation time for a link n.
- $t_{\text{MaxPacketSize}}$ is the transmission time for a maximum-sized packet P carrying (radio) data, including Ethernet preamble, start of frame delimiter, and the following interframe gap.
- t_{DUproc} is the DU processing time outside baseband-related processing.
- t_{CUproc} is the CU processing time outside baseband-related processing.

Fronthaul became topical and a heated topic in the industry after the introduction of the Centralized RAN (C-RAN) architecture [4,5,29,30]. The basic

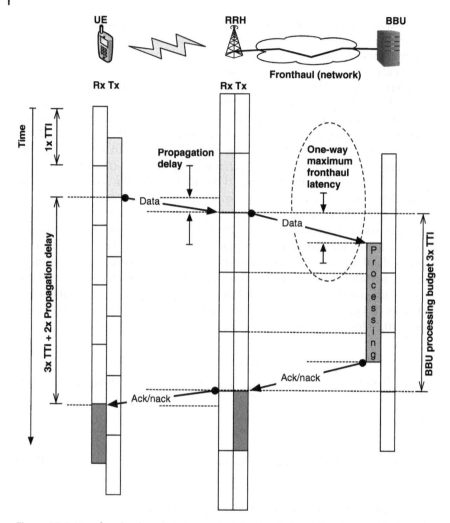

Figure 10.4 How fronthaul maximum round-trip latency is determined – example from UE initiated uplink data transmission.

architectural idea of the C-RAN is centralizing the baseband processing of several traditional distributed bases stations into a central office (CO), typically located deep in the RAN, and then connecting to a large number of RRHs. Figure 10.5 illustrates a simplified C-RAN architecture. The transition from the distributed base stations to C-RAN is not a major rearchitecting effort; more of a new way of deploying the existing technology and somehow realizing the transport between RRHs and the BBU. C-RAN is transparent to the backhaul and the mobile core network, but has a major impact on the fronthaul. Depend-

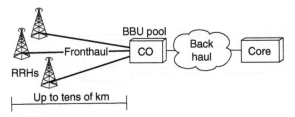

Figure 10.5 Simplified centralized RAN architecture.

ing on the deployment, RRHs can be several kilometers away from the central office. A larger central office can cover over 10 km² radio coverage area. C-RAN is mainly driven by operational cost savings. Centralizing baseband processing in one location (CO) has a significant positive impact on the following:

- Significantly reduced electricity consumption, cooling, and so on.
- Operations and management (since everything is in one place).
- Better security (both physical and network by eliminating inter-basestation communication needs).
- Less sites to rent, build, maintain, and so on.

In addition to these benefits, C-RAN specifically enables advanced coordinated radio features (e.g., various CoMP schemes) that can significantly boost the radio performance. Specifically, the coordinated radio features will be essential in the forthcoming 5G radio due to network densification and shrinking cell sizes.

Assuming the current functional split described in CPRI [23] and LTE radio numerology, a simple rule of thumb helps to estimate the bandwidth required by the fronthaul. Each 20 MHz antenna carrier requires 0.98304 Gb/s of bandwidth without line coding (25% overhead with 8b10b coding and 3.125% with 64b66b line coding). A 10 MHz carrier would be half of 20 MHz bandwidth and 100 MHz would then be five times more. These numbers are just quick and dirty tools for estimation, since they blindly assume certain LTE radio numerology and radio conditions. As an example, eight 20MHz antenna carriers, two sectors with 4 × 4 multiple-input multiple-output (MIMO) would require 7.86432 Gb/s (or 9.8304 Gb/s with 8b10b line coding) of network bandwidth. Assuming LTE Advanced, this 7.86432 Gb/s of fronthaul traffic would translate to roughly 600 Mb/s of user plane traffic.

The existing fronthaul deployment model is fiber hungry, which is a significant contributor hindering the success of C-RAN. When the radio capacity increases the need for additional fiber capacity is also evident. Figure 10.6 illustrates an example of fronthaul deployment where six point-to-point fibers (or pairs) are needed to serve a three sector (three RRHs) macrocell site with three microlayer small cells. In the worst case, each RRH or small cell could

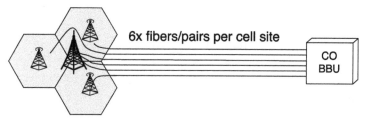

Figure 10.6 6x dark fiber (pairs) way of doing fronthaul.

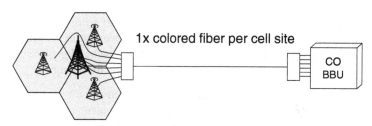

Figure 10.7 WDM way of doing fronthaul.

need its own dedicated fiber. On the other hand, Figure 10.7 illustrates the same cell site configuration, but using colored optics (WDM). Obviously, the latter requires only one fiber but more expensive optics and transceiver equipment. The predicted growth in the fronthaul traffic volumes is actually pushing WDN technology even deeper in the fronthaul networks.

10.1.4 Network Synchronization and Latency in RAN

When it comes to radio systems and specifically coordinated radio systems where multiple transmission and reception points (TRP)[2] can interfere with each other, an accurate network synchronization becomes a key building block for a proper functioning of the entire radio system. The synchronization requirements within the existing RAN can be found from multiple sources [9,19,23,31,32].

Prior TDD-LTE radio only frequency synchronization was required and often provided as a part of the TDM-based transport network. Since the gradual decommission of TDM links in favor of Ethernet and IP/MPLS, Synchronous Ethernet (SyncE) [33] has established itself as a viable frequency synchronization distribution solution. The introduction of TDD-LTE created a need for phase synchronization. Similarly, all new radio enhancing features, such as

---

2 Antenna array with one or more antenna elements available to the network located at a specific geographical location for a specific area/sector.

enhanced Inter-cell interference coordination (eICIC), CoMP, Carrier Aggregation (CA), and MIMO, also depend on an accurate phase synchronization. In IP networks, IEEE 1588 Precision Time Protocol (PTP) [34] has proved itself as a creditable solution for network synchronization. IEEE 1588 and SyncE are often used together for a better performance and shorter convergence times [35,36]. SyncE is used for the frequency reference and IEEE 1588 for time reference. Table 10.1 lists a snapshot of the LTE radio system and fronthaul network synchronization and latency requirements at the time of writing this chapter. Precise 5G NR requirements were not yet reached their final form. Therefore, trying to understand the 5G through the LTE is justifiable based on the assumption that the new radio system and RAN would not change fundamentally from the current designs. Anyway, the synchronization and the latency requirements are expected to be tighter in 5G than in the current system.

Network synchronization can obviously be provided using means other than SyncE and IEEE 1588. Global Navigation Satellite System (GNSS)-based systems like Global Positioning System (GPS) and Galileo are widely used and field-proven solutions. The expected accuracy in a typical GNSS-based installation is within the range of 100 ns. GNSS-based solutions obviously need to receive a signal from a number of satellites, which would require an additional antenna at the site and a good signal reception. This potentially is an issue, especially in indoor environments. Furthermore, mobile operators are often uneasy to depend on a technology that they cannot control/manage, specifically if that technology is provided by a foreign government institution. These reasons among others have steered the solution space toward SyncE and IEEE 1588.

SyncE is often criticized for requiring every node to be compliant, whereas IEEE 1588 can possibly bypass islands of nonsupporting nodes. However, network synchronization performance will always suffer if the path contains non-supporting nodes, and this also holds for IEEE 1588. From Table 10.1, there is no doubt that the network synchronization solution has high-performance expectations. It is "safe" to assume the 5G ready RAN requires an upgrade cycle in most of its networking nodes to prepare for the 5G bandwidth needs. Based on this "safe" assumption it makes sense to expect that all nodes in the 5G would also have required network synchronization support, at least in the nodes that are located in the fronthaul part of the RAN. In the case of IEEE 1588, it would mean supporting at minimum a highly accurate transparent clock (TC) functionality in every networking node participating in synchronization distribution within the RAN. Depending on the deployed radio features, boundary clocks (BC) and a common master (CM) may also become necessary in a close proximity to radios "under the same coordination" [32].

Figure 10.8 illustrates a high-level network synchronization overview within a packet-based fronthaul network. In the context of Table 10.1, the time accuracy (or time error) is measured against the nearest common point in the synchronization chain (e.g., a common master clock in the case of PTP). A

Table 10.1 Time synchronization requirements for TDD-LTE, and both TDM and Ethernet-based CPRI.

Technology/Feature	Frequency Accuracy	Phase/Time Accuracy	Latency
TDD-LTE	±50 ppb (wide area eNB)	±1.5 µs (radius<3 km)	Tens of ms
	±100 ppb (med range eNB)	±5 µs (radius>3 km)	
	±100 ppb (local area eNB)		
	±250 ppb (home eNB)		
eICIC		±1.5 − −5 µs	Tens of ms
CoMP (examples):			
- Dyn. point blanking		±1.5 µs	1–10 ms
- JR[a]		±1.5 µs	< 1 ms
- Coord. BF[b]		±1.5 µs	< 1 ms
- Coord. sched.		±5 µs	1–10 ms
- Noncoherent JT[c]		±5 µs	< 1 ms
Timing alignment error:			
- Positioning		≤30 ns	
- MIMO/TX diversity		≤65 ns	
- Intraband cont. CA[d]		≤130 ns	
- Intraband noncont. CA		≤260 ns	
- Interband CA		≤260 ns	
CPRI fronthaul:	±2 ppb		> 100µs
- RTT delay variation		±16.276 ns	
- DL delay accuracy		±8.138 ns	
Ethernet fronthaul [32]:	±2 ppb		< 100µs
- Category A+		±12.5 ns	
- Category A		±45 ns	
- Category B		±110 ns	
- Category C		±1.38–148 µs	
Backhaul (non-ideal)	±16 ppb	See TDD-LTE	Tens of ms
Backhaul (ideal)	±16 ppb	See TDD-LTE	2.5 µs

[a] Joint reception.
[b] Beamforming.
[c] Joint transmission.
[d] Contiguous carrier aggregation.

high-performance SyncE and IEEE 1588 combination can match and even outperform a GNSS-based solution [37]. However, the number of networking hops obviously affects the SyncE and IEEE 1588 performance. Each networking hop is a potential source and contributor to wander and phase noise. IEEE P802.1CM

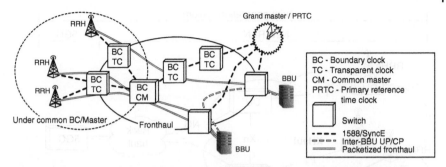

Figure 10.8 Network synchronization in an Ethernet-based and IEEE 1588-enabled fronthaul network. (Adapted from Reference [32], IEEE.)

specification [32] contains network synchronization requirements specifically targeted to Ethernet-based fronthaul network deployments [38,38,39]. Traditionally, the network synchronization requirements have been different in backhaul, midhaul, and fronthaul networks. A rule of thumb is that fronthaul synchronization requirements are a magnitude stricter than those of backhaul and midhaul. However, this might change in the future 5G RAN if and when the different "hauls" converge to the same network infrastructure. Current standardized network synchronization profiles are primarily for 4G or earlier generations [33,40] and need to be revised for 5G.

CPRI has its own synchronization solution as part of the overall protocol. Its synchronization fundamentals resemble IEEE 1588 to a large extent.

There are certain deployment consideration that are good to comprehend. Typically, each operator has their own primary reference time clock (PRTC) and a grand master (GM) that they want to base their synchronization chain on. In a shared RAN transport network environment, it could be tempting to think that the transport provider could offer time as a service to all operators that use the same network. Unfortunately, this is unlikely to be the case. As a consequence, the transport infrastructure has to prepare to serve multiple parallel grand masters and separate IEEE 1588 PTP domains (assuming IEEE 1588 would be de facto network time distribution solution). Related to distribution, the security aspects of the network time distribution have not typically been properly addressed, for example, authentication of the master or verifying the integrity of the PTP message. This oversight will be addressed in the forthcoming update of IEEE 1588 version 3.

The available latency budget for the fronthaul is a nonstatic measure, and varies on multiple nonstandard implementation-specific details like processing time within the CU and the DU. The familiar numbers seen, for example, in Table 10.1 are results of either typical implementations (see Figure 10.4) or agreed on consensus for a "reasonable value" [23,29,32]. The latency requirements are going to evolve once the new 5G ready RAN and new radio gets

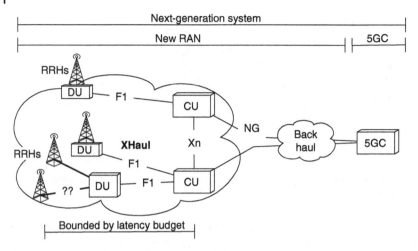

Figure 10.9 5G next-generation system and the new RAN.

deployed. The details of impacting factors will be discussed further in Section 10.2.

10.2 Evolving RAN Toward 5G Requirements

3GPP Release-14 marked the start of the specification work toward the 5G. At the time of this writing, the 3GPP terminology for 5G was not stabilized and was bound to change. However, this document uses the work in progress names: *New RAN* for the 5G RAN, *gNB* for the 5G base station, *Next Generation Core* (NGC) for the 5G packet core, and *new radio* (NR) for the 5G radio. Figure 10.9 illustrates a simplified standalone 5G system architecture. The naming of the main RAN interfaces has changed. The New RAN flavor of X2 is Xn. Similarly, NG-C and NG-U are the 5G flavors of EPS' S1 control plane and S1 user plane, respectively. The working assumption[3] is that GTP-U remains as the user data plane protocol in spite of repeated attempts to replace it. Similarly, SCTP remains as the control plane signaling transport protocol. The *new interface* between the CU and the DU is called F1 [41] and the user plane will be build on top of GTP. Traditionally, system vendors have considered the interface between the CU and the DU as a base station internal vendor-specific interface. The interface between the CU and the DU will be discussed further in Section 10.3.

---

3 As of start of 3GPP Release-14.

Although the basic new RAN architecture remains the same on most parts, there will be evolutionary changes [3]. The chief contributor is obviously the *new radio* [2,42,43]. The radio comes in three major flavors categorized by three different use cases:

Enhanced Mobile Broadband (eMBB) enables access to rich multimedia content. Its main characteristic is realistic wireless broadband (macro or hot spot coverage) with peak bandwidths exceeding 10Gb/s.

Ultrareliable Low-Latency Communications (URLLC) enables use cases such as industrial control loops and remote medical operations. URLLC has stringent latency and reliability requirements (e.g., the 1 ms round-trip time).

Massive Machine-Type Communication (mMTC) enables a massive number of devices with relatively low bandwidth and communications requirements.

Our main interest is in eMBB and to some extent in URLLC. The bandwidth requirements that eMMB require within the RAN are discussed in Section 10.2.2. A related topic is the new radio functional split, which is discussed further in Section10.2.1. The other two selected discussion topics are the latency considerations for URLLC and ultralow-latency applications, and the RAN network slicing. They are discussed in Sections 10.2.4 and 10.2.5.

Looking at the eMBB use case, Table 10.2 shows an example configuration for an envisioned dense urban area 5G base station system with both macro layer (for coverage) and an assisting micro layer (for capacity) [2]. Using LTE-Advanced and known fronthaul parameters, it is possible to calculate a rough estimate for the worst-case scenarios for a fronthaul and a backhaul network bandwidth. The fronthaul traffic volumes are roughly 2.5 Tb/s for the macrolayer and 38 Tb/s for the microlayer in the 5G cell site based on the values listed in Table 10.2.[4] On the other hand, the backhaul volumes are considerably relaxed and in scales of 200 Gb/s for the macrolayer and 2.9 Tb/s for the microlayer. These estimates naively assume fully digitalized signal feeds to each individual antenna element, which might not hold for real implementations. It is likely that large antenna arrays, for example, for beamforming are a mixture of digital and analog solutions just to keep traffic volumes within reasonable levels [44]. The presented example is for a single 5G base station with 1000 antenna carrier connections. It is expected that a central office in a RAN could host multiple such base station instances totaling thousands of

---

4 The example assumes 3GPP Functional Split Option 8.

Table 10.2 Example attributes of a dense urban area 5G cell site.

Attributes	Values and assumptions	Bandwidth assumptions
Carrier frequency and aggregated bandwidth	Around 30 GHz: up to 1 GHz (DL+UL) Around 4 GHz: up to 200 MHz (DL+UL)	One TX antenna element has one antenna carrier 20 MHz radio spectrum equals:
Layout	Two layers: -Macro layer: hexagonal grid - Micro layer: random drop Note: Both around 4 GHz and around 30 GHz may be available in macro and micro layers.	- 75Mbps end user traffic. - 1 Gbps in the fronthaul (CPRI like time domain transport) 200 MHz radio spectrum equals:
Intersystem distance	Macro layer: 200 m between transmit/ receive points (TRP). Micro layer: 3 micro TRPs per macro TRP	- 750 Mbps end user traffic. - 10Gbps in the fronthaul.
Base station antenna elements	Around 30 GHz: up to 256 TXand RX antenna elements Around 4 GHz: up to 256 TX and RX antenna elements	1 GHz equals: - 3750 Mbps end user traffic. - 50Gbps in the fronthaul.

antenna carrier connections. Calculating needed network bandwidth for each radio configuration and deployment case is a nontrivial exercise. The calculations are further complicated when different radio functional split options and other optimizations (such as compression of samples or resampling takes place). Most of the numbers presented in literature are rough estimations and heavily assume some specific radio feature set or deployment case. Unless "bit level precise" results are needed there are nice tools available to calculate user plane bandwidth even taking 3GPP-defined radio functional splits (see Section 10.2.1) – as they as described now – into account [45].

It is obvious that de facto fronthaul technology, CPRI, does not scale to the above terabit bandwidths. For example, the latest CPRI version is limited to maximum ~25 Gb/s link speeds, and ~10 Gb/s being the realistic deployable link speed at the moment (cost of 10 Gb/s fiber, and networking hardware being in place) [23]. Also, the availability of fiber capacity for each remote site is a concrete issue. Another concern is the lack of mainstream transport technologies used for the fronthaul, which immediately translates to increased cost. The bandwidth issue has already been addressed in multiple ways. Time domain I/Q sample compression and downsampling are common techniques [46]. Unfortunately, these enhancements come with a price of additional specialized hardware, software, and power consumption. Although bandwidth savings are relatively good, these enhancements are still temporary "band-aid" fixes and do not even try to address the root cause of the problem–transport of I/Q samples at the sampling rates of the used radio. Although not directly bandwidth-related issue, but closely coupled, the fact is that time domain sample transport also has

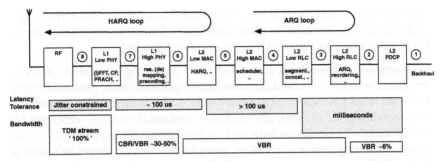

Figure 10.10 3GPP documented functional splits for 5G (and 4G) radio, (Adapted from Reference [44], IEEE.)

to meet the stringent time synchronization requirements there are for fronthaul (see Table 10.1).

The obvious way forward in the future RAN is rearchitecting it, however, without taking a clean slate approach to it. There have been several proposals on how to evolve to the new RAN for 5G and also cover the existing deployments of 4G [3,5,47–50]. They all follow more or less the same principles in the solution space.

Radio Functional Splits   Redefine the functional roles of the RRH (RE/DU) and the BBU (REC/CU). The *functional splits* have a pivotal role reducing both the fronthaul traffic volumes and relaxing synchronization/latency requirements [51–53]. 3GPP has also started to look into various functional splits and the current split categorization, as illustrated in Figure 10.10. The figure also shows the scale of change to both bandwidth and latency tolerance. The more radio stack processing is done at the RRH, the more relaxed are the fronthaul transport network requirements. The functional split point has to be carefully selected for the target radio deployment, since the different splits have varying constraints on what type of coordinated radio features they can support. Distributing the main bulk of the heavy radio signal processing to RRHs is also the key enabler for a virtualized RAN (V-RAN) paradigm, where the central office "BBU" compute platform could be entirely implemented using Data Center (DC) architectures running virtualized networking functions leveraging white boxes and general-purpose processors (GPP). Earlier this had not really been economically feasible due to the heavy signal processing needs and real-time properties of the compute tasks. Another key benefit from the reduced bandwidth and specifically synchronization/latency requirements is that the network could potentially be fully implemented using commodity off-the-shelf networking components. Specifically, Ethernet technology spun-off from data centers is a strong (if not only reasonable) candidate in this context. Various radio functional splits options and their impact to Ethernet-based networking solutions and architectures have received substantial research interest. Especially, the performance, latency, and bandwidth characteristics of transporting fronthaul traffic over Ethernet have been studied [54–56].

Convergence of Fronthaul and Backhaul   Assuming functional splits materialize, the fronthaul would not have a homogeneous traffic profile anymore. Different splits have varying characteristics ranging from the traditional backhaul to stringent CPRI-like traffic profiles. Therefore, the new 5G-enabled RAN would actually be capable of both backhaul and fronthaul category traffic. This opens interesting new opportunities, since there would no longer be a need to deploy different separate networks for fronthaul, midhaul, and backhaul. All these networks would be served by a single unified networking infrastructure and platform. The converged network is typically referred to *XHaul* or *crosshaul* in the literature [49,50,57]. In addition to the cellular, traffic originated from the same converged network would equally be capable of transporting any Internet-type traffic (since non-ideal backhaul traffic more or less equals to Internet traffic profile), which has the benefit of natural integration of Wi-Fi and enterprise traffic over the same transport infrastructure.

Softwarization of RAN   One of the goals of 5G is that the same infrastructure would serve multiple use cases and services without making compromises in the radio and the rest of the service infrastructure. The overall goal concerns not only the radio, but RAN transport network, base stations, and the core network platforms. The new cellular system has to be software reconfigurable nearly in real time (where real time is a moving target when it comes to actual timescales) [58]. Anyway, from the future RAN point of view, softwarization means more software-controlled components and also bringing more localized general compute platforms closed to the radio edge. Combined with the two earlier primary principles, softwarization of the RAN is also one of the key enablers of V-RAN and special need/purpose applications at the mobile edge (i.e., Mobile Edge Computing (MEC) [59,60]). A MEC platform can be part of the base station's local data center (V-RAN approach) or a separate local data center within RAN (e.g., located in a central office). Another prominent and well-known use case of the RAN softwarization is the RAN slicing [61]. Network slicing is a concept of creating one or more instances of end-to-end networks/services tailored to specific service requirements over a shared infrastructure utilizing its physical or virtual networking functions with strict isolation of transport, compute, storage, and radio resources.

10.2.1   New Radio Functional Splits

As already mentioned in Section 10.2, Figure 10.10 illustrates the functional splits that are studied in 3GPP Release-14 [3]. There are currently eight identified main split options with suboptions under split options 3 and 7. Within

3GPP specifications, DU is extensively used instead of RRH or RE. Similarly, CU is used instead of BBU or REC. The list below lists the functional split options and gives a short description of each:

Option 1 (1A-Like Split) The functional split in this option is similar to *option 1A architecture in dual connectivity (DC)* [62]. This split is also close to that of a distributed RAN, where each base station implements the entire radio protocol stack and Radio Resource Control (RRC) management being outside (in the EPC Mobility Management Entity (MME)). This split could be easily standardized and can be made interoperable between RAN vendors.

Option 2 (3C-Like Split) The functional split would be equivalent to the *option 3C architecture in DC*, where RRC and Packet Data Convergence Protocol (PDCP) are in the CU and the rest of the radio stack is located in the DU. This split has the base solution already standardized in 3GPP and could be made interoperable between RAN vendors.

Option 3.1 (Intra RLC Split) Part of the link control (RLC) and up is located in the CU and the rest of the radio stack in the DU. The split would allow aggregating both new radio and LTE, enjoy centralized pooling, and also have centralized Automatic Repeat Query (ARQ) robustness benefits, for example, to also recover from transport network caused packet losses. This split remotely resembles what was done for 3G (RLC between the RNC and the NB).

Option 3.2 (Intra RLC Split) The receive (in the CU) and transmit (in the DU) functions of RLC are separated.

Option 4 (RLC–MAC Split) Radio stack from media access control (MAC) below are in the DU. This split has not been considered useful enough to be explored further.

Option 5 (Intra MAC Split) This split would implement parts of the time critical MAC functions in the DU like the HARQ process. Interference coordination CoMP functions like joint processing (JP) and coordinated scheduling (CS) between DUs (RRHs) would be in the CU.

Option 6 (MAC–PHY Split) This split implements MAC layer and up in the CU and the physical layer (PHY) in the DU. The split allows centralized scheduling between DUs and, for example, Joint Transmission (JT) CoMP. There are likely to be sub-TTI delay time bounded interactions between the CU MAC and the DU PHY layers.

Option 7.1 (Intra PHY Split) This split implements FFT, cyclic prefix (CP) add and remove, and possible uplink PRACH handling in the DU. The transported data over the fronthaul is in frequency domain (marked as "I/Q (F)" in the table) reducing bandwidth requirements immediately roughly to half. In the downlink direction, reference signals and other

Table 10.3 3GPP radio functional split summary. (Adapted from Reference [3], IEEE.)

Split option	1	2	3-2	3-1	5	6	7.3	7.2	7.1	8
Baseline exists	No	DC	No							CPRI
ARQ location	DU				CU					
HARQ location	DU						CU			
CU res. pooling	Low									High
	RRC	RRC + L2				RRC + L2 + PHY				
Transport peak bw	-	Low								High
	-	Baseband bits					I/Q (F)			I/Q (T)
	-	Scales with MIMO layers..							..antenna ports	
Latency	Loose (ms scale)					tbd	Tight (< 100 μs)			
Latency tolerance	ms scale						μs scale			ns scale
Coordination	Multiple schedulers (one per DU)						Centralized scheduler (e.g., one per CU)			

repeating signals can be generated locally in the DU, thus there is no need to transport those over the network. Option(s) 7 in general allows most if not all CoMP, MIMO, and beamforming enhancement, which makes option 7 a prominent candidate for being the "sweet spot" of the functional splits.

Option 7.2 (Intra PHY Split) Similar to option 7.1 but adds PHY functions like resource (de)mapping, prefiltering, and precoding to the DU responsibilities. In a partially loaded cell this option refrains sending unused subcarrier, which reduces bandwidth requirements even further.

Option 7.3 (Intra PHY Split) A downlink enhancement, where options 7.1 and 7.2 apply, but only the encoder is located in the CU.

Option 8 (PHY-RF split) This functional split is equivalent to that of CPRI. The transported data is time domain sample (marked as "I/Q (T)" in the table). It is unlikely this split will ever interoperate between RAN vendors on any layer other than the transport network layer.

The above list of functional split options is further summarized at a high level in Table 10.3. Note that options 5–7.1 pose additional complexity because the MAC scheduler and the PHY processing are separated. For instance, the current scheduling decision accounts for the fronthaul latency, which might impact performance in the case of a short TTI and possible "long" latencies. Furthermore, option 5 also separates the MAC scheduler and the HARQ into different units, which again adds complexity to keep these functions performing optimally. In general, presenting precise maximum available network latency numbers for each split option is nontrivial. In general, options 2–5 are driven by the PDCP and RLC timers that are rather forgiving (e.g., RLC ARQ timers are in ms scale). Options 4 and 5 are driven by the TTI length-bound MAC scheduler (for LTE the TTI is currently 1ms but one cannot assume whole 1ms is available for one-way latency). Finally, options greater than 5 are bound to HARQ loop and the "leftover" time after the baseband processing, which was already discussed in more detail in Section 10.1.3.

The future RAN is likely to implement multiple functional split points in parallel. The reason is that different deployments have varying requirements and also different split options have obviously different cost structures. However, it is very likely that there will not be a fully standardized interface between the CU and the DU in addition to those that have already been standardized to some extent in 3GPP (i.e., the option 2). The best that can be expected is a unified transport network layer solution, that is, the CU and the DU are able to use a common framing protocol, commodity off-the-shelf transport network components, and use standard network synchronization solutions. The DU–CU protocol or interface is also often referred to as the *next-generation fronthaul interface* (NGFI).

3GPP is not the only organization working on the fronthaul functional splits. There are notable standardization efforts:

IEEE P1914.1 Next-generation fronthaul interface (NGFI) documenting the next-generation fronthaul networking requirements and architecture for Ethernet-based packet networks [63].

IEEE P1914.3 Radio over Ethernet (RoE) standardizing transport and framing protocol for current CPRI and future NGFI needs [64].

ITU-T SG15 Q2/15 G.RoF (Radio over Fiber) standardizing CPRI and radio transport over Optical Transport Networks (OTN) [65].

CPRI producing a new revision of CPRI with Ethernet-based transport (eCPRI) and a new split-PHY functional split for CPRI [92]. The "standardization" process for eCPRI is not open and only taking place within the CPRI cooperation members.

Functional splits are the fundamental enabler for 5G, the RAN transport network, and the V-RAN architecture. Without the new function division the RAN would face tremendous scaling challenges. Even with one of the prominent split points, such as option 7, the scaling requirements are still significant compared to the existing RAN transport network. Most likely the advanced radio enhancements are deployed only in very few selected locations. Although the details of the functional splits and their impact on radio processing are important to the radio system as whole, *the details of the splits are less relevant to the RAN transport network layer*. Looking at the *transport peak bandwidth*, *latency*, and *latency tolerance* rows in Table 10.3 and Figure 10.10 it becomes clear that there are only 2 or 3 relevant *groups of splits* from the transport network point of view. This very observation is both strikingly obvious but also means there are only a few categories of forwarding hardware that is needed

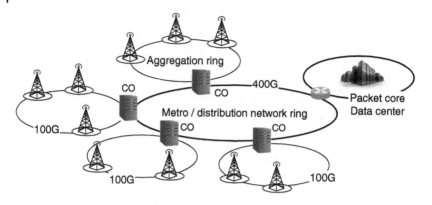

Figure 10.11 High-level architecture of the new RAN network.

to support. The key grouping factors are latency and latency tolerance, for example millisecond versus microsecond. The bandwidth needs of a forwarding hardware depend more on the deployment scale.

10.2.2 New RAN Network Architecture

Section 10.1 and Figures 10.1 and 10.3 illustrated a typical RAN network with separate fronthaul and backhaul network segments. As already mentioned earlier, such separation of networks is likely to diminish. Varying networking requirements of different functional splits (see Section 10.2.1), are the chief driver to this development. Another driver is the cost of running several parallel special-purpose networks with disjoint O&M and networking equipment. At the high level, the architecture still remains the same as illustrated in Figure 10.11. The architecture obviously has room for deployment and operator-specific variations, but the distribution of functions and introduction of local data centers in central offices (where the V-RAN compute platform is located) drives the trend. The notable aspect is the scale of networks. The distribution network between central offices and connecting to the next-generation core will mostly deal with backhaul type traffic. In the case of intercentral office traffic, for example, between V-RANs for some intercell site coordination cases, the traffic would be more closer to that of *optimal backhaul*. In any case, the envisioned aggregated traffic volumes in the distribution network are multiples of 100 Gb/s. The bandwidth need can be estimated based on the assumption that one distribution network could aggregate around 4 aggregation rings each of 100 Gb/s. The geographical area that could be covered with such a distribution network is somewhere between 15 and 20 km^2 in (dense) urban areas.

Figure 10.12 zooms in on one of the aggregation rings shown in Figure 10.11. This aggregation architecture is known by many names: *converged backhaul* and *fronthaul, XHaul* [49], *crosshaul* [50], or *next-generation fronthaul interface* (NGFI) [66]. From now on we will use XHaul as the common name for Figure 10.12 network architecture. All these proposed architectures are essentially the same and aim to the same goal: enabling the same network infrastructure transport mobile network-originated traffic over the same infrastructure and networking nodes and still being able to meet the service level agreements (SLA) for all diverse traffic types. Furthermore, the same infrastructure should be backward compatible with old style distributed base stations as well as C-RAN.

Let us look into various components of Figure 10.12 XHaul network. Bearing in mind deployment and operator-specific requirements, assuming 100 Gb/s aggregation ring capacity is a good compromise. The aggregation ring bandwidth need can be estimated based on the assumptions that it would aggregate up to 8 cell sites, and the maximum peak rates per a cell site could be around 10–20 Gb/s (based on the 5G new radio peak rates).The geographical area that could be covered with such an aggregation network is somewhere between 2 and 4 km² in (dense) urban areas.

The XHaul aggregation network needs to deal with various types of cell sites. The incumbent deployments of 4G (or even older technologies) base station cell sites are not going to change for 5G, and they are connected to the NGC through XHaul using a conventional Ethernet and IP/MPLS backhaul. The traffic volumes are moderate, and 1 Gb/s uplinks to the aggregation could suffice (assuming LTE Advanced with 4×4 MIMO reaching around 300 Mb/s per sector). LTE evolution can potentially reach backhaul bandwidths of up to 1 Mb/s per sector. A six sector cell site could still be easily served with a 10

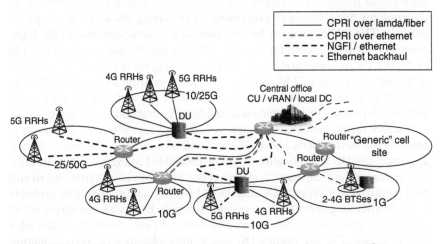

Figure 10.12 New RAN aggregation ring – the next-generation fronthaul interface network.

Gb/s uplinks. The internal structure and connectivity within an existing 4G cell site is transparent to the XHaul and would use any technology it was originally using.

XHaul is also required to support a C-RAN type of deployment, at least, during the transition period from existing deployments to V-RAN deployments. Since direct RRH connectivity using CPRI is the de facto way of doing the C-RAN RRH fanout, the XHaul would need to support "legacy CPRI" transport technology. In practice, this would imply either dedicated fibers for CPRI connections or dedicating lambdas for CPRI connections. Either way it is a wasteful use of optical resources and might have issues traversing networking nodes that do not provide optical level switching. Another alternative for CPRI interworking over the XHaul is packetizing that at the edge of the cell size, for example, in a CPRI-aware router/switch using Ethernet encapsulation methods such as radio over Ethernet (RoE) structure agnostic mapper mode [64]. Section 10.3 will elaborate on CPRI aggregation scenarios and solutions further. Straightforward CPRI over Ethernet (or packet) has the inevitable disadvantage of stringent bandwidth and especially time synchronization requirements (see Table 10.1 "Ethernet fronthaul" row). Furthermore, as stated earlier, current CPRI [23] is bounded to maximum ~25 Gb/s link option and in practice ~10 Gb/s link option is what gets deployed in numbers (due to current 10 Gb/s versus 25 Gb/s optics cost difference).

The 5G and 4G radios can also connect to the XHaul via a "centralized DU," which aggregates a number of RRHs within the cell site. The centralized DU would implement the desired functional splits and communicate with the CU using Ethernet (or packet)-based NGFI transport. The RRHs themselves within the cell site could still connect the DU using, for example, CPRI. This deployment model is prominent and the DU as an aggregation device can do the complex heavy lifting of processing while leaving the RRHs as simple as possible. The DU is likely to be and remain a system vendor-specific high-capacity networking node, for example, 10 Gb/s uplink could serve up to ~130 20 MHz LTE antenna carries (obviously depending on the selected functional split) and ~16 10 Gb/s CPRI connections. However, from an external Ethernet port I/O point of view this would still be a rather low end and low port count device with possibly few carefully selected features that makes it fit to the NGFI.

Another variation of a cell site deployment could use direct NGFI transport connectivity to the RRHs (either 5G new radios or evolved 4G radios). In this approach, the DU functionality would be embedded in every RRH. Naturally, this increases the complexity and cost of the RRH, but allows true end to end NGFI transport between the DU and the CU. Direct NGFI connectivity probably is the desired long-term (and greenfield) solution. Even with the direct NGFI transport connectivity it makes sense to deploy a centralized DU at the edge of the cell site. The aggregating DU would allow placing a radio coordination function closer to a network edge controlling multiple RRHs within the same site, and further reduces the bandwidth requirements toward the CU. The outcome of this kind of arrangement is a two-level functional split: from a RRH-DU to a cell site centralized DU to a CU. Again, it is highly probable that DUs and CUs that are connected via the NGFI transport must be from the same system vendor. This is independent of whether the transport network layer actually interoperates, but every advanced management and radio-related functionality would be specific to a system vendor. With CPRI there were attempts to break the DU–CU coupling, but without considerable success.[5]

So far there has been no mention of control plane versus user plane handling in the new RAN. The clear intention, at least in 3GPP, is to separate the control plane and user plane completely. The intention is to allow centralization of the control plane functions and management, and user plane networking nodes whose sole purpose is plain packets forwarding. This development fits well into the overall softwarization and virtualization of the networking functions. From the NGFI-enabled RAN point of view the trend is to a right direction.

10.2.3 5G RAN Migration Concerns

Migration from previous generations of the technology to a new is always a challenge. There is seldom a luxury of a clean-slate approach or a flag day swap from the old to the new. 3GPP has, however, stated that 5G does not need to care about backward compatibility with anything else than 4G EPS. 3GPP has studied multiple migration scenarios for the new RAN and NGC [67]. Unfortunately, these migration considerations typically put little effort on realities at the transport network layer. The new RAN still need to cope with incumbent cell sites and base stations' backhaul transport into the foreseen future. If we put aside truly legacy nonemulated TDM and ATM transports all these incumbent backhaul transports can safely be grouped into the same class of transport with 4G E-UTRAN backhaul.

10.2.4 Low-Latency Applications and Edge Computing

One of the 5G radio use cases was ultrareliable low-latency communication (URLLC). There are two sides to it. First, the "ultra reliable" means no packet/connectivity loss. Over the wireless link, achieving the goal can be challenging. One way of approaching it is sending the same data over multiple wireless links simultaneously in addition to other error correction and transmit diversity techniques. On the fixed transport network side reliability is easier to achieve, but in principle mechanisms are the same as over the wireless. There are multiple existing solutions on the transport/link protection area, where

---

5 The Open Radio Interface (ORI) activity, see http://www.etsi.org/technologies-clusters/technologies/ori.

the same data is delivered via multiple paths in parallel or alternative paths are preconfigured waiting for to be activated. Obviously, parallel connections consume more network capacity. The reliability solutions are discussed further in Section 10.3 in the context of Ethernet-based NGFI transport.

Second, the "low latency" might turn out particularly challenging to achieve. The 5G requirements set the level to ambitious 1 ms round trip between the application and the service, not just a network level round trip time. Effectively, the one-way maximum latency of 500 μs includes the entire software stack and the actual application processing. The requirement is ambitious because it does not, as of LTE, even fit into current radio framing and scheduling timescales. As mentioned earlier, the shortest scheduling interval is one TTI, which in LTE is already 1 ms and a size of one radio subframe.

The obvious way forward is coming up with a new numerology and also a new radio format that allows less than 500 μs long framing. 3GPP has approached the latency enhancement from multiple angles: shortened TTI, evolved (e.g., shorter) radio framing, and reducing error recovery/correction mechanisms such as FEC, ARQ, and HARQ [26,68]. On the uplink direction enhancements to uplink scheduling grant delays to send data can have a significant improvement. These extensive radio level latency optimizations are scrutinized in 3GPP for both LTE evolution and 5G NR.

How do the latency reduction schemes then affect the new RAN transport layer? Section 10.1.3 illustrated how the maximum latency for the fronthaul is determined. Now if, for example, the TTI length shrinks say to 1/7th or 1/14th of the 1 ms TTI, that immediately means there is less "available latency" for the fronthaul purposes after baseband processing. The concrete effect is that the maximum reach of the fronthaul shortens. It has been estimated that maximum fronthaul distances (bounded by the available latency) are considerably less than 10 km (meaning the available latency has reduced to 50 μs or less). This immediately sets the bounds for the geographical area a base station can cover/support. Actually, modeling distance solely based on the fiber delay ($\sim$5 μs per kilometer) does not produce a good result in the new NGFI-based RAN. The connection between the DU and the CU is likely to be multiple networking hops, thus the residence time of each hop, various queuing delays, and so on have to be accounted for as well. Due to the diminishing latency budget any improvement in the networking node residence time (i.e., how long the packet spends inside the device) increases in importance. Furthermore, some applications are also sensitive to changes in latency (i.e., jitter or frame delay variation), which further impacts the networking node implementation. It is not only the residence time, but deterministic delay and minimized jitter that counts. In some cases bounded jitter is more valuable than the delay through the networking node. These topics will be discussed further in detail in the context of Ethernet in Section 10.3. Another aspect here is that a late packet that missed its schedule is typically useless for its application. If a networking

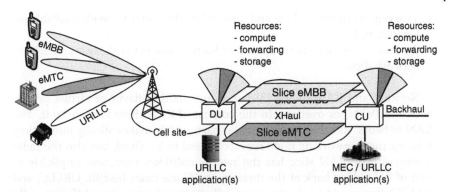

Resources:
- compute
- forwarding
- storage

Resources:
- compute
- forwarding
- storage

Figure 10.13 RAN network slicing example.

node on a path can deterministically detect a packet being late it can usually be safely silently discarded.

Another RAN concerning latency reducing solution is mobile edge computing (MEC). The basic idea is simple, bring the latency critical applications closer to the end users/devices, and thus save on the transport network originated latencies. Looking at the new RAN architecture proposals, see Section 10.2.2 and Figure 10.12, central offices are already located rather deep in the RAN and close to the radio edge. A central office would be an apparent location for a MEC compute platform, especially, it would allow sharing the same location with the CU, that is, the V-RAN local data center compute platform. Low-latency services and applications could be entirely, even those of third parties, deployed in a MEC platform in the proximity of the application/service users. Noting that centralized DUs potentially evolve toward a lightweight local data center platform architecture as a part of the overall softwarization trend, they could also host the MEC platform. Obviously, the amount of compute available for generic MEC applications/services would be less than in a central office platform. In both locations, the benefits of collocation with the actual radio processing and awareness are available, nearly real-time access to cell or even per subscriber radio level data (such as radio link condition).

10.2.5 RAN Slicing

Network slicing is another emerging concept popularized during the 5G development [61,69] and the network softwarization trend. The network slicing concept allows the infrastructure owner (e.g., a mobile operator) to create end-to-end network views, where

- slices and tenants are completely isolated from each other,
- each slice and tenant may have their own SLA,

- a varying set of networking and service functions may be made available for tenants, and
- the end-to-end slice is composed of both a core network slice and a RAN network slice.

Network and RAN slicing [3] is not just another flavor of virtual private networks (VPN) or overlays. In the context of RAN there is more to it. The RAN network slice has also the radio part, which makes slicing interesting. It is not that networking node resources need to be sliced, but also the radio resources. The RAN slice has the radio conditions awareness implicitly as part of it. Looking back at the three main 5G use cases (eMBB, URLLC, and mMTC) this would imply the same cell site could serve very different radio configurations (or rather numerologies). Since a mobile device (the UE) should also be able to attach to more than one slice simultaneously, it is likely to complicate the solution even further. Figure 10.13 illustrates a high-level view of slicing within the 5G RAN.

To the new RAN transport (i.e., the XHaul) and NGFI, slicing has a definite impact. In addition to managing a number of different functional splits, networking nodes also need to manage traffic characteristics for three completely differently parameterized radios. Combining the above, the complexity of an XHaul transport node is substantial, especially when it comes to traffic isolation (i.e., preventing each traffic flow from interfering with each other) and meeting the SLA guarantees. This can be a real challenge depending on the networking node implementation and how much different traffic types share internal resources within the node implementation. What makes slicing even more challenging is the traffic that has extremely low latency tolerance such as the CPRI over Ethernet mentioned in Section 10.2.2. That kind of traffic profile mixed with traditional backhaul traffic will require support from the forwarding hardware in order to work properly.

On the other hand, an interesting data point is that the XHaul/NGFI traffic profile depends on the selected functional split between the DU and the CU and not necessarily on the user plane traffic. Splits that are not concerned with a radio resource blocks in active use will have a constant bitrate (assuming no compression) independent of whether there is communication taking place on the user plane or not (e.g, split options 8 and 7.1). The radio is always on and sending and receiving *something*. The splits that actually send only data that is in use by the upper layers or that change as a result of radio conditions (such as MAC frames) will result in variable bitrate traffic. A further modification to this would be XON/OFF type sender policy, where nothing is sent over the XHaul/NGFI when there is no real data to send. The radio part could be fully syntonized with "noise" in the absence of actual user plane data. Such an enhancement to functional splits would result in good averaged bandwidth savings over longer observation time. However, if the XHaul/NGFI still has to service peak hour fully loaded cells without congestion, then basing the traffic profile design on the average might not be a good idea.

10.3   Ethernet-Based 5G RAN

Throughout earlier sections we have hinted about Ethernet being a strong candidate of the new 5G RAN transport network layer solution (i.e., for the XHaul and NGFI transport solutions), see Section 10.2. Naturally, we mean new network deployments, since there is seldom pressing reasons to swap existing transport networks for incumbent deployments if they work.

The premise of Ethernet in 5G new RAN originates from its *cost structure and the ecosystem around it*, not because of its technical superiority. Ethernet is also perceived as the next-generation generic connectivity solution for the different "hauls." The statistical multiplexing benefits of packet-based networking, scalability, and cost structure of the Ethernet-based unified ecosystem look unbeatable. The fact that the Ethernet ecosystem is driven by multiple market segments makes it a safe candidate. Developments in data center (DC) networking drive the terabit level scalability, low-cost optics for new interesting link speeds (e.g., 25G, 50G), and the overall cost structure. Furthermore, the ever increasing race for lower latencies has pushed Ethernet switching latencies to submicrosecond level. At the same time, other industry segments, including industrial automation and automotive, have helped drive time-sensitive networking (TSN) capabilities into Ethernet [70]. Operations and management solutions are sophisticated and within data centers designed for massive deployments, and nowadays also virtualization in mind. Virtualization drives for tenant isolation, which is a key feature needed, for example, in RAN slicing. An unified network management all the way from the remote radio heading to the packet core is worth pursuing.

Although not immediately, data centers have also found uses for time-sensitive networking features (in addition to low latency, which hardly is a TSN feature). Again, time synchronization and accurate time stamping are the key features in globally distributed data centers and their data base replication solutions [71]. Another example of the need of time synchronization and time stamping is the rise in importance of in-band telemetry and monitoring of data paths/links in data centers [72]. In the other industry segments, automotive and industrial Ethernet are the main drives of TSN feature set such as time-aware shapers (TAS), cycling queuing, preemption, and so forth. Combining these areas create a strong feature set for the 5G RAN purposes: *low-cost forwarding and switching silicon, low-cost optics, high-proven scalability, and deterministic and bounded latency.* However, another great benefit of the Ethernet technology in the RAN would be its direct *plugability* to envisioned V-RAN and MEC platforms. As we know, both V-RAN and MEC are small-scale data centers built using the main stream data center technologies, that is, Ethernet as the connectivity and transport solution.

There are still a number of details to consider. First, Ethernet is a packet technology and greatly benefits from the statistical multiplexing. Trying to repurpose Ethernet as a direct replacement of a TDM transport solution such as CPRI might not be the best way forward. Originally, Ethernet was not designed with time synchronization in mind. Second, the low cost can only be maintained if technologies used in "cellular repurposed" Ethernet solutions also find applications in other high-volume Ethernet-based ecosystems.

The rest of this section is structured as follows. Section 10.3.1 takes a quick look into Ethernet time-sensitive networking toolbox, which is believed to be the key building block for an Ethernet-based XHaul and NGFI networks. Section 10.3.2 lists and discusses some deployment considerations for Ethernet-based 5G RAN. Section 10.3.3 describes Radio over Ethernet (RoE) technology, which has been developed particularly with Ethernet-based NGFI in mind. Next-generation Ethernet-based base station architectures are elaborated in Section 10.3.4.

10.3.1 Ethernet Tools for Time-Sensitive Networking

IEEE 802.3 Ethernet is a packet-based best effort technology without any great guarantees of timeliness. An Ethernet switch may have a very low latency through the switch depending on the switch internal pipeline implementation, and ingress and egress port speeds. The standard Ethernet switching model only supports store-and-forward (SAF) architecture where a frame from an ingress port is always queued before starting to transmit it to the egress port. State-of-the-art switches have a port-to-port latency of less than 400 ns usually at port speeds of 10 Gb/s or greater.[6] These latencies are often measured using a nonstandard technology called *cut-through*, where a frame is transmitted to the egress port before it has been entirely received from the ingress port. We will discuss the cut-through further later.

Within the 5G RAN switch latency is just one component. The more pivotal features relate to network synchronization, deterministic latency (i.e., the frame delay variation (FDV) or jitter), and reliability. Looking back at the network synchronization requirements listed in Table 10.1, and functional split impact to a latency tolerance shown in Table 10.3, it becomes obvious that Ethernet could use some enhancements. IEEE 802.1TSN Task Group has developed a number of amendments to IEEE 802.1Q [73] and other related standards to

---

6 https://www.innovium.com/wp-content/uploads/2017/03/Innovium-Whitepaper.pdf
https://www.arista.com/en/products/7150-series.

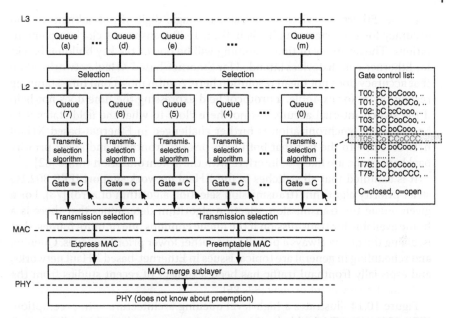

Figure 10.14 IEEE 802.1Q, queuing, and preemption. (Adapted from Reference [73], IEEE.)

make Ethernet more suitable for time-sensitive networking. This section will only look at a subset of the all available TSN amendments, namely, those that have been considered useful and discussed in the context of Ethernet-based fronthaul and XHaul networks [32]. The TSN features we look into are as follows:

IEEE 802.3br and IEEE P802.1Qbu The Frame Preemption [74,75].

IEEE P802.1Qbv Enhancements for Scheduled Traffic, also known as the time-aware shapers (TAS) [76].

IEEE P802.1CB Frame Replication and Elimination for Reliability [77].

In addition to the features listed above, IEEE P802.1Qci (Per-Stream Filtering and Policing) [78] and IEEE P802.Qch (Cyclic Queuing and Forwarding) [79] are briefly discussed in the context of IEEE P802.1Qbv.

Network synchronization is also an integral part of TSN and time-sensitive application. Therefore, we assume the presence of relevant network synchronization protocols such as IEEE P802.1AS [80] and/or IEEE P1588 (version 2) [34]. For a frequency reference, the presence of ITU-T Synchronous Ethernet (SyncE) can be assumed [33]. The accuracy of the network synchronization distribution has an extremely important role, specifically in time distribution networks with multiple networking hops [81]. There is not much to say here except the more accurate time stamping of packets the Ethernet device can have, the better performing network synchronization solutions can be achieved. For example, Ethernet devices should aim to single digit, if not subnanosecond accuracy for time stamping in their IEEE 1588 transparent clock implementations. The accuracy of time stamping will be one of the key building blocks for Ethernet switch devices intended for Ethernet-based XHaul networks. Note that accuracy does not mean resolution here. A subnanosecond time stamping resolution in, for example, a protocol field expressing the time does not help if the switch internal accuracy is nowhere close to what the field states it to be. Network synchronization is further challenged in Ethernet-based XHaul networks due to the fact that both high-priority fronthaul and lower priority backhaul traffic are likely to interfere with the synchronization traffic [82].

It is assumed that all switches in the XHaul network support IEEE 802.1Q strict priority algorithm (transmission selection algorithm of zero) [73]. For a given queue the transmission selection algorithm determines that there is a frame available for transmission if the queue contains one or more frames, that is, giving the queue always a priority over other lower priority queues. Queuing and scheduling in general are topical issues in Ethernet-based XHaul networks, and especially fronthaul traffic has been a subject of recent studies from the FDV and latency point of view [83–85].

Figure 10.14 illustrates a high-level queuing architecture with preemption. IEEE P802.1Qbv TAS added a time-aware *gate control list*, which allows controlling the output gates' open and close states associated with each queue according to a specific repeating schedule. Based on the schedule programmed in the gate control list, frames are selected for transmission. TAS allows building complex time-based schedules relative to a known timescale. As a result, TAS requires a fully synchronized and managed network to work. Each node in the network has to adhere to the network-wide calculated schedules, which are known to be a real management burden for any larger and nontrivial topology.

While TAS provides a deterministic time slot for a selected traffic (i.e., a sender knows when it can transmit, and a receiver knows when to expect a certain type of traffic to arrive), a single interfering frame from another queue, but using the same egress port can break the schedule. Once the transmission of a frame has started the other traffic has to wait until the transmission has completed. This can cause the scheduled traffic to miss the schedule and add unwanted FDV. There are ways to mitigate the issue. For example, the schedules could contain a *guard band* period, which has to be the length of the largest possible interfering frame. No frame is scheduled during the guard band period, which ensures the egress port is always available for the scheduled traffic when its time slot starts. The obvious price to pay for the guard bands is the waste of bandwidth. The schedules would now contain periods when no traffic can be sent.

Another approach to mitigate the impact of interfering traffic is the frame preemption. The frame preemption consists of two standards: IEEE 802.3br Interspersing Express Traffic (IET) that amends the IEEE 802.3 Ethernet MAC

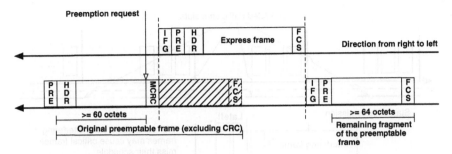

Figure 10.15 Ethernet frame preemption.

layer and IEEE P802.1Qbu that amends the IEEE 802.1Q queuing. As shown in Figure 10.14, there are two MAC instances instead of one: the *Express MAC* and the *Preemptable MAC*. The layers above these MACs see no difference and the *MAC merge sublayer* makes sure the PHY sees no difference. A frame arriving from a higher priority Express MAC can always preempt a frame from a lower priority Preemptable MAC in the MAC merge sublayer. Effectively, the preemptable traffic gets fragmented as many times as needed to make sure the express traffic gets through. Figure 10.15 illustrates how the preemption works.

There are a few restrictions to fragmenting, though. Preemption can only take place if at least 60 octets of the preemptable frame have been transmitted and at least 64 octets including the frame check sequence (FCS) remain to be transmitted. The minimum preemptable frame size is therefore 124 octets, which is also the maximum FDV introduced by the frame preemption to the express traffic. In practice, the minimum preemptable packet size is a bit higher (155 octets being the minimum, see the calculation in Reference [32]). Fragmentation also adds some overhead. Every time a frame gets fragmented an additional CRC has to be added to that fragment, and each remaining fragment has an additional inter frame gap (IFG) and preamble to add. These additional octets have to be taken into account when calculating link budgets.

TAS had the issue with guard bands wasting bandwidth. A combination of TAS and the frame preemption allows reducing the size of the guard band to the minimum of the frame preemption supported fragment size. Figure 10.16 illustrates what this means in practice. This solution has been shown to achieve almost zero FDV [86] to the high-priority express traffic. While the solution looks promising there are concerns associated with it, specifically in the context of fronthaul [87,88]:

- One of the main concerns with TAS is the operation and management burden associated with defining network-wide schedules (gate control lists). TAS also requires a fully synchronized network, which is not desirable. The complexity

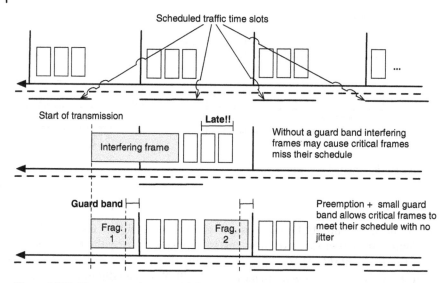

Figure 10.16 Time-aware shapers with frame preemption.

of the operation is spread throughout the entire network and not, for example, to fewer nodes at the network edges.

- The interval that a switch uses to traverse the gate control list can be too coarse for XHaul and NGFI applications. For example, current IEEE 802.1Q limits the measurement interval to only 8 kHz, that is, 125 μs, which may turn out to be an issue. In general, many TSN features, like TAS, were designed with link speeds of 1 Gb/s or much less in mind, which are not adequate for XHaul and NGFI.
- While guard bands can get small with the help of the frame preemption, they cannot be avoided, which in larger high capacity aggregation switches and networks can be significant.
- The frame preemption usefulness can be questioned on high-speed links (e.g., 100 Gb/s). Transmitting a 1500 octet frame at 100 Gb/s takes only ~120 ns. However, whether to use the frame preemption should always be evaluated through the potential traffic profiles. Even at 100 Gb/s link speeds, one 9000 octet jumbo frame takes ~720 ns, which depending on the case can already cause too high FDV.

Queuing at the network edges combined with a playout buffering (or egress buffering) can be used to achieve zero FDV without any sophisticated TSN enhancement. However, the required queue sizes might cause queuing latencies that are excessive for a given deployment. The rule of thumb is that the less FDV there is in the network the less buffering there has to be at the end stations (or other edge nodes) to eliminate the FDV. Combining centralized control over the entire network (software defined networking (SDN) approach) with state-of-the-art TSN features has been shown to achieve nearly zero FDV and, therefore, reducing the sizes of the buffering at the egress [89]. However, this comes with the cost of tight centralized control and programming TDM-like circuits over the packet network.

IEEE 802.1Qci and IEEE 802.1Qch both extend the data structures developed for IEEE 802.1Qbv. IEEE 802.1Qci could be used to block excess traffic at the ingress side of a switch (e.g., some node has gone wild/rogue) and prevent the offending traffic from entering to the rest of the network. However, similar policing can also be achieved using conventional existing policing tools without requiring the rest of the TSN toolbox. IEEE 802.1Qch cyclic queuing has been brought up as a possible solution to guarantee a deterministic latency for traffic through the network (XHaul in this case). However, this solution shares the same issues as TAS in general.

When frame loss ratio (FLR) on the XHaul side is minimal, IEEE 802.1CB Frame Replication and Elimination for Reliability (FRER) could be used. This could be the case, for example, for the envisioned 5G URLLC use cases and deployments. The FRER principle is "simple." Identify an important stream for reliability, tag the stream, and then use replication of the stream data to transport multiple copies of the desired data through the network using alternative paths. At some point in the network or at the receiver, merge the streams, discard duplicates, and deliver a single (original) stream to the end application. This sounds straightforward, which it actually is. The criticism against FRER is also straightforward: at minimum it duplicates the bandwidth requirements. Excess bandwidth requirements might be a showstopper, for example, in the case of XHaul that is already struggling with its envisioned traffic volumes. As was the case with the frame preemption, the usefulness of the FRER depends on the use case and the deployment. There might be deployment cases, for example, some large factory automation/control system, where duplication of the network bandwidth can be justified for better reliability. Another important data point regarding IEEE 802.1CB is that its stream identification function is reused by other TSN standards. For example, an implementation of IEEE 802.1Qci relies on IEEE 802.1CB stream identification for its per-stream filtering and policing.

Last, the cut-through technology keeps showing up in fronthaul-related discussion every now and then. Cut-through is not a TSN feature. Actually, it is not even a standardized solution and is likely to remain a vendor-specific "hack." However, cut-through is still widely implemented and known to reduce latencies through Ethernet switches, which makes it a tempting technology for latency bound XHaul where every means to reduce latency is seen as beneficial. However, the 5G RAN network is going to be somewhat more complex than a simple control loop ring topology with three port switches, where cut-through would be an excellent choice to reduce latency (e.g., consider a switch with one

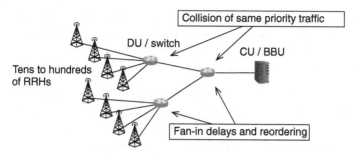

Figure 10.17 RAN aggregation and many-to-one issues (fan-in).

ingress and one egress to the ring, and then one host port where the traffic going through the switch is multicasted "for free"). The following list summarizes the criticism against (any form of) cut-through in any nontrivial topology, and specifically in the 5G RAN network context:

- MAC client interface is a packet-based interface. There is no formal way to describe cut-through without rewriting the interface and all the MACs.
- Bit errors may have changed Ethernet frame header and/or tag fields (source and destination addresses, priority, VLAN, and so forth), therefore, the cut-through may send a frame to the wrong output queue making data visible on a port/link (security issue). Also, in this case the erroneous cut-through traffic uses bandwidth that has been allocated for other traffic (delaying or breaking schedules for TSN traffic).
- Cut-through is only useful if the ingress and egress links have the same speed. This makes cut-through questionable in aggregation cases.
- Cut-through is likely to interfere with both frame preemption and scheduled traffic, since they assume to know the frame size before transmission decisions, which is not the case with cut-through.
- Cut-through works only for the first frame that goes to the egress port. Other traffic destined to the same egress port has to be queued, which reduces the usefulness of cut-through in the case of aggregation.

Summarizing, the IEEE 802.1 TSN toolbox offers a wide range of usable tools, specifically to minimize FDV and give guarantees that the high priority traffic gets through a switch in due time. These are important measures to meet the XHaul (for a given functional split and radio deployment) latency budgets, specifically if a network edge has to do "de-jitter buffering" (which shows as additional latency). On the other hand, all the TSN features essentially engineer the Ethernet network toward a TDM-like circuit-switched deployment. The benefit of statistical multiplexing start is to diminish at that point, and the question arises: why not use circuit-switched (TDM) transport technologies instead of trying to make Ethernet behave like one?

10.3.2   NGFI and XHaul Deployment and Implementation Considerations

In Section 10.3.1, we mentioned aggregation in the network several times. As an example, consider the RAN architecture shown in Figure 10.17. Assume that the switches next to radios aggregate 10 Gb/s links to a 40 or 50 Gb/s uplinks. The next level of switches aggregate 40 or 50 Gb/s links to a 100 Gb/s uplink. Unless all the sources (RRHs) and the entire network is perfectly synchronized, which is unlikely the case, there are bound to be collisions with equal priority traffic. Since frames have to be serialized to the single uplink port, there will be reordering and queuing of the frames. The reordering subsequently adds FDV for a given flow of frames. This is called the *fan-in delay* or *many-to-one issue*. Figure 10.18 illustrates the fan-in delay behavior.

The fan-in delay could be avoided if each source would be perfectly synchronized with the uplink switch and its internal ingress ports serving as an "arbiter" that selects a port at a time for input. This is very hard to achieve. Assuming the network implements TAS, increasing the scheduled traffic slot time and the guard band beyond what is actually needed to take the uncertainty or jitter into account in aggregating nodes would help. However, this would be again wasting bandwidth. This is not a TAS-only problem but an overall synchronization issue in a network that is supposed to make use of statistical multiplexing. Probably, the easiest way around the fan-in delay and reordering originated issue is adding a "de-jitter and reordering buffer" at the edge (or end stations) of the network instead of trying to battle all network-wide schedules to be perfect [86,89]. The perfectly scheduled network would essentially be a circuit-switched TDM-network, not a packet network. Furthermore, adding new sources and nodes would always mean recomputing the entire network-wide schedule. The "de-jitter buffers" have the downside of adding unwanted latency. Here, if the FDV throughout the network is small then the size of the de-jitter buffers are also smaller (thus less buffering imposed latency).

In Section 10.2.2, we mentioned that the XHaul network has to cope with existing CPRI deployments. Dedicating fibers or lambdas for CPRI connections is not a forward-looking solution. Figure 10.19 illustrates a simple transition solution for these incumbent CPRI deployments. For example, a (super) macro

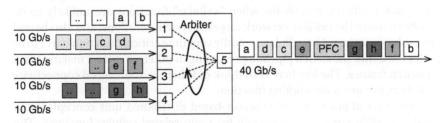

Figure 10.18 Reordering of frames due to fan-in.

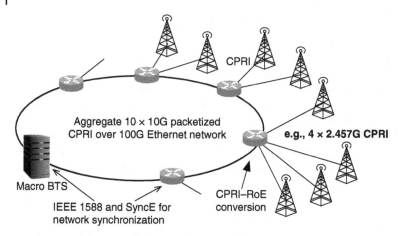

Figure 10.19 Aggregation of multiple legacy CPRI-enabled cell sites over a 100 Gb/s Ethernet-based ring.

base station and cell sites could be part of a 100 Gb/s aggregation ring. At each cell site and also at the proximity of the macro base station would resize a "CPRI-to-Ethernet-to-CPRI" translation function (a node, a switch, form factor is not important). Instead of "CPRI trunking," a normal XHaul and NGFI-capable Ethernet-based aggregation network could be used. The CPRI-over-Ethernet trunking is actually a very prominent use case, and would help conserve optical transport resources. For example, aggregating 10 option 7 CPRI links over 100 Gb/s fiber would require only 4 × 25 Gb/s or 1 × 100 Gb/s lambdas/fibers instead of 10 × 10 Gb/s or 40 × 2.5 Gb/s individual lambdas/fibers. Radio over Ethernet (RoE) puts a lot of effort into enabling a structure agnostic "CPRI-over-Ethernet" trunking use case, also in multioperator environment where each CPRI link can have its view of time, that is, the clock has to be recovered separately for each CPRI link (see Section 10.3.3 for further discussion).

Section 10.2.1 described a number of functional splits and their topical role in the 5G RAN. However, most of the split-related processing and complexity have practically nothing to do with Ethernet and transport networks. As pointed out in Section 10.3, it is hard to justify non-Ethernet or networking features in a switch silicon, specifically when "cellular" functions are unlikely to be usable outside the cellular network targeted system solutions, and also bound to evolve over time. The cellular, typically very time critical, functions have to be integrated into the switch pipeline processing in other ways than making them a switch feature. The key here is to think of Ethernet as universal connectivity solution, not just a networking function.

Figure 10.20 illustrates an Ethernet-based distributed unit concept design with a flexible integration approach for radio-related cellular functions. The integration interface is also based on standard Ethernet. The benefits of the

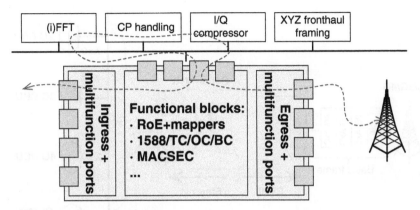

Figure 10.20 A concept of an Ethernet-based distributed unit with flexible integration of "cellular features and accelerators."

unified technology are obvious. Whether a packet is circulated through one or more "cellular features and accelerator" functions is just another switching decision, and can be placed anywhere in the system the packet can be switched to. The functions can be internal to the switch system on a chip package, part of the board, behind a network, or any combination of the previous – the same framework still applies.

The integration of other functions is primarily limited by the port speeds and possible latency considerations. This highlights several concerns in the concept illustrated in Figure 10.20: Flow characteristics change after going through various external functions. For instance, a flow bandwidth may vary significantly from what it was earlier after exiting compression or fast furier transform (FFT) functions. Furthermore, the same flow may traverse the switch pipeline multiple times both adding latency and consuming internal bandwidth/resources, and the list goes on. Details like these have to be fully understood when designing and dimensioning a product.

10.3.3 Radio over Ethernet

IEEE P1914.3 Radio over Ethernet (RoE) is a standard protocol and encapsulation specification for radio data over Ethernet transport [64] and standardized in IEEE 1914 NGFI Working Group. The RoE specification defines a common header format for both data payload and control information transport. RoE specifically concentrates on providing the minimal required building blocks for the following:

- Encapsulation for transporting time-sensitive radio data, including the related control information.

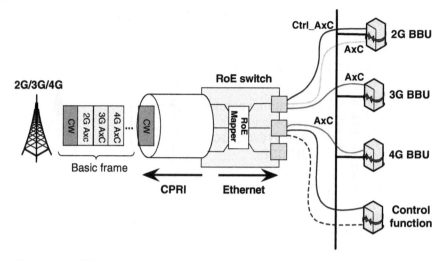

Figure 10.21 CPRI to structure-aware RoE mapping, and individual switching of the component flows.

- Multiplexing of flows that is independent of the underlying transport solution.
- Synchronization and sequencing of the flows.

It needs to be emphasized that RoE is not concerned with how the radio application uses the payload it carries. All RoE are interested in transporting the radio-related data from a RoE endpoint A to a RoE endpoint B in a format that an Ethernet switch can easily process and deliver to the applications. Furthermore, RoE does not provide any specific tools for better network synchronization, latency reduction, or meeting the schedule for a frame delivery. These functions belong to the underlying transport network and network synchronization protocols. For example, for an Ethernet-based XHaul and NGFI network, RoE implementations could assume that the underlying Ethernet network adheres to the IEEE 802.1CM network profile.

RoE is designed for multiple use cases and deployment scenarios. The use cases listed in the standard include three transition cases, and the native use of RoE when both endpoints are RoE capable. The transition cases are for deployments where the other endpoint is RoE enabled and the other one is not. The interoperability in this case is handled by a specific mapper function that we will explain later. A combination of using mappers at the both ends of the RoE connection allows interconnecting existing (legacy) radio framing protocols over an Ethernet transport network. RoE is not tied to a specific network topology. Any common networking topologies apply star, tree, ring, mesh, or even a daisy chain.

One of the powerful features of the RoE are mappers. The role of a mapper is to describe how to encapsulate, decapsulate, and handle specific radio data. The mappers at the time of writing this chapter are as follows:

Structure Agnostic Mapper   This mapper encapsulates an existing radio framing protocol into the RoE without caring about the content and the internal structure of the transported framing protocol (i.e., the "guest payload"). Examples of a structure agnostic mapper are CPRI or OBSAI transport over RoE. In practice, the agnostic mapper is not and cannot be 100% agnostic to its guest payload. The mapper has to know at least the frame boundaries and the synchronization requirements of the guest payload. The latter is specifically important when the receiver of the RoE flow has to be able to recreate the clock for its guest payload [90,91]. The "major" synchronization event acknowledged by the RoE specification is the start of the radio frame (which is, for example, 10 ms for current LTE radio) [93]. The specification goes to great lengths describing how these synchronization events and points are encoded and discovered. The structure agnostic mapper also knows how to strip and/or recreate the PHY line coding of its guest payload. Both 8b10b (25% overhead) and 64b66b (~3% overhead) line codings are supported. This feature is important, for example, when transporting option 7 CPRI guest payload over a 10 Gb/s Ethernet link. With line coding there is no way to fit ~9.8 Gb/s CPRI guest payload and the additional RoE encapsulation overhead to a 10 Gb/s link.

Structure-Aware Mapper   This mapper is only defined for CPRI V7.0. The mapper is aware of the inner details of the CPRI basic frames (BF), hyper frames (HF), and how control words are structured. The chief benefit of the structure-aware mapper is that it can break the CPRI flow into individual component flows: one or more antenna carrier flows (AxC), Slow C&M flow, and a control word flow. The Fast C&M encoded Ethernet frames are transported as-is without any RoE encapsulation over the real Ethernet link. This typically means that the mapper must, for instance, redo the line coding. The mapper also defines a "description language," which is used to describe how the component flows are extracted or reassembled. The component flow concept is powerful and allows individual switching of antenna carrier flows and control word information. Figure 10.21 shows how a CPRI flow that actually contains several antenna carriers for 2/3/4G radio is broken into component flows and individually switched to the corresponding baseband units, and the control information is switched to a dedicated control function in the network. The structure-aware mode of operation is close to what is needed in data center style base stations (e.g., a V-RAN). The obvious drawback of the structure-aware mapper is that it cannot interpret vendor proprietary use of the CPRI information (a common way to

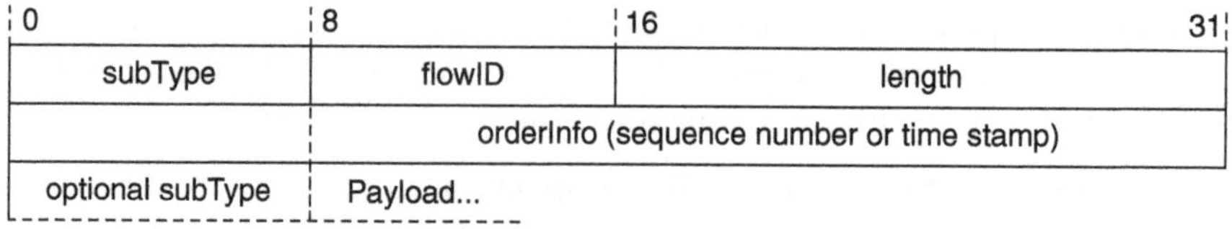

0	8	16	31
subType	flowID	length	
	orderInfo (sequence number or time stamp)		
optional subType	Payload...		

Figure 10.22 IEEE P1914.3 Radio over Ethernet generic header format.

guarantee a vendor lock between a CPRI RE and REC), which limits the usability of the RoE structure agnostic mapper.

Tunneling Mapper   This mapper is a simplification of the structure agnostic mapper. It is used to transport just a stream of octets. Even the guest payload line coding remains unchanged. This mode is inefficient compared to the structure agnostic mapper.

Native Mapper   A common name for a mapper to new radio transports and modes. Current specification work intends to specify an encapsulation and transport of frequency domain samples [94]. This basically means transporting the frequency domain samples needed by the (I)FFT function. The transport of the frequency domain samples is far more bandwidth-friendly and less time critical than the transport of the time domain samples (see Section 10.2.1 and split option 7 as an example).

Control Words   This is technically not a mapper, but still an integral part of the structure-aware mapper. RoE also contains a simple description language just for CPRI control words. Based on the description, control packets containing information extracted from CPRI control words are transported to and from the RoE endpoint.

New mappers may be specific in the future amendment to the IEEE P1914.3 RoE specification.

Figure 10.22 illustrates the common RoE header layout. The common header is intentionally minimal with a size of just 64 bits, and without optional and extendable fields. The fields are always in a networking byte order. The RoE common header is designed to be easy to process in the hardware fast path. The common header contains the following fields:

subType (8 bits) Describes the RoE packet type. The subType 0x00 is reserved for control packets, and the other subTypes are for various data packets. RoE extensibility is also handled through subTypes. A "new" subType can define how the payload following the common header (i.e., after the orderInfo) is interpreted and structured. Control packets are an example of adding new headers into the payload area, that is, the opCode field.

flowID (8 bits) Used for multiplexing RoE flows in a typical case (8 bits). The flowID 0xff is reserved for the NIL value, which is used to mark

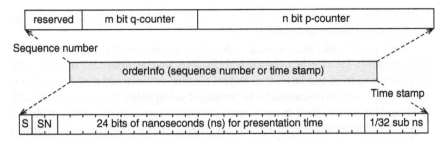

Figure 10.23 IEEE P1914.3 orderInfo fiels details–time stamp or sequence number formats.

that the flowID is not in use. In a case of control packets, the flowID can be used to signal other information (there is no multiplexing of control packets between two RoE nodes in the current specification). FlowIDs are always unique between two RoE end points. The flowID number space is managed so that the end point that receives a RoE flow assigns the flowID number, that is, a RoE endpoint can never receive overlapping flowIDs, but on the other hand can send overlapping flowIDs if they are at different destinations.

length (16 bits) The length of the payload field is in octets, that is, the octets following immediately after the common RoE header. If the payload field contains additional extension headers, they are also counted as part of the length.

orderInfo (32 bits) Either a sequence number or a presentation time time stamp. The orderInfo is used mainly for synchronizing streams. Technically, it can be used for reordering RoE packets if that is needed. The orderInfo is a complex field and will be described further later in this section. Whether the orderInfo contains a sequence number or a time stamp depends on the RoE flow setup time configuration. See Figure 10.23 for an illustration of the orderInfo field.

opCode (8 bits) (optional) Currently, defined only for control packets, and details the actual control packet type. There are control packets for transporting part of the CPRI control word information and for informing the binding between a specific sequence number, flowID, and a time stamp.

payload (0 or more octets) The transported guest payload and/or new header extensions.

RoE endpoints need to agree on a common set of parameters that describe the RoE flow. There is no RoE-specific protocol defined for this purpose, rather it is assumed that a system management does it "somehow." Actually, the management side in the standard is rather scarce. This is expected, since the

RoE specification concentrates mainly on describing how packets are to be constructed, parsed, and transported.

Like it is with all radio-related technologies, synchronization is important. RoE common header has the orderInfo field that is designed to help with synchronization of the flows it transports. It should be noted that RoE expects the network synchronization to be arranged using other means such as IEEE 1588 and/or SyncE. RoE is not a time synchronization distribution protocol. Figure 10.23 illustrates the orderInfo field in more detail; both sequence number and time stamp modes are shown.

The sequence number is divided into three fields: the p-counter, the q-counter, and the reserved bits. All these fields are flexible in size as long as their total sum equals 32 bits. Both p-counter and q-counter have their own maximum values, incrementing and overflow properties that need to be programmed when setting up the RoE connection. The idea is to allow complex sequence numbers where possible synchronization-related information can be embedded into and retrieved from it. The rollover of the p-counter has a specific meaning in this context: it indicates the guest payload radio frame boundary (typically in 10 ms intervals). Let us consider a made up example based on CPRI. The p-counter could count from 0 to 150 * 256 (150 hyper frames and 256 basic frames in each hyper frame). Whenever the p-counter becomes 0, it indicates the radio frame boundary, and causes an increment of the q-counter. The q-counter would then count Node B Frame Numbers (BFN). If both RoE endpoints are synchronized, recovering the clock based on the sequence number is then rather straightforward.

In the time stamp mode, the orderInfo field actually contains a presentation time, when the guest payload should be presented to the radio (or rather to the mapper). There are enough bits to represent time ~16 ms to the future. In practice, some amount of time is reserved to detect whether the time stamp is in the past (late), so the time stamp is actually a signed number whose zero point is not exactly at the middle, but configurable. The time stamp has a sub-nanosecond resolution up to 1/32 nanoseconds. The time stamp also has two bits of sequence number (the SN field) that are supposed to be the two lowest bits of the internally maintained p-counter. The intention of this "mini" sequence number is to detect up to three consecutive lost or reordered packets.

The interesting piece in the time stamp is the "S" bit, that is, the Start of Frame indication. As the name hints, it is used to indicate the start of the guest payload's start of the radio frame in a similar manner to the rollover of the p-counter does. This is a powerful feature. It specifically allows differential time stamping of individual RoE flows, and then recovering the clock for each flow even if view of time is different, and also independent of the transport network view of time. The use of time stamp and individual clock recovery is discussed in length specifically in the context of the structure-aware mapper, and CPRI

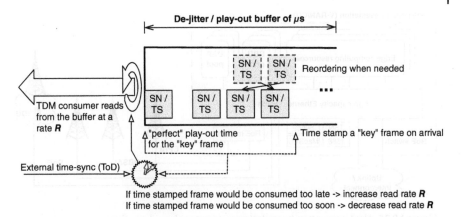

Figure 10.24 Recovering TDM stream from RoE packets using de-jitter/play-out buffer.

trunking over an Ethernet transport network segment with its own clock source that is different from any of the transported CPRI flow clocks [90,91,93].

If the start of the radio frame event and indication is so important, it begs the question: why is there no dedicated flag for it that would cover both sequence number and time stamp? Good question. There are two main reasons: There are no spare bits left in the header for this purpose, and the sequence number has the same information implicitly built-in. The sequence number is expected to be used more frequently than time stamp. This is not the greatest design, though.

As a generic example of using the orderInfo for the RoE flow synchronization, Figure 10.24 illustrates a made up RoE buffer (also used for de-jittering), and a clock recovery implementation on the receiver side. The basic idea is that the mapper knows when a specific sequence number or time stamp should be read out from the buffer by the consumer (e.g., radio). The packets are time stamped when they arrive from the RoE port (actually not all packets need to be time stamped, only the key packets, for example, those carrying the start of the frame information). By comparing the information found in the orderInfo and the mapper internal view of time, the mapper knows whether the RoE packets' guest payload is consumed too fast or too slow. Based on this information, the consumption rate is adjusted to be higher or lower.

Finally, although the RoE common header is presently specified for Ethernet transport, there is no binding between the RoE layer and the transport layer. That was, for example, the reason for seemingly redundant information in the RoE header: the length and the flowID fields. Both the payload length and the multiplexing could be realized in some way from the Ethernet transport layer. Actually, RoE can be transported over any transport technology. IP and MPLS are strong candidates as future transport layers for RoE.

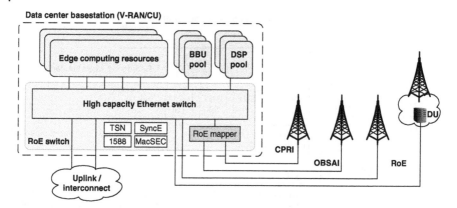

Figure 10.25 Next-generation base station architecture based on ethernet.

10.3.4 Next-Generation Ethernet-Based Base Stations

Next-generation 5G V-RAN base stations are essentially (small scale) data centers, and designed to scale to the requirements of dense and diverse radio deployments (see Table 10.2). Ethernet has gained momentum replacing existing base station internal interconnection technologies such as RapidIO. Figure 10.25 illustrates a high level and greatly simplified Ethernet-based base station design for a C-RAN and a V-RAN. A high-capacity switch (or multiple of those) is in a central role providing connectivity between various functions such as the baseband processing units, the general-purpose compute platforms (e.g., for mobile edge computing and selected packet core functions), and the remote radio heads. The baseband units and the remote radios can be any permutation of 2G to 5G radio technologies. In the 3GPP architecture, the next-generation base stations also serve the function of the centralized unit (CU).

The same platform has to handle the technology transition as well as the connectivity to a local compute, storage, and services. This is one specific area where the challenges reside. The base station is not a simple radio-specific processing entity anymore. The base station must handle arbitrary applications, transition scenarios between different generations in a distributed manner, and deal with diverse types of traffic profiles within the system. Some traffic is extremely time sensitive, and some traffic is inherently unpredictable. Furthermore, varying application needs and transition scenarios imply different transport protocols: plain Ethernet, Multiprotocol Label Switching (MPLS), plain IP, and nonstandard protocols like CPRI may all terminate or get switched in the same system. Specifically, the base station internal heterogeneous traffic mixture is a challenge from a time-sensitive networking and a latency point of view. The internal traffic could be highly engineered, and packet sizes controlled by software in order to guarantee that there are no time synchronization or latency issues within the system. However, like it was with Ethernet replacing CPRI in the fronthaul, trying to mimic an old technology/solution directly using Ethernet is not a good design choice. Instead, the new solution should leverage the strong features of the "new" technology and architecture. In this specific case, using Ethernet's scaling could be the answer: *significant overprovisioning and use of high port speeds for internal communication.* This would ensure low latencies and lower the FDV. Sophisticated TSN features could be used when deemed absolutely necessary.

In addition to handling the transition from legacy radio technologies such as 2G/3G all the way to 5G, the connectivity between the base station and the antenna sites is also a challenge. While native Radio over Ethernet (RoE) technology is a natural choice for the 5G remote radio connectivity, there remains a significant deployed base of old remote radio heads using, for example, CPRI or OBSAI fronthauls. Radio over Ethernet is a potential replacement for the legacy fronthaul technologies in new deployments, and as a standard protocol it could be implemented in merchant silicon targeted to cellular network applications.

A straightforward upgrade of the legacy radios to Radio over Ethernet transport is unlikely as already noted in Section 10.2.2. There are several strategies for the upgrade path. First, the internal communication of the next-generation base station evolves to native Ethernet. Then, the radio (in various representation forms depending, for example, on the processing stage or radio functional split within the system) can also be transported within the system using Ethernet encapsulation. A solution that looks promising is to add a conversion function, a CPRI or OBSAI mapper block, into the Ethernet switch. These mappers, like those specified for RoE, would convert, for example, from CPRI to Radio over Ethernet and vice versa. A mapper allows legacy remote radios to be connected directly into the base station without any changes or upgrades in the legacy fronthaul network, while keeping the base station internal communication entirely based on Ethernet. The mapper can be a simple structure agnostic mapper that is not concerned with the detail of the legacy radio format it deals with, except making it more bandwidth-friendly by removing, for example, redundant line coding (see Section 10.3.3). On the other hand, the mapper can be a more intelligent structure-aware mapper that is able to decompose the legacy radio framing format into different component flows, thus allowing both more bandwidth-friendly transport and individual switching of now separate component flows to their respective receivers (which again allows greater flexibility within the base station to separate processing to different dedicated compute functions).

10.4 Summary

This chapter presented the evolution of the current radio access networks towards Ethernet-based 5G ready radio access network. The chief challenge is to meet the bandwidth, latency, and network synchronization requirements of the 5G. Repurposing data center-driven Ethernet-based technologies into the new 5G RAN have strong potential, however, direct replacement of the current technology is likely to be a wrong approach. This leads to evolving the key building blocks in the 5G RAN: the RAN architecture, the transport network, the radio, and the next-generation base station systems. Distribution of functions, virtualization, and flexible radio functional splits are the key concepts. Especially, the new radio functional splits have a pivotal role to rationalize both bandwidth and time synchronization requirements in the 5G RAN transport network to levels that are commercially deployable. This chapter also took a look into a potential Ethernet-based standardized solution for Radio over Ethernet transport.

References

1 NGMN Alliance, 5G White Paper, ver. 1.0, Next Generation Mobile Networks, 2015. Available at https://www.ngmn.org/uploads/ media/NGMN_5G_White_Paper_V1_0.pdf.

2 3GPP, Technical Specification Group Radio Access Network, Study on Scenarios and Requirements for Next Generation Access Technologies (Release 14), Technical report 38.913 v14.0.0, 2016. Available at http://www.3gpp.org/DynaReport/38913.htm.

3 3GPP, Technical Specification Group Radio Access Network, Study on New Radio Access Technology; Radio Access Architecture and Interfaces (Release 14), Technical report 38.801 v0.7.0, 2016. Available at http://www.3gpp.org/DynaReport/38801.htm.

4 C. Chen, *Some View on Next Generation Radio Interface*, 2015. Available at http://www.ieee1904.org/3/meeting_archive/2015/02/tf3_1502_clark_1a.pdf.

5 C.-l. I, Y. Yuan, J. Huang, S. Ma, C. Cui, and R. Duan, "Rethink fronthaul for soft RAN," *IEEE Commun. Mag.*, vol. 53, no. 9, pp. 82–88, 2015.

6 T. Pfeiffer, "Next generation mobile fronthaul and midhaul architectures invited," *IEEE/OSA J. Opt. Commun. Netw.*, vol. 7, no. 11, pp. 38–45, 2015.

7 K.M.S. Huq and J. Rodriguez, *Backhauling/Fronthauling for Future Wireless Systems*, 1st Ed., Chichester: John Wiley & Sons, Ltd, 2017.

8 3GPP, Technical Specification Group Radio Access Network, Evolved Universal Terrestrial Radio Access (E-UTRA) and Evolved Universal Terrestrial Radio Access Network (E-UTRAN); Overall description; Stage

2 (Release 14), Technical report 36.300 v14.0.0, 2016. Available at http://www.3gpp.org/DynaReport/36300.htm.

9 3GPP, Technical Specification Group Radio Access Network, Evolved Universal Terrestrial Radio Access Network (E-UTRAN); Architecture description (Release 13), Technical report 36.410 v13.2.0, 2016. Available at http://www.3gpp.org/DynaReport/36401.htm.

10 3GPP, Technical Specification Group Services and System Aspects; General Packet Radio Service (GPRS) enhancements for Evolved Universal Terrestrial Radio Access Network (E-UTRAN) access (Release 14), Technical report 24.401 v14.1.0, 2016. Available at http://www.3gpp.org/DynaReport/23401.htm.

11 3GPP, Technical Specification Group Radio Access Network, Evolved Universal Terrestrial Radio Access Network (E-UTRAN); S1 General Aspects and Principles (Release 13), Technical report 36.410 v13.0.0, 2015. Available at http://www.3gpp.org/DynaReport/36420.htm.

12 3GPP, Technical Specification Group Radio Access Network, Evolved Universal Terrestrial Radio Access Network (E-UTRAN); S1 Data Transport (Release 13), Technical report 36.414 v13.0.0, 2015. Available at http://www.3gpp.org/DynaReport/36414.htm.

13 3GPP, Technical Specification Group Radio Access Network, Evolved Universal Terrestrial Radio Access Network (E-UTRAN); S1 Layer 1 (Release 13), Technical report 36.411 v13.0.0, 2015. Available at http://www.3gpp.org/DynaReport/36411.htm.

14 J. Salmelin and E. Metsala, *Mobile Backhaul*, 1st Ed., Chichester: John Wiley & Sons, Ltd, 2012.

15 MEF, Implementation Agreement, MEF 22.2 Mobile Backhaul Phase 3, 2016. Available at https://www.mef.net/Assets/Technical_Specifications/PDF/MEF_22.2.pdf.

16 MEF, Implementation Agreement, MEF 22.1 Mobile Backhaul Phase 2, 2012. Available at https://www.mef.net/Assets/Technical_Specifications/PDF/MEF_22.1.pdf.

17 R. Webb, Macrocell mobile backhaul equipment market tracker – regional – H2 2016, Market research. 2015. Available at http://www.ieee1904.org/3/meeting_archive/2015/02/tf3_1502_clark_1a.pdf .

18 S. Bryant and P. Pate, Pseudo wire emulation edge-to-edge (PWE3) architecture, RFC 3985 (Informational), 2005. Available at http://www.ietf.org/rfc/rfc3985.txt (updated by RFC 5462).

19 3GPP, Technical Specification Group Radio Access Network, Evolved Universal Terrestrial Radio Access (E-UTRA); Base Station (BS) radio transmission and reception (Release 14), Technical specification 36.104 v14.0.0, 2016. Available at http://www.3gpp.org/DynaReport/36104.htm.

20 J. Korhonen, T. Savolainen, and J. Soininen, *Deploying IPv6 in 3GPP Networks*, 1st Ed., John Wiley & Sons, Ltd, Chichester, 2013.

21 NGMN Alliance, Guidelines for LTE Backhaul Traffic Estimation, Guideline 0.4.2, Next Generation Mobile Networks, 2011. Available at https://www.ngmn.org/uploads/media/NGMN_Whitepaper_Guideline_for_LTE_Backhaul_Traffic_Estimation.pdf.

22 3GPP, Technical Specification Group Radio Access Network, Scenarios and Requirements for Small Cell Enhancements for E-UTRA and E-UTRAN (Release 13), Technical report 36.932 v14..0, 2015. Available at URL http://www.3gpp.org/DynaReport/36932.htm.

23 CPRI, Interface Specification, *Technical report V7.0, CPRI Cooperation*, 2015. Available at http://www.cpri.info/spec.html.

24 OBSAI, Reference Point 3 Specification, Version 4.2, Open Base Station Architecture Initiative, 2010. Available at http://www.obsai.com/specifications.htm.

25 3GPP, Technical Specification Group Radio Access Network, Evolved Universal Terrestrial Radio Access (E-UTRA); Medium Access Control (MAC) protocol specification (Release 14), Technical specification 36.321 v14.0.0, 2016. Available at http://www.3gpp.org/DynaReport/36104.htm.

26 3GPP, Numerology for new radio interface, pCR R1-162386, 2016. Available at http://www.3gpp.org/ftp/tsg_ran/WG1_RL1/TSGR1_84b/Docs/R1-162386.zip.

27 C. Y. Chang, N. Nikaein, and T. Spyropoulos, "Impact of packetization and scheduling on C-RAN fronthaul performance," in IEEE Global Communications Conference (GLOBECOM), 2016, pp. 1–7.

28 P. Assimakopoulos, M. K. Al-Hares, and N. Gomes, "Switched ethernet fronthaul architecture for cloud-radio access networks," *IEEE/OSA J. Opt. Commun. Netw.*, vol. 8, no. 12, pp. 135–146, 2016.

29 NGMN Alliance, Fronthaul Requirements for C-RAN, Requirements 1.0, Next Generation Mobile Networks, 2015. Available at https://www.ngmn.org/uploads/media/NGMN_RANEV_D1_C-RAN_Fronthaul_Requirements_v1.0.pdf.

30 A. Checko, H. Christiansen, Y. Yan, L. Scolari, G. Kardaras, M. Berger, and L. Dittmann, "Cloud RAN for mobile networks – a technology overview," *IEEE Commun. Surv. Tutor.*, vol. 17, no. 1, pp. 405–426, 2015.

31 3GPP, Technical Specification Group Radio Access Network, Evolved Universal Terrestrial Radio Access (E-UTRA); Requirements for Support of Radio Resource Management (Release 14), Technical specification 36.113 v14.1.0. Available at http://www.3gpp.org/DynaReport/36113.htm .

32 IEEE, Draft Standard for Local and Metropolitan Area Networks – Time-Sensitive Networks for Fronthaul, IEEE Draft Std P802.1CM/D0.5, 2016.

33 ITU-T, Timing and synchronization aspects in packet networks, Recommendation T-REC-G.8261, v3.0, 2013.

34 IEEE, IEEE Standard for a Precision Clock Synchronization Protocol for Networked Measurement and Control Systems, IEEE Std 1588-2008, 2008.

35 Y. Stein, A. Geva, and G. Zigelboim, "Delivering better time-of-day using synchronous ethernet and 1588," in the *5th International Telecoms Sync Forum, ITSF*, 2007, pp. 1–23.

36 M. Lipinski, T. Wlostowski, J. Serrano, and P. Alvarez, "White rabbit: a PTP application for robust sub-nanosecond synchronization," in *IEEE International Symposium on Precision Clock Synchronization for Measurement, Control and Communication*, 2011, pp. 25–30.

37 CMCC and Broadcom, *CMRI-Broadcom Radio over Ethernet Validation Testing Report*, Test report, 2015.

38 CPRI, CPRI functional decomposition requirements, Presentation, Common Public Radio Interface, 2016. Available at http://www.ieee802.org/1/ files/public/docs2016/cm-CPRI-functional-decomposition-requirements-0516-v01.pdf.

39 S. Ruffini, P802.1CM Synchronization Considerations, Presentation, Ericsson, 2016. Available at http://www.ieee802.org/1/files/public/docs2016/cm-ruffini-synchronization-considerations-0516-v01.pdf.

40 ITU-T, Precision time protocol telecom profile for phase/time synchronization with full timing support from the network, Recommendation T-REC-G.8275.1, v1.1, 2015.

41 3GPP, Technical Specification Group Services and System Aspects, NG Radio Access Network (NG-RAN); f1 Data Transport (Release 15), TS 38.474 V0.0.0 (R3-171323), 2017. Available at http://list.etsi.org/scripts/wa.exe?A2=ind1704C L=3GPP_TSG_RAN_WG3 F= S= P=6142.

42 3GPP, New SID Proposal: Study on New Radio Access Technology, SID RP-160671, 2016. Available at http://www.3gpp.org/ftp/tsg_ran/ TSG_RAN/TSGR_71/Docs/RP-160671.zip.

43 3GPP, Technical Specification Group Radio Access Network, Study on New Radio (NR) Access Technology Physical Layer Aspects (Release 14), Technical report 38.802 V1.0.0, 2016. Available at http://www.3gpp.org/ DynaReport/38802.htm.

44 D. Chen, *Fronthaul Bandwidth Analysis and Latency Constraint Considerations*, Presentation, Nokia, 2016. Available at http://www.ieee802.org/1/files/ public/docs2016/cm-chen-fronthaul-bandwidth-analysis-0716-v01.pdf.

45 A. Checko and M. Hoegdal, Fronthaul throughput dimensioning tool, 2017. Available at http://sites.ieee.org/sagroups-1914/files/2017/04/FHDimensioning_v0_3.xlsx.

46 B. Guo, W. Cao, A. Tao, and D. Samardzija, "LTE/LTE – a signal compression on the CPRI interface," *Bell Labs Tech. J.*, vol. 18, no. 2, pp. 117–133, 2013.

47 iJOIN, 2015, http://www.ict-ijoin.eu/.

48 iCIRRUS, 2016, http://www.icirrus-5gnet.eu/.

49 5G-XHaul, 2016, http://www.5g-xhaul-project.eu/.

50 5G-CrossHaul, 2016, http://5g-crosshaul.eu/.

51 U. Dotsch, M. Doll, H. P. Mayer, F. Schaich, J. Segel, and P. Sehier, "Quantitative analysis of split base station processing and determination of advantageous architectures for LTE," *Bell Labs Tech. J.*, vol. 18, no. 1, pp. 105–128, 2013.

52 A. Maeder, M. Lalam, A. D. Domenico, E. Pateromichelakis, D. Wubben, J. Bartelt, F. Richard, and P. Rost, "Towards a flexible functional split for cloud-ran networks," in *Networks and Communications (EuCNC)*, 2014, pp. 1–5.

53 D. Wobben, P. Rost, J. Bartelt, M. Lalam, V. Savin, M. Gorgoglione, A. Dekorsy, and G. Fettweis, "Benefits and impact of cloud computing on 5G signal processing: flexible centralization through cloud-ran," *IEEE Signal Process. Mag.*, vol. 31, no. 6, pp. 35–44, 2014.

54 N. Shibata, T. Tashiro, S. Kuwano, N. Yuki, Y. Fukada, J. Terada, and A. Otaka, "Performance evaluation of mobile front-haul employing Ethernet-based TDM-PON with IQ data compression (Invited)," *IEEE/OSA J. Opt. Commun. Netw.*, vol. 7, no. 11, pp. 16–22, 2015.

55 S. Nishihara, S. Kuwano, K. Miyamoto, J. Terada, and A. Ootaka, "Study on protocol and required bandwidth for 5G mobile fronthaul in C-RAN architecture with MAC-PHY split," in *22nd Asia-Pacific Conference on Communications (APCC)*, 2016, pp. 1–5.

56 C. Y. Chang, R. Schiavi, N. Nikaein, T. Spyropoulos, and C. Bonnet, "Impact of packetization and functional split on C-RAN fronthaul performance," in *IEEE International Conference on Communications (ICC)*, 2016.

57 X. Costa-Perez, A. Garcia-Saavedra, X. Li, T. Deiss, A. de la Oliva, A. di Giglio, P. Iovanna, and A. Moored, "5G-Crosshaul: an SDN/NFV integrated fronthaul/backhaul transport network architecture," *IEEE Wirel. Commun.*, vol. 24, no. 1, pp. 38–45, 2017.

58 V. Jungnickel, M. Parker, J. F. Riera, C. Bock, V. Marques, D. Levi, D. Schulz, J. Hilt, K. Habel, L. F. del Rosal, R. Freund, and S. Walker, "Performance evaluation of mobile front-haul employing Ethernet- based TDM-PON with IQ data compression (Invited)," *IEEE International Conference on Communications Workshops (ICC)*, 2016, pp. 360–366.

59 ETSI Mobile-Edge Computing (MEC), 2016. Available at http://www.etsi.org/technologies-clusters/technologies/mobile-edge-computing.

60 P. Mach and Z. Becvar, "Mobile edge computing: a survey on architecture and computation offloading," *IEEE Commun. Surv. Tutor.*, vol. 19, no. 3, pp. 1628–1656, 2017.

61 NGMN Alliance, Description of Network Slicing Concept, Technical report 1.0, Next Generation Mobile Networks, 2016. Available at https://www.ngmn.org/uploads/media/160113_Network_Slicing_v1_0.pdf.

62 3GPP, Technical Specification Group Radio Access Network, Study on Small Cell Enhancements for E-UTRA and E-UTRAN; Higher Layer Aspects (Release 12), Technical report 36.842 V12.0.0, 2013. Available at http://www.3gpp.org/DynaReport/36842.htm.

63 IEEE, Draft Standard for Packet-Based Fronthaul Transport Networks, IEEE Draft Std P1914.1/D0.0, 2016. Available at http://sites.ieee.org/sagroups-1914/p1914-1/ieee-p1914-1-draft-specifications/.

64 IEEE, Draft Standard for Radio over Ethernet Encapsulations and Mappings, IEEE Draft Std P1914.3/D2.0, 2016. Available at http://sites.ieee.org/sagroups-1914/p1914-3/ieee-p1914-3-draft-specifications/.

65 ITU-T, G.Sup55 : *Radio-over-fibre (RoF) Technologies and Their Applications*, 2016. Available at https://www.itu.int/rec/T-REC-G.Sup55-201507-I/en.

66 CMCC, NGFI (next generation fronthaul interface), White Paper Version 1.0, China Mobile (CMCC), 2015.

67 3GPP, 5G architecture options full set, pCR RP-161266, 2016. Available at http://www.3gpp.org/ftp/tsg_ran/TSG_RAN/TSGR_72/joint_RAN_SA/RP-161266.zip.

68 3GPP, Technical Specification Group Radio Access Network, Evolved Universal Terrestrial Radio Access (E-UTRA); Study on Latency Reduction Techniques for LTE (Release 14), Technical report 36.881 V14.0.0, 2016. Available at http://www.3gpp.org/DynaReport/36881.htm.

69 3GPP, Technical Specification Group Services and System Aspects, Study on Architecture for Next Generation System (Release 14), Technical report 23.799 V2.0.0, 2016. Available at http://www.3gpp.org/DynaReport/23799.htm.

70 IEEE 802.1 Time-Sensitive Networking Task Group, 2016. Available at http://www.ieee802.org/1/pages/tsn.html.

71 Google, *Spanner: Google's Globally-Distributed Database*, Research publications, 2012. Available at https://research.google.com/archive/spanner.html.

72 F. Brockners et. al, *Data Fields for In Situ OAM*, 2017. Available at https://tools.ietf.org/html/draft-brockners-inband-oam-data-03 , work in progress.

73 IEEE, IEEE Standard for Local and Metropolitan Area Networks – Bridges and Bridged Networks, IEEE Std 802.1Q-2014, 2014.

74 IEEE, IEEE Standard for Ethernet – Amendment for Interspersing Express Traffic (IET), IEEE Std 802.3br-2016, 2016.

75 IEEE, IEEE Standard for Local and Metropolitan Area Networks – Bridges and Bridged Networks – Amendment: Frame Preemption, IEEE Draft Std P802.1Qbu/D3.0, IEEE Computer Society, 2015.

76 IEEE, IEEE Standard for Local and Metropolitan Area Networks – Bridges and Bridged Networks – Amendment 25: Enhancements for Scheduled Traffic, IEEE Std 802.1Qbv-2015, 2015.

77 IEEE, Draft Standard for Local and Metropolitan Area Networks – Frame Replication and Elimination for Reliability, IEEE Draft Standard for P802.1CB/D2.6, 2016.

78 IEEE, Draft Standard for Local and Metropolitan Area Networks – Bridges and Bridged Networks – Amendment: Per-Stream Filtering and Policing, IEEE Draft Standard for P802.1Qci/D2.1, 2016.

79 IEEE, Draft Standard for Local and Metropolitan Area Networks – Bridges and Bridged Networks – Amendment: Cyclic Queuing and Forwarding, IEEE Draft Standard for P802.1Qch/D2.0, 2016.

80 IEEE, Draft Standard for Local and Metropolitan Area Networks – Timing and Synchronization for Time-Sensitive Applications, IEEE Draft Standard for P802.1AS-Rev/D4.2, 2016.

81 A. Checko, A.C. Juul, H. Christiansen, and M. Berger, "Synchronization challenges in packet-based cloud-RAN fronthaul for mobile networks," in *IEEE International Conference on Communication Workshop (ICCW)*, 2015, pp. 113–118.

82 I. Freire, I. Sousa, A. Klautau, I. Almeida, C. Lu, and M. Berg, "Analysis and evaluation of end-to-end PTP synchronization for Ethernet-based fronthaul," in *IEEE Global Communications Conference (GLOBECOM)*, 2016.

83 D. Chitimalla, K. Kondepu, L. Valcarenghi, M. Tornatore, and B. Mukherjee, "Switched ethernet fronthaul architecture for cloud-radio access networks," *IEEE/OSA J. Opt. Commun. Netw.*, vol. 9, no. 2, pp. 172–182, 2017.

84 M. K. Al-Hares, P. Assimakopoulos, S. Hill, and N. Gomes, "The effect of different queuing regimes on a switched Ethernet fronthaul," in *18th International Conference on Transparent Optical Networks (ICTON)*, 2016.

85 P. Assimakopoulos, M. K. Al-Hares, S. Hill, A. Abu-Amara, and N. Gomes, "Statistical distribution of packet inter-arrival rates in an Ethernet fronthaul," in *IEEE International Conference on Communications Workshops (ICC)*, 2016, pp. 140–144.

86 J. Farkas and B. Varga, *P802.1CM Simulation Results for Profiles A & B*, Presentation, Ericsson, 2016. Available at http://www.ieee802.org/1/files/public/docs2016/cm-farkas-profiles-A-and-B-0316-v01.pdf.

87 J. Korhonen, *802.1CM and 802.1Qbv Considerations*, Presentation, Broadcom, 2016. Available at http://www.ieee802.org/1/files/public/docs2016/cm-jik-tasconsiderations-0316-v00.pdf.

88 D. Chen and E. Ryytty, *Qbv Optional for Fronthaul over Ethernet*, Presentation, Nokia, 2016. Available at http://www.ieee802.org/1/files/public/docs2016/cm-chen-Qbv-Optional-for-Fronthaul-over-Ethernet-0316-v02.pdf.

89 T. Wan, B. McCormick, Y. Wang, and P. Ashwood-Smith, ZeroJitter: an SDN based scheduling for CPRI over Ethernet, in *IEEE Global Communications Conference (GLOBECOM)*, 2016, pp. 1–7.

90 R. Tse, J. Korhonen, and B. Li, *RoE Use Cases*, Presentation, 2016. Available at http://grouper.ieee.org/groups/1904/3/meeting_archive/2016/04/tf3_1604_tse_use_cases_1.pdf.

91 R. Tse, J. Korhonen, and B. Li, *Data Path vs Control Path for Timing of Radio Data*, Presentation, 2016. Available at http://grouper.ieee.org/groups/1904/3/meeting_archive/2016/04/tf3_1604_tse_datapath_1.pdf.

92 CPRI, eCPRI Interface Specification. Technical report V1.0, CPRI Cooperation, 2017. Available at http://www.cpri.info/spec.html

93 J. Korhonen, *RoE Structure Agnostic Mapper and Start of Frame*, Presentation, 2016. Available at http://grouper.ieee.org/groups/1904/3/meeting_archive/2016/06/tf3_1606_korhonen_sof_2.pdf.

94 R. Maiden, *Native Packet Types – Handling Frequency Domain I/Q Encoding*, Presentation, 2016. Available at http://sites.ieee.org/sagroups-1914/files/2016/11/tf3_1610_maiden_native-packet-types_2.pdf.

Jouni Korhonen Ph.D, is a Principal R&D engineer with Nordic Semiconductor. Previously he was with Broadcom and was active in Ethernet-based base station architectures, Ethernet-based fronthaul networks, and time-sensitive networking. Dr. Korhonen was instrumental forming the IEEE P1914.3 Radio over Ethernet Task Force (formerly IEEE P1904.3 RoE TF) and chaired the TF and heavily contributed to the standard development from the beginning. Previously, he was heavily involved with IPv6, DNS, and other core network signaling matters in 3GPP. He also held multiple leadership positions within IETF along the years and is still an active contributor with 38 published RFCs to date. His research interests include Cellular IoT and 3GPP system architecture evolution.

11

Energy-Efficient 5G Networks Using Joint Energy Harvesting and Scheduling

Ahmad Alsharoa,[1] Abdulkadir Celik,[2] and Ahmed E. Kamal[1]

[1] Electrical and Computer Engineering department, Iowa State University (ISU), Ames, IA, USA
[2] Computer, Electrical, Mathematical Sciences & Engineering Division, King Abdullah University of Science and Technology (KAUST), Kingdom of Saudi Arabia

11.1 Introduction

Prospective demands of next-generation wireless networks are ambitious and will require cellular networks to support thousand times higher data rates and ten times lower round-trip latency [1]. While this data deluge is a natural outcome of the increasing number of mobile devices with data-hungry applications and the Internet of Things (IoT), the low latency demand is required by the future interactive applications such as "tactile internet," virtual and enhanced reality, and online internet gaming. Overall mobile data traffic is expected to grow to 49 exabytes per month by 2021, a sevenfold increase over 2016 [2]. Mobile data traffic will grow at a compound annual growth rate of 47% from 2016 to 2021 as shown in Figure 11.1. Furthermore, the cellular infrastructure currently contributes approximately 2% of carbon footprint and 3% of worldwide energy consumption, as a result of more than three million base stations (BSs) worldwide [3]. Also, noting that the carbon emission of information and communication technologies (ICT) is predicted to increase from 170 metric tons in 2014 to 235 metric tons by 2020 [4]; these statistics led telecom industry, governmental institutions, and researchers to initiate *green* measures.

With the increasing number of mobile broadband data users and bandwidth-intensive services, the demand for radio resources has increased tremendously. One of the methods used by mobile operators to meet this challenge is to deploy additional low-powered BSs, such as smallcell BSs (SBSs) and microcell BSs (MSBs), in areas of high demand, as shown in Figure 11.2. The resulting

5G Networks: Fundamental Requirements, Enabling Technologies, and Operations Management, First Edition.
Anwer Al-Dulaimi, Xianbin Wang, and Chih-Lin I.
© 2018 by The Institute of Electrical and Electronics Engineers, Inc. Published 2018 by John Wiley & Sons, Inc.

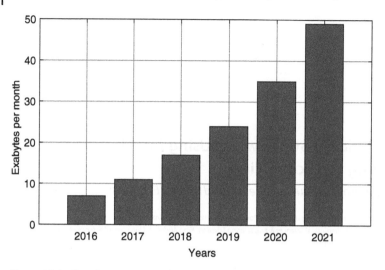

Figure 11.1 Cisco forecasts 49 exabytes per month of mobile data traffic by 2021. (Reproduced with permission from Reference [3]. Copyright 2013, IEEE.)

networks, referred to as heterogeneous networks (HetNets), help in maintaining the quality of service (QoS) for a larger number of users by reusing the spectrum [5,6]. Furthermore, HetNets have already been considered as a promising solution in which SBSs and MBSs are deployed to boost network coverage and capacity while reducing operational and capital expenditures of mobile operators. However, with the densification of these HetNets, energy consumption and the carbon footprint have significantly raised. Therefore, conserving energy while meeting the users QoS requirements has been the focus of the green communications researchers.

On the other hand, most of the wireless data usage is in indoor environments such as offices, residential buildings, shopping malls, where the users may face difficulties in achieving high-data rates while connecting to the macrocell BSs. This is mainly due to the penetration loss incurred by the wireless signals inside the buildings. Therefore, to increase the capacity of the network in these hot spots, SBSs are deployed in close proximity to the buildings [7].

11.1.1 Sleeping Strategy

In general, MBSs and SBSs provide increased coverage and network capacity during peak times. However, they might not be very useful under light traffic load scenarios. Instead, they might be underutilized or completely redundant leading to inefficient use of energy and communication resources. Hence, dynamic BS on/off switching, which is known as BS sleeping strategy, is shown to be highly useful in reducing energy consumption of cellular HetNets [8,9].

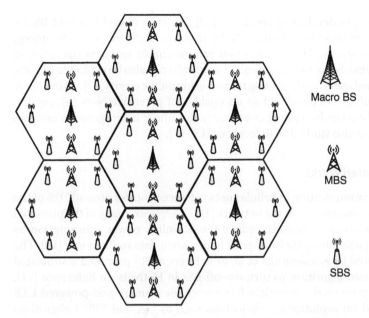

Figure 11.2 Heterogenous networks.

The BSs are turned off during periods of low traffic and the small number of active users are offloaded to a nearby BS. As a result, the power consumption of lightly loaded BSs can be reduced or completely eliminated depending on the state of the turned off BSs.

11.1.2 Energy Harvesting

Energy harvesting (EH) has been considered as one of the most effective and robust solutions to protract the lifetime and sustainability of wireless networks [10], where harvesting energy from ambient renewable energy sources is considered a promising eco-friendly solution. Many promising practical applications that use EH nodes have been discussed recently, such as, emerging ultradense small cell deployments, point-to-point sensor networks, and far-field microwave power transfer [11].

Depending on geographical and environmental characteristics, renewable energy (RE) sources such as solar radiation and/or wind energy are powerful candidates for alternative energy solutions [12]. Based on the availability and quality of the energy resources, BSs can be classified as *off-grid* and *bad-grid*, where the former stands for BSs without grid power supply (e.g., diesel-powered BSs in rural areas) while the latter describes BSs connected to a grid supply but with frequent power outages, loss of phase, or fluctuating voltages, which is mostly seen in developing countries [13]. As off-grid and bad-grid BS are expected to increase by 22 and 13% by the year 2020, respectively, energy harvesting solutions play an important role to enable seamless operation of cellular systems. Solar-powered mobile smallcells can also be exploited to serve huge crowds of users for short/long terms during public events [14]. As cognitive radio technology is considered as an enabler of green HetNets [6], energy-efficient and energy-harvesting cognitive radios under heterogeneous network conditions are also studied in References [15–17].

11.1.3 Related Works

Downlink communication in cellular networks accounts for around 70% of the total energy consumption in the network [18]. Therefore, many of the proposed works in the literature tried to reduce the downlink power consumption by switching off BSs during their off-peak hours when data traffic is low [19]. The work presented by Koudouridis et al. in Reference [20] proposed a simulated annealing-based algorithm to turn on–off BSs in HetNets. In Reference [21], the authors presented a complete framework for a smart-grid-powered LTE system based on evolutionary algorithms such as GA and BPSO algorithm; these heuristic switching on/off approaches were proposed under equal power distribution scenario. In Reference [22], the impact of turning off macrocell BSs on the energy efficiency of the HetNet is studied while keeping the SBSs active. Several robust and efficient schemes for BS on/off switching have been proposed in the literature [23,24]. For instance, in Reference [23], three different approaches for SBS switching in HetNets are discussed. The on/off status of the SBSs is controlled by the detection of active users by the SBSs, or wake-up signals by the core network, or wake-up signals by the users. In Reference [24], the authors have introduced two switching modes that operate on intermediate and fast timescales in order to cater for the short and long idle periods of the users. It is shown that dense HetNets can be used to achieve higher capacity and performance while simultaneously reducing energy consumption.

Most of promising solutions of energy efficiency with EH in HetNets are based on RE-based EH technique to power cellular networks [25–30]. The benefit of using RE-based EH technique in HetNets has been recently discussed in literature [31–33]. RE-based EH technique has shown to yield a significant carbon dioxide (CO_2) reduction by reducing the reliance on traditional electricity supplies [34]. In Reference [35], an energy-efficient resource allocation scheme for distributed antenna system with energy recycling via electromagnetic radiation is proposed. The authors developed a low complexity algorithm to find the subchannel and power allocation in order to maximize the energy efficiency utility.

One of the limitation of the RE-based EH is the discontinuity of the power generation that affects reliability of service. In Reference [36], the authors develop a tractable model based on discrete-time Markov chain to analyze the performance of downlink heterogeneous cellular networks with both power-grid-connected BS and energy harvesting SBSs. Each SBS forms a personal cell that is active only when its own priority user requests service and its battery contains sufficient energy to transmit. In Reference [37], the authors consider hybrid powering BSs connected to different microgrids that cooperate to minimize the total power cost by optimizing their resources allocation. The authors assume that each microgrid can purchase backup power from the main grid when needed, in order to ensure a reliable service to users. A hybrid energy-sharing framework is presented in Reference [38], where the BSs are powered by smart grid and have RE generation capabilities. In addition to that, physical power lines infrastructure between BSs is proposed to share energy between BSs when needed.

11.1.4 Contributions

Motivated by the above discussions, we consider the design and optimization of hybrid-powered HetNets where traditional microgrid is not exploited as long as the BSs can sustain their power demands from harvested and stored renewable energy. Taking power consumption and spectrum resources as the optimization variables, we aim to minimize the carbon footprint of the network subject to transmission power constraints, energy storage capacity, users' QoS requirements, and bandwidth limitations. In this chapter, we consider a down-link EH HetNet system where each BS is equipped to harvest from wireless and renewable sources. Moreover on/off switching strategy is used to reduce the total energy consumption. The contribution of this work can be summarized as follows:

- Consisting of green (renewable) and traditional microgrid, a hybrid power supply is considered such that traditional microgrid is not exploited as long as the BSs can meet their power demands from harvested and stored green energy.
- A generic optimization problem is formulated with the objective of minimizing the network-wide energy consumption over a given time horizon. The goal is to optimize the BS sleeping and user–cell association variables under BS's maximum power constraint, maximum BS's storing energy constraint, and user's QoS constraint.
- Depending on the knowledge level about future RE generation, two cases are investigated:

    1) The zero knowledge case: In this case, future RE generation statistics are unknown for the mobile operator. A binary linear programming (BLP) problem is formulated to optimize the BS sleeping status and user–cell association.

2) The perfect knowledge case: This case assumes that all future statistics of the network are perfectly known and estimated.

- Based on BPSO algorithm, a near optimal and low-complexity green optimization approach is developed and compared with the well-known evolutionary genetic algorithm (GA) [39].

11.1.5 Organization

The remainder of the chapter is organized as follows. Section 11.2 presents the EH HetNets system model. The problem formulation is given in Section 11.3. Low-complexity algorithms are proposed in Section 11.4. Section 11.5 discusses the selected numerical results. Conclusions and future research directions are given in Section 11.6.

11.2 System Model

We consider a half-duplex downlink transmission of three-tiered HetNets consisting of a macrocell tier, microcell tier, and smallcell tier with a total of $L + 1$ BSs (i.e., a single macrocell BS and L_M MBSs, and L_S SBSs, where $L = L_M + L_S$). The locations of all BSs are modeled by an independent homogeneous Poisson point process (PPP). The hybrid power supply microgrid sources consisting of a green grid (GG) and a traditional grid (TG) is considered. The former uses renewable sources to generate the electric power, while the latter uses classical sources to generate the electric power. Each BS is connected to the GG so as to provide help in energy when needed. The GG has the ability to purchase a backup power from the TG that is controlled by a control unit when needed, as shown in Figure 11.3.

Denoting U^b as the total number of users in the network during time slot b, we denote $\bar{U}_l$ as the maximum number of users that can be served by a BS l, where index $l = 0$ for macrocell BS and $l \geq 1$ for other BS tiers, such that $\bar{U}_l \ll \bar{U}_0$. These numbers reflect the BSs' capacities due to available number of frequency carriers and/or hardware and transmit power limitations. In order to avoid the cochannel interference, all the channels are assumed to share the spectrum orthogonally between the BS. Finally, it is assumed that each user is served by at most one BS (either macrocell BS, MBS, or SBS).

In general, we assume that the communication channel between two nodes x and y at time slot b is given as follows

$$h_{xy}^b = \sqrt{d_{xy}^{-\varpi}} \tilde{h}_{xy}^b, \tag{11.1}$$

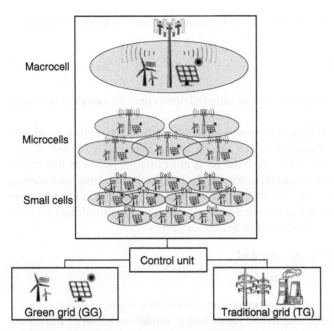

Figure 11.3 System model of HetNets hybrid EH.

where d_{xy} is the Euclidean distance between the nodes x and y, ϖ is a path loss exponent, and $\tilde{h}^b_{xy}$ is a fading coefficient with a coherence time slot T_b sec. Without loss of generality, all channel gains are assumed to be constant during T_b.

11.2.1 Base Station Power Model

Since the energy arrivals and energy consumption of the BSs are random and their energy storage capacities are finite, some BSs might not have enough energy to serve users at a particular time. Under such scenario, it is preferred that some of the BSs are kept off and allowed to recharge while their load is handled by the neighboring BSs that are on. On the other hand, dynamic BS switching on/off can help in ensuring power saving of HetNets by reducing the traditional (nonrenewable) power consumption of BSs that have a heavy energy usage mainly during low-traffic period.

Each BS can be set in either of two operational modes: active mode (AM) or sleep mode (SM). The decision to toggle the operational state from one to another is taken centrally (i.e., the decision is taken by some central entity based on the current load offered to the network). In the AM, the BS is serving a certain number of users, thus, the BS-radiated power can be expressed as

$$P_l^{\mathrm{BS}} = \sum_{u=1}^{U_l} P_{l,u}, \tag{11.2}$$

which corresponds to the sum of the radiated power over all users U_l connected to a certain BS l.

In the SM, the BS l consumes power equal to γ_l. The sleep mode is a reduced power consumption state in which the BS in not completely turned off and can be readily activated. Although the BS is not radiating power in this mode, elements such as power supply, baseband digital signal processing, and cooling are still active. Therefore, the BS keeps consuming power unless it is in a state of complete shutdown. For simplicity, the total power consumption of BS l can be approximated by a linear model as follows [40]:

$$P_l = \begin{cases} \alpha_l P_l^{\mathrm{BS}} + \beta_l, & \text{for AM,} \\ \gamma_l, & \text{for SM,} \end{cases} \tag{11.3}$$

where α_l corresponds to the power consumption that scales with the radiated power due to amplifier and feeder losses and β_l models an offset of site power that is consumed independently of the average transmit power.

Let ϵ^b denotes a binary matrix of size $(L+1) \times U$. Its entries $\epsilon_{l,u}^b$ is given by

$$\epsilon_{l,u}^b = \begin{cases} 1, & \text{if user } u \text{ is allocated to BS } l \text{ during time slot } b, \\ 0, & \text{otherwise.} \end{cases} \tag{11.4}$$

On the other hand, a dynamic on/off switching mechanism is considered to turn off redundant MBSs and SBSs whenever it is possible. More specifically, BS l can be turned off during low-traffic periods and the small number of active users are offloaded to nearby BSs. A binary vector π^b of size $L \times 1$ is introduced to indicate the status of each BS l. Its entries π_l^b is given by and is given as

$$\pi_l^b = \begin{cases} 1, & \text{if BS } l \text{ in AM during time slot } b. \\ 0, & \text{otherwise.} \end{cases} \tag{11.5}$$

Note that in order to ensure that the users can not be connected to a BS in the SM, then, the following condition should be respected

$$\epsilon_{l,u}^b \le \pi_l^b, \quad \forall l = 1,.,L, \forall u = 1,.,U, \forall b = 1,.,B. \tag{11.6}$$

The constraint given in Eq. (11.6) enforces $\epsilon_{l,u}^b = 0, \forall u$ when π_l^b in the SM (i.e., $\pi_l^b = 0$). In this chapter, we always keep the macrocell BS active (i.e., $\pi_0^b, \forall b = 1,.., B$) to ensure coverage and minimum connectivity in this typical HetNet (i.e., one macrocell BS surrounded by multiple of MBSs and SBSs). In the case of multiple macrocell BSs covering a bigger geographical area, macrocell BSs could be turned off and cell breathing mechanisms can be employed to ensure connectivity [41].

11.2.2 Energy Harvesting Model

It is assumed that each BS can harvest from RE in both AM and SM. We model the RE stochastic energy arrival rate as a random variable Φ (Watts) defined by a probability density function (pdf) $f(\varphi)$. For example, for photovoltaic energy, Φ can be interpreted as the received amount of energy per time unit with respect to the received luminous intensity in a particular direction per unit solid angle. In general, the energy consumption of the BS l during time slot b can be expressed as

$$E_0^b = T_b \left(\alpha_0 \sum_{u=1}^{U} \epsilon_{0,u}^b P_{0,u} + \beta_0 \right), \quad l = 0. \tag{11.7}$$

$$E_l^b = T_b \left(\pi_l^b \left[\alpha_l \sum_{u=1}^{U} \epsilon_{l,u}^b P_{l,u} + \beta_l \right] + (1 - \pi_l^b)\gamma_l \right), \quad l \geq 1, \tag{11.8}$$

By using Eq. 11.6, we can rewrite Eq. 11.8 as follows

$$E_l^b = T_b \left(\alpha_l \sum_{u=1}^{U} \epsilon_{l,u}^b P_{l,u} + \pi_l^b \beta_l + (1 - \pi_l^b)\gamma_l \right), \quad l \geq 1. \tag{11.9}$$

The harvested energy in BS l and GG at the end of time slot b, are given, respectively, by

$$H_l^b = T_b \eta_l \varphi_l^b, \tag{11.10}$$

$$H_g^b = T_b \eta_g \varphi_g^b, \tag{11.11}$$

where η_l and η_g are the energy conversion efficiency coefficient of the RE at BS l and GG, respectively, where $0 \leq \eta_l, \eta_g \leq 1$. Notice that the current stored energy in BS l and GG depends on both the current harvested energy during slot time b and the previously stored energy during previous slots. Therefore, the stored energy in BS l at the end of time slot b is given by

$$S_l^b = \left[S_l^{b-1} + H_l^b - E_l^b - E_{\text{le}} \right]^+, \tag{11.12}$$

where E_{le} is the leakage energy during T_b.

11.3 Problem Formulation and Solution

In this section, we formulate and solve optimally two problems, based on the knowledge level of the RE generation, aiming to minimize the networks energy consumption during the B time slots. The first optimization problem corresponds to the zero knowledge case where the mobile operator manages its BSs time slot by time slot without any prior information about the future RE generation. The second one corresponds to the perfect knowledge case with full information about the future RE generation where all the decision variables are simultaneously optimized for the B time slots. The perfect knowledge case is a not realistic case. In this study, it is used as a benchmark scenario for comparison with zero knowledge case or as an approximation of the case where RE energy uncertainty is almost negligible. The achievable data rate of user u served by BS l at a given time b is given by

$$R_{l,u}^b = \log_2\left(1 + \frac{P_{l,u} |h_{l,u}^b|^2}{N_0}\right), \tag{11.13}$$

where N_0 is the noise power density.

11.3.1 Zero Knowledge Case

In this case, we assume that the mobile operator is not aware about the future RE generation (i.e., φ_l^b and φ_g^b are known during b only). Therefore, the optimization problem that aims to minimize the total consumed energy from the grid at each time slot b is formulated as follows:

$$\underset{\pi_l^b, \epsilon_{l,u}^b \geq 0}{\text{minimize}} \quad E_c^b = \sum_{l=0}^{L} E_l^b(\pi_l^b, \epsilon_{l,u}^b) - S_l^b(\pi_l^{b-1}, \epsilon_{l,u}^{b-1}), \tag{11.14}$$

subject to:

$$\sum_{u=1}^{U} \epsilon_{l,u}^b P_{l,u} \leq \bar{P}_l, \quad \forall l = 0, .., L, \tag{11.15}$$

$$R_0 \leq \sum_{l=0}^{L} \epsilon_{l,u}^b R_{l,u}^b, \quad \forall u = 1, .., U, \tag{11.16}$$

$$S_l^{b-1}(\pi_l^b, \epsilon_{l,u}^b) + H_l^b \leq \bar{S}_l, \quad \forall l = 0, .., L, \tag{11.17}$$

$$\sum_{u=1}^{U} \epsilon_{l,u}^b \leq \bar{U}_l, \quad \forall l = 0, .., L, \tag{11.18}$$

$$\sum_{l=0}^{L} \epsilon_{l,u}^b \leq 1, \quad \forall u = 1, .., U, \tag{11.19}$$

$$\epsilon_{l,u}^b \leq \pi_l^b, \quad \forall l = 1, .., L, \forall u = 1, .., U, \tag{11.20}$$

where constraint 11.15 and 11.16 represent the maximum allowable transmit energy of BS l and user QoS, respectively. Constraint 11.17 forces the total energy stored in the battery of a BS l during the time slot b to be less than the battery capacity denoted by $\bar{S}_l$. Constraints 11.18 and 11.19 are to satisfy the backhauling condition and to ensure that each user is served by at most one BS, respectively. Notice that, this optimization problem will be solved at the beginning of each time slot. Hence, the optimal solutions for such a problem can be determined using an integer programming interface, for example, Gurobi/CVX [42,43].

11.3.2 Perfect Knowledge Case

In this case, we assume that the mobile operator can perfectly predict the future RE generation ahead of time. This case can be considered as a useful benchmark to compare with the zero knowledge case. Therefore, the objective function becomes the minimization of the total consumed energy from the grid during all B time slots.

$$\underset{\pi_l^b, \epsilon_{l,u}^b \geq 0}{\text{minimize}} \quad E_c = \sum_{b=1}^{B} \sum_{l=0}^{L} E_l^b(\pi_l^b, \epsilon_{l,u}^b) - S_l^b(\pi_l^{b-1}, \epsilon_{l,u}^{b-1}), \tag{11.21}$$

subject to:

$$\sum_{u=1}^{U} \epsilon_{l,u}^b P_{l,u} \leq \bar{P}_l, \quad \forall l = 0, .., L, \forall b = 1, .., B, \tag{11.22}$$

$$\sum_{l=0}^{L} \epsilon_{l,u}^b R_{l,u}^b \geq R_0, \quad \forall u = 1, .., U, \forall b = 1, .., B, \tag{11.23}$$

$$S_l^{b-1}(\pi_l^b, \epsilon_{l,u}^b) + H_l^b \leq \bar{S}_l, \quad \forall l = 0, .., L, \forall b = 1, .., B, \tag{11.24}$$

$$\sum_{u=1}^{U} \epsilon_{l,u}^b \leq \bar{U}_l, \quad \forall l = 0, .., L, \forall b = 1, .., B, \tag{11.25}$$

$$\sum_{l=0}^{L} \epsilon_{l,u}^b \leq 1, \quad \forall u = 1, .., U, \forall b = 1, .., B, \tag{11.26}$$

$$\epsilon_{l,u}^b \leq \pi_l^b, \quad \forall l = 1, .., L, \forall u = 1, .., U, \forall b = 1, .., B. \tag{11.27}$$

Notice that the constraints 11.22–11.27 are similar to the constraints 11.15–11.20 except that they have to be satisfied for all time slots $b = 1, .., B$. The perfect knowledge problem can also be solved using an integer programming solver, for example, Gurobi/CVX interface [42,43].

11.3.3 Cost Utility

After solving the optimization problem, the total cost of the nonrenewable energy consumption is equal to the cost of the energy consumed by all BSs that exceed the available harvested energy stored at time b and given by

$$C^b = \left[\sum_{l=0}^{L} [E_l^b - S_l^{b-1}]^+ - S_g^{b-1} \right]^+, \tag{11.28}$$

where S_g^{b-1} is the stored energy at the GG at the end of time slot $b - 1$. Therefore, the total cost over multiple time slots is given by

$$C = \sum_{b=1}^{B} C^b. \tag{11.29}$$

11.3.4 Special Case

The communication channel is assumed to be a block fading channel with a coherence time T_c second. Therefore, the scheduling and user–cell association can be assumed to be taken over a short timescale. The operational state of the switching on/off of the BSs can be taken over a long timescale, where each long time slot consists of multiple short slots. Hence, the problem can be solved by optimizing only $\epsilon_{l,u}^b$ at the beginning of the short time slot and optimizing both π_l^b and $\epsilon_{l,u}^b$ at the beginning of the long time slot.

11.4   Low-Complexity Algorithm

The formulated BLP optimization problems given in Section 11.3 are considered as NP-hard problem due to the existence of the binary variables; hence, we propose a metaheuristic algorithm, namely, BPSO. Then, we compare its performances with the well-known evolutionary GA [39].

11.4.1   Binary Particle Swarm Optimization (BPSO)

The BPSO starts by generating N particles $\mathcal{N} = [\pi_1^1, .., \pi_L^B, .., \epsilon_{1,1}^1, .., \epsilon_{L,U}^B]$; $n = 1, .., N$ of size $(L + (L + 1)U) \times 1$ for zero knowledge case (solved for each time slot b) and $(LB + (L + 1)UB) \times 1$ for perfect knowledge case to form an initial population S. Then, it determines the minimum energy consumed by each particle that satisfies the QoS by solving the optimization problem. Then, it finds the particle that provides the best solution for this iteration, denoted by $\mathcal{N}^{\text{best}}$. In addition, for each particle n, it saves a record of the position of its previous best performance, denoted by $\mathcal{N}^{(n,\text{local})}$. Then, at each iteration q, BPSO updates its velocity $v_j^{(n)}$ and particle positions $\mathcal{N}_j^{(n)}$, respectively, as follows:

$$
\begin{aligned}
v_{l,b}^{(n)}(q + 1) = & \psi_0 v_{l,b}^{(n)}(q) + \psi_1(q) \left(\mathcal{N}_j^{(n,\text{local})}(q) - \mathcal{N}_j^{(n)}(q) \right) \\
& + \psi_2(q) \left(\mathcal{N}_j^{\max}(q) - \mathcal{N}_j^{(n)}(q) \right),
\end{aligned}
\tag{11.30}
$$

$$
\mathcal{N}_j^{(n)}(q + 1) = \begin{cases} 1, & \text{if } r_{\text{rand}} < \Psi_{\text{BPSO}} \left(v_{l,b}^{(n)}(q + 1) \right), \\ 0, & \text{otherwise,} \end{cases}
\tag{11.31}
$$

where ψ_0 is the inertia weight used to control the convergence speed ($0.8 \leq \psi_0 \leq 1.2$). ψ_1 and ψ_2 are two random positive numbers generated for iteration q ($\psi_1, \psi_2 \in [0, 2]$) [44]. r_{rand} is a pseudorandom number selected from a uniform distribution in [0, 1]. Ψ_{BPSO} is a sigmoid function for transforming the velocity to probabilities and is given as

$$
\Psi_{\text{BPSO}}(x) = \frac{1}{1 + e^{-x}}.
\tag{11.32}
$$

These steps are repeated until reaching convergence by either attaining the maximum number of iterations or stopping the algorithm when no improvement is noticed. Details of the proposed optimization approach based on BPSO are given in Algorithm 1.

Algorithm 1 Proposed solution using BPSO algorithm

1: $q = 1$.

2: Generate an initial population S composed of N random particles $\mathcal{N}^{(n)}$, $n = 1 \cdots N$.

3: **while** not converged **do**

4:    **for** $n = 1 \cdots N$ **do**

5:       Compute the corresponding consumed utility function $E_c^{(n)}(q)$.

6:    **end for**

7:    Find $(n_m, q_m) = \arg\min_{n,q} E_c^{(n)}(q)$ (i.e., n_m and q_m indicate the index and the position of the particle that results in the minimum energy consumption).

8:    Set $E_c^{\text{best}} = E_c^{(n_m)}(q_m)$ and $\mathcal{N}^{\text{best}} = \mathcal{N}^{(n_m)}(q_m)$.

9:    Find $q_n = \arg\min_q E_c^{(n)}(q)$ for each particle n (i.e., q_n indicates the position of the particle n that results in best local utility).

10:   Set $\mathcal{N}^{(n,\text{local})} = \mathcal{N}^{(n)}(q_n)$.

11:   Adjust velocities and positions of all particles using 11.30 and 11.31.

12:   $q = q + 1$.

13: **end while**

11.4.2 Genetic Algorithm (GA)

The performances of the proposed BPSO algorithm is compared to those of the well-know GA. In our genetic-based approach, we generate randomly N particles $\mathcal{N}^{(n)}$, $n = 1, \ldots, N$ of size $(L + (L + 1)U) \times 1$ for zero knowledge case (solved for each time slot b) and $(LB + (L + 1)UB) \times 1$ for perfect knowledge case to form an initial population S. Then, it determines the minimum energy consumed by each particle that satisfies the QoS by solving the optimization problem. After that, the algorithm selects $\tau (1 \leq \tau \leq N)$ strings that provide the minimum consumed energy and keeps them to the next population while the $N - \tau$ remaining strings are generated by applying crossovers and mutations to the τ survived parents as shown in Figure 11.4.

Crossovers consist in cutting two selected random parent strings at a corresponding point that is chosen randomly. The obtained fragments are then swapped and recombined to produce two new strings. Then, mutation (i.e., changing a bit value of the string randomly) is applied with a probability p [45]. This procedure is repeated until reaching convergence or reaching the maximum number of iterations. Details of the proposed optimization approach based on BPSO are given in Algorithm 2.

Algorithm 2 Proposed solution using GA

1: $q = 1$.
2: Generate an initial population S composed of N random particles $\pi^{(n)}$, $n = 1 \cdots N$.
3: **while** not converged **do**
4:    **for** $n = 1 \cdots N$ **do**
5:       Compute the corresponding consumed utility function $E_{\mathrm{c}}^{(n)}(i)$.
6:    **end for**
7:    Find $(n_m, q_m) = \arg\min_{n,q} E_{\mathrm{c}}^{(n)}(q)$ (i.e., n_m and i_m indicate the index and the position of the particle that results in the minimum energy consumption).
8:    Set $E_{\mathrm{c}}^{\mathrm{best}} = E_{\mathrm{c}}^{(n_m)}(q_m)$ and $\mathcal{N}^{\mathrm{best}} = \mathcal{N}^{(n_m)}(q_m)$.
9:    Find $q_n = \arg\min_{q} E_{\mathrm{c}}^{(n)}(q)$ for each particle n (i.e., q_n indicates the position of the particle n that results in best local utility).
10:   Set $\mathcal{N}^{(\mathrm{n,local})} = \mathcal{N}^{(n)}(q_n)$.
11:   Keep the best τ strings providing the highest data rates to the next population.
12:   From the survived τ strings, generate $N - \tau$ new strings by applying crossovers and mutations to generate a new population set.
13:   $q = q + 1$.
14: **end while**

11.5 Simulation Results

In this section, selected numerical results are provided to evaluate the performance of the EH HetNets systems. Selected BSs transmit their messages periodically every T_b sec. All the fading channel gains adopted in the framework are assumed to be i.i.d Rayleigh fading gains. The efficiency transmission and conversion ratios are set to $\eta_l = \eta_g = 0.3$, respectively. The target data rate user (R_0), the number of MBSs and SBSs are 10 bits/s/Hz, 4 and 8, respectively, unless otherwise stated. The noise power is taken to be $N_0 = \bar{N}_0 W$, where $\bar{N}_0 = -174$ dBm/Hz and $W = 180$ KHz. The power consumption parameters are selected according to the energy aware radio and network technologies (EARTH) model for macrocell BS, MBSs, SBSs, are given, respectively [40], as follows: $\alpha_l = \{4.7, 2.6, 4\}$ W and $\beta_l = \{130, 56, 6.8\}$ W. The other power consumption parameters for MBSs and SBSs are given, respectively, by $\gamma_l = \{39, 2.9\}$ W. The maximum transmit power levels for the macrocell BS, MBSs, SBSs, are set, respectively, to $\bar{P}_l = \{46, 38, 20\}$ dBm.

At each BS, RE is assumed to be generated following gamma distributions $\Gamma(20, 2)$, $\Gamma(12, 2)$, and $\Gamma(3, 1)$ for macrocell BS, MBSs, and SBS, respectively, where in $\Gamma(x, y)$, x is the shape parameter and y the scale parameter. While for

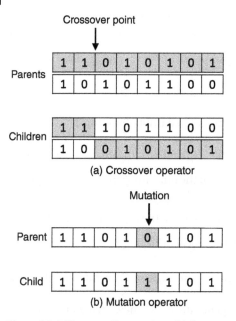

Figure 11.4 Two genetic operators. (a) Crossover operator. (b) Mutation operator.

GG, RE is assumed to be generated following a gamma distribution $\Gamma(25, 2)$. The total stored energy at macrocell BS, MBSs, and SBSs cannot exceed $\bar{S}_l = \{50, 12, 6\}$ kJ, respectively, and the battery leakage is set to be $E_{le} = 1$ mJ every T_b. The BPSO is executed with the following parameters: $N = 20$ and $\psi_0 \in [0, 1]$ is a linear decreasing function of the BPSO iterations expressed as follows: $\psi_0 = 0.9 - \frac{n(0.9-0.2)}{I_{max}}$, where $I_{max} = 200$ is the maximum number of iterations.

Figure 11.5 plots the total average energy cost, which is equal to $\frac{C}{B}$, for $B = 20$ versus number of users ($U^b, \forall b = 1, .., B$), for zero knowledge case. This figure investigates the impact of RE with two scenarios: (1) With the proposed EH (i.e., hybrid of RE and TG energy) and (2) without EH(the energy depends on the TG energy only). It also investigate the impact of the sleeping strategy (i.e., optimizing π) on the system performance. We can see that the proposed scheme (with EH and with sleeping strategy) offers a significant amount of energy saving switching over the other scenarios. It should be noted that the sleeping strategy is very useful specially for low-traffic period with a considerable energy cost gap. Indeed, for $U^b = 100$ users, the average energy cost can be decreased by around 30% for the EH scenario by going from 13.5 kJ to around 9.5 kJ. However, this gap reduces when number of users increases. This can be justified by the fact that when the number of users are relatively high, most of BSs should be in the AM in order to satisfy the user QoS.

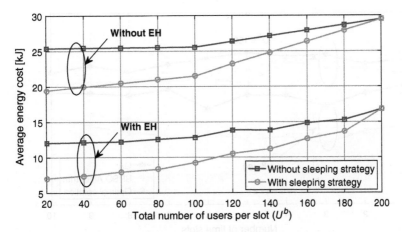

Figure 11.5 Average energy cost of $B = 20$ time slots versus total number of users.

Table 11.1 MBSs and SBSs status during multiple time slots.

Number of Users per b	Active MBSs				Active SBSs			
	m_1	m_2	m_3	m_4	s_1	s_2	s_3	s_4
$U^1 = 100$	×	-	-	×	×	-	×	-
$U^2 = 40$	×	-	-	×	×	-	-	×
$U^3 = 200$	×	×	×	×	×	×	×	-
$U^4 = 80$	-	×	-	×	-	-	×	×
$U^5 = 140$	×	-	-	×	×	×	-	×
$U^6 = 220$	×	×	×	×	×	×	×	×
$U^7 = 80$	×	-	-	×	×	-	×	-
$U^8 = 160$	×	×	×	-	-	×	-	×
$U^9 = 160$	×	×	-	×	×	×	×	-
$U^{10} = 60$	-	-	×	-	×	×	×	×

Table 11.1 confirms the sleeping strategy results in Figure 11.5. In general, it can be noted that activating the MBSs and SBSs essentially depends on the traffic and BS's battery level. For example, as shown in Table 11.1, during low-traffic periods, for example, $b = \{2, 4, 7, 10\}$ (i.e., $U^2 = 40, U^4 = 80, U^7 = 80, U^{10} = 60$), the sleeping strategy activates some of BSs and keeps the others in the SM in order to harvest some energy. On the other hand, when the network is more congested, for example, during slots $b = \{3, 6, 8, 9\}$ (i.e., $U^3 = 200, U^6 = 220, U^8 = 160, U^9 = 160$), most of the BSs are in AM.

Under the same setup of Figure 11.5, Figure 11.6 compares between the optimal solution (obtained by solving the BLP using Gurobi/CVX) with BPSO

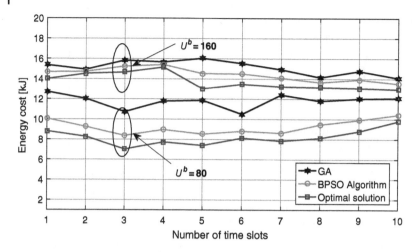

Figure 11.6 Comparison between optimal solution with BPSO algorithm and GA. Energy cost versus number of time slot.

algorithm and the well known GA for different total number of users $U^b = \{80, 160\}$. It can be seen that the BPSO achieves better performance than GA and close to the optimal solution in both low- and high-traffic periods. We can notice that both algorithms are close to the optimal when the network is more congested. This can be explained by knowing that during high-traffic period, the network needs to keep most of the BSs in AM, hence, optimizing only the association variable (i.e., ϵ). It is also worth to note that optimizing π has more weight in saving energy than optimizing ϵ due to the high values of offset power parameter β_l compared to the amplified power parameter α_l.

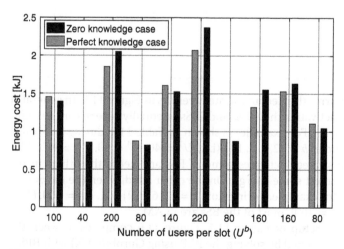

Figure 11.7 Comparison between zero knowledge and perfect knowledge cases.

Finally, Figure 11.7 compares the zero knowledge case to a benchmark case (i.e., perfect knowledge case). Figure 11.7 plots the total energy cost of the network for both cases versus different numbers of users. Since activating the BSs depends on their battery levels and the traffic status, the perfect knowledge case can manage the available resources globally and more efficiently. For example, during $b = 7$ (i.e., $U^7 = 80$), the perfect knowledge case consumes more energy by forcing some BSs to be in SM and activating them where the network is more congested, that is, $U^8 = U^9 = 160$. Although it consumes more energy than the zero knowledge case, which is around 0.1 kJ, when $b = 7$, the perfect knowledge case saves more energy, which is around 0.6 kJ, during the next two time slots $b = 8$ and $b = 9$.

11.6 Chapter Summary

11.6.1 Conclusion

In this chapter, the planning and allocation problem of downlink EH in HetNets using hybrid power sources is proposed. All the BSs are equipped with a harvested source and can get some energy from green grid or/and traditional grid when needed. We formulated a BLP aims to minimize the consumed energy over multiple time slots. The problem was solved optimally and compared with two low-complexity algorithms. After solving the problem, we investigated, via numerical results, the behavior of the proposed scheme versus various system parameters. Finally, the effects of sleeping strategy to the system average energy cost were discussed.

11.6.2 Possible Future Works

11.6.2.1 Massive MIMO

Massive (multi-input multi-output) MIMO technology is introduced recently to improve the system performance and achieve high data rate, where it refers to the idea of equipping cellular BSs with a very large number of antennas. It has been shown to potentially allow for orders of magnitude improvement in spectral and energy efficiency using relatively simple (linear) processing. [46,47].

In massive MIMO each BS is equipped with more than 100 antennas. The channel characteristic between the BSs and the set of active users are quasiorthogonal. Therefore, for the same set of users, the increase of the number of antennas in the BS will help eliminate the noise and the interference. This elimination is caused by the effect that the interference is inversely proportional to the number of antennas in the base station. Moreover, the number of users per cell are independent of the size of the cell, and the required transmitted energy per bit vanishes as the number of antennas in a MIMO cell grows to infinity. Furthermore, simple linear signal processing approaches, such as matched filter (MF) precoding/detection, can be used in massive MIMO systems to achieve these advantages [48].

One of the possible extensions is to formulate a downlink optimization problem for massive MIMO HetNets that aim to maximize the energy efficiency (EE) of the network taking into account the power budget of the BSs, the interference between neighboring BSs, and respecting a QoS for each served user.

11.6.2.2 NOMA

Having its root in multiuser detection (MUD), nonorthogonal multiple access (NOMA) has recently received attention with its ability to multiplex multiple users in the same radio resources in spectral, temporal, or code domains [49]. In the code domain, NOMA can be realized using either sparse or low-density spreading codes. On the other hand, power-domain NOMA serves users at different power levels such that high channel gain users with low-power levels can cancel interfering signals of low channel gain users with high-power levels before decoding their own signal [50]. Therefore, these inherent features of NOMA establishes the fairness among the users either by allowing the interference cancellation or higher power levels. Even though it is an emerging topic with its promising spectral efficiency, EE aspects of NOMA is still an open research in the realm of green communications.

References

1 J. G. Andrews, S. Buzzi, W. Choi, S. V. Hanly, A. Lozano, A. C. K. Soong, and J. C. Zhang, "What will 5G be?," *IEEE J. Select. Areas Commun.*, vol. 32, pp. 1065–1082, 2014.

2 Cisco VNI Mobile, Cisco visual networking index: Global mobile data traffic forecast update, 2016–2021, 2017.

3 H. Bogucka and O. Holland, "Multi-layer approach to future green mobile communications," *IEEE Intell. Transp. Syst. Mag.*, vol. 5, pp. 28–37, 2013.

4 V. Chamola and B. Sikdar, "Solar powered cellular base stations: current scenario, issues and proposed solutions," *IEEE Commun. Mag.*, vol. 54, pp. 108–114, 2016.

5 A. Ghosh, N. Mangalvedhe, R. Ratasuk, B. Mondal, M. Cudak, E. Visotsky, T. A. Thomas, J. G. Andrews, P. Xia, H. S. Jo, H. S. Dhillon, and T. D. Novlan, "Heterogeneous cellular networks: from theory to practice," *IEEE Commun. Mag.*, vol. 50, pp. 54–64, 2012.

6 A. Alsharoa and A. E. Kamal, "Green downlink radio management based cognitive radio LTE HetNets," in *Proc. of the IEEE Global Communications Conference (GLOBECOM)*, San Diego, CA, USA, Dec. 2015, pp. 1–6.

7 E. Yaacoub, "Green communications in LTE networks with environmentally friendly small cell base stations," in *Proc. of the 2012 IEEE Online Conference on Green Communications (GreenCom)*, Sept. 2012, pp. 110–115.

8 C. Liu, B. Natarajan, and H. Xia, "Small cell base station sleep strategies for energy efficiency," *IEEE Trans. Veh. Technol.*, vol. 65, no. 3, pp. 1652–1661, 2016.

9 A. Alsharoa, H. Ghazzai, E. Yaacoub, and M. S. Alouini, "On the dual-decomposition-based resource and power allocation with sleeping strategy for heterogeneous networks," in *Proc. of the 81st Vehicular Technology Conference (VTC Spring), Glasgow, Scotland*, May 2015, pp. 1–5.

10 J. Xu and R. Zhang, "Throughput optimal policies for energy harvesting wireless transmitters with non-ideal circuit power," *IEEE J. Slect. Areas Commun.*, vol. 32, no. 2, pp. 322–332, 2014.

11 H. Tabassum, E. Hossain, A. Ogundipe, and D. I. Kim, "Wireless-powered cellular networks: key challenges and solution techniques," *IEEE Commun. Mag.*, vol. 53, no. 6, pp. 63–71, 2015.

12 V. Raghunathan, S. Ganeriwal, and M. Srivastava, "Emerging techniques for long lived wireless sensor networks," *IEEE Commun. Mag.*, vol. 44, no. 4, pp. 108–114, 2006.

13 IFC, Green power for mobile bi annual report, 10th biannual report, Aug. 2014.

14 Nokia Bell Labs, *F-Cell technology from Nokia Bell Labs revolutionizes small cell deployment by cutting wires, costs and time*, Oct. 2016.

15 A. Celik and A. E. Kamal, "Multi-objective clustering optimization for multi-channel cooperative spectrum sensing in heterogeneous green crns," *IEEE Trans. Cogn. Commun. Netw.*, vol. 2, pp. 150–161, 2016.

16 A. Celik and A. E. Kamal, "Green cooperative spectrum sensing and scheduling in heterogeneous cognitive radio networks," *IEEE Trans. Cogn. Commun. Netw.*, vol. 2, pp. 238–248, 2016.

17 A. Celik, A. Alsharoa, and A. E. Kamal, "Hybrid energy harvesting-based cooperative spectrum sensing and access in heterogeneous cognitive radio networks," *IEEE Trans. Cogn. Commun. Netw.*, vol. 3, pp. 37–48, 2017.

18 G. Fettweis and E. Zimmermann, "ICT energy consumption trends and challenges," in *Proc. of the 11th International Symposium on Wireless Personal Multimedia Communications (WPMC 2008), Oulu, Filand*, Sept. 2008.

19 E. Oh, K. Son, and B. Krishnamachari, "Dynamic base station switching-on/off strategies for green cellular networks," *IEEE Trans. Wirel. Commun.*, vol. 12, no. 5, pp. 2126–2136, 2013.

20 G. P. Koudouridis, H. Gao, and P. Legg, "A centralised approach to power on–off optimisation for heterogeneous networks," in *Proc. of the IEEE Vehicular Technology Conference (VTC Fall 2012), Quebec City, Canada*, Sept. 2012, pp. 1–5.

21 H. Ghazzai, E. Yaacoub, M. S. Alouini, and A. Abu-Dayya, "Optimized smart grid energy procurement for LTE networks using evolutionary algorithms," *IEEE Trans. Veh. Technol.*, vol. 63, no. 9, pp. 4508–4519, 2014.

22 Y. S. Soh, T. Q. S. Quek, M. Kountouris, and H. Shin, "Energy efficient heterogeneous cellular networks," *IEEE J. Select. Areas Commun.*, vol. 31, no. 5, pp. 840–850, 2013.

23 I. Ashraf, F. Boccardi, and L. Ho, "Sleep mode techniques for small cell deployments," *IEEE Commun. Mag.*, vol. 49, no. 8, pp. 72–79, 2011.

24 L. Falconetti, P. Frenger, H. Kallin, and T. Rimhagen, "Energy efficiency in heterogeneous networks," in *IEEE Online Conference on Green Communications (GreenCom' 12)*, Sept. 2012.

25 H. Liang, A. K. Tamang, W. Zhuang, and X. S. Shen, "Stochastic information management in smart grid," *IEEE Commun. Surv. Tutor.*, vol. 16, pp. 1746–1770, 2014.

26 J. Leithon, S. Sun, and T. J. Lim, "Energy management strategies for base stations powered by the smart grid," in *Proc. of the IEEE Global Communications Conference (GLOBECOM), Atlanta, GA, USA*, Dec. 2013, pp. 2635–2640.

27 T. Han and N. Ansari, "On optimizing green energy utilization for cellular networks with hybrid energy supplies," *IEEE Trans. Wirel. Commun.*, vol. 12, pp. 3872–3882, 2013.

28 H. A. H. Hassan, A. Pelov, and L. Nuaymi, "Integrating cellular networks, smart grid, and renewable energy: analysis, architecture, and challenges," *IEEE Access*, vol. 3, pp. 2755–2770, 2015.

29 T. Han and N. Ansari, "Powering mobile networks with green energy," *IEEE Wirel. Commun.*, vol. 21, pp. 90–96, 2014.

30 V. Chamola and B. Sikdar, "Solar powered cellular base stations: current scenario, issues and proposed solutions," *IEEE Commun. Mag.*, vol. 54, pp. 108–114, 2016.

31 J. Leithon, T. J. Lim, and S. Sun, "Online energy management strategies for base stations powered by the smart grid," in *2013 IEEE International Conference on Smart Grid Communications (SmartGridComm)*, Oct. 2013, pp. 199–204.

32 S. Bu, F. R. Yu, Y. Cai, and X. P. Liu, "When the smart grid meets energy-efficient communications: green wireless cellular networks powered by the smart grid," *IEEE Trans. Wirel. Commun.*, vol. 11, pp. 3014–3024, 2012.

33 J. Xu, L. Duan, and R. Zhang, "Cost-aware green cellular networks with energy and communication cooperation," *IEEE Commun. Mag.*, vol. 53, pp. 257–263, 2015.

34 V. K. Bhargava and A. Leon-Garcia, "Green cellular networks: a survey, some research issues and challenges," in *2012 26th Biennial Symposium on Communications (QBSC)*, May 2012, pp. 1–2.

35 Y. Dong, H. Zhang, M. J. Hossain, J. Cheng, and V. C. M. Leung, "Energy efficient resource allocation for OFDMA full duplex distributed antenna systems with

energy recycling," in *2015 IEEE Global Communications Conference (GLOBE-COM), San Diego, CA, USA*, Dec. 2015, pp. 1–6.

36 P. S. Yu, J. Lee, T. Q. S. Quek, and Y. W. P. Hong, "Traffic offloading in heterogeneous networks with energy harvesting personal cells-network throughput and energy efficiency," *IEEE Trans. Wirel. Commun.*, vol. 15, pp. 1146–1161, 2016.

37 M. B. Ghorbel, T. Touzri, B. Hamdaoui, M. Guizani, and B. Khalfi, "Joint user-channel assignment for efficient use of renewable energy in hybrid powered communication systems," in *Proc. of the IEEE Global Communications Conference (GLOBECOM 2015), San Diego, CA, USA*, Dec. 2015, pp. 1–6.

38 M. J. Farooq, H. Ghazzai, A. Kadri, H. ElSawy, and M. S. Alouini, "A hybrid energy sharing framework for green cellular networks," *IEEE Trans. Commun.*, vol. 65, pp. 918–934, 2017.

39 M. Mitchell, *An Introduction to Genetic Algorithms*. Cambridge, MA: MIT Press, 1998.

40 G. Auer, O. Blume, V. Giannini, I. Godor, M. Imran, Y. Jading, E. Katranaras, M. Olsson, D. Sabella, P. Skillermark, et al., "Energy efficiency analysis of the reference systems, areas of improvements and target breakdown," *INFSOICT-247733 EARTH (Energy Aware Radio and NeTwork TecHnologies), Technical report*, Dec. 2010.

41 Z. Hasan, H. Boostanimehr, and V. Bhargava, "Green cellular networks: a survey, some research issues and challenges," *IEEE Commun. Surv. Tutor.*, vol. 13, no. 4, pp. 524–540, 2011.

42 Gurobi optimizer reference manual, 2016. Available at http://www.gurobi.com/.

43 M. Grant and S. Boyd, *CVX: Matlab software for disciplined convex programming, version 2.1 cvxr.com/cvx*, June 2015.

44 J. Kennedy and R. Eberhart, "A discrete binary version of the particle swarm algorithm," in *Proc. of the IEEE International Conference on Systems, Man, and Cybernetics. Computational Cybernetics and Simulation*, vol. 5, Orlando, FL, USA, Oct. 1997, pp. 4104–4108.

45 D. Beasley, D. R. Bull, and R. R. Martin, "An overview of genetic algorithms: part 2, research topics," *Univ. Comput.*, vol. 15, no. 4, pp. 170–181, 1993.

46 Z. Gao, L. Dai, D. Mi, Z. Wang, M. Imran, and M. Shakir, "Mmwave massive-MIMO-based wireless backhaul for the 5G ultra-dense network," *IEEE Wirel. Commun.*, vol. 22, no. 5, pp. 13–21, 2015.

47 B. Panzner, W. Zirwas, S. Dierks, M. Lauridsen, P. Mogensen, K. Pajukoski, and D. Miao, "Deployment and implementation strategies for massive MIMO in 5G," in *Globecom Workshops (GC Wkshps), 2014*, Dec. 2014, pp. 346–351.

48 L. Lu, G. Li, A. Swindlehurst, A. Ashikhmin, and R. Zhang, "An overview of massive MIMO: benefits and challenges," *IEEE J. Select. Top. Signal Process.*, vol. 8, no. 5, pp. 742–758, 2014.

49 L. Dai, B. Wang, Y. Yuan, S. Han, C. l. I, and Z. Wang, "Non-orthogonal multiple access for 5G: solutions, challenges, opportunities, and future research trends," *IEEE Commun. Mag.*, vol. 53, pp. 74–81, 2015.

50 M. S. Ali, H. Tabassum, and E. Hossain, "Dynamic user clustering and power allocation for uplink and downlink non-orthogonal multiple access (NOMA) systems," *IEEE Access*, vol. 4, pp. 6325–6343, 2016.

Ahmad Alsharoa (S'14, M'18) received the B.Sc degree (with honors) from Jordan University of Science and Technology (JUST), Irbid, Jordan, in January 2011 and the M.Sc. degree from King Abdullah University of Science and Technology (KAUST), Thuwal, Saudi Arabia in May 2013, both in electrical engineering. He received the Ph.D. degree with co-majors in electrical engineering and computer engineering from Iowa State University (ISU), Ames, IA, USA, in May 2017.

He was the recipient of the ISC International Scholarships Award, Research Excellence Award, and GPSS Research Award from Iowa State University in 2015, 2016, and 2017, respectively.

He was a visiting researcher at Centre Tecnologic de Telecomunicacions de Catalunya (CTTC), Barcelona, Spain, from May 2012 to August 2012. He was a postdoctoral research associate at ISU from July 2017 to August 2017 before joining University of Central Florida (UCF), Orlando, FL, USA, in September 2017 as a postdoctoral research associate.

His current research interests include: mmWave communications, visible light communications (VLC), UAV-based communications, Internet of Things (IoT), energy harvesting, green wireless communications, cognitive radio networks, cooperative relay networks, self-healing communications, and MIMO communications.

Abdulkadir Celik (S'14-M'16) received the M.S. degree in electrical engineering in 2013, the M.S. degree in computer engineering in 2015, and the Ph.D. degree in co-majors of electrical engineering and computer engineering in 2016, all from Iowa State University, Ames, IA, USA. Dr. Celik is currently a postdoctoral research fellow at Communication Theory Laboratory of King Abdullah University of Science and Technology (KAUST). His current research interests include but not limited to cognitive radio networks, green communications, non-orthogonal multiple access, D2D communications, heterogeneous networks, and optical wireless communications for data centers and underwater sensor networks.

 **Ahmed E. Kamal** (S'82, M'87, SM'91, F'12) is a professor of electrical and computer engineering at Iowa State University in the USA. He received the B.Sc. (distinction with honors) and M.Sc. degrees from Cairo University, Egypt, and the M.A.Sc. and Ph.D. from the University of Toronto, Canada, all in electrical engineering. He is a fellow of the IEEE and a senior member of the Association of Computing Machinery. He was an IEEE Communications Society Distinguished Lecturer for 2013 and 2014. His research interests include cognitive radio networks, optical networks, wireless sensor networks, and performance evaluation. He received the 1993 IEE Hartree Premium for papers published in Computers and Control in IEE Proceedings, and the best paper award of the IEEE Globecom 2008 Symposium on Ad Hoc and Sensors Networks Symposium. He chaired or cochaired Technical Program Committees of several IEEE-sponsored conferences, including the Optical Networks and Systems Symposia of the IEEE Globecom 2007 and 2010, the Cognitive Radio and Networks Symposia of the IEEE Globecom 2012 and 2014, and the Access Systems and Networks track of the IEEE International Conference on Communications 2016. He is on the editorial boards of the *IEEE Communications Surveys and Tutorials*, the *Computer Networks* journal, and the *Optical Switching and Networking* journal.

Part III

5G Network Interworking and Core Network Advancements

12

Characterizing and Learning the Mobile Data Traffic in Cellular Network

Rongpeng Li, Zhifeng Zhao, Chen Qi, and Honggang Zhang

College of Information Science and Electronic Engineering, Zhejiang University, Hangzhou, China

Traffic characterization, learning, and prediction in cellular networks, which is a classical yet still appealing field, yields a significant number of meaningful results. From a macroscopic perspective, it provides the commonly believed result that mobile Internet will witness a thousandfold traffic growth in the next 10 years [1], which is acting as a crucial anchor for the design of next-generation cellular network architecture and embedded algorithms. On the other hand, the fine traffic learning on a daily, hourly, or even minutely basis could contribute to the optimization and management of cellular networks like energy savings [2–4], opportunistic scheduling [5], and network anomaly detection [6]. In other words, the precise knowledge of cellular network traffic plays an important role in future cellular networks. In this chapter, we take advantage of a large amount of practical traffic records from one of the largest operators in China and start with the characterization over cellular network traffic. Afterward, we delve into the traffic predictability and exploit the information theory. Finally, we talk about one learning framework to exploit the aforementioned traffic characteristics and yield appealing prediction accuracy.

12.1 Understanding the Traffic Nature: A Revisiting to α-Stable Models

Instantaneous messaging (IM) services, longly running on PC platforms for personal and business communications, have recently flourished in mobile devices and quickly generated significant amount of traffic loads within cellular

5G Networks: Fundamental Requirements, Enabling Technologies, and Operations Management, First Edition.
Anwer Al-Dulaimi, Xianbin Wang, and Chih-Lin I.

networks. However, compared with traditional voice and messaging services within cellular networks, these newly emerging mobile IM (MIM) services distinguish themselves with the inborn packet switching nature and its accompanied keep-alive (KA) mechanisms, which imply to consume only a small amount of core network bandwidth but considerable radio resources of mobile access networks. Moreover, due to the KA mechanism to keep mobile users in touch with servers, MIM services fundamentally affect the stabilization and reliability of cellular networks [7,8], and could become a huge burden on the network operators [9]. Therefore, it is meaningful to carefully examine the traffic nature of MIM services, so as to design MIM service-oriented protocols to overcome their induced negative influence to cellular networks. In this section, we start with the characteristic analyses over the widely booming MIM service "WeChat/Weixin," which allows over six hundred million mobile users to exchange text messages and multimedia files like voices, pictures, and videos with each other via smartphones [10], in China as well as around the world.

In this section, we make an intensive study on the fundamental traffic nature of MIM service through a large amount of "Wechat/Weixin" traffic observations from operating cellular networks. Based on these practical measurements, we aim to find the precise model for the traffic of MIM services, and try to explain the reasons behind the somewhat conflicting results in previous studies. Interestingly, our measurements reveal that the distribution of individual message's length and inter arrival time could be better fitted using a power-law distribution and a lognormal distribution, which are completely different from the recommended models in 3GPP [11]. Instead, the aggregated traffic of these individual messages in a specific base station (BS) of cellular access networks obeys α-stable models, which usually characterize the statistical patterns of the summation of lots of independently identically distributed random variables [12]. Besides, we build up a theoretical explanation to the evolution from power-law distributed individual message length to aggregated traffic within one BS. In a word, this section contributes to the comprehensive understanding of traffic nature of MIM services, by analyzing practical traffic records of MIM services and further building models for traffic characteristics. Consequently, the related research results are able to benefit the MIM services-related traffic prediction and network protocol design.

12.1.1 MIM Working Mechanisms and Dataset Description

MIM services, which solely rely on mobile Internet to exchange information, have quite distinct working mechanisms from traditional short messaging services. One of the prominent differences is that born with standard protocols [13], traditional short messaging services could leverage well-designed mobility management schemes to conveniently fulfill timely information delivery and provision of "always-online" service. On the contrary, MIM services completely

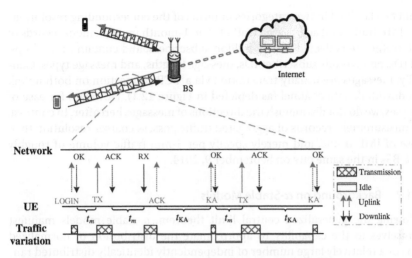

Figure 12.1 An illustration of mobile instantaneous messaging activities.

rely on the transmission control protocol/Internet protocol (TCP/IP) to send or receive text, voice, and multimedia content. However, for mobile Internet in packet switching domain, a TCP connection would release itself if exceeding a TCP inactivity timer. Therefore, as depicted in Figure 12.1, besides transmitting (TX) and receiving (RX) normal packets after logging onto a server, MIM services commonly take advantage of keep-alive mechanisms to send packets containing little information periodically and maintain a long-lived TCP connection. In particular, once the user equipment (UE) logins the network and successfully transmits messages, the network has to acknowledge the message using the OK packet. Similarly, when one UE successfully receives some messages, the UE also needs to acknowledge the successful delivery. Besides, when there exists no packet transmission between the UE and the network, the UE will periodically trigger some keep-alive (KA) message to maintain the TCP connection and the network will respond the OK packet. Hereinafter, a *message* refers to a series of packets transmitted between the user equipment (UE) and the servers of service provider on application layer. Therefore, the messages delivered on every TCP connection constitute the fundamental elements of MIM services, and are named as *individual message level* (IML) traffic in this chapter. Comparatively, when the messages are transmitted through one BS, they become accumulated and could be regarded as the *aggregated traffic* from a slightly more macroscopical perspective.

In order to build primary models, we collect measurements of the MIM traffic from the on-operating cellular networks. Our datasets collected from the Gb and Gn interfaces [9], covering about 15000 GSM and UMTS BSs of China Mobile in an eastern provincial capital within a region of 3000 km², could be classified in two categories in terms of the corresponding resolutions (i.e., IML traffic and aggregated traffic). The 1-month measurement records of IML traffic are collected from 7 million subscribers, and contain timestamps, cell IDs, anonymous subscriber IDs, message lengths, and message types. Generally, messages are usually transmitted via a TCP connection on both uplink and downlink data channel (as depicted in Figure 12.1). However, for ease of analyses, we do not distinguish the directions of messages hereafter. In contrast, the measurement records of aggregated traffic possess coarser resolution than those of IML traffic, and merely specify per 5-min traffic volume of roughly 6000 BSs in the same city on September 9, 2014.

12.1.2 Background on α-Stable Models

Following the generalized central limit theorem, α-stable models manifest themselves in the capability to approximate the distribution of normalized sums of a relatively large number of independently identically distributed random variables [12]. Besides, α-stable models[1] produce strong bursty results with properties of heavy-tailed distributions and long-range dependence. Therefore, they arose in a natural way to characterize the traffic in fixed broadband networks [14,15] and have been exploited in resource management analyses [16,17].

α-Stable models, with few exceptions, lack a closed-form expression of the PDF (probability distribution function), and are generally specified by their characteristic functions.

Definition 12.1   A random variable X is said to obey α-stable models if there are parameters $0 < \alpha \leq 2$, $\sigma \geq 0$, $-1 \leq \beta \leq 1$, and $\mu \in \mathcal{R}$ such that its characteristic function is of the following form:

$$
\begin{aligned}
\Phi(\omega) &= E(\exp j\omega X) \\
&= \begin{cases} \exp\left\{-\sigma|\omega|\left(1 + j\frac{2\beta}{\pi}(\mathrm{sgn}(\omega))\ln|\omega|\right) + j\mu\omega\right\}, \alpha = 1 \\ \exp\left\{-\sigma^{\alpha}|\omega|^{\alpha}\left(1 - j\beta(\mathrm{sgn}(\omega))\tan\frac{\pi\alpha}{2}\right) + j\mu\omega\right\}, \alpha \neq 1. \end{cases}
\end{aligned} \tag{12.1}
$$

Here, the function $E(\cdot)$ represents the expectation operation with respect to a random variable. α is called the characteristic exponent and indicates the index of stability, while β is identified as the skewness parameter. α and β together determine the shape of the models. Moreover, σ and μ are called scale and shift parameters, respectively. Specifically, if $\alpha = 2$, α-stable models reduce to

---

1 In this section, the term α-stable models is interchangeable with α-stable distributions.

Gaussian distributions. Furthermore, for an α-stable-modeled random variable X, there exists a linear relationship between the parameter α and the function $\Psi(\omega) = \ln\{-\text{Re}\,[\ln\,(\Phi(\omega))]\}$ as

$$\Psi(\omega) = \ln\{-\text{Re}\,[\ln\,(\Phi(\omega))]\} = \alpha\ln(\omega) + \alpha\ln(\sigma), \qquad (12.2)$$

where the function $\text{Re}(\cdot)$ calculates the real part of the input variable.

Usually, it is challenging to prove whether a dataset follows a specific distribution, especially for α-stable models without a closed-form expression for their PDF. Therefore, when a dataset is said to satisfy α-stable models, it usually means the dataset is consistent with the hypothetical distribution and the corresponding properties. In other words, the validation needs to first estimate parameters of α-stable models from the given dataset, and then compare the real distribution of the dataset with the estimated α-stable model [14]. Specifically, the corresponding parameters in α-stable models can be determined by quantile methods or sample characteristic function methods [14,15].

12.1.3 The Statistical Pattern and Inherited Methodology of MIM Services

12.1.3.1 IML Traffic

In order to understand the IML traffic nature of MIM services, we first calculate the PDF of message length, and then fitting them to common heavy-tailed distributions listed in Table 12.1. Specifically, during the fitting procedures, we obtain the unknown parameters in candidate distribution functions (except α-stable models) using maximum likelihood estimation (MLE) methodology. For α-stable models, we estimate the relevant parameters using quantile methods [18], correspondingly build the models to generate some random variable, and finally compare its induced PDF with the exact (empirical) one.

In Figure 12.2, we provide the corresponding results after fitting candidate distribution functions to the empirical PDF of message length v_i. Recalling the statements in Section 12.1.2, if the simulated dataset generated by one distribution has the same or approximately same PDF as the real one, the distribution of empirical dataset could be coarsely determined. Interestingly, Figure 12.2 demonstrates that instead of geometric distribution function recommended by 3GPP [11] power-law distribution (i.e., $0.347x^{-2.407}$) could most accurately approximate the empirical PDF of message length. Furthermore, a root mean square error (RMSE)[2], a larger one of which reflects a lower degree of fitting

---

2 Our previous study in Reference [13] provides a similar fitting preciseness result after performing an Akaike information criterion (AIC) test [19]. However, AIC test is not applicable for α-stable models. Here, in consideration of its simplicity, RMSE is exploited as the uniform criterion for gauging the fitting preciseness. In Section 12.1.4, the Kolmogorov–Smirnov (K-S) goodness-of-fit (GoF) test [20,21] is further exploited to examine the fitting accuracy.

Table 12.1 The accuracy measured by RMSE after fitting empirical data to candidate distributions.

Distribution	PDF	RMSE		
		$\overline{W}_{v_i}$	W_t	$\overline{W}_{v_a}$
Power-law	ax^{-b}	9.76e-5	9.25e-5	0.0357
Geometric	$(1-a)^x a$	607e-5	48.0e-5	0.0258
Exponential	ae^{-bx}	56.0e-5	22.9e-5	0.0899
Weibull	$abx^{b-1}e^{-ax^b}$	65.8e-5	8.08e-5	0.0470
Lognormal	$1/(\sqrt{2\pi}bx)\exp(-\frac{(\ln x-a)^2}{2b^2})$	34.0e-5	7.44e-5	0.0491
α —Stable	—	790e-5	170e-5	0.0144

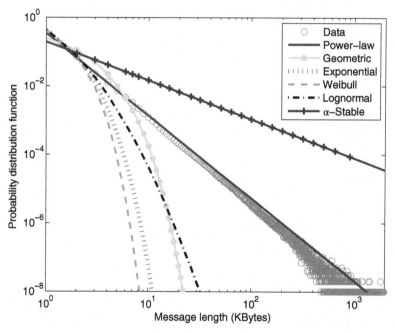

Figure 12.2 The fitting results of MIM activities' message length by candidate distribution functions.

preciseness, is also applied to quantitatively find the fittest distribution function. The RMSE results in column W_{v_i} of Table 12.1 also show that the PDF of message length is most appropriately to be modeled by power-law distribution function.

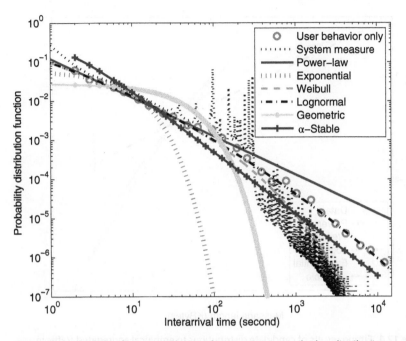

Figure 12.3 The PDF of inter arrival time of MIM messages and other distribution functions with MLE parameters estimation.

On the other hand, according to the timestamps of messages, we calculate inter arrival time t between consecutive messages on the order of seconds, and examine the fitting preciseness of MLE-estimated candidate distribution functions to the corresponding PDF. Figure 12.3 depicts the related fitting results compared to the empirical data (see legend: User behavior only) with the inter arrival time from 2nd to 3000th second. Similarly, column W_t in Table 12.1 shows the results in terms of RMSE. Compared with the exponential distribution function recommended in Reference [11], Figure 12.3 and Table 12.1 show lognormal distribution function (i.e., $\frac{1}{\sqrt{2\pi} \times 2.975x} e^{-\frac{(\ln x - 2.36)^2}{2 \times 2.975^2}}$) exhibits superior fitting preciseness for the inter arrival time of MIM messages. Notably, Figure 12.3 shows that consecutive packets arrive with very small inter arrival time, the probability for which more than 20 s is smaller than 1%. Meanwhile, the peaks in Figure 12.3 with the legend: system measure are incurred by KA messages from various UE's operating systems and diverse MIM versions. Usually, KA cycles are quite different but appear at multiples of 30 s. Moreover, the PDF of the inter arrival time decreases sharply when t_m is larger than the maximal KA (300 s) and the percentage of messages with inter arrival time larger than the maximal KA period accounts for only 2%, implying that MIM

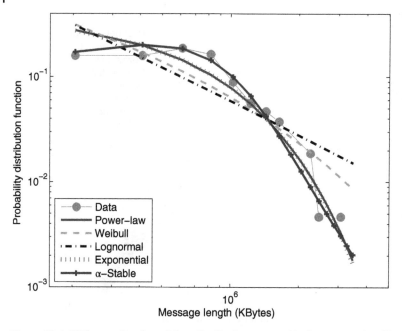

Figure 12.4 Fitting results of candidate distributions to empirical aggregated traffic in one randomly selected BS.

has to send the KA messages if it has been out of touch with the servers for KA period unless the connection to the network is abnormal.

12.1.3.2 Aggregated Traffic

In this part, from the perspective of one whole BS, we examine the fitting results of aggregated traffic within one BS to candidate distributions. Figure 12.4 presents the corresponding PDF comparison between the simulated results and the real aggregated traffic in one randomly selected BS. By taking advantage of a similar methodology to analyze IML traffic, Figure 12.4 implies the traffic records in these selected areas could be better simulated by α-stable models. Similarly, column W_{v_a} in Table 12.1 shows that α-stable models lead to better fitting accuracy in terms of RMSE. Furthermore, Figure 12.5a–d verifies the fitting preciseness of empirical data to α-stable models in another four randomly selected BSs. Figure 12.5e shows the cumulative distribution function (CDF) of preciseness error for all the cells after fitting $\Psi(\omega)$ with respect to $\ln(\omega)$ to a linear function, and demonstrates that there merely exists minor fitting errors. In other words, according to the statements in Section 12.1.2, Figure 12.5e implies that the aggregated traffic possesses the property of α-stable models. Given the previous results, it safely comes to the following conclusion: α-stable

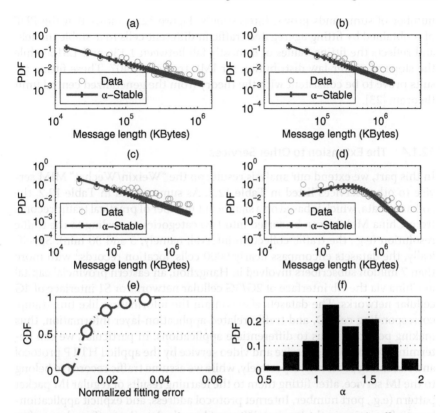

Figure 12.5 (a–d): Fitting results of α-stable models to empirical aggregated traffic in another two randomly selected BSs. (e) The preciseness error CDF for all the cells after fitting $\Psi(\omega)$ with respect to $\ln(\omega)$ to a linear function. (f) The PDF of α estimated for aggregated traffic in different cells.

models are proven to accurately model the aggregated traffic from cellular access networks to core networks.

On one hand, the universal existence of α-stable models implies and contributes to understanding the intrinsic self-similarity feature in MIM traffic [22]. On the other hand, the reasons that MIM traffic universally obeys α-stable models can be explained as follows. Previously, we unveiled that the length of one individual MIM message follows a power-law distribution. Meanwhile, the distribution of aggregated traffic within one BS can be regarded as the accumulation of lots of IM messages from diverse UEs. Moreover, the analysis results of inter arrival time imply frequent packet transmission. Therefore, according to the generalized central limit theorem [23], the sum of a number of random variables with power-law distributions decreasing as $|x|^{-\alpha-1}$ where $0 < \alpha < 2$ (and therefore having infinite variance) will tend to be an α-stable model as the number of summands grows. Interestingly, Figure 12.5f shows that the PDF of α obtained by fitting aggregated traffic in different cells to α-stable models, and reflects the fitting values of α mostly fall between 1.136 and 1.515, while the slope of power-law distribution for IML traffic is 2.407. These fitting results prove to be consistent with the theory from the generalized central limit theorem [23].

12.1.4 The Extension to Other Services

In this part, we extend our analysis results on the "Weixin/Wechat" MIM service to other services listed in Table 12.2. As summarized in Table 12.2, the collected data, which is based on a significant number of practical traffic records from China Mobile, can be sorted into two categories according to the traffic resolution (e.g., the traffic collection interval, namely, 5 and 30 min). Specifically, the datasets encompass nearly 6000 cells' location records[3] with more than 7 million subscribers involved in Hangzhou, an eastern provincial capital in China via the Gb interface of 2G/3G cellular networks or S1 interface of 4G cellular networks. The datasets also contain the information like timestamp, corresponding cell ID, and traffic-related application-layer information, thus making possible for us to differentiate applications. In particular, we can determine web browsing service and video service by the applied HTTP protocol and streaming protocol, respectively, while we assume traffic records to belong to the IM service, after fitting them to the learning results of regular IM packet pattern (e.g., port number, Internet protocol address, and explicit application-layer information in the header). We can also calculate the traffic volume after aggregating packets to each influx base station. Finally, the traffic records in Dataset 1 are collected on September 9, 2014 with Weixin/Wechat, HTTP Web Browsing, and QQLive Video[4] selected as the representatives of these three service types, while those in Dataset 2 are from July 14 to July 27, 2014 with QQ[5], HTTP Web Browsing, and QQLive Video as the representatives, respectively.

Based on the datasets in Table 12.2, Figure 12.6 illustrates the traffic variations generated by these applications in the randomly selected cells. Indeed, the phenomena in Figure 12.6 universally exist in other individual cells, that is, different services exhibit distinct traffic characteristics. IM and HTTP web browsing services frequently produce traffic loads; while distinct from them,

---

3 Indeed, at one specific location, there might exist several cells operating on different frequencies or modes. For simplicity of representation, in the following analyses, we merge the information for different cells at the same location into one.

4 QQLive Video is a popular live streaming video platform in China.

5 QQ is another instant messaging service developed by Tencent Inc. with more than 800 million active users. Due to some practical reasons, per 30-minute Weixin traffic records are unavailable.

Table 12.2 Traffic dataset description for characteristics analyses.

	MIM (Weixin)	Web Browsing (HTTP)	Video (QQLive)
Dataset 1 Under Study			
Traffic resolution (collection interval)	5 min	5 min	5 min
Duration	1 day	1 day	1 day
No. of active cells[6]	2292	4507	4472
Location info. (latitude and longitude)	Yes	Yes	Yes
Dataset 2 Under Study			
	MIM (QQ)	Web Browsing (HTTP)	Video (QQLive)
Traffic resolution (collection interval)	30 min	30 min	30 min
Duration	2 weeks	2 weeks	2 weeks
No. of active cells	5868	5984	5906
Location info. (latitude and longitude)	Yes	Yes	Yes

Table 12.3 The parameter fitting results in the α-stable models.

Name	Parameters				K–S Test	
	α	β	σ	μ	GoF	95% Thres.
Dataset 1 Under Study						
MIM	1.61	1	188.67	221.83	0.0576	0.0800
Web browsing	1.60	1	32.33	42.75	0.0434	0.0800
Video	0.51	1	1×10^{-10}	0	0.0382	0.0800
Dataset 2 Under Study						
MIM	0.70	1	26.32	−100.69	0.0483	0.0524
Web browsing	2	1	2.03×10^3	2.01×10^3	0.0504	0.0524
Video	0.51	1	136.52	−341.15	0.0237	0.0524

video service with more sporadic activities may generate more significant traffic loads.

In this section, we examine the results of fitting the application-level dataset to α-stable models. First, in Figure 12.6, we list the parameter fitting results using quantile methods [18], when we take into consideration the traffic records in three randomly selected cells (each for one service type in one dataset) and quantize the volume of each traffic vector into 100 parts.

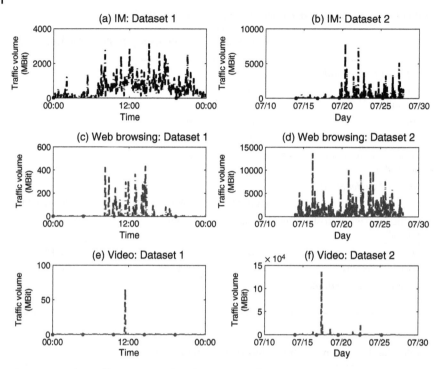

Figure 12.6 The traffic variations of applications in different service types in the randomly selected (single) cells.

Afterward, we use the α-stable models, produced by the aforementioned estimated parameters, to generate some random variable and compare the induced quantized cumulative distribution function (CDF) with the real quantized one. Figure 12.7 presents the corresponding comparison between the simulated results and the real ones. Notably, due to the quantization in the real and estimated CDF, if the PDF at the first quantized value does not equal to zero (e.g., 0.3 for Figure 12.7a and Figure 12.7b), the corresponding CDF will start from a positive value. As stated in Section 12.1.2, if the simulated dataset has the same or approximately same distribution as the real one, the empirical dataset could be deemed as α-stable modeled. Therefore, Figure 12.7 indicates that the traffic records in these selected areas could be simulated by α-stable models. We also perform the Kolmogorov–Smirnov (K–S) goodness-of-fit (GoF) test [20,21]. Table 12.3 summarizes the K–S GoF values and the 95% confidence thresholds. From the table, it is observed that the GoF values are smaller than the thresholds, which further validates the conclusion drawn from Figure 12.7.

On the other hand, recalling the statements in Section 12.1.2, for an α-stable-modeled random variable X, there exists a linear relationship between

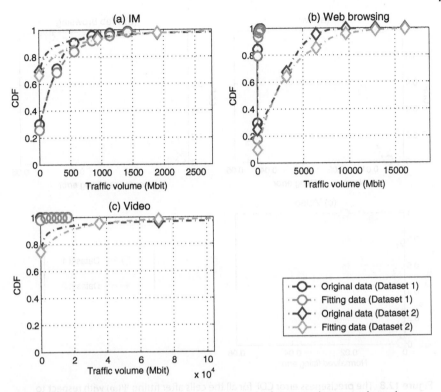

Figure 12.7 For different service types, α-stable model fitting results versus the real (empirical) ones in terms of the CDF.

the parameter α and the function $\Psi(\omega) = \ln\{-\text{Re}\,[\ln(\Phi(\omega))]\}$. Thus, we fit the estimated parameter α with the computing function $\Psi(\omega)$ and provide the preciseness error CDF for all the cells in Figure 12.8. According to Figure 12.8, the normalized fitting errors for 80% cells in both datasets are less than 0.02. Therefore, the practical application-level traffic records follow the property of α-stable models (in Eq. 12.2) and further enhances the validation results as in Figure 12.7. Moreover, different application-level traffic exhibits different fitting accuracy. In that regard, the video traffic in Figure 12.8c has the minimal fitting error, while the fitting error of the web browsing traffic in Figure 12.8b is the largest. But, the fitting error quickly decreases along with the increase in traffic resolution, since a larger traffic resolution means a confluence of more application-level traffic packets and could better demonstrate the accumulative property of α-stable models. In a word, due to their generality, α-stable models are suitable to characterize the application-level traffic loads in cellular networks, even though it might not be the most accurate one. In fact, the universal existence of α-stable models also implies the self-similarity of application-level

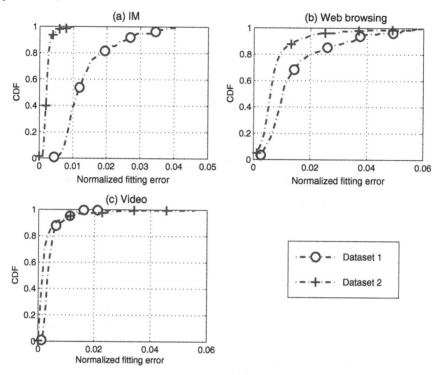

Figure 12.8 The preciseness error CDF for all the cells after fitting $\Psi(\omega)$ with respect to $\ln(\omega)$ to a linear function.

traffic [22]. Hence, in the following sections, it is sufficient to only present and discuss the results from Dataset 1 in Table 12.2.

Additionally, data traffic in wired broadband networks [24] and voice and text traffic in circuit switching domain of cellular networks [2] prove to possess the spatio temporal sparsity characteristic. Indeed, the application-level traffic spatially possesses this sparse property as well. Figure 12.9 depicts the traffic density at 10a.m. and 4p.m. in randomly selected dense urban areas. Here, the traffic density is achieved by dividing the cell traffic of each BS by the corresponding Voronoi cell area [25]. When the derived traffic density in one cell is comparatively larger than that in others, it is depicted as a red "hot spot." As shown in Figure 12.9, there appears a limited number of traffic hot spots and the number of "hot spots" change in both temporal and spatial domain. This spatially clustering property is also consistent with the findings in Reference [26] and proves the traffic's spatial sparsity. It can also be observed that the locations of "hot spots" are also service specific. In other words, different services have distinct requirements on bandwidth, thus leading to various types

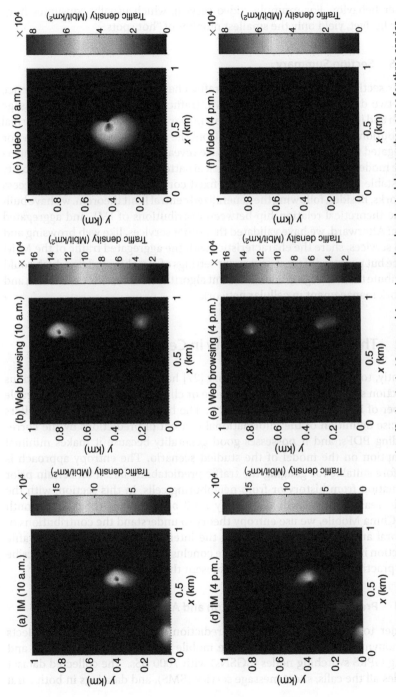

Figure 12.9 The application-level cellular network traffic density in 10a.m. and 4p.m. in randomly selected dense urban areas for three service types of applications. The area for IM, Web browsing, and Video contains 23, 39, and 35 active cells, respectively.

of user behavior. For example, video service, which usually consumes huge traffic budget, yield only the smallest number of "hot spots."

12.1.5 Section Summary

In this section, we have investigated traffic characteristics of the MIM service from two different viewpoints. For IML traffic, we have shown that message length and inter arrival time better follow power-law distribution and lognormal distribution, which are quite different from the recommendation by 3GPP. For aggregated traffic within one BS, we have revealed the accuracy of applying α-stable models to characterize this statistical pattern, and extended the suitability of α-stable models for traffic in both fixed core networks and cellular access networks. Besides, following the generalized central limit theorem, we have built up the theoretical relationship between distributions of IML and aggregated traffic. Afterward, we have validated that other services, like web browsing and video services, share the characteristics with the aggregated traffic of the MIM service but possess different parameter settings of α-stable models, which could contribute to the design of more efficient algorithms for resource allocation and network management in cellular networks.

12.2 The Traffic Predictability in Cellular Networks

Recently, tools from information theory [27] have been introduced in various prediction scenarios, such as atmosphere or climate, and given a considerable number of intuitive conclusions [28,29]. The basic idea is that *entropy* offers a precise definition of the informational content of predictions by the corresponding PDFs; and it possesses good generality because it makes minimal assumption on the model of the studied scenario. The entropy approach is therefore suitable for gauging the traffic predictability based on certain prior information from history or from neighboring cells. In this section, with the help of real traffic records of roughly 7000 base stations (BSs) in 1 month from China Mobile, we use entropy theory to understand the contributions of temporal and spatial dimensions and the inter service relationship to traffic prediction in CRANs and provide some conclusions. Furthermore, we describe some practical prediction means and present the relevant performance.

12.2.1 Prediction Dataset Description and Analysis Methodology

In order to smoothly perform the prediction analysis, this section collects the anonymous traffic records of nine mobile switching centers (MSCs) and serving GPRS switching nodes (SGSNs) with 7000 BSs. The collected dataset includes all the calls, short message service (SMS), and data logs in both rural

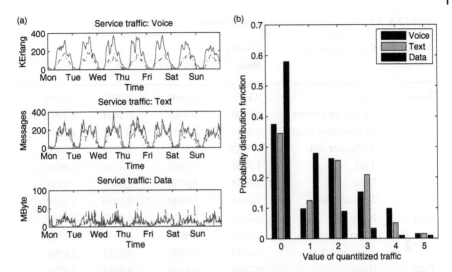

Figure 12.10 Two typical cells' traffic in one week with different entropies. The random entropies of voice, text, and data for the cell in the solid line are 2.4034, 2.1177, and 2.0415 while the counterparts for the cell in the dash line are 2.1834.

and urban areas of around 780 km^2, serving about 3 million subscribers. The duration of dataset spans from March 2012 to April 2012. The dataset also contains the fields such as timestamp and cell ID to record when and where one call/SMS/session appears. For the services of voice and data, call duration and transmitted volume are also incorporated in each record.

After obtaining the massive dataset, a preprocessing procedure is conducted to sort the traffic records by time and cell ID in the first place, and then compute voice, text, and data traffic according to the number of voice (e.g., calls), the count of texts (e.g., SMSs), and the volume of transmitted data during a certain period (e.g., 30 min) within the same cell, respectively. To ease the following analysis, the traffic during a certain period i within a cell is quantized into Q levels, based on $\text{QuantitizedTraffic}(i) = Q \times \frac{\text{RealTraffic}(i) - \text{MinimumTraffic}}{\text{MaximumTraffic} - \text{MinimumTraffic}}$. Here, the MaximumTraffic and MinimumTraffic are the maximal and minimal traffic values of the cell in the time span of interest. It should be noted here that the quantization error will affect the calculation accuracy of random entropy at some extent, but it would be reduced when we later calculate the conditional entropy using the operation of subtraction. Thus, with the quantized traffic values of every period within each cell, the corresponding traffic distributions can be obtained. For example, Figure 12.10b depicts the PDF in one cell with respect to the quantized traffic and shows that both voice and text traffic are medium while data traffic is lower.

Table 12.4 A brief summarization of entropy values.

Service	Conditioned on	Entropy			
		Mean	Variance	Min	Max
Voice	None	2.1429	0.0680	1.0936	2.4539
	2 preceding hours	0.5880	0.0107	0.2051	0.7982
	12 preceding hours	0.0228	0.0013	0	0.1399
	3 adjacent cells	0.9043	0.0581	0.4517	1.7000
	Text	1.3966	0.0851	0.5794	1.9836
	Data	2.0068	0.0730	1.0841	2.4142
	Text and data	1.2956	0.0920	0.5030	1.9339
Text	None	1.8796	0.1552	0.3053	2.4572
	12 preceding hours	0.0353	0.0023	0	0.1775
	Voice and data	1.0549	0.0833	0.2743	1.6790
Data	None	0.7785	0.2436	0.0283	2.0535
	12 preceding hours	0.0845	0.0037	0	0.1806
	Voice and text	0.5640	0.1242	0.0060	1.5095

The entropy, which Shannon utilizes to measure the uncertainty of events [27], is defined by a discrete random variable X with possible values $\{x_1, \ldots, x_n\}$ and the corresponding PDF $P\left(X = x_i\right) = p(x_i)$,

$$H(X) = -\sum_{i=1}^{n} p(x_i) \log_b p(x_i), \tag{12.3}$$

where b is the base of the logarithm and commonly takes the value of 2, when the unit of entropy is 1 bit. According to this definition of entropy, traffic distributions, which heavily depend on the specific location characteristics (e.g., residential or central business districts) and the related user behaviors, lead to distinct entropy values. Therefore, it is possible to use entropy to describe the uncertainty of traffic in CRANs. For example, Figure 12.10a depicts the traffic variations of two typical cells in one week. For the cell with the dash line, the entropies for voice, text, and data traffic are 2.1834, 2.1733, and 1.5472, respectively. In contrast, the other cell with the solid line has a comparatively larger data traffic entropy of 2.0415, which implies a more volatile traffic variation in this cell.

Meanwhile, the traffic variations illustrated in Figure 12.10a implies that the traffic of the voice service varies more dramatically than that of the others. Indeed, this phenomenon applies for other cells. Figure 12.11 plots the cumulative distribution functions (CDFs) with respect to the traffic entropies in the

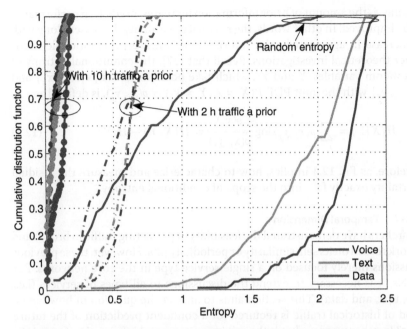

Figure 12.11 The cumulative density functions with respect to the corresponding entropies of traffic under the cases: no prior information, 2 h information, and 10 h information.

cells. Besides, Table 12.4 lists a brief summarization of the entropy calculation results. Both of them express that among all the services, voice traffic has the largest entropy with mean of 2.1429 and minimum of 1.0936, and thus its traffic distribution is more uniform. Comparatively, data traffic is relatively stable since the entropy of data traffic is the smallest and it is lower than 1 bit for more than 70% of cells.

Similarly, traffic predictability can be examined from different perspectives if we depend on the following different prior information cases and calculate the conditional entropies:

1) Temporal conditional entropy of traffic based on a certain preceding duration
2) Spatial conditional entropy of traffic based on adjacent cells
3) A specific service conditional entropy based on traffic of other services

Section 12.2.2 will address these in detail.

12.2.2 Prediction Analysis: To What Extent Is the Prior Information Required?

Traffic prediction relies on the periodical similarity of the traffic itself and requires a certain quantity of prior information to reduce uncertainty. Generally speaking, as the amount of prior information increases, prediction performance will be improved. In other words, the residual uncertainty decreases along with the increase in aggregated prior information. However, this rationale needs further theoretical investigations. Recall that [27], the conditional entropy of two random variables X and Y, which take possible values $\{x_1, \ldots, x_n\}$ and $\{y_1, \ldots, y_n\}$ with the joint PDF $P(X = x_i, Y = y_j) = p(x_i, y_j)$, is defined as

$$H(X|Y) = \sum_{i,j} p(x_i, y_j) \log \frac{p(y_j)}{p(x_i, y_j)} = H(X, Y) - H(Y). \tag{12.4}$$

Therefore, as Eq. 12.4 implies, how to characterize and measure the residual uncertainty exactly falls into the scope of conditional entropy.

12.2.2.1 Temporal Dimension

Researchers have demonstrated the feasibility of predicting traffic in broadband networks based on its self-similarity or periodicity [30]. However, these previous discussions merely focused on a single service type in the core networks. For CRANs, it is necessary to extend the discussions to all types of services (i.e., voice, text, and data). This section aims to answer the question of how long a period of historical traffic is required for a confident prediction of the future traffic? In other words, what is the minimum temporal information for a given conditional entropy constraint?

Figure 12.11 plots the CDFs of the conditional entropies in the cells with the previous 2 and 10 h traffic information. As it demonstrates, by introducing preceding traffic information, the uncertainty can be reduced effectively. Moreover, with equal preceding hours of traffic information, the conditional entropies of voice traffic decrease most rapidly, even though the random entropy of voice is comparatively large. For example, voice's temporal conditional entropy with 2 h historical information decreases from 2.1429 to 0.5880 bit on average, which indicates that it would become much easier to trace the variation of voice traffic, given 2 h temporal information. Table 12.4 further states that when the historical time increases to 12 h, voice's conditional entropy shrinks to less than 0.1 bit, a reduction of over 80% compared to the case without any prior information. As a result, the more historical traffic information is provided, the more precisely and easily the quantized traffic can be predicted. Additionally, compared to voice and text services, the temporal conditional entropy of data traffic degrades slowest when the information of the preceding hours is adopted. Therefore, it is more challenging to predict data traffic based on historical traffic knowledge.

12.2.2.2 Spatial Dimension

To guarantee a full coverage over the region of interest and ensure quality of experience all the time, mobile operators deploy coverage-overlapped adjacent

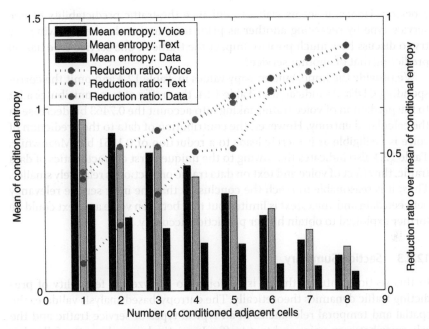

Figure 12.12 The mean of conditional entropies versus the number of conditioned adjacent cells.

cells, which in turn lead to some similarities between traffic in the adjacent cell. Meanwhile, user mobility behavior adds to the spatial relevancy in the traffic. In this part, we attempt to measure this spatial relevancy and describe how much information adjacent cells could provide.

Figure 12.12 demonstrates how the mean of spatial conditional entropy varies with the number of considered adjacent cells. As Figure 12.12 illustrates, the traffic knowledge from adjacent cells can enhance the predictability. For example, in Table 12.4, the mean of spatial conditional entropy with voice traffic information from three adjacent cells reduces to 0.9043 bit. As the number of adjacent cells exploited for traffic information rises, the mean of spatial conditional entropy continues to decline. On the other hand, slightly less than the contribution of temporal relevancy to traffic prediction, the spatial factor decreases the entropy mean to 30% with the traffic information from six adjacent cells. Similar to the effect of temporal information, Figure 12.12 implies that the relevant spatial information exhibits a larger contribution to the predictability of voice and text traffic than that of data traffic.

12.2.2.3 Interservice Relationship

As mentioned already, traffic of the three typical service types (i.e., voice, text, and data) is influenced by several common factors, such as idle/busy time of a person. Hence, it seems viable to enhance the traffic predictability of one service type by regarding another as prior information. In the following, we try to discuss how much positive impact the traffic of one service type has on predicting that of another service?

We provide the important entropy values in Table 12.4 and omit the corresponding CDFs. As indicated in Table 12.4, the traffic of text contributes a lot to the prediction of voice traffic, taking into account the 0.7463 bit decrease in the calculated entropy. However, the contribution of data to the prediction of voice is negligible as it merely leads to a reduction of 0.1361 bit. Meanwhile, Table 12.4 also indicates that owing to the unique burst characteristics of data traffic, the effect of voice and text on data traffic prediction is relatively smaller. Thus, it is reasonable to reach the conclusion that the inter service relevancy between data and voice/text is limited but that between voice and text could be further exploited to obtain higher prediction accuracy.

12.2.3 Section Summary

In this section, entropy theory is exploited to analyze the feasibility of predicting traffic dynamics theoretically. The entropy-based analysis validates the spatial and temporal relevancies in all typical types of service traffic and the relevancy between voice and text traffic. In particular, we have the following observations: (1) The temporal relevancy is the dominant contributing factor to traffic predictability. (2) The contribution of spatial relevancy is much less, but it also makes sense to the traffic prediction. (3) The inter service relevancy between voice and text can be applied to the prediction problem while the one between data and the other two types cannot. (4) Data traffic prediction can only depend on temporal and spatial relevancies.

12.3 The Prediction of Application-Level Traffic

In Section 12.2, we have demonstrated the microscopic traffic predictability in cellular networks for circuit switching's voice and short message service and packet switching's data service. However, compared to the more accurate prediction performance for voice and text service in circuit switching domain, the state-of-the-art research in packet switching's data service is still not satisfactory enough. Furthermore, the fifth-generation (5G) cellular networks, which is under the standardization and assumed to be the key enabler and infrastructure provider in the information communication technology industry, aim to cater different types of services such as enhanced mobile broadband (eMBB) with bandwidth-consuming and throughput-driving requirements and ultra reliable low latency service (URLLC). Moreover, one type of service (e.g., URLLC) has a higher priority than another one (e.g., eMBB). Hence, if we can detect the coming of the service with higher priority, we can timely reserve and specifically configure the resources (e.g., shorter transmission time interval) to guarantee the service provisioning. In a word, a learning and prediction study over application-level data traffic might contribute to understanding data service's characteristics and performing finer resource management in the 5G era [27]. In this section, we investigate on the prediction accuracy for application-level traffic generated by three popular service types (i.e., instantaneous message (IM), web browsing, and video).

12.3.1 Sparse Representation and Dictionary Learning

In recent years, sparsity methods or the related compressive sensing (CS) methods have been significantly investigated [32–34]. Mathematically, sparsity methods aim to tackle this sparse signal recovery problem in the form of

$$\min |s|_0 \, , \text{s.t. } y = Ds, \tag{12.5}$$

or

$$\min |s|_0 \, , \text{s.t. } |y - Ds| \leq \iota. \tag{12.6}$$

Here, s denotes a sparse signal vector while y denotes a measurement vector based on a transform matrix or dictionary D. Moreover, $|s|_0$ is an l_0-norm, counting the number of non-zero entries in s, while an l_p-norm $|s|_p$, $p \geq 1$ of a $1 \times n$ vector $s = (s_1, \cdots, s_n)$ is defined by $\sqrt[p]{\sum_i^n |s_i|^p}$. Besides, ι is a predefined integer indicating the sparsity. By leveraging the embedded sparsity in the signals, sparsity methods could successfully recover the sparse signal with a high probability, depending on a small number of measurements fewer than that required in Nyquist sampling theorem. Basis pursuit (BP) [35], one of typical sparsity methods, solves the problem in terms of maximizing a posterior (MAP) criterion by relaxing the l_0-norm to an l_1-norm. On the other hand, orthogonal matching pursuit (OMP) [36] greedily achieves the final outcome in a sequential manner, by computing inner products between the signal and dictionary columns and possibly solving them using the least square criterion.

For sparsity methods above, there usually exists an assumption that the transform matrix or dictionary D is already known or fixed. However, in spite of their computation simplicity, such prespecified transform matrices like Fourier transforms and overcomplete wavelets might not be suitable to lead to a sparse signal [37]. Consequently, some researchers proposed to design D based on learning [37,38]. In other words, during the sparse signal recovery procedure, machine learning and statistics are leveraged to compute the vectors in D from the measurement vector y, so as to grant more flexibility to get a sparse representation s from y. Mathematically, dictionary learning methods would

The framework

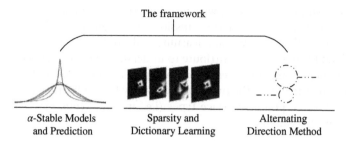

| α-Stable Models and Prediction | Sparsity and Dictionary Learning | Alternating Direction Method |

Figure 12.13 The traffic learning framework.

yield a final transform matrix by alternating between a sparse computation process based on the dictionary estimated at the current stage and a dictionary update process to approach the measurement vector.

12.3.2 The Traffic Prediction Framework

12.3.2.1 Problem Formulation

In this section, we aim to fully take advantage of the traffic modeling results in Section 12.1 and propose a new framework in Figure 12.13 to predict the traffic. The proposed framework consists of three modules. Among them the "α-Stable Model and Prediction" module would take advantage of the already known traffic knowledge to learn and distill the parameters in α-stable models and provide a coarse prediction result. Meanwhile, the "Sparsity & Dictionary Learning" module imposes constraints to make the final prediction results satisfy the spatial sparsity. But, these two modules inevitably add multiple parameters unknown *a priori* and thus need specific mathematical operations to obtain a solution. Hence, the proposed framework also contain an "Alternating Direction Method" module to iteratively process the other modules and yield the final result.

For simplicity of representation, we introduce a traffic vector x, whose entries archive the volume of traffic in one given cell at different moments. Furthermore, by augmenting the traffic vectors for different cells, we refer to a traffic matrix X to denote the traffic records in an area of interest. Then, every row vector of traffic matrix indicates traffic loads at one specific cell with respect to the time while every column vector reflects volumes of traffic of several adjacent cells at one specific moment. Specifically, for a traffic resolution Δt, $X(i, t)$ in a traffic matrix X denotes traffic loads of cell i from t to $t + \Delta t$. Mathematically, traffic prediction can be regarded as the procedure to obtain a column vector $\hat{x}_p = \hat{X}(:, t)^T$ at a future moment t, based on the already known traffic records. Each entry $\hat{x}_p^{(i)}$ in $\hat{x}_p$ corresponds to the future traffic for cell i. Here, $(\cdot)^T$ denotes a transpose operation of a matrix or vector.

Temporal Modeling Component. Benefiting from the substantial body of works toward α-stable model-based linear prediction [14,61], coarse prediction results can be achieved by computing linear prediction coefficients in terms of the least mean square error criterion, the minimum dispersion criterion, or the covariation orthogonal criterion [54]. Due to its simplicity and comparatively low variability, the covariation orthogonal criterion [54,55] is chosen in this section to demonstrate the α-stable-based linear prediction performance.

Without loss of generality, assume that there exist N cells in the area of interest. For a cell $i \in N$ with a known n-length traffic vector $x^{(i)} = (x^{(i)}(1), \cdots)$, $\hat{x}_\alpha^{(i)}$ in α-stable models-based predicted traffic vector $\hat{x}_\alpha = \left(\hat{x}_\alpha^{(1)}, \cdots \right)$ is approximated by

$$\tilde{x}_\alpha^{(i)} = \sum_{j=1}^{m} a^{(i)}(j) x^{(i)}(n + 1 - j), \tag{12.7}$$

with $1 < m \leq n$, where $a^{(i)} = (a^{(i)}(1), \cdots, a^{(i)}(m))$ denotes the prediction coefficients by α-stable models-based linear prediction algorithms. For example, in order to make the 1-step-ahead linear prediction $\tilde{x}_\alpha^{(i)}$ covariation orthogonal to $x^{(i)}(t)$, $\forall t \in \{1, \cdots, n\}$, coefficient $a^{(i)}(h)$, $\forall h \in \{1, \cdots, m\}$ should be given as [55]

$$
\begin{aligned}
a^{(i)}(h) \\
= \sum_{l=1}^{m} \Bigg[\sum_{j=\max(h,l)}^{n} x^{(i)}(j - l + 1) \left(x^{(i)}(j - h + 1) \right)^{<\alpha-1>} \\
\times \sum_{j=l+k}^{n} x^{(i)}(j) \left(x^{(i)}(j - k - l + 1) \right)^{<\alpha-1>} \Bigg].
\end{aligned}
\tag{12.8}
$$

Here, the signed power $v^{<\alpha-1>} = |v|^{(\alpha-1)} \mathrm{sgn}(v)$, while $\mathrm{sgn}(x)$ with respect to $x \in \mathcal{R}$ is defined as $\mathrm{sgn}(x) = x/|x|$ when $x \neq 0$; and $\mathrm{sgn}(x) = 0$ when $x = 0$. For simplicity of representation, the terminology "$(n = 36, m = 10, k = 1)$-linear prediction" is used to denote a prediction method, which first utilizes $n = 36$ consecutive traffic records in one randomly selected cell, then calculates $m = 10$ prediction coefficients, and finally predicts the traffic value at the next (i.e., $k = 1$) moment.

Noise Component. For any prediction algorithm, there dooms to exist some prediction error. Therefore, final traffic prediction vector $\hat{x}_p$ is approximated by $\hat{x}_\alpha$ plus Gaussian noise z. Here, there are two reasons leading to the assumption that noise is Gaussian distributed. First, Gaussian distributed noise is widely used to characterize the fitting error between models and practical data. Second, we have conducted an experiment to examine the prediction performance of a simple $(n = 36, m = 10, k = 1)$-linear prediction procedure and found that the prediction procedure could well predict the traffic trend. However, there would exist some gap between the real traffic trace and the predicted one. But, Figure 12.14 indicates that the prediction error can be approximated by the Gaussian distribution. The K–S test further shows the GoF statistics for the IM, web browsing, and video services are 0.0319, 0.0657, and 0.0437, respectively, smaller than the 95% confidence threshold (i.e., 0.3507). So, we model the prediction error by Gaussian distribution. Combining the temporal modeling and noise components, $\hat{x}_p$ could be achieved by

$$
\min_{\hat{x}_p, \hat{x}_\alpha, z} |\hat{x}_\alpha - \tilde{x}_\alpha|_2^2 + \lambda_1 |z|_2^2,
$$

$$
\text{s.t.} \hat{x}_p = \hat{x}_\alpha + z,
$$

$$
\tilde{x}_\alpha = \left(\tilde{x}_\alpha^{(1)}, \cdots, \tilde{x}_\alpha^{(N)} \right), \tag{12.9}
$$

$$
\tilde{x}_\alpha^{(i)} = \sum_{j=1}^{m} a^{(i)}(j) x^{(i)}(n + 1 - j), \forall i \in \{1, \cdots, N\}.
$$

For simplicity of representation, we omit the last two constraints in Eq. 12.9 in the following statements.

Spatial Sparse Component. In Section 12.1.4, application-level traffic is shown to exhibit the spatial sparsity. Therefore, $\hat{x}_p$ could be further refined by minimizing the gap between $\hat{x}_p$ and a sparse linear combination (i.e., $s \in \mathcal{R}^{K \times 1}$) of a dictionary $D \in \mathcal{R}^{N \times K}$, namely,

$$
\min_{\hat{x}_p, D, s} \left| \hat{x}_p - Ds \right|_2^2, \text{ s.t. } |s|_0 \leq \epsilon. \tag{12.10}
$$

Notably, in Figure 12.9, we observe sparse application-level cellular network traffic density. In other words, there merely exist few traffic spots with significantly large traffic volume. On the other hand, in the area of sparse representation, an l_0-norm, which counts the number of non-zero elements in the vector, is often used to characterize the sparse property. Therefore, in Eq. 12.10, we use an l_0-norm to add the sparse constraint to the final optimization problem. Moreover, the exact representation of the dictionary, which the previous sparsity analyses do not mention, remains a problem and would be solved later.

Therefore, it is natural to consider the original dataset as a mixture of these effects and propose a new framework to combine these two components together to get a superior forecasting performance.

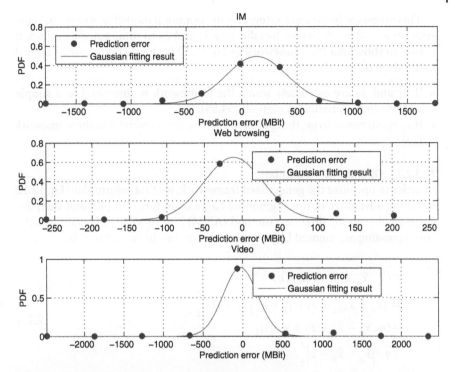

Figure 12.14 The result by fitting the prediction error to a Gaussian distribution, after an α-stable model-based $(36, 10, 1)$-linear prediction method.

In order to capture the temporal α-stable-modeled variations while keeping the spatial sparsity, a new framework is proposed as follows:

$$\min_{\widehat{x}_p, \widehat{x}_\alpha, z, D, s} \left| \widehat{x}_\alpha - \widetilde{x}_\alpha \right|_2^2 + \lambda_1 |z|_2^2 + \lambda_2 \left| \widehat{x}_p - Ds \right|_2^2,$$
$$\text{s.t.} \widehat{x}_p = \widehat{x}_\alpha + z, \ |s|_0 \le \epsilon. \tag{12.11}$$

Due to the nonconvexity of l_0-norm, the constraints in Eq. 12.11 are not directly tractable. Thanks to the sparsity methods discussed in Section 12.3.1, an l_1-norm relaxation is employed to make the problem convex while still preserving the sparsity property [39]. Therefore, Eq. 12.11 can be reformulated as

$$\min_{\widehat{x}_p, \widehat{x}_\alpha, z, D, s} \left| \widehat{x}_\alpha - \widetilde{x}_\alpha \right|_2^2 + \lambda_1 |z|_2^2 + \lambda_2 \left| \widehat{x}_p - Ds \right|_2^2,$$
$$\text{s.t.} \widehat{x}_p = \widehat{x}_\alpha + z, \ |s|_1 \le \varepsilon \epsilon, \tag{12.12}$$

where ε is a predefined constraint, similar to ϵ.

This proposed framework integrates the temporal modeling and spatial correlation together. Moreover, by adjusting λ_1 and λ_2 to some extreme values, it is easy to show that the framework in Eq. 12.12 is closely tied to some typical methods in other references.

- If λ_1 and λ_2 are extremely small, the framework is simplified to a simple α-stable linear prediction method [54,55].
- If λ_2 is extremely large, the spatial sparsity factor dominates in the framework [24].

12.3.2.2 Optimization Algorithm

In order to optimize the generalized framework, we first reformulate Eq. 12.12 by taking advantage of the augmented Lagrangian function [40] and then develop an alternating direction method (ADM) [59] to solve it. Specifically, the corresponding augmented Lagrangian function can be formulated as

$$
\begin{aligned}
&\mathcal{L}(\hat{x}_p, \hat{x}_\alpha, z, D, s, m, \gamma, \eta) \\
&\stackrel{\Delta}{=} \left|\hat{x}_\alpha - \tilde{x}_\alpha\right|_2^2 + \lambda_1 \left|z\right|_2^2 + \lambda_2 \left|\hat{x}_p - Ds\right|_2^2 \\
&+ \langle m, \hat{x}_p - \hat{x}_\alpha - z \rangle + \gamma \cdot |s|_1 \\
&+ \eta \cdot \left|\hat{x}_p - \hat{x}_\alpha - z\right|_2^2 .
\end{aligned}
\tag{12.13}
$$

Besides, m and γ are the Lagrangian multipliers, while η is a factor for the penalty term. Essentially, the augmented Lagrangian function includes the original objective, two Lagrange multiplier terms (i.e., the third line in Eq. 12.13), and one penalty term converted from the equality constraint (i.e., the forth line in Eq. 12.13). The operation $\langle x, y \rangle$ denotes the summation operation of element-wise multiplication in x and y with the same size. Specifically, introducing Lagrange multipliers conveniently convert an optimization problem with equality constraints into an unconstrained one. Moreover, for any optimal solution that minimizes the (augmented) Lagrangian function, the partial derivatives with respect to the Lagrange multipliers must be zero [41]. Additionally, the penalty terms enforce the original equality constraints. Consequently, the original equality constraints are satisfied. Besides, by including Lagrange multiplier terms as well as the penalty terms, it is not necessary to iteratively increase η to ∞ to solve the original constrained problem, thereby avoiding ill-conditioning [40].

The ADM algorithm progresses in an iterative manner. During each iteration, we alternate among the optimization of the augmented function by varying each one of $(\hat{x}_p, \hat{x}_\alpha, z, D, s, m, \gamma, \eta)$ while fixing the other variables. Specifically, the ADM algorithm involves the following steps:

1) Find $\hat{x}_\alpha$ to minimize the augmented Lagrangian function $\mathcal{L}(\hat{x}_p, \hat{x}_\alpha, z, D, s, m, \gamma, \eta)$ with other variables fixed. Removing the fixed items, the objective turns into

$$\arg\min_{\hat{x}_\alpha} |\hat{x}_\alpha - \tilde{x}_\alpha|_2^2 + \langle m, \hat{x}_p - \hat{x}_\alpha - z \rangle + \eta \cdot \left|\hat{x}_p - \hat{x}_\alpha - z\right|_2^2, \quad (12.14)$$

which can be further reformulated as

$$\arg\min_{\hat{x}_\alpha} \frac{1}{\eta} \cdot |\hat{x}_\alpha - \tilde{x}_\alpha|_2^2 + \left|\hat{x}_\alpha - (\hat{x}_p - z + \frac{m}{2\eta})\right|_2^2. \quad (12.15)$$

Letting $J_{\hat{x}_\alpha} = \hat{x}_p - z + \frac{m}{2\eta}$ and setting the gradient of the objective function in Eq. 12.15 to be zero, it yields

$$\hat{x}_\alpha = \frac{1}{\eta + 1} \cdot (\tilde{x}_\alpha + \eta \cdot J_{\hat{x}_\alpha}). \quad (12.16)$$

2) Find z to minimize the augmented Lagrangian function $\mathcal{L}(\hat{x}_p, \hat{x}_\alpha, z, D, s, m, \gamma, \eta)$ with other variables fixed. The corresponding mathematical formula is

$$\arg\min_{z} \lambda_1 |z|_2^2 + \langle m, \hat{x}_p - \hat{x}_\alpha - z \rangle + \eta \cdot \left|\hat{x}_p - \hat{x}_\alpha - z\right|_2^2. \quad (12.17)$$

Similarly, it can be reformulated as

$$\arg\min_{z} \frac{\lambda_1}{\eta} \cdot |z|_2^2 + \left|z - (\hat{x}_p - \hat{x}_\alpha + \frac{m}{2\eta})\right|_2^2. \quad (12.18)$$

Letting $J_z = \hat{x}_p - \hat{x}_\alpha + \frac{m}{2\eta}$ and setting the gradient of the objective function in Eq. 12.18 to be zero, it yields

$$z = \frac{1}{\lambda_1/\eta + 1} \cdot J_z. \quad (12.19)$$

3) Find $\hat{x}_p$ to minimize the augmented Lagrangian function $\mathcal{L}(\hat{x}_p, \hat{x}_\alpha, z, D, s, m, \gamma, \eta)$ with other variables fixed. It gives

$$\arg\min_{\hat{x}_p} \lambda_2 \left|\hat{x}_p - Ds\right|_2^2 + \langle m, \hat{x}_p - \hat{x}_\alpha - z \rangle + \eta \cdot \left|\hat{x}_p - \hat{x}_\alpha - z\right|_2^2. \quad (12.20)$$

That is

$$\arg\min_{\hat{x}_p} \frac{\lambda_2}{\eta} \cdot \left|\hat{x}_p - Ds\right|_2^2 + \left|\hat{x}_p - (\hat{x}_\alpha + z - \frac{m}{2\eta})\right|_2^2. \quad (12.21)$$

Define $J_{\hat{x}_p} = \hat{x}_\alpha + z - \frac{m}{2\eta}$ and set the corresponding gradient in Eq. 12.21 to be zero. It becomes

$$\hat{x}_p = 1/(\frac{\lambda_2}{\eta} + 1) \cdot (\frac{\lambda_2}{\eta} Ds + J_{\hat{x}_p}). \tag{12.22}$$

4) Find D and s to minimize the augmented Lagrangian function $\mathcal{L}(\hat{x}_p, \hat{x}_\alpha, z, D, s, m, \gamma, \eta)$ with other variables fixed. In fact, the objective function turns into

$$\arg\min_{D, s} \lambda_2 \left| \hat{x}_p - Ds \right|_2^2 + \gamma \cdot |s|_1. \tag{12.23}$$

Obviously, this optimization problem in Eq. 12.23 is exactly the sparse signal recovery problem without the dictionary *a priori* in Section 12.3.1. Inspired by the dictionary learning methodology (namely, the means to learn the dictionary or basis sets of large-scale data) in Reference [38], the corresponding solution alternatively determines D and s and thus involves two sub procedures, namely, online learning algorithm [38] and LARS–Lasso algorithm [42]. Algorithm 12.1 provides the skeleton of this solution.

In order to update the dictionary in Eq. 12.1, the proposed sparse signal recovery algorithm utilizes the concept of stochastic approximation, which is first introduced and mathematically proved convergent to a stationary point in Reference [38].

On the other hand, based on the learned dictionary, the concerted effort to recover a sparse signal could be exploited. As mentioned already, the well-known LARS–Lasso algorithm [42], which is a forward stage-wise regression algorithm and gradually finds the most suitable solution along a equiangular path among the already known predictors, is used here to solve the problem in Eq. 12.1. Meanwhile, it is worthwhile to note that other compressive sensing algorithms [36] could also be used here.

5) Update estimate for the Lagrangian multiplier m according to steepest gradient descent method [43], namely, $m \leftarrow m + \eta \cdot (\hat{x}_p - \hat{x}_\alpha - z)$. Similarly, update estimate γ by $\gamma \leftarrow \gamma + \eta \cdot |s|_1$.
6) Update $\eta \leftarrow \eta \cdot \rho$.

In Algorithm 12.2, we summarize the steps during each iteration. Notably, without loss of generality, consider a known traffic vector $x(0, \cdots, t)$ of a given cell at different moments $(0, \cdots, t)$. Then, we could estimate the α-stable-related parameters according to maximum likelihood methods, quantile methods, or sample characteristic function methods in References [15] and [55]. Afterward, we could conduct Algorithm 12.2 to predict the traffic volume at moment $t + 1$. Similarly, we need to estimate the α-stable-related parameters according to methods in References [15] and [55], in terms of the traffic vector $x(0, \cdots, t + 1)$,

Algorithm 12.1 The sparse signal recovery algorithm without a predetermined dictionary.

Initialize the dictionary D as an input dictionary $D^{(0)}$ (which could be the dictionary learned in last calling this Algorithm), the number of iterations for learning a dictionary as T, two auxiliary matrices $A^{(0)} \in \mathcal{R}^{K \times K}$ and $B^{(0)} \in \mathcal{R}^{K \times K}$ with all elements therein equaling zero.

1: **For** $t = 1$ to T **do**

2: Sparse coding: computing $s^{(t)}$ using LARS-Lasso algorithm [42] to obtain

$$s^{(t)} = \arg\min_s \lambda_2 \left| \hat{x}_p - D^{(t-1)} s \right|_2^2 + \gamma \cdot |s|_1 . \tag{12.23}$$

3: Update $A^{(t)}$ according to

$$A^{(t)} \leftarrow A^{(t)} + s^{(t)}(s^{(t)})^T . \tag{12.23}$$

4: Update $B^{(t)}$ according to

$$B^{(t)} \leftarrow B^{(t)} + \hat{x}_p(s^{(t)})^T . \tag{12.23}$$

5: Dictionary Update: computing $D^{(t)}$ online learning algorithm [38] to obtain

$$\begin{aligned} D^{(t)} &= \arg\min_D \lambda_2 \left| \hat{x}_p - Ds^{(t)} \right|_2^2 + \gamma \cdot |s^{(t)}|_1 \\ &= \arg\min_D \text{Tr}(D^T D A^{(t)}) - 2\text{Tr}(D^T B^{(t)}). \end{aligned} \tag{12.23}$$

6: **End For**

7: **Return** the learned dictionary $D^{(t)}$ and the sparse coding vector $s^{(t)}$.

---

and perform Algorithm 12.2 to predict the traffic volume at moment $t + 2$. It can be observed that, compared to Algorithm 12.1, which is an application of the lines in Reference [38], Algorithm 12.2 is made up of some additional iterative procedures to procure the parameters unknown *a priori*. Besides, most steps involved in Algorithm 12.2 are deterministic vector computations and thus computationally efficient. Therefore, the whole framework could effectively yield the traffic forecasting results.

12.3.3 Performance Evaluation

We validate the prediction accuracy improvement of our proposed framework in Algorithm 12.2 relying on the practical traffic dataset. Specifically, we choose the traffic load records of these three service types of applications generated in 113 cells within a randomly selected region from Dataset 1 in Table 12.2. Moreover, we intentionally divide the traffic dataset into two parts. One is used to learn and distill the parameters related to traffic characteristics, and the other part is to conduct the experiments to verify and validate the accuracy of the proposed framework in Algorithm 12.2. Specifically, we compare our prediction $\hat{x}_p$ with the ground truth x in terms of the normalized mean absolute

Algorithm 12.2 The dictionary learning-based alternating direction method

---

Initialize $\hat{x}_p, \hat{x}_\alpha, z, s, m, \gamma, \eta$ according to $\hat{x}_p^{(0)}, \hat{x}_\alpha^{(0)}, z^{(0)}, D^{(0)}, s^{(0)}, m^{(0)}, \gamma^{(0)}, \eta^{(0)}$, and the number of iterations T. Compute $\tilde{x}_\alpha$ according to α-stable model based linear prediction algorithms [14,61].

1: **For** $t = 1$ to T **do**

2: Update $\hat{x}_\alpha$ according to $\hat{x}_\alpha^{(t)} \leftarrow \frac{1}{\eta^{(t-1)}+1} \cdot \left(\tilde{x}_\alpha + \eta^{(t-1)} \cdot \left(\hat{x}_p^{(t-1)} - z^{(t-1)} + \frac{m^{(t-1)}}{2\eta^{(t-1)}} \right) \right)$.

3: Update z according to $z^{(t)} \leftarrow \frac{1}{\lambda_1/\eta^{(t-1)}+1} \cdot \left(\hat{x}_p^{(t-1)} - \hat{x}_\alpha^{(t)} + \frac{m^{(t-1)}}{2\eta^{(t-1)}} \right)$.

4: Update $\hat{x}_p$ according to $\hat{x}_p^{(t)} \leftarrow 1/(\frac{\lambda_2}{\eta^{(t-1)}} + 1) \cdot \left(\frac{\lambda_2}{\eta^{(t-1)}} D^{(t-1)} s^{(t-1)} + \hat{x}_\alpha^{(t)} + z^{(t)} - \frac{m^{(t-1)}}{2\eta^{(t-1)}} \right)$.

5: Update D and s according to sparse signal recovery algorithm (i.e., Algorithm 12.1). In particular, use two sub-procedures namely online learning algorithm [38] and LARS-lasso algorithm [42] to update D and s, respectively.

6: Update m according to $m^{(t)} \leftarrow m^{(t-1)} + \eta^{(t-1)} \cdot (\hat{x}_p^{(t)} - \hat{x}_\alpha^{(t)} - z^{(t)})$.

7: Update γ by $\gamma^{(t)} \leftarrow \gamma^{(t-1)} + \eta^{(t-1)} \cdot \left| s^{(t)} \right|_1$.

8: Update η by $\eta^{(t)} \leftarrow \eta^{(t-1)} \cdot \rho$, here ρ is an iteration ratio.

9: **End For**

10: **Return** the predicted traffic vector $\hat{x}_p$.

---

error (NMAE) [59], which is defined as

$$\text{NMAE} = \frac{\sum_{i=1}^{N} \left| \hat{x}_p(i) - x(i) \right|}{\sum_{i=1}^{N} |x(i)|}. \tag{12.24}$$

As described in Eq. 12.24, most of the parameters could be set easily and tuned dynamically within the framework. Therefore, we can benefit from this advantage and only need to examine the performance impact of few parameters, namely, λ_1, λ_2, γ, and η by dynamically adjusting them. By default, we set $\lambda_1 = 10$, $\lambda_2 = 1$, $\gamma = 1$, and $\eta = 10^{-4}$ and the number of iterations in Algorithm 12.2 and sparse signal recovery algorithm (i.e., Algorithm 12.1) to be 20 and 3, respectively. Besides, we impose no prior constraints on D, s, and z, and set them as zero vectors.

Figure 12.15 gives the performance of our proposed framework in terms of NMAE, by taking advantage of the (36,10,1)-linear prediction algorithm in Section 12.3.2 to provide the "coarse" prediction results $\tilde{x}_\alpha$. In other words, we would exploit traffic records in the last three hours to train the parameters of α-stable models and predict traffic loads in the next 5 min. In order to provide a more comprehensive comparison, the simulations run in both busy moments (i.e., 9a.m., 12p.m., and 4p.m.) and idle ones (i.e., 7a.m. and 9p.m.)

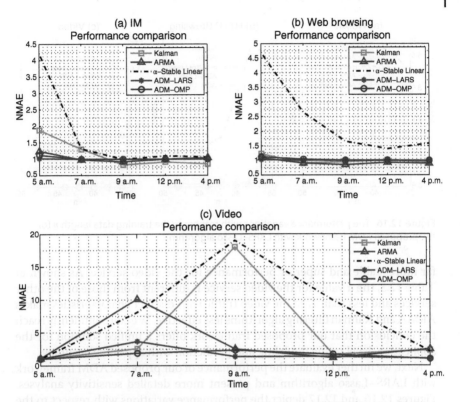

Figure 12.15 The performance comparison between the proposed ADM framework with different sparse signal recovery algorithms (i.e., LARS–Lasso and OMP), and the α-stable model-based (36,10,1)-linear prediction algorithm.

of one day. We first examine the corresponding performance improvement of the proposed ADM framework with different sparse signal recovery algorithm (i.e., LARS–Lasso algorithm [42] and OMP algorithm [36]). It can be observed that in most cases, different sparse signal recovery algorithm has little impact on the prediction accuracy. Therefore, the applications of the proposed framework could pay little attention to the involved sparsity methods. Afterward, we can find that the proposed framework significantly outperforms the classical α-stable model-based (36,10,1)-linear prediction algorithm. In particular, the NMAE of the proposed framework can be 12% small (e.g., prediction for 12p.m. video traffic) as that for the classical linear algorithm. This performance improvement can be interpreted as the gain by exploiting the embedded sparsity in traffic and taking account of the originally existing prediction error of linear prediction. Furthermore, we also compare the proposed framework with ARMA and Kalman filtering algorithms and show that our solution can achieve competitive performance for IM and web browsing services and yield

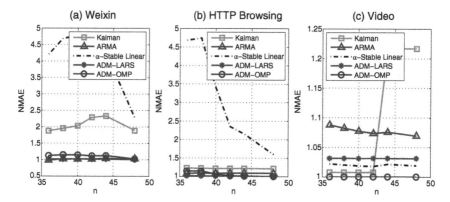

Figure 12.16 The performance variations with respect to the training data length n for the proposed ADM framework with LARS–Lasso algorithm.

far more stable and superior performance for the video service. As shown in Table 12.3, the α value for the video service is different from those of the other services and less than 1, so the video traffic with distinct characteristics makes ARMA and Kalman filtering algorithms less effective. We can confidently reach the conclusion that our proposed framework offers a unified solution for the application-level traffic modeling and prediction with appealing accuracy.

Next, we further evaluate the performance of our proposed ADM framework with LARS–Lasso algorithm and present more detailed sensitivity analyses. Figures 12.16 and 12.17 depict the performance variations with respect to the training data length n and the number of prediction coefficients m, respectively. From the figures, the increase of n or m contributes to improving the prediction accuracy for all types of applications especially the video service, which is consistent with our intuition. As depicted in Figure 12.18, similar result is applicable to the case where we increase the number of iterations in Algorithm 12.2. However, Figure 12.18 implies that the loss in prediction accuracy is rather small when the number of iterations decreases from 20 to 4. Hence, if we initialized the prediction process with 20 iterations, we can stop the iterative process whenever the results between two consecutive iterations become sufficiently small, so as to reduce the computational complexity.

Figure 12.19(a), Figure 12.19(b) and Figure 12.19(c)a–c shows that the prediction accuracy nearly stays the same irrespective of λ_1. This means that the noise component has limited contribution to the corresponding performance. It also implies that the choice of λ_1 could be flexible when we apply the framework in practice. Figure 12.19d demonstrates that the influence of λ_2 is comparatively more obvious and even diverges for different service types. Specifically, a larger λ_2 has a slightly negative impact on predicting the traffic loads for IM and web browsing service, but it contributes to the prediction of video service. Recalling the sparsity analyses in Section 12.1.4, video service demonstrates the strongest sparsity. Hence, by increasing λ_2 it implies to put more emphasis on

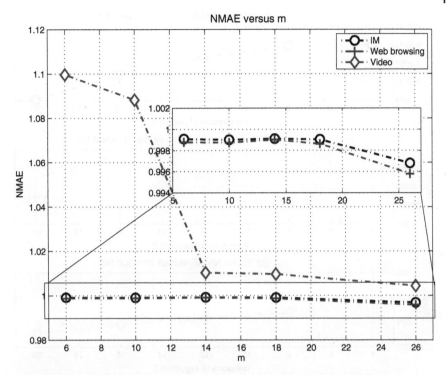

Figure 12.17 The performance variations with respect to the number of prediction coefficients m for the proposed ADM framework with LARS–Lasso algorithm.

the importance of sparsity and results in a better performance for video service. It is worthwhile to note here that in Eq. 12.13, λ_2 and γ are coupled together as well and should have inverse performance impact. Therefore, due to the space limitation, the performance impact of γ is omitted here. Figure 12.19e depicts the performance variation with respect to η, which is similar to that with respect to λ_2. Since a larger η has a positive impact on predicting the traffic loads for IM and web browsing service, it degrades the prediction performance of video service. This phenomenon is potentially originated from the very distinct characteristics of these three services types (e.g., different α-stable models' parameters and different sparsity representation) and needs a further careful investigation. However, it safely comes to the conclusion that the proposed framework provides a superior and robust performance than the classical linear algorithm.

12.3.4 Section Summary

In this section, we have collected the application-level traffic data from one operator in China. With the aid of this practical traffic data, we have confirmed

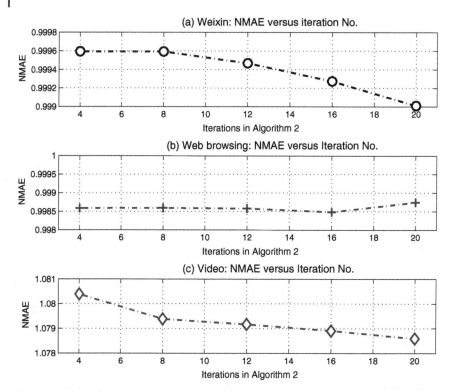

Figure 12.18 The performance variations with respect to the number of iterations in Algorithm 12.2

several important statistical characteristics like temporally α-stable-modeled property and spatial sparsity. Afterward, we have proposed a traffic prediction framework, which takes advantage of the already known traffic knowledge to distill the parameters related to aforementioned traffic characteristics and forecasts future traffic results bearing the same characteristics. We have also developed a dictionary learning-based alternating direction method to solve the framework, and manifested the effectiveness and robustness of our algorithm through extensive simulation results.

12.4 Related Works

Due to its apparent importance to the protocol design and performance evaluation of telecommunications networks, there have already existed some former works toward modeling the traffic in various networks. In fixed broadband networks, researchers showed that aggregate traffic traces demonstrate strong

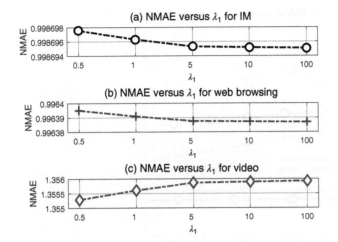

Figure 12.19 The performance variations with respect to λ_1, λ_2, and η, for the proposed ADM framework with LARS–Lasso algorithm.

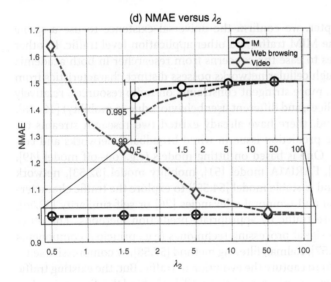

Figure 12.19 *continued.*

burstiness and could be modeled with α-stable models [14,15]. On the other hand, the investigation over traffic characteristics of IM in wired Internet revealed heavy-tailed distribution phenomena in services like AIM (AOL Instant Messenger) and Windows Live Messenger [44,45]. Therefore, it is natural to investigate which one of the aforementioned models is more suitable for MIM

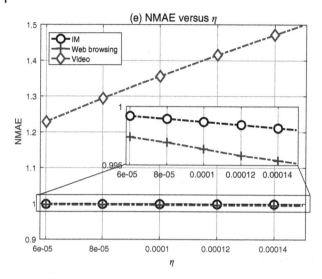

Figure 12.19 *continued.*

traffic? In this chapter, we confirm the universal existence to use α-stable models to model the MIM traffic and other application-level traffic. In other words, it contributes to ease the concerns from researcher in both academia and industry, although cellular networks possess distinct characteristics from fixed networks (e.g., more stringent constraints on radio resources, relatively expensive billing polices, and different user behaviors due to mobility) [46–48].

On the other hand, there have already existed two research streams toward the fine traffic prediction issue in wired broadband networks and cellular networks [46]. One is based on fitting models (e.g., on–off model [49], ARIMA model [50], FARIMA model [51], mobility model [52,53], network traffic model [53], and α-stable model [54,55]) to explore the traffic characteristics, such as spatial and temporal relevancies [26] or self-similarity [22,56], and obtain the future traffic by appropriate prediction methods. The other is based on modern signal processing techniques (e.g., principal components analysis method [24,57], Kalman filtering method [57,58], or compressive sensing method [2,24,59]) to capture the evolution of traffic. But, the existing traffic prediction methods in this microscopic case still lag behind the diverse requirements of various application scenarios. For example, most of them still focus on the traffic of all data services [5] and seldom shed light on a specific type of services (e.g., video, web browsing, and IM). Besides, the existing prediction methods usually follow the analysis results in wired broadband networks like the α-stable models [61,62] or the often accompanied self-similarity [22] to forecast future traffic values [51,55,56]. Therefore, it is still meaningful to further investigate the prediction accuracy in cellular networks. In this chapter, we take advantage of the advancement in both signal processing and statistical modeling and provide a uniform prediction framework.

12.5 Conclusion

In this chapter, we have discussed about the characterization and learning of the cellular network traffic. Based on practical traffic records, we have shown that α-stable models with different parameters could accurately characterize different service types (e.g., MIM, web browsing, video) of traffic aggregated in one BS, which could be regarded as the summation of lots of individual packets and further explained by the generalized central limit theorem. It implies that α-stable models are the uniform models to characterize both wired networks and cellular networks, in spite of significant differences in the application-level protocol design.

Afterward, based on the entropy theory, we have proved that compared to the spatial and inter service relevancy, the temporal relevancy has the dominant effect on traffic prediction. Hence, it will be more appealing to build the temporal series and derive the short- or long-range dependence, when we want to predict the future traffic loads.

Finally, we have designed one unified learning framework to mine the embedded traffic characteristics and predict the future traffic loads for three typical types of applications/services. Our extensive simulation results have also validated its appealing traffic prediction accuracy. In this regard, future works might include how to balance the prediction accuracy and the corresponding computation resource cost, and how to construct on-demand prediction mechanism according to the requirements of different services.

References

1 Cisco, Cisco visual networking index: global mobile data traffic forecast update, 2012–2017, 2013. Available at http://www.cisco.com/en/US/solutions/collateral/ns341/ns525/ns537/ns705/ns827/white_paper_c11-520862.html (accessed Jan. 20, 2014).

2 R. Li, Z. Zhao, X. Zhou, and H. Zhang, "Energy savings scheme in radio access networks via compressive sensing-based traffic load prediction," *Trans. Emerg. Telecommun. Technol. (ETT)*, vol. 25, no. 4, pp. 468–478, 2014.

3 Z. Niu, Y. Wu, J. Gong, and Z. Yang, "Cell zooming for cost-efficient green cellular networks," *IEEE Commun. Mag.*, vol. 48, no. 11, pp. 74–79, 2010.

4 Z. Niu, "TANGO: traffic-aware network planning and green operation," *IEEE Wirel. Commun.*, vol. 18, no. 5, pp. 25–29, 2011.

5 U. Paul, M. M. Buddhikot, and S. R. Das, "Opportunistic traffic scheduling in cellular data networks," in *Proc. IEEE DySPAN 2012*, Bellevue, WA, USA, 2012.

6 P. Romirer-Maierhofer, M. Schiavone, and A. D'Alconzo, Device-specific traffic characterization for root cause analysis in cellular networks, in *Traffic Monitoring and Analysis*, M. Steiner, P. Barlet-Ros, and O. Bonaventure (eds.), Springer International Publishing, 2015, pp. 64–78.

7 M. Donegan, Android signaling storm rises in Japan, *Light Reading*, 2012. Available at http://www.lightreading.com/mobile/device-operating-systems/android-signaling-storm-rises-in-japan/a/d-id/693138 (accessed Feb. 11, 2015).

8 Reuters Staff, O2 says iPhone demand strained its London network, *Reuters*, Dec. 2009.

9 X. Zhou, Z. Zhao, R. Li, Y. Zhou, J. Palicot, and H. Zhang, "Understanding the nature of social mobile instant messaging in cellular networks," *IEEE Commun. Lett.*, vol. 18, no. 3, pp. 389–392, 2014.

10 Tencent Inc., *WeChat: the new way to connect*, 2011. Available at http://www. wechat.com/en/.

11 3GPP, GERAN study on mobile data applications, 3GPP TR 43.802, May 2011.

12 G. Samorodnitsky, Stable Non-Gaussian Random Processes: Stochastic Models with Infinite Variance. New York: Chapman and Hall/CRC, 1994.

13 European Telecommunications Standards Institute, Services and Facilities to be Provided in the GSM System, GSM Doc 28/85, Jun. 1985.

14 X. Ge, G. Zhu, and Y. Zhu, "On the testing for alpha-stable distributions of network traffic," *Comput. Commun.*, vol. 27, no. 5, pp. 447–457, 2004.

15 J. R. Gallardo, D. Makrakis, and L. Orozco-Barbosa, "Use of alpha-stable self-similar stochastic processes for modeling traffic in broadband networks," *Perform. Eval.*, vol. 40, pp. 71–98, 2000.

16 W. Song and W. Zhuang, "Resource reservation for self-similar data traffic in cellular/WLAN integrated mobile hotspots," in *Proc. IEEE ICC 2010*, Cape Town, South Africa, 2010.

17 J. C.-I. Chuang and N. R. Sollenberger, "Spectrum resource allocation for wireless packet access with application to advanced cellular Internet service," *IEEE J. Select. Area. Comm.*, vol. 16, no. 6, pp. 820–829, 1998.

18 J. H. McCulloch, "Simple consistent estimators of stable distribution parameters," *Commun. Stat. Simul. Comput.*, vol. 15, no. 4, pp. 1109–1136, 1986.

19 K. P. Burnham and D. R. Anderson, "Multimodel inference understanding AIC and BIC in model selection," *Sociol. Methods. Res.*, vol. 33, no. 2, pp. 261–304, 2004.

20 M. A. Stephens, "EDF statistics for goodness of fit and some comparisons," *J. Am. Statist. Assoc.*, vol. 69, no. 347, pp. 730–737, 1974.

21 M. Vidyasagar, Fitting data to distributions (Lecture 4), *Quantitative Introduction to Risk and Uncertainty in Business*, 2016. Available at http://www.utdallas.edu/m.vidyasagar/Fall-2014/6303/Lect-4.pdf (accessed Oct. 26, 2014).

22 M. E. Crovella and A. Bestavros, "Self-similarity in World Wide Web traffic: evidence and possible causes," *IEEE/ACM Trans. Netw.*, vol. 5, no. 6, pp. 835–846, 1997.

23 A. N. Kolmogorov, K. L. Chung, and B. V. Gnedenko, Limit Distributions for Sums of Independent Random Variables, Revised ed. Reading, MA: Addison-Wesley, 1968.

24 Y. Zhang, M. Roughan, W. Willinger, and L. Qiu, "Spatio-temporal compressive sensing and internet traffic matrices," in *Proc. ACM SIGCOMM 2009*, Barcelona, Spain, 2009.

25 D. Lee, S. Zhou, X. Zhong, Z. Niu, X. Zhou, and H. Zhang, "Spatial modeling of the traffic density in cellular networks," *IEEE Wirel. Commun.*, vol. 21, no. 1, pp. 80–88, 2014.

26 M. Z. Shafiq, L. Ji, A. X. Liu, J. Pang, and J. Wang, "Geospatial and temporal dynamics of application usage in cellular data networks," *IEEE Trans. Mob. Comput.*, vol. 14, no. 7, pp. 1369–1381, 2014.

27 T. M. Cover and J. A. Thomas, Elements of Information Theory. Wiley Inter-Science, 2006.

28 R. Kleeman, A. J. Majda, and I. Timofeyev, "Quantifying predictability in a model with statistical features of the atmosphere," *Proc. Natl. Acad. Sci. USA*, vol. 99, no. 24, pp. 15291–15296, 2002.

29 T. Schneider and S. M. Griffies, "A conceptual framework for predictability studies," *J. Clim.*, vol. 12, no. 10, pp. 3133–3155, 1999.

30 A. Adas, "Traffic models in broadband networks," *IEEE Commun. Mag.*, vol. 35, no. 7, pp. 82–89, 1997.

31 N. Bui, M. Cesana, S. A. Hosseini, Q. Liao, I. Malanchini, and J. Widmer, "Anticipatory networking in future generation mobile networks: a survey," *arXiv:1606.00191 [cs]*, Jun. 2016.

32 D. Donoho, "Compressed sensing," *IEEE Trans. Inf. Theory*, vol. 52, no. 4, pp. 4036–4048, 2006.

33 J. Romberg and M. Wakin, "Compressed sensing: a tutorial," in *Proc. IEEE SSP Workshop 2007*, Madison, Wisconsin, 2007.

34 R. G. Baraniuk, "Compressive sensing (lecture notes)," *IEEE Signal Process. Mag.*, vol. 24, no. 4, pp. 118–121, 2007.

35 S. S. Chen, D. L. Donoho, and M. A. Saunders, "Atomic decomposition by basis pursuit," *SIAM J. Sci. Comput.*, vol. 20, no. 1, pp. 33–61, 1998.

36 Y. C. Pati, R. Rezaiifar, and P. S. Krishnaprasad, "Orthogonal matching pursuit: recursive function approximation with applications to wavelet decomposition," in *Proc. ACSSC 1993*, Pacific Grove, CA, USA, 1993.

37 M. Aharon, M. Elad, and A. Bruckstein, "K-SVD: an algorithm for designing overcomplete dictionaries for sparse representation," *IEEE Trans. Signal Process.*, vol. 54, no. 11, pp. 4311–4322, 2006.

38 J. Mairal, F. Bach, J. Ponce, and G. Sapiro, "Online learning for matrix factorization and sparse coding," *J. Mach. Learn. Res.*, vol. 11, pp. 19–60, 2010.

39 R. Fang, T. Chen, and P. C. Sanelli, "Towards robust deconvolution of low-dose perfusion CT: sparse perfusion deconvolution using online dictionary learning," *Med. Image Anal.*, vol. 17, no. 4, pp. 417–428, 2013.

40 Wikipedia, *Augmented Lagrangian method*, 2014. Available at http://en.wikipedia.org/w/index.php?title=Augmented_Lagrangian_method (accessed Oct 26, 2014).

41 S. Boyd and L. Vandenberghe, Convex Optimization. Cambridge, UK: Cambridge University Press, 2004.

42 B. Efron, T. Hastie, I. Johnstone, and R. Tibshirani, "Least angle regression," *Ann. Stat.*, vol. 32, pp. 407–499, 2004.

43 R. S. Sutton and A. G. Barto, Reinforcement Learning: An Introduction. Cambridge University Press, 1998.

44 J. Leskovec and E. Horvitz, "Planetary-scale views on a large instant-messaging network," in *Proc. WWW 2008*, Beijing, China, 2008.

45 Z. Xiao, L. Guo, and J. Tracey, "Understanding Instant Messaging Traffic Characteristics," in Proc. *IEEE ICDCS 2007*, Toronto, ON, Canada, 2007.

46 R. Li, Z. Zhao, X. Zhou, J. Palicot, and H. Zhang, "The prediction analysis of cellular radio access network traffic: from entropy theory to networking practice," *IEEE Commun. Mag.*, vol. 52, no. 6, pp. 238–244, 2014.

47 F. Qian, Z. Wang, A. Gerber, Z. M. Mao, S. Sen, and O. Spatscheck, "Characterizing radio resource allocation for 3G networks," in *Proc. ACM SIGCOMM 2010*, New York, NY, USA, 2010.

48 F. P. Tso, J. Teng, W. Jia, and D. Xuan, "Mobility: a double-edged sword for HSPA networks: a large-scale test on Hong Kong mobile HSPA networks," in *Proc. ACM Mobihoc 2010*, 2010.

49 IEEE 802.16 Boradband Wireless Access Working Group, IEEE 802.16m evaluation methodology document, 2008. Available at http://ieee802.org/16.

50 B. Zhou, D. He, Z. Sun, and W. H. Ng, "Network traffic modeling and prediction with ARIMA/GARCH," in *Proc. HET-NETs Conference*, Ilkley, UK, 2005.

51 O. Cappe, E. Moulines, J.-C. Pesquet, A. P. Petropulu, and Y. Xueshi, "Long-range dependence and heavy-tail modeling for teletraffic data," *IEEE Signal Process. Mag.*, vol. 19, no. 3, pp. 14–27, 2002.

52 F. Ashtiani, J. A. Salehi, and M. R. Aref, "Mobility modeling and analytical solution for spatial traffic distribution in wireless multimedia networks," *IEEE J. Select. Area. Comm.*, vol. 21, no. 10, pp. 1699–1709, 2003.

53 K. Tutschku and P. Tran-Gia, "Spatial traffic estimation and characterization for mobile communication network design," *IEEE J. Select. Area. Comm.*, vol. 16, no. 5, pp. 804–811, 1998.

54 L. Xiang, X. Ge, C. Liu, L. Shu, and C. Wang, "A new hybrid network traffic prediction method," in *Proc. IEEE Globecom 2010*, Miami, Florida, USA, 2010.

55 X. Ge, S. Yu, W.-S. Yoon, and Y.-D. Kim, "A new prediction method of alpha-stable processes for self-similar traffic," in *Proc. IEEE Globecom 2004*, Dallas, Texas, USA, 2004.

56 W. E. Leland, M. S. Taqqu, W. Willinger, and D. V. Wilson, "On the self-similar nature of Ethernet traffic," *IEEE/ACM Trans. Netw.*, vol. 2, no. 1, pp. 1–15, 1994.

57 A. Soule, A. Lakhina, and N. Taft, "Traffic matrices: balancing measurements, inference and modeling," in *Proc. ACM SIGMETRICS 2005*, Banff, Alberta, Canada, 2005.

58 M. C. Falvo, M. Gastaldi, A. Nardecchia, and A. Prudenzi, "Kalman filter for short-term load forecasting: an hourly predictor of municipal load," in *Proc. IASTED ASM 2007*, Palma de Mallorca, Spain, 2007.

59 Y.-C. Chen, L. Qiu, Y. Zhang, G. Xue, and Z. Hu, "Robust network compressive sensing," in Proc. *ACM Mobicom 2014*, Maui, Hawaii, USA, 2014.

60 J. Mairal, F. Bach, J. Ponce, and G. Sapiro, "Online learning for matrix factorization and sparse coding," *J. Mach. Learn. Res.*, vol. 11, pp. 19–60, 2010.

61 J. B. Hill, *"Minimum dispersion and unbiasedness: 'Best' linear predictors for stationary arma a-stable processes,"* Discussion Papers in Economics Working Paper No. 00-06, University of Colorado at Boulder, Sep. 2000.

62 A. Karasaridis and D. Hatzinakos, "Network heavy traffic modeling using alpha-stable self-similar processes," *IEEE Trans. Commun.*, vol. 49, no. 7, pp. 1203–1214, 2001.

Rongpeng Li received his Ph.D and B.E. from Zhejiang University, Hangzhou, China and Xidian University, Xian, China, in June 2015 and June 2010, respectively, both as "Excellent Graduates." He is now a postdoctoral researcher in College of Computer Science and Technologies, Zhejiang University, Hangzhou, China. Prior to that, from August 2015 to September 2016, he was a researcher in Wireless Communication Laboratory, Huawei Technologies Co. Ltd., Shanghai, China. He was a visiting doctoral student in Supelec, Rennes, France from September 2013 to December 2013, and an intern researcher in China Mobile Research Institute, Beijing, China from May 2014 to August 2014. His research interests currently focus on applications of artificial intelligence, data-driven network design, resource allocation of cellular networks (especially full-duplex networks). He was granted by the National Postdoctoral Program for Innovative Talents, which had a grant ratio of 13% in 2016. He serves as an editor of China Communications.

Zhifeng Zhao is an associate professor at the Department of Information Science and Electronic Engineering, Zhejiang University, China. He received the Ph.D. degree in communication and information system from the PLA University of Science and Technology, Nanjing, China, in 2002. Prior to that, he received the master's degree in communication and information system in 1999 and bachelor's degree in computer science in 1996 from the PLA University of Science and Technology, respectively. From September 2002 to December 2004, he was a postdoctoral researcher at the Zhejiang University, where his researches were focused on multimedia NGN (next-generation networks) and soft-switch technology for energy efficiency. From January 2005 to August 2006, he was a senior researcher at the PLA University of Science and Technology, Nanjing, China, where he performed research and development on advanced energy-efficient wireless router, ad hoc network simulator, and cognitive mesh networking testbed. His research area includes cognitive radio, wireless multi hop networks (Ad Hoc, Mesh, WSN, etc.), wireless multimedia network, and green Communications.

Dr. Zhao is the symposium co-chair of ChinaCom 2009 and 2010. He is the TPC (Technical Program Committee) co-chair of IEEE ISCIT 2010 (10th IEEE International Symposium on Communication and Information Technology).

Qi Chen received her B.S. in communication engineering from Zhejiang University in 2014. Currently, she is a Ph.D. candidate in the College of Information Science and Electrical Engineering, Zhejiang University, China. Her research interests include data mining of wireless networks and green communications.

Honggang Zhang is a full professor with the College of Information Science and Electronic Engineering, Zhejiang University, Hangzhou, China. He is an honorary visiting professor at the University of York, York, U.K. He was the international chair professor of excellence for Universite Europenne de Bretagne (UEB) and Supelec, France. He is currently active in the research on green communications and was the leading Guest Editor of the *IEEE Communications Magazine* special issues on "Green Communications". He is associate Editor-in-Chief (AEIC) of China Communications as well as the series editor of IEEE Communications Magazine for its Green Communications and Computing Networks Series. He served as the Chair of the Technical Committee on Cognitive Networks of the IEEE Communications Society from 2011 to 2012. He was the co author and an editor of two books: *Cognitive Communications: Distributed Artificial Intelligence (DAI), Regulatory Policy and Economics, Implementation* (John Wiley & Sons, Ltd) and *Green Communications: Theoretical Fundamentals, Algorithms and Applications* (CRC Press), respectively.

13

Network Softwarization View of 5G Networks

Takashi Shimizu, Akihiro Nakao, and Kohei Satoh

The Fifth Generation Mobile Communications Promotion Forum (5GMF), Tokyo, Japan

13.1 Introduction

This chapter describes a view of 5G networks based on the outcome of the study conducted in the 5GMF in Japan. The 5GMF, the fifth-generation mobile communications promotion forum, was established on September 30, 2014, according to the roadmap reported by the Radio Policy Vision Council of Ministry of Internal Affairs and Communications in Japan. It aimed at giving strategic guidance on R&D, standardization, international cooperation, and public relations. This forum has provided the venue for operators, manufacturers, academics, and government organizations in Japan together to exchange ideas and views on 5G to draw up the plans for overall R&D activities through Industry-Academic-Government Cooperation. In May 2016, the 5GMF completed the initial study and published the white paper [1], which clarified the key concepts and key technologies of 5G. The white paper especially emphasized the industry trend, "network softwarization and slicing," to facilitate harmonization of enabling technologies for 5G networks, and illustrated the network softwarization view of 5G networks.

In the following section, the key concepts and key technologies of 5G are revisited, and the views of 5G networks are described in detail with the brief history for the industry reaching the idea of network softwarization and slicing. The outstanding issues for further R&D activities are discussed with the latest activity and deliverables in ITU-T SG13 [2,3].

5G Networks: Fundamental Requirements, Enabling Technologies, and Operations Management, First Edition. Anwer Al-Dulaimi, Xianbin Wang, and Chih-Lin I.

13.2   Key Concept of 5G[1]

The 5GMF identifies two key concepts for 5G: "Satisfaction of end-to-end (E2E) quality" and "extreme flexibility." "Satisfaction of E2E quality" means providing every user satisfactory access to any application, anytime, anywhere, and under any circumstance. "Extreme flexibility" is the feature of communications systems that will allow 5G to always achieve E2E quality.

The demands by users and needs by applications for E2E quality in the 5G era will be much more diverse when compared to previous generation systems. The dynamic ranges fluctuated by the temporal and spatial factors will also expand more dramatically. These changes determine a major requirement for 5G that is completely different from previous generation systems, for which providing best quality was sufficient. Additionally, when thinking about radio access and network coordination concerning constraints of temporal factors, the minimum latency is determined by the path length between servers and terminals or controllers and controlled equipment in the network. Thus, extreme flexibility cannot be realized by an individual radio access or core network on its own. Rather, extreme flexibility through the coordination of networks will be needed in order to provide the dynamically diverse and fluctuating E2E quality that users will demand.

In previous generation systems up to 4G, radio access networks were regarded as the dominant bottleneck in determining E2E quality of mobile applications and services, since the performance of radio access networks were limited by a number of constraints, including radio propagation characteristics, available bandwidth, handset power, and mobility. In the 5G era, however, it is expected that most of these constraints will be greatly relaxed by the advancement of radio technologies, meaning the performance of radio access networks alone no longer the sole bottleneck. The performance of core networks will also be taken into account to satisfy E2E quality. Therefore, the technologies for radio access and core networks should be studied jointly and developed on an equal basis in order to realize "extreme flexibility."

The 5GMF also identifies two key technologies necessary to support "extreme flexibility" and thus the wide range of use cases expected in the 5G era: "Advanced Heterogeneous Network" and "Network Softwarization and Slicing." It is emphasized that through the use of these two key technologies, 5G standards will allow wired and wireless networks to handle the growing demand for larger capacity and higher speeds as previous mobile communication systems have been able to do so in the past.

---

1 This section is derived, in part, from an article published as the 5GMF white paper, "5G Mobile Communications Systems for 2020 and Beyond," Available at http://5gmf.jp/en/whitepaper/.

13.3 Network Softwarization View of 5G Networks[2]

Network softwarization is an overall transformation trend about designing, implementing, deploying, managing, and maintaining network equipment and/or network components through software programming. It exploits the ability of the software, such as flexibility and rapidity, to manage the lifecycle of network functions and services. It will enable redesigning of network and service architectures, in order to optimize processes and expenditure, self-management, and added value to an infrastructure [2].

The term "network softwarization" was first introduced at the academic conference, NetSoft 2015, the first IEEE Conference on Network Softwarization. It encompasses broader ideas in the industry, including network virtualization, network functions virtualisation (NFV) [4], SDN (software-defined networking) [5,6], mobile edge computing (MEC) [7], cloud/IoT technologies, and so forth.

The term "network softwarization" is introduced to describe the view of 5G networks with the notion of programmable software-defined infrastructure.

The basic capability provided by "network softwarization" is "slicing," as defined in Refs [8,9]. Slicing allows logically isolated network partitions (LINP) to exist in an infrastructure. Considering the wide variety of application domains to be supported by 5G, it is necessary to extend the concept of slicing to cover a wider range of use cases than those targeted by NFV/SDN technologies, and a number of issues are to be addressed on how to compose and manage slices created on top of the infrastructure.

Figure 13.1 illustrates the network softwarization view of 5G networks, which consists of a couple of slices created on a physical infrastructure and a "network management and orchestration" box. A slice is a collection of virtualized or physical network functions connected by links, and it constitutes a networked system. In this figure, the slice A consists of a radio access network (RAN), a mobile packet core, an UE (user equipment)/device, and a cloud, each of which is a collection of virtualized or physical network functions. Note that the entities in Figure 13.1 are described symbolically: Links are not described for simplicity. The box "network management and orchestration" manages the life cycle of slices: creation, update, and deletion. It also manages the physical infrastructure and virtual resources via abstraction of physical ones. The physical infrastructure consists of computation and storage resources that include UEs/devices (e.g., sensors) and data centers, and network resources that include RATs (Radio Access Technologies), MFH (Mobile FrontHaul), MBH (Mobile BackHaul), and transport. It should be noted that both computation/storage resources and network resources are distributed and are available for virtualized network functions wherever required.

---

2 This section is derived, in part, from a report published as, "FG IMT-2020: Report on Standards Gap Analysis." Available at http://www.itu.int/en/ITU-T/focusgroups/imt-2020/Documents/T13-SG13-151130-TD-PLEN-0208%21%21MSW-E.docx.

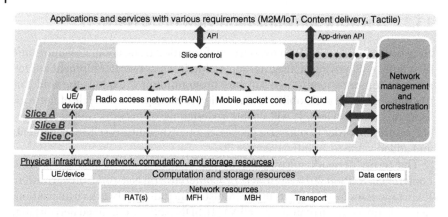

Figure 13.1 Network softwarization view of 5G networks.

In addition, virtualized network functions and other functions assigned to a slice are controlled by the "slice control." It oversees the overall networked system by configuring its entities appropriately. It may include network layer control and service/application layer control. In some cases, it makes a part of infrastructure being service-aware. It depends on the requirements presented for the networked system, for example, a slice to provide the support of information-centric networks (ICN).

Orchestration is defined as the sequencing of management operations. For example, a customer can send a request to the "network management and orchestration" box with his or her own requirements of an end-to-end service and other related attributes. The request is handled in the box to create a network system with network programmability functions, if they exist, in fronthaul/backhaul or transport networks. This involves support for on-demand composition of network functions and capabilities and enforcement of required capability, capacity, security, elasticity, adaptability, and flexibility where and when needed.

Step 1) **Creating a slice**

Based on a request, the "network management and orchestration" creates virtualized or physical network functions and connects them as appropriate and instantiates all the network functions.

Step 2) **Configuring the slice**

The slice control takes over the control of all the network functions and network programmability functions, if they exists, and configures them as appropriate to start an end-to-end service.

13.4   Brief History of Network Softwarization and Slicing

Network slicing is the major feature provided by the overall trend expressed by the network softwarization. It is the topic of discussion in research communities and industries, as well as standards developing organizations (SDOs) such as 3GPP (The Third-Generation Partnership Project) and ITU-T. This is because, in the context of 5G, it allows mobile operators to enable different classes of communications, for example, for providing eMBB (enhanced mobile broadBand), mMTC (massive machine-type communications), URLLC (Ultrareliable and low-latency communications), and so forth.

The definition of "network slicing" generally means an isolated collection of resources and functions implemented through software programs on top of the resources to enable QoS guarantee for the network requirements as well as in-network processing along end-to-end communications. The "network slicing" has been first introduced in the overlay network research efforts, such as PlanetLab [10,11], in 2002. At that time, a slice was defined as an isolated set of network bandwidth and computational/storage resources allocated for a group of users that "program" network functions and services over their overlay network overlaid across "the planet." Later, various network virtualization testbed efforts such as PlanetLab EU [12], GENI [13,14], VNode [15,16], FLARE [17], and Fed4Fire [18] were made, each inheriting the concept of the slice as a basis of its infrastructure, as a set of programmable resources to tailor new network services and protocols.

The concept of network virtualization has been discussed and developed in the AKARI project in Japan. Then, VNode [16], a Japanese research project, started in 2008, aimed at implementing a platform to support the practical requirements for operators to utilize virtualization in their networks. This project generated the VNode architecture, a building component for a network virtualization platform. The architecture is based on the definition of "network slicing" in PlanetLab, and it extended the details so as to be utilized in operators' networks. The outcome of the project not only affected the development of other virtualization testbeds, such as GENI but also contributed to the standardization activities in ITU-T, where we could find the recommendations on the framework [8], requirements [9], and functional architectures [19], in addition to a number of industries' efforts of NFV (Network Functions Virtualization) and SDN (Software Defined Networking), including the large-scale experimental evaluation of virtualized EPC (Evolved Packet Core) [20].

In 2014, NGMN (Next-Generation Mobile Networks) Alliance published the White Paper [21] and identified "network slicing" as the major feature for networking for 5G. It is intended to embrace flexible functions and capabilities of overall systems by leveraging the developments relating to NFV/SDN. Subsequently, both academia and industry have acknowledged the essential value of network slicing and the 5G promotion forum, such as 5GMF, has been guiding a number of efforts to leverage and extend the feature, under the umbrella of "network softwarization." Such efforts are aiming to address practical issues to provide end-to-end quality, which are elaborated in the following sections. In terms of standardization, the mobile part of the overall architecture has been studied in 3GPP [22], and has been completed in 2018. The specifications will allow industry players to create business cases for 5G, when utilizing the plenty of experience and knowledge earned by long-term academic research and development worldwide.

13.5   Issues for Slicing Towards 5G[3]

13.5.1   Horizontal Extension of Slicing

In 5G, to satisfy end-to-end quality is an important requirement. Especially, as wireless technologies are expected to advance, networking technologies should support as appropriate to sustain end-to-end quality of communications. Therefore, it is natural to consider extending the slicing concept to cover end-to-end context, that is, from UE to cloud. Issues in extending slices have to then be addressed, not only the software defined infrastructure in a limited part of a network, but also the entire end-to-end path.

The scope of the current SDN technology primarily focuses on the portions of the network such as within data centers or transport networks. In 5G, it is necessary to consider end-to-end quality. Therefore, there exists a gap between the current projection of SDN technology development and the requirement for end-to-end quality. It is desired that an infrastructure for 5G will support end-to-end control and management of slices and the composition of multiple slices, especially with consideration of slicing over wireless and wireline parts of end-to-end paths.

Figure 13.2 shows the breakdown of the end-to-end latency in the current mobile network. The figure shows the network architecture to allow latency-aware deployment of network functions and services in order to make the most of wireless latency reduction (targeted from 10 to 1 ms) described in ITU-R IMT Vision [23].

3GPP carried out latency studies for 3G, which are documented in specifications. 3GPP has carried out studies for future network service requirements, which are documented in TR22.891 [24]. Operators are building LTE networks

---

3 This section is derived, in part, from a report published as, "FG IMT-2020: Report on Standards Gap Analysis," Available at http://www.itu.int/en/ITU-T/focusgroups/imt-2020/Documents/T13-SG13-151130-TD-PLEN-0208%21%21MSW-E.docx.

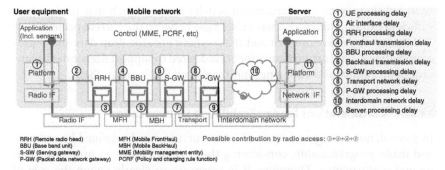

Figure 13.2 Breakdown of end-to-end latency of the current mobile network.

to meet the latency budget provided in the 3GPP specification. Latency studies carried out on many LTE-deployed networks demonstrate that the 3GPP specifications provide adequate guidelines [25-28]. Actual LTE network performances varied, however, due to a variety of variables as well as adjacent ecosystems.

For 5G, an extensive latency study should be carried out in order to provide guidelines for a number of latency-critical services. In order to structure the latency study framework, it is suggested to use breakdown of latency according to Figure 13.2.

13.5.2   Vertical Extension of Slicing: Data Plane Enhancement

5G networks may support various communication protocols, even those that have not yet been invented, for services such as Internet of Things (IoT) and content delivery provided by information-centric networking (ICN) and content-centric networking (CCN). Advanced infrastructure may need the capability of data-plane programmability and associated programming interfaces, which we could call the vertical extension of slicing. The current SDN technology primarily focuses on the programmability of control-plane, and only recently the extension of programmability to data-plane has been discussed in the research community and in ITU-T SG13 without well-defined use cases. For 5G, there are several use cases for driving invention and introduction of new protocols and architectures especially at the edge of networks. For instance, the need for redundancy elimination and low latency access to contents in content distribution drives ICN at mobile backhaul networks. Protocol agnostic forwarding methods such as protocol oblivious forwarding (POF) discuss the extension to SDN, addressing forwarding with new protocols. In addition, protocols requiring large cache storage such as ICN needs new enhancement. A few academic research projects such as P4 [29] and FLARE [30] discuss the possibility of deeply programmable data-plane that could implement new protocols such as

ICN, but there is no standardization activity to cover such new protocols to sufficient extent. Therefore, there exists a gap between the current projection of SDN technology development and the requirements for deep data-plane programmability. The infrastructure for 5G is desired to support deeper data-plane programmability for defining new protocols and mechanisms [31,32].

13.5.3 Considerations for Applicability of Softwarization

In general, not every component of infrastructure may be defined by software and made programmable, considering the trade-off between programmability and performance. Therefore, it is necessary to clearly define the role of hardware and software according to the potential use cases when softwarizing infrastructure.

SDN is primarily motivated by reduction of operating and capital expenditure and flexible and logically centralized control of network operations. Operators might be motivated to softwarize everything everywhere possible to meet various network management and service objectives.

In 5G, some applications have stringent performance requirements such as ultra low latency and high data rate, while others may require cost-effective solutions. A range of solutions exists from application-driven software-based solutions executed on virtualization platform with hypervisor, container, or bare metals to complete hardware-assisted solutions. The former may need performance enhancement enabled by hardware-assisted solutions, while the latter may be facilitated by software-based solutions. The infrastructure for 5G may need to support traffic classification performed not only by flow basis but also by other metrics and bundles such as per-device and per-application basis so as to apply software-/hardware-based solutions appropriately tailored for individual use cases. Therefore, there exists a gap between the current projection of SDN technology development and the requirements for applicability of softwarization.

13.5.4 End-to-End Reference Model for Scalable Operation

Softwarized systems should have sufficient levels of scalability in various aspects of functions, capabilities, and components. First, the target range of the number of instances should be considered, for example, service slices to be configured and to be in operation concurrently. The number of clients and service providers accommodated by each service slice is also an important metric for the practical deployment of the systems. The main constraints for scalability would be the dynamic behavior of each slice and control granularity of physical resources. The communication session established by a mobile packet core, however, would be challenging, because it requires a dedicated system for such an extraordinary multiple-state and real-time control, especially for mobility handling. The coordination and isolation between these systems should be clearly defined. Nevertheless, scalability for other types of sessions would also be an issue concerning architectural modeling, including application services, system operation, or advanced network services.

In addition to the dimensions and dynamics of the systems, further research is required from the perspective of resiliency and intersystem coordination. For resiliency, some new aspects might be considered other than traditional mean-time-between-failure (MTBF) type faulty conditions. In case of disaster, for example, fault localization, analysis, and recovery of softwarized systems could be more complicated. Traditional operation architecture also finds it difficult to cope with misbehaviors caused by human factors because of the indirectness arisen when operating softwarized systems.

The intersystem coordination architecture should be clearly structured and modeled for efficient standardization and for scalability evaluation of softwarized systems. There might be two categories of the coordination: horizontal and vertical. The horizontal coordination is for between slice, cloud systems, and UE-the end-to-end system coordination. Vertical coordination can be distinguished in two ways. One way is for slice and service provider through APIs and the other way is for virtual and physical resource coordination aimed to efficient resource handling through policy and analytics.

In summary, softwarized systems should have sufficient levels of scalability as follows:

- The number of instances/service slices to be supported
- Series of capabilities provided by service slices
- The number of service sessions to be handled concurrently
- Dynamic behavior of the instances and slices
- Granularity of resource management, especially for policy control and/or analytics
- Resiliency for various faulty conditions
- Intraslice coordination among end-to-end resources, and
- Interslice coordination, specifically with various external systems.

Intensive studies are required on both the dimension and the dynamic behavior of softwarized systems, since such systems will have an enormous number of instances and their reactions are not easy to extrapolate from the current physical systems.

Virtual resource handling must be an essential part of the scalable and novel operation architecture, which potentially improves conventional network operations and possibly even up to the level of supporting disaster recovery by using network resiliency and recovery of/with the systems both in a single domain and in multiple domains.

The end-to-end quality management is a key capability required for 5G. However, this capability will be established on the complex interaction among softwarized systems including UEs, cloud systems, applications, and networks. An appropriate end-to-end reference model and architecture should be intensively investigated for such complex systems.

13.5.5 Coordinated APIs

It may be useful to define APIs so that applications and services can program network functions directly bypassing control and management to optimize the performance, for example, to achieve ultra low latency applications.

Discussions on the capabilities of the programmable interface should be objective based, for example, accommodating a variety of application services easily, enabling higher velocity of service deployment, and operation and efficient physical resource utilization. Users or developers who utilize the APIs can be categorized according to their roles. Application service providers will enable value-added services over the end-to-end connectivity through the APIs. Advanced network service providers will add some sophisticated functions to communications sessions, such as security and reliability, in order to facilitate faster application service deployment by the aforementioned application service providers. Network management operators will also utilize the APIs for more efficient and agile resource handling.

Information modeling should be the most significant issue for API definitions. It should include virtual resource characteristics, relationships between various resources, operational models, and so on. Levels of abstraction should be carefully investigated, so that the model and APIs should be human-readable and machine/system-implementable at higher performance simultaneously. Since considerations on software development methodologies will have an impact on the development model, the choice of the proper methodology for each capability will be important.

The system control and coordination architecture is another issue that will affect the achievement of scalable and agile APIs. Not only the traditional provisioning/configuration or distributed control of networking systems but also automatic and autonomic system control should be the main target. The closed-loop control architecture might be the most innovative enhancement from the traditional networking systems even for the APIs.

The robustness and fault tolerance are absolutely necessary for open systems controlled through the APIs by various providers. Isolation over virtual resources should be carefully structured with the APIs' functionalities and constraints.

In summary, discussions on the programmable interface capabilities should embrace the following:

- Level of abstraction sufficient for both system operations and customization of the capability provided by the interfaces,

- Modeling for virtual/abstracted resources in a multiple-technology environment,
- Ease of programming for service and operation velocity,
- Technologies for automatic and/or autonomic operations, and
- Provisioning of classified functional elements suitable for a range of system developers such as supplication service providers, network service providers, and network management operator.

13.6 Information-Centric Network (ICN) Enabled by Network Softwarization[4]

13.6.1 General Characteristics

13.6.1.1 Overview

One of the aims of 5G is the provision of the emerging network paradigm that fits social requirements. ICN is a promising candidate, with a variety of R&D activities ongoing worldwide. ICN has several merits, including the following:

1) Content access by its name
2) Traffic reduction by in-network caching
3) Provisioning of in-network data processing
4) Content security
5) Robustness to network failures by multipath routing

Details of these aspects are described in the following section.

This architecture, however, adopts a new data forwarding mechanism different from the current Internet. Therefore, it is necessary to have data-plane programmability.

13.6.1.2 Content Access by Its Name

The prime difference between ICN and the current Internet is how content is accessed. Content is accessed on the Internet by knowing where the content server is located on the network. ICN, on the other hand, content is accessed by submitting a request of the name of the content on the network. The network will then route the request to the appropriate network node that is storing or caching the named content. The capability to access content by finding "named content" is the basis of ICN. It is emphasized that a named content is

---

4 This section is derived, in part, from a report published as, "FG IMT-2020: Report on Standards Gap Analysis," Available at http://www.itu.int/en/ITU-T/focusgroups/imt-2020/Documents/T13-SG13-151130-TD-PLEN-0208%21%21MSW-E.docx.

stored dynamically and moves to the node where the content is most frequently requested and therefore is more efficiently served to the end user. This can also apply to in-network data processing services. Accessing named content also makes it easier to support consumer mobility by making the ability to serve content more efficiently, as well as improving human readability of content requests.

13.6.1.3 Traffic Reduction by In-Network Caching

Another feature of ICN is in-network caching. ICN network nodes are equipped with a content cache server that caches content going through a particular node. The server will then autonomously select which content to cache based on the need of the users accessing the node. Generally, despite different use cases, content will generally move toward the network edge node where the specific named content is frequently requested. Once the most popular content is cached at the network edge node, subsequent content requests will be served at the particular network edge node, with future communication being terminated at this edge, resulting in a total reduction of the network traffic and lessening the overall server load.

13.6.1.4 Provisioning of In-Network Data Processing

In-network data processing will provide network nodes to do network-wide data processing and provide application services on network nodes. The current configuration will need a basic structural change to handle the increase of video traffic and the expansion of IoT, as well as to provide shorter response times. Currently, data processing is done at a remote data center and the network functions only as a data pipe. In 5G, data processing for application services will be provided with the aim of reducing network congestion as well as shortening response time when necessary. Two typical examples of in-network data processing are ICN, which reduces traffic congestion and response time through the use of a network cache, and edge computing, which provides data processing and service provisioning at the network edge. In-network processing can be considered generally an expanded form of edge computing, where data processing and service provisioning will be provided dynamically any place on a network that is appropriate. Due to the dynamic nature of service and data processing points, ICN's basic mechanism of accessing requested content by name rather than location is especially suitable to provide in-network data processing. Edge computing is efficient in terms of shortening response times and reducing network congestion when the target data for computing is close to an edge node area. Some IoT use cases will, however, have target data needed for processing across many edge node areas; therefore, the inner node of a network will be more appropriate for processing. Another example is on-path data processing, where data processing is applied in tandem on a transmission path.

This is frequently used in big data processing. There are also some use cases in which the inner network node is better suited to perform data processing, for example, when users for a particular service are few in number and yet distributed across several edge nodes.

13.6.1.5 Content Security
In some ICN architecture such as CCN and NDN (Named Data Networking), content security is provided as a basic function. Since security is a key concern in several systems like content delivery and IoT, having a built-in security mechanism is a very attractive point of ICN.

13.6.1.6 Robustness to Network Failures by Multipath Routing
To enable the content access by name, ICN routing/forwarding is capable of multipath routing, because the contents once cached in certain node will not be available at the next chance. In ICN multipath routing, when the response does not come back from the direction the interest is sent out, the node will automatically issue the same request to another direction. This mechanism is very helpful when a part of the network fails, such as during disaster, and makes the network robust to failures.

13.6.2 Applications of ICN

13.6.2.1 Networking in a Disaster Area
This service scenario describes ICN as a communication architecture that provides an efficient and resilient data dissemination in a disaster area.

A provider using ICN will be able to directly disseminate emergency data to specific individuals or groups. In this use case, consumers in advance will express their interest in a specific type of emergency data and information, which ICN will deliver when available. Providers will also be able to directly disseminate emergency data to its users in case of an emergency regardless of any prior requests for this service as well.

A provider can push emergency data to the cache or storage of ICN nodes, and then the ICN nodes can indirectly deliver the emergency data from their cache or storage to specific individuals and groups as well as to a larger population using the network on a case-by-case basis.

ICN nodes have sufficient storage capacity and so they can hold emergency data for a long time. Both providers and consumers can use the ICN storage system as intermediate devices to share any emergency data with others during a disaster period.

Providers will be able to efficiently and resiliently disseminate emergency data in a disaster area due to the forwarding and caching functions of ICN, in which emergency data are forwarded to an intermediate ICN node where the data are kept (cached) first and then sent to the final destination or to another intermediate ICN node. The caching data can be served for other consumers (efficient data dissemination) even when the original provider is not available due to a temporal network partition (resilient data dissemination).

A consumer can retrieve emergency data even from an intermittently connected network during a disruption or disaster period. ICN follows a receiver (consumer)-driven communication model where receivers can regulate if and when they wish to receive segments of data, and so continuous data retrieval from multiple ICN caching points is possible without regard to end-to-end session.

Operators will benefit from the reductions in system construction costs related to protecting their networks in case of a disaster. Due to name-based communication, there is no clear, functional boundary between the network and end devices in ICN. This means ICN nodes can act on behalf of end devices by recognizing and responding to user requests. For example, all ICN nodes will be able to respond to all consumers using its storage capability to share information in a disaster area. This will be a particularly useful feature at times when it is impossible to predict which parts of a network will not be accessible during a disaster.

13.6.2.2   Advanced Metering Infrastructure (AMI) on a Smart Grid

This service scenario involves smart meters, communications networks, and data management systems that provide two-way communication between utility companies and their customers. Customers will be given assistance through devices such as in-home displays and power management tools. On the communication network, ICN nodes can be installed in order to keep a copy of these data in its cache, which can then be used to present the data in a desired format for the convenience of both the consumer as well as the utility company. The ability to quickly retrieve use pattern data of a particular service is very important in order to efficiently plan and consume services provided by utilities to consumers.

By using the ICN nodes in the network, efficient resources usage and effective load control is possible. Besides an ICN approach for AMI systems in smart grid, they can efficiently control network congestion, support mobility, and ensure security.

Since with ICN it is possible to secure data, customer will feel more comfortable with the smart grid infrastructure based on the ICN. Furthermore, the operator (utility company) can manage the data more cost-effectively. It also adds value in the scalability issue.

13.6.2.3   Proactive Caching

This service scenario involves people who will access the Internet through their portable device, such as a smartphone or laptop, while they are the passengers

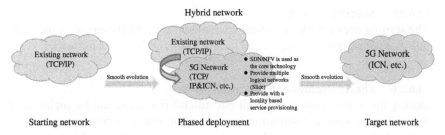

Figure 13.3 Phased migration.

of a moving vehicle, such as trains, cars, and buses. A certain passenger wants to watch a video-on-demand on his or her smartphone. If this passenger is on a commuter train, the desired video will be proactively cached in every train station's ICN node according to the scheduler, which decides how much video content should be proactively cached according to video and transportation information. If the user is a vehicle passenger, such as a car, the vehicle mobility information, accessed from the navigation system, can be used to choose an ICN node where content/video will be cached proactively.

The quality of video delivery can be significantly improved by using proactive caching integrated with ICN nodes. Since an ICN node fetches a data object in advance, data objects requested by the mobile user will be immediately available after changing the point of attachment. The delay will be minimized due to the reduction of number of hops taken during data transmission. In addition, since the ICN node will maintain a cache of a particular data object, all subsequent requesters of the same data object will reuse the data already cached by the ICN node.

Network operators will benefit as well. First, bandwidth consumption will decrease due to caching and data reuse. Second, energy consumption will be reduced since data objects accessed from ICN nodes through Wi-Fi, reducing traffic on 3G/4G networks. The reduction of transmission delays will also allow providers to offer enhanced user experiences for their customers.

13.6.2.4 Migration Scenario

5G will coexist with legacy network equipment and be compatible with existing network technologies. In other words, it should work in a hybrid manner: It may be composed of classical physical network appliances and softwarized appliances during the intermediate phase toward full deployment. Therefore, migration from the starting network to the target will gradually be accomplished by using a hybrid deployment model, as shown in the following three-steps-migration path (Figure 13.3).

13.6.2.5   Starting Network
The starting network phase utilizes current and state-of-the-art network technologies (existing technologies), including LTE- and IP-based networks.

13.6.2.6   Phased Deployment: Intermediate Phase
During the intermediate phase, all end-to-end resources can be maintained through conventional communication means in order to communicate with each other. However, the softwarization mechanisms allow existence of the migrated end-to-end resources deployed in conjunction with existing ones. It enhances the overall migration process feasibility by enabling gradual deployment while maintaining current communication networks simultaneously during migration period.

Migration scenarios from the early stage include the following:

- Locality-based service provisioning mechanisms and architecture, for example, mobile edge computing and local area computing.
- In-network data processing/service provisioning capability, where each network node carries out some data processing and service provisioning - a feature especially useful for the efficient management of IoT devices and big data.
- Adoption of emerging network technology widely.

Possible technological directions include the following:

- Application of network softwarization, such as SDN and NFV, as a core technology.
- Adoption of multiple logical network partitions (slices), each having different architecture that fits to the services provided on the slices, such as IP, ICN, IoT, and low latency.
- Having a clear API to provide for the development and distribution of a variety of applications and services.
- Providing in-network data processing capabilities, whereby each network node carries out some data processing and service provisioning. This feature will allow handling IoT devices and big data efficiently.

13.6.2.7   Target Network
The target network will benefit operators in three ways. First, the bandwidth consumption will be low due to caching and data reuse. Second, energy consumption will be reduced by accessing data objects from the ICN node through Wi-Fi, reducing the 3G/4G traffic. Since the transmission delays will be minimized, operators will be able to provide an enhanced user experience as well.

13.7 Studies in ITU-T SG13 Focus Group on IMT-2020

In April 2015, ITU-T SG13 launched the focus group (FG) on IMT-2020 in order to accelerate the study of the nonradio aspects of IMT-2020. IMT-2020 is the term in ITU to identify the systems for the year 2020, possibly 5G. During the first year, FG conducted gap analysis of existing technologies against the requirements of IMT-2020. The areas of the analysis include network softwarization, overall architecture, QoS, fronthaul/backhaul, and emerging technologies, such as ICN (Information Centric Networking) and CCN (Content Centric Networking). This activity, called Phase-1, delivered the reports [2], which illustrated the network softwarization as the central concept and detailed gaps in those areas in December 2015. Based on the outcome of the Phase-1, during the second year term, called Phase-2, FG continued the study with the goal of compiling new standardization items to address the requirements for IMT-2020, clarified by the gap analysis in Phase-1. This activity was completed in December 2016 and the report was delivered [3], detailing the latest updates and clarifying the issues to be addressed in the industry for 5G networks. Further study of those issues for standardization has been continued as the work of ITU-T SG13 in the study period 2017-2020.

13.8 Conclusion

Network softwarization view of 5G networks have provided the conceptual basis for utilizing network slicing feature to enable different classes of communications, such as eMBB, mMTC, and URLLC. Based on the discussion of 5GMF, ITU-T FG IMT-2020, and other organizations, the definition and views of network softwarization and the issues and challenges toward the realization of 5G networks has been reviewed with focus on the pioneering activities in academia. A number of field trials, together with the standardization effort, are expected to form a consolidated architecture and associated system developments, which will accelerate the early deployment of 5G in many regions in the world.

References

1 5GMF, *5G Mobile Communications Systems for 2020 and Beyond.*, Version 1.01, July 4, 2016. Available: at http://5gmf.jp/en/whitepaper/.

2 ITU-T SG13 FG IMT-2020, *FG IMT-2020: Report on Standards Gap Analysis*, ITU, 2015. Available: at http://www.itu.int/en/ITU-T/focusgroups/imt-2020/Documents/T13-SG13-151130-TD-PLEN-0208%21%21MSW-E.docx.

3 ITU-T SG13 FG IMT-2020, *Nine Deliverables prepared Prepared in 2016*, ITU, 2016. Available: at https://www.itu.int/en/ITU-T/focusgroups/imt-2020/Documents/FG_IMT-2020_Deliverables_2016.zip

4 ETSI ISG NFV. *Network Functions Virtualisation: An introductionIntroduction, Benefits, Enablers, Challenges & Call for Action*, ETSI, 2012. Available: at https://portal.etsi.org/NFV/NFV_White_Paper.pdf

5 ONF, *Open Networking Foundation*, 2011. Available: at https://www.opennetworking.org/.

6 IEEE SDN, 2015. Available at: http://sdn.ieee.org/.

7 ETSI ISG MEC, *Mobile-Edge Computing: Introductory Technical White Paper*, ETSI, 2014. Available: at https://portal.etsi.org/portals/0/tbpages/mec/docs/mobile-edge_computing_-_introductory_technical_white_paper_v1% 2018-09-14.pdf

8 ITU-T. Y.3011, *Framework of Network Virtualization for Future networksNetworks*, Geneva, ITU, 2012. Available from:at http://www.itu.int/rec/T-REC-Y.3011-201201-I.

9 ITU-T Y.3012, *Requirements of Network for Future networksNetworks*, Geneva, ITU, 2014. Available: at https://www.itu.int/rec/T-REC-Y.3012.

10 L. Peterson and T. Roscoe, "The design principles of PlanetLab," *ACM SIGOPS Operating Syst. Rev.*, vol. 40, no. 1, pp. 11–16, 2006.

11 PLANETLAB, *An Open Platform for Developing, Deploying, and Accessing Planetary-Scale Services*, 2012. Available: at http://www.planet-lab.org.

12 PLANETLAB Europe. *An Open Platform for Developing, Deploying, and Accessing Planetary-Scale Services*, 2012. Available: at https://www.planet-lab.eu.

13 M. Berman, J. Chase, L. Landweber, A. Nakao, M. Ott, D. Raychaudhuri, et al., "GENI: a federated testbed for innovative network experiments," *Comput. Netw.*, vol. 61, pp. 5–23, 2014.

14 GENI, *An Open Infrastructure for At-Scale Networking and Distributed Systems Research and Education That Spans the US*, 2007. Available from:at https://www.geni.net.

15 A. Nakao, "VNode: a deeply programmable network testbed through network virtualization," in: 3rd IEICE Technical Committee on Network Virtualization, 2012.

16 K. Yamada, Y. Kanada, K. Amemiya, A. Nakao, and Y. Saida, "VNode infrastructure enhancement: deeply programmable network virtualization," in: 21st Asia-Pacific Conference on Communications (APCC), IEEE, 2015, pp. 244–249.

17 A. Nakao, *FLARE Open Deeply Programmable Network Node Architecture*. 2012. Available: at http://netseminar.stanford.edu/10_18_12.html

18 Fed4Fire, *Federation for Future Internet Research and Experimentation*. Available: at https://www.fed4fire.eu.

19 ITU-T. Y.3015, *Functional architecture Architecture of Network Virtualization for Future Networks*. Geneva, ITU, 2016. Available: at https://www.itu.int/rec/T-REC-Y.3015.

20 T. Shimizu, T. Nakamura, S. Iwashina, W. Takita, A. Iwata, M. Kiuchi, et al., "An experimental evaluation of dynamic virtualized networking resource control on an evolved mobile core network: a new approach to reducing massive traffic congestion after a devastating disaster," in: IEEE Region 10 Humanitarian Technology Conference, 2013.

21 NGMN, *NGMN 5G White Paper*, Version 1.0., February 17, 2015. Available: at https://www.ngmn.org/5g-white-paper.html.

22 3GPP. TR23.799, *Study on Architecture for Next Generation System.*

23 ITU-R, M.2083-0 *IMT-Vision Framework and Overall Objectives of the Future Development of IMT for 2020 and Beyond*, Geneva, ITU. 2015. Available: at https://www.itu.int/rec/R-REC-M.2083

24 3GPP. TR22.891, *Study on New Services and Markets Technology Enablers.*

25 3GPP. TR25.912, *Feasibility Study for Evolved Universal Terrestrial Radio Access (UTRA) and Universal Terrestrial Radio Access Network (UTRAN).*

26 3GPP. TR25.913, *Requirements for Evolved UTRA (E-UTRA) and Evolved UTRAN (E-UTRAN).*

27 3GPP. TR36.912, *Feasibility Study for Further Advancements for E-UTRA (LTE-Advanced).*

28 3GPP. TR36.913, *Requirements for Further Advancements for Evolved Universal Terrestrial Radio Access (E-UTRA) (LTE-Advanced).*

29 P. Bosshart, D. Daly, M. Izzard, N. McKeown, J. Rexford, and C. Schlesinger, et al. *Programming Protocol-Independent Packet Processors*, Available from;at http://arxiv.org/abs/1312.1719.

30 A. Nakao, "Software-defined data plane enhancing SDN and NFV," *IEICE Trans. Commun.*, vol. E98.B, no. 1, pp. 12–19, 2015.

31 A. Nakao, "Network virtualization as foundation for enabling new network architectures and applications," *IEICE Trans. Commun.*, vol. E93-B, no. 3, pp. 454–457, 2010.

32 A. Nakao, "End-to-end network slicing for 5G mobile networks," *J. Inform. Process.*, vol. 25, no. 2, pp. 153–163, 2017.

Akihiro Nakao received B.S. (1991) in Physics and M.E. (1994) in Information Engineering from the University of Tokyo. He was at IBM Yamato Laboratory, Tokyo Research Laboratory, and IBM Texas Austin from 1994 to 2005. He received M.S. (2001) and Ph.D. (2005) in Computer Science from Princeton University. He has been teaching as an associate professor (2005-2014) and as a professor (2014 to present) in Applied Computer Science at Interfaculty Initiative in Information (III) Studies, Graduate School of Interdisciplinary Information Studies, the University of Tokyo. He is a chairperson of the department at III. He is also a chairman of the Network Architecture

Committee of 5G Mobile Communication Promotion Forum (5GMF) since September 2014.

Kohei Satoh joined the Electrical Communication Laboratories, Nippon Telegraph and Telephone Public Corporation (NTT), Japan, in 1975 and transferred to NTT Mobile Communications Network, Inc. (NTT DOCOMO) in 1992. Since 1975, he has been engaged in research on radio propagation for satellite communication systems, and R&D of mobile satellite communications and mobile communications systems. After 1985, he has also been engaged in standardization activities of mobile satellite services in ITU-R and international alliance activities for IMT-2000. In July 2002, he moved from NTT DoCoMo to the Association of Radio Industries and Businesses (ARIB). He is now an executive manager on Standardization of ARIB, and his current job is to promote R&D and standardization activities for the Fifth-Generation Mobile Communications System. He is also Chairman of APT Wireless Group and Secretary General of the Fifth-Generation Mobile Communications Promotion Forum (5GMF). Dr. Satoh is a member of the IEEE and the Institute of Electronics, Information, and Communication Engineers of Japan, and a Fellow of Wireless World Research Forum.

Takashi Shimizu is a senior research engineer, a supervisor of NTT Network Innovation Laboratories. Shimizu received his doctoral degree in electrical engineering from Tokyo Institute of Technology in 1996, and joined NTT Laboratories. In 2004, he spent a year as a visiting scholar at Stanford University. Since 2011, he has been affiliated with the research laboratory in NTT DOCOMO INC., and has been engaged in the research and development for network functions virtualization of mobile systems. In 2014, he returned to NTT Network Innovation Laboratories and has engaged in the research on advanced wireless systems for future IoT applications. He is a member of IEEE, ACM, and IEICE.

14

Machine-Type Communication in the 5G Era: Massive and Ultrareliable Connectivity Forces of Evolution, Revolution, and Complementarity

Renaud Di Francesco and Peter Karlsson

Sony Europe Research and Standardisation Department, Sony Mobile, Lund, Sweden

14.1 Overview

This chapter assumes a commonly adopted framework of use for Mobile Connectivity in the 5G era.

Such use can be categorized as follows:

- *Human broadband use*, where human users equipped with a device can browse the Internet, stream content, use applications and servers, and communicate with other people.
- *Machine use*, where human interactions are not needed at every step, and machines or devices communicate with other machines to fulfill tasks and missions.

We will detail what is intended under these two categories of use. Since our focus in this chapter is specifically on the machine aspect, we will further analyze the requirements for such use.

We will also review the standardization path so far, and the expected objectives and characteristics in the 5G phase, learning from best practice and bottlenecks alike, in the development of Mobile Standards in previous and current generations: 2G, 3G, and 4G (LTE).

This chapter takes a problem-solving approach: starting with a problem inventory, leading to an inventory of methods used so far for solution development, and then addressing the possibilities to advance the state of the art into 5G, referring as much as possible to promising paths opened by the current research and development reported in the technical literature. We will conclude

5G Networks: Fundamental Requirements, Enabling Technologies, and Operations Management, First Edition.
Anwer Al-Dulaimi, Xianbin Wang, and Chih-Lin I.
© 2018 by The Institute of Electrical and Electronics Engineers, Inc. Published 2018 by John Wiley & Sons, Inc.

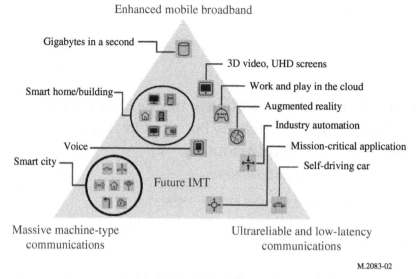

Figure 14.1 The "5G golden triangle of use" as produced by ITU-R for IMT2020 [1]

by sketching perceived opportunities and challenges for 5G in the machine communications domain.

14.2 Introduction

The 5G golden triangle (Figure 14.1) of use cases has three summits: the top edge standing for the broadband use by humans, as expected in a human-centric society, and the bottom two edges are about machine-type communication use case categories. These machine-type communication categories are massive connectivity for machines and machines operating in real-time mission critical environment.

Massive connectivity for machines (or devices) requires light communication client software and hardware. It is assumed in such application that the volume of data to be transmitted is typically on the low side, and that the frequency of such data transmission may probably be low. Conversely, such use case would require strong minimization of energy consumption, and in most cases a battery operation with a very long autonomy time before recharge if ever.

Machines operating in real-time, and mission-critical, environments, face tight constraints on latencies, feedback loops, complex and demanding cycles, reliability needs, and potentially the ability to ingest or output significant amounts of data at high speed. This can typically be considered under

Supervisory control and data acquisition (SCADA) type of remote supervision during any operation.

For engineers, machines have always been fascinating – they are the raison-d'etre of engineers as early as Archimedes of Syracuse and Heron of Alexandria.

In the same way, machines gave human muscular strength a multiplier order of magnitude (still remembered in automotive engines referring to horse power), digital engineering combined with physical engineering can achieve further upscales of efficiency and disruptive solutions.

Naturally, the digital leverage borrows a set of technologies where networks are only one part in a wider set of information and communications technologies. However, networks are always on the critical path of digitally powered operation.

For instance, the use of cloud architectures is increasingly frequent to support Enterprise Software and applications. This makes it essential to consider robustness of the end-to-end software operation with regard to network vulnerabilities. In this context, the energy consumption and the environmental efficiency have become visible, and make it necessary to review the allocation of weights between the cloud and the edge (these very machines) so that on-site preprocessing might alleviate the transmission volume and energy burden, while not overconsuming resources locally if they are constrained. We will discuss this optimization between centralized and distributed execution of tasks in the light of distributed cloud/mobile edge computing advances [2], processing, storage, and services for specific use cases of machine-type communications. It is important to explore a potential tuning to match legacy network resource as well as the future 5G resource as it becomes available.

How transmission requirements are addressed needs to be revisited for the 5G era, so that all feasible optimization can occur. This entails access to the network by the machine (device) and the possible capillarity architecture [3] designed to accommodate the edge demand efficiently, with a certain degree of traffic aggregation and handling of cycles in multiple access protocols for local access.

Machines are also mostly supervised, whenever possible, by SCADA systems. We will discuss specifically the supervision requirements, and look at these from the edge as well as from the Control Center, so that everything connects and operates as reliably as possible within the cost and technical resource budgets allocated to this supervised operation.

This will lead us to a second step of architectural discussion, motivated by open system semantics, whereas the first discussion and optimization step considers load balancing between edge and centralized cloud in a closed system.

The demand analysis will be followed by a top-down review of the standardization path so far, and how it expands into the 5G phase.

For instance, one could revisit the end-to-end development of 3G (UMTS) and assess how such a development occurred, keeping all identified use categories within the scope, with voice, data, and messaging carried further from the 2G era into the 3G era, while the innovation on the service side had been a significant uptake of data use for the early stage of mobile Internet.

As opposed to this, 4G picked up and focused on addressing only the mobile broadband (MBB) data need, and left voice support to be solved by adjacent technologies (OTT,* IMS,** or CSFB***), or in practice to tag it as out of scope for the radio access network.

This will lead us to identify some best practice, and to discuss a few of the bottlenecks encountered. Recommendations for the path to follow into the 5G era will be extracted from this analysis of recent standardization history.

The 5G candidate solution space will be explored from a wide sample of recent articles coming from the Research and Development community aiming at paving the way to 5G with innovative approaches and technical contributions. Conclusions will be proposed on what seems most promising and on what seems to need further assessment and investigation.

14.3 Demand Analysis

14.3.1 Machines Serving Humans

5G has a more stringent dual purpose, compared to previous mobile network generations, with the objectives of serving both humans and machines.

Humans will benefit from higher Internet speed.

The identified requirement is to provide network speed, compute speed, store/retrieval speed pulled by a 5G access speed increase, and supported by a wide coverage across territories where humans wish high performance Internet access.

Machines, need to leverage on the Internet of Things (IoT) with ubiquitous, massive, and ultrareliable connectivity.

Here the criteria are likely to include coverage across geographies, and indoor/outdoor professional and other environments of use, capillarity of network, device to device, data aggregation features, granularity (in time, and space) as required by the particular professional and other context of use.

---

* Over the top.

** IP Multimedia Subsystem.

*** Circuit Switched FallBack; in other terms, transferring to a 3G or 2G voice service the call initiated under 4G.

Machines, at their origin, helped multiply human strength. Beyond this multiplier effect, digital technologies including networking further upscale machine efficiency and increase precision and maneuverability.

Physics of information is leveraged upon, and energy is saved along the way wherever a substitution of physical activity by digital (virtual) activity is possible: electrons and photons carrying information travel more easily than atoms and molecules required by physical movement and transport at macroscopic scale (human travel, logistics of goods and commodities). Remote and virtual inspection and maintenance, for instance, saves on-site visits and inspections.

Within the machine realm, or the Internet of Things, what types of architectural orientations [4] can be taken?

When a number of devices or sensors and actuators are considered, a choice opens between centralizing or distributing digital processing [5] for data gathering and analysis and decision making.

The recent uptake of cloud architecture makes big data analytics affordable, by means of gathering data in a compute resource, and performing digital processing efficiently after data transmission, "in the Cloud." More recently still, the ETSI Industry Specification Group (ISG) initially named Mobile Edge Computing (MEC), now renamed "Multi-access* Edge computing [6]," has shown that for many cases, using a degree of distributed processing resulted in energy, network, or other resource saving [7], compared to a purely centralized architecture as in the pure cloud model. This underlines the dimension of multiple access technologies in distributed computing, and it has harnessed within an end-to-end architecture, which encompasses the power- and spectrum-constrained mobile segment.

Whenever designing a new service to support an Internet of Things application, a first analysis of load allocation of processing capabilities should be performed. It is likely that a mix of centralization and distribution in the cloud with Edge Computing or similar distributed processing as well as some degree of local processing in the device or machine will match the needs best. There is a long path of evolution in computing, always oscillating between further centralization and further distribution. A good illustration of the semantics behind such architecture has been offered by the systolic architectures [8].

14.3.2 Eyes and Hands to Control Industrial Systems: SCADA

Supervisory control and data acquisition is the overall system and service architecture supporting monitoring and action from a control room on a set of machines and technical infrastructure. Its dependence on the underlying net-

---

* "Multi-access edge computing" is an evolution in name and scope of the group initially formed as "Mobile Edge Computing" Industry Specification Group (ETSI ISG-MEC) at ETSI.

work connecting the parts of the supervised system has been analyzed in detail [9] for Industrial Control Systems.

14.3.2.1 Description of SCADA

The user interface of SCADA is classically a control room with a number of visual displays, enhanced by audio alarms and additional communication modalities as required. Operators monitor, supervise, and control remotely the operation of a machine system and infrastructure, such as a power plant, a transport, communication, or other utility network.

Images of space launches have captured the imagination of everyone, with real-time broadcast coverage showing the control center (NASA, ESA, etc.) as well as the launch pad and the rocket being launched, to carry satellites or other vessels to outer space.

The SCADA paradigm, shown in it as the control center, is progressively expanding to more and more categories of use, with more people put in the driving monitoring and ultimately controlling seat.

Mobile phones can be tracked by their owners, and many other objects and assets are following. To meet this market need of tracking objects and devices of all shapes and technical capabilities, services have been developed using several radio access technologies, including 3GPP-based systems and those operating in unlicensed spectrum. The latter are mainly local area networks such as Bluetooth and Wi-Fi and also Low Power Wide Area Networks [10] such as SigFox, LoRA, Weightless, with a more recent candidate L-four from Sony, added. The common challenge address is to fulfill the expected coverage area and trade off the long reach with throughput needs and low energy consumption in the remote device.

14.3.2.2 Mobile Networks Support for SCADA

Through their universal and federating purpose, Mobile Networks incorporating 5G "machine-friendly" developments will further extend the reach of monitoring and control.

In industrial environments, technical operators are usually very skilled and expert in their field of activity. When moving away from a professional skilled arena, the SCADA functionalities undergo a redesign making them more universally usable and scalable.

With SCADA-based monitoring, the challenge is not to raise more alarms but better alarms and ensure that the support to decision making and issue resolution does not leave the user down.

Error messages like "error 231" may not help anyone but the system developers, if error 231 is defined with an inward focus of what it means for internal processes involved in the execution of software or hardware action. In operation, alarms and issues should be raised with a view to be understood and acted upon when needed.

Humans remain in many cases at critical decision points, and their psychology influences how they react.

The analysis of air crashes has shown how humans ignoring deliberately alarm systems while under stress could contribute to aggravate an issue, when an actual detection is estimated as false detection by humans, who override the automation.

Conversely, lack of detection by the machines with no alarm/intervention triggered from humans can lead to catastrophes.

14.3.2.3 Data Processing in SCADA Systems

Data processing in SCADA systems is best described looking at the input signals, and then at an expert (automated) processing and at resulting output commands that will be delivered for execution.

Input Signals Input signals describe the status of system elements, and raise alarms when needed. They are sent by the nodes to the control center. In the particular instance shown in illustration, sensors generate the input signals.

Expert Processing An expert processing system establishes diagnosis and proposes actions to fulfill operational goals. In the illustration, this function is the Controller.

Output Commands Output commands are distributed to the nodes, to be executed remotely by each system element as required. These are shown as Actuators in the illustration

The three functions in the illustration, as sensors, actuators, and controller, are connected by networks. The SCADA system overall consists here of two categories of locations: a control center and multiple field sites, as shown in Figure 14.2. Locally elements are connected as the site allows, with local connectivity. Then the field sites and the control center are connected by reliable links, over networks that may benefit from wireless connectivity. Since the supervised system or process has its own requirements, the network used has to meet the corresponding latency and other reliability and performance constraints. Hence the industrial interest for the ultrareliable and low-latency 5G network.

14.3.2.4 National Electricity Grid Example

A practical illustration of such a supervised automated system is given by implementations of supervision and control for the operation of national electricity transport grids. Status information is visible from the control center, and commands can be sent in an automated or a semiautomated way to each remote node.

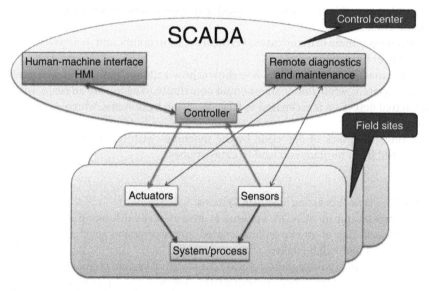

Figure 14.2 SCADA controller and filed sites [11].

In such systems the time is critical: Status descriptors need to be consistently time-stamped and commands need to be executed with time consistency and synchronicity where required across all nodes.

In a previous practical instance, one of the authors had to use a satellite-based communication network to ensure power network resilience to natural and other terrestrial disasters, and a specific clocking system for the nodes, to match synchronized response needs beyond the capabilities of a communication network at the time of implementation. This long confidential project has been described and published since then [12].

14.3.3 Machines and 5G

14.3.3.1 Digital Transformation of the Machines

The digital transformation wave is now reaching a further step in automation and industrial processes: Machines are connected to the network(s), and can be remotely supervised and controlled, leading to a new step of automation, making full benefit of networks and, in many cases, of sensors. The recent description of this as Industrial IoT (IIoT) is gaining further momentum.

This is leading to the engineering of solutions across the physical and the digital world, with, for instance, real-time goals and tasks in the physical world, combined with status and context awareness processed in the digital domain.

Such mixed physical and digital systems are often called cyber-physical systems. Requirements for the most demanding automation use cases are described within the 5G ultrareliable low-latency communications (URLLC) area. Such use cases require extremely low latency, very high reliability, and availability in harsh environments with innovative services and applications related to, for example, automation and industrial control in cyber-physical systems.

14.3.3.2 Cyber-Physical System Requirements

The system and service requirements for a distributed system relying on connected machines comprise 5G requirements defined in 3GPP for new services and markets technology enablers (SMARTER) specify reliability >99.999% for most of the URLLC use cases and in some specific applications like industry control the requirement is as high as 99.9999999%.

Granularity Requirements   The dimensioning factors can exist for any physical parameter, although two of them are usually the most central: time and space.

Particularly stringent demands occur in the time domain, with the need to respond quickly to an event, leading to different latency constraints:

- Latency for extreme industrial control <1 ms
- Latency for typical industrial control <8 ms
- Latency for industrial automation <25 ms

Similar constraints in the space domain depend on the ability to determine the position of an asset with a required precision.

Physical Resource Need   The need to access and use a physical resource determines the possibility for the system to reach its physical goals

In the physical domain, system reliability may also be complemented by health and safety requirements for humans involved in the use of the system, or who may come into its area of action. This leads to interesting developments in human–robot cooperation:

- *Digital Resource Need*

The performance of tasks goes through the access and use of digital assets such as the capacities to network, compute, and store.

- The digital parts of the execution of tasks implies associated information security requirements, and wherever humans are involved, privacy requirements.

- *Energy*

Each particular activity exhibits energy consumption patterns with specifically a battery autonomy requirement before recharge/change.

14.3.3.3 Vertical Use Case Examples

Any domain of the economy can be in scope. However, the following categories may have the most pressing need and ability to transform. We will highlight a few areas where fast digital growth is being observed: energy, health care, and smart cities. After this practical analysis, a more complete view for the digital transformation of the whole economy will be discussed.

- *Energy*

Energy in general, and in the current electronic era mostly electricity, is the first resource to assume for any digital activity to take place: No energy, no computing. Energy sometimes comes before water, depending on the context, as demonstrated in the large tsunami in Japan in 2011, where disaster recovery reactivating the water supply started with reactivating the power supply to water pumps, as was extensively described by the IEEE Robotics Society [13], with IEEE putting in practice daily its tagline "Advancing Technology for Humanity."

The high-voltage electrical grid is the energy transport backbone of the electrical utilities, with infrastructure typically at 400 kV. It is supervised using SCADA, with real-time response for command and control [14] required at large scale (country and even multi-country scale, because grids are interconnected [15]) closer to homes and workplaces. Smart meters allow fine-grain measurement of consumption and anticipation of local demand. The distribution voltage is typically 63 kV, down to 220 V at the home or office plug. Smart meters gather consumption data, and transmission patterns may be sparse packets every now and then.

- *Health*

Human life can be at stake, in most health care services. The demand for health care services scales up rapidly, with a special further acceleration caused by age, in regions with demographics of an aging population (Europe, Northern America, Japan, etc.). To meet this growing demand, several approaches are commonly being combined. A centralized approach is built around the hospital. Core health care can be concentrated in centers of clinical excellence to maximize the efficient use of rare resource: equipment and expert consultants in particular. It requires the transport and hosting of patients, with complex processes from admission, sorting by level of urgency required in treatment, to potentially bed allocation, surgical intervention, condition monitoring, and discharging. The complexity of patient, medical staff, and medical and other hospital asset management benefits from intelligent Information Systems and ubiquitous and responsive connectivity. The dynamic hospital environment with variable time constraints and indoor mobility of all elements and processes (with people and equipment, resources) is likely to benefit from an improved connectivity [16] of sensors collecting data. This will allow for supervision and command and efficient data analysis systems, with ubiquitous rendering for medical staff in charge. Wearable devices, indoor mobility, and positioning might help further increase care efficiency, with automation where appropriate and support of valuable but rare human expertise and wisdom. A distributed approach complements the above, with home support of patients.

Hospital stays are costly, and hospital beds should be saved for those really needing them. Furthermore, elderly patients [17] live longer and better when they are given the possibility to stay at home, with care coming to them rather the other way.

Active-assisted living (AAL), supporting people in their homes, is currently under study at the International Electrotechnical Committee (IEC)*, prior to standards development supporting future services. This committee has already assessed use cases [18], highlighting in particular the data and connectivity requirements to support such use cases.

• *Smart Cities*

The population of the world is increasingly living in cities and suburban areas. Facing daily transport and resource demand peaks, smart cities have strong incentives to use digital technologies for flow management and synchronization. They are places where demand and supply for resource can fluctuate from extreme density to very sparse density between peak and off-peak time and location, depending on where the population concentrates or spreads, with different objectives and interests (work, leisure, shopping, commuting between home and work, dining out, attending a game, concert or show, etc.).

Physical flows of people, goods, and other resources can be optimized, and peak management in particular is a common challenge faced. As the granularity of the network grows with thorough seamless indoor-outdoor coverage [19] and data are analyzed in-depth to understand and address phenomena involving people, asset, and process, related (private life, professional life, production, consumption, etc.) multifaceted needs are likely to appear and existing ones to evolve and grow.

Among services and functions designed to meet such needs, let us highlight three that are essential to the operation of a smart city: transport, energy, and communication. All of these use a networked infrastructure with multiple nodes (distributed needs in time and space), which can be supervised and impacted seasonally, be it daily with peak hours or yearly with peak season.

---

* International Electrotechnical Commission (IEC), *Mandate of System Committee on Active Assisted Living (IEC SyC AAL)*.

- *Transport*

Transport planning for both suppliers of infrastructure and service and users of these services starts ahead of journey and passenger flows to/through/from the smart city.

It continues as passengers and drivers live their journeys and experience navigation needs, congestion, rerouting, parking, or changing over to another transport modality during comprehensive physical transport sequences.

A second aspect of digital services along with transport experience is the continuation of digital services for users boarding a vehicle or entering a station or an airport. Such services include entertainment, communication, leisure, or work-related applications, possibly health and other ones too. Transport gives a pulse of how a city lives.

- *Energy*

The smart city with its smart infrastructures, including transport and buildings in particular, is subject to fluctuations of occupancy as population moves in and out of the locations of the smart city. Energy demand and supplies can be optimized accordingly. Energy is another pulse of how life is carried out in the city.

- *Communication*

As for transport and energy infrastructures, the high density of population in some locations at certain times creates peak effects on the demand for communication. This is both a challenge and an opportunity for communications services providers and network infrastructure operators. A massive surge in customer demand is observed, and value can be created if the resource to meet such demand has been adapted adequately. Resource dimensioning and provisioning and traffic management policies come into play to ensure fair access and value, as well as efficient resource utilization from the perspective of the resource and infrastructure owner. Communication in the smart city poses challenges, which can be alleviated by efficient edge computing [20].

5G communication can play a central role in enabling the following:

- *Coverage:* Beyond the canyon effects of streets on wave propagation, and the deep indoor coverage challenges.
- *Speed:* Ensuring that users get access to sufficient transmission speed for applications and use forms to meet the expected quality of experience from the users, be they humans or machine types.

- *Physical Implementation on the Busy Streets of the Smart City*

By nature, the smart city faces demand peaks for any resource. An important category of such resource is the street space. Let us highlight particularly two aspects of use of the street space: lighting systems and street furniture.

Lighting on the streets requires installing poles and other lighting systems, which are physically positioned as a particular case of street furniture, usually in adequate locations where people and machines are expected to pass by in significant flows and densities. Hence colocating radio and other network equipment with lighting equipment is usually efficient, from three viewpoints for efficiency maximization.

The demand for light is aligned with a demand for communication (humans and machines). In some cases the same lamp post can host lighting and radio functions.

Construction is costly and heavy. Civil engineering for light and other purposes (e.g., communications) allows a single inconveniency to be met by multiple benefits: lighting and communication.

Efficient maintenance and supervision is then obtained, after efficiency in the installation phase and efficient use of the resource by design.

Other street furniture with benefits similar to the lighting poles in use, installation, maintenance, and supervision include bus stops, traffic lights and street crossings, dustbins, street-level outlets and cabinets for water supply, and optical fiber in the near underground of the city.

Beyond the use categories addressed in detail above as energy, health care, and smart cities, the unavoidable impact of digital transformation is changing irreversibly economy and society. The vertical markets or sectors that are considered as fairly digitized may include: day-to-day human consumer markets, manufacturing lines, Energy sectors, Governments and Legal, telecommunication and transportation.

The list above is not the only view on the verticals. Let us also quote the more concise categorization made by the authoritative World Economic Forum:

- Automotive
- Consumer
- Electricity
- Health care
- Logistics
- Media
- Aviation and travel
- Chemistry and advanced materials
- Professional services
- Telecommunications
- Mining and minerals
- Oil and gas

The World Economic Forum [21] observes the importance of cross-industry themes such as the digital enterprise, digital consumption, societal implications, and introducing digital value to society. It lists a selection of key technologies for the purposes of digital transformation, as 3D printing and manufacturing on demand, artificial intelligence, autonomous vehicles, big data analytics and the cloud, the Internet of Things, and connected devices, robots, and drones.

An observed challenge is the possible tension between the providers of digital transformation, mostly with digital skills, and the current owners of the sector to transform. Both the target scenario and the transition path will orient the state of play after transformation. The transformation can be seen as hostile, in the case of a new entrant introducing a digital platform logic where incumbents had none, causing disruption and deintermediation. It can also be seen as an opportunity when the companies and organizations proactively shape their future [22], creating open standards and efficiencies for all stakeholders [23].

14.3.3.4 Machines and Humans

Whereas the previous sections addressed the needs of systems and organizations, let us now consider more directly the level of an individual, in their daily life, and in their work life.

In our daily life, as individual, we are owners, users, and beneficiaries of the operation of machines around us.

A smooth and reliable connectivity is of paramount importance, especially when the user does not want to be bothered with low-level operational matters, which in turn may be essential for the operation of the machine. This is a matter of quality of experience for the human user, and reliability and dependability of the machine system. This domain is often discussed as the smarter home [24].

Our work may focus on using machines to enhance our human efficiency in specific tasks, or organize the operation of machines in multiple ways. The overall efficiency gain matters to the organization employing workers and machines. It is also critical that the machine operation is safe, suitable, and acceptable for the human worker and user of the machines. An illustration is the human–robot cooperation [25]. An important case is the factory environment where the robot has to operate safely, reliably, and has to be trusted by its human coworker [26].

14.4 Reviewing the Standardization Path* So Far

14.4.1 Overview: From 3G to 4G

Lessons learnt from developing 3G and 4G will be summarized here, looking specifically on what can be gained as a criterion or as a method for machine-type communication in the 5G era.

---

* At 3GPP, mainly.

14.4.1.1 From "Voice-Mainly"** to "IP Focus"

3G chose an integrated approach to develop data and keep legacy telecom services voice and SMS.

4G developed high-speed data with focus on IP performance and left circuit switched (CS) voice outside. This made it possible to handle the tremendous growth of mobile IP-based data traffic. The voice calls triggered with smart-phone in 4G coverage had to be handled initially with CSFB (Circuit Switch Fall-Back) transferring the call back to a 3G network or similar, until IMS*** -based telephony was outlined as "VoLTE" (Voice over LTE) in a dedicated profile similar to a patch and fixed the voice issue.

14.4.1.2 Machine-Type Communication

4G kickstarted the machine-type communication and low-power wide area IoT connectivity with dedicated LTE categories. The first 3GPP initiative to specify LTE-based cellular IoT was taken in Release 12 with the UE (user equipment) category 0 (Cat-0). The UE specification according to Cat-0 had a clear target of IoT use cases, with lower data rate, reduced complexity, and minimized power consumption. The Cat-0 peak data rate is 1 Mbps, for both uplink and downlink and the complexity could be reduced by up to 50% compared to Rel-8 Cat-1 UEs, since the requirements include only one receiver antenna and support of half-duplex operation.

There has however not been any market traction for Cat-0 and instead more efforts were spent in the standardization to make further complexity and power reductions in Release 13. Two cellular IoT versions, the MTC Cat M1 and the Narrow Band IoT Cat NB1, have been defined in Release 13. These are both meeting the specific IoT requirements on UE complexity and power reduction, mainly due to a significantly lower RF BW compared to the 20 MHz in Cat-1 and Cat-0.

Another key challenge for the machine-type communication use cases is to achieve excellent coverage in small and low-cost devices. Such use cases include, for example, small tracking devices and smart meters, the latter that could be placed stationary in poor coverage (e.g., a basement) thus always remain in the same bad position. Coverage enhancement was addressed in Rel-13 MTC and NB-IOT. The goal was formulated as increasing maximum possible coupling loss (MCL) between the base station and the terminal by 15 dB (Cat M1) and 20 dB (Cat NB1) compared to legacy systems (LTE and GPRS, respectively). The method used is mainly based on repetition (up to 2048 times) and the solutions

---

** Voice support is provided through VoLTE (Voice over LTE), or through the voice communi-cation component in OTT (Over the Top) applications.

*** IMS stands for IP Multimedia Subsystem(IMS). It is a core network solution to delivermedia-rich real-time communication services.

	Cat 1	Cat 0	Cat M1	Cat NB1	[Cat M2[a]]	[CAT NB2]
3GPP release	Rel-8	Rel-12	Rel-13	Rel-13	Rel-14	Rel-14
UE RF BW (MHz)	20	20	1.4	0.2	[1.4-5]	[0.2]
Peak rate (Mbps)	10	1	0.8	0.1	[1-3]	[0.2]
Max power (dBm)	23	23	20/23	20/23	[20/23]	[14/20/23]

[a]3GPP Rel.14 standardization efforts with further improvements continued into Rel-15. Specifications were finalized in April 2017.

now show that both Cat M1 and Cat NB1 achieve more than 20 dB coverage enhancement.

There are further improvements to be made and Rel-14 has addressed some MTC features such as support for positioning and higher data rates for a wider range of use cases. But it has also reduced output power to meet system on-chip implementation aspects as well as battery current consumption constraints. Key parameters of the cellular IoT standards are given in the following table:

14.4.2 The 5G Path Ahead

This section focuses on communication and interaction with machines mainly and the expected evolution and further enhancements toward the 5G massive MTC that will continue in 3GPP Rel-15 during 2017 described as even further enhanced MTC (efeMTC), still based on the fundamentals of LTE. Machine-type communication will thus continuously evolve toward 5G and beyond. The related deployment scenarios and key performance indicators (KPIs) are described in 3GPP requirements [27]. The expected deployment scenarios are dense urban and urban coverage for massive connection.

Some of the challenging KPIs for machine-type communications are the target for UE battery life for massive MTC should be beyond 10 years, 15 years is desirable. The power and stamina targets also include demands for operating the devices by means of coin cell batteries where the peak current drain must be significantly lower than in existing cellular devices. The target for connection density is set to manage up to 1 000 000 device/km^2 in urban environment. The 5G new radio (NR) being defined for evolved Mobile Broadband and ultrareliable low-latency communications (URRLC) use cases in Rel-15 is a potential candidate to evaluate the progress on extended use cases, connection density, and reduced device power consumption in Rel-16. It is thus essential that mMTC in 5G is equipped with all the existing state-of-the-art features identified as well as potentially new technical solutions in order to fulfill all possible 5G KPIs and requirements.

As 3GPP specifications progress, from a large package called a release to the next, the different domains of use benefit from enriched functionalities. Let us review an important subset of these below:

- *Critical Communications and Public Safety*

This domain [28] is of increasing interest, following unfortunate events, happening in a dense sequence since September 11, 2001, and causing reinforced State Security aiming at the protection of the public.

According to ETSI, which has developed the set of TETRA standards, TErrestrial Trunked RAdio (TETRA) is a digital trunked mobile radio standard developed to meet the needs of traditional Professional Mobile Radio(PMR) user organizations. Potential users are Public Safety,Transportation, Utilities, Government, Military, PAMR, Commercial & Industry, Oil and Gas. The replacement of the TETRA trunk telephony system is being actively considered at 3GPP, with functions now necessary, which had not been considered technically available when the TETRA trunk telephony for civil emergency forces had been defined. These include video, critical IoT, remote control, and monitoring of machines and robotics illustrated by interventions in nuclear sites such as the Fukushima plant [29].

- *Internet of Things: The Republic of the Machines*

The Internet of Things may provide a huge advantage in many cases, where humans "are not in the way," and therefore the risk of human error does not exist in automated tasks. However, a key criterion of the Internet of Things is that it serves the humans as enterprise or organization, or as individuals. This leads to use cases and scenarios for the matching connectivity in the 5G era, with the work context or the daily life context aggregating most cases.

Then, within the Republic of the Machines, the handling of time constraints, with task scheduling, and the use of data, especially those regarding humans directly or indirectly, require special care, first for safety and second for privacy protection, and probably a number of criteria for technology acceptance and endorsement. Energy and action, with sensors, decisions, commands, and execution, constitute the segment of goals achievement and delivery of objectives in the Republic of the Machines. This is the rationale for machines being put in operation. To such effect, overall architectures will be defined, sector by sector, and within each sector, problem by problem. These architecture will define network requirements in space and time, quality, and other parameters. Sensor data will be gathered, actuator commands will be distributed for execution, with the needed capillarity, on suitable networks, with matching latencies, energy consumptions, ranges, and protocols. Within these constraints and needs, 5G will definitely play an enabling part, either at sensor/actuator level or at an aggregated multisensor/multiactuator level.

Whereas the commercial deployment of cloud systems is in full swing as we write this chapter, the edge computing [30] and hybrid combination of edge and cloud for different digital roles of computing, storing, and networking is already actively explored and in many cases implemented [31].

14.4.3   5G Candidate Solution Space

A main contributor to 3GPP, Ericsson, has shared in a white paper the following assessment of the sequencing of feature releases under 5G [32]:

"With 5G research progressing at a rapid pace, the standardization process has started in 3GPP. As the most prevalent mobile broadband communication technology worldwide, LTE constitutes an essential piece of the 5G puzzle. As such, its upcoming releases (Rel-14 and Rel-15) are intended to meet as many 5G requirements as possible and address the relevant use cases expected in the 5G era.

Let us highlight areas of rapid growth, either as generic improvement or as use case domains broadening the scope of mobile systems. They take the shape of technology streams consolidated within 3GPP's relevant committees.

- *Energy Efficiency Improvement*

5G has set itself as a goal to improve energy efficiency. Concepts such as device sleep, named "micro-sleep" in an article have been considered. Further savings may be achieved through robustness to discontinuities in signal availability/level for transmitting or receiving, since the possibility of such discontinuity occurs, when the device is moving or any other event causing the signal power to fluctuate. As in other consumer electronics systems, the functionality for waking-up the receiver may contribute additional energy savings.

An overview paper reviewing such techniques [33] claims achievement of 20% energy savings as compared to LTE, thanks to "micro-sleep," and up to 80%, thanks to robustness to discontinuous receive/send phase, and even up to 90% for some traffic type with scheduling, using the receiver wake-up trigger.

- *Critical Communication and Public Safety*

One of the features required in a current mobile system used for civil emergency forces (ambulance, fire department, and police), TETRA, is a direct push-to-talk functionality whereby a device can call another device without relying on a base station to relay this communication.

Therefore, there is sustained interest in having 5G to support (direct) device-to-device (D2D) communication. This interest is also materializing in the MTC area, with potentially different objectives:

- Robustness to infrastructure availability and "signal power" as in critical conditions
- Energy saving
- Reducing data transfer delay

An interesting article [34] discusses the possibility to trade off energy consumption and data transfer delay through the possibility to cluster multiple devices before transmitting to eNodeBs. The results obtained hint at strategies to match specific constraints occurring in MTC.

- Location

With the advent of always more accurate and ubiquitous positioning systems, using Global Navigation Satellite Systems (GNSS) enhanced and complemented by other radio techniques, for outdoors, but more critically for indoors or where satellites are not visible, positions can be determined. Using the spatial information about location can lead to improved communications in each of different layers: Physical, MAC, Network and transport.

Developments using location information to improve the efficiency of operation in each of these layers have been consolidated and analyzed in an article [35], where remaining technical challenges for practical use in 5G are also discussed.

- *Ultrareliable Communication*

Ultrareliable communication is required for the implementation of the Advanced Driver Assistance systems and the Automotive Industry Roadmap toward Autonomous Driving also called Automated Driving by the German Automotive Association VDA (Verband der Automobilindustrie) [36].

The corresponding technologies need to support Low-Latency Communications to vehicles to maintain instant location updates [37].

14.5 Conclusion on Machine-Type 5G

Machine-type 5G will serve two domains, with consistent expectations formulated in slightly different ways:

- *Humans* using machine assets, either because it is their job to do so at work or because such assets serve them in their private life for convenience and a variety of needs (home, car, appliances, any other asset).
- *Systems Combining Machines Within a Process for a Service Typically*

Systems aim at operational goals. They have requirements such as end-to-end health and safety and information security protection for its infrastructure and assets. Humans tend to emphasize usability requirements, and privacy protection understood as supporting the freedom to be what they want, and as they want, within the task they want to achieve (in particular at work). Enterprise solutions delivered to organizations, with 5G technology elements or network segments, will have to address the criteria above. Services delivered to individuals, in support of their own environment, may have common path to the enterprise ones, but need to match potentially different priorities as described above. Gaining acceptance of the individual worker remains essential on the path to providing an efficient solution to an enterprise. Accompanying measures could include training where needed.

References

1 Recommendation ITU-R M.2083-0 (09/2015) "IMT Vision: framework and overall objectives of the future development of IMT for 2020 and beyond," M Series, Mobile, radiodetermination, amateur and related satellite services.

2 Y. C. Hu, M. Patel, D. Sabella, N. Sprecher, and V. Young, ETSI White Paper No. 11: *Mobile Edge Computing: A Key Technology Towards 5G*, 1st edition, ETSI, 2015. Available at http://www.etsi.org/images/files/ETSIWhitePapers/etsi_wp11_mec_a_key_technology_towards_5g.pdf (accessed June 3, 2018).

3 O. Novo, N. Beijar, M. Ocak, J. Kjallman, M. Komu and T. Kauppinen, Capillary Networks: Bridging the Cellular and IoT Worlds, *2015 IEEE 2nd World Forum on Internet of Things (WF-IoT)*, Milan, 2015, pp. 571–578.

4 J. Laassiri and S. D. Krit, "Internet of Things: architecture and concepts in ODP information language," in *2016 International Conference on Engineering & MIS (ICEMIS)*, Agadir, 2016, pp. 1-5.

5 K. Hwang, J. Dongarra, and G.C. Fox, *Distributed and Cloud Computing: From Parallel Processing to the Internet of Things.* Morgan Kaufmann, 2013.

6 M. Patel et al., Mobile-edge computing: Introductory technical white paper, ETSI White Paper, Sep. 2014.

7 A. Cocana-Fernandez, J. Ranilla, and L. Sanchez, "Energy-efficient allocation of computing node slots in HPC clusters through parameter learning and hybrid genetic fuzzy system modelling," J. Supercomput., vol. 71, no. 3, pp. 1163–1174, 2015.

8 K.-H. Lee, H.-J. Lee, and Y.-M. Kim, "Design and implementation of a systolic architecture for low power wireless sensor network," *J. Korea Inst. Electron. Commun. Sci.*, vol. 10, no. 6, pp. 749–756, 2015.

9 European Union Agency for Network and Information Security, Report on "Communication Network Dependencies for ICS/SCADA Systems," December 2016. Available at https://www.enisa.europa.eu/publications/ics-scada-dependencies (accessed June 3, 2018).

10 U. Raza, P. Kulkarni and M. Sooriyabandara, "Low power wide area networks: an overview," in *IEEE Commun. Surv. Tutor.*, vol. 19, no. 2, pp. 855–873, 2017.

11 K. Stouffer, J. Falco, and K. Kent, "Guide to Supervisory Control and Data Acquisition (SCADA) and Industrial Control Systems Security", *National Institute of Standards and Technology*, Technical REPORT, Sept. 2006.

12 C. Bouneau, L. Jacques, A. Jean-Yves, B. Christophe, C.Richard, D. Bernard, F. Claude, L. Andre, and P. Jacques, *Le systeme nerveux du rvseau franais detransport d'electricite (1946-2006) 60 annee de contrle lectrique*, editions EDF R&D (See Sections 7.2–7.4).

13 M. Kumagai, *IEEE Spectrum: Japan Earthquake Diary: Suddenly I Heard a Roar from the Ground* May 3, 2011.

14 L. Papangelis, M. S. Debry, P. Panciatici and T. Van Cutsem, "Coordinated supervisory control of multi-terminal HVDC grids: a model predictive control approach," in *IEEE Trans. Power Syst.*, vol. 32, no. 6, pp. 4673–4683, 2017.

15 C. W. Gellings, *Let's Build a Global Power Grid: With a Little DC Wizardry and a Lot of Cash, We Could Swap Power Across Continents*, IEEE Spectrum, July 28, 2015.

16 C. Thuemmler, A. Paulin, and A. K. Lim, "Determinants of next generation e-health network and architecture specifications," *2016 IEEE 18th International Conference on e-Health Networking, Applications and Services (Healthcom)*, Munich, 2016, pp. 1–6.

17 World Health Organisation (WHO), "Ageing and health," Fact Sheet No. 404, September 2015.

18 IEC document, PNW TS AAL-60 ED active assisted living (AAL) use cases.

19 T. Han, X. Ge, L. Wang, K. S. Kwak, Y. Han, and X. Liu, "5G converged cell-less communications in smart cities," in *IEEE Commun. Mag.*, vol. 55, no. 3, pp. 44–50, 2017.

20 T. Taleb, S. Dutta, A. Ksentini, M. Iqbal, and H. Flinck, "Mobile edge computing potential in making cities smarter," in *IEEE Commun. Mag.*, vol. 55, no. 3, pp. 38–43, March 2017.

21 World Economic Forum, Digital Transformation Initiative (DTI): Telecommunications Industry, White Paper, Jan 2017. Available at http://reports.weforum.org/digital-transformation/wp-content/blogs. dir/94/mp/ files/pages/files/white-paper-dti-2017-telecommunications.pdf (accessed June 3, 2018).

22 N. Roedder, D. Dauer, K. Laubis, P. Karaenke, and C. Weinhardt, "The digital transformation and smart data analytics: an overview of enabling developments and application areas," *2016 IEEE International Conference on Big Data (Big Data)*, Washington, DC, 2016, pp. 2795–2802.

23 MIT Sloan Management Review, Reframing Growth Strategy in a Digital Economy, in *How to Go Digital:Practical Wisdom to Help Drive Your Organization's Digital Transformation*, 1, MIT Press, 2018, pp. 224.

24 K. Pretz, *IEEE Provides the Keys to a Smarter Home: Members Are Making It Possible for All Your Appliances and Devices to Communicate with Each Other*, IEEE "The Institute" magazine, 2015. Avaiulable at http://theinstitute.ieee.org/technology-topics/smart-technology/ieee-provides-the-keys-toa-smarter-home (accessed June 3, 2018).

25 T. Salmi, I. Marstio, T. Malm and J. Montonen, "Advanced safety solutions for human-robot-cooperation," *Proceedings of ISR 2016: 47th International Symposium on Robotics*, Munich, Germany, 2016, pp. 1–6.

26 T. Meneweger, D. Wurhofer, V. Fuchsberger, and M. Tscheligi, "Working together with industrial robots: experiencing robots in a production environment," *2015 24th IEEE International Symposium on Robot and Human Interactive Communication (RO-MAN)*, Kobe, 2015, pp. 833–838.

27 Third Generation Partnership Project. Technical Specification Group Radio Access Network, Study on Scenarios and Requirements for Next Generation Access Technologies (Release 14), TR 38.913 V14.1.0 (2016).

28 A. Kumbhar, F. Koohifar, I. Guven and B. Mueller, "A Survey on legacy and emerging technologies for public safety communications," in *IEEE Commun. Surv. Tutor.*, vol. 19, no. 1, pp. 97–124, 2017.

29 E. Ackerman, Another Robot to Enter Fukushima Reactor, and We Wish It Were Modular, IEEE Spectrum, 2015. Available at http://spectrum.ieee.org/automaton/robotics/industrial-robots/heres-whywe-should-be-using-modular-robots-to-explore-fukushima (accessed June 3, 2018).

30 O. Salman, I. Elhajj, A. Kayssi, and A. Chehab, "Edge computing enabling the Internet of Things," *2015 IEEE 2nd World Forum on Internet of Things (WF-IoT)*, Milan, 2015, pp. 603–608.

31 W. Shi, J. Cao, Q. Zhang, Y. Li, and L. Xu, "Edge computing: vision and challenges," in *IEEE Internet of Things Journal*, vol. 3, no. 5, pp. 637–646, 2016.

32 O. Teyeb, G. Wikstrom, M. Stattin, T. Cheng, S. Faxr, and H. Do, *Evolving LTE to Fit the 5G Future*. Available at https://www.ericsson.com/publications/ericsson-technology-review/archive/2017/evolvinglte-to-fit-the-5g-future (accessed June 3, 2018).

33 M. Lauridsen, G. Berardinelli, F. M. L. Tavares, F. Frederiksen, and P. Mogensen, "Sleep modes for enhanced battery life of 5G mobile terminals," *2016 IEEE 83rd Vehicular Technology Conference (VTC Spring)*, Nanjing, 2016, pp. 1–6.

34 M. Bagaa, A. Ksentini, T. Taleb, R. Jantti, A. Chelli, and I. Balasingham, "An efficient D2D-based strategies for machine type communications in 5G mobile systems," *2016 IEEE Wireless Communications and Networking Conference*, Doha, 2016, pp. 1–6.

35 R. D. Taranto, S. Muppirisetty, R. Raulefs, D. Slock, T. Svensson. and H. Wymeersch, "Location-aware communications for 5G networks: how location information can improve scalability, latency, and robustness of 5G," in *IEEE Signal Process. Mag.*, vol. 31, no. 6, pp. 102–112, 2014.

36 VDA, Automation: From Driver Assistance Systems to Automated Driving, White Paper, Berlin, Sept. 2015. Available at https://www.vda.de/dam/vda/publications/2015/automation.pdf (accessed June 3, 2018).

37 S. A. A. Shah, E. Ahmed, M. Imran, and S. Zeadally, "5G for vehicular communications," in *IEEE Commun. Mag.*, vol. 56, no. 1, pp. 111–117, 2018.

Renaud Di Francesco is a telecommunications engineer with degrees from Ecole Polytechnique Paris, Telecom Paris (ENST), a Ph.D. from Telecom-Paris-Tech, and has worked with AT&T Bell Labs, France Telecom, the European Commission DG Information Society (DG Connect), and Sony. His career took him through different aspects of the digital and digitally powered technologies transforming today's economy, from energy to user interface through networks and multimedia, service design, and deployment. The dominance of big data in information technology led him recently to develop the critical view that data could be the hero instead of the information technology, resulting in three books published at Amazon/Kindle Edition: *Big Data, Driving Decisions* on the scope shift from semistatic to real time; *Big Data Economics, towards Market Places* on the pricing/nonpricing of source data, and economic mechanisms around the raw value of data; and *Microeconomics for Big Data*, expressing the value carried by data, in words and equations.

Peter Karlsson obtained his Ph.D. in Applied Electronics from Lund Institute of Technology in 1995 with a thesis on indoor radio wave propagation. He then joined Telia Research and had different research and management positions in the wireless and mobile communications systems area. In 2000 Peter had a 1 year postdoc and research fellow position in the Center for Communications Research at the University of Bristol. Peter joined Sony Ericsson corporate technology office 2007 and was leading the network Technology Lab in the new Sony Mobile organization in 2012. He is now head of the Research and Standardisation Department at Sony Mobile, Research and Incubation, Lund, focusing on 5G and IoT topics. Dr. Karlsson has written and coauthored some 75 conference and journal papers in the mobile and wireless communications area.

Part IV

Vertical 5G Applications

15

Social-Aware Content Delivery in Device-to-Device Underlay Networks

Chen Xu,[1] Caixia Gao,[1] Zhenyu Zhou,[1] Shahid Mumtaz,[2] and Jonathan Rodriguez[2,3]

[1]*School of Electrical and Electronic Engineering, North China Electric Power University, Beijing, China*
[2]*Instituto de Telecomunicações, Aveiro, Portugal*
[3]*University of South Wales, Pontypridd, UK*

15.1 Introduction

With the popularity of high-performance intelligent terminals and the emergence of new mobile multimedia services, the demand on wireless high data rate has been growing continuously [1]. The contradiction between the growing service demands of users and the limited network bandwidth has become increasingly prominent [2]. The existing wireless network architecture needs to be upgraded [3].

Researchers in academia and industry therefore attempt to explore new valuable communication technologies that can improve the system capacity by spectrum reuse. Device-to-device (D2D) communication is one of the key solutions for future 5G system. In D2D communication, mobile devices transmit data signals using cellular resources over local peer-to-peer links instead of through a traditional infrastructure, that is the base station (BS) of cellular network, which differs from the traditional communication mode that data signals need to be forwarded through the BS. Thus, repeated transmissions are not necessary for all mobile devices that request the same content. By reusing cellular spectrum resources under the control of the BS, D2D communication can dramatically increase the spectrum efficiency and network capacity [4,5]. Moreover, because of the proximity effect of direct connections, D2D is expected to enhance the data transmission rates and promote new applications [6,7].

5G Networks: Fundamental Requirements, Enabling Technologies, and Operations Management, First Edition.
Anwer Al-Dulaimi, Xianbin Wang, and Chih-Lin I.
© 2018 by The Institute of Electrical and Electronics Engineers, Inc. Published 2018 by John Wiley & Sons, Inc.

Hence, D2D communication can significantly improve spectrum efficiency and reduce transmission latency compared with the traditional communication mode, due to the reusing gain, proximity gain, and hop gain. Furthermore, the cost brought by enabling D2D communication is unnecessary to be considered here owing to the fact that the link establishment and resource management of D2D communication are controlled by the BS.

According to the analytical data results [1], a large amount of data traffic are generated from hot spots, where the distribution of mobile users is extremely dense, such as a subway train, a concert hall, and other public places. From another perspective, the users located in the hot spots may have "relationships" with other ones, which can be obtained from their social data on the social platforms. The social relationship, generally speaking, includes real friend relations and virtual relations associated with interests in similar contents. In practice, multiple users may request for the same content, while the BS has to transmit the content to these users by multiple repeated transmissions in the traditional way. For this case, it is reasonable to apply D2D technique to push or share the same content from the content holder to users with tight social relationships. It is worth noting that the appropriate cellular spectrum resources need to be reused by the D2D links. As a result, the heavy data traffic is offloaded from the cellular infrastructure, and at the same time the spectrum efficiency is increased.

There exist some works on cellular data offloading by integrating D2D communications with social networks [8–11]. In such a scenario, the BS pushes the content to a set of seed users, who then transmit the content to other users in proximity by D2D links. The relationships between the seed and nonseed users are defined as "social ties," which reflect the similarity of users' preferences on the content. The core objective is to spread the popular content in as short a time period as possible. However, in practice, there is another situation that users in the hot spots may not be interested in the same content. For instance, two passengers on a subway train are using Facebook to browse pictures posted by their mutual friend, while another passenger has just downloaded a video that other passengers around him may be interested in. Therefore, in order to implement effective content delivery and achieve good user satisfactions, it is necessary to consider the different preferences of users on the contents based on historical data obtained from the social platforms.

The above consideration brings challenges to the system. First, the social relationships that reflect the close degree of users, that is the consistency degree of preferences on similar contents, are required for determining the transmitter and receiver of D2D communication, which can be regarded as a process of peer discovery. Second, since the D2D transmitters push contents to the receivers by reusing the cellular spectrum, the cochannel interference cannot be ignored, which requires an efficient resource management to optimize the system performance and guarantee the quality of service (QoS) as well. Combining these two aspects, the strategy of content delivery should consider the system status information from both social layer and physical layer. On the one side, the content delivered to the user is expected to be what he just wants; on the other side, the reused spectrum is hoped to be the best choice for maximizing the system sum rate.

This chapter focuses on implementing effective content delivery by employing a joint peer discovery and resource allocation approach with the objective of maximizing the system sum rate weighted by the intensity of users' social relationships, and at the same time guaranteeing the QoS of both cellular and D2D links [12]. Due to the uncontrollability and uncertainty of users' activities in social network [13], we utilize the probabilities of selecting similar contents, which can be estimated by Bayesian nonparametric models [14], to obtain the social relationships among users. Considering the different preferences of users on the contents and spectrum resources, we focus on solving the joint optimization problem by matching theory [15], which attempts to describe the formation of mutually beneficial relationships. Matching theory has been already employed in some works to allocate limited spectrum resources to users to maximize spectrum efficiency [16–19], and energy-efficient resource management schemes based on matching theory have been proposed in some works for D2D communications [20,21]. Note that in our problem, the matching between D2D transmitters and receivers, and the matching between D2D pairs and resource blocks (RBs), should be jointly considered. Thus, we propose a three-dimensional matching process to achieve the coordinated allocation of users, contents, and spectrum resources, based on the social layer and physical layer information. The main contributions of this chapter are summarized as follows:

- We comprehensively introduce and summarize the related works and existing progress associated with the research direction of resource allocation in D2D communications, including social-unaware and social-aware techniques. The related literatures are categorized according to the motivations and application scenarios. Meanwhile, depth discussions and analysis of the contributions of the researched literatures, as well as the application scenarios, general assumptions, merits and drawbacks, and future research directions are provided.
- We propose a social network-based content delivery approach to offload the cellular data traffic by D2D links. Specifically, we define the intensity of two users' social relationship as the normalized correlation of the probabilities of selecting similar contents that are estimated by the Bayesian nonparametric models. Moreover, a joint peer discovery and spectrum resource allocation problem, which involves the matching between content providers (transmitters) and content consumers (receivers), and the matching between D2D links and spectrum resources, respectively, is proposed and formulated as a

three-dimensional matching that maximizes the system sum rate weighted
by the intensity of social relationships.

- Due to its combinatorial nature, the joint allocation problem is intractable
  and belongs to the class of NP-hard problems. We simplify the problem
  based on pricing strategy and propose a three-dimensional matching algorithm, which can approach the performance of the exhaustive optimal
  algorithm with a much lower complexity. In the algorithm, we also consider
  the power control for D2D transmissions to avoid excessive interference
  to cellular users. Then we provide theoretical discussion and analysis for
  convergence, stability, optimality, and complexity of the proposed three-
  dimensional matching algorithm.
- In the simulation, we compare the proposed matching algorithm with the
  exhaustive optimal and random matching algorithm in terms of the achieved
  weighted sum rate for D2D communications under different scenarios. Numerical results show that our proposed scheme can achieve a considerable
  performance gain, and the satisfaction of users on the shared contents is
  substantially improved with the consideration of social relationships.

The rest of this chapter is organized as follows: Section 15.2 provides a brief
review of the related works. Section 15.3 introduces the system model that consists of physical layer and social layer. Section 15.4 introduces the formulation of
the social network-based content delivery problem. Section 15.5 describes the
three-dimensional matching algorithm with relevant theoretical concepts and
analysis. Section 15.6 presents the simulation results and discussions. Section
15.7 concludes the chapter.

15.2 Related Works

As a result of the limited spectrum resources and battery capacity, new challenges have been indeed brought for resource allocation optimization in D2D
communication underlay cellular networks. One main line of works optimizing
resource allocation for D2D communications is to maximize the spectrum efficiency (SE). Comprehensive investigations of resource allocation management
associated with SE for D2D communications were summarized in Refs [5,22].
And in Ref. [23], authors analyzed the optimum SE performances for various
resource sharing modes under practical constraints. A three-step scheme including admission control, power allocation, and partner matching is proposed
in Ref, [24] to maximize the SE performance of the system. In Ref. [25], authors
proposed a reverse iterative combinatorial auction-based resource allocation
scheme to optimize the system sum rate. A distributed approach based on game
theory was proposed in Ref. [26] to optimize SE performance, where a virtual
pricing mechanism for signaling was employed to make each D2D user play

a best response. In Ref. [27], the resource allocation problem was formulated into an infinite-horizon constrained Markov decision process (CMDP) with consideration of the dynamic data arrival. Besides, spectrum-efficient schemes have been proposed to solve resource allocation problems under various scenarios, for instance, software-based heterogeneous networks [28,29], wireless video networks [30], relay-assisted networks [22,31,32], capacity-constrained networks [33], energy harvesting networks [34,35], cloud radio access networks [36], and intelligent transportation systems [37].

Despite that the above works can achieve significant gains on SE performance, energy consumption of mobile user equipment is ignored in the design of resource allocation schemes. In other words, the energy efficiency (EE) performance of users is neglected, and thus users need to continuously raise the transmission power to achieve the stringent QoS requirements in interference-limited systems. However, the increase of transmission power would lead to rapid consumption of battery energy and cause more interference to other mobile users. Hence, energy-efficient resource allocation schemes need to be designed to improve the system performance. There exist some works investigating resource allocation schemes associated with EE for D2D communications. In Refs [38,39], a joint spectrum and power allocation scheme was proposed to optimize the EE performance of user equipment, where an iterative price-based auction game was introduced. While an auction-based energy-efficient resource allocation scheme was proposed in Ref. [39] for a cooperative system, where mobile users with high-level battery would help carry the data of mobile users with low-level battery. In Ref. [40], the authors proposed a joint mode selection and resource management scheme based on coalition game model to improve the system performance. In Ref. [41], a modified genetic-based algorithm was studied to optimize EE performance under a certain throughput insurance for the scenario with multiple resource pool multiplexing rather than the one-time reusing system. Two novel resource allocation mechanisms were proposed in Ref. [42], in which the first one aimed at mitigating the interference and the second one managed the available resources to optimize EE. Energy-efficient resource allocation scheme integrated with noncooperative game theory for cloud RAN-based networks was designed in Ref. [43]. Theoretical analysis of trade-off between SE performance and EE performance was analyzed in detail in Refs [44,45].

Our previous works mainly focused on the resource allocation problem and provided a theoretical analysis on the trade-off between energy efficiency and spectrum efficiency [46,47]. However, the peer discovery problem has not been taken into consideration. In comparison, this chapter aims to solve the joint peer discovery and resource allocation problem with power control in D2D communications underlaying cellular networks by exploring both social and physical layer information. In Ref. [48], authors proposed a centralized D2D discovery method that can adaptively allocate spectrum resources for the discovery to avoid the underutilization of resource blocks based on the random access procedure in LTE-A system by employing the location information of users. A social-aware peer discovery scheme for D2D users based on an established paradigm was proposed in Ref. [49], where mobile users were divided into groups by using the social information including location, background, and interest. In Ref. [50], authors proposed a code-based discovery scheme to realize proximity-based services, where the discovery code containing the compressed information of mobile applications was utilized to find the nearby devices that have interests in the mobile applications. The above works mainly solve the peer discovery issue of D2D communications considering physical location information, social information, interests in mobile applications, and so on.

In addition, social information in social network is utilized to enhance various performance metrics of D2D communications. For instance, authors in Ref. [51] proposed a clustering scheme with an admission policy to increase system rate, while in Ref. [52] it was proposed to improve the system throughput and energy efficiency by employing Chinese Restaurant Process (CRP). Mode selection of content downloading for D2D users and relay selection for social-trust-based and social-reciprocity-based cooperative D2D communications were studied in Refs [8],[53], respectively. In Ref. [9], authors proposed a sharing strategy using social relationship with consideration of minimum delay. While in Ref. [10], a sharing strategy utilizing social information was proposed with consideration of formation of a practical network. In Ref. [11], the authors proposed a proactive caching framework by exploiting social structure of the network under storage constraints to relieve peak data demands and offload heavy data traffic.

Besides, resource allocation issues for D2D pairs with consideration of the social relationship were studied in different scenarios, such as a single community in Refs [54–56], cooperative communities in Ref. [57], and a slotted system in Ref. [58]. The comprehensive summary of the classifications of social-aware resource allocation schemes is shown in Table 15.1. In a single community, D2D pairs can simply reuse the RBs occupied by the cellular users that are in the same community; while in the scenario of cooperative communities, D2D pairs can reuse the RBs of the cellular users that are in the community coalition, namely, the aggregation of the cooperative communities. Due to the human mobility in a slotted system, a D2D link can be considered for resource allocation only when the two users encounter and the contact time is long enough to complete a meaningful transmission. Focus on different methods, the allocation of RBs to D2D pairs was solved by utilizing the matching game in Ref. [54], two-step coalitional game in Ref. [57], and other maximization games in Refs [55–58] with different objective functions. Resource allocation problem can be modeled as a two-sided matching problem using matching theory. As in Ref. [54], authors formulated the resource allocation problem as a matching game in which D2D pairs and RBs rank one another based on the utility functions considering both

Table 15.1 A comprehensive summary of the classifications of social-aware resource allocation schemes.

Categories	Application scenarios	Optimization goals	Variables	Solution methods
Social-aware resource allocation	Single community [54–56]	Maximizing the utility function of achievable rates and social ties [54]	Achievable rate, social tie	Matching theory
		Maximizing the required transmission slots [55]	Decision variable	Maximization method
		Maximizing social group utility of each D2D user [56]	Channel rate, social tie, decision variable	Maximization method
	Cooperative communities [57]	Maximizing the satisfied D2D requests of the coa-lition	Channel rate, allocation strategy	Coalitional game
	Slotted system [58]	Maximizing the aggregated throughput of all the users	Channel rate, contact time	Maximization method

social and physical metrics. Also, matching theory has been utilized to solve resource allocation problems considering two-dimensional matching with mutual preferences in D2D communications [20,21,32,59,60]. In Ref. [20], a relay selection scheme was designed to reduce the energy consumption of mobile devices and extend the lifetime of the system for the scenario, in which multiple D2D pairs exploit full duplex relays to achieve communication utilizing directional antennas. Authors in Ref. [21] employed a game-theoretic method and the Gale-Shapley (GS) algorithm to address the resource allocation problem, which is formulated as a one-to-one matching problem to optimize EE performance. In Ref. [59], resource allocation scheme based on matching theory was proposed for a heterogeneous cellular network composed of macro users, small cell users, and D2D users, where the matching problem between spectrum resource blocks and users was solved. Afterward the work was extended to the relay-assisted D2D communication with the consideration of uncertainties of wireless links, where ellipsoidal uncertainty sets were utilized to model the uncertainties based on robust optimization theory and the matching theory was utilized to develop a distributed resource allocation scheme [32]. In Ref. [60], the idea of cheating was introduced into the process of matching to optimize system performance while satisfying QoS requirements of users simultaneously, where the network stability was ensured by implementing matching theory. In Ref. [61], an energy-efficient resource allocation scheme was proposed integrated with game theory and matching theory, where interactions among users were analyzed by adopting noncooperative game model. In addition, matching theory has been employed in heterogeneous cellular networks [16,18], cognitive radios [17], and so on.

However, the previous works have not employed social information to solve the joint peer discovery and resource allocation problem, which actually involves a three-dimensional matching among D2D transmitters, D2D receivers, and RBs in the content delivery process.

15.3   System Model

We consider a cellular network with one BS and multiple users involving traditional cellular user equipment (CUEs) and potential D2D pairs. Each user can receive data from either the BS or the another user through potential D2D links. In this chapter, the mode selection problem is left out of consideration, and thus we assume that there exist some users satisfying the physical requirement of D2D, such as the constraint of transmission distance. Once it is found that two users can be matched to form a D2D pair, the content holder transmits signals to the requester. Here, we focus on two key problems: (1) How to match the content transmitter (TX) with the receiver (RX) so that the RX would be satisfied with the received content. (2) How to design an efficient resource allocation scheme for D2D pairs to maximize the system performance.

An illustration of social-aware D2D underlay network is shown in Figure 15.1. The architecture can be divided into two layers consisting of social layer and physical layer. In the social layer, users' behaviors in social network reflect their real social connections, which can be obtained from social platforms such as Microblog, Facebook, Twitter, and so on. Thus, we can derive the real close degree of user relationships by exploring their behaviors in such platforms. In the physical layer, the establishment of D2D links is mainly determined by transmission distance between two mobile nodes, namely, smart terminals, such as smartphones and tablets. For each user in the social layer, there exists a corresponding terminal in the physical layer. To achieve successful message pushing or content sharing through D2D links, both the social relations and the physical locations need to be taken into account.

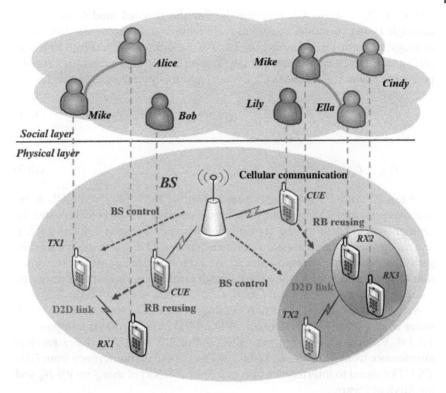

Figure 15.1 System model of social network-based content delivery in D2D underlay cellular networks.

In general, if two users have a stronger social relationship, the probability of establishing direct link between them would be higher, which is because their content preferences are more similar than that of users with weak social connections. Meanwhile, a better channel quality between users promotes an establishment of D2D link. In this section, we introduce the system model of social-aware D2D underlay network. The physical transmission model is first described, and then, the social relationship between users is quantified.

15.3.1 Physical Layer Model

In the system, we assume that D2D links share uplink (UL) resource blocks (RBs) occupied by cellular users, and for simplicity, one RB is allocated to one CUE and can be reused by at most one D2D pair. Furthermore, we assume that there are N D2D TXs (content providers) and N D2D RXs, which are denoted by the set $\mathcal{N}_{\mathcal{T}} = \{1, \cdots, i, \cdots, N\}$ and $\mathcal{N}_R = \{1, \cdots, j, \cdots, N\}$, respectively. K RBs and the corresponding cellular users are denoted by the set $\mathcal{N}_K = \{N_1, \cdots, N_k, \cdots, N_K\}$

and $\mathcal{K}=\{1,2,\cdots,k,\cdots,K\}$, respectively. For the channel model, we use the Rayleigh fading to model the small-scale fading, and employ the free space propagation path-loss to model the large-scale fading. The received power of D2D link between transmitter $i \in \mathcal{N}_T$ and receiver $j \in \mathcal{N}_R$, and the received power of cellular link between CUE $k \in \mathcal{K}$ and the BS, can be expressed as

$$P_{r,j} = P_i^D h_{ij}^2 = P_i^D d_{ij}^{-\alpha} h_{0,ij}^2, \tag{15.1}$$

$$P_{r,k} = P_k^C h_k^2 = P_k^C d_k^{-\alpha} h_{0,k}^2, \tag{15.2}$$

where P_i^D and P_k^C are the transmit power of D2D TX i and CUE k, respectively. h_{ij} and h_k represent the channel response of the D2D link and the cellular link. d_{ij} denotes the transmission distance between TX i and RX j, and d_k denotes the transmission distance between CUE k and the BS. α is the path-loss exponent corresponding to the large-scale fading of the transmission channel, and $h_{0,ij}$, $h_{0,k}$ are the Rayleigh channel coefficient that obeys the complex Gaussian distribution $\mathcal{CN}(0,1)$.

As a result of uplink spectrum reusing, both D2D receivers and the BS suffer from cochannel interference. When D2D pair D_{ij} that is composed of TX $i \in \mathcal{N}_T$ and RX $j \in \mathcal{N}_R$ reuses the uplink RB $N_k \in \mathcal{N}_K$, RX j receives interference from CUE $k \in \mathcal{K}$, and the BS is exposed to interference from D2D TX i. The signal to interference plus noise ratio (SINR) of user j on RB N_k and the SINR of BS are

$$\gamma_{D_{ij},k} = \frac{P_i^D h_{ij}^2}{P_k^C h_{kj}^2 + N_0} = \frac{P_i^D d_{ij}^{-\alpha} h_{0,ij}^2}{P_k^C d_{kj}^{-\alpha} h_{0,kj}^2 + N_0}, \tag{15.3}$$

$$\gamma_{k,i} = \frac{P_k^C h_k^2}{P_i^D h_{iB}^2 + N_0} = \frac{P_k^C d_k^{-\alpha} h_{0,k}^2}{P_i^D d_{iB}^{-\alpha} h_{0,iB}^2 + N_0}. \tag{15.4}$$

Here, h_{kj} and h_{iB} denote the channel response of the interference links between CUE k and D2D RX j, and between D2D TX i and the BS, respectively. N_0 is the one-sided power spectral density of the additive white Gaussian noise (AWGN) at the receivers. Based on the above expressions, the channel rate of D2D pair D_{ij} reusing RB N_k and the rate of cellular link between k and the BS are obtained by

$$r_{D_{ij},k} = \log_2\left(1 + \frac{P_i^D h_{ij}^2}{P_k^C h_{kj}^2 + N_0}\right), \tag{15.5}$$

$$r_{k,i} = \log_2\left(1 + \frac{P_k^C h_k^2}{P_i^D h_{iB}^2 + N_0}\right). \tag{15.6}$$

15.3.2 Social Layer Model

In social network, users' behaviors reflect the close degree of their relationships. Therefore, it is extremely important to analyze users' social behaviors during the process of social layer modeling. However, it is hard to find an appropriate model to describe the properties of social behaviors due to their uncontrollability and uncertainty. Thus, we utilize the probability of selecting similar contents to represent the similarity of users' behaviors, which determines the intensity of their social relationships. Bayesian model is an efficient model that apply probability and statistics into complex area to handle the uncertainty reasoning. Integrating the prior information and sample information, it is easy to obtain the posterior probability distribution. It means that the system can obtain the probability distributions of users' content selections by integrating the history records collected from different social network platforms using the Bayesian technique [62–64]. After that, the intensity of social relationship, that is the consistency degree of preferences on similar contents, can be quantified.

Statistical modeling is a useful tool that models the process as a stochastic variable with a correlative probability density function (pdf) in a feature space. A particular statistical distribution, which is supposed to approximate the practical distribution with the parameters estimated from the sample, is used to represent the pdf parametrically. In this process, we have to find an appropriate model that approximates the actual distribution to estimate the associated parameters. However, Bayesian nonparametric models can estimate the pdf directly from the samples without making any assumptions for the underlaying distribution so as to avoid the parameter estimation process and the accuracy of the estimation would be improved as more data are observed. Dirichlet processes [65,66], which are a family of stochastic processes, are often used in Bayesian nonparametric statistics since the prior and posterior distributions in Bayesian nonparametric models are stochastic processes rather than parametric distributions. In the following paragraph, we will introduce the process to build the social relationship among users in details. Readers interested in the theoretical basis of the process can refer to Ref. [12] for more theoretical details.

15.3.2.1 Estimation of Probability Distribution

We assume that the users in our system are denoted by the set C. For a certain user $c \in C$, q observation sets that involve the probabilities of selecting the similar contents can be obtained from social network platforms in several time periods. And we denote the q observation sets as the set Q. At a certain time, for observation set $Q \in \mathcal{Q}$, user c selects the similar contents with the probability p_{Qc}. Hence, the value of p_{Qc} is a random variable with a pdf $P_{Qc}(p_{Qc})$ over the state space $\Theta = [0, 1]$. In each observation set $Q \in \mathcal{Q}$, Z_{Qc} observations are performed, which are denoted by $\mathcal{Z}_{Qc} = \{p_{Qc}^1, p_{Qc}^2, \cdots, p_{Qc}^{Z_{Qc}}\}$, $\forall c \in C$, $Q \in \mathcal{Q}$. Employing the DP, the predictive pdf of the next observation $p_{Qc}^{Z_{Qc}+1}$ can be obtained by using the following formula based on the observation set $\mathcal{Z}_{Qc}$:

$$
\begin{aligned}
& P_{Qc}(p_{Qc}^{Z_{Qc}+1} \in \mathcal{E} \mid p_{Qc}^1, p_{Qc}^2, \cdots, p_{Qc}^{Z_{Qc}}) \\
&= \frac{1}{\varpi + Z_{Qc}}(\varpi G(\mathcal{E}) + \Sigma_{z=1}^{Z_{Qc}} \omega_{p_{Qc}^z}(\mathcal{E})),
\end{aligned} \tag{15.7}
$$

where $\mathcal{E}$ is a measurable partition of Θ. G is the base distribution as the prior and ϖ is viewed as the strength associated with the prior base distribution for the estimation of the posterior. With the DP marginalized out, the predictive distribution of the next observation $p_{Qc}^{Z_{Qc}+1}$ conditioned on the observation set $\mathcal{Z}_{Qc}$ can be expressed as

$$
p_{Qc}^{Z_{Qc}+1} \mid p_{Qc}^1, p_{Qc}^2, \cdots, p_{Qc}^{Z_{Qc}} \sim \frac{1}{\varpi + Z_{Qc}}(\varpi G + \Sigma_{z=1}^{Z_{Qc}} \omega_{p_{Qc}^z}). \tag{15.8}
$$

When the base distribution G and the concentration parameter ϖ of the DP are unknown, we express the predictive pdf of the next observation $p_{Qc}^{Z_{Qc}+1}$ as follows (Eq. 15.7),

$$
P_{Qc}(p_{Qc}^{Z_{Qc}+1} \in \mathcal{E} \mid p_{Qc}^1, p_{Qc}^2, \cdots, p_{Qc}^{Z_{Qc}}) = \frac{\Sigma_{z=1}^{Z_{Qc}} \omega_{p_{Qc}^z}(\mathcal{E})}{Z_{Qc}}. \tag{15.9}
$$

$\omega_{p_{Qc}^z}$ is the point mass located at p_{Qc}^z and $\omega_{p_{Qc}^z}(\mathcal{E}) = 1$ when $p_{Qc}^z \in \mathcal{E}$; $\omega_{p_{Qc}^z}(\mathcal{E}) = 0$ otherwise. Then we use kernel to smooth out the distribution drawn from the DP to get the continuous estimate $\tilde{P}_{Qc}$ of P_{Qc}. However, as the number of the available observations Z_{Qc} is small, we consider another approach to improve the estimates.

For user $c \in C$, given the subset $\mathcal{W} \subseteq \mathcal{Q}$ and the observation set $Q \in \mathcal{Q}$, we denote the rest observation sets in subset $\mathcal{W}$ except Q as $\mathcal{W}_{Qc} = \mathcal{W} \setminus \{Q\}$, which represents the priors. Then, we can integrate the observation set Q with the set of validated priors $\mathcal{W}_{Qc}$ to derive the pdf of any new observation $p_{Qc}^{Z_{Qc}+1}$ using the following expression:

$$
P_{Qc}^{w} = \varphi_Q \tilde{P}_{Qc}(\mathcal{E}) + \sum_{L \in \mathcal{W}_{Qc}} \varphi_L \tilde{P}_{Lc}(\mathcal{E}). \tag{15.10}
$$

The contribution of the observation set Q for the generation of the pdf P_{Qc}^{W} is quantified by φ_Q while that of $L \in W_{Qc}$ is quantified by φ_L. In practice, we set the weights φ_Q and φ_L to be proportional to the number of observations, which are expressed as

$$\varphi_Q = \frac{Z_{Qc}}{\sum_{V \in W} Z_{Vc}}, \varphi_L = \frac{Z_{Lc}}{\sum_{V \in W} Z_{Vc}}, \forall L \in W_{Qc}. \tag{15.11}$$

With the consideration of the equal availability of observation sets, we define that $P_c = P_{Qc}^{W}$.

15.3.2.2 Intensity of Social Relationship

Due to the fact that the social relationship at close degree of any two users is measured by the similarity of their selection on contents, the probability corresponding to the selection of similar contents is utilized to derive the normalized correlation that indicates the intensity of the social relationship. For D2D TX $i \in \mathcal{N}_T$ and RX $j \in \mathcal{N}_R$, the intensity of their social relationship can be expressed as

$$\rho_{ij} = (corr(p_i, p_j) + 1)/2, \tag{15.12}$$

where $p_i \sim P_i(p)$ and $p_j \sim P_j(p)$. P_i and P_j represent the estimated correlative pdfs. And ρ_{ij} varies from 0 to 1, namely, $\rho_{ij} \in [0, 1]$.

15.4 Problem Formulation

The purpose of our work is to achieve content delivery with high satisfaction of users by employing social-aware D2D techniques, while at the same time maximizing the transmission sum rate of D2D links. Hence, we need to consider an optimization problem involving both the social layer and the physical layer. Furthermore, we formulate the objective function as a weighted channel rate, that is the rate weighted by the intensity of social relationship. The weighted rate of the link between D2D TX i and RX j when reusing RB N_k can be obtained by

$$R_{D_{ij},k} = I(\rho_{ij})\rho_{ij}r_{D_{ij},k}. \tag{15.13}$$

In practice, TX i is approved to share contents with RX j only when the intensity of social relationship between them is no less than a threshold δ, that is to say, it is potential for i and j to form a D2D link when $\rho_{ij} \geq \delta$. Hence, we define $I(\rho_{ij})$ as an indicator function of ρ_{ij} that $I(\rho_{ij}) = 1$ when $\rho_{ij} \geq \delta$; $I(\rho_{ij}) = 0$ otherwise.

To maximize the weighted sum rate of all the D2D pairs, we need to design an efficient mechanism for pairing the content provider (TX) with the content consumer (RX) and allocating the spectrum resource to the transmission link. In other words, it is an issue of joint peer discovery and resource allocation for D2D communication. To avoid excessive interference to cellular links, power control for D2D TX should also be taken into account. We use a set of binary variables $X = \{x_{i,j,k}\}$ to formulate the user pairing and resource allocation. $x_{i,j,k} = 1$ denotes that a D2D link is established between TX i and RX j reusing RB N_k. Accordingly, we jointly design the binary decision variables $\{x_{i,j,k}\}$ and the continuous power variables P_i^D to optimize the system performance. A mixed integer programming problem is formulated as

$$
\max_{\{X, P_i^D\}} \quad \sum_{k=1}^{K}\sum_{j=1}^{N}\sum_{i=1}^{N} x_{i,j,k} R_{D_{ij},k}
$$

$$
\text{s.t.} \quad C1 : \ 0 \leq P_i^D \leq P_{\max},
$$

$$
C2 : \ x_{i,j,k} \in \{0,1\}, \forall i \in \mathcal{N}_T, j \in \mathcal{N}_R, N_k \in \mathcal{N}_K,
$$

$$
C3 : \ \sum_{j \in \mathcal{N}_R, N_k \in \mathcal{N}_K} x_{i,j,k} \leq 1, \forall i \in \mathcal{N}_T,
$$

$$
\sum_{i \in \mathcal{N}_T, N_k \in \mathcal{N}_K} x_{i,j,k} \leq 1, \forall j \in \mathcal{N}_R,
$$

$$
\sum_{i \in \mathcal{N}_T, j \in \mathcal{N}_R} x_{i,j,k} \leq 1, \forall k \in \mathcal{N}_K,
$$

$$
C4 : \ r_{D_{ij},k} \geq r_{\min}^d, \forall i \in \mathcal{N}_T, j \in \mathcal{N}_R, N_k \in \mathcal{N}_K,
$$

$$
C5 : \ r_{k,i} \geq r_{\min}^c, \forall i \in \mathcal{N}_T, j \in \mathcal{N}_R, N_k \in \mathcal{N}_K. \tag{15.14}
$$

Here, constraint $C1$ gives the transmit power range of D2D TXs, which ensures the power would not exceed the maximum $P_{\max}$. The three inequalities in $C3$ ensures that each TX can only be paired with at most one RX and vice versa, while each RB can only be assigned to at most one D2D pair and vice versa. $C4$ and $C5$ guarantee the QoS requirements of D2D links and cellular links, respectively.

15.5 Social Network-Based Content Delivery Matching Algorithm for D2D Underlay Networks

In this section, we investigate a three-dimensional matching approach to solve the mixed integer programming problem (Eq. 15.14). First, we introduce some concepts of matching theory that are the basis of our algorithm. Then, we give the establishment process of the preference list, which is the critical component of matching model. The preference list is mainly based on maximizing the weighted channel rate, which is coupled with a power control problem. Afterward we introduce a pricing strategy to simplify the three-dimensional matching problem, and propose an iterative algorithm to derive a stable matching among D2D TXs, D2D RXs, and RBs. Finally, the properties of the proposed matching approach, including convergence, stability, optimality, and complexity, are analyzed in detail.

15.5.1 Matching Concepts

In our system, we attempt to solve the problem (Eq. 15.14) by employing the three-dimensional matching that pairs D2D TXs, D2D RXs, and RBs with each other. For its high complexity, we transform it to a two-sided matching. First, we define a RX–RB unit, which is composed of one RX and one RB. Due to the assumption that there is one CUE on each RB, we then rewrite the RX–RB unit as RX–CUE (RC) unit. Owing to the existence of N RXs and K CUEs, there are $N \times K$ different RC units, denoted by $RC = \{RC_{j,k}\}_{j=1,k=1}^{j=N,k=K}$. Thus, the three-dimensional matching problem can be simplified to a two-sided matching with N TXs on one side and $N \times K$ RC units on the other side. We have the definition as below:

Definition 15.1   A matching Φ is a one-to-one correspondence $\mathcal{N}_T \cup RC \rightarrow \mathcal{N}_T \cup RC \cup \{\emptyset\}$ and such that $\Phi(i) = RC_{j,k}$ means that TX i is matched with the unit $RC_{j,k}$ consisting of RX j and CUE k.

Because of the constraint that the matching among TXs, RXs, and RBs is a three-dimensional one-to-one correspondence, when $\Phi(i) = RC_{j,k}$, for $\forall i' \in \mathcal{N}_T \setminus \{i\}, \Phi(i') = \{RC \setminus \{RC_{j,k}\}\} \cup \{\emptyset\}$. The matching Φ is stable when there is no blocking pair, that is to say, there is no pair consisting of TX i and RC unit $RC_{j,k}$ that is not matched with each other but prefer each other to be their mates under matching Φ.

15.5.2 Preference Establishment

In a matching process, individuals on one side propose to establish pairs with ones on the other side based on their own preference lists. Since the three-dimensional matching problem is transformed into a two-sided matching problem with N TXs on one side and NK RC units on the other side, the essential issue is to find the preference lists of TXs on RC units. For TX i, when paired with different RC units, it can achieve different channel rates and different content satisfactions of RX, due to the different physical and social layer information. Therefore, the preference of TX on RC units can be formulated as the weighted channel rate (Eq. 15.13) with the optimization of power variables P_i^D. In the process of preference lists establishment, we need to temporar-

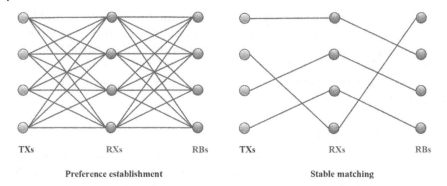

TXs RXs RBs TXs RXs RBs

Preference establishment Stable matching

Figure 15.2 Graphical expressions of preference establishment and a stable three-dimensional matching.

ily pair each TX ($\forall i \in \mathcal{N}_T$) with each RC units ($\{RC_{j,k}\}_{j=1,k=1}^{j=N,k=K}$), and thus to obtain the weighted channel rate corresponding to each three-dimensional combination TX-RX–CUE with the transmit power of TX being restricted to meet the QoS of CUE. Let $T_i = \{t_1, t_2, \cdots, t_{N \times K}\}$ denote the achieved maximum weighted rate of TX i paired with each RC units in descending order, and $O_i = \{o_1, o_2, \cdots, o_{N \times K}\}$ denote the corresponding RC units, which can be defined as the preference list of TX i. Then, we define $\mathcal{T} = \{T_1, T_2, \cdots, T_i, \cdots, T_N\}$ as the weighted rate set of all the TX-RC pairs, $\mathcal{O} = \{O_1, O_2, \cdots, O_i, \cdots, O_N\}$ as the preference list set of TX i, $\forall i \in \mathcal{N}_T$ on $\mathcal{RC}$ corresponding to $\mathcal{T}$. To obtain the maximum weighted channel rate for each TX-RC pair, we formulate the following problem:

$$\max_{\{P_i^D\}} \quad R_{D_{ij},k}$$

$$\text{s.t.} \quad C1 : 0 \le P_i^D \le P_{\max}$$

$$C2 : r_{D_{ij},k} \ge r_{\min}^d,$$

$$C3 : r_{k,i} \ge r_{\min}^c. \tag{15.15}$$

Thus, the preference list of D2D TX i on RC units $RC_{j,k}$ can be derived by solving problem (Eq. 15.15), and a detailed preference establishment algorithm is summarized in Algorithm 15.1, which constitutes the basis of the matching algorithm. An illustration of the preference lists establishment and a stable matching that we expected is shown in Figure 15.2.

15.5.3 Three-Dimensional Matching Algorithm

Based on the established preference lists, TXs could propose toward the RC units in their own first order. However, there exists a situation that more than one TX propose toward the same RC unit. Here, we propose a pricing strategy to decide the winner. The proposed matching algorithm is described briefly as follows.

- First of all, we introduce the concept of *price* for each RC unit that represents the matching cost for each TX. These prices are set to be zero at the beginning and they are virtual money without any physical significance. Let $\mathcal{CR}=\{CR_1, \cdots, CR_j, \cdots, CR_N\}$, $\forall j \in \mathcal{N}_R$, and $\mathcal{CK}=\{CK_1, \cdots, CK_k, \cdots, CK_K\}$, $\forall k \in \mathcal{K}$ denote the price sets of RXs and CUEs, respectively. The prices of RC units are denoted by $C=\{C_{j,k}\}_{j=1,k=1}^{j=N,k=K}$ where the price $C_{j,k}$ of $RC_{j,k}$ is the sum of RX j's price CR_j and CUE k's price CK_k.

- The proposed algorithm proceeds iteratively. In each iteration, any TX i that has not been matched with any RC unit would propose to its most preferred RC unit in O_i based on its payoff, which is equal to the achieved maximum weighted rate minus the matching cost, that is the current price of the RC unit. If any RX or CUE receives request from only one TX, the requested RC units would be directly matched with the TXs that initiate requests, and thus form a stable matching.

- Otherwise, the conflicting elements set consisting of RXs and CUEs that have received requests from more than one TX is denoted by Ω. Then, the elements in Ω would raise their prices with the price step s, which is determined by the minimum of the differences between any two adjacent values in the ordered weighted rate set. Accordingly, each TX proposed updates its preference list and renews its request. The process of rising prices continues until there is only one request received for the RC units.

- The algorithm would end if there exists no new request from TXs, that is all the TXs are matched when $K \geq N$ or all the CUEs are matched when $N \geq K$.

The above steps can lead to a stable matching that is proved in Section 15.5.4. We summarize the proposed three-dimensional matching algorithm in Algorithm 15.2.

In D2D underlay cellular network, the BS is the controller of resource management and link establishment, and thus the global channel state information (CSI) should be available at the BS for the matching approach. However, it is unnecessary for D2D users to obtain the global CSI, but just to feedback detected CSI by receiving detection signals at each terminal to the BS.

Algorithm 15.1

Preference Establishment Algorithm {

      Input: $\mathcal{N}_T, \mathcal{N}_R, \mathcal{N}_K, \mathcal{K}, \rho_{ij}, r_{\min}$.

      Output: $\{P_i^D\}, \mathcal{O}, \mathcal{T}$.

      for each neuron $i \in \mathcal{N}_T$

      {

            for each neuron $j \in \mathcal{N}_R$

            {

            for each neuron $k \in \mathcal{K}$

            {

                Calculate the maximum weighted rate $R_{D_{ij},k}$ [Equation

(15.15)] with the optimization of transmit power P_i^D.

                }

            }

            }

      for each neuron $i \in \mathcal{N}_T$

      {

            Obtain T_i by sorting the achieved maximum

weighted rates $R_{D_{ij},k}, \forall j \in \mathcal{N}_R, k \in \mathcal{K}$ in descending order.

            Establish the preference list O_i of TX i on RC

units by sorting each RC unit $RC_{j,k}$ in descending order based on T_i.

      }

}

15.5.4 Properties of the Three-Dimensional Matching Algorithm

In this section, the properties involving convergence, stability, optimality, and complexity of the proposed three-dimensional matching algorithm are analyzed in detail. All proofs are omitted here to save space and can be found in Ref. [12].

15.5.4.1 Convergence

We define the achieved maximum weighted rate $R_{D_{ij},k}$ as valuation $v_{i,j,k}$ of $RC_{j,k}$ for TX i, and the price $C_{j,k}$ of $RC_{j,k}$ as the matching cost for TXs. Then the payoff of TX i being matched with $RC_{j,k}$ can be written as $v_{i,j,k} - C_{j,k}$. In addition, it is denoted that there exists contention among TXs when any RX or CUE receives requests from more than one TX. At the start of each contention, the prices of RXs and CUEs are set to be zero and they would gradually increase by the step size s in the process of the contention. Any $i \in \mathcal{N}_T$ that has proposed to the conflicting elements would change its choice

Algorithm 15.2

The Three-Dimensional Matching Algorithm {

          Input: $\mathcal{N}_T$, $\mathcal{N}_R$, $\mathcal{K}$, $\mathcal{O}$, $\mathcal{T}$, $C\mathcal{R}$, $C\mathcal{K}$, C, Ω, s.

          Output: Φ, $\{x_{i,j,k}\}$.

          Initialization:

          Every TX $i \in \mathcal{N}_T$ builds its preference list on $\mathcal{R}C$ by using Algorithm

1.

          Set:

          $\Phi = \emptyset$, $\Omega = \emptyset$, $s = 0.1$.

          while ($\exists \Phi(i) = \emptyset$)

          {

            if ($O_i \neq \emptyset$)

            {

               for each neuron $i \in \mathcal{N}_T$

               {

                  TX i which has not been matched proposes to its most preferred RC unit in updated O_i.

               }

               Count the amount of RXs and CUEs that have received requests and put the conflicting elements that have received more than one request into Ω.

            if ($\Omega = \emptyset$)

            {

               Match the RC unit with its requestor TX directly.

            }

            if ($\Omega \neq \emptyset$)

            {

               for each neuron $RC_{j,k} \in \mathcal{R}C$

               {

                  if (CUE k receive requests from more than one TX)

                  {

                    RX j and CUE k in Ω increase their prices CR_j and CK_k with the price step s, and then TXs would update their preference lists and change their choices on RC units according to the price $C_{j,k}$. After this process, RX j and CUE k would be matched with the last remaining TX i that proposes to them, which is denoted by $\Phi(i) = RC_{j,k}$.

Algorithm 15.2

                              }
                         }
                    }
              Update:
                    Update $\mathcal{O}$ and $\mathcal{T}$ by deleting the RC units involving the matched
              RX j or CUE k and the corresponding achieved weighted rate respectively.
        Set $C\mathcal{R} = \{0\}, C\mathcal{K} = \{0\}, C = \{0\}$.
                         else
                         {
                              break
                         }
                    }
              }
        }

---

with the increase of the prices, which is based on its current maximum payoff:

$$(j, k) = \arg \max_{j \in \mathcal{N}_R, k \in K} (v_{i,j,k} - C_{j,k}) \tag{15.16}$$

The matching rules from Algorithm 15.2 show that the conflicting elements would be assigned to the TX that is the last one remaining in the request queue with the increase of the conflicting elements' prices. Assuming that TX i is matched with the conflicting RC unit $RC_{j,k}$, the contention must come to an end within $v_{i,j,k}/s$ steps. Hence, we can conclude the matching process within finite iterations.

15.5.4.2 Stability

Theorem 15.1 The proposed Algorithm 15.2 can converge to a two-sided stable matching Φ in finite iterations.

15.5.4.3 Optimality

Theorem 15.2 The content delivery one-to-one matching Φ is weak Pareto optimal for D2D transmitters on combinations of D2D receivers and spectrum resources.

15.5.4.4 Complexity
In the process of preference establishment, the computational complexity for any TX $i \in \mathcal{N}_T$ to obtain the preferences is $\mathcal{O}(NK)$ since each TX has to find

Algorithm 15.3 Simulation Parameters.

Simulation Parameter	Value
Cell radius R	200 m
Radius of the hot zone r	30 m
Max D2D transmission distance d_{max}	50 m
Pathloss exponent α	4
Max transmission power of D2D TXs P_{max}	23 dBm
Transmission power of cellular users P_k^C	23 dBm
Noise power N_0	-114 dBm
Number of D2D transmitters and receivers N	$1 \sim 6$
Number of resource blocks and cellular users K	$1 \sim 6$
QoS requirement r_{min}	0.5 bit/(s Hz)
Step size s	0.1

its preference value for each RC unit, which is corresponding to the achieved weighted rate. The computational complexity to derive the preference list by sorting the preference values for each TX is $\mathcal{O}(NK \log(NK))$. In Algorithm 15.2, the complexity of each process, in which TXs do not have matched propose to their most preferred RC units, is $\mathcal{O}(N^{\text{loop}})$ [67]. N^{loop} is the required number of iterations in the process of rising prices based on the step size s, that is during N^{loop} iterations, the assignment of the conflicting elements are finished when $\Omega \neq \emptyset$. We have $N^{\text{loop}} = 1$ when $\Omega = \emptyset$. Then, the computational complexity of the matching process is $\mathcal{O}(N N^{\text{loop}})$ $(N \geq K)$ or $\mathcal{O}(K N^{\text{loop}})$ $(K \geq N)$.

For the centralized exhaustive search, the total number of possible matching results is $N! \times K!$. The complexity of the algorithm can be written as $\mathcal{O}(N! \times K!)$. It is obvious that the proposed matching algorithm results in a much lower complexity for sufficient large values of N and K.

15.6 Numerical Results

In this section, the performance of the proposed iterative matching algorithm and impacts of the social relationships on D2D receivers' satisfactions are validated through simulations. The simulation parameters are summarized in Table 15.2 [10,21,46,47]. A single cellular network with a radius of $R = 200$ m is considered, where K CUEs are randomly distributed. N D2D transmitters and N receivers are randomly deployed in a circular hot spot area with the radius of $r = 30$ m. Figure 15.3 shows a snapshot of UEs' locations with $K = N = 6$. In the circular hot spot area represented by the inner dotted circle, D2D TXs and RXs

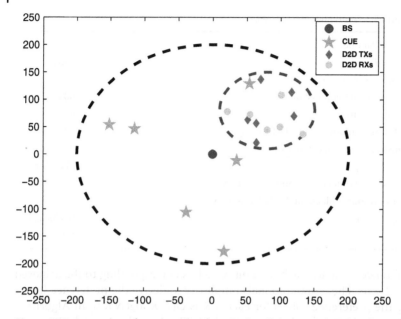

Figure 15.3 A snapshot of user locations for a single cellular network with K CUEs, N D2D TXs, and N D2D RXs ($K = 6$, $N = 6$, $d_{max} = 50$ m, the cell radius = 200 m, and the size of the spot hot is 30 m, respectively).

that satisfy both the physical and social requirements of D2D communication can form a D2D pair to directly exchange contents.

15.6.1 Convergence

The proposed algorithm is compared with two heuristic algorithms, that is the exhaustive and random matching algorithms. In particular, the exhaustive matching algorithm that examines every possible solution to find the optimum one is used to serve as an upper performance benchmark, while the random matching algorithm is used to serve as a lower performance benchmark. The convergence of the proposed matching algorithm is shown in Figure 15.4, which represents the weighted sum rate of D2D pairs versus the matching iterations. In Algorithm 15.2, we denote that at least one TX-RX–CUE pair would be formed in each iteration, thus we can derive that the number of the iterations required for the proposed algorithm to converge is related with the number of the TXs, RXs, and CUEs. Given $K = 6$, we can see that it only takes four and six matching iterations for the proposed algorithm to converge when $N = 4$ and $N = 6$, respectively. Moreover, it can be seen that the performance of matching is quite close to that of the exhaustive algorithm after the convergence.

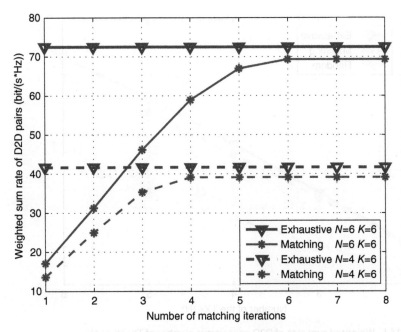

Figure 15.4 Weighted sum rate of D2D pairs versus number of matching iterations ($N = 6$).

15.6.2 Weighted Sum Rate

Figure 15.5 shows the weighted sum rate of all D2D pairs versus the number of TXs, while Figure 15.6 shows the weighted sum rate of all D2D pairs versus the number of CUEs. It is observed that the performance gaps between the proposed algorithm and the optimum exhaustive matching algorithm in Figures 15.5 and 15.6 are small. For instance, in Figure 15.5, the proposed algorithm is able to achieve 94.92% of the optimum performance, and outperforms the random matching algorithm by as much as 74.89% when $N = 5$ and $K = 6$. In Figure 15.6, the corresponding values of the performance compared with the optimum performance and the random performance are 93.33% and 74.61%, respectively, when $N = 6$ and $K = 5$. On the other hand, the computational complexity of the proposed matching algorithm is an order of magnitude lower than that of the exhaustive algorithm. For example, when $N = K = 6$, it takes 5.184×10^5 iterations for the exhaustive matching algorithm to find the optimum solution, while the proposed algorithm only requires 600 iterations, which reduces the complexity by nearly a thousand times. Compared with the exhaustive matching, the proposed algorithm does not need to achieve every possible matching result, which significantly reduces the computational complexity.

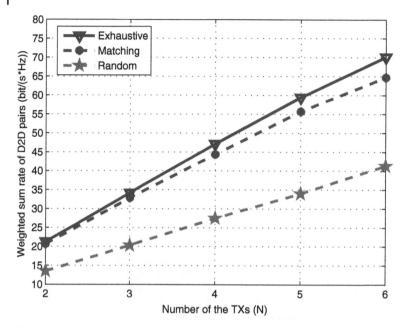

Figure 15.5 Weighted sum rate of D2D pairs versus number of TXs ($K = 6$).

From another perspective, we can find that the weighted sum rate raises up with both the number of D2D TXs and the number of CUEs (RBs) increasing. On one hand, when the amount of RBs is fixed, more D2D pairs contribute to a higher sum rate of D2D links. On the other hand, as the amount of RBs increases, the system supports to establish more D2D links. In Figure 15.6, it is obvious that the increment of weighted sum rate decreases continuously as the number of RBs increases. The reason is that the probability of accessing to the most preferred RB for a D2D pair becomes lower as more D2D pairs access to the network.

15.6.3 User Satisfaction

Figure 15.7 shows the cumulative distribution functions (CDFs) of the satisfactions for D2D RXs, namely, the similarity of users' preferences on the content that is reflected by the intensity of social relationships between the mutually matched D2D TXs and RXs. To evaluate the impacts of the social relationships on D2D RXs' satisfactions, both the social-aware and social-unaware matching algorithms are compared by varying the threshold of social relationships. Simulation results show that for the social-unaware algorithm, the proportion of D2D RXs whose satisfaction is greater than 0.8 is 15%, while the corresponding proportions achieved by the proposed algorithm are much higher, that is 41,

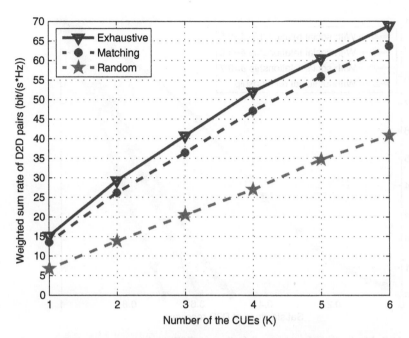

Figure 15.6 Weighted sum rate of D2D pairs versus number of CUEs ($N = 6$).

47, and 65% for $\delta = 0.5$, $\delta = 0.6$, and $\delta = 0.7$, respectively. It is noted that when the threshold δ decreases, the satisfaction performance also becomes worse. The reason is that it is much easier for D2D TXs and RXs with weak intensity of social relationship to form a D2D pair when the threshold is lower, which in turn degrades the satisfaction performance.

15.7 Conclusions

In this chapter, we studied the content delivery problem in social network-based D2D communications with uplink spectrum reusing. Both the social layer and the physical layer information were exploited in the optimization of the matching among users, contents, and spectrum resources. First, we modeled the social relationship between two users as the probability of selecting similar contents, which was estimated by using Bayesian nonparametric models. Then, we proposed a three-dimensional iterative matching algorithm to maximize the sum rate of D2D pairs weighted by the intensity of social relationships while guaranteeing the quality of service requirements of both cellular and D2D links simultaneously. Finally, the proposed algorithm was validated through simulations and compared with exhaustive optimal and random matching algorithms.

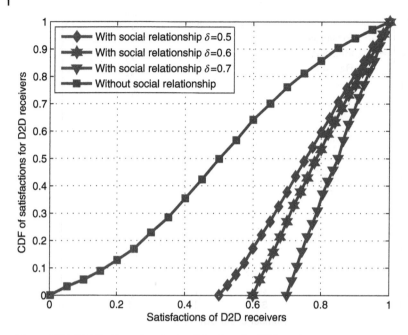

Figure 15.7 Distribution of content satisfactions for D2D receivers.

Simulation results demonstrated that the performance of the proposed iterative matching algorithm is much better than that of the random matching algorithm, and is very close to that of the optimum exhaustive matching but with a much lower computational complexity. Furthermore, the content satisfactions of D2D receivers are dramatically improved if social layer information is considered during the matching process. In future works, we will focus on the design of social-aware resource allocation algorithms for D2D communication by incorporating distributed caching schemes.

References

1 Cisco, "Cisco visual networking index: global mobile data traffic forecast update, 2015–2020," White Paper, 2016.
2 N. Zhang, N. Cheng, A. Gamage, K. Zhang, J. W. Mark, and X. Shen, "Cloud assisted HetNets toward 5G wireless networks," *IEEE Commun. Mag.*, vol. EDL-53, no. 6., p. 59, 2015.
3 S. Zhang, N. Zhang, S. Zhou, J. Gong, Z. Niu, and X. Shen, "Energy-aware traffic offloading for green heterogeneous networks," *IEEE J. Sel. Areas Commun.*, vol. EDL-34, no. 5, p. 1116, 2016.

4 K. Doppler, M. Rinne, C. Wijting, C. B. Ribeiro, and K. Hugl, "Device-to-device communication as an underlay to LTE-advanced networks," *IEEE Commun. Mag.*, vol. EDL-47, no. 12, p. 42, 2009.

5 C. Xu, L. Song, and Z. Han, *Resource Management for Device-to-Device Underlay Communication*, Springer Briefs in Computer Science, Springer, 2013.

6 G. Fodor, E. Dahlman, G. Mildh, S. Parkvall, and N. Reider, "Design aspects of network assisted device-to-device communications," *IEEE Commun. Mag.*, vol. EDL-50, no. 3, p. 170, 2012.

7 L. Wei, R. Hu, Y. Qian, and G. Wu, "Enable device-to-device communications underlaying cellular networks: challenges and research aspects," *IEEE Trans. Commun.*, vol. EDL-52, no. 6, p. 90, 2014.

8 Y. Cai, D. Wu, and W. Yang, "Social-aware content downloading mode selection for D2D communications," in *Proc. IEEE ICC'15*, p. 2931, 2015.

9 Z. Zheng, T. Wang, L. Song, Z. Han, and J. Wu, "Social-aware multi-file dissemination in device-to-device overlay networks," in *Proc. IEEE INFOCOM'14*, p. 219, 2014.

10 T. Wang, Y. Sun, L. Song, and Z. Han, "Social data offloading in D2D-enhanced cellular networks by network formation games," *IEEE Trans. Wirel. Commun.*, vol. EDL-14, no. 12, p. 7004, 2015.

11 J. Iqbal and P. Giaccone, "Social and spatial proactive caching for mobile data offloading," in *Proc. IEEE ICC'14*, p. 581, 2014.

12 C. Xu, C. Gao, Z. Zhou, Z. Chang, and Y. Jia, "Social network-based content delivery in device-to-device underlay cellular networks using matching theory," *IEEE Access*, vol. EDL-PP, no. 99, p. 1, 2016.

13 T. Ma, J. Zhou, M. Tang, Y. Tian, A. Al-Dhelaan, M. Al-Rodhaan, and S. Lee, "Social network and tag sources based aug menting collaborative recommender system," *IEICE Trans. Inform. Syst.*, vol. EDL-E98-D, no. 4, p. 902, 2015.

14 G. W. Corder and D. I. Foreman, *Nonparametric Statictics for NonStatisticians: A Step-by-Step Approach*. New York: John Wiley & Sons, Inc., 2009.

15 A. Roth and M. Sotomayor, *Two-Sided Matching: A Study in GameTheoretic Modeling and Analysis*. Cambridge University Press, 1992.

16 Y. Gu, W. Saad, M. Bennis, M. Debbah, and Z. Han, "Matching theory for future wireless networks: fundamentals and applications," *IEEE Commun. Mag.*, vol. EDL-53, no. 5, p. 52, 2015.

17 X. Feng, G. Sun, X. Gan et al., "Cooperative spectrum sharing in cognitive radio networks: a distributed matching approach," *IEEE Trans. Commun.*, vol. EDL-62, no. 8, p. 2651, 2014.

18 A. M. EI-Hajj, Z. Dawy, and W. Saad, "A stable matching game for joint uplink/downlink resource allocation in OFDMA wireless networks," in *Proc. IEEE ICC'12*, p. 5354, 2012.

19 B. Di, S. Bayat, L. Song, and Y. Li, "Radio resource allocation for full-duplex OFDMA networks using matching theory," in *Proc. IEEE INFOCOM'14*, p. 197, 2014.

20 B. Ma, H. S. Mansouri, and V. W. S. Wong, "A matching approach for power efficient relay selection in full duplex D2D networks," in *Proc. IEEE ICC'16*, p. 1, 2016.

21 Z. Zhou, M. Dong, K. Ota, and C. Xu, "Energy-efficient matching for resource allocation in D2D enabled cellular networks," *IEEE Trans. Veh. Technol.*, vol. EDL-PP, no. 99, p. 1, 2016.

22 J. Liu, N. Kato, J. Ma, and N. Kadowaki, "Device-to-device communication in LTE-advanced networks: a survey," *IEEE Commun. Surv. Tutor.*, vol. EDL-17, no. 4, p. 1923, 2015.

23 Y. Chia, K. Doppler, C. B. Ribeiro, and O. Tirkkonen, "Resource sharing optimization for device-to-device communication underlaying cellular networks," *IEEE Trans. Wirel. Commun.*, vol. EDL-10, no. 8, p. 2752, 2011.

24 D. Feng, L. Lu, Y. Wu, G. Y. Li, G. Feng, and S. Li, "Device-to-device communications underlaying cellular networks," *IEEE Trans. Commun.*, vol. EDL-61, no. 8, p. 3541, 2013.

25 C. Xu, L. Song, Z. Han, and Q. Zhao, "Efficiency resource allocation for device-to-device underlay communication systems: a reverse iterative combinatorial auction based approach," *IEEE J. Sel. Areas Commun.*, vol. EDL-31, no. 9, p. 348, 2013.

26 Q. Ye, M. Al-Shalash, C. Caramanis, and J. G. Andrews, "Distributed resource allocation in device-to-device enhanced cellular networks," *IEEE Trans. Commun.*, vol. EDL-63, no. 2, p. 441, 2015.

27 L. Lei, Y. Kuang, N. Cheng, and X. Shen, "Delay-optimal dynamic mode selection and resource allocation in device-to-device communications – Part I: optimal policy," *IEEE Trans. Veh. Technol.*, vol. EDL-65, no. 5, p. 3474, 2016.

28 J. Liu, S. Zhang, N. Kato, H. Ujikawa, and K. Suzuki, "Device-to-device communications for enhancing quality of experience in software defined multi-tier LTE-A networks," *IEEE Net. Mag.*, vol. EDL-29, no. 4, p. 46, 2015.

29 J. Liu, H. Nishiyama, N. Kato, and J. Guo, "On the outage probability of device-to-device-communication-enabled multichannel cellular networks: an RSS-threshold-based perspective," *IEEE J. Sel. Areas Commun.*, vol. EDL-34, no. 1, p. 163, 2016.

30 N. Golrezaei, P. Mansourifard, A. F. Molisch, and A. G. Dimakis, "Base-station assisted device-to-device communications for high-throughput wireless video networks," *IEEE Trans. Wirel. Commun.*, vol. EDL-13, no. 7, p. 3665, 2014.

31 B. Zhou, H. Hu, S. Huang, and H. Chen, "Intracluster device-to-device relay algorithm with optimal resource utilization," *IEEE Trans. Veh. Technol.*, vol. EDL-62, no. 5, p. 2315, 2013.

32 M. Hasan and E. Hossain, "Distributed resource allocation for relay-aided device-to-device communication under channel uncertainties: a stable matching approach," *IEEE Wirel. Commun.*, EDL-63, no. 10, p. 3882, 2015.

33 Y. Cheng, Y. Gu, and X. Lin, "Combined power control and link selection in device-to-device enabled cellular systems," *IET Commun.*, vol. EDL-7, no. 12, p. 1221, 2013.

34 X. Zhang, Z. Zheng, Q. Shen, J. Liu, X. S. Shen, and L. L. Xie, "Optimizing network sustainability and efficiency in green cellular networks," *IEEE Trans. Wirel. Comm.*, vol. EDL-13, no. 2, p. 1129, 2014.

35 S. Gupta, R. Zhang, and L. Hanzo, "Energy harvesting aided device-to-device communication underlaying the cellular downlink," *IEEE Access*, vol. EDL-PP99, p. 1, 2016.

36 J. Liu, M. Sheng, T. Q. S. Quek, and J. Li, "D2D enhanced co-ordinated multipoint in cloud radio access networks," *IEEE Trans. Wirel. Comm.*, vol. EDL-15, no. 6, p. 4248, 2016.

37 X. Cheng, L. Yang, and X. Shen, "D2D for intelligent transportation systems: a feasibility study," *IEEE Trans. Intell. Transp. Syst.*, vol. EDL-16, no. 4, p. 1784, 2015.

38 F. Wang, C. Xu, L. Song, Q. Zhao, X. Wang, and Z. Han, "Energy-aware resource allocation for device-to-device underlay communication," in *Proc. IEEE ICC'13*, p. 6076, 2013.

39 F. Wang, C. Xu, L. Song, and Z. Han, "Energy-efficient resource allocation for device-to-device underlay communication," *IEEE Trans. Wirel. Comm.*, vol. EDL-14, no. 4, p. 2082, 2015.

40 D. Wu, J. Wang, R. Q. Hu, Y. Cai, and L. Zhou, "Energy-efficient resource sharing for mobile device-to-device multimedia communications," *IEEE Trans. Veh. Technol.*, vol. EDL-63, no. 5, p. 2093, 2014.

41 C. Yang, X. Xu, J. Han, and X. Tao, "Energy efficiency-based device-to-device uplink resource allocation with multiple resource reusing," *Electron. Lett.*, vol. EDL-51, no. 3, p. 293, 2015.

42 S. Mumtaz, K. M. S. Huq, A. Radwan, J. Rodriguez, and R. L. Aguiar, "Energy-efficient interference-aware resource allocation in LTE-D2D communication," in *Proc. IEEE ICC'14*, p. 2826, 2014.

43 Z. Zhou, M. Dong, K. Ota, G. Wang, and L. T. Yang, "Energy-efficient resource allocation for D2D communications underlaying cloud RAN-based LTE$_i$"A networks," *IEEE Internet Things J.*, vol. EDL-3, no. 3, p. 428, 2015.

44 D. Wu, L. Zhou, Y. Cai, R. Q. Hu, and Y. Qian, "The role of mobility for D2D communications in LTE-advanced networks: energy vs. bandwidth efficiency," *IEEE Wireless Commun.*, vol. EDL-21, no. 2, p. 66, 2014.

45 L. Wei, R. Q. Hu, Y. Cai, and G. Wu, "Energy-efficiency and spectrum-efficiency of multi-hop device-to-device communications underlaying cellular networks," *IEEE Trans. Veh. Technol.*, vol. EDL-65, no. 1, p. 367, 2016.

46 Z. Zhou, M. Dong, K. Ota, J. Wu, and T. Sato, "Energy efficiency and spectral efficiency tradeoff in device-to-device D2D communications," *IEEE Wirel. Commun. Lett.*, vol. EDL-3, no. 5, p. 485, 2014.

47 Z. Zhou, M. Dong, K. Ota et al., "A game-theoretic approach to energy-efficient resource allocation in device-to-device underlay communications," *IET Commun.*, vol. EDL-9, no. 3, p. 375, 2015.

48 K. W. Choi and Z. Han, "Device-to-device discovery for proximity-based service in LTE-advanced system," *IEEE J. Sel. Areas Commun.*, vol. EDL-33, no. 1, p. 55, 2015.

49 B. Zhang, Y. Li, D. Jin, P. Hui, and Z. Han, "Social-aware peer discovery for D2D communications underlaying cellular networks," *IEEE Trans. Wirel. Commun.*, vol. EDL-14, no. 5, p. 2426, 2015.

50 K. W. Choi, D. T. Wiriaatmadja, and E. Hossain, "Discovering mobile applications in cellular device-to-device communications: hash function and bloom filter-based approach," *IEEE Trans. Mobile Comput.*, vol. EDL-15, no. 2, p. 336, 2016.

51 L. Wang, H. Tang, and M. ierny, "Device-to-device link admission policy based on social interaction information," *IEEE Trans. Veh. Technol.*, vol. EDL-64, no. 9, p. 4180, 2015.

52 C. Cao, L. Wang, and M. Song, "Joint social and physical clustering scheme for device-to-device communications," in *Proc. IEEE IIKI'14*, p. 175, 2014.

53 X. Chen, B. Proulx, X. Gong, and J. Zhang, "Exploiting social ties for cooperative D2D communications: a mobile social networking case," *IEEE Trans. Netw.*, vol. EDL-23, no. 5, p. 1471, 2015.

54 O. Semiari, W. Saad, S. Valentin, M. Bennis, and H. V. Poor, "Context-aware small cell networks: how social metrics improve wireless resource allocation," *IEEE Trans. Wirel. Commun.*, vol. EDL-1411, p. 5927, 2015.

55 Y. Cao, T. Jiang, X. Chen, and J. Zhang, "Social-aware video multicast based on device-to-device communications," *IEEE Trans. Mobile Comput.*, vol. EDL-15, no. 6, p. 1528, 2016.

56 Y. Zhao, Y. Li, Y. Cao, T. Jiang, and N. Ge, "Social-aware resource allocation for device-to-device communications underlaying cellular networks," *IEEE Trans. Wirel. Commun.*, vol. EDL-1412, p. 6621, 2015.

57 F. Wang, Y. Li, Z. Wang, and Z. Yang, "Social community aware resource allocation for D2D communications underlaying cellular networks," *IEEE Trans. Veh. Technol.*, vol. EDL-65, no. 5, p. 3628, 2016.

58 L. Wang, L. Liu, X. Cao, X. Tian, and Y. Cheng, "Sociality-aware resource allocation for device-to-device communications in cellular networks," *IET Commun.*, vol. EDL-9, no. 3, p. 342, 2015.

59 M. Hasan and E. Hossain, *Distributed resource allocation in 5G cellular networks*, 2014. Available at http://arxiv.org/abs/1409.2475

60 Y. Gu, Y. Zhang, M. Pan, and Z. Han, "Matching and cheating in device to device communications underlying cellular networks," *IEEE J. Sel. Areas Commun.*, vol. EDL-3310, p. 2156, 2015.

61 Z. Zhou, G. Ma, C. Xu, Z. Chang, and T. Ristaniemi, "Energy-efficient resource allocation in cognitive D2D communications: a game-theoretical and matching approach," in *Proc. IEEE ICC*, p. 1, 2016.

62 Z. Han, R. Zheng, and H. V. Poor, "Repeated auctions with Bayesian nonparametric learning for spectrum access in cognitive radio networks," *IEEE Trans. Wirel. Commun.*, vol. EDL-10, no. 3, p. 1536, 2011.

63 B. Gu, X. Sun, and V. S. Sheng, "Structural minimax probability machine," *IEEE Trans. Neural Netw. Learn. Syst.*, vol. EDL-PP99, p. 1, 2016.

64 B. Gu and V. S. Sheng, "A robust regularization path algorithm for V-support vector classification," *IEEE Trans. Neural Netw. Learn. Syst.*, vol. EDL-PP99, p. 1, 2016.

65 L. Ding, A. Yilmaz, and R. Yan, "Interactive image segmentation using Dirichlet process multiple-view learning," *IEEE Trans. Image Process.*, vol. EDL-21, no. 4, p. 2119, 2012.

66 Y. W. Teh, M. I. Jordan, M. J. Beal, and D. M. Blei, "Hierarchical Dirichlet processes," *J. Am. Stat. Assoc.*, vol. EDL-101, no. 476, p. 1566, 2006.

67 G. O'Malley, "Algorithmic aspects of stable matching problems," Ph.D. dissertation, University of Glasgow, 2007.

Chen Xu received the B.S. degree from Beijing University of Posts and Telecommunications, in 2010, and the Ph.D. degree from Peking University, in 2015. She is now a lecturer in School of Electrical and Electronic Engineering, North China Electric Power University, China. Her research interests mainly include wireless resource allocation and management, game theory, optimization theory, heterogeneous networks, and smart grid communication. She received the best paper award in International Conference on Wireless Communications and Signal Processing (WCSP 2012), and she is the winner of IEEE Leonard G. Abraham Prize 2016.

Caixia Gao is currently working toward the M.S. degree at North China Electric Power University, China. Her research interests include resource allocation, interference management, and energy management in D2D communications.

Zhenyu Zhou received his M.E. and Ph.D degree from Waseda University, Tokyo, Japan in 2008 and 2011, respectively. From April 2012 to March 2013, he was the chief researcher at Department of Technology, KDDI, Tokyo, Japan. From March 2013 to now, he is an Associate Professor at School of Electrical and Electronic Engineering, North China Electric Power University, China. He is also a visiting scholar with Tsinghua-Hitachi Joint Lab on Environment-Harmonious ICT at University of Tsinghua, Beijing from 2014 until now. He served as workshop co-chair for IEEE ISADS 2015, session chair for IEEE Globecom 2014, and TPC member for IEEE Globecom 2015, ACM Mobimedia 2015, IEEE AFRICON 2015, and so on. He received the "Young Researcher Encouragement Award" from IEEE Vehicular Technology Society in 2009. His research interests include green communications and smart grid. He is a member of IEEE, IEICE, and CSEE.

Shahid Mumtaz received the M.Sc. degree from the Blekinge Institute of Technology, Sweden, and the Ph.D. degree from the University of Aveiro, Portugal. He is currently a Senior Research Engineer with the Instituto de Telecomunicaes, Aveiro, Portugal, where he is involved in EU-funded projects. His research interests include MIMO techniques, multi-hop relaying communication, cooperative techniques, cognitive radios, game theory, energy-efficient framework for 4G, position information-assisted communication, and joint PHY and MAC layer optimization in LTE standard. He has authored several conferences, journals, and books publications.

Jonathan Rodriguez received his Master's and Ph.D degree in Electronic and Electrical Engineering from the University of Surrey (UK), in 1998 and 2004 respectively. In 2005, he became a researcher at the Institut? de Telecomunica?es and Senior Researcher in the same institution in 2008 where he established the 4TELL Research Group (http://www.av.it.pt/4TELL/) targeting the next-generation mobile networks with key interests on energy-efficient design, cooperative strategies, and security and electronic circuit design. He has served as project coordinator for major international research projects (Eureka LOOP, FP7 C2POWER), while acting as the technical manager for FP7 COGEU and FP7 SALUS. He is currently leading the H2020-ETN SECRET project, a European Training Network on 5G communications. Since 2009, he has been invited as the Assistant Professor at the Universidade de Aveiro, and granted as the Associate Professor in 2015. He has authored more than 350 scientific works, including nine book editorials. Since 2017, he is Professor of Mobile Communications at the University of South Wales, UK.

16

Service-Oriented Architecture for IoT Home Area Networking in 5G

Mohd Rozaini Abd Rahim, Rozeha A. Rashid, Ahmad M. Rateb, Mohd Adib Sarijari, Ahmad Shahidan Abdullah, Abdul Hadi Fikri Abdul Hamid, Hamdan Sayuti, and Norsheila Fisal

Faculty of Electrical Engineering, Universiti Teknologi Malaysia, Johor, Malaysia

16.1 Introduction

Internet of Things (IoT) is a network that interconnects objects from everyday life to create smarter homes, cities, transport, and health care systems [1]. The number of IoT devices is expected to range between 30 and 75 billion devices by 2020 [2–5], and hence they will represent the majority of the 5G network terminals. This large population of machines will use the 5G network to communicate with each other, with the human user, and to send data collected by embedded sensors to the cloud for analysis and processing. As a result, IoT devices have acquired a significant share of the 5G systems design effort, where several components of the 5G network are being designed to handle the scalability, heterogeneity, power, and cost requirements of the IoT [6,7].

Smart cities and health monitoring are among the most prevailing vertical applications of 5G networks [8]. Smart cities will involve a highly dense clusters of wireless sensors that enable a large set of services and applications such as household electrical power consumption monitoring to enable power grid optimization, environmental services, access control, security system, and pollution monitoring. These applications demand highly efficient utilization and management of the sensor resources at the bottom level in order to secure stable operation and traffic intensity within the network that interconnects these nodes.

5G Networks: Fundamental Requirements, Enabling Technologies, and Operations Management, First Edition.
Anwer Al-Dulaimi, Xianbin Wang, and Chih-Lin I.
© 2018 by The Institute of Electrical and Electronics Engineers, Inc. Published 2018 by John Wiley & Sons, Inc.

On the other hand, health services is going to be revolutionized by 5G due to enabling of real-time health monitoring through wearables, in addition to enabling distant intervention by health specialist in cases of emergencies. Enabling a highly reliable operation for these critical services requires viable guarantees on system connectivity and latency, which might be hurdles due to network congestion and inefficient management of sensors.

A large portion of IoT devices resides within the vicinity of our homes, such as personal smartphones, wearable health monitors, smart light bulbs, smart locks, security cameras, smart air conditioners. The interconnection between all these devices creates a smart IoT home area network (HAN). Devices within the IoT HAN enable smart home applications such as smart lighting, heating, security, remote health care, entertainment [9–11]. Efficient implementation of IoT HAN for smart home applications faces several challenges, among which we mention the following:

1) *Inefficient Utilization and Management of Sensor Nodes*: The strong coupling between application and hardware mandates that one or more sensor nodes are allocated for every smart home application. For example, room temperature control and fire alarm applications mainly sense the same quantity, which is temperature. This introduces redundancy in the sensors and the data they transmit due to limited reusability of sensor nodes that adds cost and shortens node battery life. In addition, implementation of new smart home applications will essentially imply installing more sensor nodes, and hence the transmission of all sensed data from these nodes will represent an overload on the network, and leads to increasing its latency due to the increased probability of collision. Conversely, 5G networks demand stringent latency specifications [6].

2) *Interoperability in Heterogeneous Networks*: IoT smart home applications involve installing a large number of sensing nodes from various manufacturers with essentially different wireless access technologies such as ZigBee, Bluetooth Low Energy (BLE), and Wi-Fi [7]. Two nodes employing different technologies cannot achieve direct machine-to-machine (M2M) communications and thus are incompatible [12,13]. For example, in a home security application, a motion sensor that uses Bluetooth technology will not be able to trigger a camera that uses Wi-Fi technology to capture a picture of an intruder. A straightforward solution to this problem is to employ a gateway server to connect these nodes, however, this solution has drawbacks such as cost, difficulty of usage, and maintenance [14].

3) *Application Development*: The conventional approach for developing smart home applications is usually tailored for a specific sensor node type, and requires the developer to be well aware of low-level details of node operating system and programming, sensor operation, and its wireless access technology. It becomes even more challenging to develop an application that

supports interoperability of heterogeneous sensors that use different transmission technology and/or run on different operating systems [12,15,16]. As a result, the overall application development time and cost are high.

In this chapter, we introduce our proposed solution for the already mentioned problems. We present a novel service-oriented architecture that relies on the concept of sensor virtualization, where the low-level functionalities of sensor nodes are abstracted into a group of services that are made available to the application developer, eliminating the need for the developer to be aware of low-level details. This is achieved through development of highly scalable middleware that establishes interoperability between heterogeneous sensor nodes, and optimizes resource utilization. The proposed solution has the following state-of-the-art contributions:

- Smart management and utilization of WSN resources, where the redundancy in sensing data acquisition and transmission are minimized, and hence sensing node battery life is maximized.
- Ensures minimum network latency by limiting the data packets transmitted by sensor nodes, which in turn minimizes the probability of packet collision.
- Enables implementing new smart home applications without adding new sensor nodes by reusing services provided by the existing nodes.
- Hides low-level details of sensor node functionality and connectivity from the application developer, which reduces application development time, effort, and cost.
- Enables seamless M2M communication between nodes that use different wireless technologies without using a gateway.
- Improves overall network reliability, where a data from a faulty sensor used by one application can be automatically replaced by data from another sensor used by another application in close proximity.

The proceeding section will describe the basic service-oriented architecture, followed by related works in the implementation of service-oriented architecture in wireless sensor network application. Next, the proposed service-oriented architecture for home area network will be explained in Section 16.4 with its performance analysis presented in Section 16.5. Finally, the conclusion is drawn in Section 16.6.

16.2 Service-Oriented Architecture

Service-oriented architecture is a distributed software architecture that consists of multiple autonomous services. Services are distributed such that they can execute on different nodes with different service providers [17]. A service is comprised of a set of functions and each function defines the operation that

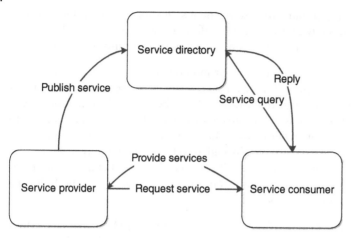

Figure 16.1 Basic operation of service-oriented architecture.

can be executed by the service and whether these operations are provided or required [18].

Figure 16.1 shows the basic operation of service-oriented architecture. Service-oriented architecture is divided into three elements: service provider, service directory, and service consumer. The service provider is responsible to design and develop the service. The developed service information will be published into the service directory. The services inside the service directory will publish the service information into the network. When the service consumer requires the service, service consumer requests the service information from the service directory and binds with the service provider to invoke the services.

The advantage of using service-oriented middleware is modularity, autonomous operation, and well-defined interface by which the service can be described, published, invoked, and discovered over network. The development of service-oriented middleware should provide the following features [16,19,20]:

Abstraction. The middleware system should hide as much as possible of the heterogeneity of underlying environment from the developer.

Interoperability. Services should have ability to interact with different service providers and wireless technologies.

Scalability. The middleware should have the capability to support a huge number of sensor nodes or users in term of accessibility, and also provides efficient query optimization.

Resource Utilization. The service may receive a large number of similar requests in a relatively short period of time. By using mechanisms, such as request similarity detection and caching, the service may be able to answer most of the requests without requesting the same data again.

Cost Reduction. The services are expected to be reusable and often generic in order to reduce application development cost.

Adaptability. The system should be able to adapt to network or environment changes to maintain continuity of the communication system.

Topology. To provide a mechanism to support the ever-changing network topology and guarantee its robustness.

Extensibility. To be able to add new sensor types or new data operators to the software installed at the network nodes.

Programmability. It should be flexible, allowing for configuration or reconfiguration of its features and functionality in the real time.

The mentioned features form the basis of the proposed service-oriented architecture, specifically designed for a home area network that consists a large number of heterogeneous devices and various applications. The main contributions of the proposed architecture include a better utilization of available devices, higher energy efficiency, and reduced latency. Consequently, the proposed work offers a promising solution for device and sensor management in a residential setting.

16.3 Related Work

Recently, several efforts have been carried out on the implementation of service-oriented architecture into wireless sensor network middleware for various applications. Table 16.1 shows the summary of existing service-oriented middleware features.

A Web Service Middleware for Ambient Intelligence (aWESoME) [21] is a middleware developed to address the issues of universal, homogenous access to the system function and fulfill functional and nonfunctional requirements from the system architecture. It is based on ambient intelligence environment and consumes low power without compromise, with reliability, and fast response time. Service broker has been used for registering and providing the service.

USEME [22] middleware allows the combination between microprogramming and node-centric programming to develop real-time and efficient application for wireless sensor and actuator network. It provides the real-time specification between services, uses group network structure, and supports dependent-free platform. The USEME middleware allows the developer to create application without knowing the low-level detail and repetitive task in heterogeneous wireless sensor network. It is suitable for various wireless sensor network application developments.

HERA [23] is an upgraded version of services layer over light physical device (SYLPH), which is based on a distributed platform and implemented using an SOA approach into heterogeneous wireless sensor network. The HERA archi-

Table 16.1 Summary of the existing service-oriented middleware features.

Middleware	Main Features	Disadvantage
aWESoME	• Supports universal and homogenous access to system function • Supports functional and nonfunctional requirements • Able to expose function and data from various devices	• Only tackles interoperability problem • Does not support code reusability • Uses third-party software, therefore, modifying the programming at gateway is not possible • Centralized topology results in everything being done at the base station. Node just sends the data to the base station
USEME	• Allows combination of macro-programming and node centric programming. • Supports and allows the specification of real-time constrain between services • Permits use of groups to structure the network platform independently	• Requires cluster head to manage the group • The cluster head will store all the information about the node member
HERA	• Allows different wireless technologies of the device to work together in a distributed way • The platform is specially designed to implement hardware agent • Capable of recovering from error and more flexible to adjust its behavior in execution time	• High latency during the initialization of HERA agent • Requires many initial packet for HERA initialization • More focus on the service delivery
μSMS	• Applies a virtual sensor service method using in-network agentbased services • Uses event model and publishes/subscribes for service discovery	• Sensor information will be sent to the gateway before service formation • One sensor for one sensor node
KASO	• Integrates the WSN into cloud service • Offers advanced and enriched pervasive service to everyone connected to Internet • Supports event-based and on-demand service	• Requires cluster head to operate • Only cluster head be able to communicate to the sink node • Two-level service composition at the sensor node and cluster head

Table 16.1 Continued

Middleware	Main Features	Disadvantage
OASIS	• Provides abstractions object centric • Supports ambient aware • Provides tracking application	• Only focuses on developing abstraction for object-centric application • Does not describe the method of data collection from multiple sensors on a single node and create the service • Service-oriented architecture design provides the standard data representation, information of service interface, and service discovery
TinySOA	• Independent programming language	• Does not support multiple sensors on a single node

tecture can operate into various sensor nodes platform with independent wireless technology, architecture, and programming language. It also uses reactive agents with case-based planning features to recover from error by considering the previous history. Generally, HERA can work together with wireless devices from different technologies in a distributed way.

A micro-subscription management system (μSMS) [24] middleware has been proposed for smart infrastructure over wireless sensor network by using service-oriented architecture as a software architecture. The approach of this middleware is to specify and develop the notion of virtual sensor services created for smart environment via sensor network from tiny in-network services by using agent-based technology. The developed middleware provided medical status monitoring, perimeter surveillance, and location tracking. Generally, the main purpose of this middleware focuses on translating the wireless sensor network architecture into Internet-based architecture for world smart digital ecosystem.

Knowledge-Aware and Service-Oriented (KASO) [25] middleware tries to integrate the wireless sensor and actuator network with service cloud. The main key element of KASO is to offer advance and enrich pervasive services to everyone connected to Internet. It will implement mechanism and protocol that allows managing the knowledge generated in pervasive embedded networks in order to disclose it to Internet user in a readable way. The energy consumption, memory, and bandwidth are considered in developing of KASO middleware.

OASIS [26] middleware provides abstraction for object-centric, service-oriented sensor network application, and ambient aware. It also provides location tracking to track the heat source. The programming framework for OASIS enables the program developer not to deal with low-level system and network issues and also provides well-defined model. It also decomposes specified application behavior and produces the suitable node-level code for placement into the sensor network. The development of middleware based on function block programming abstraction in Reference [19] provides reprogramming of sensor node with new applications by injecting the base station with a mobile agent. It also facilitates flexibility and adaptability features by making their operation more complete and efficient even in dynamic environment and at the same time keeping the complexity and overhead programming low. Wiring concept has been used to link between function blocks in order to develop the application. This method will hide the underlying hardware and software complexity from the application developer.

TinySOA [28] is a service-oriented architecture middleware that lets programmers access WSN from their application based on a simple service-oriented API over their own language programming. The approach does not take into the account the nonfunctional capability. It consists of four components such as gateway, node, server, and registry. The scope of TinySOA is to cover monitoring and visualization application. The general target of TinySOA is to facilitate the technique developer control and access of wireless sensor network and incorporate them into application implementation.

In summary, most of the existing service-oriented middleware developments focus on improving the abstraction and reducing the code's complexity in order to simplify the application programmer in developing and integrating the applications. In addition, many of the current service-oriented middleware works are centralized in nature where the data collected by the sensor node will be sent to the cloud. The sensor data gathered by the sensor node is not able to be transmitted directly to the base station. It needs to go through the cluster head before being forwarded to base station. The focus also concentrates more on the service delivery. By focusing on the service delivery, more packet needs to be introduced in order to ensure the service reaches the destination. This will result in an increase in latency.

16.4 Service-Oriented Architecture for Home Area Network (SoHAN)

In this section, a more detailed description of the proposed SoHAN architecture is described.

16.4.1 SoHAN Network

Figure 16.2 shows the proposed SoHAN network. The SoHAN network architecture is organized into four main parts, including home area network (HAN), 5G network, server, and monitoring devices.

The components of HAN are divided into two parts: sensor nodes and home gateway. Wireless sensor nodes are responsible to collect and process data sensed from the environment and send them to the gateway through other

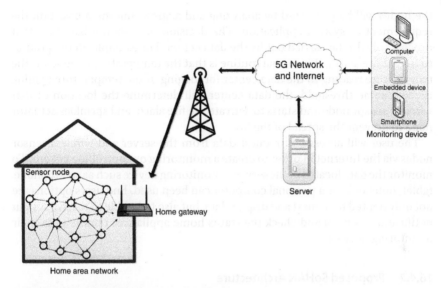

Figure 16.2 Proposed SoHAN network solution.

sensor nodes within the HAN. The typical sensor node consists of one or more sensors that sense the environment, a processing unit such as microcontroller to analyze the sensor data, and a wireless communication module to send the processed data to the gateway. Each sensor node is equipped with a sensor and an actuator to perform sensing and control operations, respectively. The sensing operation includes temperature, humidity, fire, physiological signals, and light in the home environment, while controling operation includes air conditioner, washing machine, security system, television, and electric kettle. The wireless sensor node in the proposed SoHAN middleware is implemented within mesh network. In the mesh network, each sensor node is capable of relaying the message from the origin to destination by using a routing technique or flooding technique. All the sensor nodes are located inside the home.

The home gateway is responsible for enabling communication between the sensor node and server via the 5G network. The gateway relays the message from the sensor node to the server. At the same time, the gateway is also responsible for channeling the message from the server to the destination node within the HAN. The gateway in the proposed SoHAN is located within the house premises.

A server plays a main role in the SoHAN system architecture. The server contains a repository that consists of detailed information about the HAN applications, such as temperature sensing data, wireless sensor node status, wireless sensor node location, network status, home appliances status, services, and the elderly health status. The huge amount of information received from the nodes will be processed by analyzing and aggregating them based on the requirement of system application. The decision or response based on that process will also be carried out by the data center. For example, in the process to trigger fire alarm, the normal routine is that the temperature node sends the temperature reading to the data center for logging. If the temperature reading exceeds some threshold, the data center will determine the location of that wireless sensor node and starts to activate the fire alarm and sprinkler actuator node to prevent the spread of the fire.

The user will access and request data from the server and wireless sensor nodes via the Internet in order to create a monitoring or controlling system. To monitor the data located at the server, a monitoring device such as smart phone, tablet, notebook, or a personal computer can been used. End users options are not only limited to request and display data, but also to be able to receive system notification, control and check the status home appliances remotely via their monitoring devices.

16.4.2 Proposed SoHAN Architecture

The proposed SoHAN architecture is organized into three main sections: sensors, services, and applications as shown at Figure 16.3. Sensors section represents the number, type, and capability of the sensors available within the SoHAN architecture. The services section shows the number and type of possible services that can be generated based on the sensors attached to the sensor nodes. The combination of the services existing in the SoHAN architecture will create the application and resides at the application section.

Sensor section consists of the sensors or actuators attached to the sensor node. It represents the existence of the various heterogeneous sensors in the HAN network. All the sensor nodes are independent. The sensor data are the values generated by either the sensors or by the actuators such as actuator status. Typical examples of sensors include gap sensor, smoke sensor, temperature sensor, motion sensor, and camera. On the other hand, examples of actuators include magnetic door lock, siren controller, and fire sprinkler system controller. The number of sensors and actuators attached to the sensor node can be more than one sensor and one actuator.

The service section contains the services offered by the SoHAN architecture. A service is a combination of the functionalities of sensors and actuators residing on a single sensor node. The maximum possible number of services is dependent on the number of sensor and actuators attached to a single node. For example, let us say we have one sensor node with a gap sensor and a magnetic door lock. The gap sensor is responsible to detect the door status either open or closed, and the magnetic door lock operation is to lock and unlock the door. The total number of services that can be generated from this sensor node is only three: door status service that checks whether the door is open or closed, door

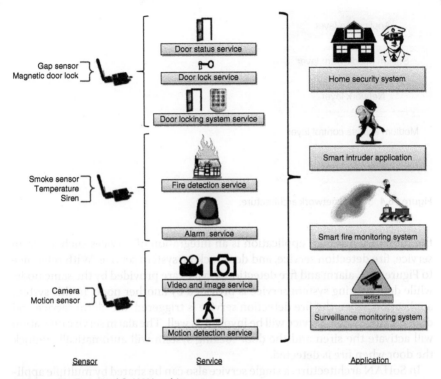

Figure 16.3 Proposed SoHAN architecture.

lock service that enables locking/unlocking the door, and door locking system service that combines the functionalities of both aforementioned services. The services can be created by the combination of sensors/actuators residing on the same sensor node only. Services are provided by node sharing data generated by the sensors attached to the same node.

For example, with reference to Figure 16.3, consider the two services operating on the same node: video and image and motion detection services. The latter only requires data from the motion sensor, while the former also requires the motion data to activate the security camera. As a result, two services used data from the same sensor without requiring separate sensors. Therefore, the SoHAN middleware features provide interoperability between the sensor data and service when one sensor data can be utilized by two or more services at one time.

Application section consists of the number of possible applications that can be formed based on the services available at the service section. The applications are designed by integrating the multiple services offered by HAN nodes. Application can be composed of more than one service. Unlike services, applications can utilize services provided by different nodes. For example, the smart

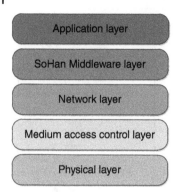

Figure 16.4 SoHAN network architecture.

fire monitoring system application is an integration of services such as alarm service, fire detection service, and door locking system service. With reference to Figure 16.3, alarm and fire detection services are provided by the same node, while door locking system service is provided by another node. In this system application, when the fire detection service is triggered, the alarm service and door locking system service will be invoked as well. The alarm service operation will activate the siren and the door locking system will automatically unlock the door when fire is detected.

In SoHAN architecture, a single service also can be shared by multiple applications at the same time. For example, home security system application and smart fire monitoring system application. Home security system application integrates motion detection service, video and image service, alarm service, and door locking system service. Here we can see that both applications share the alarm service. This feature makes SoHAN architecture highly flexible, where the developer does not need to develop new services or acquire a new sensor node if the service is provided by an existing node. Figure 16.4 illustrates the position of SoHAN middleware layer compared to the OSI reference model. SoHAN middleware layer is located between the network layer and application layer. The SoHAN middleware layer consists of two sublayers, namely, sensor-dependent sublayer and service-dependent sublayer. The sensor-dependent sublayer mainly focuses on managing the sensor operation, while the service-dependent sublayer focuses on managing the service in SoHAN architecture. Both proposed sublayers are independent in terms of operation and implementation.

16.4.3 The Proposed SoHAN Middleware Framework

Figure 16.5 shows the proposed SoHAN middleware structure. SoHAN middleware structure mainly consist of two layer: sensor dependent layer and

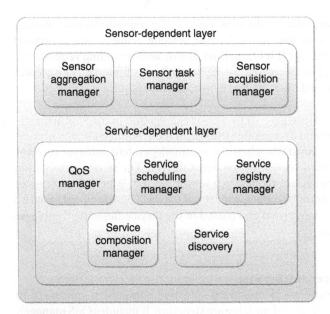

Figure 16.5 SoHAN middleware framework.

service-dependent layer. Sensor-dependent layer is responsible to manage the sensor operation and service-dependent layer is responsible to manage the service in SoHAN middleware structure.

16.4.3.1 Sensor-Dependent Sublayer

There are several component managers involved in the sensor-dependent sublayer to manage the sensor activity in SoHAN middleware structure, including sensor aggregation manager, sensor task manager, and sensor acquisition manager.

Sensor Acquisition Manager The sensor acquisition manager is in charge of issuing read requests to the sensor and receiving response data. Normally, each sensor on the sensor node has three basic operations: sensor initialization, sensor data register, and sensor data request. Sensor initialization is used to set the sensor parameters before usage, such as analog-to-digital converter (ADC), digital port, USART, and I2C setting. Sensor data register will be triggered after sensor data request process is complete. Sensor data register will then pass the requested data.

Sensor Task Manager Sensor task manager is the core manager in sensor-dependent sublayer. Sensor task manager is responsible for handling the multiple read request tasks initiated from the sensor aggregation manager to multiple sensors available on the sensor node. The requested sensor reading tasks will be queued in the buffer. Sensor reading requests are processed based on a first-in-first-out (FIFO) method. If two sensor reading requests are received, the first will be processed, while the second is buffered. The second request will be processed after the first sensor reading request is complete.

Sensor Aggregation Manager   The sensor aggregation manager's main operation is to assemble and disassemble the sensor data required by each service. The dissemble operation starts when the sensor-dependent sublayer receives the service request from service-dependent sublayer. The sensor aggregation manager converts the service request to a sensor request. When the process of the sensor request is complete, the sensor result data will be assembled based on the service request frame format and passed to the service-dependent sublayer to process the service.

16.4.3.2   Service-Dependent Sublayer

Service-dependent sublayer is organized into five component managers to manage the services in SoHAN middleware structure. It includes QoS manager, service scheduling manager, service registry manager, service composition manager, and service discovery manager.

QoS Manager   QoS manager is responsible for assigning the priority level for each services contained in the SoHAN middleware structure. Service priority can be divided into three levels: high, normal, and low. The assigning of priority level is based on service type. The assigning of service based on priority is very important to make sure the service with high priority will be served first compared with lower priority services.

Service Scheduling Manager   Service scheduling manager is the main part in service-dependent sublayer. It is responsible for handling the service scheduling process. The service scheduling process will be executed based on priority level assigned by the QoS manager.

Service Registry Manager   The main operation of service registry manager is to store the available services in the SoHAN middleware structure registry. The service registry manager performs three tasks: adding a new service to the registry, deleting a service when the service is not available, and lookup for available services within the SoHAN middleware structure.

Service Composition Manager   Service composition manager is responsible for handling the multiple service combinations and for serving multiple applications in the SoHAN middleware structure. Service composition manager enables a single service to be reused by multiple applications.

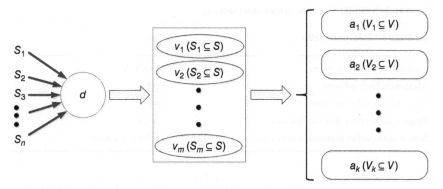

Figure 16.6 General network model for SoHAN.

Service Discovery Manager Service discovery manager is responsible for discovering the services available from the sensor nodes within the SoHAN middleware structure. It achieved this by broadcasting its services during its initialization and the maintenance phase. Service discovery manager is also responsible for updating the current status of a service in the registry.

16.5 Performance Evaluation

In this section, we evaluate the performance of our proposed architecture. We start by describing the network model, and then we proceed to describe the simulation setup, including the main parameters and reference scenarios. Finally, we present and discuss the generated results.

16.5.1 Network Model

Figure 16.6 illustrates the network model for SoHAN middleware.

It consist of a sensor node denoted by d, to which n sensors are attached denoted by $S = \{s_1, \dots, s_n\}$. This sensor node is capable of generating m services denoted by $\mathcal{V} = \{v_1, \dots, v_m\}$. Finally, k applications are constructed from the available services denoted by $\{a_1, \dots, a_k\}$. In the proposed middleware, the service can only be created using the data generated by sensors within the same node. Only the related sensor data required by the service that meet the application requirement will be sent to the base station or gateway.

16.5.2 Simulation Setup

We evaluate the performance of SoHAN middleware using a discrete event simulator developed in MATLAB software and using IEEE 802.1.5.4 MAC

Table 16.2 IEEE 802.14.5 MAC parameters setting.

IEE 802.15.4 MAC parameter	Value
Collision avoidance scheme	CSMA/CA
Maximum back-off time	5 cycles
Maximum number of retries	3 times
Header packet transmission duration	3 cycles
Sensor data packet transmission duration	7 cycles per sensor

protocol [29] as a reference standard model. Relevant parameter settings for IEEE 802.15.4 MAC protocol used in our simulation were set according to Table 16.2.

To simulate the real application of SoHAN architecture, we created two applications called a_1 and a_2. Figure 16.7 shows the proposed application scenario where a_1 is a combination between the service v_1 located at the sensor node d_1 and service v_4 located at the sensor node d_2. Service v_1 is based on data generated by sensors s_1 and s_2 and service v_4 is based on data generated by sensor s_6. The second application is a_2, where it is composed of service v_2 that resides at sensor node d_1 and service v_3 that resides at sensor node d_2. Service v_2 is based on data generated by sensor s_3 and v_3 is based on data generated by sensors s_7 and s_8. Figure 16.7 illustrates the two proposed applications. In our

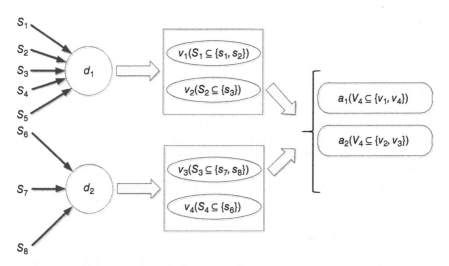

Figure 16.7 Network model for simulation.

simulation, we set the probability of executing applications a_1 and a_2 to 0.005 and 0.0025, respectively.

We evaluate and compare the performance of the proposed SoHAN framework with two other reference cases executing the same applications. We describe the two cases as follows:

Case 1: In Case 1, the sensor node will send the sensor data based on the application request. For example, if a sensor node has three attached sensors and the application only requires data from two sensors, the sensor node will send two sensor data packets, where each packet carries data from one of the sensors. The total number of packets transmitted depends on how many sensors are required to forward their data to the application.

Case 2: For Case 2, each sensor node will transmit all its sensor data together in one packet regardless of whether the application requires data from these sensors or not. For example, if a sensor node consists of four sensors and application only needs data from two sensors, the sensor node will combine all sensor data it has in one packet and send it to the application. In this case, the total number of transmitted packets does not depend on the number of sensors required to forward their data to the application. However, the transmitted packet length depends on the type and number of sensors attached to each node. Therefore, we may deduce that packet size in Case 2 is larger than or equal to that of Case 1.

16.5.3 Results

The SoHAN framework performance evaluation is presented in this section. SoHAN framework has been evaluated using four criteria, namely, packet loss, packet latency, total packet generation, and energy consumption. Figure 16.8 shows the total number of packets dropped versus time for the proposed SoHAN framework and the two references schemes, namely, Case 1 and Case 2. From the displayed results, the average packet drop rate for the SoHAN framework is calculated to be 3.17×10^{-3} packet/cycle, while for Case 1 and Case 2 is 9.5×10^{-3} and 3.15×10^{-3} packet/cycle, respectively. The packet drop rate for SoHAN framework is significantly less than Case 1 due to the fact that in Case 1 each sensor data is transmitted in a separate packet, leading to a large number of generated packets. Hence, the number of collisions increases. This can be confirmed by referring to Figure 16.8, where the number of packets generated versus time is illustrated. In Figure 16.9, the rate by which packets are generated with time for Case 1 is 2.73×10^{-3} packet/cycle compared to 1.05×10^{-3} packet/cycle for SoHAN.

On the other hand, packet drop rate for SoHAN is comparable to that of Case 2 since both follow a similar approach based on producing a single packet

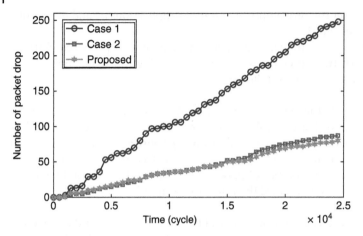

Figure 16.8 Number of drop packet versus time.

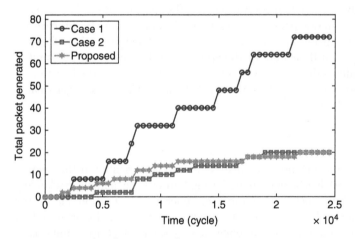

Figure 16.9 Total packet generated versus time.

upon sensor data request. From Figure 16.9, packet generation rate for Case 2 is 0.7×10^{-3} packet/cycle, which is less than SoHAN. However, packet size produced in Case 2 is larger since it includes data from all sensors. Since the probability of packet collision is directly proportional to packet generation rate and packet size [27], similar packet drop rates are observed for Case 2 and SoHAN.

Figure 16.10 shows the latency versus the number of cycles. It can be observed clearly that SoHAN achieves lower latency than both Case 1 and Case 2 over the whole time span. SoHAN achieves an average latency of 37 cycles compared to 53 cycles for Case 1 and 78 cycles for Case 2. This performance improvement

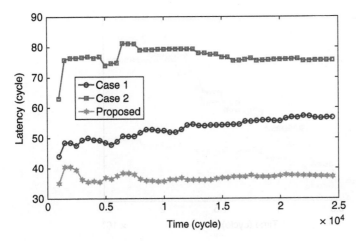

Figure 16.10 Average packet latency over time.

is referred to the fact that SoHAN transmits fewer packets than Case 1, hence avoids latency caused by increased probability of collision. On the other hand, SoHAN packet size is smaller than Case 2, hence it avoids latency caused by long packet transmission time. Furthermore, consider that one cycle is equal to one symbol duration (i.e., 16 μs) [13,29], hence, the corresponding average latencies are 0.85, 1.2 and, 0.59 ms for Case 1, Case 2, and SoHAN, respectively. It is clear that SoHAN manages to achieve a latency below the target latency for 5G systems that is, 1 ms [30].

Finally, Figure 16.11 illustrates the normalized energy consumption versus time in cycles. With reference to parameters in Table 16.3, energy consumption in packet transmission can be calculated as $E = N_p N_b E_b$. On the other hand, normalized energy consumption is calculated as $\bar{E} = E/E_b = N_p N_b$. We observe that SoHAN consumes less energy over time than both Case 1 and Case 2. SoHAN achieves average normalized power consumption of 14.23×10^{-3} cycle^{-1} compared to 27.25×10^{-3} and 21.6×10^{-3} cycle^{-1} for Case 1 and Case 2, respectively, where normalized power consumption is defined as normalized energy consumption per cycle.

Table 16.3 summarizes the performance criteria achieved in the presented case studies. We observe that packet drop rate for SoHAN is 67% less that Case 1 and only 0.6% higher than Case 2. Packet generation rate for SoHAN is 62% less that Case 1 and 50% higher than Case 2. Therefore, we can deduce that SoHAN performance in terms of the aforementioned criteria is significantly improved compared to Case 1, while basically similar to Case 2. On the other hand, for average latency and power consumption, SoHAN achieves performance improvement over both Cases, by which SoHAN average latency is 30% less than Case 1 and 52% less than Case 2, and power consumption 48% less than

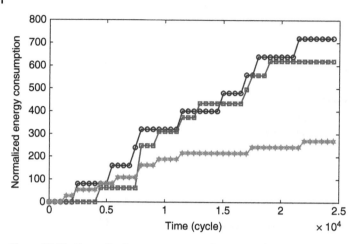

Figure 16.11 Normalized energy consumption versus time.

Table 16.3 Summary of SoHAN performance evaluation.

Criterion	Case 1	Case 2	SoHAN
Packet drop rate (packet/cycle)	9.50×10^{-3}	3.15×10^{-3}	3.17×10^{-3}
Packet generation rate (packet/cycle)	2.73×10^{-3}	0.70×10^{-3}	1.05×10^{-3}
Average latency (cycle)	53.00	76.80	36.98
Normalized power consumption ($cycle^{-1}$)	27.25×10^{-3}	21.60×10^{-3}	14.23×10^{-3}

Case 1 and 34% less than Case 2. Therefore, SoHAN achieves a significant overall performance improvement compared to the other studied Cases.

16.6 Conclusion

This chapter presents the service-oriented middleware architecture for efficient Internet of Things (IoT) home area network in 5G. The SoHAN architecture is organized into three divisions, namely, sensor, service, and application. Sensor acts as a service provider, while the service acts as a service directory and followed by the application as a service consumer. The SoHAN middleware layer consists of two sublayers, namely, sensor-dependent sublayer and service-dependent sublayer. The sensor-dependent sublayer focuses on the sensor management while the service-dependent sublayer concentrates on service operation.

The SoHAN performance evaluation was carried out by comparing two conventional methods and the result was promising in terms of reliability, latency, and power consumption. This work provides smart management and enhances resource utilization in home area network. It promotes cost efficiency by reusing the services offered by the existing sensor node in composing new application by minimizing the complexity of low-level details from application developer. The authors wish to express their gratitude to Ministry of Higher Education (MOHE), Research Management Center (RMC) for the sponsorship, and Advanced Telecommunication Technology Research Group, Universiti Teknologi Malaysia for the financial support and advice for this project. (Vot number Q.J130000.2523.19H25)

References

1 A. Al-Fuqaha, M. Guizani, M. Mohammadi, M. Aledhari, and M. Ayyash, "Internet of Things: a Survey on enabling technologies, protocols, and applications," *IEEE Commun. Surv. Tutor.*, vol. 17, no. 4, pp. 2347–2376, 2015.

2 W. Ejaz, A. Anpalagan, M. A. Imran, M. Jo, M. Naeem, S. B. Qaisar, and W. Wang, "Internet of Things (IoT) in 5G wireless communications," *IEEE Access*, vol. 4, pp. 10310–10314, 2016.

3 ABI Research, More than 30 billion devices will wirelessly connect to the Internet of Everything in 2020, 2013.

4 T. Danova, Morgan Stanley: 75 billion devices will be connected to the Internet Of Things by 2020, 2013.

5 D. Evans, The Internet of Things: How the Next Evolution of the Internet Is Changing Everything, CISCO white paper, 2011.

6 I. F. Akyildiz, S. Nie, S.-C. Lin, and M. Chandrasekaran, "5G roadmap: 10 key enabling technologies," *Comput. Netw.*, vol. 106, p. 17–48, 2016.

7 M. R. Palattella, M. Dohler, A. Grieco, G. Rizzo, J. Torsner, T. Engel, and L. Ladid, "Internet of Things in the 5G era: enablers, architecture, and business models," *IEEE J. Select. Areas Commun.*, vol. 3, no. 34, pp. 510–527, 2016.

8 G. Eastwood, *A Guide to 5G Drivers and Vertical Applications*, vol. 11, pp. 4380–4392, 2017.

9 M. S. Obaidat and P. Nicopolitidis, "Smart Cities and Homes: Key Enabling Technologies," *Elsevier*, 2016.

10 O. Vermesan and P. Friess, *Internet of Things: Converging Technologies for Smart Environments and Integrated Ecosystems*, River Publishers, 2013.

11 O. Vermesan and P. Friess, *Internet of Things Applications: From Research and Innovation to Market Deployment*, River Publishers, 2014.

12 N. Mohamed and J. Al-Jaroodi, "A survey on service-oriented middleware for wireless sensor networks," *Serv. Oriented Comput. Appl.*, vol. 2, no. 5, pp. 71–85, 2011.

13 M. A. Sarijari, A. Lo, M.S. Abdullah, S. H. D. Groot, I. G. M. M. Niemegeers, and R. A. Rashid, "Coexistence of heterogeneous and homogeneous wireless technologies in smart grid-home area network," *2013 International Conference on Parallel and Distributed Systems*, 2013, pp. 576–581.

14 M. R. B. A. Rahim, N. Fisal, R. A. Rashid, and Z. Khalid, "A service oriented middleware for smart home and ambient assisted living," in *2015 1st International Conference on Telematics and Future Generation Networks (TAFGEN)*, 2015, p. 49–53.

15 J. Leguay, M. Lopez-Ramos, K. Jean-Marie, Kathlyn and V. Conan, "An efficient service oriented architecture for heterogeneous and dynamic wireless sensor networks," *2008 33rd IEEE Conference on Local Computer Networks (LCN)*, 2008, pp. 740–747.

16 A. Rezgui and M. Eltoweissy, "Service-oriented sensor-actuator networks: promises, challenges, and the road ahead," *Comput. Commun.*, vol. 13, no. 30, pp. 2627–2648, 2007.

17 H. Gomma, *Software Modelling and Design*, Cambridge University Press, 2011.

18 E. Canete, J. Chen, M. Diaz, L. Llopis, and B. Rubio, "A service-oriented approach to facilitate WSAN application development," *Ad Hoc Netw.*, vol. 3, no. 9, pp. 430–452, 2011.

19 J. Al-Jaroodi and N. Mohamed, "Service-oriented middleware: a survey," *J. Netw. Comput. Appl.*, vol. 1, no. 35, pp. 211–220, 2012.

20 I. Chatzigiannakis, G. Mylonas, and S. Nikoletseas, "50 ways to build your application: a survey of middleware and systems for wireless sensor networks," *in IEEE Conference on Emerging Technologies and Factory Automation*, 2007.

21 T. G. Stavropoulos, K. Gottis, D. Vrakas, and I. Vlahavas, "aWESoME: a web service middleware for ambient intelligence," *Expert Syst. Appl.*, vol. 11, no. 40, pp. 4380–4392, 2013.

22 E. Caete, J. Chen, M. Diaz, L. Llopis, and B. Rubio, "USEME: a service-oriented framework for wireless sensor and actor networks," in *Eighth International Workshop (ASWN '08)*, 2008, pp. 47–53.

23 S. R. Alonso, D. I. Tapia, J. Bajo, O. Garcia, J. F. de Paz, and J. M. Corchado, "Implementing a hardware-embedded reactive agents platform based on a service-oriented architecture over heterogeneous wireless sensor networks," *Ad Hoc Netw.*, vol. 1, no. 11, pp. 151–166, 2013.

24 M. S. Familiar, J. F. Martnez, I. Corredor, and C. Garca-Rubio, "Building service-oriented smart infrastructures over wireless ad hoc sensor networks: a middleware perspective," *Comput. Netw.*, vol. 4, no. 56, pp. 1303–1328, 2012.

25 I. Corredor, J. F. Martnez, S. M. Familiar, and L. Lourdes, "Knowledge-aware and service-oriented middleware for deploying pervasive services," *J. Netw. Comput. Appl.*, vol. 2, no. 35, pp. 562–576, 2012.

26 M. Kushwaha, I. Amundson, X. Koutsoukos, N. Sandeep, and J. Sztipanovits, "OASiS: a programming framework for service-oriented sensor networks," in

2nd International Conference on Communication Systems Software and Middleware, 2007, pp. 1–8.

27 S. H. Nguyen, H. L. Vu, and L. L. H. Andrew, "Packet size variability affects collisions and energy efficiency in WLANs," in *2010 IEEE Wireless Communication and Networking Conference*, 2010, pp. 1–6.

28 E. Aviles-Lopez and J. A. Garca- Macas, "TinySOA: a service-oriented architecture for wireless sensor networks," *Serv. Oriented Comput. Appl.*, vol. 2, no. 3, pp. 99–108, 2009.

29 IEEE Computer Society, IEEE Standard for Information Technology - Telecommunications and Information Exchange Between Systems - Local and Metropolitan Area Networks Specific Requirements Part 15.4: Wireless Medium Access Control (MAC) and Physical Layer (PHY) Specifications for Low-Rate Wireless Personal Area Networks (LR-WPANs), 2003, pp. 1–670.

30 P. Schulz, M. Matthe, H. Klessig, M. Simsek, G. Fettweis, J. Ansari, S. A. Ashraf, B. Almeroth, J. Voigt, I. Riedel, A. Puschmann, A. Mitschele-Thiel, M. Muller, T. Elste, and M. Windisch, "Latency critical IoT applications in 5G: perspective on the design of radio interface and network architecture," *IEEE Commun. Mag.*, vol. 2, no. 55, pp. 70–78, 2017.

Mohd Rozaini bin Abd Rahim received the B.Eng and master's degrees from the University Teknologi Malaysia, in 2007 and 2011, respectively, both in electrical engineering. He is currently pursuing the Ph.D. degree with the Advanced Telecommunication Technology (ATT)Research Group. His research interest include wireless sensor network, service-oriented middleware, and wireless biomedical sensor network.

Rozeha A. Rashid received her B.Sc. degree in electrical and electronic engineering from the University of Michigan, Ann Arbor in 1989. She received her M.E.E. and Ph.D. degrees in telecommunication engineering from Universiti Teknologi Malaysia (UTM) in 1993 and 2015, respectively. She is a senior lecturer in the Department of Communication Engineering, Faculty of Electrical Engineering, Universiti Technologi Malaysia. Her research interests include wireless communications, sensor network, cognitive radio, and Internet of Things.

Ahmad M. Rateb received his Ph.D. in 2013 from Universiti Teknologi Malaysia (UTM), Malaysia. Currently, he is with the Advanced Telecommunication Technology research group (ATT) in UTM as a postdoctoral research fellow. His research interest covers compressed sensing, 5G communications, and cognitive radio technology.

M.A.B (Mohd Adib bin) Sarijari received his bachelor's degree in engineering (first class, and with honors) in 2007, and the Master of Science in electrical engineering in 2011, both from Universiti Teknologi Malaysia (UTM), Johor, Malaysia. In 2012, he received his Ph.D. from Delft University of Technology, the Netherlands. He is currently a senior lecturer at the Department of Communication Engineering, Faculty of Electrical Engineering, UTM. His general research interest lies in the field of communications, optimization, and system design. In particular, he is interested in cognitive radio, home area networks, wireless sensor networks, software defined radio, smart home, and smart city.

Ahmad Shahidan Abdullah received the B.Eng. degree in telecommunication engineering and Ph.D. degree in electrical engineering from the Universiti Teknologi Malaysia (UTM), Johor Bahru, Malaysia, in 2009 and 2014, respectively. In 2009, he joined the Communication Engineering Department, Faculty of Electrical Engineering, UTM, as a tutor, and later as a lecturer, in 2014. His current research interests include channel and network coding, wireless sensor networks, and IoT applications.

Abdul Hadi Fikri Abdul Hamid received the B.Eng. degree from the Universiti Teknologi Malaysia, in 2007, and the master's degree from the same university, in 2011, both in electrical engineering. He is currently pursuing the Ph.D. degree with the Advanced Telecommunication Technology (ATT) Research Group in UTM. His research interests include wireless sensor network, cognitive Radio, and embedded system.

 ***Hamdan Sayuti*** received the B.Eng. degree in computer engineering and the master's degree from the Universiti Teknologi Malaysia (UTM), Johor Bahru, Malaysia, in 2007 and 2016, respectively. He is currently a research officer at UTM. His research interests include wireless sensor network, IoT system, and cloud system design.

 ***Norsheila Fisal*** received her B.Sc. degree in electronics communication from the University of Salford, Manchester, United Kingdom, in 1984, M.Sc. degree in telecommunication technology, and Ph.D. degree in data communication from the University of Aston, Birmingham, United Kingdom, in 1986 and 1993, respectively. Currently, she is a professor with the Faculty of Electrical Engineering, Universiti Technologi Malaysia, and the head of the UTM-MIMOS Telecommunication Technology Research Group.

17

Provisioning Unlicensed LAA Interface for Smart Grid Applications

Saba Al-Rubaye[1] and John Cosmas[2]

[1] Instituto de Telecomunicações Campus Universitário de Santiago Aveiro - Portugal

[2] Department of Electronic and Electrical Engineering, Brunel University London, Uxbridge, UK

17.1 Introduction

The new era of fifth-generation (5G) telecommunications enables the necessary infrastructure to transfer big data of information for industry sectors such as power utility. In this context, Licensed Assisted Access (LAA) emerges as a new multiradio interface technology that can integrate a Wi-Fi radio interface with the Long-Term Evolution-Advanced (LTE-A). This combination enables the LAA node to access the unlicensed band as a supplementary downlink to increase throughput to the end consumers. In this chapter, we propose new algorithms for bandwidth allocation technique, which have the ability to reserve unlicensed band channels to the smart grid applications through a priority order by mapping accessible channels to the minimum uality of service (QoS) requirements. Thus, bandwidth requirement for each smart grid traffic flow can be configured based on the whole network states in real time. Finally, under the framework of current power grid bandwidth estimation technique, an evaluation method based on Poisson distribution is proposed to predict the communication bandwidth for utility, which will help to reach the optimum bandwidth calculation to guarantee the QoS requirements by considering the dynamic communication characteristics of utility services. To evaluate the system performance of the proposed method, utility service quality indexes (e.g., delay, packet loss rate) are considered with LAA node service along with

5G Networks: Fundamental Requirements, Enabling Technologies, and Operations Management, First Edition.
Anwer Al-Dulaimi, Xianbin Wang, and Chih-Lin I.

bandwidth needs to raise the utilization ratio. Simulation results proved the validation of the proposed method through comparative analysis of critical load demand and normal utility load demand parameter for bandwidth allocation method.

The high data demands and the automation signaling requirements in smart grid networks exceed the available resources of current wireless technologies employed by power utilities. Therefore, smart grids are motivated to build new communication architectures that provide more flexibility in accessing the spectrum to exchange status information and control commands [1]. Practically, this revaluation needs to adopt new technologies that can allow interoperability between multibandwidths to upgrade the communication infrastructure. To this point, the vision of using cellular operators to provide this additional bandwidth is the only way to restore reliability, resilience, and stability for smart grids. Ideally, there are two types of information streams in smart grid communications. First, the data delivered from electrical appliances to the smart meters. These data can be transmitted for short rang using communication (e.g., ZigBee). Second, data between smart meters and the utility control centers that require third-party networks [2]. The LAA technology can stimulate as new adjustments in the smart grid infrastructure and help to mitigate the obstacle of dense network deploying. From a network infrastructure perspective, LAA can support smart meter data by establishing reliable communication link between control centers and customers [3], as shown in Figure 17.1.

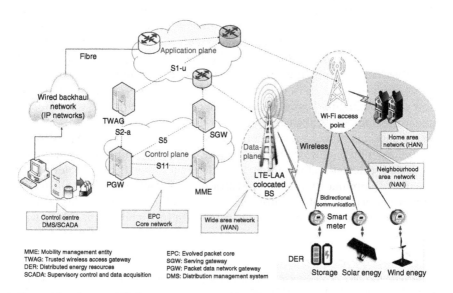

Figure 17.1 Smart grid infrastructure powered by LTE-LAA.

LAA physical layer proposed by 3rd Partnership Project (3GPP) for Release 13 and beyond focuses on the integration of unlicensed band that can facilitate similar technology of Wi-Fi [4]. In other words, engaging the LAA in the infrastructure design of the smart grid communications to enhance the communications is not necessary and can improve the spectrum access of the network domains. The major problem comes when the spectrum bandwidth is not sufficient to allocate all the power grid data transmitting, which creates a bottleneck and unreliable communication link. The challenge caused by huge amount of data generated by smart meters and utility consumers will continue to grow in the next few years [5]. Therefore, the prospective communication concept provides us with a framework modification in the Wi-Fi transmission scheme to support a new LAA spectrum access model for sensing and transmission operations.

This allows exploiting the necessary wireless resources to meet the smart grid requirements of growing sensors and control messaging. In this proposed model, the data from smart grid applications will be assigned to the free channel based on the priority of these applications. The priority values are decided upon the time delay and bit error rate (BER) thresholds that an application can tolerate before exceeding the limits of the requested QoS. In comparison to our previous research [3], this work considered LAA integrated with unlicensed band aiming to tackle the lack of bandwidth resources problems using an emerging technology of interfacing LTE and Wi-Fi platforms. This solution can deliver an optimum spectrum utilization and system reliability to the smart grid communications.

17.2 Smart Grid Architecture-Based 5G Communications

The utility providers should consider a smart monitoring system that can control the distribution generation to achieve the stability and scalability. These obligations will help to develop the data signal control in the smart grid communications system by delivering a reliable secure channel. The main source of data generation is the smart meters (SMs) used for measuring and transmitting the utility information and power consumption status. The data will be delivered using communication system to the control unit and switches; they signal content status of the load and control messages. Basically, smart power grids enable two-way communications that require to secure appropriate delivery of the data transmission. These demands require multiple radio interface technology that integrates different grid units to motivate the market and utility industry to consider the LAA node for smart grids. A smart grid consists of one or more control centers that govern a hierarchical architecture of three communication networks including home area network (HAN), neighborhood area network (NAN), and wide area network (WAN). Although LAA technology can slice the bandwidth to share the resources with the power grid, still the smart power grid requires a real-time data exchange between different levels and vendors by applying ubiquitous communication technology. For example, the new era of 5G platform can transmit the data over different channel bands with multi-interfaces technology to support smart grid infrastructure, which can be a promising technology to meet the customer utility desire and emergency request. The main architecture of smart grid communication is explained in the following sections.

17.2.1 Control Center Architecture

Control center is the main entity with ability to access and communicate with various devises in the utility domain, including actuator, smart meters, transformers, and so on, through communication technology networks. Therefore, the data centers need to deal with various power grid parameters to be able to achieve an accurate decision in reaction to the instantaneous electricity load. The control system receiving this information will start to define the best energy scheduling in more efficient way and make the right decision for data flow. Terminal device will deliver its data to the aggregate, and then the data will be forwarded to the coordinator as per the routing table. In this proposed architecture, the terminal devices are SMs) that generate a random load without any topology.

17.2.2 Home Area Network

It is part of smart grid networks with enabling ZigBee technologies that can manage the wireless connectivity between sensors, smart metering, and home appliances. HAN creates a technology platform for exchanging the grid information such as load management, demand response, power consumption, and smart metering. The transferring of each of these parameters is performed for certain goals to help the smart grid providing a power supply that meets power demand at reasonable power prices [6]. There are many requirements for the information flow in smart grid to provide a real-time and a reliable grid status such as packet loss rate, throughput, and latency. Therefore, smart grids employ their own network infrastructures to secure a two-way communications between utility control centers and consumers. The grid information system and control entities can be connected using different network infrastructure and communications technologies.

Considering HAN, the wireless coverage in the 2.4 and 5 GHz unlicensed band, industrial, scientific, and medical (ISM) radio band uses technologies such as Wi-Fi access point. However, the limited temporarily spectrum access of such technologies increases the occurrence of disconnectivity and degrades the reliability of smart grids. Therefore, the introduction of LAA technology at the

NAN may boost the current connectivity schemes with HAN through improved sensing and adjustment of transmission/reception parameters to increase the spectrum accessibility at different channels. This also enables performing optimal transmissions with lower disconnectivity and interference levels and allows increasing the number of connected households and consumers [7].

17.2.3 Neighborhood Area Network

This network can be labeled as the second layer of smart grid communications infrastructure. NAN is the network that interconnects different consumers from HAN, including homes, industries, and businesses. Also, it includes a set of utility meters that are deployed to measure community consumption and not included as part of consumers' meters. The main functionality of this network is to gather energy status information from neighborhood areas and deliver this status information to utility controllers through WANs. The obtained data are then transmitted through third-party communication broadband providers in order to meet the performance requirements of the information flow in smart grids. Enabling LAA technology in the NAN can facilitate an informal access of the HAN data flow in unlicensed band to the licensed WAN through the LAA nodes. This will reduce the complexity of current infrastructure to have different integrated technologies when transferring the information from HAN to WAN through NAN. The enhancement of the LAA unlicensed node by improving the opportunistic access to unused available spectrum and adopting new channel access scheme is another way to avoid the bottleneck at the NAN segment. However, using LAA node as a gateway can enhance the bandwidth utilization for different consumers. This improvement can achieve optimum bandwidth utilization and higher throughput and guarantee the QoS among different smart grid consumers [8].

17.2.4 Wide Area Network

It is labeled as an upper layer in the smart grid communication infrastructure that can carry the data flow between local smart grid domain and the main grid controllers [9]. The concept of smart grid access to the wireless networks to transfer their information and control messaging can be identified as a smart architecture that enables such integration [1]. Multilayer networks have various communications technique through the distribution power grid and employ an infrastructure of WAN. The utilities WAN requires its entire distribution system, along with substations and incorporated with other distribution assets. The WAN will also deliver two-way communications as required for distribution automation (DA), monitoring system, and substation levels. At the same time, WAN will support the data collection for advanced metering infrastructure (AMI) and demand response. Regarding the security considerations, the

LAA node can be the prompting machinery for such an arrangement that has ability to reserve some bandwidths for the huge data transmissions. To this end, we assume that the WAN communications are provided by the LTE radio interface with LAA unlicensed band.

17.3   Bandwidth Utilization Method

We consider a LAA network with combination of supportive smart meters that are connected to the LAA unlicensed interface node in the NAN area. The objective of this chapter is to maximize the available resources for smart grid applications using a developed LAA node along with Wi-Fi access point to exploit additional available bandwidth resources. In the context of smart grids, each consumer/application may be assigned a different level of QoS subject to the transmission requirements and availability of resources. Therefore, in order to reach a reliable system for industry purposes, the communications resources need to be assigned a priority order to maximize spectrum/bandwidth operation under the constraint that each smart grid application can meet its minimum requirements. Each smart application needs a robust bandwidth that can be requested from the network's resources through sharing the spectrum chart and the grid responsible for distributing the resources between the consumers within the unlicensed band. In such new schemes, transmissions should be precisely tuned and dynamically allocated between cooperative transmitters in order to avoid collisions through channel reuse. In this chapter, we identify the spectrum/bandwidth operation to utilize the spectrum space within a channel or to transmit the data in the suitable channel, which is available for a data transmission. An effective spectrum utilization can only be realized by showing complete transmissions without interfering with other access points. The proposed bandwidth utilization method has various functionalities and adaptive parameters that are located in the physical (PHY) layer and media access control (MAC) layers. The functionality enables instantaneous track of transmissions status between the PHY and MAC layers in order to avoid the fundamental Wi-Fi difficulty of temporarily wireless access. In this scheme, the bandwidth utilization unit can manage the channel accessibility, including the time of reserving the channel and the decision to release that channel for other consumers and/or applications. The channel availability analyses begin once a transmission request is approved in the upper layers, where a smart power grid application is set for processing. Each application will be associated with the specific time interval in order to meet QoS requirements. Hence, the applications delivered from upper layers will have particular characteristics. When the data are received by bandwidth utilization technique, the request will be sent to the sensor unit for channel availability. The channel sensing is performed using the energy detection technique at the radio frequency (RF)

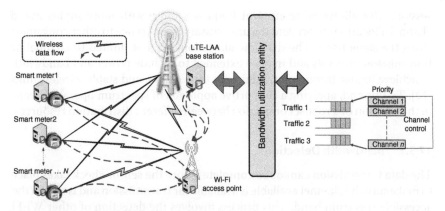

Figure 17.2 Bandwidth utilization employing LTE LAA.

observation module. The statistics data about the absence/presence of channels of other Wi-Fi consumers are calculated based on signal to noise ratio (SNR) and interference measurements. The proposed bandwidth utilization scheme allocates one smart grid application per channel in specific time interval to prevent long buffering duration for data transmission or interference afterward. Bandwidth/spectrum detection entity in the smart metering is responsible for getting channel accessibility and obtaining the transmission opportunities over the whole bandwidth band from long observation. The information about channel availability is forwarded to the bandwidth utilization unit to be matched with the QoS requirements of awaiting smart grid applications. The decision on allocating any application to a certain channel is made by the spectrum decision entity that performs the required assessments for a valid transmission action. Such decisions will always consider the time interval of channel vacancy to improve the coexistence with other Wi-Fi consumer's whether they are smart grid consumers or just normal Wi-Fi clients. The operation of bandwidth utilization powered by LTE LAA radio access is illustrated in Figure 17.2.

The free channel can provide good services for continuing data streaming in the communication link. Thus, the applications coming from sensors or electrical consumers are allocated on a priority order in specific time intervals. Spectrum decision entity considers channel characteristics such as latency and throughput. This can allow the system to use the available channels in a correct way and reduce the time buffering of the data during time intervals [10]. If the Wi-Fi access point fails to restore a transmission on the same channel, then it has to resettle to another channel. This process involves an additional time delay of approximately 224 s for each application undergoing a transmission [11]. For the best practice of the proposed model, we assume that a consumer needs to complete transmitting all the buffered data before starting another transmission session. This allows more efficient frequency reuse with more sophisticated channel allocations to prevent the unnecessary interference to other consumers along the same band. The dynamic allocations are not limited to channel or transmission intervals and may be extended to include power and bandwidth to achieve higher throughput, minimum time delays, and stable wireless links [12]. To this end, applying a supportive approach of spectrum operation leads to the creation of a robust connection between different distribution resources.

17.3.1 Bandwidth Detection

The data transmission cannot be complete before the sensor check takes place to make sure the channel available is available for transmission and is within the accessible spectrum band. This process involves the detection of other Wi-Fi consumers through different features such as interference, SNR, and energy detection [13]. The obtained detection results will be only instant estimates of the channel availability and these evaluations may change afterward due to the dynamic nature of channel availability in unlicensed band. Therefore, the spectrum decision module may assign a certain information flow/application to the chosen channel that meets the QoS requirements at the time of making the transmission decision. However, the spectrum allocation module continues to analyze the availability of that channel using detection reports or once more suitable channel becomes available for that flow. In this case, the system will be based on sensor detection results to make a decision on the prospective and current data transmissions in order to obtain reliable and high system performance. It is very important to obtain an accurate spectrum detection report to prevent any interference to other consumers and to avoid unexpected backoffs that may increase the size of buffered data. Therefore, detection must be performed using the most appropriate technology that fits with the local wireless environment. In our model, we develop a new sensing mechanism as shown in Algorithm 17.1. We show that we use the power detection to decide up on channel availability and to choose the channel that can be available for longer time intervals.

17.3.2 Interference Avoidance

Interference may occur at any time during the transmission process. For Wi-Fi, errors of misdetection are unlikely to happen. However, if two Wi-Fi consumers perform the same observation interval at the same time duration, both consumers may experience interference due to collision of transmissions over the same channel. The interference occurring in a particular condition can imitate the impact of general system performance. The possibility of interference occurrence can occur whenever any Wi-Fi access point starts to transmit after the end of the time interval [14]. Thus, a new technique for interference

Algorithm 17.1 Bandwidth detection mechanism

1:	*Transmission request arrival*
2:	*detection sessions begin*
3:	*RF detectors begin to evaluate certain channel by calculate the power*
4:	*Determine the Signal Interference*
5:	*if(interference measurement<threshold)* **then**
6:	*Estimate the channel time interval*
7:	**else**
8:	*Adjust to the new channel*
9:	Return to 3
10:	**end if**

prevention is important to reduce any retransmissions. The interference mechanism is very important in order to monitoring the data arrival of prior and existing data packets. Harm interference may cause packet damage; therefore, the validity of a new received packet is compared with the packet received before. An unsuccessful packet will be dropped and the scanning session will be resumed to look for a new packet, as shown in Algorithm 17.2.

17.3.3 Spectrum Access

Accessing the spectrum is a procedure that involves spectrum detection to identify the available free channels, evaluate the compatibility of detected free channels durations to meet the requested QoS, and decide to transmit when no interference is expected. Therefore, the spectrum access decision is categorized

Algorithm 17.2 Interference measurement

1:	*Previous packet received, current packet received*
2:	*if previous packet ends prior the new packet arrival* **then**
3:	*No collision occurred*
4:	**else**
5:	*Collide occurred: start scanning for new packet*
6:	**end if**
7:	*Calculate if the new received packet is a valid packet or noise*
8:	*if arriving packet is valid* **then:**
9:	*Evaluate interference using the previous packet and the newly arrived packet*
10:	**else**
11:	Go to 7
12:	**end if**

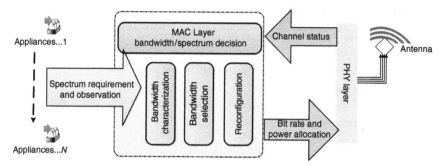

Figure 17.3 Bandwidth/spectrum decision procedure.

by successful spectrum detection where it has the ability to update it based on the dynamic changes in the channel [15], as shown in Figure 17.3.

Again, each application can be located in one channel, and the data transmission will begin once there are enough resources to meet the transmission requirements in that channel. The decision will also include some additional information, such as the suitable transmission power in order to avoid any interference with other consumers. Therefore, a spectrum access decision may be issued for a short time interval to avoid collisions in a demanding wireless environment where so many consumers are computing to access the available channels. In this case, the data transmission will occur in multichannel future LTE LAA unlicensed band technique, rather than the occupied one channel for transmission.

The procedure of accessing the spectrum is based on a set of prior operations [11]:

- **Characterization:** In this stage, the determination of channel availability is required to avoid the interference.
- **Selection:** A specific channel is selected to transmit the data application once the free time intervals offer the necessary capacity to meet the requested QoS.
- **Reconfiguration:** Communication system has the ability to change the parameter features and adapt its transmission characteristics (e.g., frequency and coding) in order to maintain a connection through the available bandwidth.

The developed decision procedure module is presented in Algorithm 17.3.

In this algorithm, we assume that each smart grid consumer can be transmitted using a different data rate to prevent disconnectivity in case of limited bandwidth capacity and spectrum availability. Therefore, various delay transmissions for data demands will be rescheduled based on the channel availability and QoS requirement to avoid retransmission of unsuccessfully data distribution.

Algorithm 17.3 Spectrum decision

1:	***Procedure***
2:	*Smart grid data arrival*
3:	*allocation to the adjacent LTE LAA base station*
4:	*utilization unit send request to the sensor*
5:	*Spectrum availability*
6:	*if (there is no channels available)* **then:**
7:	*Buffers/Queues smart grid data of different applications*
8:	*else*
9:	*Searching for free new channels*
10:	*end if*
11:	*if(the new Channel is Available)* **then:**
12:	*Check SNR and BER*
13:	*else*
14:	*Set transmission parameters (power, modulation, interference)*
15:	*end if*
16:	*if(channel transmit time interval is sufficient)*
17:	**then:**
18:	*transmit*
	else
19:	*Buffer electric data according to the load/application*
20:	*end if*
21:	*Check if there is any data from other smart grid applications*
22:	*Return to 5*

17.3.4 Bandwidth Utilization

The spectrum bandwidth has a limited resource generating the necessary requirement for an effective multipath transmission for end-to-end channels or the same channel. Spectrum allocation technique is the process of assigning bandwidth to different smart grid applications.

The LAA node is assumed to process several queuing applications. To concluded, the produce of smart grid system A_{pp} applications are arranged in order $1, 2, \dots, M$, and the consumers are located within the same transmission domain of that access point. Then, each smart grid request received by the LAA node is assigned a priority value according to the maximum delay that can be experienced by each application [11].

In this proposed framework technique, a pool of channels is assumed to be accessible to different smart grid applications. All these applications have similar delay and BER demands. The spectrum allocation module can evaluate the

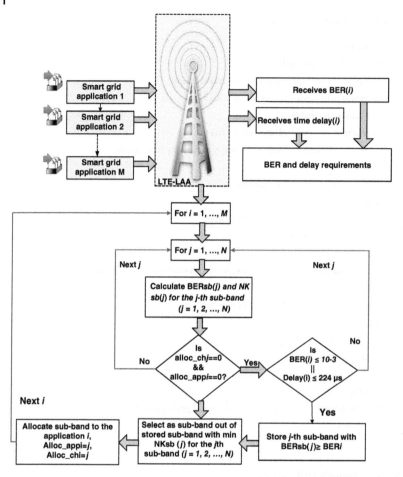

Figure 17.4 Bandwidth utilization scheme for smart grid applications.

BER requirements of all smart grid applications in the queue prior to allocating for any subband. The real-time applications requirement to get enhanced BER performance should reach minimum transmission delay ($<124\mu$). On the other hand, subbands with lower BER performance can serve all nonreal-time smart grid applications. In this case, the LAA node allocates the best available subband to a certain smart grid consumer if that channel provides the necessary resources to meet the requested QoS. The limitation of channel allocation is building based on prioritized different characteristic (e.g., bandwidth, load demand, BER, latency, and throughput) for system requirements [16]. The developed bandwidth utilization scheme is shown in Figure 17.4.

When an LAA node receives M smart grid application requests, it assign them priority values of ith request, for $i = 1, 2, \ldots, M$ considering each application requirements of BER and time delay. The LAA also determine the BER of the *j*th for accessible sub carrier, $BER_{sb}(j)$, and its number of available consumers N_k of OFDM symbol $N_{k_{sb}}(j)$. In this model, we consider two vectors channel sub-band allocation *alloc_chj* and application allocation *alloc_appi* to enumerate each of channel sub-bands that belong to the sub-band set N. Only one sub-band is assigned to a certain smart grid application. The decision block ensures that an allocated *j*th sub-band is never allocated again to a requesting smart grid consumer ($alloc_{chj} == 0$) as well as that a *i*th smart grid consumer is never allocated two or more subbands at the same time instant ($alloc\_appi == 0$). Subsequently, a subband is only allocated to a $BER_{sb}(j)$ if the later appears to have a higher or equivalent requested BER*i* of *ithith* consumers. Once a channel is allocated, the status of *j*th sub-band and *i*th consumer request are changed to duplicate allocation of sub-bands and consumer requests (alloc*chj* $== 1$ and $alloc_{appi} == 1$). This process is repeated for all M requests.

17.4 System Implementation and Simulation Platform

The proposed system model is implemented using Riverbed Modeler wireless network simulator [17]. This simulator supports large-scale network simulations that consist of many access points and end consumers with multiple connections separated by chosen distances. Modeler simulator provides the statistics obtained from various scenarios using sequential state tracking. In this case, one transitional state is analyzed during the steps of load generations and communications. For a Wi-Fi access point, the information is transmitted using IEEE 802.11 distributed coordination function (DCF) mode, while the channel assignment control messages, issued by our proposed model, are performed using a common spectrum control channel (CSCC) [16].

During the channel allocation process, a negotiation message will be established between LAA node and Wi-Fi access point, as shown in Figure 17.5. This process involves connecting the Wi-Fi to be adapted to another available channel if the choice was made to change the existing frequency. Network infrastructure and the communication link between different nodes can be effected by this change. In this chapter, we characterize an adapted interface in the Wi-Fi access point that is integrated with LTE node to develop the LAA multiradio interface. The main challenging task is to implement the bandwidth utilization entity in the new characteristic node that is able to improve the default node performance in accessing the spectrum holes.

17.4.1 Enable Career Detection for LAA Unlicensed Interface

This section provides the algorithm used for managing spectrum access for the LAA in unlicensed band. In the MAC layer, an IEEE 802.11e standard can adopt

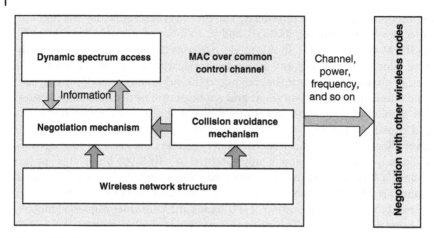

Figure 17.5 Structure of channel assignment policy.

carrier sense multiple access with collision avoidance (CSMA/CA) technique [18]. The CSMA/CA is adopted as a listen-before-talk (LBT) procedure that enables sensing the channel before any data transmission. Therefore, this chapter employs a CSMA technique in order to develop a suitable spectrum decision procedure, as shown in Algorithm 17.4. Thus, LAA node interface under unlicensed band can conduct a transmission as soon as data are transmitted. This technique will allow the smart grid system to reallocate consumers between different channels subject to their updated load status and QoS. The dynamic collaboration between channels using PHY and MAC layer sensing techniques allow additional resources that we need for smart grid.

The successful packet delivery to the end consumers at the household/enterprise locations is confirmed using Acknowledgement (ACK) packets. These packets need to be sent either way to confirm successful delivery of information/control messages without any collision. If no ACK received after certain time counter, the LA unlicensed/ Wi-Fi will retransmit the same packet to the other consumer. This process will be repeated for certain iterations before enabling Algorithm 17.1 to move to a new channel that has more reliability and less excepted collisions from other channel consumers.

17.4.2 System Performance and Analysis

The smart grid communications is simulated in a city scale network project of 2 km × 2 km. The scenario consists Wi-Fi access point along with LTE and the proposed LAA unlicensed node, 100 LAA node are distributed and connected to the consumers, 20 Wi-Fi access point are deployed for data service, and one client server application running over TCP/IP. Typical characteristics of a smart

Algorithm 17.4 Developed CSMA CA mechanism

1:	*Channel evaluation*
2:	*if channel available = CSMA/CA (0)* **then:**
3:	*Data Transmit*
4:	**else**
5:	*Check the time interval of data transmission*
6:	*Buffer and queues the data*
7:	**end if**
8:	*If time interval expires then restore transmission*
9:	*If resend packet fails*
10:	*Buffer packet and go to 2*
11:	**else**
12:	*Activate Algorithm 17.3*
13:	**end if**

meter lay within the TCP/IP application protocol. Alternatively, the parameters of power grid (e.g., current, voltage, and power factor) are evaluated in the process model using the function block, which is located in the node model of the application layer. This unit can maintain transmitting and receiving data from the smart meter to the control center and server. The simulated scenarios are configured to perform a random spectrum access at variable transmission intervals. The system bandwidth is specified as 5 MHz, and the threshold of SNR is specified as 10 dB. We assume that each consumer delivers the data traffic every one second. In each time period, the consumer delivers 500 packets, which follow a Poisson distribution, and packet size is considered as 1024 bytes. Two different types of traffic connection is considered for the data traffic of SMs with services time interval of 1 ms. Similarly, is defined as a smart meter density that can represent the number of smart meters per unit area. The distribution of a spatial Poisson process is adopted. In this way, has a uniform density and the count of the smart meters SMs has a Poisson distribution. This allows us to simulate performance of smart meters in a smart grid area similar to a realistic distribution of smart grid consumers. Moreover, a smart meter sends its readings every 30 min in duration of 1 h. During simulation scenario, the anticipated result was evaluated as an average value and set in 15 min for each case scenario. The simulation parameters were configured as shown in Table 17.1. We assumed there is no interference between LAA and Wi-Fi with consumers, which may impact the wireless channel.

The consumers are associated with two different load priorities: one is considered a critical load and the other a normal load. The consumer appliances that are connected with fridge are classified as critical load, whereas normal

Table 17.1 Smart grid communications characteristics.

Name	Value
Network size	2 km..2 km
Bandwidth	5 MHz
LTE LAA transmit power	40 dBm
Noise variance at receiver	−95.0
Path loss exponent	3
Packet size	1024 bits
Number of traffic per consumer	4
Meter Rreading Ppayload	1 KBkB
Meter Rreading Ttransmission Frequency	30 min

appliances for entertainment (e.g., TVs) are classified as normal load. Based on this priority connection, the loads are generated by consumers normally scheduled to be admission in the available channels by applying the presented transmission technique. The impact of increasing the number of utility services request on the SNR performance is considered, as shown in Figure 17.6. The proposed LAA radio interface and developed algorithms for smart grid system can select a suitable channel at each time slot for certain types of load and choose the next load in the priority queuing order. The average transmission rate is a function for the time available for each request. Due to the heavy penetration loss of the signal from LAA node and Wi-Fi access point led to boost up the SNR of the entire system with maximum of 50 dbm.

Figure 17.7 shows the latency delays for different types of loads. The latency escalated gradually with increase in the number of service request as the number of available channels decaled. The improvement of employing the proposed algorithms can be seen in all simulated scenarios. The proposed bandwidth utilization scheme was able to support more resources to the critical load flow resulting in higher performance and less delay. The longer spectrum channel availability for time intervals also results in better system performance compared to the conventional system.

Figure 17.8 shows the number of successfully delivered packets for all case studies. The proposed scheme can improve the channel assignment using LAA node for smart grid consumers by improving the number of successful packets compared to the conventional Wi-Fi scenario. This improvement is due to the proposed algorithms, which help to enable more refined channels subject to consumers load profiles. To this end, new connection routes and reliable connections are achieved between the LTE operating in the unlicensed band in the form of SMs and the network backbone for service control.

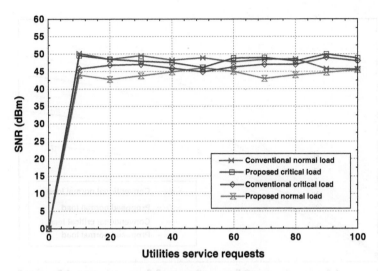

Figure 17.6 SNR versus number of utility services request.

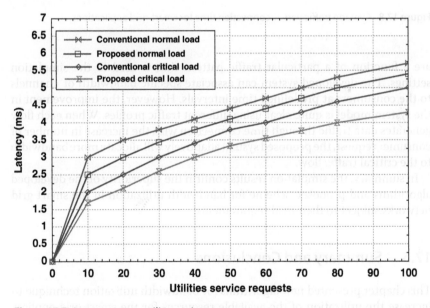

Figure 17.7 Latency versus utility services request.

It is noticeable that there is a significant enhancement in the data traffic when employing the proposed bandwidth utilization scheme compared to the conventional scheme for both critical load and normal load. The given results

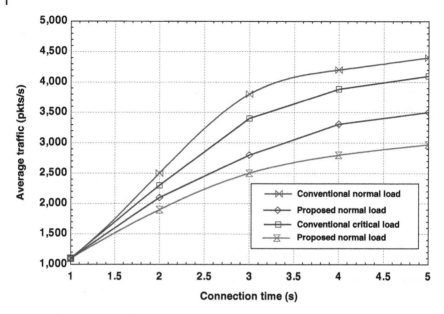

Figure 17.8 Average traffic versus connection time for smart grid applications.

are a function of a particular traffic pattern that is used in the simulation settings. The developed system can associate specific bandwidth or channels to the consumers at predefined time intervals. However, the improvement in the system model is quite clear using different traffic profiles. When each flow generates data traffic through time connection with the increase in number of consumer request, the proposed system adaptively will allocate more bandwidth to the critical traffic flow.

In summary, the simulation results confirm the superiority of the developed algorithms and proposed scheme for bandwidth utilization in the smart grid networks subject to their load requirements.

17.5 Summary and Conclusions

This chapter presented new algorithms for bandwidth utilization technique to increase the utilization of the available resources for the smart grid applications. The proposed utilization framework incorporates LAA radio interface and Wi-Fi access point in order to assign suitable utility services in appropriate available transmission opportunities. The proposed system with the developed algorithms was able to increase the bandwidth utilization of the available resources for the smart grid system and improve the overall spectrum efficiency.

In this way, the smart grid consumers can deliver their information using LAA node in unlicensed access technology. The new bandwidth utilization technique performed a priority queuing for real-time loads and allocates the critical loads to the available subband to meet the QoS requirements using the channels available for transmissions.

References

1 C. Kalalas, L. Thrybom, and J. Alonso-Zarate, "Cellular communications for smart grid neighborhood area networks: a survey," *IEEE Access*, vol. 4, pp. 1469–1493, 2016.

2 M. Fathi, "A spectrum allocation scheme between smart grid communication and neighbor communication networks," *IEEE Syst. J.*, vol. 99, pp. P 1–8, 2015.

3 S. Al-Rubaye, A. Al-Dulaimi, and J. Cosmas, "Spectrum allocation techniques for industrial smart grid infrastructure," in *2016 IEEE 14th International Conference on Industrial Informatics (INDIN)*, France, July 2016.

4 A. Mukherjee, J. F. Cheng, S. Falahati, H. Koorapaty, D. H. Kang, R. Karaki, L. Falconetti, and D. Larsson, "Licensed-assisted access LTE: coexistence with IEEE 802.11 and the evolution toward 5G," *IEEE Commun. Mag.*, vol. 54, no. 6, pp. 50–57, 2016.

5 R. Yu, C. Zhang, X. Zhang, L. Zhou, and K. Yang, "Hybrid spectrum access in cognitive-radio-based smart-grid communications systems," *IEEE Syst. J.*, vol. 8, no. 2, pp. 577–587, 2014.

6 R. C. Qui, Z. Hu, N. Guo, R. Ranganathan, S. Hou, and G. Zheng, "Cognitive radio network for the smart grid: experimental system architecture, control algorithms, security and micro grid testbed," *IEEE Trans. Smart Grid*, vol. 2, no. 4, pp. 724–740, 2011.

7 Y. Zhang, R. Yu, M. Nekovee, Y. Liu, S. Xie and S. Gjessing, "Cognitive machine-to-machine communications: visions and potentials for the smart grid," *IEEE Netw*, vol. 26, no. 3, pp. 6–13, 2012.

8 V. C. Gungor, D. Sahin, T. Kocak, S. Ergut, C. Buccella, C. Cecati, and G. P. Hancke, "Smart grid technologies: communication technologies and standards," *IEEE Trans. Ind. Inform.*, vol. 7, no. 4, pp. 529–539, 2011.

9 Standards Council of Canada, "The Canadian Smart Grid Standards Roadmap: A Strategic Planning Document," CNC/IEC Task Force on Smart Grid Technology and Standards, October 2012.

10 W. Lee and D. H. Cho, "Channel selection and spectrum availability check scheme for cognitive radio systems considering consumer mobility," *IEEE Commun. Lett.*, vol. 17, no.3, pp. 463–466, 2013.

11 S. Al-Rubaye, (a) "Radio Network Management in Cognitive LTE Systems," Ph.D. thesis, Electronic and Computer Engineering, Brunel University, London, UK, 2013. (b) W. Ikram, S. Petersen, P. Orten, and N.F. Thornhill, "Adaptive

multi-channel transmission power control for industrial wireless instrumentation," *IEEE Trans. Ind. Inform.*, vol. 10, no. 2, pp. 978–990, 2014.

12 V. C. Gungor and D. Sahin, "Cognitive radio networks for smart grid applications: a promising technology to overcome spectrum inefficiency," *IEEE Veh. Tech. Mag.*, vol. 7, no. 2, pp. 41–46, 2012.

13 U. Paul, A. Kashyap, R. Maheshwari, and S. R. Das, "Passive measurement of interference in Wi-Fi networks with application in misbehavior detection," *IEEE Trans. Mobile Comput.*, vol. 12, no. 3, pp. 434–446, 2013.

14 N. H. Tran, L. B. Le, S. Ren, Z. Han, and C. S. Hong, "Joint pricing and load balancing for cognitive spectrum access: non-cooperation versus cooperation," *IEEE J. Sel. Areas Commun.*, vol. 33, no. 5, pp. 972–985, 2015.

15 B. Vucetic, "An adaptive coding scheme for time-varying channels," *IEEE Trans. Wireless Commun.*, vol. 39, pp. P 653–663, 1991.

16 Riverbed Modeller. Available at http://www.riverbed.com/products/steelcentral/steelcentral-riverbed-modeler.html (accessed February 1, 2014).

17 IEEE Standard 802.11-1999, "Wireless LAN Medium Access Control (MAC) and Physical Layer (PHY) Specifications," November 1999.

18 B. F. Lo, "A survey of common control channel design in cognitive radio networks," *J. Phys. Commun.*, vol. 4, no. 1, pp. P 26–39 2011.

John Cosmas received his Bachelor of Engineering degree in electronic engineering from Liverpool University in 1978 and his Ph.D. degree from Imperial College, University of London, in 1986. He is currently Professor of Multimedia Systems at the College of Engineering, Design and Physical Sciences, Brunel University, London. He coleads the Wireless Networks and Communications Research Centre. He is the course director of M.Sc. Advanced Multimedia Design and 3D Technologies. He is Associate Editor of the *IEEE Transactions on Broadcasting*. His research interests are concerned with the development of multimedia systems applied to future of broadcasting, cellular communications, 2D/3D digital video/graphics media, and the synergies between these technologies toward their application for the benefit of the environment, health, and societies. He has participated in 11 EU-IST and 2 EPSRC-funded research projects since 1986 and has led 3 of these (CISMUNDUS, PLUTO, and 3D MURALE). His latest research is concerned with management of heterogeneous cellular networks, convergence of cellular and ad hoc networks, 3D MIMO, and efficient software-defined networks architectures.

 ***Saba Al-Rubaye*** received her Ph.D. degree in electrical and electronic engineering from Brunel University, London in 2013 and postdoctoral fellow in Smart Grids from Stony Brook University in 2014. She is currently working as a senior research engineer at Institute of Telecommunications, Aveiro, Portugal. Prior to her current position, Dr. Al-Rubaye was working with advisory service on several integration, analysis, and testing projects at Sustainable Technology Integration Laboratory (QT-STIL), Quanta Technology, Toronto, Canada. She has more than 10 years of experience in industry and research projects with strong track record of launching innovative technical solutions. She led and managed several projects focusing on smart grids, IIoT, cyber-physical systems, SDN platform, 5G and beyond wireless networks. Dr. Al-Rubaye served as technical program committee member for several prestigious conferences and workshops, and participated in IEEE standardization. She has published many papers on IEEE journals and conferences in the areas of smart grids and telecommunications system. She is registered as a Chartered Engineer (CEng) by British Engineering Council and recognized as Associate Fellow of the British Higher Education Academy in the United Kingdom. She is a senior member of IEEE, a voting member of IEEE 1932.1 WG Licensed/Unlicensed spectrum interoperability standard and IEEE P2784 WG of building a framework for smart city standard.

Part V

R&D and 5G Standardization

18

5G Communication System: A Network Operator Perspective

Bruno Jacobfeuerborn[1] and Frank H. P. Fitzek[2]

[1] *Deutsche Telekom AG, Berlin, Germany*
[2] *5G Lab Germany and Technical University Dresden, Dresden, Germany*

18.1 Introduction

Currently, the fifth-generation (5G) communication system is discussed in the research and industry domain. Not just the network operators and network manufactures, but also the so-called verticals, which apply the technologies later on, are interested in these discussions with the potential opening new markets in connected cars, industrial Internet, smart energy networks, health care, and even agriculture. It is assumed that 5G will address 500 billion wireless devices [1] in 2020, compared to the 10 billion mobile smartphones the networks support today. This development is a result of the current development of the Internet of Things (IoT). The trend will even intensify over the next years. As given in Figure 18.1, the first four generations of communication systems were focusing on ten billion human beings with their mobile handsets enabling ubiquitous content communication. With 5G, there is a disruptive change, not only in the number of devices, but also in the main application, from ubiquitous content communication to ubiquitous controlling and steering of devices.

At this point of time, we are able to identify a larger number of 5G use cases, which will already guarantee the success of this technology wave. But we can already assume that this list is by far not complete and that the next generation of researchers and industry will come up with even more successful use cases added to this list. It is generally accepted that 5G will not only provide better data rates, as all previous generations had to do, but it must go beyond and also improve significantly the resilience, energy consumption, security, and

5G Networks: Fundamental Requirements, Enabling Technologies, and Operations Management, First Edition.
Anwer Al-Dulaimi, Xianbin Wang, and Chih-Lin I.

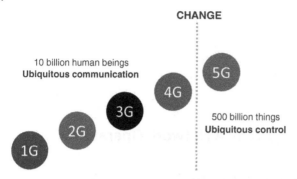

Figure 18.1 Progression of mobile communication technology

latency of the network. These massive technical requirements ask for significant changes in the network architecture. In contrast to past generations, the changes are not focusing solely on the radio access technologies and the radio access network (RAN). Instead, 5G requires a holistic approach that includes the core network and cloud infrastructure. Within this paradigm, software-defined radio (SDR), software-defined networking (SDN), and network function virtualization (NFV) will play a major role introducing fundamental changes to the network [2].

It is not the first time that we are discussing a fundamental change of our communication networks. Already in the 1960s, Paul Baran proposed to change the circuit-switched networks to a packet-switched approach [3] in order to increase resilience in the network, and to be less prone to single point failures. As given in Figure 18.2, the change from circuit switched to packet switched was referred to as a revolution at that point in time. Within this fundamental change, the resilience of the communication was achieved with the implementation of multipath communication. The Internet was built on Baran(tm)s ideas, and it deployed the packet-switched features. However, it cut down on the multi-path communication as it showed to have a negative impact on the data throughput of the networks due to the use of inefficient repetition coding. As we will show within this chapter, the technical requirements are historical trade-offs, ones against each other, and in order to break with this trade-offs, we need new technologies.

In the spirit of Paul Baran, new concepts are currently discussed for communication systems, such as information-centric networking (ICN), content delivery networks (CDN), SDN, and NFV, which comes hand in hand with a process called softwarization. ICN is a new concept that allows users to retrieve content based on identifiers and not based on physical location. ICN is currently discussed in standardization bodies, but it will, for sure, play a similar role as SDN/NFV in future communication systems such as 5G.

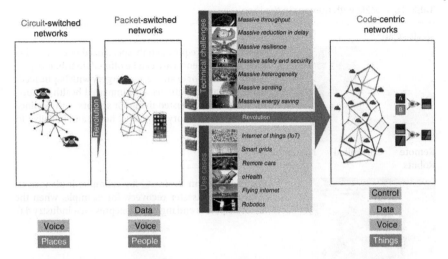

Figure 18.2 Evolution of data transmission in networks.

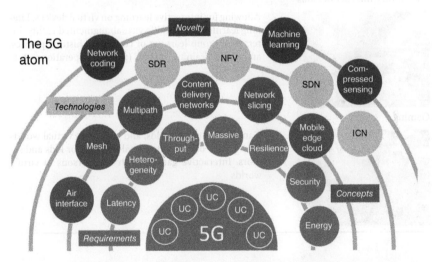

Figure 18.3 5G atom. The 5G use cases are in the center. The layers, from the center out, represent the requirements of the 5G use cases, the concepts that will allow network operators to satisfy the requirements, the technologies that enable the implementation of the concepts, and the novelties, that is, technologies that can be easily implemented due to softwarization and virtualization techniques

The ongoing discussions over 5G are summarized in Figure 18.3, which shows the *5G atom* with the 5G use cases in the core, and the technical requirements, concepts, technologies, and novelty that each use case requires represented on the different tiers. The 5G uses cases are described in full detail in Tables 18.1 and 18.2.

Table 18.1 M2M with humans in the loop.

Health Care

Tele hospitals requiring only specialized operation theaters while patients care can be offered distributed, possibly at home. For example, telesurgery with haptic feedback, telecare of patients with impaired health in nursing homes, tele-assisted living for seniors with declined cognitive and sensory abilities at home and in daily life

Remote   Controlled   Humanoid Robots

For construction business, for example, in dangerous settings. For disaster recovery, for example, when the life of helpers is endangered. Telepresence Industry 4.0

Education - Internet of Skills

Allowing for interactive learning on virtual devices. Lifelong learning. Second life virtual/augmented reality. Interactive group learning of languages, history, physics, and more. Training of skills (painting, operations, etc)

Gaming

Gaming in virtual worlds, and gaming in virtual worlds as cognitive or socioemotional training for kids and seniors. Interactive gaming with real persons in virtual worlds

Even though the use cases appear to be quite heterogeneous, the technical requirements are quite similar. Figure 18.3 lists them as latency, heterogeneity, throughput, massive, resilience, security, and energy. Latency is needed to enable the request use cases based on control and steering. The exact value of the latency varies for each use case, but currently one millisecond end to end seems to be a good choice for both, machine to machine (M2M) only as well as M2M with humans in the loop. So far, latency was not really in the focus for the previous generations of mobile communication. But already in the 1990s, there was one group underlining the need of low-latency communication, the

Table 18.2 M2M.

Autonomous and Connected Driving

Driving vehicles on roads safely and efficiently. Automatic logistics: distribution centers, for example, with fully controlled automatic fork lifts and parcel pickup systems

Factory Automation, Industry 4.0

Replacement of current data/control cables in machines and robots, which break easily, by wireless systems. Enabling mobile robots that, for example, manually assemble individualized goods. Co-working with sensitive robots for flexible automation of workshops

Agriculture

Precision-controlled fleets of harvesting vehicles, also including augmented reality by using vision -and radar-equipped drones around harvesters. Precision controlled fleets of drones for surveillance, fertilization, and pest control

Smart Grid - Smart Building - Smart City

Mobile robots as, for example, drones for packet delivery, for building maintenance and repair, and many other markets. Robots in construction industry, and in underwater construction and maintenance

Energy Grids

To enable a synchronized stable grid network with minimized reactive power, each subnet must be synchronized within 20′ phase, which results at 50Hz AC frequency in 1ms maximum synchronization offset.

gamers [4]. The gamers represent a very interesting application field as they showed one solution for low latency with omnipresent technologies, namely, so-called LAN parties. All gamers gathered at one location and played their games. This shows us one important fact we will exploit later to realize low-latency communication, and it is the proximity between communication entities. The next technical requirement is focusing on the data rate supported in one communication area for a massive number of devices. For sure, the data rates will increase, and currently, 10 Gbps are discussed for individual high-end devices. How to increase the data rates is known to the engineers, and it will be realized by increased spectrum efficiency, smaller cells, and more spectrum. In order to achieve more spectrum, 5G will target higher carrier frequencies between 3.5 and 60 GHz. Nevertheless, higher data rates are needed to support the massive number of devices we expect to see in 5G as well to support high-quality video feeds with 4k or 8k. Besides pure data rate, the massive number of devices itself will impose drastic challenges to any network. For example, reading out 10,000 sensors with low latency is a challenge and another example for a classical trade-off between two technical requirements. To enable ubiquitous steering and control, we also need resilience and security to make sure that the network is always available and guarantee that no third parties are maliciously taking over our communication. Furthermore, the energy consumption of the network as well as the devices needs to be considered when we design 5G. Several billion devices will consume a large amount of energy if the system is not designed in a tight manner. One technical requirement that is not mentioned very often when it comes to 5G is heterogeneity. This reflects the fact that devices are heterogeneous in terms of computing power, air interfaces, and functionality. Nowadays, mobile phones are more or less homogeneous, even if manufacturers want us to believe differently. The answer to this cannot be to group the devices into different heterogeneity classes, since we would lose cognition in our network [5], but instead, we need to design scalable solutions that can be run on any platform.

If we want to support the aforementioned technical requirements in our 5G networks, we will need new concepts. In Figure 18.3, the concepts listed are air interfaces, mesh, multi-path, content delivery networks, network slicing, and mobile edge cloud. Each concept will support at least one if not several technical requirements. Subsequently, new air interfaces will improve latency and data rates; an efficient support of multi path communication will improve the security, resilience, and data rates; network slicing will introduce quality of service support; the support of meshed networks will allow for a massive number of devices; and the concept of the mobile edge cloud (MEC) will improve the latency. While the first named concepts are easy to couple with the technical requirements, the need of the MEC may not be a straightforward concept to grasp for all the readers. While cloud applications are hosted in huge data centers today, they will become part of the network in future communication systems as part of the MEC concept. The fusion of storage and cloud computing with the transport network will be the core of the 5G communication system.

In order to realize the aforementioned concepts in an efficient manner, we have to rely on SDN, NFV, and ICN as well as on network coding, compressed sensing, and machine learning as the enabling technologies and novelties. For example, in order to provide low latency, the MEC is a key concept, but it is realized, as seen later, with SDN and NFV as core technolog supported by network coding (to improve the efficiency) and machine learning (to optimize the SDN/NFV settings). Therefore, we claim that SDN and NFV are key technologies for 5G communication networks. Furthermore, they open the door for further novelties, for example, network coding. In the atom, the tiers two and three, namely, concepts and technologies, are mainly open source and standardized. It is in tier 4, where the IPs and network operators have the selling point.

Now, let us shortly introduce SDN and NFV here. SDN advocates for centralized controlled network protocols replacing the state-of-the-art distributed protocols. The centralized approach plus the softwarization of the network protocols makes it easier to experiment with new ideas, allows for stronger optimizations that cannot be achieved with distributed approaches, and speeds up the deployment of new or upgraded protocols. SDN is composed out of one logical (not necessarily one physical) SDN controller that is in charge of one single network formed by several white boxes on which softwarization takes place. SDN is a result of consequent fusion of computing and communication, or software and networking. This fusion is not new, as it has been already discussed in the context of software-defined radio (SDR) or active networks (AN). Most definitions of SDN are focusing solely on the split of the management and control plane and the data plane, but SDN is far more than that. Actually, SDN follows the abstraction concepts known from computer programming languages, and uses high-level abstraction levels, for example, application programmer interface (API) instead of redefining protocols primitives or bit level concepts as given in every standardization document. SDN abstractions include packet forwarding, routing, and configuration as a computational problem instead of specialized protocols. NFV is a direct request from telecommunication operators to shorten development cycles as well as cutting costs for service deployment by replacing specialized and static hardware solutions with software on standard hardware using virtualization concepts. The softwarization allows for quick deployments of new services, while the virtualization allows for relocation, live migration, upgrade, and downgrade of services wherever and whenever they are needed. Furthermore, the softwarization will decrease the cost of exchanging and maintaining new services reducing the capital expenses (CAPEX) and operating expenses (OPEX) of network operators. The profitability of cloud and smart-phone application deployments inspired the request for shorter deployment circles for the network domain. While the deployment cycles of the former were only days, the telecommunication operators had to undergo time-consuming discussions with vendors and standardization bodies, which easily led to deployment cycles of several month or sometimes even years, as the deployment of IPv6 illustrates. The NFV framework consists of the virtualized network functions (VNF), the network function virtualization infrastructure (NFVI), and the network function virtualization management and orchestration architectural framework (NFV-MANO). The latter one is in charge of orchestration and management of the NFVI and multiple VNFs. NFVI, on the other hand, is a collection of hardware (white boxes) and software where VNFs are deployed, which in turn are the core of the NFV concept. In order to be more concrete, the standardization ETSI NFV POC group introduced nine examples for possible virtualizations in the area of cloud, mobile networks, data caters, and access [6]. Out of those, we shortly discuss the CDN example. The motivation is based on the fact that video will play a larger role in current and future networks. CDNs have the capability to reduce the traffic for video significantly if they are placed closer to the user, that is, at the edge of the network, avoiding traffic in the backbone. Currently, large content provider as network operators deploy their proprietary cache nodes into the ISP network (Netflix, OpenConnect, program, Akamai, Aura CDN), which lead to the following disadvantages: (i) high costs due to design of capacity and resilience of CDNs for peak hours, (ii) more dedicated hardware in the network operators' networks is required, and (iii) it grants access to third parties in the network operators' network.

Loosely speaking, while SDN can be seen as the network operating system, NFV is comparable to mobile apps on smartphones. The fusion of SDN/NFV is a game changer as telecommunication operators are regaining lost ground. Nowadays, network operators suffer by over the top (OTT) services. With 5G, OTT will be dead and all services, such as the mobile edge cloud, will be deployed and anchored within the network. Big players, such as Facebook and Google, are aware of this trend, and subsequently, they are answering network initiatives such as Loon, Aquila, and Google Fiber. Earlier the market was owned by the manufacturers of the phones and the cloud service providers. In the near future, the one who owns the network is the one who will controls the market.

18.2 Softwarization for the 5G Communication System

There are different new concepts that are needed in order to satisfy the requirements of the 5G use cases, as it is depicted in Figure 18.3. There are many different technologies that will enable the deployment of these concepts in the 5G networks. However, there is a clear trend. They aim for the softwarization of everything. For instance, telecommunication services that run nowadays in dedicated hardware can be designed as VNFs that can run in industry standard

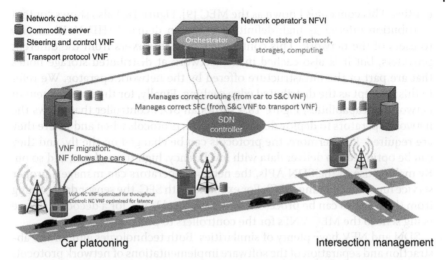

Figure 18.4 Softwarization within the context of the 5G communication system.

equipment, and can be scaled and migrated as needed [7]. Moreover, softwarization and virtualization can potentially reduce the costs of the deployment of novel technologies such as network coding and compressed sensing. The reason is that advances in software and hardware become independent of each other, and therefore, they can be upgraded separately as required. Furthermore, the acquisition of commodity servers and industry standard computing hardware will provide network operators with an infrastructure that can be offered as a service for network functions to other service providers. This infrastructure can also be extended to storage nodes. Therefore, the network operators in the future can offer network, storage, and computing resources as services.

Managing the infrastructure will become particularly important in the 5G networks. In order to satisfy the strong requirements of the 5G use cases, it will be mandatory to have a holistic approach for the quality of service (QoS). Softwarization tools are useful to perform network slicing, which is QoS for the fixed network. However, QoS is also needed in the radio access and in each involved platform, for example, white boxes in the SDN context, mobile phones, and sensor networks. The latter can be achieved with modern coding techniques such FULCRUM codes [8] and with multi core optimizations for the software.

In Figure 18.4, we show an example of a typical 5G environment. The network operator offers its infrastructure for the deployment of a steering and control (S&C) network function (NF). The NF is software that receives, as an input, information from the sensors of a group of vehicles at a highway or at a busy intersections and provides, as an output, control instructions. The NF can be virtualized and migrated to different base stations following the controllable entities. This concept is known as the MEC [9]. Figure 18.4 also shows content distribution offered as high-definition video on demand (HD-VoD) service to users of the network. This content is stored in servers from the different providers, but it is also cached in the network at distributed storage nodes that are part of the infrastructure offered by the network operator. We refer to this concept as the distributed edge caching. Finally, for the management of networking capabilities, Figure 18.4 shows an SDN controller that allows the network operators to deploy different network protocols when and where they are required. Furthermore, the protocols can be changed on the fly, and they can be optimized to deliver data with low latency, high throughput, and so on. By making use of the SDN APIs, the network operators can manage a proper service function chaining (SFC). For example, with SFC, the sensor data coming from the vehicles can be processed at the protocol VNFs for a decoding before being sent to the MEC VNFs for the controllers to process.

SDN and NFV have plenty of similarities. Both technologies aim for the abstraction and separation of the software implementations of network protocols and network functions from dedicated hardware. Consequently, protocols and functions can be dynamically instantiated and deployed with predefined life cycles when they are needed. In a typical 5G communication ecosystem as the one shown in Figure 18.4, NFV and SDN are complimentary. For instance, the SDN control plane can be implemented as pure software that runs in industry standard hardware in the data centers of the network operator. Furthermore, different communication protocols can be instantiated as network functions, and therefore, they can benefit from the flexibility and reliability of NFV.

For example, different network coding protocols, which are optimized for latency, throughput, or computational complexity, can be seen as a set of different encoders and decoders that are implemented in software. When an user or application has a specific requirement, for example, low latency, the network will deploy as a VNF, in standard hardware, the appropriate set of encoders and decoders. The SDN architecture, by making use of virtualization, will also allocate the abstract resources to the client or application [10] in order to guarantee that its requirements are satisfied. Additionally, by making use of the SDN capabilities, the network will manage the correct SFC. In this context, the packets that go from the network to the users of the application will be processed by the appropriate encoder before being sent over the air interface. Similarly, the incoming packets for that application will be processed by the corresponding decoder before being routed to the other VNFs providing the service. Now, we will introduce three scenarios where SDN/NFV will be beneficial for the 5G communication systems.

18.2.1 Network Coding as a Service

It has been long perceived that store-and-forward networks with lossy channels are characterized by high latency and low throughput. On the other hand, employing network coding guarantees maximum throughput under similar circumstances [11]. In this paradigm of store-code-forward, several packets are combined together using linear operations over finite fields, therefore, reducing latency and increasing the overall bandwidth and resilience of lossy networks.

Random linear network Coding (RLNC) has been shown to be applicable in distributed environments [12] while performing asymptotically optimally. In RLNC, the data is fragmented into smaller groups of packets called generations, which are subject to linear combination with randomly chosen coefficients to produce coded packets. The randomness of the coefficients reduces the need for coordination, and the capability to recode at intermediate nodes in the network helps RLNC to further reduce the end-to-end latency in lossy channels. Fortunately, SDN and NFV fit perfectly into the whole picture. NFV allows network coding software to be implemented as a service in the form of VNF. SDN allows for setting up end-to-end paths on demand with pre specified QoS. Thus, NC can be deployed as a service at the network position where and when it is needed. In the following sections, we will quickly describe three use-cases of network coding. These are point to point, point to multi point, and multi-hop. They demonstrate the superior properties of network coding over conventional store-and-forward paradigm.

18.2.1.1 Point to Point

This scenario demonstrates a simple yet fundamental situation when two nodes communicate over a lossy link. Conventional error control mechanisms, such as Automatic Repeat reQuest (ARQ), keep track of transmission errors by employing an acknowledgment mechanism for each transmitted packet. The sender keeps sending a new packet as long as it receives an ACK from the receiver about the previous transmission. An absence of the ACK packet triggers a retransmission even when the data packet is received but the ACK is lost, leading to unnecessary retransmission and increasing delay. With network coding, the ACK from the receiver can contain information of the decoding status. Consequently, unnecessary retransmissions are avoided because the sender is informed with more details of the decoding status. To confirm the advantage of network coding over uncoded transmissions, we perform an experiment with two nodes, source node (S) and the destination node (D). S must send 100 packets to D over a 20% lossy link. The expected latency of each packet as compared to an ideal case with no link loss is then recorded and plotted in Figure 18.5a. As soon as a packet is lost, a retransmission occurs, increasing the latency of the lost packet and also the accumulated latency at subsequent packets. Sliding window RLNC, however, reduces almost 50% of the latency due to avoiding unnecessary retransmission. Furthermore, compared with traditional block codes, network coding can reduce the per packet latency of the transmission.

18.2.1.2 Point to Multipoint

This scenario illustrates a classical reliable multicast problem, where the same information is sent to multiple destination nodes simultaneously. For example, the traffic information at the road intersection needs to be sent to all the vehicles approaching it. In an uncoded transmission, the source needs to resend individual lost packets upon request from each receiver. The other receivers receive the resent packet but discard it in most cases because it is irrelevant to them. However, in RLNC, retransmissions are useful for each receiver, since each encoded packet contains information of the whole data. To demonstrate the advantage of network coding in this scenario, we set up a topology consisting of one source node (S) and N destination nodes with a channel erasure rate of 20% in each link. We vary the number of receivers between 1 and 50, and measure the resulting total number of packets transmitted by the source for both uncoded and coded scenarios. Figure 18.5b shows that as the number of receivers increase, the number of transmitted packets increase significantly in the uncoded transmission while in the coded transmission, it remains constant. The results confirm that network coding can significantly reduce the amount of transmitted data.

18.2.1.3 Multi-Hop

This scenario illustrates a situation where two end nodes communicate over several intermediate nodes such as in a car-to-car communication. Conventional end-to-end communication results in significant delay due to retransmission from the source regardless of where the packet losses occur. RLNC allows for recoding at the intermediate relay nodes, thus, introducing redundancy to the network where it is needed, and thereby, reducing end-to-end latency. To demonstrate the advantage of network coding in this scenario, we set up a chain topology consisting of one source node (S), one destination node (D) and N relay nodes. We assume a packet loss probability of 20% at each link between two consecutive nodes. We vary the number of intermediate nodes (or hops) between 1 and 50 and measure the average latency of the destination node to receive 100 packets. We perform the experiment for both network coding and Reed–Solomon codes. As it is illustrated in Figure 18.5c, Reed–Solomon increases latency drastically as the number of hops increases. On the contrary, the latency increases slowly in case of RLNC. This demonstrates that RLNC outperforms conventional end-to-end error-correcting code.

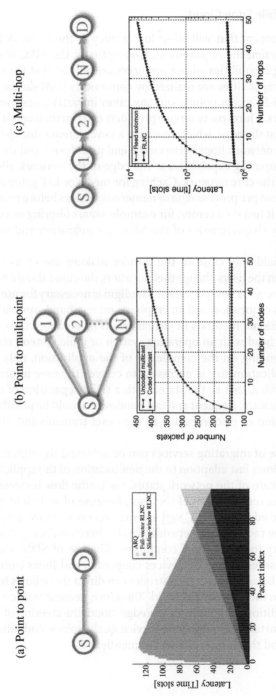

Figure 18.5 Network Coding as a Service (NCaaS) – advantages of network coding in various transmission scenarios where channels are lossy.
(a) Point to point. (b) Point to multi point. (c) Multi-hop.

18.2.2 The Mobile Edge Cloud

The MEC is a concept that will allow low latency in the network for the 5G use cases that present the requirement. Conceptually, the MEC is realized by cloud computing capabilities whose resources are allocated at the edge of the network. These capabilities are realized by commodity hardware set up at the base stations, Wi-Fi access points, and any other infrastructure owned by the network operators. Third-party service providers can run their services as VNFs on the hardware at the edge, which results in a lower latency due to the shorter distance and less network hops between the end user device and the cloud.

Moreover, computing capabilities at the edge of the network allow for decreased traffic in the core network. Caching for media or IoT gateways close to the data sources can pre process data of numerous sensors before reacting to the input or sending it to a data center, for example, smart filtering or compressed sensing. Two key characteristics of the MEC are migration and synergies of SDN/NFV.

The MEC should stay close to the devices making use of its computing capabilities. When the users change their location, the cloud should migrate and follow them. However, a new migration paradigm is necessary for use cases with frequent changes of location and an emphasis on fast handover with minimal downtime. Traditional methods in virtualization or containerization transfer a significant overhead with an operating system or a file system, respectively. Meanwhile, depending on the complexity of the application, only a fraction of the actual application size is necessary to convey the user information for providing the service, that is, the state. With a strict separation of the service framework and its state, frequent and fast handovers would be possible without extensive expansion of resources due to the exact transmission of necessary data.

An efficient use of migrating services can be achieved through deployment of SDN, which allows fast adaption to the new location of the application. This includes being aware of the network status, the traffic flow between the edge nodes, and the current position of the user. Because of sufficient computing capabilities at the edge, the MEC can leverage access to more detailed status information of the radio access network (RAN), therefore, being able to adjust the service quality to the channel conditions. The use of SDN is even more important when several users or devices cooperate, and flows from different locations have to be aggregated. A controller can direct these flows to the MEC without relying on the backbone network, therefore, keeping traffic and latency low. With the addition of mobility of the edge cloud, the steering of data flows is of utmost importance to guarantee service quality since endpoints of both the application and the devices change frequently.

18.2.3 Distributed Edge Caching and Computing

We are interested in managing the storage and computing capabilities distributed at the edge of the network in order to increase the resilience and reduce the latency. For this purpose, network coding is an ideal technology for the management of distributed edge caches [13] and computing processes. Network coding allows the management of communication and storage with a single code structure. In the 5G context represented in Figure 18.4, the network operators of the future will have infrastructure that will provide storage capabilities to service providers, either for the VNFs that require storage, or as edge caching of content. These storage nodes are susceptible to failures that can affect the reliability of the stored information. To maintain the data reliable against failures of storage or computing nodes, it is necessary to introduce some form of redundancy. The redundancy could be simple replication or coded information. For the coded redundancy, we could rely on traditional block codes or on network coding.

Once any node fails, the system must repair the lost data. This is known as the repair problem. The different redundancy schemes behave differently and have different costs when they are used to repair failing nodes. We will describe the characteristics of each redundancy scheme in a distributed storage system, their repair strategies, and elaborate on why network coding presents gains on fully distributed systems compared with other schemes.

18.2.3.1 Block Codes versus Replication

The simpler form of redundancy is replication. To maintain a certain level of reliability, the stored data must be replicated in different storage nodes. Therefore, if a storage node fails, the data can be recovered from the copies stored somewhere else. This kind of redundancy is easy to implement, but it requires high costs in storage. A benefit of replication is that when a node fails, the traffic required over the network in order to repair it is minimal. Once the failing node is replaced, the system needs to transfer only the damaged pieces of the files.

The more complex form of redundancy are traditional block codes such as the Reed–Solomon codes used in RAID-6 systems. These codes work by forming linear combinations of the original data. For example, they can produce, out of M original pieces, a set of $K > M$ linear combinations. Any subset of M linear combinations can be decoded to recover the original data. This coding scheme drastically reduces the costs of storage compared with replication. To repair a failing node, the system must transfer over the network all the information (i.e., M linear combinations), performs a decoding process to obtain the original file, and encodes it again in order to recover the lost information.

18.2.3.2   Network Coding in Distributed Storage Systems

For years, it was thought that the overhead in network traffic was unavoidable when using coding in distributed storage systems. However, it has been proven that network coded systems can operate in optimal operation points in the storage–repair traffic trade-off [14]. This is due to the ability to recode linear combinations, which can only be done in network-coded systems. If network operators want to provide scalable and distributed caching and computing capabilities for the ICN and CDN of the 5G communication system, then network coded-based systems might be the solution. Without a centralized overview of the state of the system, the stored data would degrade after each repair if traditional redundancy schemes are implemented.

The benefits of a network coded scheme over the traditional redundancy schemes are shown in the *melting ice cubes* from Figure 18.6. In Figure 18.6, we implemented a completely distributed storage system [15]. We varied different parameters such as how much data each node stored and how much transfers the system did to repair a failing node. We introduced failures, replaced the storage nodes, and performed the repairs. For the repairs, the replaced node downloaded the stored data at other nodes and stored some of these pieces or recoded versions of them. After 100 repairs, the data has practically degraded completely for the uncoded storage. For the Reed–Solomon case, the data is fully degraded for 100 and 1000 repairs unless the system invests a lot of storage and traffic resources. On the other hand, the recoding capabilities of network coding, allowed a high reliability of the information without the expensive investments needed with block codes.

18.2.3.3   Security Aspects: Algebraic and Light Weight

In terms of security, network coded distributed storage systems offer benefits over the other two schemes. It offers an inherent algebraic security [14]. In order to decode the original information, an attacker would need to break into enough storage nodes and collect enough linear combinations. Otherwise, since the information is coded, the attacker would not make sense of it. Additionally, to increase the security of the system, network coding offers the possibility of implementing a light weight encryption technique. Instead of encrypting the whole data, it is possible to encrypt only the coding coefficients used to generate the coded packets. The coefficients represent just a small fraction of the data, thus the light weight. Even if an attacker collects enough linear combinations, it would not be able to make sense of the information without cracking the coefficients.

18.3   5G Holistic Testbed

To facilitate research activities and demonstrate proof of concepts towards 5G, building testbeds is one of the fundamental steps. For this reason, the 5G

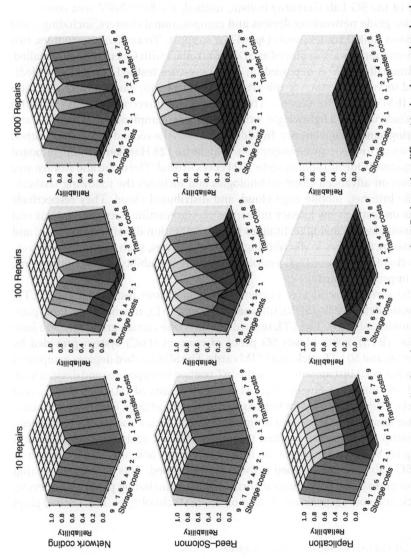

Figure 18.6 Data integrity in a distributed storage system over time. Each row is a different technology and each column represent 10, 100, and 1000 repairs correspondingly. The plots show the data reliability of the system versus the storage and traffic costs [15].

Lab Germany[1] holistic testbed is built toward the creation of an experimental network infrastructure that anticipates the 5G communication system. There are four main areas of the 5G Lab Germany testbed: SDN/NFV, cloud and storage, access network, and security with individual testbeds as illustrated in Figure 18.7. The main objective is to demonstrate and assess the level of maturity of adopted solutions in a holistic manner. The testbeds are also capable of illustrating how they can unleash the full potential in a real-world infrastructure by creating, nurturing, and supporting an innovative 5G ecosystem. As the core of the 5G Lab Germany holistic testbed, the SDN/NFV area consists of carrier-grade networking devices and computational clusters, including three Ericsson R6000 routers with OpenFlow support. To control the routers, two high-end servers are deployed as SDN controllers with OpenDaylight installed. Additionally, NFV is deployed on the many-core testbed consisting of dedicated servers with state-of-the-art CPU architectures, namely, Intel x86 Xeon and IBM's POWER8 servers. Those servers are interconnected via RapidIO switches running a lightweight switching protocol implemented on FPGA and offering nanosecond latency. In addition to that, the second many-core testbed focuses on low-energy consumption. It includes 128 Hardkernel's single-board computers (Odroid) adding up to 1024 cores in total. The cloud and storage area focuses on advanced cloud technologies and includes the following testbeds: tactile Internet, mobile edge cloud, and distributed cloud. They respectively focus on testing how latency influences the interaction between humans and technologies, accessing techniques for live migration of software services, and evaluating advances in distributing Internet services, such as storage over various clouds to achieve high performance, high availability, privacy preserving, and improved security.

The wireless access area currently consists of four testbeds, namely, LTE, software-define radio (SDR), Internet of Things (IoT), and multihop, multipath, and multicast (3M). The LTE testbed employs one carrier-grade Ericsson base station (BTS), a full-blown 5G proof of concept (PoC) system provided by Ericsson, and 50 LTE sticks with SIM cards. The SDR testbed deploys equipment from National Instruments while the IoT testbed incorporates end devices from the LoRa initiative. Finally, the 3M testbed consists of 150 high-end Huawei tablets. They can be used for both fixed and mobile settings. Currently, the tablet cluster demonstrates the synchronized video multicast use case. Nevertheless, the setup can be easily adapted to other scenarios, such as server migration using NFV/SDN to evaluate the impact of latency and jitter. The last area in the 5G Lab Germany testbed is the security testbed. This enables the testing and demonstration of various security advances, including privacy preserving, attack, and countermeasure of state-of-the-art technologies. This testbed plays

---

1 5G Lab Germany (2016). http://http://5glab.de/.

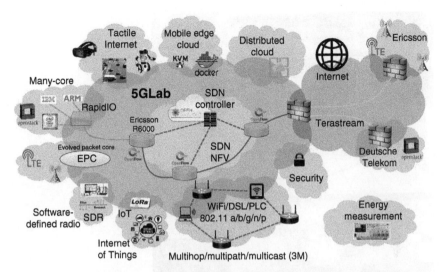

Figure 18.7 5G Lab Germany holistic testbed consisting of four main areas. (i) SDN/NFV as the core testbed with Ericsson routers and many-core platforms. (ii) Cloud and storage with tactile Internet, mobile edge cloud, and distributed cloud testbeds. (iii) Access network with LTE, SDR, IoT, and 3M testbeds.(iv) Security.

an important role in the overall 5G Lab Germany holistic testbed, especially to demonstrate service function chaining later on. The individual testbeds in the four areas of the 5G Lab Germany holistic testbed are interconnected via the SDN/NFV cluster.

The 5G Lab Germany testbed is capable of extending its connections and capabilities in several directions. First, the Ericsson BTS can connect with other Ericsson's PoC to orchestrate more sophisticated evaluations. Second, the SDN/NFV can be expanded to include Deutsche Telekom's OpenStack cluster with dedicated Tbps optical connections. Third, we also would like to evaluate our testbeds with energy measurement tools, such as N6705B DC Power Analyzer. The equipment allows for benchmarking computational gain versus consumed energy. This will be interesting to investigate the energy consumption especially from the ICT sector.

At the Mobile World Congress (MWC) 2016, the Deutsche Telekom showcased the first 5G end-to-end connection for industrial Internet applications, as shown in Figure 18.8. A robotic arm was controlled remotely to place objects in the real world. Communication systems with state-of-the-art latencies failed to place the object at the right position. Meanwhile, 5G latencies lead to successful placement of the objects. In this demonstrated setup, the low-latency communication was achieved by enabling network slicing.

Also, at the MWC 2016, Nokia together with the 5G Lab Germany demonstrated autonomous and connected cars. Six model cars, which can be seen in

Figure 18.8 Deutsche Telekom end-to-end low latency robotic demonstrator.

Figure 18.9 Nokia and 5G Lab Germany connected cars demonstrator.

Figure 18.9, were controlled by the network and the mobile edge cloud to drive over an intersection without traffic lights. The audience could even interact with the demo and stop the cars at any moment and at any point of the track. However, thanks to the mobile edge cloud, it was assured that no accidents could happen. Both demos showed the main concepts and power of 5G.

18.4   5G as Game Changer in the Value Chain

In order to understand the revolutionary aspect of 5G for the business of network operators, we have to look at the history of the different generations as well as the current situation in the 4G systems. As illustrated in Figure 18.10, the 3G and previous architectures were dominated by three main players, namely, the cloud, the network, and the devices. With the introduction of the smartphones the device market rocketed, becoming even more important. A new market segment driven by apps development was born. Later on, the apps became more powerful and capable of many more functionalities with the support of cloud technology introducing the digital layer and over the top (OTT) application. The companies developing and designing their products for cloud and devices are the well-known big players such as Google, Facebook, Amazon, Samsung, and Apple. The network operators tried to take part in the development of both entities, but their core business was limited to transporting bits between the clouds and the devices, getting more and more pressure from the OTT players. So we reached the 4G system. But this will change with the introduction of 5G networks. The most interesting services, such as the low-latency communication, can only be realized by new ideas such as mobile edge cloud and network slicing that will happen inside the network, as given by Figure 18.10b. Both ideas can only happen within the network of the operators and cannot be offered by OTT. In other words, OTT is dead for real 5G use cases. This is the main motivation driving well-established cloud service players, such as Google or Facebook, to start their initiative to build up their own networks. The main message from Figure 18.10 is that before 5G, the innovation was mainly driven by the cloud and the devices while network operators had to provide suitable networks. With 5G, the network operators will gain some ground to drive the 5G use cases. Furthermore, we see that new architectures are supported. Beside the old-fashioned end-to-end communication paradigm, we see the mobile cloud communication [16], multipath communication toward devices, and IoT-based communication.

18.5   Conclusion

5G is a disruptive technology that does not follow the linear evolutionary innovation process in the history of 2G, 3G, and 4G. Differencing it from the predecessor generations, 5G targets the ubiquitous control and steering of IoT devices and cyber-physical systems rather than ubiquitous content communi-

(a)
4G and before

(b)
5G and beyond

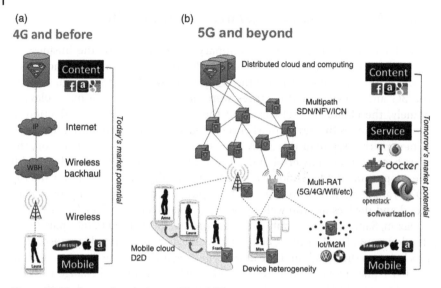

Figure 18.10 Comparison between 4G and 5G.

cation. For this reason, 5G is not only throughput driven, but it will also focus on latency, resilience, security, among others.

5G has already many defined possible use cases. Even though these might not be the final or the most successful use cases, for there are still many more to come; 5G, in comparison with previous generations, is a use case-driven communication system and not a technology-driven one. In order to realize this new system, we need new concepts and technologies like network coding and compressed sensing that break the traditional ideas such as the end-to-end paradigms or the deterministic thinking. With 5G, softwarization will play a major role. This is not only the enabler for new services, such as the mobile edge cloud, but it will also put an end to the generations as such. Instead of having updates every 10 years, we will see updates on a daily or weekly basis. Nevertheless, from the marketing perspective, 6G might be still intriguing. The position of the network operators in the markets will be dramatically strengthened in 5G, while the OTT services will be dead.

18.6 Acknowledgments

Thanks to all the partners who have contributed in the different projects of the 5G Lab Germany and to those who have helped to setup the 5G holistic testbed.

The authors would also like to acknowledge Juan A. Cabrera G., Giang Nguyen, Robert-Steve Schmoll, and Sreekrishna Pandi for their contributions to the projects and the text.

References

1 J. Chamber, *Beyond the hype: Internet of Things shows up strong at mobile world congress*, 2016. Available at http://www.pcworld.com .

2 J. Hansen, D. E. Lucani, J. Krigslund, M. Medard, and F. H. P. Fitzek, "Network coded software defined networking: enabling 5G transmission and storage networks," *IEEE Commun. Mag.*, vol. **53**, no. 9, pp. 100–107, 2015.

3 P. Baran, "On distributed communications networks," *IEEE Trans. Commun. Syst.*, vol. **12**, no. 1, pp. 1–9, 1964.

4 S. Cheshire, Bandwidth and latency: it's the latency, stupid (part 1), *TidBITS*, 1997. Available at https://tidbits.com/article/729.

5 F.H. Fitzek, and M. D. Katz, *Cognitive Wireless Networks: Concepts, Methodologies and Visions Inspiring the Age of Enlightenment of Wireless Communications*, Springer Science & Business Media, 2007.

6 ETSI Industry Specification Group (ISG), ETSI GS NFV 001 V1.1.1: Network Function Virtualisation: Use Cases, 2013.

7 R. Mijumbi, J. Serrat, J. L. Gorricho, N. Bouten, F. D. Turck, and R. Boutaba, "Network function virtualization: state-of-the-art and research challenges," *IEEE Commun. Surv. Tutor.*, vol. 18, NO. 1, PP. 236–262, 2016.

8 D. E. Lucani, M. V. Pedersen, J. Heide, and F. H. P. Fitzek, "Coping with the upcoming heterogeneity in 5G communications and storage using fulcrum network codes," in *2014 11th International Symposium on Wireless Communications Systems (ISWCS)*, 2014, pp. 997–1001.

9 U. Drolia, R. Martins, J. Tan, A. Chheda, M. Sanghavi, R. Gandhi, and P. Narasimhan, "The case for mobile edge-clouds," in *2013 IEEE 10th International Conference on Ubiquitous Intelligence and Computing and 10th International Conference on Autonomic and Trusted Computing (UIC/ATC)*, 2013, pp. 209–215.

10 Open Networking Foundation, SDN Architecture. Issue 1, 2014. Available at https://www.opennetworking.org/images/stories/downloads/sdn-resources/technical-reports/TR_SDN_ARCH_1.0_06062014.pdf.

11 R. Ahlswede, N. Cai, S.Y.R. Li, and R. W. Yeung, "Network information flow," *IEEE Trans. Inf. Theory*, vol. 46, no. 4, pp. 1204–1216, 2000.

12 M. Medard, F. Fitzek, M. J. Montpetit, and C. Rosenberg, "Network coding mythbusting: why it is not about butterflies anymore," *IEEE Commun. Mag.*, vol. 52, no. 7, pp. 177–183, 2014.

13 M. A. Maddah-Ali and U. Niesen, "Fundamental limits of caching," *IEEE Trans. Inf. Theory*, vol. 60, no. 5, pp. 2856–2867, 2014.

14 A. G. Dimakis, P. B. Godfrey, Y. Wu, M. J. Wainwright, and K. Ramchandran, "Network coding for distributed storage systems," *IEEE Trans. Inf. Theory*, vol. 56, no. 9, pp. 4539–4551, 2010.

15 F.H.P. Fitzek, T. Toth, A. Szabados, M. V. Pedersen, D. E. Lucani, M. Sipos, H. Charaf, and M. Medard, "Implementation and performance evaluation of distributed cloud storage solutions using random linear network coding," in *2014 IEEE International Conference on Communications Workshops (ICC)*, 2014, pp. 249–254.

16 F. H. Fitzek and M. D. Katz, *Mobile Clouds: Exploiting Distributed Resources in Wireless, Mobile and Social Networks*, Chichester, U.K.: John Wiley & Sons, Ltd, 2013.

Bruno Jacobfeuerborn has been CEO of DFMG Deutsche Funkturm GmbH since January 2017. In addition, he is CEO of Comfortcharge GmbH and Chairman of the Supervisory Board of 1NCE GmbH, both since January 2018. Previously, he served as Chief Technology Officer (CTO) of Deutsche Telekom AG from February 2012 to December 2017 and Director of Technology Telekom Deutschland GmbH from April 2010 to December 2016. Additionally, he was Chairman of the Next Generation Mobile Networks alliance (NGMN) as well as President of the Association for Electrical, Electronic and Information Technologies (VDE Verband der Elektrotechnik Elektronik Intormationstechnologie e.V.). Prior to that he was Director of Technology at T-Mobile Deutschland and T-Home in Germany from July 2009 to March 2010. In this double role, he was responsible for the technology business (both mobile and fixed network) in Germany. From April 2007 to July 2009, he was Managing Director of Technology, IT and Procurement at Polska Telefonica Cyfrowa. From 2002 to 2007, he was a member of the management team in his capacity as Managing Director of Technology, IT, Procurement and, until 2004, Customer Service at T-Mobile Netherlands in The Hague. Jacobfeuerborn joined what is now Deutsche Telekom AG in 1989 and has held several positions with increasing responsibility within the group.

Frank H. P. Fitzek is a professor and head of the Deutsche Telekom Chair of Communication Networks at the Technical University Dresden, Germany, coordinating the 5G Lab Germany. He received his diploma (Dipl.-Ing.) degree in electrical engineering from the University of Technology – Rheinisch-Westfalische Technische Hochschule ¤ (RWTH) – Aachen, Germany, in 1997 and his Ph.D. (Dr.-Ing.) in electrical engineering from the Technical University Berlin, Germany in 2002, and became Adjunct Professor at the University of Ferrara, Italy in the same year. In 2003, he joined Aalborg University as an associate professor and later became a professor. Dr. Fitzek cofounded several start-up companies starting with acticom GmbH in Berlin in 1999. He was selected to receive the Nokia Champion Award several times in a row from 2007 to 2011. In 2008, he was awarded the Nokia Achievement Award for his work on cooperative networks. In 2011, he received the SAPERE AUDE research grant from the Danish government and in 2012 he received the Vodafone Innovation prize. In 2015, he was awarded the honorary degree "Doctor Honoris Causa" from Budapest University of Technology and Economy (BUTE). His current research interests are in the areas of wireless and mobile 5G communication networks, mobile phone programming, network coding, cross layer, and energy-efficient protocol design and cooperative networking.

19

Toward All-IT 5G End-to-End Infrastructure

Alex Jinsung Choi,[1] Jinhyo Park,[2] Sungho Jo,[3] and Sangsoo Jeong[3]

[1] *Deutsche Telekom, T-Laboratories Innovation, Friedrich-Ebert-Allee 140, 53113 Bonn, Germany*
[2] *SK Telecom, ICT R&D Center, SK T-Tower, 65, Eulji-ro, Jung-gu, Seoul, Korea*
[3] *SK Telecom, Network Technology R&D Center, Hwangsaeul-ro, 258beon-gil, Bundang-gu, Seongnam-si, Gyeonggi-do, Korea*

19.1   Introduction

19.1.1   Background and Purpose

The innovation regarding the architecture of Telco infrastructure is being promoted to accommodate rapidly increasing mobile traffic in a cost-effective manner and to ensure flexible adoption of a variety of future services triggered by the ICT-based convergence trend and the 5G era [1–8]. Advances in IT technologies such as SDN/NFV-based virtualization technology and cloud computing have laid the basis for the innovation of the architecture of Telco infrastructure, and various reference models are being presented accordingly [9,10].

The utmost priorities of SK Telecom's infrastructure outlined in this chapter are "Scalable, Cognitive, Automated, Lean and End-to-End." The ATSCALE architecture with its key functions and implementation technologies, which allow us to achieve the stated goals are presented as well. Through this, we plan to present anticipated benefits as models of open and cooperation with ICT ecosystem. The aforementioned benefits include TCO (total cost of ownership) efficiency/operational intelligence and business innovation that allow us to quickly offer customized product/service based on an open platform.

5G Networks: Fundamental Requirements, Enabling Technologies, and Operations Management, First Edition.
Anwer Al-Dulaimi, Xianbin Wang, and Chih-Lin I.
© 2018 by The Institute of Electrical and Electronics Engineers, Inc. Published 2018 by John Wiley & Sons, Inc.

19.1.2 Evolution Trend of Telco Infrastructure

19.1.2.1 Telco Infrastructure Virtualization

The virtualization of Telco infrastructure has been expanded beyond the traditional data center to encompass even the areas of core and access network [11,12]. ETSI provides an explanation of an architecture that enables to implement a network function on a virtualized server platform. Several Telcos including AT&T and Verizon have announced their plans to shift their Telco infrastructure and services to cloud services in order to ensure efficient operations and business transformation.

19.1.2.2 Open Software and Hardware

The application scope of both open-source hardware and software technologies is being expanded gradually across the Telco infrastructure. OpenStack, an open-source project, mainly led by developers in its early days, has evolved as a representative community to develop an operating system for cloud computing services backed by the involvement of companies from a diverse range of business sectors [13]. Furthermore, an ecosystem centered around such open-source communities such as OCP (Open Compute Project) and Open ROADM MSA is being actively created, and Telcos including SK Telecom along with vendors have formed a Telecom Infrastructure Project (TIP) to share technologies for the Telco infrastructure innovation and to develop open hardware solutions [14–16].

19.1.2.3 Evolution into Platform to Allow "As-a-Service"

Future networks are expected to evolve into a programmable platform that allows modifying freely network structure/function according to a specific user/service class. Well-known cases of this include eDECOR of 3GPP, Mobile Edge Computing (MEC) of ETSI, M-CORD of ON.Lab, Network Slicing of Ericsson, and Ultra Service Platform of Cisco [17–20].

19.1.2.4 Intelligence and Operation Efficiency

The efforts that have been made by telecom service providers to transform themselves into digital service providers are being materialized through machine learning and big data analytics technology based on a virtualized infrastructure. In order to support agile services and dynamics that have been enabled by a programmable platform, an OSS (operation supporting system) is expected to evolve into an intelligent and automated operation from an end-to-end perspective. A TM forum that has been organized by Telcos and vendors to drive the standardization of BSS/OSS is currently carrying out studies and standardization. Ways to ensure an effective management in OSS triggered by the introduction of NFV/SDN and hybrid networks have been shaped through ZOOM and Catalyst Project [21,22].

19.1.3   SK Telecom's Perspective on NFV/SDN

The technology related contents described in this chapter are based on NFV/SDN, and the ATSCALE architecture that will be covered later is also designed on the basis of their paradigm.

First, NFV is a technology to implement network functions in the form of hardware-independent, software-based virtual appliances and run them on any infrastructure instead of conventional telecommunication equipment (both hardware and software) provided by vendors through separation of software from hardware [23].

The transition from hardware-centric, closed, and dedicated network equipment to software-centric, open, scalable, and programmable network functions that allow easy installation and modification will enable to realize efficient operation and rapid service deployment while serving as an underlying technology to implement a cost-effective network through a pooling of general-purpose/open hardware and resources, and flexible operation.

SDN is another enabling technology of ATSCALE that separates network node by its function and role and provides an open interface to enable communication between network nodes [24]. SDN classifies network functions and roles by plane (either control or user plane) and offers communications between planes through an open interface, and thereby allows users to purchase each function separately while offering an optimal configuration (centralized control plane and decentralized user plane), a cost-effective extension, and a flexible functional change (programmability).

Moving further beyond the traditional SDN that splits user and control plane functions, there are four planes that compose a Telco function: (i) a radio plane, (ii) a user plane (data transmission), (iii) a control plane (data delivery control logic and policy), and (iv) a data management plane. The ATSCALE Architecture outlined later is basically formed on the basis of the aforementioned four planes of SDN architecture, and the structural details by domain are described in great detail later in this chapter.

19.2   Development Status and Lesson Learned

This section will review PoC (proof-of-concept) projects that SK Telecom has promoted in each network domain for the innovation/evolution of Telcos network architecture as well as commercialization status, and aims to suggest the future direction of Telcos and the issues that are deemed necessary for them to consider while heading into such direction.

Performance monitoring

Over-the air transmission in a shield box

vRAN server

Figure 19.1 Mobile World Congress 2014 Demonstration.

19.2.1 Radio Access Network

19.2.1.1 RAN Virtualization

SK Telecom has completed a PoC to run L1/L2 modem functions conducted by the existing base stations on a GPP server as software and demonstrated the project at the MWC 2014. We were able to achieve a data transfer rate of 300 Mbps by using a 1 ms TTI for real-time data processing function and 4×4 MIMO technology in an LTE cell with 20 MHz bandwidth (February 2014).

We were also able to improve the performance of vRAN server by performing a physical layer (PHY) functionality that involves a complicated process on RU through L1–L2 function split and have applied an Ethernet-based fronthaul network instead of the existing CPRI/OBSAI. Moreover, we have launched a PoC project to set up an integrated end-to-end virtualization system that is connected to a virtualized EPC (vEPC). These efforts have resulted in considerable improvements of operational efficiency to deliver data rates of 300 Mbps based on LTE-A 2CC CA, along with scale-in/scale-out of cells based on virtual machine (VM) scaling and live migration functions, and so on (November 2014).

We carried out a field test in a multicell/multi-UE-based mobile environment using commercial smartphones and confirmed that both access success rate and handover success rate reached more than 99% and also verified a technology to strengthen risk management via autorecovery (November 2015).

19.2.1.2 Mobile Edge Service (MEC)

A PoC to develop a base-station-based edge service platform to offer services such as local contents, AR, and advertisements was conducted to verify a reduction in both required backhaul capacity and packet delivery latency, as well as the provisioning of base station cell ID-based service functions (September 2013).

Figure 19.2 Lab & Field Trial, 2014 2015.

Key Considerations

- Needs to redesign VNF in modular form on an unbundled basis.
- Needs to secure performance, stability, and capacity under commercial conditions.
- Needs to reinforce differentiated functions such as autorecovery and live migration, and so on.
- One-stop management through end-to-end network orchestrator interconnection.

SK Telecom developed and commercialized a local breakout-based private LTE technology that supports designing of an enterprise-specific network. The private LTE technology offers a high level of security for networks while making it easier to manage network traffic by building an independent network through an enterprise-specific traffic routing function (December 2015).

19.2.2 Core Network

In order to improve both flexibility and agility of a core network, several existing Telco's central offices have been partially reorganized as a data center, and an NFV technology is being applied to some core network functions in a gradual phase.

19.2.2.1 Virtualized EPC/IMS Commercialization
We have been pushing ahead with the virtualization of EPC and IMS equipment in packet core in order to handle network traffic in a cost-effective manner and to meet increasing demand in a timely way, and introduced the relevant commercial service in 2015.

The company has commercialized virtualized mobility management entity (MME) supporting up to 1 million subscribers and virtualized S/P-GW serving 1 million data session for the Internet of Things (IoT) services (August 2015). In order to cope with the rising demand for IMS services (VoLTE), SK Telecom has commercialized a virtualized CSCF (Call Session Control Function) that supports up to 10 million subscribers (September 2015), and is currently promoting the commercialization of virtualized AGW covering 4 million subscribers (November 2015) and TAS serving 2 million subscribers (February 2016) on a step-by-step basis.

19.2.2.2 NFV MANO (Management and Orchestration) Commercialization

Due to lack of standards for interoperability between VNF, VIM, and VNFM of different vendors in the ETSI NFV, SK Telecom has materialized interworking standards up to the level that allows MANO requirements, standards, and data model skeleton to be implemented and has commercialized NFVO. Heterogeneous VNFM and VIM of different vendors are being operated within the current network. At the early stage of NFV adoption, virtual infrastructure was isolated by VNF and then managed, and NFVO was also divided by operation office in order to accommodate each operation office's specific requirements.

19.2.2.3 Service Orchestration PoC

SK Telecom has noticed that the network slicing that has been discussed previously mainly concentrates on the implementation of network infrastructure per business unit such as mobile broadband and IoT that are somewhat macroscopic and relatively static. We have defined service slicing as a virtualization-based, dynamic customized network segment that is optimally designed to satisfy product/service requirements, and also developed a service orchestration technology that oversees the design, implementation, and life cycle of service slice while offering functions of distributing and establishing policies for network products and services.

19.2.2.4 SDN-Based vEPC PoC

SK Telecom has completed an SDN-based vEPC PoC that manages traffic paths in a core network using a centralized control plane – a method to split the control plane handling mobility and session and user plane responsible for data packet processing (October 2015) – and functions of charging in dispersed UP nodes and delivering QoS are currently under development.

19.2.3 Transport Network

SK Telecom is currently in the process of developing T-SDN platform to offer a software-based unified control and management of a multi-vendor/multilayer transport network, and this platform is expected to serve a wide range of equip-

---

Key Considerations

- Needs to redesign and implement a virtual network function (VNF) by taking into account a virtual environment.
- Early commercialization of vEPC, vIMS, and NFVO in a situation where standards are yet to be confirmed, can make it difficult to link them to each other.
- Needs to develop an orchestrator for operational intelligence.
- Although virtualization has led to hardware cost-savings, such cost-saving efforts should be continued through an open-hardware and open-software-based implementation, and automated operation to achieve an optimized TCO.
- Given the fact that the current orchestration technology supports up to the level of establishment and operation of a virtual network infrastructure, an evolution of orchestration technology from the view of service/product implementation is necessary.

---

ment ranging from legacy ROADM and OTN to even transport equipment to be adopted in the future.

19.2.3.1 Unified Control Function of ROADM/OTN on Commercial Network

The first phase of T-SDN platform has been commercialized in October 2015 that offers functions such as autodiscovery, management, design, and carrier application for legacy ROADM(L0)/OTN(L1) network to enhance operational efficiency and creation/deletion of OTN services, and has now moved into the second phase of developing control function of IP equipment (L3) in the LTE backhaul network. We are delivering both engineering and operation convenience functions to obtain network resource efficiency. These functions offer unified management for both OTN and ROADM.

19.2.3.2 Common Hardware Platform-Based POTN

We have developed common hardware platform-based POTN (Packet Optical Transport Network) equipment and succeeded in testing its interoperability with vendors existing PTN in January 2016. It can be used by embedding network OS of each vendor on POTN equipment with the same common hardware platform.

19.2.3.3 PTN/POTN Unified Control in Multi-Vendor Environment

We have succeeded in developing end-to-end, leased line service provisioning function through a PoC for unified control of multi-vendor PTN/POTN equipment using T-SDN platform. By applying south-bound plug-in software

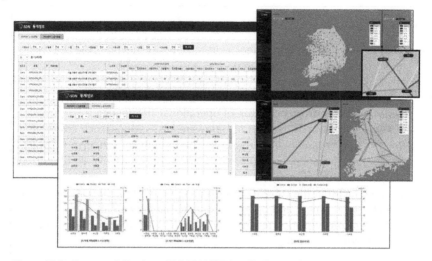

Figure 19.3 Commercialization of ROADM/OTN unified control.

Key Considerations

- Needs to redesign and implement a virtual network function (VNF) by taking into account a virtual environment.
- Open standard for network element data model and interface.
- Hardware disaggregation of transport equipment and extended use of open technologies are required [24].

---

modules developed by each vendor to OpenDaylight-based T-SDN platform, we have succeeded in testing its interoperability between this south-bound plug-in and PTN topology manager, inventory manger, and service manager in T-SDN platform and have also verified bandwidth-on-demand and bandwidth calendaring service feasibility in multi-vendor PTN/POTN network.

19.2.4 M-CORD

SK Telecom has introduced an M-CORD (Mobile-Central Office rearchitected as a data center) in collaboration with AT&T and Verizon, the members of ON.Lab, and relevant manufacturers, and demonstrated the PoC at the ONS 2016, which was held in March 2016. M-CORD is a platform that integrates disaggregated networks and by being applied to edge of mobile network, it enables to offer customized and personalized services.

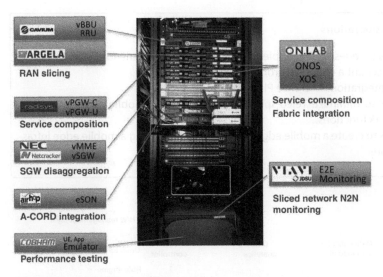

Figure 19.4 M-CORD Proto-type.

19.2.4.1 Integration of SDN/NFV Technology

M-CORD allows redesigning (or transformation) of a mobile communications service infrastructure into a data center architecture and based upon such rearchitected data center. It aims to deliver services that meet customer requirements in a cost-effective manner through speedy and economic establishment and deployment of virtualized network functions wherever and whenever it is needed. To achieve its goals with a specific use case, mobile edge computing, M-CORD adopts an SDN-based network and utilizes commercial off-the-shelf (COTS) hardware-based virtualized network functions.

19.2.4.2 RAN Virtualization/Disaggregation

By making DU functions (L2, L3, and application), to be handled in a general-purpose hardware-based, and using a disaggregated RAN function and virtualized BBU to process L1 and some portion of L2 in RRH, we were able to achieve better scalability and resource utilization. With the disaggregation of RAN function, Ethernet fronthaul has also become possible. Through a centralized coordination, the spectrum efficiency has been enhanced highly while the overhead of mobility management dropped.

19.2.4.3 EPC Virtualization/Disaggregation

We have built both MME and S/P-GW to be performed in general-purpose hardware by using a virtualized VM. Moreover, we have split the control plane and user plane of S/P-GW and used an SDN OpenFlow to allow the user plane of P-GW, in particular, to communicate with the control plane.

Key Considerations

- Needs to redesign and implement a virtual network function (VNF) by taking into account a virtual environment.
- NFV Integration Reference Platform is required.
- Needs to minimize Telco-specific requirements in mobile communications network functions.
- Needs to create a mobile edge service ecosystem using a mobile edge infrastructure.

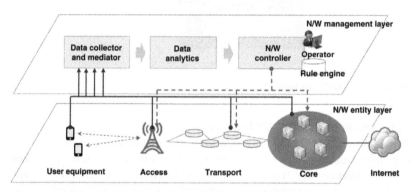

Figure 19.5 End-to-end operational intelligence.

19.2.4.4 Mobile Edge Services

M-CORD platform enables to create RAN/EPC components and applications in the form of "as-a-Service." We have demonstrated a video cache service that provides an improved QoE with a low latency in a nearby location from a customer and high-throughput by creating edge services based on a service platform located in a mobile edge.

19.2.5 Operational Intelligence

SK Telecom has defined OAM data for an intelligent/automated network management and is currently implementing a real-time monitoring of network quality and managing of customer experience. These are achieved by conducting a correlation analysis on the data collected from an end-to-end perspective. In addition, by detecting abnormal conditions of the applications and devices used by a customer, we are able to prevent the problems and respond to such issues promptly based on automated processes.

19.2.5.1   Intelligence for Network Big Data Collection/Analytics

First, APOLLO (Analytics Platform for Intelligent Operation) collects data related to wireless environment from a base station in real time, and conducts real-time management on areas with deteriorated quality. APOLLO carries out real-time anomaly detection, optimization, and proactive measures. We have employed a machine learning approach to learn network management indicators and to detect abnormalities and made the verified use case as a rule to be applied in real-time monitoring and analytics processes.

DoD (diagnostic on device) collects data related to wireless quality environments and abnormal events calculated by a users device and manages the areas where optimization is required. Network data that have been collected based upon such purpose is then analyzed by an open-source based (e.g., Flume/Kafka, Spark, and Hadoop) big data system.

T-PANI (T-Packet Analysis & Network Intelligence) receives customer's service traffic-related data from a core network and analyzes quality of customer services on an individual customer basis, and thus manages key indicators of the network management. For instance, in case the distribution of wireless environment indicators shows a meaningful difference compared to other cells and when such event is detected as an abnormal RU activity, it is possible to take measures automatically. Network optimization and investment efficiency based on the customer experience index (CEI) of a device can be achieved by analyzing traffic data flowing in the core network and monitoring the areas or applications or devices where abnormal events occurred.

19.2.5.2   Telco-Defined Network Management Indicators

We have introduced our own definition of network management index (CEI), which allows us to carry out a total customer experience quality management instead of using quality indicators provided by vendors, and through this we are promoting network improvement activities. In addition, various data linkage analysis environments are provided, which enable to easily perform a correlation analysis with the CEI by utilizing a data warehouse.

19.2.5.3   Big Data-Based Automated Operation

All of the collected and analyzed data are utilized across all areas covering network engineering, construction, and operation. Especially in the field of operation, an automated control is implemented based upon the rule optimized for a particular operator's operational environment. Moreover, wireless quality is measured automatically through which we can minimize operational expenses and improve the accuracy of measurement.

19.2.5.4   Monitoring and Management of Virtual Resources

An O&M that monitors the status of virtualized network functions and virtual resources is performed by an NFVO at base station levels and an O&M to

---

Key Considerations

- Needs to redesign and implement a virtual network function (VNF) by taking into account a virtual environment.
- Openness and standardization are required to carry out a Telco-driven data collection/analytics/control that is cost-effective.
- An environment to allow operators to directly conduct analysis/verification is essential, and thus delivering automated operation based on the rules and use cases optimized for an operator.
- A function that manages hybrid resources from an end-to-end perspective is essential in a hybrid network environment with a mixture of legacy PNF and VNF.

---

monitor the service status of each VNF is managed by an NMS (network management system) through legacy EMS (equipment management system).

19.3 Infrastructure Evolution of SK Telecom for 5G: ATSCALE

This section will present the evolution direction of network infrastructure architecture that a Telco needs to achieve in the future based on the SK Telecom's enriched experiences in technology developments and commercialization and key considerations. This section also proposes the ATSCALE architecture that SK Telecom aims to accomplish.

19.3.1 Evolution Direction

We have set the future evolution direction of Telco infrastructure mainly focusing on "scalable, cognitive, automated, lean, and end-to-end."

The meaning and objective of five evolution directions are as follows.

19.3.1.1 Scalable

Aims to achieve an architecture that delivers on-demand agility and adoptive operation. This improves the business efficiency in costs and operations, and is able to constantly respond to the ever-growing customer/service needs, volume, diversities, and new use cases through the efficient use of resources.

19.3.1.2 Cognitive

Aims to achieve an architecture that realizes closed-loop process where the "context cognition-decision–action-improvement" is conducted. Based on this,

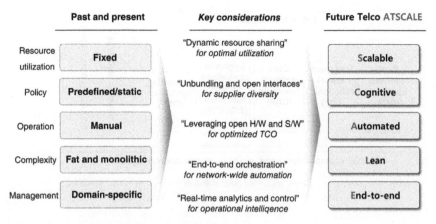

Figure 19.6 Evolution direction – ATSCALE.

visibility of end-to-end network/service as well as network optimization, automated operation, and intelligent service can be offered.

19.3.1.3 Automated

Aims to achieve an architecture that efficiently performs infrastructure deployment, network configurations, modifications, rebundling/extension of network functions, and network/service operation. This allows an operational intelligence-based automation, and thus establishes an infrastructure that allows offering new businesses and services.

19.3.1.4 Lean

Aims to maintain the TCO structure lean by using open software and hardware based technologies, and realizes a network that can be easily managed.

19.3.1.5 End-to-End

Aims to establish an architecture that allows automation and optimization at the entire network in service level while avoiding segregated/disconnected operation.

19.3.2 Telco Functions on COSMOS

Before designing the overall ATSCALE architecture, we have defined the infrastructure and resource layer on which Telco networks and IT functions run as COSMOS (composable, open, scalable, mobile-oriented system) [3].

COSMOS allows a Telco to deliver all of its services under a single, common, open, and programmable infrastructure environment based on software-defined technologies. The two building blocks of a Telco infrastructure – dis-

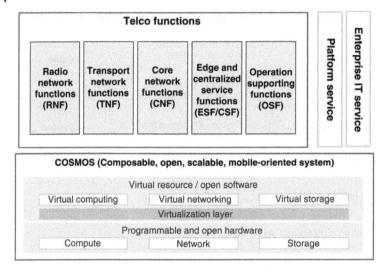

Figure 19.7 Telco and IT functions running on COSMOS.

Table 19.1 Telco functions and roles.

Telco Functions	Functions and Roles
RNF (radio network function)	A function that performs resource management, policy control, and packet management in order to provide Telco services in RAN domain
CNF (core network function)	A function that offers features ranging from mobility management to authentication, session, policy management, subscriber DB, and voice/message/data service
TNF (transport network function)	L3–L7 transport function including router, switch, and load balancer
ESF (edge service function)	Value-added service functions that need to be delivered from mobile edge
CSF (centralized service function)	In-line, proxy network functions that offer network-based value-added services
OSF (operation supporting function)	A function that offers operation and management and OSS interworking

tributed edge cloud (edge data center) and centralized cloud (centralized data center)– are built based on COSMOS, and the functions and roles that run on these two building blocks are described in the following sections.

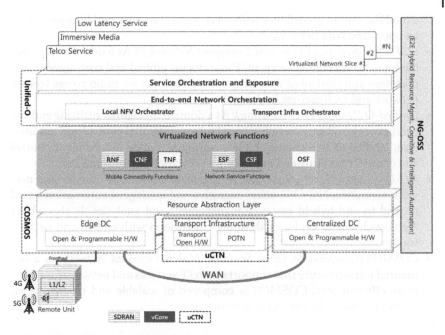

Figure 19.8 ATSCALE End-to-End Architecture

19.3.3 ATSCALE Architecture

Based on the aforementioned five evolution directions, we have defined the mandatory technology domains that allow provisioning of the existing Telco services as well as new future services: software-defined RAN (SDRAN), virtualized core (vCore), unified and converged transport network (uCTN), unified orchestration (Unified-O), NG-OSS, and COSMOS.

Explanation on each of the six technology domains is given below:

- **SDRAN (Software-defined RAN):** As a domain providing wireless connections to offer mobile communications services to customers and managing features including allocation, cancellation, and scheduling of wireless resources, SDRAN supports provisioning of edge services and operation through network virtualization based on open hardware, software, and interfaces.
- **vCore (Virtualized Core):** As a domain offering the fundamental features of mobile telecommunications services, namely, authentication, session management, mobility management, charging control, and value-added services, vCore allows unbundling (or decoupling) and redesigning traditional complex architecture by modular (or unit) function (control/user plane) to achieve simplification and rebundling/redeploying the functions divided into

a modular unit in accordance with service requirements, and thus delivering an optimized core network.

- *uCTN (Unified and Converged Transport Network)P:* As a domain responsible for network connectivity, uCTN performs unified and converged transportation and control of data across the distance from radio access to core network and IX (Internet Exchange).

- *Unified-O (Unified Orchestration):* As a domain that allows realizing end-to-end network and service agility based on an integrated control and automation, Unified-O is responsible for ensuring a vendor-independent control and consistent end-to-end policy implementation and automation.

- *NG-OSS (Next-Generation OSS):* As a domain dedicated to service and network assurance, NG-OSS offers cognitive and intelligent automation based on end-to-end hybrid resource management and analytics. It also allows zero-touch operation by establishing closed-loop with end-to-end orchestration.

- *COSMOS (Composable, Open, Scalable, Mobile-Oriented System):* As a virtualized infrastructure that supports both IT services and network functions in an efficient way, COSMOS is composed of scalable and efficient open hardware and open software components.

19.4   Detailed Architecture and Key Enabling Technology

This section will present architectural details with regard to the domains of the aforementioned ATSCALE, namely, SDRAN, vCore, uCTN, Unified-O, and NG-OSS and describe each one of them in detail from the viewpoint of implementation technology.

19.4.1   Software-Defined RAN

SDRAN aims for an open architecture based on both open hardware and software, and an open interface while offering various frequency bands in a flexible way. A technology to achieve the stated functions is defined as FCOMMA (fronthaul enhancement, CP/UP separation, open architecture, MEC and M-CORD, analytics (SON) agent).

19.4.1.1   Fronthaul Enhancement

The evolution direction of SDRAN fronthaul is to achieve bandwidth savings, flexibility improvement, and interface openness, and a structural innovation that allows designing function split of DU (digital unit) and RU (radio unit) is necessary to obtain the stated objectives.

The function split here refers to moving certain part of the digital processing functions that were exclusively handled in DU to RU, and there are three dif-

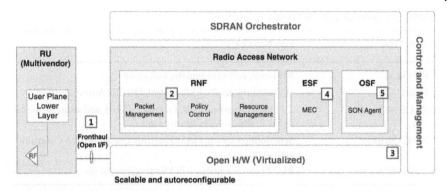

Figure 19.9 SDRAN architecture.

ferent split options depending on the borders subject to separation: (i) RF/PHY split, (ii) PHY/MAC split, and (iii) RLC/PDCP split. With the aim of evolving into a Telco DC, SDRAN has adopted the RLC/PDCP split method as a reference architecture enabling to connect DU cloud and RU of each site without any restriction in distance while supporting flexible structures. Based on the RLC/PDCP split method, SDRAN needs to support a dual connectivity function so that it can be adopted to multi-RAT (radio access technology) environments in the future such as 4G–5G Interworking.

A new interface standard is required in order to ensure a new connection between DU and RU based on a function split. A new fronthaul interface needs to adopt the Ethernet Protocol, and it is necessary to establish standard specification for an open interface.

- *O&M Interface*: Offers the O&M functions of RU such as turning on or off of various functions built in RU or changing activation parameters of the function concerned, monitoring of alarm and status of RU, and authentication and security RU.
- *Data Interface*: Allows transmitiing downlink signaling and user data that have been created in a core network and DU to RU, and forwarding the uplink signaling and user data received from UE to DU.

19.4.1.2   CP/UP Separation

SDRAN is built in a way to ensure its connectivity to a commodity UP hardware by controlling the CP function defined by a Telco based on the CP and UP separation. In addition, CP generates a policy optimized for the current network based on the information provided by a resource management function.

Three components that are needed for the separation of CP and UP in the SDRAN as well as their characteristics are as follows.

Table 19.2 Summary of O&M type and function.

O&M Type	Function List
RU O&M	RU Reset, software update, and so on
Link O&M	DU/RU Link Fault (BER, Optic), DU/RU Authentication
RF O&M	Tx Power, Antenna Path, Frequency Band, and so on RF Signal (VSWR, RSSI), RF hardware fault
MAC O&M	Scheduling Option, HARQ Setting, CA Activation Condition, and so on
RLC O&M	RLC Mode Selection, ARQ Setting, and so on

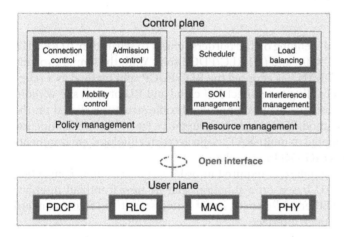

Figure 19.10 CP/UP separation.

- *Virtualized CP Function:* A function that defines a differentiated control policy in accordance with a Telcos policy and operational environment based on resource management functions such as scheduling and load balancing.
- *Standardized UP Hardware:* An UP data processing defined by the standards in use can be implemented in a cost-effective manner based on COTS hardware.
- *Open Interface:* Offers a definition of an open API that allows a smooth connection between Telco-specific CP functions and common UP hardware.

19.4.1.3 Open Hardware and Software

With the virtualization of Telco functions, decoupled hardware is expected to evolve into a common infrastructure that is irrelevant to network functions or vendors.

SDRAN is set up based on COSMOS, a common, standards-based infrastructure. Also it defines a separate function and performance to design a hardware architecture that is best optimized for Telco functions, and thereby promotes a global alliance (e.g., OCP and TIP) for the expansion of ecosystem and maximization of openness. As SDRAN is an NFV-based platform, a virtual infrastructure is deployed based on an open-source software.

19.4.1.4 MEC

As a Telco infrastructure is becoming standardized and moving toward a cloud-based data center, the hardware borders between access and core networks have been blurred that allowed the access network to effectively offer the services that were provided by the core network before. It also allows developing various specialized services centered around four areas, including mission critical, local-specific service, QoS enhancement, and traffic offloading, via an Open API.

MEC is required to use a standardized platform as it needs to allow third party applications to be built on it. SDRAN needs to be aligned with global standards such as ETSI MEC ISG while supporting an Open API to offer the services.

19.4.1.5 Analytics (SON) Agent

SON agent in SDRAN collects status information on a radio environment and radio resource as well as real-time traffic data to adjust scheduling scheme (e.g., Scheduling Algorithm and Weighting Factor) on a per-user basis and modify network parameters (e.g., CA Enable and Event Trigger Condition).

As the SON agent processes the collected data into a form agreed upon with OSS and then transmits the data, it enables to achieve the overall system optimization. For this, it needs to collect the information including network, radio, traffic, and fault status.

19.4.2 Virtualized Core (vCore)

vCore simplifies the existing complex core network and devices by function and role on a software modular basis with the aim of delivering a simplified architecture, and has allowed to minimize redundancy by integrating similar functions. For this, we have separated and simplified the built-in user plane specialized for each existing device and made it a common function (simple user plane), whereas the control plane has been rearchitected by each control function (e.g., mobility/session/ and QoS management) vCore also includes centralized service functions (CSF) and operation supporting functions (OSF). CSFs allow offering an improved customer experience (B2C/B2B) and third-party service experience, while OSFs collect data on operation, management, performance, conditions and faults, and so on.

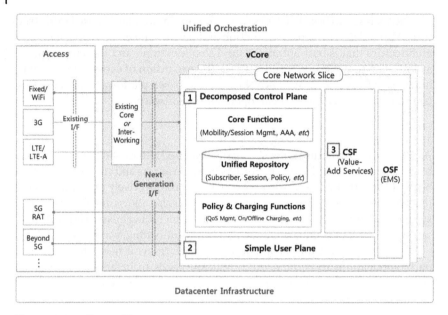

Figure 19.11 vCore architecture.

19.4.2.1 Decomposed Control Plane

In decomposed control plane, each function unit as a module (atomic service or microservice) is virtualized through a unified repository-based stateless architecture, thus allowing to select, combine, and recycle (or reuse) necessary functions and then put the right function in the right place on an on-demand basis. As a result, it allows dynamically creating networks that meet the diverse functional and performance requirements expected to arise from future networks.

- *Virtualized and Decomposed Control Plane Functions:* Avoids a complex VNF architecture that is comprised of a combination of VNF components, and performs a micro or atomic function on a VM basis through modularization. Based on a unified repository of distributed DB architecture, each micro or atomic function is built with stateless control plane architecture, and thereby ensures the consistency and availability of information for sessions and conditions.
- *Recomposed Control Plane:* Allows selecting only necessary functions and deploying them in the optimal location. Control plane unit functions as well as their detailed configurations are defined in the form of descriptor based on open standard data models.
- *Next-Generation Protocol and Interface:* Allows the evolution of core network functions as independently as possible and also offers a simplified in-

terworking architecture that allows seamless mobility and traffic aggregation between heterogeneous networks.

19.4.2.2 Simple User Plane

Simple user plane is a software function that has laid the foundation for data transmission in core networks. Simple user plane operates based on service policy control and the data paths set up by the control plane.

- *User Plane Programmability:* With the implementation of an open and programmable interface (south-bound interface) between control and user planes, User Plane Programmability function enables to establish the data transmission path of simple user plane and service policy control.
- *Core Network User Plane:* Core Network User Plane performs the user plane functions of S-GW/P-GW/TDF in 3GPP core network, namely, GTP encapsulation/decapsulation, packet forwarding, packet marking, downlink data notification, accounting, rating, and QoS enforcement function. In addition, it offers traffic classification, detection, and steering functions to enable service-function chaining.
- *Provisioning of Unlimited Capacity and Scalability:* Without being restricted by traditional unit-based capacity engineering and capacity limit, it offers a capacity expansion that is a linearly increasing function and serves as an all-active high-availability function through a scale-out at VM level

19.4.2.3 Centralized Service Functions (CSF)

In order to offer a "specialized function, differentiated quality, and reliable security level" optimized for customer's needs and product/service requirements in a network, it needs to be evolved into an architecture that supports data service delivery best suited for a customer's real-time situation and needs by being linked to a centralized policy server.

In order to keep the vCore architecture lean, we have softwarized the VAS that allows installing and realizing the operation of necessary service functions in on-demand fashion based on the plug-in architecture. It also supports 3GPP FMSS (flexible mobile service steering) and IETF SFC (service function chaining) functions [25,26].

19.4.3 Unified and Converged Transport Network (uCTN)

uCTN is a domain that provides network connectivity to enhance the effectiveness of network resource utilization by integrating multi-vendor and multilayer network control and management functions and delivers TNaaS (transport network-as-a-service).

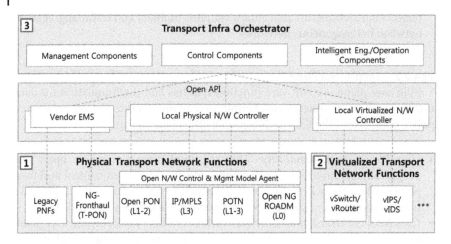

Figure 19.12 uCTN architecture.

19.4.3.1 Transport Physical Network Functions

Transport network will be simplified by migrating legacy transport physical network functions into multilayer (Layer 0-3) converged network functions.

- *POTN:* POTN, multilayer (Layer 0-3) network function, converging a packet function and an optical function and multiservices configured in each layer.
- *Open NG-ROADM:* Open NG-ROADM is open hardware based upon CDC (colorless, directionless, contention-less) technology. CDC allows it to accommodate beyond 200 Gbps optical signals and to flexibly bypass, add, and drop wavelengths in multidirections, while Open NG ROADM reduces the hardware dependency. This includes flexible photonics wherein transmission capacity and spectrum efficiency can be maximized by dynamically controlling flexible photonics considering the crosstalk on an optical signal and transmission distance.
- *NG-Fronthaul (T-PON):* With the adoption of an optical fiber-based L0 transparent architecture that does not rely on a specific protocol, NG-Fronthaul presents an architecture associated with combination of PON technology to make last-mile device into a passive device, and Wavelength-Tunable Optical Module Technology to support automatic wavelength selection method, and thus integrally accommodates wired and wireless access networks such as LTE and 5G.
- *Open PON:* Open PON allows achieving TCO reduction by converting the parts that are related to PHY performance or real-time functions into white box (pOLT), and enables setting up a flexible network in which control/management and value-added service functions can be created/deleted according to customer requirements. pOLT contains PON MAC/PHY Func-

tion, I/O Optical Interface, SDN Controller Agent, and allows using low-cost CPUs or removing L3 switches. Management applications implement management functions of network devices and network-related services on a cloud PoP in the form of a VM. Optical access network controller enables implementing a virtualized control function to handle non-real-time traffic forwarding.

- *IP/MPLS:* IP/MPLS supports OpenConfig/IETF L3 network model as well as Open APIs, including NETCONF, PCEP, and gRPC, and thereby implements a vendor-independent integrated control and management functions.
- *Open Network Control and Management Model:* Transport Infrastructure Orchestrator allows performing an integrated control of multi-vendor network equipment based on standardized control protocols (e.g., OpenFlow, NETCONF/YANG, and PCEP), and uses a common management data model to ensure the operation and management of open hardware without EMS (Element Management System).

19.4.3.2 Virtualized Transport Network Functions
Virtualized transport network functions support virtualized L7 function to provide value-added network services and with network service functions can be offered.

19.4.3.3 Transport Infrastructure Orchestrator
Transport Infrastructure Orchestrator offers a range of functions: a resource/path control engine, carrier applications to achieve network operation, and a south-bound plug-in function. The orchestrator would be based on Open-Source Projects (e.g., OpenDaylight).

- *Management Components:* A range of functions including real-time automatic information gathering, service management, quality management, and traffic analysis.
- *Control Components:* Responsible for processing unified control of transport equipment to offer functions such as life cycle management of end-to-end circuit or service, fault restoration, and QoS control by mobile media and voice service.
- *Intelligent Engineering/Operation Components:* Allow multilayer/domain path computation based on multi-path/node considerations, multilayer resource and path optimization, and provisioning of customized path to each specific network service.
- *Backhaul/Fronthaul Unified Control Architecture:* Realize an architecture where the transformation into a NaaS platform can be achieved through which agile enterprise network services (L0-L3 VPN) can easily be provided.

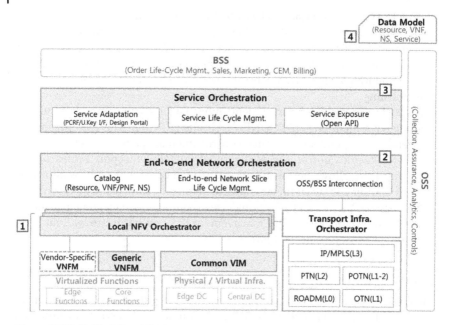

Figure 19.13 Unified-O architecture.

19.4.4 Unified Orchestration (Unified-O)

Unified-O controls and manages the entire life cycle of infrastructure resources, function, end-to-end network (network slice), and service (service slice). Unified-O empowers operators to minimize vendor dependency through orchestration technology standardization, while delivering end-to-end network automation and operation efficiency by integrating the silo-based orchestration process. To ensure the delivery of such benefits, SK Telecoms Unified Orchestration Architecture is composed of the following main building blocks.

19.4.4.1 Standardized NFV MANO Framework

NFV MANO (Management and Orchestration) manages virtual environment that contains heterogeneous/different physical/virtual resources and virtual network functions from multi-vendors. Although ETSI NFV has defined the interworking interfaces for NFV MANO, the lack of standards in the context of concrete realization has the potential to drive up operation cost and interworking complexity.

Standardized NFV MANO framework consists of the following building blocks: Local NFV Orchestrator (with end-to-end network orchestrator) that enables hierarchical orchestration structure to achieve flexibility and scalability that allows each operation office to achieve operational separation and policy

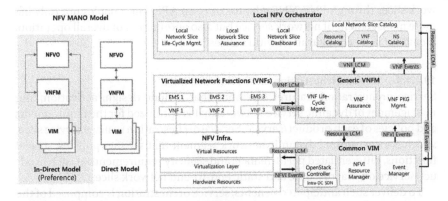

Figure 19.14 Standardized NFV MANO framework.

differentiation, generic VNFM that efficiently accommodates multi-vendor and multi-VNF, and common VIM.

- *Local NFV Orchestrator:* As an NFV orchestrator that is deployed locally in DC or DC cluster unit, Local NFV Orchestrator designs and manages virtual local network slices. Local NFV Orchestrator is composed of the following: (i) Local network slice life cycle management, (ii) local network slice assurance, (iii) local network slice dashboard, and (iv) local network slice catalogue.
- *Generic VNFM:* Performs VNF Life Cycle management functions. Although executed as a vendor-specific VNFM during the adoption phase of NFV technology, there is a need for the realization of Generic VNFM that reduces interworking complexity, enhances operation efficiency, and accommodates a wide range of VNFs. Generic VNFM is composed of the following: (i) VNF life cycle management, (ii) VNF assurance, and (iii) VNF package management.
- *Common VIM:* Manages NFV infrastructure that provides execution environment for VNF, without being bound by the type of physical hardware, virtualization layer, and OpenStack. There are three major building blocks in Common VIM: (i) OpenStack Controller, (ii) NFVI Resource Management, and (iii) Event Manager Function.

19.4.4.2   End-to-End Network Orchestration

End-to-End Orchestration integrates orchestration function in the context of end-to-end network. End-to-End Network Orchestration manages and controls the life cycle of end-to-end network slices through local NFVO, and also through interworking with Transport Infrastructure Orchestrator that manages transport network resources between them. Moreover, it allows configuration of muti-vendor and heterogeneous/different PNF/VNF within end-to-end network slices through PNF/VNF adaptor or open API. This enables not only life cycle management of network service but also automation of overall operation, including node configuration within slices.

19.4.4.3 Service Orchestration

Service orchestration features functions that include design, activation, and management of dynamic service network that is optimized for the requirements and policies of each product/service. Service orchestration realizes service slice, which is a customized dynamic network segment that satisfies policies, special functions, differentiated performance, and security requirements of each product/service on macroscopic and static end-to-end network slice.

To deliver such anticipated benefits, service orchestration is composed of the following building blocks: (i) Service slice life cycle management is responsible for managing the life cycle of dynamic network segment within end-to-end network slice. (ii) Service slice assurance is needed to collect and manage performance and fault information of service slice. (iii) Product management manages B2C/B2B products and services through interworking with BSS. (iv) Service adaptation is responsible for interworking/configuration among policy servers, subscriber DBs, and systems that perform dynamic service chaining, in the context of service realization. (v) Service slice catalogue manages product/service meta data and configuration information. (vi) Design portal and dashboard is required to design, manage, and operate service slice.

19.4.4.4 Standard Data Model

In order to achieve operation automation of orchestration and support migration and interoperability between clouds, SK Telecom, with the aim to commercialize vEPC, vIMS, and NFVO, has defined VNFD, LNSD (local NSD), and ENSD (end-to-end NSD) in XML after specifying VNF Descriptor (VNFD) and NS Descriptor (NSD) that is in compliance with ETSI NFV specification, and is accelerating the evolution into a more sophisticated NFV descriptor that supports open-sourced, automated data models such as YANG and TOSCA.

19.4.5 NG-OSS

Operation support system (OSS) is used to stabilize and optimize infrastructures and maintain quality of customer services. NG-OSS has two major building blocks: Cognitive and Intelligent Automation and End-to-End Hybrid Resource Management. Cognitive and Intelligent Automation provides dynamic assurance suitable for on-demand service and network, while forwarding information needed for service creation of Unified Orchestration. NG-OSS delivers integrated topology view and service visibility, in the end-to-end context, within hybrid network environment where legacy PNF and VNF coexist. In order to

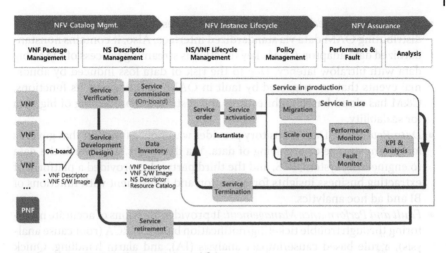

Figure 19.15 NFV MANO operation workflow.

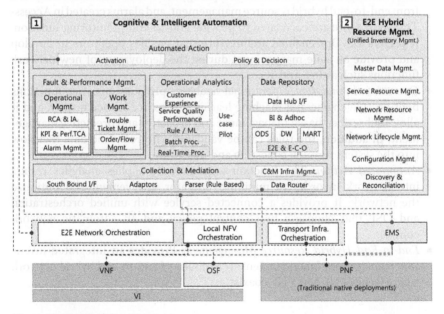

Figure 19.16 NG-OSS architecture.

deliver such benefits, End-to-End Hybrid Resource Management performs the role of integrated inventory management in the context of end-to-end network.

The major function blocks of NG-OSS have the following characteristics:

- *Collection and Mediation (C&M):* C&M is responsible for collecting and distributing OAM data via standardized interface. Also, it contains nonstandardized data adaptor. C&Ms key function is streaming process of OAM big data with ultralow latency. Due to the risk of data loss induced by abnormal events that are triggered by fault in OAM storming and its functions, C&M has cloud-based architectures and distributed processing of big data for scalability.
- *Data Repository:* Data repository provides warehouse function that governs long-term storage or archiving of data. Via data mart, it forwards the data to engineering system, BSS, and the third party, also provided a platform of extracting business insights for operation and investment with functions of BI and ad hoc analytics.
- *Fault and Performance Management:* It provides functions of accurate monitoring through trouble ticketing/notification based on RCA (root cause analysis), a rule-based cause/impact analysis (IA), and alarm handling. Quick reactive management is possible through automation of trouble ticketing and verification activities. Network topology information is loaded extracted from end-to-end hybrid resource management, and alarms created in Access-Transport-Core are correlated to provide end-to-end monitoring function. It prefers DevOps-enabled architecture that enables operators to develop and provide an agile management during the deployment of new network functions such as IoT and 5G.
- *Operational Analytics:* Operational analytics provides analytics functions built on rule engine and machine learning to secure service quality, and it allows subscriber-based management and proactive and predictive management. In addition, the rules can be validated with real-time data that flow into the system through collection and mediation.
- *Automated Action:* The function automatically applies analytics results into network in order to minimize operator involvement and optimize the network. It provides a connected service with unified orchestration and OSF that choose a best control policy and an automatic network control.
- *End-to-End Hybrid Resource Management:* Provides an integrated inventory management function to ensure visibility of end-to-end hybrid network where legacy PNF and VNF coexist.

19.4.5.1 Cognitive and Intelligent Automation

Cognitive and intelligent automation function collects and distributes data generated from the entire equipment of radio access, transport, and core network; performs real-time analysis of data; and stores data efficiently to sense and respond swiftly to new occurrences and changes within network. Policy and decision function leverages such analysis results to determine and authorize

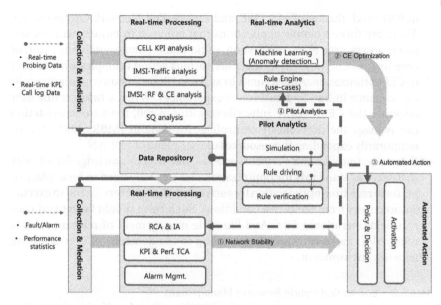

Figure 19.17 Cognitive and intelligent automation workflow

execution of actions that are optimal for a relevant situation, while minimizing operator involvement in this process.

- *Ensure Network Stability:* Information forwarded from C&M and managed at data repository is correlated and analyzed to identify abnormal conditions. Once abnormal condition is recognized, RCA and are conducted for rapid problem solving. Moreover, the overall accuracy of such analytics is enhanced by alarm management that performs clustering, filtering, and correlation of alarms.

- *Customer Experience Optimization:* NG-OSS is reinforcing its CE capabilities by moving its optimization focus from the existing cell unit to subscriber and service units. It delivers customer experience optimization by improving wireless network quality with machine learned rule. Real-time Customer Experience Index (CEI) is monitored by each subscriber and service unit. Moreover, CEI is reflected in engineering and construction process to support the design of optimized investment.

- *Automated Action:* Policy and decision determines a solution that best suits the corresponding network environment by leveraging intelligent knowledge DB. Discovery analytics turns know-how of skilled operators into policies and assists in updating the policies so that it can adapt to the rapidly changing network environment. Once the policy function determines the most suitable solution, activation automatically renders instructions and supervises the rightness of corresponding actions. The proposed actions are performed in PNF/VNF through EMS, OSF, and End-to-End Network Orchestration. There are three example applications that are used to ensure network stability: access user plane congestion, radio plane outage, and congestion of core (e.g., EPC). Meanwhile, there are various action examples with respect to CEI optimization. Adjustments to antenna tilt or Tx power of serving cells can enhance in-building customers experiences. When a heavy traffic user degrades the CE of entire subscribers within a cell, the automated action can manage the problem by controlling throughput (MBR, ABR, etc.) or temporarily expanding the resources of user plane of SDRAN.

- *Discovery Analytics and Verification (Pilot Analytics):* Knowledge-based rules that serve as standards for optimization and recovery need to be in place to perform real-time optimization for each subscriber and service and to execute autorecovery from abnormal conditions. Such rules should be verified with real-field operational data. And life cycle management of relevant rules is required to make adjustments according to the demands of ever-changing network environment.

19.4.5.2 End-to-End Hybrid Resource Management

End-to-end service assurance as well as fulfillment and unified orchestration functions need to be performed based upon reference data. Information of resources and connections of hybrid network should be offered in the form of a unified inventory. The major role of end-to-end hybrid resource management is to provide visibility of hybrid network topology that provides all the connection information of VNF and PNF and performs an integrated management of product, service, and network resources. For this, it collects Virtual Resource Descriptor and Instance Information from NFVO and gathers configuration information about PNF via EMS and Transport Infrastructure Orchestrator. Configuration Information can also be collected through OSF. By using virtual resource autosynchronization and discovery function, end-to-end hybrid resource management detects changes in each network and ensures that those changes are reflected to always manage the latest information.

- *Master Data Management:* Manages resource models and their enhancements including vendor-specific VNF and PNF and version-specific models. With respect to this legacy equipment, such as DU and EPC, Master Data Management enables to manage them along with VNF by using information modeling that splits functions from physical infrastructure.
- *Service Resource Management:* Manages information of equipment and its connection that are required for delivering services. Moreover, it allows manageing resources as well as their states that are required for service provisioning.
- *Network Resource Management:* Manages physical devices as well as VNF and the relevant connection information that construct the entire network.

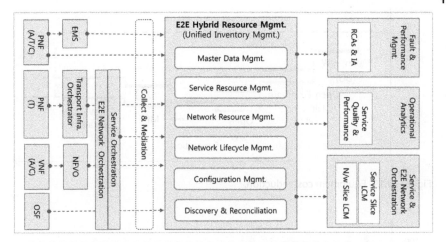

Figure 19.18 End-to-end hybrid resource management architecture.

Considering VNF, NRM manages functions and infrastructure separately and each device is managed by levels of the hierarchy container (rack), shelf, slot, and port. Port, in particular, information of physical interconnection of devices allows managing the interconnection with service resources through logical mapping of services.

- *Network Life Cycle Management:* Offers the most up-to-date information about devices and interconnections. For NFVO, it manages the resource life cycle of VNF instances.
- *Configuration Management:* Manages configuration information related to the interconnection between devices with respect to the overall end-to-end PNFs and VNFs. Both physical and logical interconnection information are generated on the basis of configuration information.
- *Discovery and Reconciliation:* Performs discovery and update resource and configuration information based on the data that have been generated by NFVO and EMS system. In case automated update is not possible, an operator may perform update action manually through business processes.

19.5 Value Proposition

19.5.1 TCO Reduction

19.5.1.1 Open-Source Hardware and Software Delivers Cost Savings
Use of standardized hardware can lead to savings on equipment costs, and commoditization of standard functions can spur competition and lead to reduction in software prices. Another effective approach is to avoid vendor lock-in and

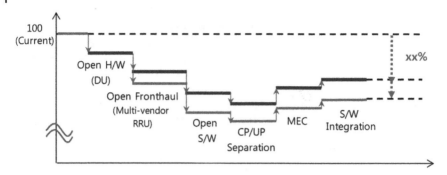

Figure 19.19 Equipment cost reduction.

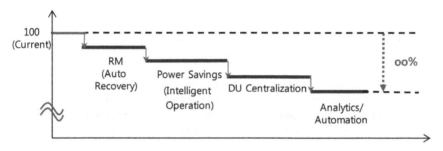

Figure 19.20 Maintenance cost reduction.

build a multi-vendor environment with open interfaces. Although this has the potential to drive up costs in the event of software integration and development of operator-specific functions, its actual impact on equipment prices is relatively small when the supply is large enough.

19.5.1.2 Optimization Based on Analytics Delivers Operation Cost Savings

Cost-savings at sites can be achieved through auto recovery function that automatically responds to faults, and also through optimization of coverage and performance based on analytics, along with base station (DU) centralization.

19.5.2 Platformization of Telco Infrastructure

Unlike IT infrastructure, Telco infrastructure has distinctive assets such as physical wired/wireless network resource, subscriber information, service use context, and location/analytics based additional information. By driving platformization of their infrastructures, Telcos are turning themselves into digital service providers. Based on a platform, Telcos can create and deliver services that meet the demand of customers and third parties.

The service that can be made available through platformization of infrastructure is categorized as TIaaS (Telco Infrastructure as a Service) and specific examples of them are stated below.

19.5.2.1 Mobile Edge Computing as a Service (MECaaS)

The goal of mobile edge computing (MEC) is to provide a better quality service by installing/deploying functions that are required to deliver the given service at Telco infrastructures located at the shortest distance from customers.

19.5.2.2 Analytics as a Service (AaaS)

AaaS enables operators to open up analytics-based information under the consent of customers and empowers platform users to leverage the information to add personalization and optimization to their services.

19.5.2.3 Policy as a Service (POaaS)

POaaS enables operators to open up policies applicable to network, thereby empowering platform users to select policies to put to use and leverage to enhance service qualities.

19.5.3 Operation Automation

19.5.3.1 Intelligent/Automated Network Operation

This system brings intelligence and automation achieved through data analytics into the entire architecture process that consists of investment, deployment, and operation. The application enables detection/prediction of customer service degradation through real-time monitoring of KPI and anomaly detection, while ensuring speedy response through its root cause analysis and impact analysis. This allows for improving customer experience quality.

19.5.3.2 Analytics and Verification with Data Analytics Capabilities

By providing data analytics platform, the system supports discovering new rules and use cases for automation. The provision of standardized life cycle management process allows rules and use cases that are verified in commercial networks to be applied in a timely fashion, while turning them into assets. Moreover, it also provides an environment where operators can manipulate data with ease and intuitively understand the meaning of data.

19.5.4 Deployment of Operator-Specific Functions

Operators seek to develop and embed unique, specialized functions into their networks to deploy networks optimized for their environment and gain competitive edge. Examples of such functions are coverage extension technology, load balancing function, wireless resource scheduler, and information collec-

Gray Key Areas	Key Functions	Enabling Technology	Values
SDRAN	Software-based open I/F and offers MEC integrated platform	Function split (DU-RU), auto recovery, and so on	- Open hardware- and software-based TCO reduction
vCore	Modularization, simplification, virtualization, and offers network-based VAS	CP/UP split (CUPS) and common user plane	- Transformation of Telco infrastructure into a platform - Analytics-based operation automation
uCTN	Multi-vendor, layer unified control and management as well as L0-3 converged equipment	T-SDN platform, POTN, NG-fronthaul (T-PON) and so on	- Service agility - Function specialized for operator
Unified-O	Multidomain, end-to-end, and unified orchestration	Standard NFV MANO, end-to-end network service orchestration	
NG-OSS	Cognitive and intelligent automation	Big data analytics, unified inventory management	
COSMOS	Open and programmable infra resources	Open hardware virtualization, container	

tion agent. Such specialized functions not attainable in the existing network architecture can be developed and deployed in equipment with ease under ATSCALE.

19.5.5 Enhanced Service Agility

19.5.5.1 Recombinable and Reusable Software Modules with Virtualization

Unlike the existing system, under the ATSCALE architecture, hardware-oriented network equipment is turned into hardware-independent software through virtualization, minimizing time, and cost required for equipment deployment, scale-in/scale-out, and relocation or even completely automating the previously mentioned processes. Moreover, the system guarantees speedy implementation of network that satisfies various functional and performance requirements of new services/products, by recomposing modularized unit functions and optimizing physical configuration.

19.6 Summary and Conclusion

SK Telecom is trying to transform itself into a platform service company. To this end, our infrastructure is required to deliver various services in a timely fashion while enabling more developers to engage in the innovative service developments. Moreover, it is essential to overcome the limitations of the current network such as its complex and closed architecture as well as the dependency on manufacturers, so as to realize a reduction in both TCO and service delivery time. A software-based IT technology is expected to serve as an enabler to drive such structural innovation of the Telco infrastructure. This chapter has presented the direction in which an innovation of Telco infrastructure shall persuade.

Initially, such changes and innovations have been detected at an SDDC using an SDx-based technology and are now being scaled out to the Telco infrastructure. To be in line with those, we released the "ATSCALE" vision, and thereby expect to lead the ecosystem and drive the innovation of Telco infrastructure. ATSCALE architecture aims at realizing a "Scalable, Cognitive, Automated, Lean, and End-to-End" Telco network, and keeping the network and TCO structure lean by using software-based technologies. This chapter has described the details on technology on each specific domain, which can be summarized as follows.

We have introduced an open-interface structure to allow manufacturers to use them as a guideline for their development work. Moreover, we have also presented the architecture allowing the automated operation and the integrated analytics for Telcos. Finally, we would like to conclude this chapter with a call for the effort from the Telco industry, because such innovation will not be possible with a small group of companies.

References

1 SK Telecom, "ATSCALE White Paper," July 2016.
2 SK Telecom, "5G White Paper: SK Telecom's View on 5G Vision, Architecture, Technology, and Spectrum," October, 2014.
3 SK Telecom, *Evolving Telco Data Center with Software-Defined Technologies – COSMOS for All IT Network*, June 2016.
4 SK Telecom, *Telecom's 5G Architecture Design and Implementation Guidelines*, October, 2015.
5 P. Marsch, I. Silva, O. Bulakci, M. Tesanovic, S. E. E. Ayoubi, T. Rosowski, A. Kaloxylos, and M. Boldi, "5G Radio Access Network Architecture: Design Guidelines and Key Considerations," *IEEE Commun. Mag.*, vol. 54, no. 11, pp. 24–32, 2016

6 A. Maeder, A. Ali, A. Bedekar, A. F. Cattoni, D. Chandramouli, S. Chandrashekar, L. Du, M. Hesse, C. Sartori, and S. Turtinen, "A scalable and flexible radio access network architecture for fifth generation mobile networks," *IEEE Commun. Mag.*, vol. 54, no. 11, pp. 16–23, 2016.

7 AT&T, "ECOMP (Enhanced Control, Orchestration, Management & Policy) Architecture White Paper," March, 2016

8 Verizon Network Infrastructure Planning, *SDN-NFV Reference Architecture*, February 2016.

9 3GPP, *3GPP TR 23.799: Study on Architecture for Next Generation System*, April, 2016.

10 ETSI NFV ISG, *ETSI GS NFV 002 v1.1.1: Network Functions Virtualization (NFV); Architectural Framework*, October 2013.

11 A. Gudipati, D. Perry, L. E. Li, and S. Katti, *SoftRAN: Software Defined Radio Access Network*, HotSDN, August 2013.

12 ON.LAB ONOS, *M-CORD: Mobile CORD; Enable 5G on CORD*, March 2016.

13 OpenStack, *Open Source Cloud Computing Software*. Available at https://www.openstack.org.

14 Open Compute Project (OCP), *Open Compute Project: Overview*, May 2013.

15 Telecom Infra Project (TIP). Available at https://telecominfraproject.com.

16 Open ROADM MSA, *Open ROADM Overview v1.0*, March 2016.

17 3GPP, *3GPP TR 23.711: Enhancements to Dedicated Core Networks Selection Mechanism*, June 2016.

18 ETSI MEC ISG, "Mobile-Edge Computing: Introductory Technical White Paper," September 2014.

19 Ericsson, *5G Systems; Enabling Industry and Society Transformation*, January 2015.

20 Cisco Systems, Inc., *Ultra Services Platform*, 2016. Available at http://www.cisco.com/c/en/us/solutions/service-provider/ultra-services-platform/index.html.

21 TM Forum, *OSS/BSS Futures Overview, Why OSS Needs to Change, IG1117, Rel. 14.5.1*, April 2015.

22 TM Forum , *Zero-Time Orchestration, Operations and Management (ZOOM)*.

23 ETSI NFV ISG, *Network Functions Virtualization (NFV)*, October 2014.

24 Open Networking Foundation (ONF), *SDN Architecture*, June 2014.

25 3GPP, *TR 22.808, Study on Flexible Mobile Service Steering (FMSS)*.

26 Internet Engineering Task Force (IETF), Principles and Strategies of Design, ActualTech Media, January 2016, Service Function Chaining Working Group. Available at http://datatracker.ietf.org/wg/sfc.

Alex Jinsung Choi is SVP of Research and Technology Innovation, Head of T-Laboratories Innovation at Deutsche Telekom AG. In addition, he is member of the Technology & Innovation management board, where he is responsible for several strategic projects.

Dr. Choi has more than 20 years of experience in the mobile telecommunication industry. Prior to joining Deutsche Telekom, he served as Chief Technology Officer (CTO) and Head of the Corporate R&D Division and the Technology Strategy Office in SK Telekom in South Korea from 2012, where he advanced key strategic and research topics. An important focus was on the development of next-generation ICT technologies such as artificial intelligence (AI), 5G, and autonomous driving. With the introduction of "NUGU," the first AI-based virtual assistant in Korea, Dr. Choi was influential in the development of AI solutions. Before SK Telekom, Dr. Choi held various positions at LG Electronics, including Head of the Mobile Core Technology Lab and the Next Generation Telecommunications Lab. He has also chaired the Korea Artificial Intelligence Industry Association. and served as Chairman of Telecom Infra Project (TIP), headed by Facebook, Deutsche Telekom, and SK Telekom.

Jinhyo Park is currently serving as CTO, Executive Vice President, and Head of the ICT R&D Center at SK Telecom. In this role, Park leads the development of next-generation 5G communications technologies as well as the latest advanced technologies in media, security, data analytics, and autonomous driving.

Having been at SK Telecom since 1997, Park has had a variety of roles in his 20+ years, including managing R&D strategy and successfully overseeing the commercialization of WCDMA and HSDPA/HSUPA networks. As leading the Access Network Lab, he is also credited with helping SK Telecom to be the first to commercialize LTE in Korea and the world's first multicarrier/HD Voice(VoLTE). For the last 5 years when he served as Senior Vice President and the Head of Network R&D Center, he has actively contributed to world-leading LTE-A, 5G, connected car, quantum cryptography communications technologies. He holds a bachelor's degree in mathematics and a master's degree in information and communications engineering from Korea University.

Sungho Jo Sungho Jo as the Senior Director, The Head of Access Network Lab is currently responsible for developing 5G, 4G, IoT Access Networks at SK Telecom R&D. He joined SK Telecom in 1997, and worked for the network operation division. From 2001, he has not only contributed commercialization of CDMA1X and EVDO but also took a responsibility for WCDMA development of SK Telecom. Since 2011, he has successfully led the Korea-first LTE launch, the world-first multi-carrier deployment, the world-first nation-wide HD Voice (VoLTE) commercialization, the world-first LTE-A (Carrier Aggregation) commercialization, and directing LTE evolution and beyond LTE-A of SK Telecom as the head of Commercialization part at Access Network Lab. He also led a successful LoRa IoT Network Commercialization in 2016.He led 5G Tech Lab and has been the head of Access Network. Lab since 2017. His current interest mainly focuses on next generation communication system like 5G, SDN/NFV and LTE-A Evolution, LPWA,IoT. He received a Bachelor degree in the department of Electronic Engineering at Kyungpook National University (KNU) and received a Master of Business Administration (MBA) degree at Korea University.

Sangsoo Jeong obtained his B.S. and Ph.D. degrees in electrical engineering and computer science from Seoul National University, Seoul, Korea, in 2003 and 2010, respectively. From 2010 to 2015, he was a researcher with the DMC R&D Center of Samsung Electronics, Korea, where he conducted research and standardization on 3GPP/LTE systems. In late 2015, he joined Network Technology R&D Center of SK Telecom, Korea, where he is currently developing new technologies for the next-generation telecommunication systems. His research interests include 5G system architecture, software-defined networking, network function virtualization, and wireless communication protocols.

20

Standardization: The Road to 5G

M. P. Galante and G. Romano

Technology Innovation Department, TIM, Torino, Italy

20.1   The Role of Standardization

If we look at recent developments of the industry, one of the main successes of standardization is mobile communications as specified by Third-Generation Partnership Project (3GPP) [1], which provided the technical specifications for UMTS and LTE. Open standards, as specified by 3GPP, allowed people to experience new services with the freedom to move all around the world, thanks to the capability of their devices to interwork with networks operated by different service providers and with network equipment from different vendors.

As such, standardization is the definition of basic aspects of the mobile system, encompassing interfaces and protocols that allow the interworking between different components. For example, standards are similar to an assembly line in a factory, where different components are assembled in a predefined manner, independent of who manufactured the specific component. Therefore, the open standards were exploited by different entities in the value chain, for example, chipset makers, original equipment manufacturers (OEMs), network vendors, network carriers, and service provides, all of them contributing to the delivery of the communication service to the mobile customer, whatever the phone he/she will use and whatever the network he/she will get service from.

Standards push competition, since they provide specifications for the interfaces, not the algorithms (which can be used to differentiate between different suppliers), nor the man–machine interface (e.g., touchscreen versus keyboard)

5G Networks: Fundamental Requirements, Enabling Technologies, and Operations Management, First Edition.
Anwer Al-Dulaimi, Xianbin Wang, and Chih-Lin I.

or the applications. Therefore, they enable differentiation and competition, rather than homologation.

An open standard is built up by a community working together and acknowledging defined rules for Intellectual Property Right (IPR) exploitation, such as licensing patents under fair, reasonable, and nondiscriminatory (FRAND) regime.

Most importantly, an open standard allows companies to implement the different interoperable components, and any supplier can get access to the product industrialization phase (FRAND regime guaranteeing a fair remuneration to the relevant patent holders). Thus, an open market is created where the competition is solely based on how good and performing the implementation is (both in technical and economic terms).

Another key factor is economy of scale and mass market production to deliver equipment and devices at a cost affordable to everybody. Nowadays, the research costs to develop and commercialize a high-technology device are incredibly high and the only way to recover the costs is to provide solutions able to work all over the world. Even large markets like China cannot afford to develop custom solutions, since the same companies will then have to make research and develop products for the rest of the world. Having different solutions for Europe, China, America, or Japan will lead to a 10-fold increase in development costs. Lack of standards and interoperability increases the costs for all actors in the value chain: end customers, service providers, network operators, and terminal and equipment vendors.

With 5G, the problem becomes even more complex, because the number of players increases outside the "usual" ICT players. According to the European Union vision, 5G is the enabler for a fully connected society by integrating many industrial vertical segments (i.e., the verticals) that will deliver applications to enable, for example, smart cities, smart grids, smart factories, public safety applications, and others on top of a common communication infrastructure.

One of the dream projects of 5G is the so-called self-driving car. Can you imagine what happens if a car manufactured in Europe is not able to "talk" with a car or infrastructure manufactured in Asia? All the components must be able to speak the same language otherwise "lost in translation" may, as a minimum, result in the failure of the dream of a 5G society!

Therefore, the key to access a global market, integrating verticals and allowing people to buy 5G devices affordable for their economic situation is to define first of all open standards.

This chapter describes how and where the standardization of 5G is carried out, by identifying the "three phases" of the process: vision, technical specifications, and policy and profiling. An overview of the main players is given, followed by an in-depth knowledge of the two major players: ITU-R [2] and 3GPP, who will shape the technical characteristics of the new system. Other authors provided an overview of the standardization process [3–9], but the idea behind this chapter is to provide a view of the process leading to success of a technology, by means of high-quality standards.

20.2 The Main Standardization Bodies

Behind the success of a technology, there is not just a single standardization body, but a number of entities strictly collaborating among them. If we look to a success story of the recent past, LTE, we can see that there are a number of players contributing to the final availability of a simple smartphone.

A first aspect that is required to ensure the success of a technology is the spectrum availability. Spectrum is the very scarce resource that is required to provide wireless communications. Without spectrum there is no wireless. And spectrum fragmentation (i.e., different frequency bands allocated in different countries or continents) is also a major hindering factor when developing products, since multiple radio frequency (RF) components are required to allow a smartphone to operate both in Europe and in the United States. The main entities that identify the spectrum needs for mobile communications and the rules for its use are ITU-R (during the World Radiocommunication Conferences (WRCs) [8]) and Regional bodies such as CEPT in Europe [10], APT [11] in Asia, and Citel [12] for the Americas. Note that the ultimate owner of the spectrum is however the national regulator, and so it will be, for example, FCC in the United States (and corresponding entities elsewhere) that will allocate and license the spectrum.

The spectrum alone is of course not sufficient to develop a successful system. First of all, common requirements and goals to be achieved by the system need to be set, and then the components of the technology need to be developed. 3GPP was the main standardization body delivering technical specifications of LTE, but the work was built upon collaboration with a number of other standards organizations in order to reuse as far as possible best-in-class standardized technologies such as IETF [13] for the IP protocol suite developed for the Internet, IEEE [14] for interworking with Wi-Fi, OMA [15] for device management, ETSI [16] for the specifications of the smart card that hosts USIM applications, and so on.

A device before being commercialized needs to be tested in order to be sure it works as expected. Therefore, 3GPP specifies testing procedures that allow certification bodies such as GCF [17] to provide the rules for certification of conformity.

GSMA [18] finally provides the business rules to ensure, for example, the roaming and interconnection procedures, fraud management, and security best practices.

In case of 5G, we can expect also new standardization bodies to enter the ecosystem, as new technologies (such as virtualization and software-defined

Technical specifications
Policy and profiling

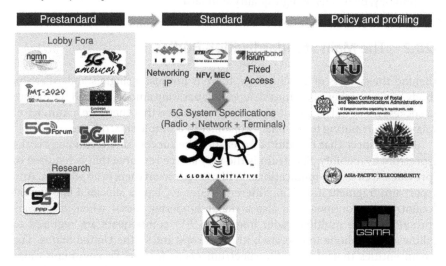

Figure 20.1 5G standardization landscape.

networking) will enter the landscape. Also, new stakeholders will enter the 5G ecosystem as new requirements from the verticals need to be developed by the relevant industry associations.

20.3 5G Standardization Process

Figure 20.1 provides an overview of the standardization process for 5G. The process can be subdivided into three phases:

- Prestandard phase, focused on industry vision building
- Technical specifications
- Policy and profiling

The setting of requirements is usually elaborated in the prestandardization phase, often by means of White Papers providing the vision of the industry and associations. Several White Papers have been published by many organizations, such as 5G Americas [19], European Union [20], 5G Forum [21], 5GMF [22], NGMN [23], and by many vendors as well. In Europe, the research program under the 5G Infrastructure Association [24] umbrella also set the basis for the vision toward 5G.

In particular, the NGMN White Paper elaborates on the new business opportunities brought by 5G, by identifying new industrial vertical segments

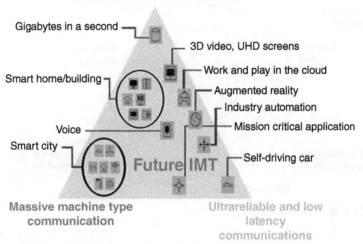

Figure 20.2 5G use cases [25].

interested in exploiting the new technology. As already explained, "vertical" is a generic name to indicate industries committed to deliver specific applications to the users, which typically falls out of the traditional Telco business. A nonexhaustive list of such verticals comprises automotive (connected cars and self-driving cars), manufacturing industry (mobile robots, Automated Guided Vehicles, closed-loop control in process automation), public safety (mission-critical services to police, fire brigades), e-Health (remote diagnosis and treatments, remote surgery), smart cities (street lighting optimal usage, waste management, smart parking, etc.), railway companies (communication between trains and infrastructure), and many others.

Taking into account 5G use cases analyses from the already cited organizations, it is generally recognized that they can be categorized into three main classes (see Figure 20.2):

- *extreme Mobile Broadband (eMBB)*: encompassing all the services derived from the evolution of traditional Telco Services toward an enhanced user experience (e.g., 3D video, augmented reality, and 50 Mbps everywhere);
- *massive Machine-Type Communications (mMTC)*: encompassing all the communications established between billions of devices and a cloud, which will create the new Internet of Everything (e.g., ultralow power, low complexity sensors such as wearables and utility meters).
- *Ultrareliable and Low-Latency Communications (URLLC)*: encompassing the capabilities to communicate and manage the status of remote objects (e.g., robots, actuators) in a very reliable way and with very low latency (e.g.,

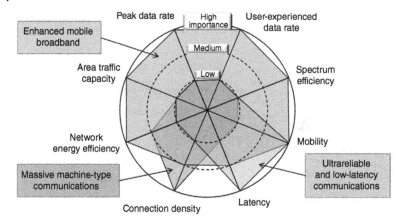

Figure 20.3 KPIs for 5G use cases [25].

for remote surgery, remotely controlled vehicles, drone delivery, and robot control in factory automation). This new communication paradigm, which enables not only internetworking among objects but also remote control over such objects, as if they were physically located nearby the controller, is often referred to as "Tactile Internet."

These three classes of use cases are often represented at the extreme corners of a triangle, which should encompass all the different 5G services, some of them even yet unknown today.

Each use case is characterized by key performance indicators (KPIs) representing the target behavior the system must exhibit to satisfy the use case. As is shown in Figure 20.3, each main category is characterised by distinct extreme KPI values. In the case of mMTC, connection density will be the critical factor to achieve a scalable network as throughput is foreseen to be quite limited for such services. On the contrary, eMBB is expected to adhere to challenging KPIs at least for user experienced data rate and mobility. Finally, URLLC applications will need challenging KPIs in the latency and mobility dimensions more than in other ones.

Also, most White Papers address the spectrum and IPR aspects, to ensure a complete ecosystem is taken care of for the new system. Spectrum is the fuel to radio communications, and it must be allocated in a suitable way with sufficient resources to satisfy the capacity needs. A clear policy for IPRs is on the other hand a must to ensure that nobody is discriminated and new companies are attracted by the 5G ecosystem, contributing to innovation with their R&D efforts and with their products.

The subsequent phase, as depicted in Figure 20.1, is the definition of technical standards. This is not a job for a single organization, as illustrated above, but most likely the result of the collaboration between different SDOs (Standard Development Organizations), each providing a subset of the full system. The expectation is that ITU-R will define the requirements that the new IMT-2020 radio family will have to meet and 3GPP will be the SDO playing the master role in the definition of the whole 5G system and radio aspects, complying with ITU-R requirements. Similar to the LTE story case, some features need to be developed by other entities, such as IETF (e.g., for IP protocols), ETSI (e.g., for NFV and management and orchestration solutions, for Smart Card Platform solutions), BBF (e.g., for integration with fixed access), IEEE (for Wi-Fi integration), and ITU-T (e.g., for transport capabilities). The next section will focus on the work of ITU-R and 3GPP to provide an insight on what their plans and milestones are.

The final leg of the standardization process is represented by policy and profiling activities. Some examples of these activities are the identification of the spectrum to be used for 5G applications and the definition of the rules on how to use such spectrum. This activity is mainly carried out by National Regulators and ITU-R (in the WRC). In particular, WRCs try to harmonize the spectrum worldwide, therefore minimizing the market fragmentation. Once 3GPP 5G system specifications will be finalized, other bodies like the GSMA are expected to define on top of those the minimum set of features (profiles) to increase interoperability among UEs and networks (e.g., like it was for the IMS-based VoLTE, Voice over LTE), and to define proper business references to operators and international carriers for the cooperative delivery of 5G services to their subscribers (e.g., international roaming models, interoperator accounting, and interconnection models).

20.4 ITU-R

As done for the previous generations of mobile systems, ITU-R defines the process for the definition of IMT-2020. The key group is Working Party 5D "IMT Systems" and the final result will be a set of recommendations containing the technical specifications of IMT-2020 (e.g., technical characteristics and out of band emissions). The process is not different from the one used for the definition of IMT-Advanced, which led to the creation of Recommendations M.2012 (detailed specifications of the terrestrial radio interfaces of International Mobile Telecommunications Advanced (IMT-Advanced) [26]), M.2070 (generic unwanted emission characteristics of base stations using the terrestrial radio interfaces of IMT-Advanced [27]), and M.2071 (generic unwanted emission characteristics of mobile stations using the terrestrial radio interfaces of IMT-Advanced [28]).

The overall process (see Figure 20.4) can be subdivided into several phases: a preparation phase, aimed to define the vision of IMT beyond 2020 and therefore

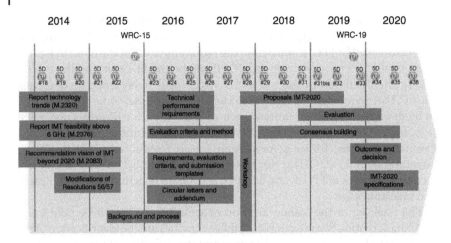

Figure 20.4 Detailed timeline and process for IMT-2020 in ITU-R [29].

start the discussion on the identification of suitable spectrum during WRC 2015; the definition of KPIs and evaluation criteria (in 2016–2017); and finally a call for proposals, their evaluation, and decision on which radio access technologies can be labeled as IMT-2020. Another important milestone will be WRC 2019, which plans to identify spectrum for IMT applications beyond 6 GHz.

The main outcome of the first phase is Recommendation M.2083 (IMT Vision "Framework and overall objectives of the future development of IMT for 2020 and beyond" [25]), which poses the basis for the definition of what should be expected in terms of new services by IMT-2020. The document identifies the three usage scenarios (eMBB, mMTC, and URLLC), similar to those identified by NGMN, and the capabilities of IMT-2020.

The second phase started with the publication of the Circular Letter 5/LCCE/59 [30], which invites the submission of Radio Interface Technologies (RIT) or Set of Radio Interface Technologies (SRIT) to be recognized as IMT-2020. The Circular Letter addresses only the terrestrial component of IMT-2020 (in scope of WP5D), while the satellite component is in the scope of other working groups. In particular, an RIT is a single technology that satisfies the performance criteria, while an SRIT may be composed by different RITs, each addressing different performance criteria (e.g., a radio interface optimized for machine type communications and a solution optimized for mobile broadband, interworking with each other). The radio interfaces are developed outside ITU and should be submitted to ITU-R according to a submission template to demonstrate that the proposal is able to fulfill the minimum technical performance requirements and evaluation criteria. The submission template must contain a self-evaluation and may be integrated by any relevant information the proponent may consider useful to better evaluate the proposal. Finally, the proponents must indicate their compliance with the ITU policy on IPRs [31].

After the submission, candidate RITs or SRITs will be assessed by organizations registered as evaluators in the IMT-2020 web page [32]. An independent evaluation may be done by ITU-R members, standards organizations, and other organizations, such as universities and research projects. The first evaluator to register was the 5G Infrastructure Association [24]. The evaluation report, based on agreed methodologies in ITU-R WP5D will be made available in the same web page [32].

Based on the different evaluations, WP5D will assess if the proposal(s) meets the minimum technical performance requirements and evaluation criteria of the IMT-2020. Based on the evaluation results, modifications to the proposals may be required, and in case of multiple candidates, a phase of consensus building will start to harmonize as much as possible for the different solutions.

Finally, a number of recommendations will be developed within ITU-R, sufficiently detailed to enable worldwide compatibility of operation and equipment, including roaming.

20.5 3GPP

The work in 3GPP is not organized directly according to the OSI layers, but it somehow reflects such a classification. 3GPP is subdivided into three main areas: System Aspects (SA), Core Network and Terminals (CT), and Radio Access Network (RAN). Each area is governed by a Technical Specification Group (TSG) Plenary, which defines the workplan and approves the technical specifications developed by the respective working groups. RAN focuses on the Media Layers (1-3); RAN3 defines the radio architecture and interfaces between radio nodes and the Core Network. RAN4 defines the performance of the radio access technologies and RAN5 defines the test procedures to ensure a device is compliant to 3GPP specifications. SA defines the services and the system aspects. SA1 specifies the requirements for the new services both from a user perspective and from a network perspective; SA2 defines the network architecture, while the security aspects are defined by SA3. SA4 focuses on codecs and SA5 on telecom management. SA6 deals with applications related to critical communications and related verticals (e.g. public safety). Finally, CT defines the protocols (layers 3–5) to enable communications inside the mobile network and the interconnectivity with external networks. The specific contribution of each working group to the 5G work plan will be detailed later in this chapter.

Following the ITU-R procedure, 3GPP decided that its solutions will be submitted to ITU-R to become part of IMT-2020. In particular, one key requirement is that LTE and the new radio (NR) must be tightly integrated, part

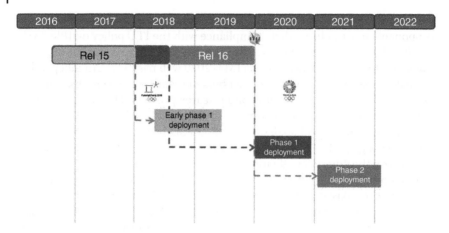

Figure 20.5 3GPP workplan.

of the 3GPP system. The new radio access technology will have ultimately the goal to satisfy all the requirements set by ITU-R, but 3GPP will submit both LTE and the new radio for inclusion in IMT-2020.

Figure 20.5 represents the 3GPP workplan as defined in September 2016. This picture may be not definitive, since 3GPP is always striving to satisfy the different market requirements. As a consequence, the roadmap has been already modified twice in 2016 to ensure that at least a subset of the expected full-blown 5G technical specifications is available already at the end of 2017 for operators planning commercial launches in 2018.

In general, the workplan is based on a phased approach. In Release 14, 3GPP studied the feasibility of 5G solutions (SA1 identified the service requirements, SA2 the system architecture, and RAN the radio access technology). Based on the study phase, technical specifications are derived in two phases. Phase 1 is delivered within the Release 15 time frame (with completion date June 2018). This phase mainly focuses on the eMBB use case and provide technical specifications for the new radio access technology and the foundations of the next-generation Core Network which are going to be enhanced in the Release 16, planned for December 2019. Therefore, the solutions specified in Release 15 must allow Release 16 to be built on such foundations set in 2018 (this is indicated within 3GPP as forward compatibility). The Release 16 specifications must fulfill all the requirements and be ready for incorporation in the ITU-R technical description of IMT-2020.

Finally, a number of operators indicated the willingness to anticipate in 2018 the commercial launch of 5G services. As a consequence, it was decided to anticipate around the end of 2017 a preliminary set of specifications based on the reuse of the 4G Core Network and a dual-connectivity approach (see

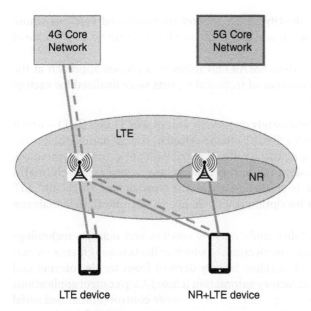

Figure 20.6 LTE-assisted approach.

Figure 20.6). The new radio access technology is mainly used for capacity enhancements of current LTE networks (with the possibility to operate the new radio on new bands, for example, 28 GHz). No modifications are required in the 4G Core Network (EPC) apart from the capability to handle greater throughputs than today. In fact, the EPC is connected to an LTE base station (eNB) via the current S1 interface. The new radio base station is connected to the LTE eNB by exploiting the "dual connectivity" feature [33], and the 5G device connects to both base stations. The LTE connection carries the signaling flow exchanges with the Core Network (e.g. mobility management, paging – dotted line in Figure 20.6), while the user data are carried by both the new radio and LTE (continuous line in Figure 20.6). This approach requires "only" the definition of the low layers of the new radio, with no functional change to EPC, and therefore it was quicker to specify and commercialize. Note that 3GPP ruled out the possibility for a new radio base station alone to attach to the LTE CN. A new radio base station in stand-alone deployment (i.e., not used in "dual connectivity" with LTE eNB) will connect only to the next-generation core.

SMARTER (New Services and Markets Technology Enablers) was the name of the SA WG1 project that developed high-level use cases and identified the related high-level potential requirements to enable 5G.

The study aimed at identifying the market segments and verticals whose needs 3GPP should focus on and that could not be met with LTE/EPS state of the technology.

The Release 14 Study Item SMARTER followed a phased approach at the conclusions of which a number of technical reports were finalized for each of the following macro areas:

- TR 22.861 [34] Feasibility study on massive Internet of Things (IoT) – which collects requirements for the IoT characterized by a large number of devices that may experience long battery life, high reliability (e.g., Smart wearables), and low complexity (sensors). In this context, also a new approach to SIM remote management is studied, such as the provisioning of "blank" IoT devices with 3GPP subscription where change subscription/credentials can happen over the air.
- TR 22.862 [35] Feasibility study on new services and markets technology enablers for critical communications – which collects requirements including high reliability and ultralow latency derived from tactile Internet and other use cases such as factory automation (closed-loop control applications running over 5G short-range radio), UAV remote control (unmanned aerial vehicles, drones), cars collision avoidance, and mission critical services.
- TR 22.863 [36] Feasibility study on new services and markets technology enablers for enhanced mobile broadband – collecting requirements from office and dense urban scenarios (e.g., real-time video meeting with very high data rates) and fast moving devices (car and trains).
- TR 22.864 [37] Feasibility study on new services and markets technology enablers for network operation - which describes the required system capabilities in terms of flexibility (e.g., network slicing, efficient user plane allocation, and exposure to third parties), scalability, mobility support, efficient content delivery, non-3GPP and 3GPP access integration, migration, and security.

Together with the previously listed SMARTER studies, SA1 has also developed a study on the enhancement of 3GPP support for 5G V2X services (TR 22.886 [38]), which has been conceived as the natural evolution of Release 14 support for LTE-based V2X.

At the end of the Release 14 5G study phase, the result of all such TRs has built the basis for the corresponding normative activity for Phase 1 (Release 15), which is documented in the Technical Specification TS 22.261 "Service requirements for next generation new services and markets" [39] (completed in March 2017).

In December 2015, SA approved the SA2 study item for the next-generation 3GPP system architecture to achieve a simple, flexible, scalable, and extensible architecture (NextGen) with two distinct characteristics with respect to previous ones:

- A "service based" design, to take advantage of the latest virtualization and software technologies
- A high flexibility in deploying multiple networks with different characteristics serving various users and service needs (network slices)

It is worth underlying again that the overall NextGen system work is split into two phases:

- Phase 1 (Release 15) - Developing a baseline NextGen system including the NextGen Core Network
- Phase 2 (Release 16) - Delivering a complete and feature rich NextGen system on top of Phase 1

The corresponding Release 15 TR 23.977 "Study on Architecture for Next Generation System" [40] (also referred to as NextGen) contains the agreed high-level principles and documents solutions to a number of key issues in which the architectural work has been decomposed, such as network slicing, QoS framework, mobility framework, session management, support for session and service continuity and efficient user plane paths, policy framework, support for IMS, just to mention some.

The architecture has been developed with the following non-exhaustive list of operational efficiency and optimization characteristics:

- Ability to handle the rapid growth in mobile data traffic/device numbers resulting from existing and new communication services in a scalable manner.
- Allow independent evolution of core and radio networks.
- Support techniques (e.g., Network Function Virtualization and Software-Defined Networking) to reduce total cost of ownership, improve operational efficiency, energy efficiency, and simplicity in and flexibility for offering new services.

From the conclusions of the TR 23.977 (December 2016), the normative work item for Phase 1 of the NextGen architecture has been derived and completed within the Release 15 framework.

In March 2016, SA3 started to study preliminary threats, requirements, and solutions for the security of next-generation mobile networks. TR 33.899 [41] captures the output of this study.

Finally, SA5 also started some NextGen-related work to understand how the network management should evolve, for example, how to satisfy the operational and management requirements and the role and location of the management functionalities, how to investigate use cases and requirements for management and orchestration of network slicing, how to define management and orchestration architecture to support network operational features such as real-time, on demand, automation as well as vertical applications (e.g., eV2X).

KPI	Value		KPI	Value
Peak data rate	20 Gbps DL 10 Gbps UL		Extreme coverage	100–400 km voice/low data
Peak spectral efficiency	30 bps/Hz – 15 bps/Hz		UE battery life	mMTC > 10 years (target 15 years)
Control plane latency	10 ms from idle to active mode		Cell/transmission point/ TRP spectral efficiency	3 × IMT-A
User plane latency	URLLC: 0.5 ms UL and DL eMBB: 4 ms UL and DL		Area traffic capacity	To be derived area capacity (bps/m2) = site density (site/m2) × bandwidth (Hz) × spectrum efficiency (bps/Hz/site)
Latency for infrequent small packets	UL: 10 s for a 20 byte application packet @ e maximum coupling loss of 164 dB			
Mobility interruption time	0 ms		User experienced data rate	TBD - NGMN: 50 Mbps everywhere (2 Gbps indoor hotspot)
Intersystem mobility	With other IMT systems		Fifth percentile user spectrum efficiency	3 × IMT-A
Reliability	URLLC: P = 10–5 in 1 ms		Connection density	mMTC 1 M device/km²
Coverage	mMTC 164 dB @ 160 bps		Mobility	500 km/h

Figure 20.7 3GPP radio KPIs for new radio.

For this purpose, SA5 developed for the Release 15 the following studies, before starting any normative activity:

- TR 28.802, Study on Management Aspects of Next-Generation Network Architecture and Features [42]
- TR 28.800, Study on Management and Orchestration Architecture of Next-Generation Network and Service [43]
- TR 28.801, Study on Management and Orchestration of Network Slicing for Next-Generation Network [44]

The studies define the relationship between network slice management and orchestration concepts developed in SA5 and the management and orchestration concepts defined by ETSI NFV.

In September 2016, RAN approved the radio requirements for the new radio [45]. This document provides the KPIs for the radio interface (see Figure 20.7) and the deployment scenarios to be used to verify the KPIs are met. The document also provides a number of requirements on architecture, supplementary services (MBMS [46], positioning, critical communications), and operational requirements.

From the architecture perspective, some new aspects have been introduced, such as splitting the RAN architecture (Cloud RAN), network function virtualization and SDN, and network slicing.

One of the big novelties for the radio is the approach to spectrum: In order to achieve very high-throughput, it is necessary to explore new spectrum.

Therefore, 3GPP decided to develop solutions able to operate up to 100 GHz. The challenge to provide cellular service at these frequency ranges is very high, but trials and literature indicate it as feasible [47].

References

1 3rd Generation Partnership Project (3GPP). Available at http://www.3gpp.org/about-3gpp.

2 ITU Radiocommunication Sector. Available at http://www.itu.int/en/ITU-R/Pages/default.aspx.

3 A. Osseiran, J. F. Monserrat, and P. Marsch, *5G Mobile and Wireless Communications Technology*, Cambridge University Press, 2016.

4 A. N. Barreto et al., "5G – wireless communications for 2020," *J. Commun. Inform. Syst.*, vol. 31, no. 1, 146–163, 2016.

5 H. Tullberg et al., "METIS research and standardization: a path towards a 5G system," *2014 IEEE Globecom Workshops (GC Wkshps)*.

6 V. Frascolla et al., "MmWave use cases and prototyping: a way towards 5G standardization," *2015 European Conference on Networks and Communications (EuCNC)*.

7 V. Frascolla et al., "Challenges and opportunities for millimeter-wave mobile access standardisation," *2014 IEEE Globecom Workshops (GC Wkshps)*.

8 World Radiocommunication Conferences (WRC). Available at http://www.itu.int/en/ITU-R/conferences/wrc/Pages/default.aspx.

9 E. Dahlman et al., *4G, LTE-Advanced Pro and the Road to 5G*, 3rd edition, Academic Press, 2016. ISBN:0128045752 9780128045756.

10 European Conference of Postal and Telecommunications Administrations. Available at http://www.cept.org/.

11 Asia-Pacific Telecommunity (APT). Available at http://www.aptsec.org/.

12 Inter-American Telecommunication Commission (CITEL). Available at https://www.citel.oas.org/en/Pages/default.aspx.

13 CAU The Internet Engineering Task Force (IETF). Available at https://www.ietf.org/.

14 Institute of Electrical and Electronic Engineers. Available at https://www.ieee.org/standards/index.html.

15 Open Mobile Alliance. Available at http://openmobilealliance.org/.

16 CAU European Telecommunications Standards Institute. Available at http://www.etsi.org/.

17 Global Certification Forum (GCF). Available at http://www.globalcertificationforum.org/.

18 GSM Association. Available at http://www.gsma.com/.

19 White Paper, "4G Americas' Recommendations on 5G Requirements and Solutions." Available at http://www.5gamericas.org/en/resources/white-papers/.

20 "Towards 5G". Available at https://ec.europa.eu/digital-single-market/en/towards-5g.

21 5G Forum (Korea). Available at https://www.5gforum.org/eng.

22 Fifth Generation Mobile Communication Promotion Forum (5GMF) (Japan). Available at http://5gmf.jp/en/.sss

23 "NGMN 5G White Paper." Available at http://ngmn.org/5g-white-paper/5g-white-paper.html.

24 5G Infrastructure Association. Available at https://5g-ppp.eu/.

25 Recommendation M.2083-0 (09/2015) IMT Vision, "Framework and overall objectives of the future development of IMT for 2020 and beyond." Available at https://www.itu.int/rec/R-REC-M.2083.

26 Recommendation M.2012-2 (09/2015), "Detailed specifications of the terrestrial radio interfaces of International Mobile Telecommunications Advanced (IMT-Advanced)." Available at https://www.itu.int/rec/R-REC-M.2012.

27 Recommendation M.2010-1 (02/2017), "Generic unwanted emission characteristics of base stations using the terrestrial radio interfaces of IMT-Advanced." Available at https://www.itu.int/rec/R-REC-M.2070/en.

28 Recommendation M.2071-1 (02/2017), "Generic unwanted emission characteristics of mobile stations using the terrestrial radio interfaces of IMT-Advanced." Available at https://www.itu.int/rec/R-REC-M.2071/en.

29 ITU towards "IMT for 2020 and beyond." Available at http://www.itu.int/en/ITU-R/study-groups/rsg5/rwp5d/imt-2020/Pages/default.aspx.

30 Radiocommunication Bureau, Circular Letter 5/LCCE/59, "Invitation for submission of proposals for candidate radio interface technologies for the terrestrial components of the radio interface(s) for IMT-2020 and invitation to participate in their subsequent evaluation." Available at http://www.itu.int/md/R00-SG05-CIR-0059.

31 Common Patent Policy for ITU-T/ITU-R/ISO/IEC. Available at http://www.itu.int/en/ITU-T/ipr/Pages/policy.aspx.

32 IMT-2020 submission and evaluation process. Available at http:// www.itu.int/en/ITU-R/study-groups/rsg5/ rwp5d/imt-2020/Pages/ submission-eval.aspx.

33 NTT DOCOMO, Inc., NEC Corporation, "Revised Work Item Description: Dual Connectivity for LTE," RP-141266. Available at http://www.3gpp.org/ftp/tsg_ran/tsg_ran/TSGR_65/Docs/.

34 3GPP TR 22.861 "FS_SMARTER: massive Internet of Things." Available at http://www.3gpp.org/DynaReport/22-series.htm.

35 3GPP TR 22.862, "Feasibility study on new services and markets technology enablers for critical communications." Available at http://www.3gpp.org/DynaReport/22-series.htm.

36 3GPP TR 22.863, "Feasibility study on new services and markets technology enablers for enhanced mobile broadband." Available at http://www.3gpp.org/DynaReport/22-series.htm.

37 3GPP TR 22.864, "Feasibility study on new services and markets technology enablers for network operation." Available at http://www.3gpp.org/DynaReport/22-series.htm

38 3GPP, "TR 22.886 study on enhancement of 3GPP support for 5G V2X services." Available at http://www.3gpp.org/DynaReport/22-series.htm.

39 3GPP TS 22.261, "Service requirements for next generation new services and markets." Available at http://www.3gpp.org/DynaReport/22-series.htm.

40 3GPP TR 23.977, "Study on architecture for next generation system." Available at http://www.3gpp.org/DynaReport/23-series.htm.

41 3GPP TR 33.899, "Study on the security aspects of the next generation system." Available at http://www.3gpp.org/DynaReport/33-series.htm.

42 3GPP TR 28.802, "Study on management aspects of next generation network architecture and features." Available at http://www.3gpp.org/DynaReport/28-series.htm.

43 3GPP TR 28.800, "Study on management and orchestration architecture of next generation network and service." Available at http://www.3gpp.org/DynaReport/28-series.htm.

44 3GPP TR 28.801, "Study on management and orchestration of network slicing for next generation network." Available at http://www.3gpp.org/DynaReport/28-series.htm.

45 3GPP TR 38.913, "Study on scenarios and requirements for next generation access technologies." Available at http://www.3gpp.org/DynaReport/38-series.htm.

46 J. Calabuig et al., C "5th generation mobile networks: a new opportunity for the convergence of mobile broadband and broadcast services," *IEEE Commun. Mag.*, vol. 53, no. 2, 198–205, 2015.

47 T. S. Rappaport et al., "Millimeter Wave Mobile Communications for 5G Cellular: It Will Work!", *IEEE Access*, vol. 1, May 10, 2013.

Giovanni Romano is currently coordinating TIM activities on technical standards related to radio access and spectrum. The role implies the identification of requirements to be transferred to standardization fora such as 3GPP to define new technical solutions for the radio interfaces of GSM, UMTS, LTE, 5G/IMT2020 and their evolutions, and the evaluation of technical solutions under development.

In October 2016, Giovanni was nominated Alternate Board Director, representing TIM in NGMN.

Since 1996, Giovanni has been active in several international standardization fora, and he is currently attending 3GPP RAN and NGMN. He is currently coordinating the exchange of information between 3GPP and ITU-R.

Giovanni has served as Vice-Chairman of 3GPP Technical Specification Group RAN for the period 2013–2017.

Until 2005, Giovanni was project leader of several activities within the R&D center of Telecom Italia, including UMTS performance evaluation, quality of service verification, standardization, field trials, and testing.

Maria Pia Galante works at TIM, where she is mainly involved in the coordination of TIM activities on standards for network architecture and services. Her roles include identifying requirements and solutions that best suit the operator strategy and promoting them accordingly, in synergy with relevant stakeholders, so that effective and market respondent technical standards are timely elaborated by 3GPP for the ever-evolving mobile communications services. She graduated from Polytechnic University of Turin and joined Telecom Italia Group (formerly CSELT) in 1998, dealing with control layer innovation technologies for third-generation systems in several international projects. She has represented Telecom Italia in 3GPP SA WG2 (Network Architecture) for the period 2000–2008, and since 2008, she is TIM delegate to 3GPP TSG SA (Services and Systems Aspects).

INDEX

5G Networks: Fundamental Requirements, Enabling Technologies, and Operations Management, First Edition.
Anwer Al-Dulaimi, Xianbin Wang, and Chih-Lin I.
© 2018 by The Institute of Electrical and Electronics Engineers, Inc. Published 2018 by John Wiley & Sons, Inc.